RETINAL DEGENERATIVE DISEASES

ADVANCES IN EXPERIMENTAL MEDICINE AND BIOLOGY

Recent Volumes in this Series

Volume 564
GLYCOBIOLOGY AND MEDICINE: PROCEEDINGS OF THE 7TH JENNER
GLYCOBIOLOGY AND MEDICINE SYMPOSIUM
Edited by John S. Axford

Volume 565
SLIDING FILAMENT MECHANISM IN MUSCLE CONTRACTION:
FIFTY YEARS OF RESEARCH
Edited by Haruo Sugi

Volume 566
OXYGEN TRANSPORT TO TISSUE XXVI
Edited by Paul Okunieff, Jacqueline Williams, and Yuhchyau Chen

Volume 567
THE GROWTH HORMONE-INSULIN-LIKE GROWTH FACTOR AXIS
DURING DEVELOPMENT
Edited by Isabel Varela-Nieto and Julie A. Chowen

Volume 568
HOT TOPICS IN INFETION AND IMMUNITY IN CHILDREN II
Edited by Andrew J. Pollard and Adam Finn

Volume 569
EARLY NUTRITION AND ITS LATER CONSEQUENCES: NEW OPPORTUNITIES
Edited by Berthold Koletzko, Peter Dodds, Hans Akerbloom, and Margaret Ashwell

Volume 570
GENOME INSTABILITY IN CANCER DEVELOPMENT
Edited by Erich A. Nigg

Volume 571
ADVANCES IN MYCOLOGY
Edited by J.I. Pitts, A.D. Hocking, and U. Thrane

Volume 572
RETINAL DEGENERATIVE DISEASES
Edited by Joe Hollyfield, Robert Anderson, and Matthew LaVail

RETINAL DEGENERATIVE DISEASES

Edited by

Joe G. Hollyfield

Cole Eye Institute
The Cleveland Clinic Foundation
Cleveland, Ohio

Robert E. Anderson

Dean A. McGee Eye Institute
University of Oklahoma Health Sciences Center
Oklahoma City, Oklahoma

and

Matthew M. LaVail

Beckman Vision Center
University of California, San Francisco
San Francisco, California

Joe G. Hollyfield
Cleveland Clinic Foundation
Cleveland, Ohio 44195
USA
hollyfj@ccf.org

Robert E. Anderson
Dean Mcgee Eye Institute
University of Oklahoma
Oklahoma City, OK 73104
robert-anderson@ouhsc.edu

Matthew M. LaVail
University of California, San Francisco
San Francisco, CA 94143
USA
MMLV@hsa.ucsf.edu

Library of Congress Control Number: 2005931128

ISBN-10: 0-387-28464-8
ISBN-13: 978-0387-28464-4

Printed on acid-free paper.

Printed in the United States of America.   (BS/MVY)

9 8 7 6 5 4 3 2

springer.com

Paul A. Sieving, M.D., Ph.D.

This book is dedicated to Paul A. Sieving, M.D., Ph.D., in recognition of his dedication to the advancement of basic and clinical sciences focused on understanding the causes, treatment and elimination of blinding diseases; and for his leadership as Director of the National Eye Institute, National Institutes of Health, Bethesda, Maryland.

PREFACE

A Symposium on Retinal Degenerations has been held in conjunction with the biennial International Congress of Eye Research (ICER) since 1984. These Retinal Degeneration Symposia have allowed scientists and clinicians from around the world to convene and present their new research findings. The Symposia have been organized to allow sufficient time for discussions and one-on-one interactions in a relaxed atmosphere, where international friendships and collaborations would be fostered.

The XI International Symposium on Retinal Degeneration (also known as RD2004) was held from August 23-28, 2004 in Perth, Western Australia. The meeting brought together 151 scientists, retinal specialists in ophthalmology and trainees in the field from all parts of the world. In the course of the meeting, 36 platform and 80 poster presentations were given, and a majority of these are presented in this proceedings volume. New discoveries and state of the art findings from most research areas in the field of retinal degenerations were presented. The RD2004 meeting was highlighted by three special lectures. The first was given by Dr. Paul Sieving, Director of the National Eye Institute, National Institutes of Health, Bethesda, Maryland, USA. Dr. Sieving discussed the evolution of understanding ocular diseases. The second lecture by Professor Ian Constable, Director of the Lions Eye Institute, Nedlands, Western Australia, covered clinical imperatives in retinal degeneration research. The third was by Professor Thaddeus Dryja, Harvard Medical School, Boston, Massachusetts, USA, who discussed the progress to date on understanding the etiology of RP and allied diseases.

We want to give special acknowledgement to the local organizers of the Symposium, Drs. Elizabeth P. Rakoczy, Sasha Pendal and Brian King from Perth. Mrs. Stacey Scaffardi, Administrative Assistant to Professor Rakoczy, worked tirelessly for months preparing an extraordinarily smooth meeting, a wonderful social program and excursion. We also thank the staff of the Burswood International Resort Hotel, the venue of the meeting, for all their help.

The Symposium received international financial support from a number of organizations. We are particularly pleased to thank The Foundation Fighting Blindness, Owings Mills, Maryland, for its continuing support of this and the previous biennial Symposia, without which we could not have held these important meetings. In addition, for the second time, the National Eye Institute of the National Institutes of Health contributed to the meeting. This additional funding allowed us to provide 20 Travel Awards to young investigators and trainees working in the field of retinal degenerations. The Foundation Fighting Blindness also contributed to the Travel Awards program providing 14 Travel Awards. The response to the Travel Awards program was extraordinary, with 74 applicants competed for the 34 Awards.

We also acknowledge the diligent and outstanding efforts of Ms. Holly Whiteside, who carried out most of the administrative aspects of the RD2004 Symposium, designed and maintained the meeting website. Holly is the Administrative Manager of Dr. Anderson's laboratory at the Oklahoma Health Sciences Center, and she has become the permanent Coordinator for the Retinal Degeneration Symposia. Her dedicated efforts with the Symposia since RD2000 have provided continuity heretofore not available, and we are deeply indebted to her.

We thank Ms. Laura Hogan, Administrative Assistant in Dr. Hollyfield's program in the Cole Eye Institute at The Cleveland Clinic Foundation, for her help in assembling this volume. Thanks also go to Springer Science+Business Media for its publication.

Joe G. Hollyfield
Robert E. Anderson
Matthew M. LaVail
Cleveland, Ohio
April 2005

ABOUT THE EDITORS

Joe G. Hollyfield, Ph.D., is Director of Ophthalmic Research and Professor of Molecular Medicine in the Cole Eye Institute at The Cleveland Clinic Foundation, Cleveland, Ohio. He received a Ph.D. from the University of Texas at Austin and was a postdoctoral fellow at the Hubrecht Laboratory in Utrecht, The Netherlands. He has held faculty positions at Columbia University College of Physicians and Surgeons in New York City and at Baylor College of Medicine in Houston, TX. He was Director of the Retinitis Pigmentosa Research Center while in Houston from 1978 until his move to The Cleveland Clinic Foundation in 1995. He is currently Director of the Foundation Fighting Blindness Research Center at The Cleveland Clinic Foundation. Dr. Hollyfield has published over 170 papers in the area of cell and developmental biology of the retina in health and disease. He has edited twelve books, eleven on retinal degenerations and one on the structure of the eye. Dr. Hollyfield received the Marjorie W. Margolin Prize (1981, 1994), the Sam and Bertha Brochstein Award (1985) and the Award of Merit in Retina Research (1998) from the Retina Research Foundation; the Olga Keith Wiess Distinguished Scholars' Award (1981), two Senior Scientific Investigator Awards (1988, 1994) from Research to Prevent Blindness; an award for Outstanding Contributions to Vision Research from the Alcon Research Institute (1987); the Distinguished Alumnus Award (1991) from Hendrix College, Conway, Arkansas; and the Endre A. Balazs Prize (1994) from the International Society for Eye Research (ISER). He is currently Editor-in-Chief of the journal, *Experimental Eye Research* published by Elsevier Press. Dr. Hollyfield has held elected leadership roles in the Association for Research in Vision and Ophthalmology (ARVO) serving on the Program Committee, as a Trustee representing Retina Cell Biology, and as President. He is also a past President and former Secretary of the International Society of Eye Research. He currently serves on the Scientific Advisory Boards of The Foundation Fighting Blindness, Research to Prevent Blindness, The Helen Keller Eye Research Foundation, The Knights Templar Eye Foundation, the Macular Degeneration program of the American Health Assistance Foundation, Retina South Africa Fighting Blindness, and is Co-Chairman of the Medical and Scientific Advisory Board of Retina International.

Robert E. Anderson, M.D., Ph.D., is Professor and Chair of Cell Biology, Dean A. McGee Professor of Ophthalmology, and Adjunct Professor of Biochemistry & Molecular Biology and Geriatric Medicine at The University of Oklahoma Health Sciences Center in

Oklahoma City, Oklahoma. He is also Director of Research at the Dean A. McGee Eye Institute. He received his Ph.D. in Biochemistry (1968) from Texas A&M University and his M.D. from Baylor College of Medicine in 1975. In 1968, he was a postdoctoral fellow at Oak Ridge Associated Universities. At Baylor, he was appointed Assistant Professor in 1969, Associate Professor in 1976, and Professor in 1981. He joined the faculty of the University of Oklahoma in January of 1995. Dr. Anderson has published over 200 research articles in the areas of lipid metabolism in the retina and biochemistry of retinal degenerations. He has edited twelve books, eleven on retinal degenerations and one on the biochemistry of the eye. Dr. Anderson has received the Sam and Bertha Brochstein Award for Outstanding Achievement in Retina Research from the Retina Research Foundation (1980), the Dolly Green Award (1982) and two Senior Scientific Investigator Awards (1990 and 1997) from Research to Prevent Blindness, Inc. He received an Award for Outstanding Contributions to Vision Research from the Alcon Research Institute (1985), and the Marjorie Margolin Prize (1994). He has served on the editorial boards of *Investigative Ophthalmology and Visual Science*, *Journal of Neuroscience Research*, *Neurochemistry International*, *Current Eye Research* and *Experimental Eye Research*. Dr. Anderson has been involved in leadership roles in the Association for Research in Vision and Ophthalmology (ARVO) and is a former trustee representing the Biochemistry and Molecular Biology section. He has served on the Vision Research Program Committee and Board of Scientific Counselors of the National Eye Institute and the Board of the Basic and Clinical Science Series of The American Academy of Ophthalmology. Currently he is a member of the Macular Degeneration grant review panel of the American Health Assistance Foundation. Dr. Anderson is currently the President of the International Society for Eye Research and has served as a past Councilor and Treasurer of this society.

Matthew M. LaVail, Ph.D., is Professor of Anatomy and Ophthalmology at the Beckman Vision Center of the University of California, San Francisco School of Medicine. He received his Ph.D. degree in Anatomy (1969) from the University of Texas Medical Branch in Galveston and was subsequently a postdoctoral fellow at Harvard Medical School. Dr. LaVail was appointed Assistant Professor of Neurology-Neuropathology at Harvard Medical School in 1973. In 1976, he moved to UCSF, where he was appointed Associate Professor of Anatomy. He was appointed to his current position in 1982, and in 1988, he also became director of the Retinitis Pigmentosa Research Center at UCSF, later named the Kearn Family Center for the Study of Retinal Degeneration. Dr. LaVail has published extensively in the research areas of photoreceptor-retinal pigment epithelial cell interactions, retinal development, circadian events in the retina, genetics of pigmentation and ocular abnormalities, inherited retinal degenerations, light-induced retinal degeneration, and pharmaceutical and gene therapy for retinal degenerative diseases. He has identified several naturally occurring murine models of human retinal degenerations and has developed transgenic mouse and rat models of others. He is the author of more than 140 research publications and has edited eleven books on inherited and environmentally induced retinal degenerations. Dr. LaVail has received the Fight for Sight Citation (1976); the Sundial Award from the Retina Foundation (1976); the Friedenwald Award from the Association for Research in Vision and Ophthalmology (ARVO, 1981); two Senior Scientific Investigators Awards from Research to Prevent Blindness (1988 and 1998); a MERIT Award from the National Eye Institute (1989); an Award for Outstanding Contributions to Vision Research from the Alcon Research Institute (1990); the Award of Merit from the Retina Research Foundation (1990); the first John A. Moran Prize for Vision Research from the University of Utah (1997); and the first Trustee

Award from The Foundation Fighting Blindness (1998). He has served on the editorial board of *Investigative Ophthalmology and Visual Science* and is currently on the editorial board of *Experimental Eye Research*. Dr. LaVail has been an active participant in the program committee of ARVO and has served as a Trustee (Retinal Cell Biology Section) of ARVO. He has been a member of the program committee and a Vice President of the International Society for Eye research. He has also served on the Scientific Advisory Board of the Foundation Fighting Blindness since 1973.

CONTENTS

PREFACE vii

ABOUT THE EDITORS ix

PART I   MOLECULAR GENETICS AND CANDIDATE GENES

1. GENETIC FACTORS MODIFYING CLINICAL EXPRESSION OF AUTOSOMAL DOMINANT RP 3

Stephen P. Daiger, Suma P. Shankar, Alice B. Schindler, Lori S. Sullivan,
Sara J. Bowne, Terri M. King, E. Warick Daw, Edwin M. Stone, and
John R. Heckenlively

2. DISEASE-ASSOCIATED VARIANTS OF THE ROD-DERIVED CONE VIABILITY FACTOR (RdCVF) IN LEBER CONGENITAL AMAUROSIS. Rod-derived cone viability variants in LCA 9

Sylvain Hanein, Isabelle Perrault, Sylvie Gerber, Hélène Dollfus,
Jean-Louis Dufier, Josué Feingold, Arnold Munnich, Shomi Bhattacharya,
Josseline Kaplan, José-Alain Sahel, Jean-Michel Rozet, and Thierry Leveillard

3. LEBER CONGENITAL AMAUROSIS: SURVEY OF THE GENETIC HETEROGENEITY, REFINEMENT OF THE CLINICAL DEFINITION AND PHENOTYPE-GENOTYPE CORRELATIONS AS A STRATEGY FOR MOLECULAR DIAGNOSIS. Clinical and molecular survey in LCA 15

Sylvain Hanein, Isabelle Perrault, Sylvie Gerber, Gaëlle Tanguy,
Jean-Michel Rozet, and Josseline Kaplan

4. **A FIRST LOCUS FOR ISOLATED AUTOSOMAL RECESSIVE OPTIC ATROPHY (ROA1) MAPS TO CHROMOSOME 8q21–q22** 21
Fabienne Barbet, Sylvie Gerber, Sélim Hakiki, Isabelle Perrault, Sylvain Hanein, Dominique Ducroq, Gaëlle Tanguy, Jean-Louis Dufier, Arnold Munnich, Josseline Kaplan, and Jean-Michel Rozet

5. **RCC1-LIKE DOMAIN AND ORF15: ESSENTIALS IN RPGR GENE** 29
Zi-Bing Jin, Mutsuko Hayakawa, Akira Murakami, and Nobuhisa Nao-i

6. **CHOROIDAL NEOVASCULARIZATION IN PATIENTS WITH ADULT-ONSET FOVEOMACULAR DYSTROPHY CAUSED BY MUTATIONS IN THE *RDS/PERIPHERIN* GENE** 35
Darius M. Moshfeghi, Zhenglin Yang, Nathan D. Faulkner, Goutam Karan, Sukanya Thirumalaichary, Erik Pearson, Yu Zhao, Thomas Tsai, and Kang Zhang

7. **BIOCHEMICAL CHARACTERISATION OF THE *C1QTNF5* GENE ASSOCIATED WITH LATE-ONSET RETINAL DEGENERATION. A genetic model of age-related macular degeneration** 41
Xinhua Shu, Brian Tulloch, Alan Lennon, Caroline Hayward, Mary O'Connell, Artur V. Cideciyan, Samuel G. Jacobson, and Alan F. Wright

8. **BIETTI CRYSTALLINE CORNEORETINAL DYSTROPHY ASSOCIATED WITH *CYP4V2* GENE MUTATIONS** 49
Makoto Nakamura, Jian Lin, Koji Nishiguchi, Mineo Kondo, Jiro Sugita, and Yozo Miyake

PART II   DIAGNOSTIC, CLINICAL, CYTOPATHOLOGICAL AND PHYSIOLOGIC ASPECTS OF RETINAL DEGENERATION

9. **FUNDUS APPEARANCE OF CHOROIDEREMIA USING OPTICAL COHERENCE TOMOGRAPY** 57
Bradley J. Katz, Zhenglin Yang, Marielle Payn, Yin Lin, Yu Zhao, Erik Pearson, Shan Duan, Shin Kamaya, Goutam Karan, and Kang Zhang

10. **A2E, A FLUOROPHORE OF RPE LIPOFUSCIN, CAN DESTABILIE MEMBRANE** 63
Janet R. Sparrow, Bolin Cai, Young Pyo Jang, Jilin Zhou, and Koji Nakanishi

11. AMINO-RETINOID COMPOUNDS IN THE HUMAN RETINAL PIGMENT EPITHELIUM 69
Heidi R. Vollmer-Snarr, McKenzie R. Pew, Mary L. Alvarez, D. Joshua Cameron, Zhibing Chen, Glenn L. Walker, Josh L. Price, and Jeffrey L. Swallow

12. ANNEXINS IN BRUCH'S MEMBRANE AND DRUSEN 75
Mary E. Rayborn, Hiro Sakaguchi, Karen G. Shadrach, John W. Crabb, and Joe G. Hollyfield

PART III   ANIMAL MODELS OF RETINAL DEGENERATION

13. MOLECULAR MECHANISMS OF PHOTORECEPTOR DEGENERATION IN RP CAUSED BY IMPDH1 MUTATIONS 81
Aileen Aherne, Avril Kennan, Paul F. Kenna, Niamh McNally, G. Jane Farrar, and Pete Humphries

14. BIOCHEMICAL FUNCTION OF THE LCA LINKED PROTEIN, ARYL HYDROCARBON RECEPTOR INTERACTING PROTEIN LIKE–1 (AIPL1). Role of AIPL1 in retina 89
Matthew L. Schwartz, James B. Hurley, and Visvanathan Ramamurthy

15. CHARACTERIZATION OF MOUSE MUTANTS WITH ABNORMAL RPE CELLS 95
Chun-hong Xia, Haiquan Liu, Meng Wang, Debra Cheung, Alex Park, Yang Yang, Xin Du, Bo Chang, Bruce Beutler, and Xiaohua Gong

16. ROD AND CONE PIGMENT REGENERATION IN $RPE65^{-/-}$ MICE 101
Baerbel Rohrer and Rosalie Crouch

17. INITIAL OBSERVATIONS OF KEY FEATURES OF AGE-RELATED MACULAR DEGENERATION IN $APOE$ TARGETED REPLACEMENT MICE 109
Goldis Malek, Brian Mace, Peter Saloupis, Donald Schmechel, Dennis Rickman, Patrick Sullivan, and Catherine Bowes Rickman

18. ALTERED RHYTHM OF PHOTORECEPTOR OUTER SEGMENT PHAGOCYTOSIS IN β5 INTEGRIN KNOCKOUT MICE 119
Emeline F. Nandrot and Silvia C. Finnemann

19. LIGHT/DARK TRANSLOCATION OF ALPHATRANSDUCIN IN MOUSE PHOTORECEPTOR CELLS EXPRESSING G90D MUTANT OPSIN 125

Zack A. Nash and Muna I. Naash

20. SLOWED PHOTORESPONSE RECOVERY AND AGE-RELATED DEGENERATION IN CONES LACKING G PROTEIN-COUPLED RECEPTOR KINASE 1 133

Xuemei Zhu, Bruce Brown, Lawrence Rife, and Cheryl M. Craft

21. TRANSGENIC ANIMAL STUDIES OF HUMAN RETINAL DISEASE CAUSED BY MUTATIONS IN PERIPHERIN/*RDS* 141

Xi-Qin Ding and Muna I. Naash

22. TRANSGENIC EXPRESSION OF LEUKEMIA INHIBITORY FACTOR INHIBITS BOTH ROD AND CONE GENE EXPRESSION. Gp130 regulates cone gene expression 147

John D. Ash and Dianca R. Graham

23. A ROLE FOR BHLH TRANSCRIPTION FACTORS IN RETINAL DEGENERATION AND DYSFUNCTION 155

Mark E. Pennesi, Debra E. Bramblett, Jang-Hyeon Cho, Ming-Jer Tsai, and Samuel M. Wu

24. CHARACTERISATION OF A MODEL FOR RETINAL NEOVASCULARISATION. VEGF MODEL CHARACTERISATION 163

Pauline E. van Eeden, Lisa Tee, Wei-Yong Shen, Sherralee Lukehurst, Chooi-May Lai, P. Elizabeth Rakoczy, Lyn D. Beazley, and Sarah A. Dunlop

25. A TWO-ALTERNATIVE FORCED CHOICE METHOD FOR ASSESSING MOUSE VISION 169

Yumiko Umino, Bridget Frio, Maryam Abbasi, and Robert Barlow

26. CONDITIONAL GENE KNOCKOUT SYSTEM IN CONE PHOTORECEPTORS 173

Yun-Zheng Le, John D. Ash, Muayyad R. Al-Ubaidi, Ying Chen, Jian-Xing Ma, and Robert E. Anderson

27. REGULATION OF TIGHT JUNCTION PROTEINS IN CULTURED RETINAL PIGMENT EPITHELIAL CELLS AND IN VEGF OVEREXPRESSING TRANSGENIC MOUSE RETINAS 179

Reza Ghassemifar, Chooi-May Lai, and P. Elizabeth. Rakoczy

28. **PATHOLOGICAL HETEROGENEITY OF VASOPROLIFERATIVE RETINOPATHY IN TRANSGENIC MICE OVEREXPRESSING VASCULAR ENDOTHELIAL GROWTH FACTOR IN PHOTORECEPTORS** 187
Wei-Yong Shen, Yvonne K.Y. Lai, Chooi-May Lai, Nicolette Binz,
Lyn D. Beazley, Sarah A. Dunlop, and P. Elizabeth Rakoczy

29. **LASER PHOTOCOAGULATION: OCULAR RESEARCH AND THERAPY IN DIABETIC RETINOPATHY** 195
Caroline E. Graham, Nicolette Binz, Wei-Yong Shen, Ian J. Constable, and
Elizabeth P. Rakoczy

30. **APPLYING TRANSGENIC ZEBRAFISH TECHNOLOGY TO STUDY THE RETINA** 201
Ross F. Collery, Maria L. Cederlund, Vincent A. Smyth, and
Breandán N. Kennedy

31. **BMI1 LOSS DELAYS PHOTORECEPTOR DEGENERATION IN *RD1* MICE. Bmi1 loss and neuroprotection in *Rd1* mice** 209
Dusan Zencak, sylvain V. Crippa, Meriem Tekaya, Ellen Tanger,
Daniel F. Schorderet, Francis L. Munier, Maarten van Lohuizen, and
Yvan Arsenijevic

32. **TRANSCRIPTIONAL AND POST-TRANSCRIPTIONAL REGULATION OF THE ROD cGMP-PHOSPHODIESTERASE β-SUBUNIT GENE. Recent advances and current concepts** 217
Leonid E. Lerner, Natik Piri, and Debora B. Farber

PART IV   GENE THERAPY AND NEUROPROTECTION

33. **DOWN-REGULATION OF RHODOPSIN GENE EXPRESSION BY AAV-VECTORED SHORT INTERFERING RNA** 233
Jacqueline T. Teusner, Alfred S. Lewin, and William W. Hauswirth

34. **ASSESSING THE EFFICACY OF GENE THERAPY IN *Rpe65$^{-/-}$* MICE USING PHOTOENTRAINMENT OF CIRCADIAN RHYTHM** 239
Chris W. Stoddart, Meaghan J.T. Yu, Matthew T. Martin-Iverson, Dru M. Daniels,
Chooi-May Lai, Nigel L. Barnett, T. Michael Redmond, Kristina Narfström, and
P. Elizabeth Rakoczy

35. LENTIVIRAL VECTORS CONTAINING A RETINAL PIGMENT EPITHELIUM SPECIFIC PROMOTER FOR LEBER CONGENITAL AMAUROSIS GENE THERAPY. Lentiviral gene therapy for LCA 247

Alexis-Pierre Bemelmans, Corinne Kostic, Dana Hornfeld, Muriel Jaquet, Sylvain V. Crippa, William W. Hauswirth, Janis Lem, Zhongyan Wang, Daniel F. Schorderet, Francis L. Munier, Andreas Wenzel, and Yvan Arsenijevic

36. GENE DELIVERY TO THE RETINA USING LENTIVIRAL VECTORS 255

Kenneth P. Greenberg, Edwin S. Lee, David V. Schaffer, and John G. Flannery

37. POTENTIAL USE OF CELLULAR PROMOTER(S) TO TARGET RPE IN AAV-MEDIATED DELIVERY. Cellular promoters and RPE-targeting 267

Erika N. Sutanto, Dan Zhang, Yvonne K.Y. Lai, Wei-Yong Shen, and P. Elizabeth Rakoczy

38. CYTOKINE-INDUCE RETINAL DEGENERATION: ROLE OF SUPPRESSORS OF CYTOKINE SIGNALING (SOCS) PROTEINS IN PROTECTION OF THE NEURORETINA 275

Charles E. Egwuagu, Cheng-Hong Yu, Rashid M. Mahdi, Maire Mameza, Chikezie Eseonu, Hiroshi Takase, and Samuel Ebong

39. DISEASE MECHANISMS AND GENE THERAPY IN A MOUSE MODEL FOR X-LINKED RETINOSCHISIS 283

Laurie L. Molday, Seok-Hong Min, Mathias W. Seeliger, Winco W.H. Wu, Astra Dinculescu, Adrian M. Timmers, Andreas Janssen, Felix Tonagel, Kristiane Hudl, Bernhard H. F. Weber, William W. Hauswirth, and Robert S. Molday

40. MOLECULAR MECHANISMS OF NEUROPROTECTION IN THE EYE 291

Colin J. Barnstable and Joyce Tombran-Tink

41. RETINAL DAMAGE CAUSED BY PHOTODYNAMIC THERAPY CAN BE REDUCED USING BDNF 297

Jacque L. Duncan, Daniel M. Paskowitz, George C. Nune, Douglas Yasumura, Haidong Yang, Michael T. Matthes, Marco A. Zarbin, and Matthew M. LaVail

42. **CONTROLLING VASCULAR ENDOTHELIAL GROWTH FACTOR: THERAPIES FOR OCULAR DISEASES ASSOCIATED WITH NEOVASCULARIZATION** 303
Robert J. Marano and P. Elizabeth Rakoczy

43. **INTRAVITREAL INJECTION OF TRIAMCINOLONE ACETONIDE FOR MACULAR EDEMA DUE TO RETINITIS PIGMENTOSA AND OTHER RETINAL DISEASES** 309
Jianbin Hu, Paul S. Bernstein, Michael P. Teske, Marielle Payne,
Zhenglin Yang Chumei Li, David Adams, Jennifer H. Baird, and Kang Zhang

44. **CONE SURVIVAL: IDENTIFICATION OF RdCVF** 315
Olivier Lorentz, José Sahel, Saddek Mohand-Saïd, and Thierry Leveillard

45. **NEUROPROTECTION OF PHOTORECEPTORS IN THE RCS RAT AFTER IMPLANTATION OF A SUBRETINAL IMPLANT IN THE SUPERIOR OR INFERIOR RETINA** 321
Machelle T. Pardue, Michael J. Phillips, Brett Hanzlicek, Hang Yin,
Alan Y. Chow, and Sherry L. Ball

46. **GLUTAMATE TRANSPORT MODULATION: A POSSIBLE ROLE IN RETINAL NEUROPROTECTION** 327
Nigel L. Barnett, Kei Takamoto, and Natalie D. Bull

47. **ACTIVATION OF CELL SURVIVAL SIGNALS IN THE GOLDFISH RETINAL GANGLION CELLS AFTER OPTIC NERVE INJURY** 333
Yoshiki Koriyama, Keiko Homma, and Satoru Kato

PART V   USHER SYNDROME

48. **ROLES AND INTERACTIONS OF USHER 1 PROTEINS IN THE OUTER RETINA** 341
Concepción Lillo, Junko Kitamoto, and David S. Williams

49. **MOLECULAR ANALYSIS OF THE SUPRAMOLECULAR USHER PROTEIN COMPLEX IN THE RETINA. Harmonin as the key protein of the Usher Syndrome** 349
Jane Reiners and Uwe Wolfrum

PART VI   STEM CELLS, TRANSPLANTATION AND RETINAL REPAIR

**50. LIMITED NEURAL DIFFERENTIATION OF RETINAL
PIGMENT EPITHELIUM** 357
Ryosuke Wakusawa, Toshiaki Abe, Yoko Saigo, and Makoto Tamai

**51. RETINAL PIGMENT EPITHELIAL CELLS FROM THERMALLY
RESPONSIVE POLYMER-GRAFTED SURFACE REDUCE
APOPTOSIS** 363
Toshiaki Abe, Masayoshi Hojo, Yoko Saigo, Masahiko Yamato, Teruo Okano,
Ryosuke Wakusawa, and Makoto Tamai

**52. RETINAL TRANSPLANTATION. A treatment strategy for retinal
degenerative disease** 367
Biju B. Thomas, Robert B. Aramant, SriniVas R. Sadda, and Magdalene J. Seiler

**53. MICROARRAY ANALYSIS REVELAS RETINAL STEM CELL
CHARACTERISTICS OF THE ADULT HUMAN EYE.
For contribution volumes** 377
B Brigitte Angénieux, Lydia Michaut, Daniel F. Schorderet, Francis L. Munier,
Walter Gehring, and Yvan Arsenijevic

**54. USING STEM CELLS TO REPAIR THE DEGENERATE RETINA.
Stem cells in the context of retinal degenerations** 381
Christine M. Hall, Anthony Kicic, Chooi-May Lai, and P. Elizabeth Rakoczy

**55. OPTIC NERVE REGENERATION: MOLECULAR
PRE-REQUISITES AND THE ROLE OF TRAINING.
Restoring vision after optic nerve injury** 389
Lyn D. Beazley, Jennifer Rodger, Carolyn E. King, Carole A. Bartlett,
Andrew L. Taylor, and Sarah A. Dunlop

**56. RETINAL GANGLION CELL REMODELLING IN
EXPERIMENTAL GLAUCOMA** 397
James E. Morgan, Amit V. Datta, Jonathan T. Erichsen, Julie Albon, and
Michael E. Boulton

PART VII   INDUCED RETINAL DEGENERATIONS

**57. NEURAL PLASTICITY REVEALED BY LIGHT-INDUCED
PHOTORECEPTOR LESIONS** 405
Bryan W. Jones, Robert E. Marc, Carl B. Watt, Dana K. Vaughan, and
Daniel T. Organisciak

58. **FACTORS UNDERLYING CIRCADIAN DEPENDENT SUSCEPTIBILITY TO LIGHT INDUCED RETINAL DAMAGE** 411
Ruby Grewal, Daniel T. Organisciak, and Paul Wong

59. **SPACE FLIGHT ENVIRONMENT INDUCES DEGENERATION IN THE RETINA OF RAT NEONATES** 417
Joyce Tombran-Tink and Colin J. Barnstable

60. **TOXICITY OF HYPEROXIA TO THE RETINA: EVIDENCE FROM THE MOUSE** 425
Scott Geller, Renata Krowka, Krisztina Valter, and Jonathan Stone

61. **TREATMENT WITH CARBONIC ANHYDRASE INHIBITORS DEPRESSES ELECTRORETINOGRAM RESPONSIVENESS IN MICE** 439
Yves Sauvé, Goutam Karan, Zhenglin Yang, Chunmei Li, Jianbin Hu, and Kang Zhang

62. **INJURY-INDUCED RETINAL GANGLION CELL LOSS IN THE NEONATAL RAT RETINA** 447
Kirsty L. Spalding, Qi Cui, Arunasalam M. Dharmarajan, and Alan R. Harvey

PART VIII   BASIC SCIENCE UNDERLYING RETINAL DEGENERATION

63. **ARRESTIN TRANSLOCATION IN ROD PHOTORECEPTORS** 455
W. Clay Smith, James J. Peterson, Wilda Orisme, and Astra Dinculescu

64. **BINDING OF N-RETINYLIDENE-PO TO ABCA4 AND A MODEL FOR ITS TRANSPORT ACROSS MEMBRANES** 465
Robert S. Molday, Seelochan Beharry, Jinhi Ahn, and Ming Zhong

65. **THE CHAPERONE FUNCTION OF THE LCA PROTEIN AIPL1. AIPL1 chaperone function** 471
Jacqueline van der Spuy and Michael E. Cheetham

66. **CRALBP LIGAND AND PROTEIN INTERACTIONS** 477
Zhiping Wu, Sanjoy K. Bhattacharya, Zhaoyan Jin, Vera L. Bonilha, Tianyun Liu, Maria Nawrot, David C. Teller, John C. Saari, and John W. Crabb

67. **FUNCITONAL STUDY OF PHOTORECEPTOR PDEδ** 485
Houbin Zhang, Jeanne M. Frederick, and Wolfgang Baehr

68. LOCALIZATION OF THE INSULIN RECEPTOR AND PHOSPHOINOSITIDE 3-KINASE IN DETERGENT-RESISTANT MEMBRANE RAFTS OF ROD PHOTORECEPTOR OUTER SEGMENTS 491

Raju V.S. Rajala, Michael H. Elliott, Mark E. McClellan, and Robert E. Anderson

69. MERTK ACTIVATION DURING RPE PHAGOCYTOSIS *IN VIVO* REQUIRES αvβ5 INTEGRIN 499

Silvia C. Finnemann and Emeline F. Nandrot

70. PHOTORECEPTOR RETINOL DEHYDROGENASES. An attempt to characterize the function of Rdh11 505

Anne Kasus-Jacobi, David G. Birch, and Robert E. Anderson

71. PIGMENT EPITHELIUM-DERIVED GROWTH FACTOR INHIBITS FETAL BOVINE SERUM STIMULATED VASCULAR ENDOTHELIAL GROWTH FACTOR SYNTHESIS IN CULTURED HUMAN RETINAL PIGMENT EPITHELIAL CELLS 513

Piyush C. Kothary, Rhonda Lahiri, Lynn Kee, Nitin Sharma, Eugene Chun, Angela Kuznia, and Monte A. Del Monte

72. THE RETINAL PIGMENT EPITHELIUM APICAL MICROVILLI AND RETINAL FUNCTION 519

Vera L. Bonilha, Mary E. Rayborn, Sanjoy K. Bhattacharya, Xiarong Gu, John S. Crabb, John W. Crabb, and Joe G. Hollyfield

73. UPREGULATION OF TRANSGLUTAMINASE IN THE GOLDFISH RETINA DURING OPTIC NERVE REGENERATION 525

Kayo Sugitani, Toru Matsukawa, Ari Maeda, and Satoru Kato

74. SURVIVAL SIGNALING IN RETINAL PIGMENT EPITHELIAL CELLS IN RESPONSE TO OXIDATIVE STRESS. Significance in retinal degenerations 531

Nicolas Bazan

INDEX 541

PART I

MOLECULAR GENETICS AND CANDIDATE GENES

CHAPTER 1

GENETIC FACTORS MODIFYING CLINICAL EXPRESSION OF AUTOSOMAL DOMINANT RP

Stephen P. Daiger[1], Suma P. Shankar[1], Alice B. Schindler[1],
Lori S. Sullivan[1], Sara J. Bowne[1], Terri M. King[2], E. Warick Daw[3],
Edwin M. Stone[4], and John R. Heckenlively[5]

1. INTRODUCTION

Factors modifying clinical expression of inherited diseases are likely to be complex, involving genetic factors, environmental factors and stochastic effects. One way to reduce the complexity is to focus on individuals who share a dominant mutation identical by descent, thus eliminating variability in the underlying mutation and variation in *cis* to the mutation. A further simplification is to limit analysis to a single, extended family, which may reduce, though not eliminate, environmental effects.

Autosomal dominant retinitis pigmentosa (adRP) offers a number of opportunities for such studies. We are investigating factors which modify clinical features consequent to an Arg677ter stop mutation in the RP1 gene, a mutation which causes adRP with variable age-of-onset, progression and end stage consequences, that is, variable severity (Berson et al., 2002; Jacobson et al., 2000; Sullivan et al., 1999).

The RP1 locus was mapped by linkage testing to 8q13 in a large, extended adRP family, RP01, largely situated in southeastern Kentucky (Blanton et al., 1991; Field et al., 1982). Subsequently, the RP1 gene was identified by positional cloning and mutation screening in RP01 and additional adRP families (Guillonneau et al., 1999; Pierce et al., 1999; Sullivan et al., 1999). Mutations in the RP1 gene account for approximately 4% of adRP families in the United States. The RP1 Arg677ter mutation, found in RP01 and other families, accounts for approximately $\frac{1}{2}$ of the total (Bowne et al., 1999).

The RP1 gene codes for a novel protein, 2,156 amino acids in length, with high sequence similarity over the first 10% of the amino terminus to doublecortin, a protein

[1] Human Genetics Ctr and Dept. of Ophthalmology, Univ. of Texas, Houston, TX, USA; [2] Dept. of Internal Medicine, Univ. of Texas, Houston, TX, USA; [3] Dept. of Epidemiology, Univ. of Texas MD Anderson Cancer Center, Houston, TX, USA; [4] Dept. of Ophthalmology, Univ. of Iowa, IA, USA; [5] Kellogg Eye Ctr, Univ. of Michigan, Ann Arbor, MI, USA. Corresponding author: S. P. Daiger, E-mail: stephen.p.daiger@uth.tmc.edu.

involved in cortical folding, but with very little sequence similarity elsewhere. The protein is photo-receptor-specific; localizes to the rod interconnecting cilium; stabilizes disc architecture in outer segments; and is most likely involved in ciliary matrix assembly and function (Gao et al., 2002; Liu et al., 2002; Liu et al., 2003; Liu et al., 2004; Pierce et al., 2004).

The Arg677ter mutation is in the final, 4th exon of the gene and probably escapes nonsense-mediated decay. If so, the product of the mutant message would be a severely truncated protein, retaining the doublecortin domain, but lacking most of the remaining sequence. Thus the pathophysiology probably involves a dominant negative or gain of function effect, rather than haploinsufficiency.

Factors modifying simple mendelian diseases, such as retinitis pigmentosa, may be just as complex as factors contributing to multifactorial diseases such as age-related macular degeneration (Haider et al., 2002; Nadeau, 2003). However, for dominant-acting diseases one factor is particularly likely: the "wild type" allele in *trans* to the mutation. For example, alleles in *trans* to alpha-spectrin mutations modify the severity of elliptocytisis (Gratzer, 1994; Randon et al., 1994). Also, of direct relevance to retinitis pigmentosa, variability in expression of the allele in *trans* to disease-causing mutations in PRPF31 modify penetrance of the RP11 form of adRP (McGee et al., 1997; Vithana et al., 2003). This is particularly relevant to RP1 because several polymorphic amino acid substitutions exist at the locus which result in at least 6 distinct amino acid haplotypes, i.e., distinct proteins, in Caucasian populations (Bowne et al., 1999). These protein haplotypes in *trans* are potential modifiers of clinical expression of the RP1 Arg677ter mutation.

2. CLINICAL AND MOLECULAR CHARACTERIZATION OF RP01

A published genealogy traces retinitis pigmentosa in the RP01 family to a single individual born in 1803 (Breeding, 1982). Over 100 living, affected family members have been identified by pedigree reconstruction (Blanton et al., 1991). The family has been the subject of clinical and genetic studies for more that 30 years, including field studies in Kentucky and ophthalmic evaluations at the Jules Stein Eye Institute, UCLA (Field et al., 1982; Heckenlively et al., 1982; Lehmer et al., 1992).

RP01 has classical type 2 adRP with late onset of night blindness, usually by the third decade of life, and slow progression. Characteristic findings include diffuse retinal pigmentation, progressive decrease in recordable ERGs, and concentric visual field loss. Fundoscopic findings include retinal atrophy, bone spicule-like pigment deposits, and vascular attenuation.

The family also shows substantial within-family variability. For example, three members, examined in their 50s, had only small areas of regional pigmentation while two of these had children with typical RP. Early changes in the children included focal depigmented spot atrophy with pigmentation of the edge and abnormal RPE granularity. By contrast, typical late changes in adults vary from diffuse atrophy with pavingstone-type changes to typical bone spicule pigmentation. Further, at least two instances of a "skipped generation" have been documented in RP01.

We have DNA samples, and in many cases, lymphoblastoid cell lines, from 96 members of the family. Of these, 66 have the Arg677ter mutation. Of the 66, we have at least minimal clinical data on 44 and detailed clinical records, including fundus photographs, on 35. For linkage and segregation analysis we use a pedigree structure with 150 individuals, including two inbreeding loops (Blanton et al., 1992).

3. DO GENETIC FACTORS PLAY A ROLE?

To answer this question we assigned an age-adjusted affectation status to the 44 individuals with at least minimal records and tested for segregation of severity using the non-parametric pedigree analysis program Loki (Daw et al., 1999; Daw et al., 1999a; Heath, 1997). Loki uses an iterated, Monte Carlo Markoff-chain approach to estimate the extent to which genetic differences play a role in variation of a trait (severity in this case) and the potential number of contributing genes. Based on this evaluation, 36% of the differences in severity in RP01 can be explained by variation at the RP1 locus (other than the mutation itself) and 18% can be explained by unlinked loci. For the unlinked loci, Loki estimated that there is an 80% probability that 6 or fewer loci are involved, and that one of these loci may account for at least 9% of the total variation (Figure 1.1; unpublished). These results are provisional, given the "soft" nature of the clinical phenotype, but they support the likelihood that genetic differences play a role in the variability of adRP caused by the Arg677ter mutation.

4. POLYMORPHIC AMINO ACID HAPLOTYPES AT THE RP1 LOCUS

Numerous non-pathogenic amino acid substitutions have been reported at the RP1 locus (Berson et al., 2002; Bowne et al., 1999; Sohocki et al., 2001; Sullivan et al., 1999). Of

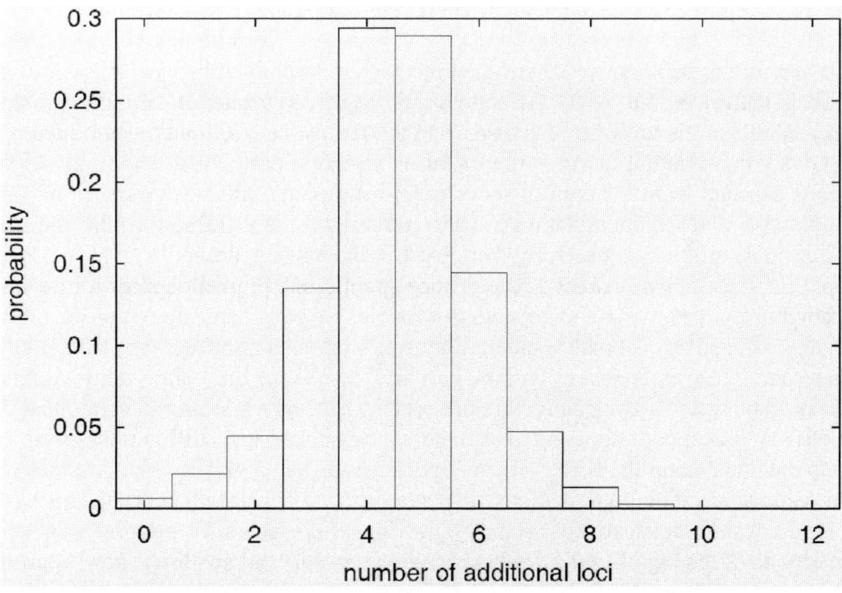

Figure 1.1. Number of genes estimated to contribute to RP1 Arg677ter severity based on Loki analysis in RP01.

Table 1.1. Polymorphic amino acid substitutions and haplotypes in RP1.

Substitution	Frequencies
Arg872His (CGT → CAT)	75%, 25%
Asn985Tyr (AAT → TAT)	54%, 46%
Ala1670Thr (GCA → ACA)	78%, 22%
Ser1691Pro (TCT → CCT)	77%, 23%
Cys2033Tyr (TGT → TAT)	46%, 54%

Haplotype	Frequencies
1: Arg – Tyr – Ala – Ser – Tyr	41%
2: Arg – Asn – Ala – Ser – Cys	30%
3: His – Asn – Thr – Pro – Cys	26%
4: Arg – Tyr – Ala – Ser – Cys	1%
5: His – Tyr – Ala – Ser – Tyr	1%
6: His – Asn – Ala – Pro – Cys	1%

these, at least 5 have high heterozygosity, that is, the lesser allele has a frequency of 10% or greater (Table 1.1). Although there is considerable linkage disequilibrium between these alleles, 6 distinct amino acid haplotypes were found in 100 unrelated CEPH parents, using offspring genotypes to reconstruct haplotypes, as shown in Table 1.1 (Sohocki et al., 2001; and unpublished). These polymorphic protein haplotypes are potential modifiers, in *trans*, of the Arg677ter mutation.

5. DO THE HAPLOTYPES IN *TRANS* PLAY A ROLE?

To test this possibility, we determined the protein haplotype in *trans* in each of the 66 individuals carrying the Arg677ter mutation. Since a sequence of approximately 3 megabases on 8q13 is tracking with disease in the family, i.e., without recombination, there is effectively no variation in *cis* to the mutation (Blanton et al., 1992). (Not all Arg677ter mutations descend from a common ancestor [Bowne et al., 2002; Schwartz et al., 2003]). The haplotype in *cis* to the mutation in RP01 is haplotype 2 in Table 1.1; thus the allele in *trans* can be determined deductively from each individual's genotype.

For this analysis we evaluated fundus photographs and clinical findings for the 33 individuals with detailed records. As dependent variables we considered age-of onset, other age-based landmarks, professionally graded fundus photographs and a composite, age-adjusted severity score. The independent variables in this case were the polymorphic amino acid alleles in *trans*, tested individually, and the protein haplotypes. Analytic methods included parametric significance testing and analysis of variance, and Kaplan-Meier survival modeling, implemented using the SAS statistical package (Allen, 1995).

In summary, a few combinations of dependent variables with independent variables show significant association, but are not significant if corrected for multiple comparisons. However, one of the significant associations, based on survival analysis, shows a protective effect of haplotype 3 in females (Figure 1.2). Because of the preliminary nature of the analysis, this should be taken as simply suggestive, but it is consistent with the Loki analysis and justifies further research.

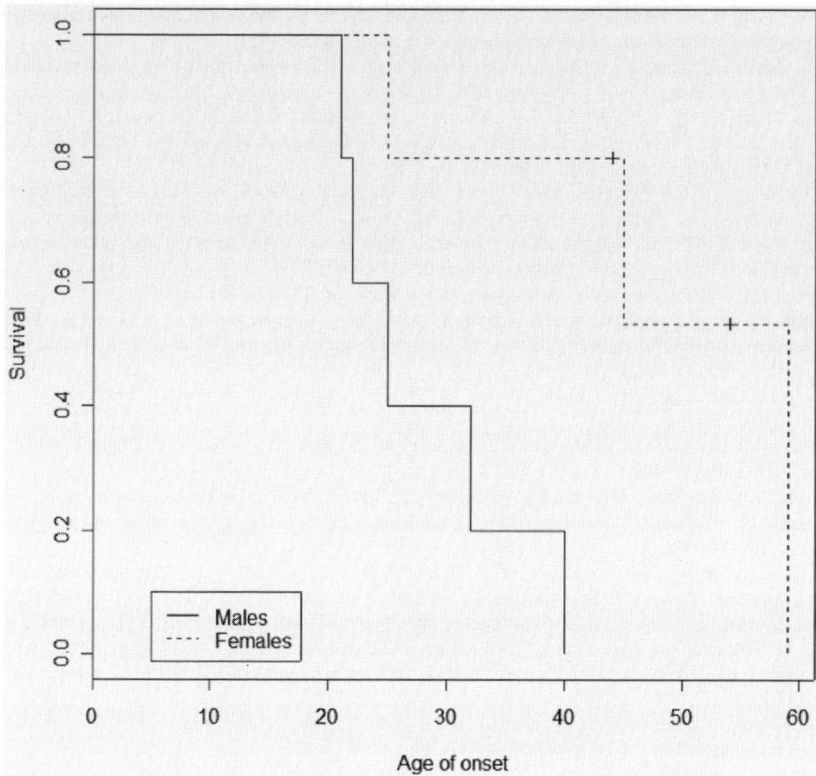

Figure 1.2. Kaplan-Meier survival analysis, RP1 protein haplotype 3 versus severity in males and females.

6. ACKNOWLEDGMENTS

Supported by grants from the Foundation Fighting Blindness and the Hermann Eye Fund, and by grants EY07142 and EY14170 from the National Eye Institute - National Institutes of Health.

7. REFERENCES

Allen, P. D., 1995, "Survival Analysis Using the SAS System: A Practical Guide", SAS Publishing, New York.

Berson, E. L., Grimsby, J. L., Adams, S. M., McGee, T. L., Sweklo, E., Pierce, E. A., Sandberg, M. A., Dryja, T. P., 2002, Clinical features and mutations in patients with dominant retinitis pigmentosa-1 (RP1), *Invest. Ophthalmol. Vis. Sci.* **42**:2217-2224.

Breeding, C. "Partial Ison Genealogy, 1650-1982". Private printing, copyright 1982, C. Breeding, Charleston, IN, USA.

Blanton, S. H., Heckenlively, J. R., Cottingham, A. W., Friedman, J., Sadler, L. A., Wagner, M., Friedman, L. H., Daiger, S. P., 1991, Linkage mapping of autosomal dominant retinitis pigmentosa (RP1) to the pericentric region of human chromosome 8, *Genomics* **11**:857-873.

Bowne, S. J., Daiger, S. P., Hims, M. W., Sohocki, M. M., Malone, K. A., McKie, A. B., Heckenlively, J. R., Birch, D. R., Inglehearn, C. F., Bhattacharya, S. S., Bird, A., Sullivan, L. S., 1999, Mutations in the RP1 gene causing autosomal dominant retinitis pigmentosa, *Hum. Mol. Genet.* **11**:2121-2128.

Daw, E. W., Heath, S. C., Wijsman, E. M., 1999, Multipoint oligogenic analysis of age-at-onset data with applications to Alzheimer disease pedigrees. *Am. J. Hum. Genet.* **64**:839-851.

Daw, E. W., Kumm, J., Snow, G. L., Thompson, E. A., Wijsman, E. M., 1999a, Monte Carlo Markov chain methods for genome screening, *Genet. Epidemiol.* **17**:133-138.

Field, L. L., Heckenlively, J. R., Sparks, R. S., Garcia, C. A., Farson, C., Zedalis, D., Sparkes, M. C., Crist, M., Tideman, S., Spence, M. A., 1982, Linkage analysis of five pedigrees affected with typical autosomal dominant retinitis pigmentosa, *J. Med. Genet.* **19**:266-270.

Gao, J., Cheon, K., Nusinowitz, S., Liu, Q., Bei, D., Atkins, K., Azimi, A., Daiger, S. P., Farber, D. B., Heckenlively, J. R., Pierce, E. A., Sullivan, L. S., Zuo, J., 2002, Progressive photoreceptor degeneration, outer segment dysplasia and rhodopsin mis-localization in mice with targeted disruption of the retinitis pigmentosa-1 (Rp1) gene, *Proc. Natl Acad. Sci. USA* **99**:5698-5703.

Gratzer, W., 1994, Human genetics. Silence speaks in spectrin. *Nat.* **372**:620-621.

Guillonneau, X., Piriev, N. I., Danciger, M., Kozak, C. A., Cideciyan, A. V., Jacobson, S. G., Farber, D. B., 1999, A nonsense mutation in a novel gene is associated with retinitis pigmentosa in a family linked to the RP1 locus, *Hum. Mol. Genet.* **8**:1541-1546.

Haider, N. B., Ikeda, A., Naggert, J. K., Nishina, P. M., 2002, Genetic modifiers of vision and hearing, *Hum. Mol. Genet.* **10**:1195-1206.

Heath, S. C., 1997, Markov chain Monte Carlo segregation and linkage analysis for oligogenic models, *Am. J. Hum. Genet.* **61**:748-760.

Heckenlively, J. R., Pearlman, J. T., Sparkes, R. S., Spence, M. A., Zedalis, D., Field, L., Sparkes, M., Crist, M., Tideman, S., 1982, Possible assignment of a dominant retinitis pigmentosa gene to chromosome 1, *Ophthalmic Res.* **14**:46-53.

Jacobson, S. G., Cideciyan, A. V., Iannaccone, A., Weleber, R. G., Fishman, G. A., Maguire, A. M., Affatigato, L. M., Bennett, J., Pierce, E. A., Danciger, M., Farber, D. B., Stone, E. M., 2000, Disease expression of RP1 mutations causing autosomal dominant retinitis pigmentosa, *Invest. Ophthalmol. Vis. Sci.* **41**:1898-1908.

Lehmer, J. M., Heckenlively, J. R., Stone, E. M., Kimura, A. E., Blanton, S. H., Daiger, S. P., 1992, Clinical characterization of chromosome 8 autosomal dominant retinitis pigmentosa (UCLA-RP01), *Invest. Ophthal. Vis. Sci.*, **33**:1396.

Liu, Q., Lyubarsky, A., Skalet, J. H., Pugh Jr, E. N., Pierce E. A., 2003, RP1 is required for the correct stacking of outer segment discs, Invest. Ophthalmol. Vis. Sci. **44**:4171-4183.

Liu, Q., Zhou, J., Daiger, S. P., Farber, D. B., Heckenlively, J. R., Smith, J. E., Sullivan, L. S., Zuo, J., Milam, A. H., Pierce, E. A., 2002, Identification and subcellular localization of the RP1 protein in human and mouse photoreceptors, *Invest. Ophthal. Vis. Sci.* **43**:22-32.

Liu, Q., Zuo, J., Pierce, E. A., 2004, The retinitis pigmentosa 1 protein is a photoreceptor microtubule-associated protein, *J. Neurosci.* **24**:6427-6436.

McGee, T. L., Devoto, M., Ott, J., Berson, E. L., Dryja, T. P., 1997, Evidence that the penetrance of mutations at the RP11 locus causing dominant retinitis pigmentosa is influenced by a gene linked to the homologous RP11 allele, *Am. J. Hum. Genet.* **61**:1059-1066.

Nadeau J., 2003, Modifier genes and protective alleles in humans and mice, *Curr. Opin. Genet. Dev.* **3**:290-295.

Pierce, E. A., Quinn, T., Meehan, T., McGee, T. L., Berson, E. L., Dryja, T. P., 1999, Mutations in a gene encoding a new oxygen-regulated photoreceptor protein cause dominant retinitis pigmentosa, *Nat. Genet.* **22**:48-254.

Randon, J., Boulanger, L., Marechal, J., Garbarz, M., Vallier, A., Ribeiro, L., Tamagnini, G., Dhermy, D., Delaunay, J., 1994, A variant of spectrin low-expression allele alpha-LELY carrying a hereditary elliptocytosis mutation in codon 28, *Brit. J. Haemat.* **88**:534-540.

Schwartz, S. B., Aleman, T. S., Cideciyan, A. V., Swaroop, A., Jacobson, S. G., Stone, E. M., 2003, De novo mutation in the RP1 gene (Arg677ter) associated with retinitis pigmentosa, *Invest. Ophthalmol. Vis. Sci.* **44**:3593-3597.

Sohocki, M. M., Daiger, S. P., Bowne, S. J., Rodriquez, J. A., Northrup, H., Heckenlively, J. R., Birch, D. G., Mintz-Hittner, H., Ruiz, R. S., Lewis, R. A., Saperstein, D. A., Sullivan, L. S. 2001, Prevalence of mutations causing retinitis pigmentosa and other inherited retinopathies, *Hum. Mutat.* **17**:42-51.

Sullivan, L. S., Heckenlively, J. R., Bowne, S. J., Zuo, J., Hide, W. A., Gal, A., Denton, M., Inglehearn, C. F., Blanton, S. H., Daiger, S. P., 1999, Mutations in a novel retina-specific gene cause autosomal dominant retinitis pigmentosa, *Nat. Genet.* **22**:248-251.

Vithana, E. N., Abu-Safieh, L., Pelosini, L., Winchester, E., Hornan, D., Bird, A. C., Hunt, D. M., Bustin, S. A., Bhattacharya, S. S., 2003, Expression of PRPF31 mRNA in patients with autosomal dominant retinitis pigmentosa: a molecular clue for incomplete penetrance?, *Invest. Ophthalmol. Vis. Sci.* **44**:4204-4109.

CHAPTER 2

DISEASE-ASSOCIATED VARIANTS OF THE ROD-DERIVED CONE VIABILITY FACTOR (RdCVF) IN LEBER CONGENITAL AMAUROSIS

Rod-derived cone viability variants in LCA

Sylvain Hanein[1], Isabelle Perrault[1], Sylvie Gerber[1], Hélène Dollfus[2], Jean-Louis Dufier[3], Josué Feingold[1], Arnold Munnich[1], Shomi Bhattacharya[4], Josseline Kaplan[1], José-Alain Sahel[5], Jean-Michel Rozet[1], and Thierry Leveillard[5]

1. INTRODUCTION

Leber congenital amaurosis (LCA) is the most early and severe form of all inherited retinal dystrophies, responsible for congenital blindness. The genetic heterogeneity of LCA has been accepted for a long time but it turned out to be largely higher than all odds. So far, 11genes have been mapped on human chromosomes and eight identified. *i*) the retinal specific guanylate cyclase gene (GUCY2D, retGC1; 17p13.1; LCA1; MIM 600179), *ii*) the gene encoding the 65-kD protein specific to the retinal pigment epithelium (RPE65; 1p31; LCA2; MIM180069), *iii*) the cone-rod homeobox-containing gene (CRX; 19q13.3; LCA7; MIM 60225), *iv*) the gene encoding the arylhydrocarbon receptor interacting protein-like 1 (AIPL1; 17p13.1; LCA; MIM 604392), *v*) the gene encoding the retinitis pigmentosa GTPase regulator-interacting protein 1 (RPGRIP1; 14q11; LCA6; MIM 605446), *vi*) the human homologue of the *drosophila melanogater* crumbs gene (CRB1; 1q31; LCA8; MIM 604210), *vii*) the gene encoding the tubby-like protein 1 (TULP1; 6q21.3; LCA10; MIM 602280), *viii*) the retinol dehydrogenase 12 (RDH12; 14q24; LCA11; MIM 608830), *ix*) LCA3 (14q24; MIM 604232), *x*) LCA5 (6q11-16; MIM 604537) and *xi*) LCA9 (1p36; MIM608553).

[1]Unité de Recherches sur les Handicaps Génétiques de l'Enfant. Hôpital Necker - Enfants Malades, 149 rue de Sèvres, 75743 Paris Cedex 15, France. Email: kaplan@necker.fr; 2 Clinique Ophtalmologique, Hopitaux Universitaires de Strasbourg, Strasbourg, France; Service d'ophtalmologie, Hôpital Necker, France; 4 Departments of Molecular Genetics and Visual Science, Institute of Ophthalmology, London, United Kingdom. Moorfields Eye Hospital, London, United Kingdom; 5 Laboratoire de physiopathologie cellulaire et moléculaire de la rétine Inserm U592, Paris, France.

Interestingly, all LCA genes hitherto identified are involved in strikingly different physiologic pathways resulting in a wide physiopathologic variety.

So far, 50% of all cases have been related to a known disease-causing gene. If in 60% of these cases, the affection develops like a congenital severe cone-rod dystrophy, in about 40% of them, the disease appear to develop as a severe yet progressive rod-cone dystrophy and may represent the extremity of a spectrum of severity of retinitis pigmentosa (RP), (Hanein et al., 2004).

On the other hand, it has been shown that factors secreted from rods are an essential requirement for cone viability (Mohand-Saïd et al., 1998). One such trophic factor has just been identified by expression cloning and named rod-derived cone viability factor (RdCVF). RdCVF is a novel protein specifically expressed by photoreceptors. By immunodepletion with specific antibodies it has been demonstrated that RdCVF is required for cone rescue in mouse retinal cultured explants. The injection of recombinant RdCVF is able to prevent 40% of cones from degeneration in the *rd1* mouse over a period of two weeks. These results provide a biochemical basis for the previously described paracrine interaction between rod and cone photoreceptors that appears to play a key role in maintaining cone cell viability.

The aim of the present study was to look for mutations in LCA patients in RdCVF, by screening the thioredoxin-like 6 gene which encode this novel protein (TXNL6; Leveillard et al., 2004).

2. MATERIAL AND METHODS

2.1. Patients Panel

A total of 200 unrelated patients were either seen at the Ophthalmo-Genetics Center of Necker–Enfants Malades Hospital or sent to the laboratory by referent ophthalmologists or geneticists from France or other countries worldwide.

Our inclusion criteria were: *i*) severe impairment of visual function detected at birth or during the first months of life with pendular nystagmus, roving eye movements, eye poking, inability to follow light or objects and normal fundus, *ii*) extinguished ERG, *iii*) exclusion of ophthalmologic or systemic diseases sharing features with LCA. Detailed clinical data were required for each patient *ie i*) age and mode of onset, *ii*) light behaviour since birth, *iii*) natural history of the visual impairment since the first months of life including the subjective impressions of parents, *iv*) refraction data, *v*) ophthalmologic findings (anterior chamber and fundus), *vi*) visual acuity (if measurable) and *vii*) electrophysiology recordings. The course of the disease was determined by interviewing the patients or their parents and a pedigree was established.

Genomic DNA was extracted from whole blood or immortalized lymphoblast cell lines of patients using standard methods. When a mutation was identified we examined the parents and other family members when available in both sporadic and familial cases.

2.2. Controls Panel

Genomic DNA obtained from 125 unrelated healthy individuals and a cohort of 55 patients affected with Stargardt (STGD) disease were used as a control panel for molecular studies.

Table 2.1. Sequences of forward and reverse primers used for the mutational screening of the TXNL6 gene.

Exon number	Forward sequence (5'-3')	Reverse sequence (5'-3')
1	GAGAGGAGCCAGTCAGCAGA	TGGATGCTTCACTTTCAGCG
2	TCAGCATCAGGGATGTGGAT	TGGAGGTTCATCAACAAACC

2.3. Mutational Screening of TXNL6

Mutational screening of TXNL6 gene was performed on genomic DNA from the patients using primers designed to flank the splice junctions of each coding exon (Table 2.1). After standard PCR amplification (conditions available on request), products were screened for mutations using denaturing high-pressure liquid chromatography (DHPLC). heteroduplex formation was induced by heat denaturation of PCR products at 94°C for 10 min, followed by gradual reannealing from 94°C to 25°C over 30 min. DHPLC analysis was performed with the WAVE DNA fragment analysis system [Transgenomic, Cheshire, UK]. PCR products were eluted at a flow rate of 0.9 ml/min with a linear acetonitrile gradient. the values of the buffer gradients (buffer A, 0.1 M triethylammoniumacetate; buffer B, 0.1 M triethylammoniumacetate/25% acetonitrile), start and end points of the gradient, and melting temperature predictions were determined by the WAVEMAKER software (Transgenomic, Cheshire, UK). Optimal run temperatures were empirically determined.

PCR fragments displaying DHPLC abnormal profiles were further sequenced using the Big Dye Terminator Cycle Sequencing Kit v3 (ABI Prism, Applied Biosystems, Foster City, USA on a 3100 automated sequencer).

2.4. Mutation Nomenclature

We have chosen to number the A of the start codon (ATG) of the TXNL6 cDNA sequence (Genbank accession number BC014127) as nucleotide 1.

2.5. Statistical Test

Comparison of the genotype and allele frequencies in patients and controls were performed by the Fisher's exact test (two sided).

3. RESULTS

3.1. Mutational Screening of the TXNL6 Gene

Sequence analysis of the 2 TXNL6 exons allowed to identify eleven variant alleles (9/11 different) in 7/200 unrelated LCA patients excluding all eight known LCA genes and in 2/56 other patients for whom the genetic screening of LCA genes is still ongoing. Single or compound heterozygosity for non-conservative amino acid substitutions was identified in seven patients (3 consanguineous) and two patients (1 consanguineous), respectively (Table 2.2). For both compound heterozygous patients, the variants were inherited from healthy parents and co-segregated with the disease.

Table 2.2. Sequence changes in patients affected with LCA.

Family	Exon	Base Change	ALLELE 1 Predicted	Exon	Base Change	ALLELE 2 Predicted Change	Control Panel
*91 F	Intrron2	c.327-9G>A	"?"		"?"	"?"	0/180
94 S	2	c.485 A>G	p.Asn162Ser	2	c.533 C>T	p.The178Ile	0/180
*103 F	2	c.533 C>T	p.The178Ile		"?"	"?"	0/180
105 S	2	c.334 G>T	p.Gly112Trp		"?"	"?"	0/180
*110 S	2	c.485 A>G	p.Asn162Ser		"?"	"?"	0/180
211 S	1	c.189 G>A	p.Glu64Lys		"?"	"?"	0/180
247 F	1	c.282 G>C	p.Met94Ile		"?"	"?"	0/180
*284 S	1	c.46 G>A	p.Asp18Asn	2	c.533 C>T	p.The178Ile	0/180
285 S	1	c.275 A>G	p.Lys92Arg		"?"	"?"	0/180

*Indicate consanguinity of the parents of LCA patients. S: sporadic case, F: familial case. A of the start codon (ATG) of the cDNA sequences of TXNL6 (Genbank accession numbers BC014127) as nucleotide 1.

Table 2.3. Sequence changes in 125 control individuals and 55 STGD patients.

Exon	Base change	Predicted change	Frequency
2	c.461 A>G	p.Glu154Val	33%
1	c.83 G>C	p.Arg28Pro	30%
1	c.93 G>C	p.Glu31Asp	2%
Intron 1	c.326+7A>C	"?"	2%
1	c.108 G>A	p.Leu91Leu	1%
2	c.457 G>A	p.Gln153Lys	1%

A of the start codon (ATG) of the TXNL6 cDNA sequence (Genbank accession number BC014127) as nucleotide 1.

All 11 variant alleles resulted from non-conservative amino acid substitutions. None of them was found either in 125 unaffected control individuals or in 55 patients affected with typical Stargardt disease (Table 2.3).

3.2. TXNL6 Variants in LCA

The proportions of TXNL6 variant alleles in LCA patients *versus* controls and STGD patients were compared by the Fisher exact test. A statistically significant difference in TXNL6 genotype frequencies was evidenced between LCA patients and control individuals: P_{LCA} = 9/200 *vs* $P_{controls}$ = 0/125, p = 0.024. This difference was even more significant when STGD patients were added to the control populations: P_{LCA} = 10/200 *vs* $P_{controls+STGD\ patients}$ = 0/180, p = 0.006. TXNL6 allele frequencies were also compared: P_{LCA} = 9/400 *vs* $P_{controls}$ = 0/250, p = 0.009 and P_{LCA} = 10/400 *vs* $P_{controls+STGD\ patients}$ = 0/360, p = 0.0008.

4. DISCUSSION

We report here the identification of compound or single heterozygosity for TXNL6 variant alleles in 2 and 7 unrelated LCA patients, respectively.

None of the 9 different variants identified in LCA patients was found in a control population of 125 healthy individuals and 55 patients affected with Stargardt disease supporting the involvement of these alterations in LCA.

The identification of compound heterozygous variants in a LCA patient born to non-consanguineous parents was consistent with a recessive inheritance. However, four patients born to consanguineous were either compound heterozygous (1/4) or single heterozygous (3/4) for TXNL6 substitutions. This observation ruled out a simple recessive transmission and supported the view that TXNL6 variants may act as modifiers of the phenotype or may be disease-causing mutations in a multiallelic mode of inheritance.

The absence of variants in control individuals and STGD patients makes the hypothesis of a modifier role of TXNL6 in LCA less likely than a possible multiallelic inheritance. From this point of view, one has to mention that a triallelism have been demonstrated in Bardet-Biedl syndrome by the identification of several patients harbouring two mutations in the BBS2 gene and one mutation in the BBS6 gene and some asymptomatic individuals carrying two BBS2 gene mutations (Katsanis et al., 2002). These data could be related to the microarray-based mutation analysis of all LCA genes (>260 mutations) in large cohorts of LCA patients showing that *i)* more than two (expected) variants in a substantial fraction of patients and that *ii)* the third allele segregated with a more severe disease phenotype in several families (Allikmets et al., 2004). Along the same lines, it is worth noting that all nine patients harbouring TXNL6 variants are affected with a severe form of congenital cone-rod dystrophy according to the description recently by Hanein et al. (2004).

In conclusion, our data suggest that TXNL6 variants may be associated to 7.5% of LCA patients in our series. Further experiments are now necessary to confirm this hypothesis by showing, for instance, that the TXNL6 variants identified in LCA patients are responsible for a significant reduction of the capacity of the RdCVF protein to maintain cone cell viability.

6. ACKNOWLEDGEMENTS

This work was supported by the Associations Retina France and Valentin Haüy. We thank the National Eye Institute for the travel award provided to SH.

7. ELECTRONIC DATA BASES

Online Mendelian inheritance in Man: http://www4.ncbi.nlm.nih.gov/OMIM/

8. REFERENCES

Hanein, S., Perrault, I., Gerber, S., Tanguy, G., Barbet, F., Ducroq, D., Calvas, P., Dollfus, H., Hamel, C., Lopponen, T., Munier, F., Santos, L., Shalev, S., Zafeiriou, D., Dufier, J.L., Munnich, A., Rozet, J.M., Kaplan, J., 2004, Leber congenital amaurosis: comprehensive survey of the genetic heterogeneity, refinement of the clinical definition, and genotype-phenotype correlations as a strategy for molecular diagnosis. Hum Mutat **23**(4):306-17.

Kaplan, J., Bonneau, D., Frezal, J., Munnich, A., Dufier, J.L., 1990, Clinical and genetic heterogeneity in retinitis pigmentosa. Hum Genet **86**:635-42.

Katsanis, N., Eichers, E.R., Ansley, S.J., Lewis, R.A., Kayserili, H., Hoskins, B.E., Scambler, P.J., Beales, P.L., Lupski, J.R., 2002. BBS4 is a minor contributor to Bardet-Biedl syndrome and may also participate in trial-lelic inheritance. Am J Hum Genet **71**(1):22-9.

Leveillard, T., Mohand-Said, S., Lorentz, O., Hicks, D., Fintz, A.C., Clerin, E., Simonutti, M., Forster, V., Cavusoglu, N., Chalmel, F., Dolle, P., Poch, O., Lambrou, G., Sahel, J.A., 2004, Identification and characterization of rod-derived cone viability factor. Nat Genet **36**(7):755-9.

Mohand-Said, S., Deudon-Combe, A., Hicks, D., Simonutti, M., Forster, V., Fintz, A.C., Leveillard, T., Dreyfus, H., Sahel, J.A., 1998, Normal retina releases a diffusible factor stimulating cone survival in the retinal degeneration mouse. Proc Natl Acad Sci U S A 7;**95**(14):8357-62.

Allikmets, R.L., Zernant, J., Perrault, I., den Hollander, A., Dharmaraj, S., F.P., Cremers, M., Kaplan, J., Koenekoop, R.K., Maumenee., I., 2004, Multiallelic inheritance and/or modifier alleles in leber congenital amaurosis: analysis with the lca disease chip, ARVO Meeting (2444/B79).

CHAPTER 3

LEBER CONGENITAL AMAUROSIS: SURVEY OF THE GENETIC HETEROGENEITY, REFINEMENT OF THE CLINICAL DEFINITION AND PHENOTYPE-GENOTYPE CORRELATIONS AS A STRATEGY FOR MOLECULAR DIAGNOSIS

Clinical and molecular survey in LCA

Sylvain Hanein, Isabelle Perrault, Sylvie Gerber, Gaëlle Tanguy, Jean-Michel Rozet, and Josseline Kaplan*

1. INTRODUCTION

Leber congenital amaurosis (LCA, MIM 204000) is the earliest and most severe form of all hereditary retinal dystrophies, responsible for congenital blindness. Its frequency, estimated until recently to 5% of all inherited retinal dystrophies[1], has been re-evaluated as some LCA cases might represent the extreme end of a spectrum of severity of retinal dystrophies.[2-4]

LCA is a genetically heterogeneous condition.[3,4] To date, 11 genes have been mapped on human chromosomes and eight identified which encode proteins involved in strikingly different physiological pathways (Table 3.1). Mutations in these genes are consistent with autosomal recessive inheritance with the exception of extremely rare autosomal dominant CRX mutations.[5]

The genetic heterogeneity of LCA that could largely increase in the coming years represents an obstacle to the molecular diagnosis in patients. Thus we performed the mutational screening of the eight hitherto identified LCA genes in a series of 195 unrelated LCA patients in search for genotype-phenotype correlations, mutational hot spots and founder effects as criteria that could help to guide genetic studies in LCA.

---

*Unité de Recherches sur les Handicaps Génétiques de l'Enfant. Hôpital Necker - Enfants Malades, 149 rue de Sèvres, 75743 Paris Cedex 15, France. Email: kaplan@necker.fr.

Table 3.1. The genetic heterogeneity of Leber congenital amaurosis.

Locus (localisation)	Gene (symbol, number of exons and MIM number)	Physiological pathways (subtissular expression)
LCA1 (17p13.1)	Retinal guanylate cyclase (GUCY2D; 20 exons; MIM600179)	transduction cascade (photoreceptors)
LCA2 (1p31)	65-kD RPE-specific protein (RPE65; 14 exons; MIM180069)	visual cycle (retinal pigment epithelium)
LCA3 (14q24)	Not identified (MIM604232)	
LCA4 (17p13.1)	Arylhydrocarbon interacting protein like 1 (AIPL1; 6 exons; MIM604392)	protein chaperoning (photoreceptors)
LCA5 (6q11-16)	Not identified (MIM604537)	
LCA6 (14q11)	Retinitis Pigmentosa GTPase regulator-interacting protein 1 (RPGRIP1; 24 exons; MIM605446)	transport along connecting cilia (photoreceptors)
LCA7 (19q13.3)	Cone-rod homeobox-containing gene; (CRX; 3 exons; MIM602225)	photoreceptor development (photoreceptors)
LCA8 (1q31)	Homologue of the drosophila crumbs gene (CRB1; 12 exons; MIM604210)	polarity of photoreceptors (retina, brain)
LCA9 (1p36)	Not identified (MIM608553)	
LCA10 (6q21.3)	Tubby-like protein 1 (TULP1; 14 exons; MIM602280)	transport of rhodopsin (photoreceptors)
LCA11 (14q24)	Retinol dehydrogenase 12 (RDH12; 8 exons; MIM608830)	visual cycle (photoreceptors)

2. MATERIAL AND METHODS

The minimal criteria for inclusion of patients (n = 195) were a severe impairment of visual function since birth or the first months of life, a normal fundus and an unrecordable ERG.

Among the families included in this study, 141/195 were simplex (29 consanguineous) and 54/195 were multiplex (26 consanguineous). Most of them hailed from Europe and North Africa.[3]

Prior to the mutational screening, the linkage status of familial cases and sporadic consanguineous patients was determined using a set of highly polymorphic markers specific to all known LCA loci. LCA genes were subsequently screened using combination of denaturing high-pressure liquid chromatography (DHPLC) and direct sequencing as described elsewhere.[3,4]

3. RESULTS

3.1. Spectrum of LCA Genes Mutations

In most cases, the genetic studies were not informative. Hints of linkage pointed to a LCA locus in only 13/52 familial cases and 6/27 consanguineous sporadic cases. One hundred seventy-four disease alleles were identified in 93/195 unrelated patients. The frequency of mutations varied greatly from gene to another, the GUCY2D gene being by far the most frequently involved in the disease. In this gene, most mutations were clustered in exons 2, 15 and 17 and two were shown to result from a founder effect: the 2000-3000 years old c.387delC mutation identified in 6/15 North African families and the 3000 years old c.2943 delG mutation found in the 5 apparently unrelated Finnish family.[3,6]

In CRB1, most mutations reside in exons 7 and 9 and several were recurrent. Conversely, in the RPE65, RPGRIP1, RDH12, AIPL1 and TULP1 genes, the mutations were spared through out the sequences and were family-specific mutations in all cases but one (AIPL1 p.Trp278X mutation). Finally, in the CRX gene, only one heterozygous 1bp de *novo* deletion was identified in a family affected with a dominant form of LCA.[3,5]

3.2. Phenotype-Genotype Correlations

We previously reported that LCA could be divided into two types a congenital severe and stationary cone-rod dystrophy and a less severe and progressive rod-cone dystrophy.[2] The study of the eight LCA genes in an enlarged cohort of patients confirmed this subdivision of the disease and even allowed to recognize two subgroups for the first type and three subgroups for the second (Figure 3.1).

On the other hand, it is worth noting that mutations in all LCA genes were shown to account for other retinal dystrophies inherited as autosomal recessive or dominant traits (Table 3.2). In GUCY2D, genotype-phenotype correlations have been proposed: homozygosity for null alleles or missense mutations encoding non functional cyclases constantly results in LCA[7,8] while homozygosity for alleles encoding a partially functional cyclase may be responsible for early-onset severe retinitis pigmentosa[10] and finally, single heterozygosity for a missense mutation located in the dimerisation domain results in autosomal dominant cone-rod dystrophy (CORD6, MIM601777). With regard to the other LCA genes, no genotype-phenotype correlations have been reported. However, no haploidentity has ever

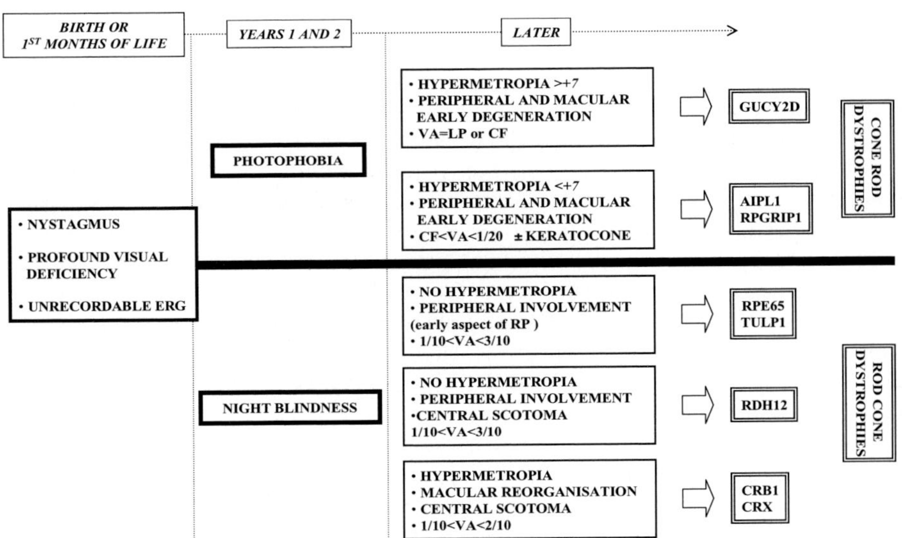

Figure 3.1. Clinical flowchart. LCA and early-onset RP are divided on the basis of light behaviour, the refraction data, the aspect of the retina, and the visual acuity (VA). CF, counting fingers; LP, light perception. This flowchart allows to direct the molecular analysis of selected LCA genes.

Table 3.2. LCA genes involved in other retinal dystrophies.

		Genotype-phenotype correlations
GUCY2D	CORD6 (MIM601777)	+++ (R838C, R838H, E837D, R838S)
	Early-onset severe RP[8]	+++ (insACCA)
RPE65	RP20 (MIM180069)	
CRX	CORD2 (MIM120970)	LCA & [RP or CRD] patients
CRB1	RP12 (MIM600105)	1/2 disease allele in common = Yes
TULP1	RP14 (MIM600132)	2/2 disease alleles in common = No
RPGRIP1	CORD9 (MIM608194)	

been evidenced between LCA and other retinal dystrophies. Furthermore, a high prevalence of null alleles has been evidenced in our series of LCA patients: (in decreasing order, CRX non included) 83.3% in RPGRIP1, 66.7% in AIPL1, 62.7% in GUCY2D, 57.9% in RPE65, 50% in RDH12, 36.1% in CRB1 and 33.3% in TULP1. Therefore, we propose that the LCA phenotype is consistently accounted for by the knocking-out of one of these genes while moderate mutations are responsible for less severe progressive retinitis pigmentosa or cone-rod dystrophy with onset ranging from infancy to adulthood.

4. DISCUSSION

To date, mutations in 11 genes encoding proteins involved in strikingly different physiologic pathways have been shown to cause LCA. The genotyping of 195 unrelated LCA patients enabled us to determine the prevalence of each genetic subtype. Mutations were identified in 93/195 patients: GUCY2D (21.2%), CRB1 (10%), RPE65 (6.1%), RPGRIP1 (4.5%), RDH12 (4.1%), AIPL1 (3.4%), TULP1 (1.7%) and CRX (0.6%).

The high prevalence of GUCY2D mutations in our series might be explained by the identification of several population-specific mutations accounting for 19/38 (50%) of patients with mutation in this gene[3,6] (c.387delC, p.Phe565Ser, 620delC and p.Ser448X in 14 North African families; c.2943delG in 5 Finish families).

Whatever the gene (CRX apart), most patients (86/93, >92.5%) were found to be homozygous (n = 50) or compound heterozygous (n = 34) for mutations. Only seven patients were single heterozygous (RPGRIP1, n = 3; RPE65, n = 3; GUCY2D, n = 1; respectively). It is likely that the second mutation lie in unscreened regions of the genes (promoter region, intragenic sequences, 3' untranslated regions).

For about 48% the patients the disease gene remained to be identified. Considering that genome wide search for homozygosity in large multiplex and consanguineous families failed to identify a major locus, it is likely that many disease genes accounting for a small proportion of patients have to be identified.

The growing number of LCA genes leads to growing difficulties in genotyping patients. However, genotyping remains essential prior to any therapeutic approach and it is thus necessary to identify criteria to direct genetic analyses. The survey of the molecular pathology in LCA enabled us to identify hot spots mutations as well as population-specific mutations that might be search in first intention. But more importantly, this study allowed identifying sound genotype-phenotype correlations as a main criterion to select genes to screen in priority. Indeed, we not only confirmed the subdivision of LCA into two main forms (types I

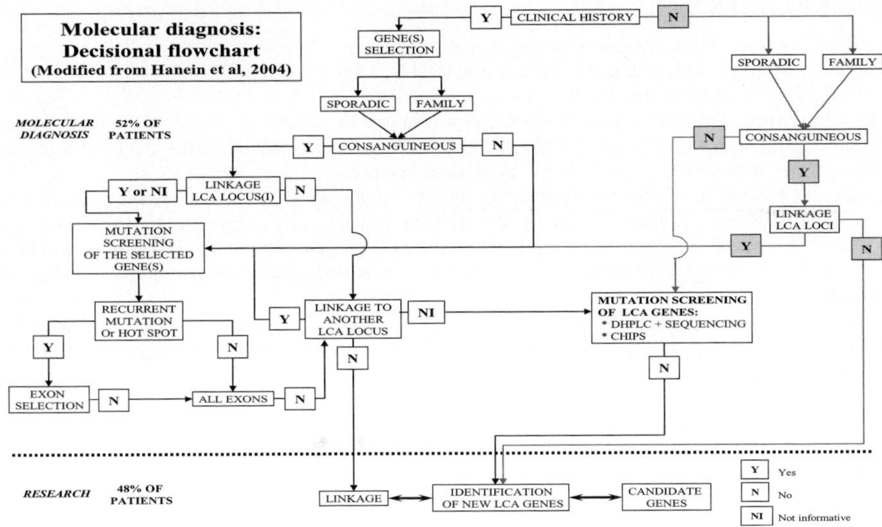

Figure 3.2. Decision-making flowchart for the molecular diagnosis of LCA.

and II) but we also subdivided each of them into distinct clinical subtypes on the basis of the progression course the disease, the refraction error, the severity of the visual deficiency, and the aspect of the retina. Each clinical subtype was specific to one or two LCA genes.

The clinical description of the congenital severe stationary cone-rod dystrophy form of the disease (LCA type I) appeared to be consistent with the traditional definition of LCA. Conversely, the boundary between LCA and early onset severe retinal dystrophy was unclear when the second subtype of the less severe and progressive rod-cone dystrophy form of the disease was considered (LCA type II, subtype II; Figure 3.1). This notion led to the idea that some LCA cases might represent the extreme end of severity in the clinical spectrum of RP.

Altogether these findings allowed us to draw a decisional flowchart to direct the genotyping of selected LCA genes (Figure 3.2). This flowchart is undoubtedly helpful to lighten the molecular diagnosis in a remarkably genetically heterogeneous condition in which linkage analyses turned out to be little useful to guide the molecular diagnosis in patients (genome identity often found by random, several homozygous loci in consanguineous cases despite informative markers).

5. ELECTRONIC DATABASE INFORMATION

OMIM: http://www.ncbi.nlm.nih.gov/entrez/query.fcgi?db=OMIM

6. ACKNOWLEDGEMENTS

This work was supported by the Associations Retina France and Valentin Haüy. We thank the National Eye Institute for the Travel Award provided to SH.

7. REFERENCES

1. Kaplan, J., Bonneau, D., Frezal, J., Munnich, A., Dufier, J.L., 1990, Clinical and genetic heterogeneity in retinitis pigmentosa. Hum Genet **86**:635-42.
2. Perrault, I., Rozet, J.M., Ghazi, I., Leowski, C., Bonnemaison, M., Gerber, S., Ducroq, D., Cabot, A., Souied, E., Dufier, J.L., Munnich, A., Kaplan, J. 1999. Different functional outcome of retGC1 and RPE65 gene mutations in Leber congenital amaurosis. Am. J. Hum. Genet **64**:1225-8.
3. Hanein, S., Perrault, I., Gerber, S., Tanguy, G., Barbet, F., Ducroq, D., Calvas, P., Dollfus, H., Hamel, C., Lopponen, T., Munier, F., Santos, L., Shalev, S., Zafeiriou, D., Dufier, J.L., Munnich, A., Rozet, J.M., Kaplan, J., 2004, Leber congenital amaurosis: comprehensive survey of the genetic heterogeneity, refinement of the clinical definition, and genotype-phenotype correlations as a strategy for molecular diagnosis. Hum Mutat **23**(4):306-17.
4. Perrault, I., Hanein, S., Gerber, S., Barbet, F., Ducroq, D., Dollfus, H., Hamel, C., Dufier, J.L., Munnich, A., Kaplan, J., Rozet, J.M., 2004, Retinal dehydrogenase 12 (RDH12) mutations in leber congenital amaurosis. Am J Hum Genet **75**(4):639-46.
5. Perrault, I., Hanein, S., Gerber, S., Barbet, F., Dufier, J-L., Munnich, A., Rozet, J-M., Kaplan, J., 2003, Evidence of autosomal dominant Leber congenital amaurosis (LCA) underlain by a CRX heterozygote null allele. J Med Genet **40**:E90.
6. Hanein, S., Perrault, I., Olsen, P., Lopponen, T., Hietala, M., Gerber, S., Jeanpierre, M., Barbet, F., Ducroq, D., Hakiki, S., Munnich, A., Rozet, J.M., Kaplan, J., 2002, Evidence of a founder effect for the RETGC1 (GUCY2D) 2943DelG mutation in Leber congenital amaurosis pedigrees of Finnish origin. Hum Mutat **20**:322-3.
7. Rozet J.M., Perrault I., Gerber S., Hanein S., Barbet F., Ducroq D., Souied E., Munnich A., Kaplan J. Complete abolition of the retinal-specific guanylyl cyclase (retGC-1) catalytic ability consistently leads to leber congenital amaurosis (LCA). Invest Ophthalmol Vis Sci. 2001 May;**42**(6):1190-2.
8. Perrault, I., Hanein, S., Gerber, S., Lebail, B., Vlajnik, P., Barbet, F., Dufier, J-L., Munnich, A., Kaplan, J. and Rozet, J-M., 2005, A novel mutation in the GUCY2D gene responsible for an early onset severe RP different from the usual GUCY2D-LCA phenotype. Hum Mutat *in press*.

CHAPTER 4

A FIRST LOCUS FOR ISOLATED AUTOSOMAL RECESSIVE OPTIC ATROPHY (ROA1) MAPS TO CHROMOSOME 8q21-q22

Fabienne Barbet[1], Sylvie Gerber[1], Sélim Hakiki[2], Isabelle Perrault[1], Sylvain Hanein[1], Dominique Ducroq[1], Gaëlle Tanguy[1], Jean-Louis Dufier[2], Arnold Munnich[1], Josseline Kaplan[1], and Jean-Michel Rozet[1]

1. INTRODUCTION

Genetically determined optic atrophies (OA) affect the retinal ganglion cells, the retinal fibre layer or the intra-ocular portion of the optic nerve. Autosomal dominant optic atrophies (DOA) are the most common form of hereditary optic neuropathy (prevalence 1 : 50,000). The genetic heterogeneity of DOA has been demonstrated. Three loci have been reported: OPA1 [3q28-q29; MIM 165500], OPA4 [18q12.2-q12.3; MIM 605293, and OPA5 (22q12.1-q13.1).[1] OPA1 which accounts for about 90% of DOA is due to mutations in the Msp1 protein [MIM 605290].

In sharp contrast to DOA, in which the optic atrophy is usually an isolated event, the recessive optic atrophies (ROA) are often multisystemic diseases involving the central nervous system and other organs. Only three loci (2/3 genes identified) of syndromic ROA have been reported: OPA3 (19q13.2-q13.3; 3-methyl-glutaconicaciduria type III; MIM606580), WFS1 (4p16.1; Wolfram syndrome; MIM 606201) and WFS2 (4q22-q24; MIM 604928).

Compared to syndromic ROA, isolated autosomal recessive optic atrophies are uncommon (MIM 258500).[2,3] These cases have been described as congenital or early infantile total OA, *i.e.* with alteration of both macular and peripheral bundles of the optic nerve resulting in the perturbation of both central and peripheral areas of the visual field before the age of 3. The visual impairment is severe: a nystagmus and a severe dyschromatopsia close to achromatopsia are usually noted. A profound alteration in visual acuity is almost always

---

[1]Unité de Recherches sur les Handicaps Génétiques de l'Enfant. Hôpital Necker-Enfants Malades, 149 rue de Sèvres, 75743 Paris cedex 15, France. E mail: kaplan@necker.fr, 2 Service d'Ophtalmologie, Hôpital Necker-Enfants Malades, Paris, France.

reported. Since their description, these observations have been largely dismissed or over-looked.[2,3] In 1992, Moller opened a debate about the existence of these isolated recessively inherited simple optic atrophy and concluded that: "a very clear-cut well documented pedigree has yet to be published".[1]

Here, we report a large multiplex consanguineous family segregating a true isolated autosomal recessive optic atrophies and the mapping of the disease-causing gene on chromosome 8q21-q22 at the ROA1 locus.

2. MATERIALS AND METHODS

A large multiplex consanguineous family of French origin affected with isolated autosomal ROA was ascertained through the Ophthalmologic Consultation of the "Hôpital des Enfants Malades" of Paris (Figure 4.3). All family members underwent general and ophthalmologic examinations and the mitochondrial DNA was analysed to exclude the diagnosis of Leber Optic Neuropathy (LHON [MIM 535500]).

A genome-wide search for homozygosity was undertaken with 382 pairs of fluorescent oligonucleotides of the Genescan Linkage Mapping Set, Version II (Perkin Elmer Cetus) under conditions recommended by the manufacturer. The polymorphic markers have an average spacing of 10 cM. Amplified fragments were electrophoresed and analysed on an automatic sequencer (ABI 3100, Applied Biosystems, Foster City, USA). Linkage analyses were performed using M-LINK and LINKMAP of the 5.1 version of the Linkage program.[5,6] (gene frequency ~1/1000, penetrance = 1; allele frequencies available from the CEPH database).

All candidate genes at the ROA1 locus were screened by direct sequencing.

3. RESULTS

3.1. Clinical Evaluation of Isolated ROA1 Family

Four out of the five sibs of the ROA1 family (Figure 4.3) were affected with an isolated, early-onset but slowly progressive optic neuropathy. Between 2 and 6 years of age, all patients complained of a visual impairment that could not be corrected by glasses. At age 10, fundus examination, fluorescein angiography, visual field testing, colour vision analysis and electrophysiological recordings of individual V2, led to the diagnosis of optic atrophy without retinal degeneration. The three younger sibs displayed the same phenotype (Figures 4.1 and 4.2). In all affected sibs, the progression of the disease was very slow (visual acuity ranging from 1/10 to 2/10 for distant vision at 36 < age of patients < 45 years old). Moderate photophobia and dyschromatopsia with red-green confusion was noted. None of the four patients had nystagmus. Both parents (IV1 and IV2), and V1 underwent complete ophthalmologic exploration and no symptom of optic neuropathy was noted. A normal ocular pressure was measured for all members of the family, as well as a normal cup/disc ratio of the optic disk. Neurological examination failed to reveal any developmental failure, pyramidal or extra pyramidal signs, ataxia or hearing loss. Furthermore, no alopecia, diabetes or any malformation was noted, allowing the exclusion of all known syndromes associated with ROA.

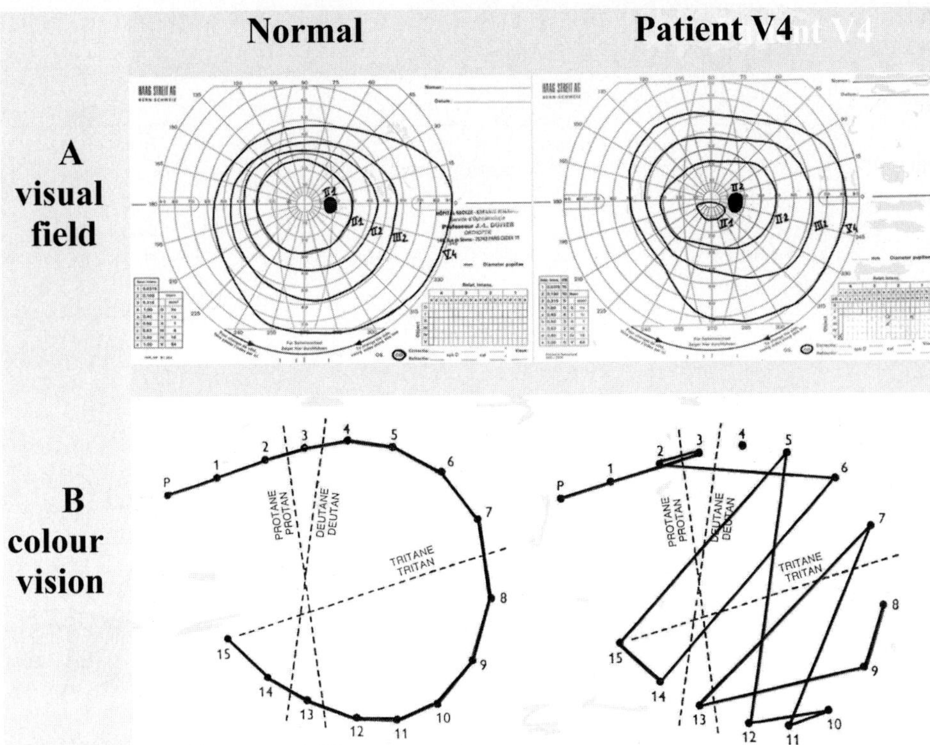

Figure 4.1. Visual field and colour vision testings in normal control and patient V4 (age = 32 years). A) Dynamic perimetry using a Goldman apparatus to test the visual field. The visual field of the patient is almost normal, except for the existence of a relative and partial scotoma at II1. B) Colour vision was tested using the D-15 Farnsworth's panel. The circling line on the left indicates that the normal individual was able to harmoniously classify the D colour chips 15 while the patient made a disorganized classification with confusion in the red-green axis.

The four patients were born to healthy second-cousins (Figure 4.3). This consanguinity associated with *i)* the existence of a common affected ancestor, *ii)* the absence of any affected individual in the common branch, ruling out autosomal dominant inheritance with incomplete penetrance, *iii)* the transmission of the disease gene through four healthy men (II2, III2, III3, and IV3), ruling out X-linked inheritance and *iv)* the affected male-female ratio of 1, strongly support the conclusion of autosomal recessive inheritance.

3.2. Primary Mapping of the Disease-Causing Gene

The 10-cM genome-wide search in the ROA1 multiplex family showed only one chromosomal region in which informative markers were found to be homozygous in all four patients. Indeed, all patients shared a common homozygote haplotype on chromosome 8q21-q22. One recombination event detected in V2 as well as a loss of homozygosity detected in all patients allowed to define a critical homozygote region between D8S1702 and D8S1794

A

B

C

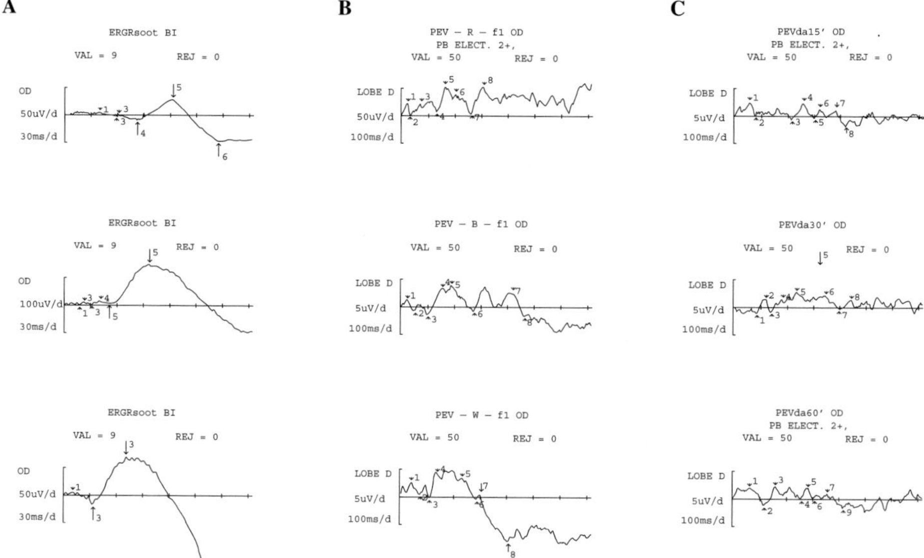

Figure 4.2. Electroretinogram (ERG) and visual evoked response (PEV) in patient V4 (age = 32 years).
A) ERG recordings showing a normal retinal response. B) Flash visually evoked response are detectable but display atypical morphology for white, red and blue light-components. C) The responses of VER derived from alternating checkerboard are badly detectable. Note the reduced amplitude measurements for the 15′, 30′ and 60′ check size, respectively.

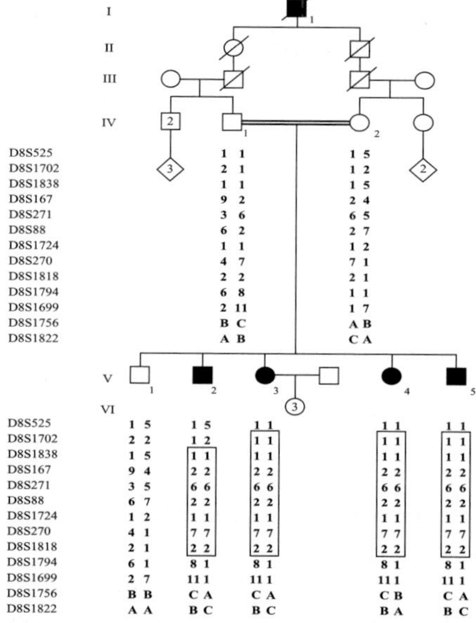

Figure 4.3. ROA1 pedigree and haplotype at the 8q21-q22 region. The homozygote haplotype is squared. A recombination event in V2 and loss of homozygosity in all patients defined a critical homozygous region between D8S1702 and D8S1794. Between D8S1818 and D8S1756 all patients were haploidentical.

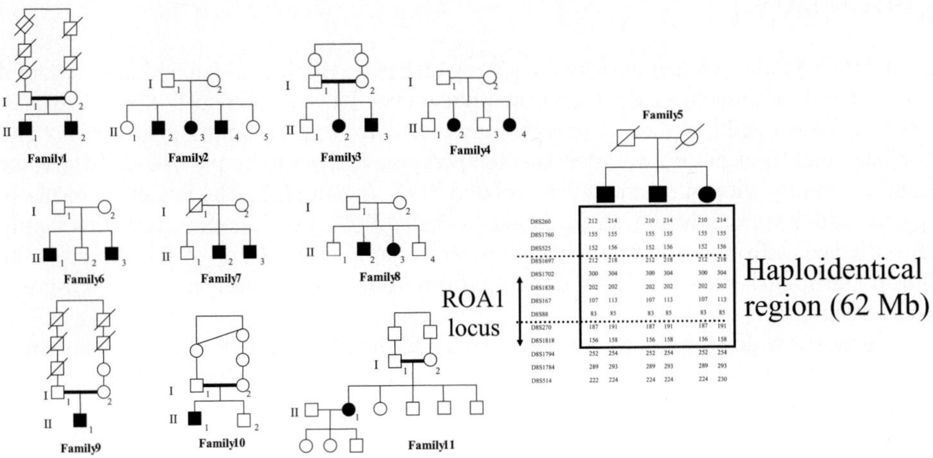

Figure 4.4. Pedigree of eleven families with an isolated ROA and haplotype of family 5 at the 8q region. The three patients are hapoidentical in a 62 Mb region containing the ROA1 locus.

(Zmax = 3.41 at θ = 0 at the D8S270 locus; Figure 4.3). Moreover two markers were haploidentical for the four affected sibs allowing to define a 2,5 Mb haploidenticial region between the D8S1818 and D8S1756 loci.

3.3. Genetic Heterogeneity of ROA

Among the 11 other families of isolated ROA available in the laboratory, only one may be linked to the ROA1 locus (3/3 affected children haploidentical in a 62 Mb region including the ROA1 locus; Figure 4.4).

3.4. Candidate Genes Study at the ROA1 Locus

The [D8S1702-D8S1794] 12 Mb region contained 25 known genes and >100 predicted genes (UCSC Human Genome Project Working Draft database, April 2003 release). The study of all known and of several hypothetical genes of the homozygous region failed to identify the disease-causing mutation. These genes encode: *i)* the β subunit of the cone cGMP gated channel, responsible for achromatopsia, *ii)* the mitochondrial 2,4-dienoyl CoA reductase 1 and protein phosphatase 2C, *iii)* the potentially mitochondrial (MITOPROT and PSORT) carbonic anhydrase III, 38 kDa lysosomal ATPase H+ transporting, V0 subunit d isoform 2, copine III, EF hand calcium binding protein 1, core-binding factor, runt domain, alpha subunit 2, CGI-90, FL35802; AL136588; AL834364; CBFA2T1 and LOC137392, *iv)* the E2F transcription factor 5, p130-binding protein, carbonic anhydrases II and XIII, protein serine kinase H2, solute carrier family 7 member 13; WW domain containing E3 ubiquitin protein ligase 1, matrix metalloproteinase 16, receptor-interacting serine-threonine kinase 2, Nijmegen breakage syndrome 1, calbindin 1, solute carrier family 26 member 7 and CGI-77 encoding genes and *v)* the KIAA1764; FLJ35775; C8ORF1 and DKFZp7620076 hypothetical proteins.

4. DISCUSSION

In 1992, Moller pointed out that the survey on causes of blindness in children reported by Fraser and Friedmann in 1967[7] and the review over 18 years of registration of visually impaired Danish children did not mention any recessive form of simple optic atrophy and concluded that "a clear-cut, well-documented pedigree had yet to be published".[4] Here, we describe a family with an unambiguous isolated ROA. In contrast to the few cases formerly reported with a very early and severe form of isolated ROA, the four patients of this family were affected with an early-onset but moderately progressive form of the disease. The clinical phenotype was strikingly different from AOD (no central scotoma, red-green dyschromatopsia).

A genome-wide search for homozygosity led to the identification of a single homozygous region on chromosome 8q21-q22 (ROA1 locus). This result excluded linkage and thus allelism of ROA1 with all known syndromic ROA (see OMIM).

Among the genes of the [D8S1702-D8S1794] 12 Mb physical interval, CNGB3 was first considered. Indeed, although the phenotype of all patients was strikingly different from that described in total colour blindness, the report of severe dyschromatopsia close to achromatopsia in the few known cases of isolated ROA[2,3] prompted us to search for mutations in this gene but no mutation was found.

Subsequently, we screened all genes encoding mitochondrial proteins owing to the view that known hereditary optic neuropathies genes are either mitochondrial (LHON) or encode proteins with high mitochondrial targeting (OPA1, OPA3, TIMM8A [MIM 300356]). Thus we first decided to screen genes known to encode mitochondrial protein such as DECR1 and PDP and proteins potentially targeted to the mitochondria, using the prediction programs MITOPROT and PSORT. The exclusion of these candidate genes prompted us to screen all known genes lying within the interval and subsequently several hypothetical genes. Since no mutation was found, we decided to screen several genes lying within the haploidenticial region defined by the D8S1794 and D8S1699 loci following the hypothesis that, in this family, the common ancestor might be affected with ROA and could carry two different mutations segregating through each branch of the family. No mutation was found and further studies are now required *i)* to identify the causative gene in this family and *ii)* to delineate the role of this locus in small non-consanguineous families affected with ROA in which linkage analysis can not be conclusive.

Finally, one can expect that the identification of the ROA1 gene and other genes involved in ROA will allow the dissection of molecular mechanisms underlying optic neuropathies to aid genetic counselling and to develop rational therapeutic tools.

5. ACKNOWLEDGEMENTS

This work was supported by the Associations Retina France and Valentin Haüy. We thank the National Eye Institute for the Travel Award provided to FB.

6. ELECTRONIC DATABASE INFORMATION

MITOPROT, Prediction of mitochondrial targeting sequences; http://ihg.gsf.de/ihg/ mitoprot.html OMIM, Online Mendelian Inheritance in Man; http://www.ncbi.nlm.nih.gov/

entrez/query.fcgi?db=OMIMPSORT, Prediction of Protein Sorting Signals and Localization Sites in Amino Acid Sequences; http://psort.nibb.ac.jp/; UCSC, Human Genome Project Working Draft, http://genome.ucsc.edu

7. REFERENCES

1. F. Barbet, S. Hakiki, C. Orssaud, S. Gerber, I. Perrault, S. Hanein, D. Ducroq, J-L. Dufier, A. Munnich, J. Kaplan., and J-M. Rozet, A third locus for dominant optic atrophy on chromosome 22q, *J Med Genet.* **42**:e001 (2005)
2. J. François, Affections du nerf optique; in Masson (eds), *L'hérédité en ophtalmologie.* Paris, 581-606 (1958)
3. S. Merin, Inherited diseases of the optic nerve, Dekker M (eds): *Inherited eye diseases, Diagnosis and clinical management.* New York, 323-344 (1991)
4. H. U. Moller, Recessively inherited, simple optic atrophy–does it exist? (Letter), *Ophthalmic Paedia Genet.* **13**:31-32 (1992)
5. G. M. Lathrop, and J. M. Lalouel, Easy calculations of lod scores and genetic risks on small computers, *Am J Hum Genet.* **36**:460-5 (1984)
6. E. S. Lander, and D. Botstein, Homozygosity mapping: a way to map human recessive traits with the DNA of inbred children, *Science.* **236**:1567-70 (1987)
7. G. R. Fraser., and A. I. Friedmann, The Causes of Blindness in Childhood. A Study of 776 Children with Severe Visual Handicaps, *Johns Hopkins Press* 1 (1967)

CHAPTER 5

RCC1-LIKE DOMAIN AND ORF15: ESSENTIALS IN RPGR GENE

Zi-Bing Jin[1], Mutsuko Hayakawa[2], Akira Murakami[2], and Nobuhisa Nao-i[1]*

1. INTRODUCTION

Retinitis Pigmentosa (RP) is a group of disease with progressive degeneration of photoreceptor. Patients present with night blindness, progressive loss of peripheral vision, and pigmentary alterations in the retina with the appearance of "bone-spicules". X-Linked Retinitis Pigmentosa (XLRP) accounts for about 15-30% of all RP cases,[1-4] and produces severe symptoms, with early onset and rapid deterioration. Five XLRP-related loci, RP2, RP3, RP6, RP23 and RP24, were mapped on the X chromosome through linkage analyses. RP3 and RP2 account for 56-90% and 10-20% of XLRP, respectively.[5-9] The two major loci, RP2 and RPGR (RP3) genes, have been cloned successfully.[10-13] Retinitis Pigmentosa GTPase Regulator (RPGR), locus on the X chromosome (Xp21.1), accounted for about 30-60% of XLRP cases in molecular screening studies.[13,14] Positional cloning of the RPGR gene originally revealed a 2784-nucleotide (nt) ubiquitously expressed transcript which include 19 exons coding for 815 amino acids. Exon ORF15 which was identified later, encodes 567 amino acids, and was shown to be a mutation hot spot. Disease-causing mutations reported so far are localized in 5-prime exons and ORF15, while none have been reported in exons 16-19. The N-terminal sequence of RPGR spanning exons 1-11 shows homology to the regulator of chromatin condensation (RCC1),[10] a nuclear protein that catalyzes guanine nucleotide exchange for the small GTPase Ran and regulates nuclear import and export.[15] On the basis of RCC1 crystal structure, the RPGR protein is predicted to have a seven-blade beta-propeller structure and to function as a GEF for a small GTP-binding protein. The mutations exclusively affect the RCC1-like domain (RLD) of RPGR and ORF15, which suggests that these regions are contribute to the physiological role of RPGR in the retina. Although a growing number of novel mutations of RPGR are being reported, its function in vivo is still unclear. Through a yeast two-hybrid screens, two proteins, the delta subunit

---

[1] Department of Ophthalmology, Miyazaki Medical College, University of Miyazaki, Japan; [2] Department of Ophthalmology, Juntendo University School of Medicine, Japan. *Nobuhisa.Nao-i. Author, Miyazaki Medical College, Kihara 5200, Kiyotake, Miyazaki, 889-1692 Japan.

of rod cyclic GMP phosphodiesterase (PDEδ) and RPGR interaction protein (RPGRIP1), were identified to interact with the RLD of RPGR.[16-18]

2. MUTATION SPECTRUM

The RPGR gene was previously identified in the RP3 region of Xp21.1. Mutational screening studies showed it was mutated in only 10-20% of patients with XLRP, however, this differed with the results from linkage analysis which suggested a 56-90% frequency in RP3.[19-23] Because of the possibility of undiscovered exons in RPGR, Veroort et al sequenced a 172kb region containing the entire gene and disclosed a new 3′terminal exon that was mutated in 60% of XLRP patients examined.[13] This exon, named ORF15, encodes 567 amino acids, with a repetitive domain rich in glutamic acid residues. Many mutations in the RPGR gene have been found through further mutational screening studies. To date (through Medline searching and conventional information), at least 300 patients had been identified with RPGR-ORF15 mutations, and at least 185 mutations were identified; 88 mutations in RPGR, and 97 in ORF15 (data not shown). Over 90% of the patients identified with mutations were from North America and Europe. Only 3% were from other continents; 6 were Asian, 1 South American and 1 South African. The mutations include nonsense and frameshift mutations, missense and inframe deletions, splice mutations, and deletions reported previously Frameshift mutations always lead to an aberrant amino acid sequence and early termination. Figure 5.1 shows the previously reported frameshift, nonsense, missense mutations and inframe deletions on RPGR-ORF15. No mutations have been reported in exons 15a, 15b1, 15b2 and 16-19 so far. It was suspected that these conserved regions are necessary for normal function in the retina, and targeted mutagenesis or transgenic experiments may help to verify this.

2.1 Mutation and Function

Several mutation-screening studies have revealed a high rate of mutations in ORF15 and exons 1-15 of RPGR. Phenotype-genotype relations were analyzed between the patients with mutations in the RP2 and RPGR genes,[24] but it would be interesting to find if a phenotypic difference presents in mutations in RLD and exon ORF15. In vivo studies to find interacting proteins in the retina cDNA library revealed two proteins, the delta subunit of rod cyclic GMP phosphodiesterase (PDEδ) and RPGR interaction protein (RPGRIP1),

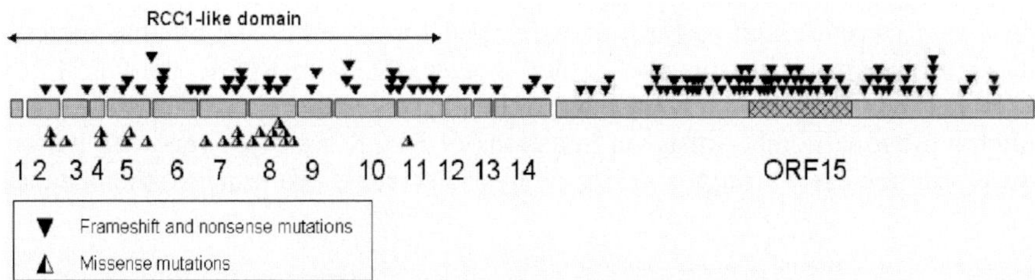

Figure 5.1. Schematic diagram of the RPGR gene containing exon ORF15 indicating the sites of mutation in XLRP to date. The exons are numbered; the sparse indicate the purine-rich domain of ORF15.

interacting with RPGR. Furthermore, using RCC1-like domain of RPGR as bait, two-hybrid screening suggested that the RLD is a key region for this interactions. A RPGR-deficient mouse model for XLRP was created by gene knockout with homologous recombination.[25] In that study, exons 6-8 were knocked out, and the authors did not detect any RPGR expression in variant organs of the knockout mice through RT-PCR or Northern blot. It indicated that transcripts lacking of exons 6-8 within RCC1-like domain was insufficient for the normal function of RPGR in the retina.

3. MACRODELETION OF RCC1-LIKE DOMAIN IN XLRP

So far, six deletions larger than one kilobase within RPGR gene were reported[10,11,13,26-28] (Table 5.1). Before the cloning of the RPGR gene, Roepman et al screened 30 unrelated patients with XLRP and identified one deletion spinning exons 15, ORF15 and exon 15a. Meindl et al identified another deletion from IVS13 to IVS15 which the exon 14-15 were absent. Other studies confirmed two deletions within RCC1-like domain, exons 8-10 and part of exon 8. Here, we describe a XLRP family with novel macrodeletion of the RPGR gene spanning from exon 1 to exon 11. Two affected males had severe symptoms with early onset and rapid deterioration. In the beginning, exons 1-11 could not be amplified by PCR while other exons were amplified successfully. To determine the suspected deletion, southern blot analysis was performed with a probe from the normal PCR products of exon 2 to exon 11, respectively. However, the result appeared normal. A Macrodeletion less than 2M base was suspected and the two sides of the suspected deletion were amplified with designed primer sets. The deletion spanning the RCC1-like domain was confirmed with direct PCR-sequencing. The deletion start-point was located 79 bp in front of exon 1, and the end-point was 42 bp downstream of exon 11 (Figure 5.2). We did not detect the pattern of the expression in the two patients, but it suggested the deletion of the RCC1-like domain could be involved in Japanese XLRP. It is suspected that transcripts lacking RLD would lose combining sites with PDE-delta or RPGRIP1, and that RLD is essential for the normal function of RPGR in the human retina.

4. SUMMARY

Clinical research into mutations of the RPGR gene showed that lack of either the RCC1-like domain of the ORF15 causes X-linked retinitis pigmentosa. Thus, the ORF15 and

Table 5.1. Deletions in RPGR Gene.

Deletions	Localization breakpoints	Reference
g.EX8+57_IVS8	EX8+58 and IVS8+1317	Vervoort et al. 2000
EX8_EX10del	IVS7 and IVS10	Buraczynska et al. 1997; Andreasson et al. 1997
EX14_EX15del	IVS13 and IVS15	Meindl et al. 1996
ORF15_EX15Adel	IVS15+1.5kb and IVS15+7.9k	Roapman et al. 1996
MO deletion	RPGR IVS10 and SRPX IVS1	Kirschner et al. 1999
SB deletion	RPGR IVS9 and telomoric of XK	Kirschner et al. 1999
EX1_EX11del	EX1-79 and EX11+42	described in present

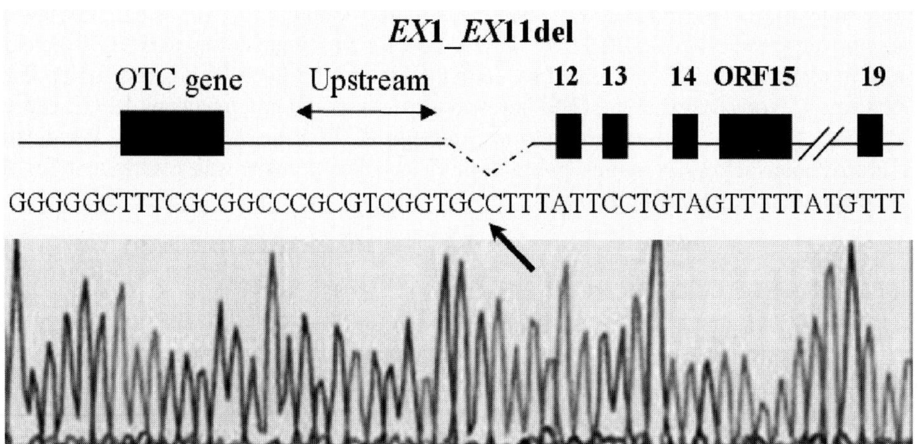

Figure 5.2. The novel deletion spans 30204 base pairs absence of exon 1 to exon 11. The dashed line indicates the deletion. The mutant allele sequence is shown at the bottom, and the arrowhead indicates the break point.

RCC1-like domain play a crucial role in the human retina. Further studies on the role of the RCC1-like domain in the visual Cascade and additional findings of related proteins in the retina or even other organs, will give us a more precise understanding of this protein.

5. REFERENCES

1. Fishman GA, 1978, Retinitis pigmentosa. Genetic percentages. Arch Ophthalmol, **96**:1185-1188.
2. Boughman JA, Conneally PM, Nance WE, 1980, Population genetic studies of retinitis pigmentosa. Am J Hum Genet, **32**:223-235.
3. Jay M, 1982, Figures and fantasies: the frequencies of the different genetic forms of retinitis pigmentosa. Birth Defects Orig Artic Ser, **18**:167-173.
4. Bunker CH, Berson EL, Bromley WC, Hayes RP, Roderick TH, 1984, Prevalence of retinitis pigmentosa in Maine. Am J Ophthalmol, **97**:357-365.
5. Musarella MA, nson-Cartwright L, Leal SM, Gilbert LD, Worton RG, Fishman GA, Ott J, 1990, Multipoint linkage analysis and heterogeneity testing in 20 X-linked retinitis pigmentosa families. Genomics, **8**:286-296.
6. Ott J, Bhattacharya S, Chen JD, Denton MJ, Donald J, Dubay C, Farrar GJ, Fishman GA, Frey D, Gal A, 1990, Localizing multiple X chromosome-linked retinitis pigmentosa loci using multilocus homogeneity tests. Proc Natl Acad Sci U S A, **87**:701-704.
7. Teague PW, Aldred MA, Jay M, Dempster M, Harrison C, Carothers AD, Hardwick LJ, Evans HJ, Strain L, Brock DJ, 1994, Heterogeneity analysis in 40 X-linked retinitis pigmentosa families. Am J Hum Genet, **55**:105-111.
8. Bergen AA, Van den Born LI, Schuurman EJ, Pinckers AJ, Van Ommen GJ, Bleekers-Wagemakers EM, Sandkuijl LA, 1995, Multipoint linkage analysis and homogeneity tests in 15 Dutch X-linked retinitis pigmentosa families. Ophthalmic Genet, **16**:63-70.
9. Fujita R, Buraczynska M, Gieser L, Wu W, Forsythe P, Abrahamson M, Jacobson SG, Sieving PA, Andreasson S, Swaroop A, 1997, Analysis of the RPGR gene in 11 pedigrees with the retinitis pigmentosa type 3 genotype: paucity of mutations in the coding region but splice defects in two families. Am J Hum Genet, **61**:571-580.
10. Meindl A, Dry K, Herrmann K, Manson F, Ciccodicola A, Edgar A, Carvalho MR, Achatz H, Hellebrand H, Lennon A, Migliaccio C, Porter K, Zrenner E, Bird A, Jay M, Lorenz B, Wittwer B, D'Urso M, Meitinger T,

Wright A, 1996, A gene (RPGR) with homology to the RCC1 guanine nucleotide exchange factor is mutated in X-linked retinitis pigmentosa (RP3). Nature Genetics, **13**:35-42.

11. Roepman R, Bauer D, Rosenberg T, van Duijnhoven G, van de Vosse E, Platzer M, Rosenthal A, Ropers H, Cremers F, Berger W, 1996, Identification of a gene disrupted by a microdeletion in a patient with X-linked retinitis pigmentosa (XLRP). Hum Mol Genet, **5**:827-833.

12. Schwahn U, Lenzner S, Dong J, Feil S, Hinzmann B, van DG, Kirschner R, Hemberger M, Bergen AA, Rosenberg T, Pinckers AJ, Fundele R, Rosenthal A, Cremers FP, Ropers HH, Berger W, 1998, Positional cloning of the gene for X-linked retinitis pigmentosa 2. Nat Genet, **19**:327-332.

13. Vervoort R, Lennon A, Bird AC, Tulloch B, Axton R, Miano MG, Meindl A, Meitinger T, Ciccodicola A, Wright AF, 2000, Mutational hot spot within a new RPGR exon in X-linked retinitis, pigmentosa. Nature Genetics, **25**:462-466.

14. Breuer DK, Yashar BM, Filippova E, Hiriyanna S, Lyons RH, Mears AJ, Asaye B, Acar C, Vervoort R, Wright AF, Musarella MA, Wheeler P, MacDonald I, Iannaccone A, Birch D, Hoffman DR, Fishman GA, Heckenlively JR, Jacobson SG, Sieving PA, Swaroop A, 2002, A Comprehensive Mutation Analysis of RP2 and RPGR in a North American Cohort of Families with X-Linked Retinitis Pigmentosa. Am J Hum Genet, **70**:1545-1554.

15. Gorlich D, Mattaj IW, 1996, Nucleocytoplasmic Transport. Science, **271**:1513-1519.

16. Linari M, Ueffing M, Manson F, Wright A, Meitinger T, Becker J, 1999, The retinitis pigmentosa GTPase regulator, RPGR, interacts with the delta subunit of rod cyclic GMP phosphodiesterase. PNAS, **96**:1315-1320.

17. Hong D-H, Yue G, Adamian M, Li T, 2001, Retinitis Pigmentosa GTPase Regulator (RPGR)-interacting Protein Is Stably Associated with the Photoreceptor Ciliary Axoneme and Anchors RPGR to the Connecting Cilium. J Biol Chem, **276**:12091-12099.

18. Boylan JP, Wright AF, 2000, Identification of a novel protein interacting with RPGR. Hum Mol Genet, **9**:2085-2093.

19. Musarella MA, nson-Cartwright L, Leal SM, Gilbert LD, Worton RG, Fishman GA, Ott J, 1990, Multipoint linkage analysis and heterogeneity testing in 20 X-linked retinitis pigmentosa families. Genomics, **8**:286-296.

20. Ott J, Bhattacharya S, Chen JD, Denton MJ, Donald J, Dubay C, Farrar GJ, Fishman GA, Frey D, Gal A, 1990, Localizing multiple X chromosome-linked retinitis pigmentosa loci using multilocus homogeneity tests. Proc Natl Acad Sci U S A, **87**:701-704.

21. Teague PW, Aldred MA, Jay M, Dempster M, Harrison C, Carothers AD, Hardwick LJ, Evans HJ, Strain L, Brock DJ, 1994, Heterogeneity analysis in 40 X-linked retinitis pigmentosa families. Am J Hum Genet, **55**:105-111.

22. Bergen AA, Van den Born LI, Schuurman EJ, Pinckers AJ, Van Ommen GJ, Bleekers-Wagemakers EM, Sandkuijl LA, 1995, Multipoint linkage analysis and homogeneity tests in 15 Dutch X-linked retinitis pigmentosa families. Ophthalmic Genet, **16**:63-70.

23. Fujita R, Buraczynska M, Gieser L, Wu W, Forsythe P, Abrahamson M, Jacobson SG, Sieving PA, Andreasson S, Swaroop A, 1997, Analysis of the RPGR gene in 11 pedigrees with the retinitis pigmentosa type 3 genotype: paucity of mutations in the coding region but splice defects in two families. Am J Hum Genet, **61**:571-580.

24. Sharon D, Sandberg MA, Rabe VW, Stillberger M, Dryja TP, Berson EL, 2003, RP2 and RPGR Mutations and Clinical Correlations in Patients with X-Linked Retinitis Pigmentosa. American Journal of Human Genetics, **73**:1131-1146.

25. Hong D-H, Pawlyk BS, Shang J, Sandberg MA, Berson EL, Li T, 2000, A retinitis pigmentosa GTPase regulator (RPGR)- deficient mouse model for X-linked retinitis pigmentosa (RP3). PNAS, **97**:3649-3654.

26. Buraczynska M, WWFRBKPEASBJBDFGHDIGJSMMSPSA, 1997, Spectrum of mutations in the RPGR gene that are identified in 20% of families with X-linked retinitis pigmentosa. Am J Hum Genet, **61**:1287-1292.

27. Andreasson S, PVAMEBWWFRBMSA, 1997, Phenotypes in three Swedish families with X-linked retinitis pigmentosa caused by different mutations in the RPGR gene. Am J Ophthalmol, **124**:95-102.

28. Kirschner R, Rosenberg T, Schultz-Heienbrok R, Lenzner S, Feil S, Roepman R, Cremers F, Ropers H, Berger W, 1999, RPGR transcription studies in mouse and human tissues reveal a retina- specific isoform that is disrupted in a patient with X-linked retinitis pigmentosa. Hum Mol Genet, **8**:1571-1578.

CHAPTER 6

CHOROIDAL NEOVASCULARIZATION IN PATIENTS WITH ADULT-ONSET FOVEOMACULAR DYSTROPHY CAUSED BY MUTATIONS IN THE *RDS/PERIPHERIN* GENE

Darius M. Moshfeghi[1], Zhenglin Yang[2], Nathan D. Faulkner[2], Goutam Karan[2], Sukanya Thirumalaichary[2], Erik Pearson[2], Yu Zhao[2], Thomas Tsai[3], and Kang Zhang[2]

1. INTRODUCTION

Adult-onset foveomacular dystrophy (AOFMD) was first described as a peculiar foveomacular dystrophy in 1974 (Gass, 1974). A mutation in the *RDS/peripherin* gene (Pro-210-Arg) was identified in this particular kindred (Gorin et al., 1994). Subsequently, Feist and coworkers reported a case of choroidal neovascularization associated with AOFMD in a patient with the Pro-210-Arg mutation (Feist et al., 1994). To our knowledge, CNV in AOFMD is rare as demonstrated by only two other descriptions of it in the literature: 1) Vine and Schatz described three instances in two patients, neither of whom had an identified mutation (Vine et al., 1980); and 2) Battaglia Parodi and coworkers described a case of subfoveal CNV in a vascularized pigment epithelial detachment in a patient with AOFMD (Battaglia et al., 2000). Recently, an A to G change, predicting a Tyr-141-Cys substitution in the RDS/peripherin gene has been described that results in AOFMD which is dominantly transmitted (Yang et al., 2004). In addition, a frameshift mutation in exon 1 of the RDS/peripherin gene, that results in a guanine deletion at nucleotide position 112, leads to a premature termination of the gene product at amino acid 38 and has been implicated in the genesis of AOFMD in a 13-family member kindred (Yang et al., 2003). We present three cases of subfoveal CNV in patients with AOFMD, which were caused by RDS/peripherin gene mutations.

---

[1]Department of Ophthalmology, Stanford University Head of Ophthalmic Oncology, [2]Moran Eye Center, Department of Ophthalmology and Visual Science, and Program in Human Molecular Biology & Genetics, Eccles Institute of Human Genetics, University of Utah, Salt Lake City, UT, [3]Ohio Retinal Associates, Parma, OH.

2. METHODS

Patients with AOFMD and CNV underwent ophthalmoscopic examination and fluorescein angiography. They also donated peripheral venous blood and underwent mutational screening to detect RDS/peripherin gene mutations using standard techniques (Zhang et al., 2001).

3. RESULTS

3.1. Case 1

A 68-year-old Caucasian female presented to an outside ophthalmologist with a complaint of diminished visual acuity in her right eye. Her best-corrected visual acuity was 20/40 OD and 20/20⁻ OS. Slit-lamp biomicroscopy demonstrated mild pseudophakic bullous keratopathy changes and posterior chamber intraocular lenses in each eye. She had a central distortion on Amsler grid testing in her right eye. Dilated fundus examination revealed a subfoveal choroidal neovascularization (CNV) with surrounding hemorrhage and lipid in the right eye (Figure 6.1A) and geographic atrophy with retinal pigment epithelial mottling in the left eye (Figure 6.1B). Fluorescein angiography demonstrated subfoveal CNV with a predominantly classic pattern (Figure 6.1C,D). She was offered verteporfin OPT and underwent treatment at that time. On two subsequent three month follow-up exams, her visual acuity was stable at 20/40 OD, and her fluorescein angiogram demonstrated mild leakage (Figure 6.1E,F). She underwent repeat verteporfin OPT at those visits. A family history survey revealed that she belonged to a family with nine relatives diagnosed with AOFMD transmitted in an autosomal dominant pattern. DNA analyses showed that all affected individuals in this family inherited a Tyr-141-Cys mutation. She has had no evidence of recurrence of her CNV either by fundus examination or as demonstrated by leakage on fluorescein angiography (data not shown), and her vision has remained stable at 20/40 OD.

3.2. Case 2

An 80-year-old Caucasian female presented to an outside ophthalmologist with a complaint of diminished visual acuity in the right eye. Her best-corrected visual acuity was $20/60^{-2}$ OD and $20/100^{+}$ OS. Slit-lamp biomicroscopy demonstrated 1+ cortical changes and nuclear sclerosis in each eye. She had a central distortion on Amsler grid testing in her right eye. Dilated fundus examination revealed a subfoveal choroidal neovascularization with surrounding hemorrhage and lipid in the right eye (Figure 6.2A) and geographic atrophy involving the center of the macula of the left eye (Figure 6.2B). Fluorescein angiography demonstrated a minimally classic, subfoveal CNV (Figure 6.2C,D). She has been observed for over 1 year without change in her visual status.

3.3. Case 3

A 53-year-old Caucasian female (daughter of Case 2) presented with a complaint of diminished visual acuity in her right eye. She had previously undergone laser photocoagu-

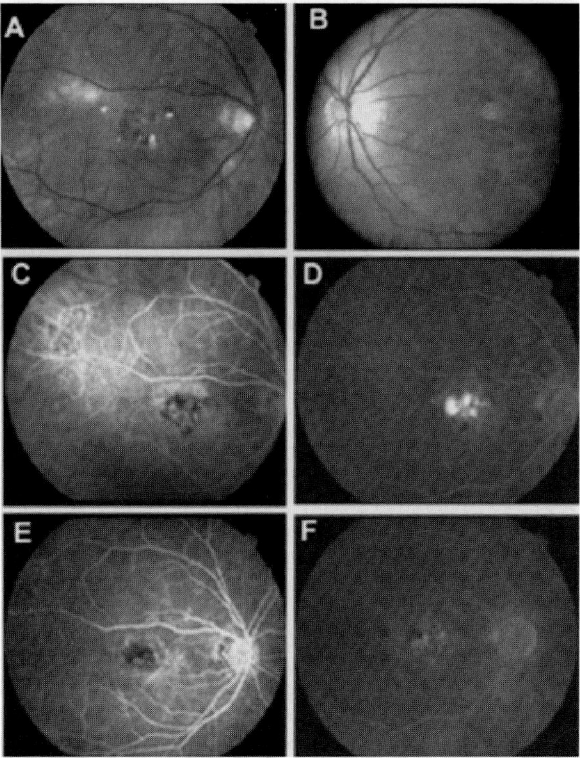

Figure 6.1. Case 1. A 68-year-old Caucasian female with adult-onset foveomacular dystrophy. Note the choroidal neovascularization in the right eye with surrounding lipid and hemorrhage (Figure 6.1A). The left fundus demonstrates a central area of retinal pigment epithelial atrophy (Figure 6.1B). Early fluorescein angiogram of the right eye demonstrates a predominantly classic, subfoveal choroidal neovascular membrane with surrounding blocking defect from hemorrhage (Figure 6.1C). On late frames, there is extensive fluorescein leakage (Figure 6.1D). Six months after PDT treatment, the right eye demonstrates early hyperfluorescence (Figure 6.1E) with mild late leakage (Figure 6.1F).

lation for a subfoveal choroidal neovascular membrane in her left eye two years earlier. Her best-corrected visual acuity was 20/30⁻ OD and 20/400 OS. Slit-lamp biomicroscopy demonstrated 1+ nuclear sclerosis in each eye. She had a central distortion on Amsler grid testing in the right eye and a large central scotoma in the left eye. Dilated fundus examination revealed a subfoveal choroidal neovascularization with surrounding hemorrhage and lipid in the right eye (Figure 6.3A) and a laser photocoagulation scar involving the center of the macula of the left eye (Figure 6.3B). Fluorescein angiography demonstrated a fibrovascular pigment epithelial detachment involving the fovea of the right eye (Figure 6.3C,D). She has been observed for 12 months without change in her visual status.

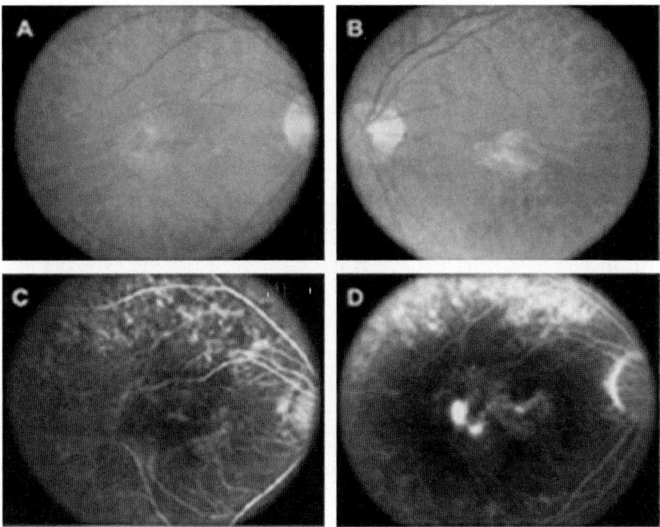

Figure 6.2. Case 2. An 80-year-old Caucasian female with adult-onset foveomacular dystrophy. In the right eye there is a choroidal neovascular membrane juxtafoveally situated with hemorrhage, elevation of the retinal pigment epithelium (RPE), and lipid exudation (Figure 6.2A). The left eye demonstrated RPE atrophy as well as subretinal yellow deposits (Figure 6.2B). Fluorescein angiogram of the right eye showed patchy early fluorescence with blocking defect corresponding to areas of hemorrhage (Figure 6.2C), as well as intense hyperflourescence and late staining (Figure 6.2D).

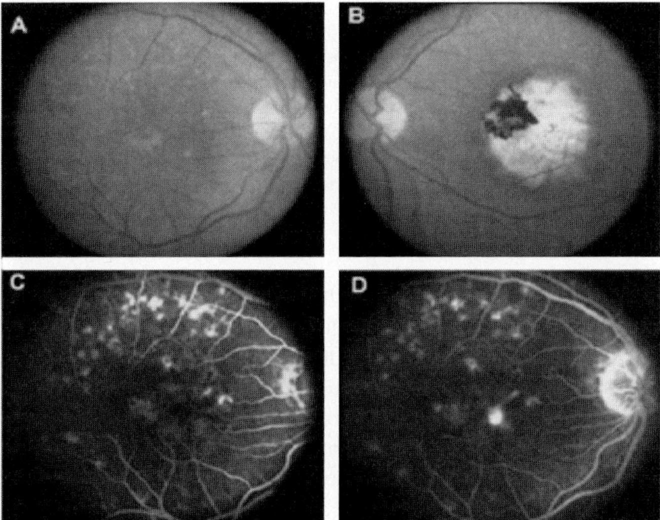

Figure 6.3. Case 3. A 53-year-old Caucasian female with adult-onset foveomacular dystrophy (daughter of case 2). Fundus examination of the right eye showed diffuse elevation of the RPE with gray coloring and a hyperpigmented spot (Figure 6.3A). Examination of the left eye demonstrated a previous laser photocoagulation scar with RPE hyperplasia, temporal RPE atrophy, and fibrosis overlying the fovea (Figure 6.3B). Fluorescein angiogram in the right eye demonstrated patchy, granular early hyperfluorescence (Figure 6.3C), that hyperfluoresced and stained intensely in late frames (Figure 6.3D).

4. DISCUSSION

This case series describes 3 patients with AOFMD, subfoveal CNV, and a mutation of the RDS/Peripherin gene. One patient had a Tyr-140-Cys substitution in the RDS/Peripherin gene and two patients (mother and daughter) had a single guanine base deletion at position 112 of the RDS/Peripherin gene. The Tyr-141-Cys mutation is seen with both butterfly-shaped pattern dystrophy and AOFMD, and is associated with moderate visual loss (Yang, 2004). The guanine base deletion at position 112 is associated with advanced macular degeneration and poor visual acuity (Payne, 2004). One patient was successfully treated with a course of verteporfin OPT, with stabilization of visual acuity in the treated eye. This represents the first documented case of a choroidal neovascular membrane secondary to a known genetic mutation treated with photodynamic therapy. The remaining two patients have been observed for over a year without precipitous decline in visual acuity.

5. CONCLUSION

Further understanding of genotype/phenotype correlations between the mutations of the RDS/peripherin gene and CNV due to AOFMD may be useful to provide prognostic information and determine which patients with AOFMD and subfoveal CNV may be candidates for treatment.

6. ACKNOWLEDGEMENTS

This research was supported by National Institutes of Health Grants R01EY14428, R01EY14448 and GCRC M01-RR00064, the Ruth and Milton Steinbach Fund, Ronald McDonald House Charities, the Macular Vision Research Foundation, the Research to Prevent Blindness, Inc., Knights Templar Eye Research Foundation, Grant Ritter Fund, American Health Assistance Foundation, the Karl Kirchgessner Foundation, Val and Edith Green Foundation, and the Simmons Foundation.

7. REFERENCES

Gass, J. D. M., 1974, A clinicopathologic study of a peculiar foveomacular dystrophy, *Trans Am Ophthalmol Soc.* **72**:139-156.

Vine, A. K., Schatz, H., 1980, Adult-onset foveomacular pigment epithelial dystrophy, *Am J Ophthalmol.* **89**:680-691.

Gorin, M. B., Jackson, K. E., Ferrell, O. D., et al., 1994, A peripherin/retinal degeneration slow mutation (Pro-210-Arg) associated with macular and peripheral retinal degeneration, *Ophthalmology* **102**:246-255.

Feist, R. M., White, M. F., Skalka, H., Stone, E. M., 1994, Choroidal neovascularization in a patient with adult foveomacular dystrophy and a mutation in the retinal degeneration slow gene (Pro 210 Arg), *Am J Ophthalmol.* **118**:259-260.

Battaglia, P. M., Di Crecchio, L., Ravalico, G., 2000, Vascularized pigment epithelial detachment in adult-onset foveomacular vitelliform dystrophy, *Eur J Ophthalmol.* **10**:266-269.

Zhang, K., et al., 2001, A 5-bp deletion in ELOVL4 is associated with two related forms of autosomal dominant macular dystrophy, *Nat Genet.* **27**(1):89-93.

Yang, Z., et al., 2003, A novel mutation in the RDS/Peripherin gene causes Adult-onset Foveomacular Dystrophy, *Am. J. Ophthalmol.* **135**:213-218.

Yang, Z., Jiang, L., Karan, G., Moshfeghi, D., O'Connor, S. Z., Lewis, H., Zack, D., Jacobsen, S., Zhang, K., 2004, A novel RDS/peripherin gene mutation associated with diverse macular phenotypes, *Ophthalmic Genet.* **25**(2):133-145.

Payne, M., et al., 2004, Dominant optic atrophy, sensorineural hearing loss, ptosis, and ophthalmoplegia: a syndrome caused by a missense mutation in OPA1, *Am. J. Ophthalmol.* **138**(5):749-755.

CHAPTER 7

BIOCHEMICAL CHARACTERISATION OF THE *C1QTNF5* GENE ASSOCIATED WITH LATE-ONSET RETINAL DEGENERATION

A genetic model of age-related macular degeneration

Xinhua Shu[1], Brian Tulloch[1], Alan Lennon[1], Caroline Hayward[1], Mary O'Connell[1], Artur V. Cideciyan[2], Samuel G. Jacobson[2], and Alan F. Wright[1,*]

1. INTRODUCTION

Age-related macular degeneration (AMD) is the commonest cause of severe vision loss in adults, affecting up to 30% of the elderly population and accounting for 50-60% of new blind registration in western countries (Green and Enger, 1993; Seddon, 2001). It is characterised by a late-onset degeneration of the retinal macula and represents the advanced stage of a more common disorder, age-related maculopathy. There are two clinical subtypes of AMD, one is a "dry" form characterised by geographic atrophy, the other a "wet" form characterised by choroidal neovascularisation (CNV). This "wet" form represents only 10% of cases but accounts for about 90% of registered blindness (Ferris et al., 1984). The important early pathological features of AMD are the presence of both focal (drusen) and diffuse extracellular (basal) deposits in the macula, between the retinal pigment epithelium (RPE) and inner collagenous layer of Bruch's membrane, a pentalaminar structure bounded by the basement membranes of RPE and choroidal capillary [1]endothelium. These deposits lead to dysfunction and later death of RPE and associated photoreceptors. The nature of the proteins within the diffuse extracellular deposits have not been elucidated but the focal deposits (drusen) include >100 proteins, together with esterified and non-esterified cholesterol and other lipids and glycosaminoglycans (Crabb et al., 2002; Malek et al., 2003). Risk factors for AMD include age, sex, family history, *APOE* genotype, smoking, ethnicity and cardio-

[1] MRC Human Genetics Unit, Western General Hospital, Edinburgh, UK; [2] Department of Ophthalmology, Scheie Eye Institute, University of Pennsylvania, Philadelphia, PA, USA. *Corresponding author: Alan Wright, E-mail: alan.wright@hgu.mrc.ac.uk.

vascular disease (Seddon, 2001). Genetic factors are implicated in AMD on the basis of twin and family studies but it appears to be a genetically complex disorder (Hammond et al., 2002).

An important means of elucidating diseases mechanism in genetically complex disorders such as AMD, is to take advantage of information from simple, often rare genetic abnormalities associated with these disorders. Late-onset retinal degeneration (L-ORD) is a rare autosomal dominant disorder characterised by onset in the fifth to sixth decade with punctate drusen-like deposits in the posterior pole of the retinal fundus, followed by macular degeneration and a diffuse chorioretinal atrophy with, in late stages, CNV and disciform scarring (Jacobson et al., 2001; Kuntz et al., 1996; Milam et al., 2000). A major and probably feature of the disease is a thick extracellular sub-RPE deposit similar to, but more extensive than that seen in AMD (Kuntz et al., 1996; Milam et al., 2000). This disorder is an excellent model for the most severe "wet" form of AMD. The causal gene in L-ORD was identified by positional cloning, which identified a Ser163Arg mutation in the *C1QTNF5* short-chain collagen gene in affected members of 7 out of 14 L-ORD families, suggesting genetic heterogeneity (Hayward et al., 2003). The *C1QTNF5* gene is strongly expressed in RPE cells and may be involved with adhesion between RPE and Bruch's membrane. The protein is predicted to contain an N-terminal secretory signal, a short helical collagen repeat and a C-terminal globular complement 1q (gC1q) domain concerned with trimerisation. The functional consequences of the Ser163Arg mutation in the gC1q domain appear to be destabilisation and aggregation of the protein as a result of an abnormal surface charge (Hayward et al., 2003). To better understand the function of C1QTNF5, we investigated the biochemical properties of its gC1q domain and found that the native protein is capable of oligomerisation into both trimeric and hexameric forms which are unstable under denaturing conditions.

2. MATERIALS AND METHODS

2.1. Preparation of C1QTNF5 gC1q Domain Constructs

C1QTNF5 gC1q domain was cloned by amplification using primers C1q N1:-5′-GTGC-CTCCGCGATCCGCCTTC - and C1q C1:-5′-AGCAAAGACTGGGGAGCTGTGCCA - using Human Retina Marathon-Ready cDNA (Clontech) as template. The amplification was carried out using Expand High Fidelity PRC System (Roche) with cycling conditions of 95°C for 30 sec, 56°C for 1 min and 72°C for 30 sec over 35 cycles. The amplified fragment was purified and ligated into pBAD/TOPO ThioFusion vector (Invitrogen). Positive clones were plasmid-purified and sequenced to determine the correct orientation of the insert. To construct the *C1QTNF5* gC1q Ser163Arg mutant, one PCR was carried using primers C1q N1 and C1qmut C1:-5′-CAGATCAAACTGCAGCCTGGCCCGGTAGACGGT.

The second PCR was carried out using primer C1qmut N1:-5′-ACCGTCTAC CGGGCCAGGCTGCAGTTTGATCTG - and C1q C1, the two reactions used Human Retina Marathon-Ready cDNA (Clontech) as templates. The two PCR products were purified, mixed and used as template in the overlapping PCR with primers C1q N1 and C1q C1. The conditions of the three separate PCR were as mentioned above. The recombinant mutant PCR fragment was purified and ligated into the pBAD/TOPO ThioFusion vector

(Invitrogen) and the plasmids from positive clones were sequenced to ensure the correct mutation was introduced and that there were no PCR errors.

2.2. Expression and Purification of the Recombinant gC1q Domain

Escherichia coli LMG194 cells harbouring the above recombinant plasmids were grown in LB medium containing 100 µg of ampicillin at 37°C with agitation at 220 rpm. The expression of the recombinant gC1q domain (wildtype and mutant) were induced at 18°C overnight by addition of 0.02% arabinose when the absorbance at 600 nm (A_{600}) reached 0.4-0.6. Cells were harvested by centrifugation and resuspended in 50 mM Tris-HCl buffer, pH8.0, containing 150.mM NaCl and 10% glycerol with proteinase inhibitors (Roche). The cells were disrupted by a French press and centrifuged at 13000 rpm for 30 min to collect the supernatant. The fusion proteins were affinity purified with Ni-NTA superflow (QIAGEN) according to the manufacturer's instructions.

2.3. Sucrose Gradient Sedimentation

150 µl of purified gC1q domain protein was applied to 4.2 ml 5-20% sucrose gradient in 50 mM Tris-HCl buffer (pH8.0) containing 150 mM NaCl. Protein standards were applied to parallel gradients: chymotrypsinogen A (25 kDa), ovalbumin (43 kDa), bovine serum albumin (67 kDa), catalase (250 kDa) and urease (272 kDa). The gradients were spun for 22 hr at 4°C in a Sorvall TH-641 rotor at 50,000 rpm and 23 fractions were sequentially collected from the bottom of the gradient and analysed by their UV absorbance at 280 nm and western blotting.

2.4. Determination of the Native Molecular Size of gC1q Domain

Purified gC1q domain and non-denatured protein markers of known molecular size (Sigma) were separated in 6, 8, 10 and 12% polyacrylamide (PAGE) gels under non-denaturing conditions. Gels were stained with Coomassie Brilliant Blue R250. The Relative Mobility (R_f) of each protein compared to the tracking dye was determined on a set of gels with various polyacrylamide concentrations; $100[\text{Log}(R_f \times 100)]$ values were plotted against the gel concentration as percentages and the negative slopes of the plots were plotted against the known molecular weights of the standards. This produced a linear plot and the molecular size of the unknown protein was determined by its negative slope.

2.5. SDS-polyacrylamide Gel Electrophoresis (PAGE) and Blotting

The samples were loaded on a 12% SDS-PAGE gel for electrophoresis. Samples were boiled for 5 min before loading as required. Gels were stained with Coomassie Brilliant Blue R250 or transferred to nitrocellulose membranes. For the blotting, membranes were blocked in 5% non-fat dried milk in PBS and primary antibody (anti-His tag, Invitrogen) was used at 1:300 dilution. Anti-mouse HRP-conjugated secondary antibody (Amersham) was used at 1:5000 dilution. Bound antibody was visualised using ECL kit (Amersham).

3. RESULTS

3.1. Expression and Purification of the C1QTNF5 gC1q Domain

cDNA fragments of both wildtype (163Ser) and mutant (163Arg) *C1QTNF5* gC1q domain expressed a major 32kDa hexa-histidine (His) tagged fusion protein when induced with arabinose (Fig. 7.1a,b) and separated by polyacrylamide gel electrophoresis (PAGE) under denaturing conditions in the presence of 0.1% SDS. This represents the monomeric form of the proteins. Wildtype gC1q and mutant expressed at similar levels *in vitro* but the mutant gC1q was less soluble (Fig. 7.1a,b). The wildtype but not mutant gC1q protein run under denaturing conditions also showed a weak band of about 64kDa, representing the dimer and indicating a strong multimeric association (Fig. 7.1a,b). However, if five-fold larger amounts of wildtype and mutant gC1q were loaded onto SDS-PAGE gels, both proteins showed a major monomeric band, and progressively weaker bands at 64kDa (dimer) and 96kDa (trimer), although the oligomeric bands were weaker in the mutant, probably because of its instability (Fig. 7.1c). These data show that both the 163Ser wildtype and 163Arg mutant gC1q domains are capable of oligomerisation, even under denaturing conditions.

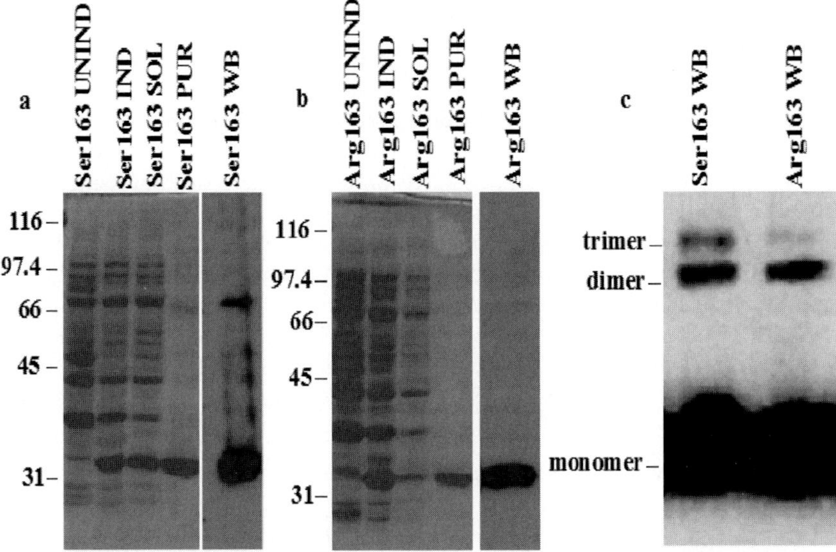

Figure 7.1. Expression and purification of CTRP5 gC1q domain. Expression of wildtype (Ser163) (**a**) and mutant (Arg163) (**b**) gC1q domain fusion protein induced (IND) or uninduced (UNIND) with arabinose. The soluble fractions (SOL) and purified (PUR) extract, obtained with a nickel affinity resin are shown. The proteins were also detected with anti-hexa-Histidine epitope tag antibody by western blot (WB). In (**c**), monomer, dimer and trimer of the fusion protein are detectable when more protein is loaded (5-fold the amounts loaded in a,b). The position of molecular weight (M_r) markers are shown on the left in kiloDaltons.

3.2. Assembly of the C1QTNF5 gC1q Domain

The C1QTNF5 gC1q domain shows significant homology to several other members of the gC1q/TNF superfamily, such as the complement C1q component, ACRP30, Emilin, and type VIII and X short-chain collagens. To further investigate whether the gC1q domain of C1QTNF5 can form homo-multimers, similar to other members of the C1q/TNF superfamily, the purified gC1q domain was analysed by sucrose gradient sedimentation. The results showed that two distinct peaks as judged by sucrose gradient sedimentation, corresponding to gC1q domains with the molecular mass of gC1q trimers (~96 kDa, major peak) and hexamers (~192 kDa, minor peak) (Fig. 7.2).

Systematic examination of the mobility of native wildtype gC1q domain protein using non-denaturing PAGE gels of different porosity (6, 8, 10 and 12% acrylamide concentration), compared with molecular weight standards (α-lactalbumin, carbonic anhydrase, chicken egg albumin, bovine serum albumin, urease), again showed molecular sizes consistent with a native gC1q trimer and hexamer, plus a high molecular weight aggregate which was not resolved (Fig. 7.3). This further confirms the observations using sucrose gradient sedimentation analysis.

3.3. C1QTNF5 gC1q Domain is Unstable Under Denaturing Conditions

To test the stability of the wildtype gC1q multimers to denaturation, wildtype His-tagged gC1q fusion protein was subjected to increasing concentrations of SDS (0-2%) prior to analysis by SDS-PAGE and western blot using anti-His tag antibody (Fig. 7.4a). In each case, the C1QTNF5 gC1q domain remains detectable as a major monomer and a weak dimer. Under non-denaturing conditions, multimerisation of the gC1q domain is seen to be sensitive to 8 M Urea and 10% trichloroacetic acid (especially at 60°C), all of which enhance the denaturation of the gC1q domain (Fig. 7.4b).

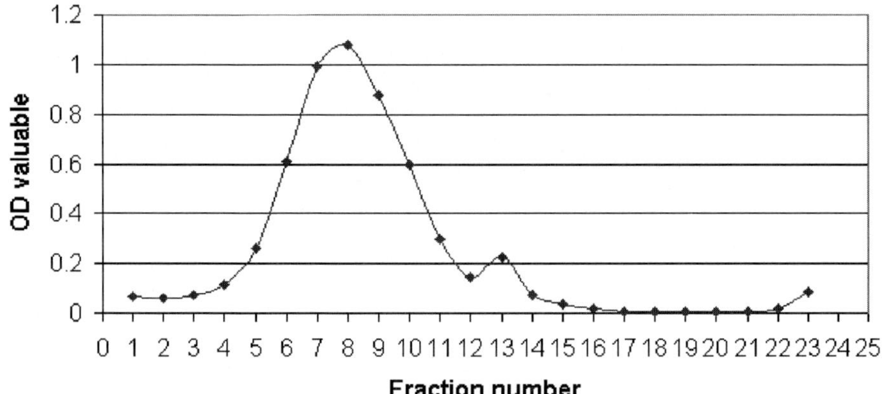

Figure 7.2. Sedimentation analysis of CTRP5 Ser163 gC1q domain. Purified Ser163 gC1q domain and size markers were applied to 5-20% sucrose gradients and detected by UV absorbance. This shows a major peak corresponding to the native trimer (~96 kDa) and a small peak consistent with a hexamer (~192 kDa).

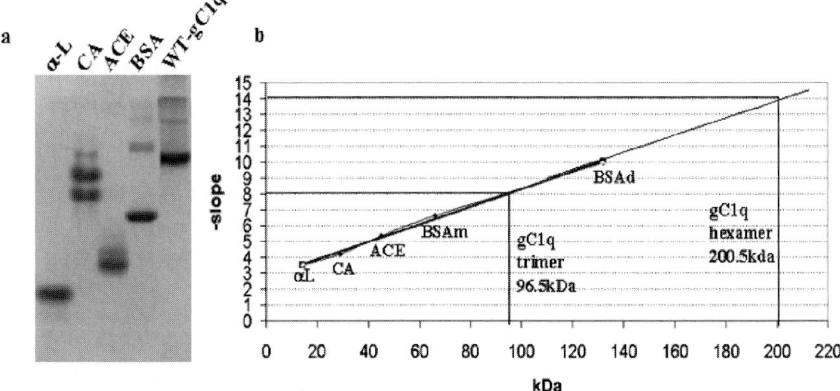

Figure 7.3. Determination of the molecular weight of CTRP5 163Ser gC1q domain (WT-gC1q) by non-denaturing polyacrylamide gel electrophoresis. **a**, an example of the separation shown for an 8% polyacrylamide gel run under native conditions. Size standards include α-lactalbumin (α-L), carbonic anhydrase (CA), chicken egg albumin (ACE), bovine serum albumin (BSA, monomer (m) or dimer (d)) and urease (not shown). **b**, the calculated sizes of the gC1q trimer and hexamer are shown.

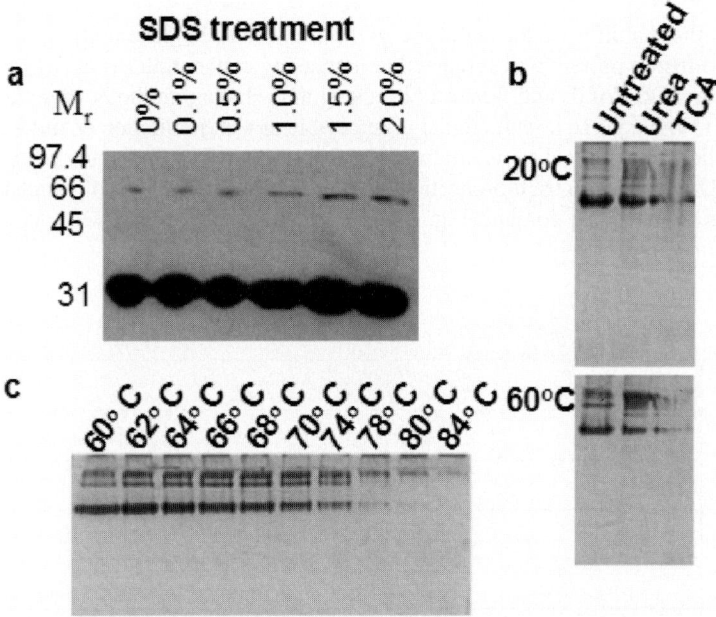

Figure 7.4. a, Bacterially expressed C1QTNF5 Ser163 gC1q domain was treated with SDS (from 0% to 2.0%) at room temperature for 10 min, separated by denaturing SDS-PAGE and detected by western blotting with anti-His antibody. The strong lower band represents gC1q monomer and the weak upper band gC1q dimer. band gC1q dimer. Molecular weight (Mr) size standards in kiloDaltons are shown on the left. **b**, Ser163 gC1q domain was treated with urea (8 M) or 10% trichloacetic acid (TCA) for 10 min at 20°C or 60°C, then separated by non-denaturing PAGE and stained for protein with Coomassie Blue. **c**, Thermostability of Ser163 gC1q domain subjected to heating at indicated temperatures for 10 min under non-reducing conditions before loading on a 12% non-denaturing PAGE gel and staining for proteins with Coomassie Blue.

In a thermostability analysis of C1QTNF5 gC1q domain, the wildtype (Ser163) gC1q domain was incubated at different temperatures (60°C-84°C) in sample buffer for 10 min and run on a 12% non-denaturing polyacrylamide gel electrophoresis (PAGE) gel. The wild-type gC1q domain is stable at temperatures up to 70°C, but starts to degrade or aggregate at higher temperatures, so that most has degraded at 80°C (Fig. 7.4c).

4. DISCUSSION

The short-chain collagen C1QTNF5, like other members of this protein family, such as collagen VIII and X, includes both a short-chain collagen and a gC1q domain, which is thought to be necessary for trimerisation and "zippering" of the collagen trimer. The gC1q domain contains a group of highly conserved aromatic amino acids which are proposed to be involved in this trimerisation step (Brass *et al.*, 1992). In the related protein collagen X, this aromatic motif is critical for the interaction of the gC1q domains (Chan *et al.*, 1999). Deletion of the same aromatic motif in C1QTNF5 also shows loss of trimerisation (unpublished data). Collagen X exists as a major trimer and as a high molecular weight multimer (Frischholz *et al.*, 1998), while EMILIN initially assembles into trimers but also forms large multimers (Mongiat et al., 2000). The results presented here suggest that the native form of C1QTNF5 gC1q domain also exists as a trimer but a hexameric form is also detectable so that the possibility of higher order multimers or a hexagonal lattice, as found in some other members of this protein family, is not excluded. C1QTNF5 is predicted to be secreted by retinal pigment epithelial (RPE) cells so that one possible role in late-onset retinal degeneration is in cell adhesion to the underlying Bruch's membrane. Further work is required to establish whether or not the 163Arg mutation compromises such cell adhesion by assay in transfected RPE cells.

Why is there a close resemblance between L-ORD and age-related macular degeneration? The most likely explanation is the common occurrence of a thick sub-RPE deposit in both conditions, which impairs transport of nutrients between choroid and RPE and/or adhesion between RPE and its basement membrane. The molecular basis of C1QTNF5 Ser163Arg pathogenicity could firstly be due to lack of wildtype protein (haploinsufficiency), since the mutant protein is unstable and readily aggregates *in vitro* (Hayward et al., 2003), so that a 50% reduction in functional C1QTNF5 is likely. Alternatively, this study shows that multimerisation is apparently normal in the 163Arg mutant, so that formation of heteromultimers of wildtype and mutant monomers, which are found *in vitro* (data not shown) could substantially reduce the amount of normal multimer. For example, assuming equal expression and random oligomerisation of subunits, the amount of functional trimer could be as little as 12.5% of normal, if the mutant monomer de-stabilises or disrupts the function of all heterotrimers. The mutation could therefore exert a dominant-negative effect on C1QTNF5 function. Further work is required to distinguish between these alternative disease models.

5. ACKNOWLEDGEMENTS

We gratefully acknowledge an RD2004 Young Investigator Travel Award from National Eye Institute, NIH to Dr. Xinhua Shu, and the financial support of the Foundation Fighting

Blindness, Macula Vision Research Foundation, Macular Disease Foundation, NIH/NEI, and Medical Research Council.

6. REFERENCES

Brass, A., Kadler, K.E., Thomas, J.T., Grant, M.E., Boot-Handford, R.P., 1992, The fibrillar collagens, collagen VIII, collagen X and the C1q complement proteins share a similar domain in their C-terminal non-collagenous regions, *FEBS* **303**:126-128.

Chan, D., Freddi, S., Weng, Y.M., Bateman, J.F., 1999, Interaction of collagen α1 (X) containing engineered NC1 mutations with normal α1 (X) *in vitro, J. Biol. Chem.* **274**:13091-13097.

Crabb, J.W., Miyagi, M., Gu, X., Shadrach, K., West, K.A., Sakaguchi, H., Kamei, M., Hasan, A., Yan, L., Rayborn, M.E., Salomon, R.G., Hollyfield, J.G., 2002, Drusen proteome analysis: an approach to the etiology of age-related macular degeneration, *Proc Natl Acad Sci U S A.* **99**:14682-14687.

Ferris, F.L., Fine, S.L., Hyman, L., 1984, Age related macular degeneration and blindness due to neovascular maculopathy, *Arch. Ophthalmol.* **102**:1640-1642.

Green, W.R., Enger, C., 1993, Age-related macular degeneration histopathologic studies: the 1992 Lorenz E. Zimmerman Lecture, *Ophthalmology* **100**:1519-1535.

Frischholz, S., Beier, F., Girkontaite, I., Wagner, K., Poschl, E., Turnay, J., Mayer, U., von der Mark, K., 1998, Characterization of human type X procollagen and its NC-1 domain expressed as recombinant proteins in HEK293 cells, *J Biol. Chem.* **273**:4547-4555.

Hammond, C.J., Webster, A.R., Snieder, H., Bird, A.C., Gilbert, C.E., Spector, T.D., 2002, Genetic influence on early age-related maculopathy: a twin study, *Ophthalmology.* **109**:730-736.

Hayward, C., Shu, X., Cideciyan, A.V., Lennon, A., Barran, P., Zareparsi, S., Sawyer, L., Hendry, G., Dhillon, B., Milam, A.H., Luthert, P.J., Swaroop, A., Hastie, N.D., Jacobson, S.G., Wright, A.F., 2003, Mutation in a short-chain collagen gene, C1QTNF5, results in extracellular deposit formation in late-onset retinal degeneration: a genetic model for age-related macular degeneration, *Hum Mol Genet.* **12**:2657-2667.

Jacobson, S.G., Cideciyan, A.V., Wright, E., Wright, A.F., 2001, Phenotypic marker for early disease detection in dominant late-onset retinal degeneration, *Invest Ophthalmol Vis Sci.* **42**:1882-1890.

Kuntz, C.A., Jacobson, S.G., Cideciyan, A.V., Li, Z.Y., Stone, E.M., Possin, D., Milam, A.H., 1996, Sub-retinal pigment epithelial deposits in a dominant late-onset retinal degeneration, *Invest Ophthalmol Vis Sci.* **37**:1772-1782.

Malek, G., Li, C.M., Guidry, C., Medeiros, N.E., Curcio, C.A., 2003, Apolipoprotein B in cholesterol-containing drusen and basal deposits of human eyes with age-related maculopathy. *Am J Pathol.* **162**:413-425.

Milam, A.H., Curcio, C.A., Cideciyan, A.V., Saxena, S., John, S.K., Kruth, H.S., Malek, G., Heckenlively, J.R., Weleber, R.G., Jacobson, S.G., 2000, Dominant late-onset retinal degeneration with regional variation of sub-retinal pigment epithelium deposits, retinal function, and photoreceptor degeneration, *Ophthalmology.* **107**:2256-2266.

Mongiat, M., Mungiguerra, G., Bot, S., Mucignat, M.T., Giacomello, E.G., Doliana, R., Colombatti, A., 2000, Self-assembly and supramolecular organization of EMILIN, *J Biol Chem.* **275**:25471-25480.

Seddon, J.M., 2001, Age related macular degeneration. In Ryan, S.J. (ed), *Retina* (3rd edn), Mosb, St Louis, MO, pp. 1039-1050.

CHAPTER 8

BIETTI CRYSTALLINE CORNEORETINAL DYSTROPHY ASSOCIATED WITH *CYP4V2* GENE MUTATIONS

Makoto Nakamura, Jian Lin, Koji Nishiguchi, Mineo Kondo, Jiro Sugita, and Yozo Miyake*

1. SUMMARY

Bietti crystalline corneoretinal dystrophy (BCD) is an autosomal recessive chorioretinal dystrophy characterized by progressive night blindness, tiny, yellowish, glistening retinal crystals, choroidal sclerosis, and crystalline deposits in the peripheral cornea. Recent studies have demonstrated that the *CYP4V2* gene which encodes a CYP450 family protein is the causative gene of the disease. We have identified a homozygous mutation in the *CYP4V2* gene in 8 separate Japanese patients with BCD and conclude that mutations in the *CYP4V2* gene are the major cause of BCD. The IVS6-8_c.810del/insGC mutation is found at a higher frequency in the Asian populations suggesting a founder effect.

2. INTRODUCTION

Bietti crystalline corneoretinal dystrophy (BCD) is an autosomal recessive chorioretinal dystrophy characterized by progressive night blindness, tiny, yellowish, glistening retinal crystals, choroidal sclerosis, and crystalline deposits in the peripheral cornea. Fluorescein angiography shows varying degrees of RPE atrophy and choriocapillaris loss. The atrophy of the retinal pigment epithelium (RPE) and choroidal sclerosis are progressive.[1-9] The patients complain of progressive night blindness associated with a reduction of visual acuity and loss of visual field.[8] Some patients show crystalline deposits in superficial limbal cornea as well as in circulating lymphocytes.[1,3-5,8] Patients with BCD have been reported to be more common in Asia especially in Japan and China.[4,7,8,10]

---

*Department of Ophthalmology, Nagoya University Graduate School of Medicine, 65 Tsuruma-cho, Showa-ku, Nagoya 466-8550, Japan.

A recent study has revealed that mutations of the *CYP4V5* gene, which encodes a member of the cytochrome p450 (CYP450) family proteins and is expressed abundantly in the retina, are the cause of this disease.[11] The protein encoded by *CYP4V5* is suggested to play a role in the lipid processing pathways,[11] because some other members of the CYP450 proteins are implicated in lipid metabolism, and because cultured lymphocytes from patients with BCD showed abnormally high levels of triglyceride and cholesterol and absence of two fatty acid-binding proteins.[12]

We have examined the *CYP4V5* gene in 8 separate Japanese patients with BCD and have identified a mutation in this gene in all. This would indicate that defects in this gene are the major cause of BCD.[13] The patients had characteristic clinical features of BCD.[13]

3. PATIENTS AND METHODS

The procedures used in this study conformed to the tenets of the Declaration of Helsinki, and informed consent was obtained from each patient after an explanation of this study. Eight Japanese patients with BCD were examined. None of the other family members was affected, and to the best of our knowledge, the families were not related. Three unrelated patients were the offsprings of three consanguineous marriages. All individuals examined have been followed in the Departments of Ophthalmology, Nagoya University. Each patient received a complete ophthalmologic examination including best-corrected visual acuity, slit-lamp and fundus examination, fundus photography, Goldmann kinetic perimetry, fluorescein angiography, and electroretinography.

Genomic DNA was extracted from peripheral leukocytes. Exons 1 through 11 with flanking intron splice sites of the *CYP4V5* gene were individually amplified by polymerase chain reaction (PCR), and the PCR products were purified and directly sequenced.

Standardized electroretinograms (ERGs) were elicited by ganzfeld stimuli after 30 minutes of dark-adaptation. The rod (scotopic) ERGs were elicited by a blue light at an intensity of 5.2×10^{-3} cd/m^2/sec. The mixed rod:cone single flash ERGs were elicited by a white stimulus at an intensity of 44.2 cd/m^2/sec. The cone ERGs and the 30 Hz flicker ERGs were elicited by a white stimulus at an intensity of 4 cd/m^2/sec and 0.9 cd/m^2/sec, respectively, on a white background of 68 cd/m^2.

4. RESULTS

A homozygous mutation in the *CYP4V5* gene was found in the 8 patients with BCD.[13] Seven patients had an IVS6-8_c.810del/insGC mutation with 17-bp deletion and 2-bp insertion that affected the IVS6 splice acceptor site probably resulting in an in-frame skipping of exon 7.[11,13] The remaining patient had an L173W mutation in the gene.[13]

The clinical characteristics of the BCD patients are summarized in Table 8.1. The ages of the subjective onset of symptoms were between 35- and 54-years-old. Four patients complained of night blindness. The first symptoms were nyctalopia in 3 patients, a reduction of central vision in 2, a disturbance of peripheral visual fields in 2, and 1 patient had no symptoms and his disease was detected during a health examination. The visual acuities of the patients ranged between 0.2 and 1.5 (Table 8.1).

Table 8.1. Clinical and genetic findings of patients with BCD.

Case	Age (y)	Sex	Visual acuity OD, OS	Goldmann visual field	Corneal deposits	*ERG amplitudes R,L (µV)	Mutation
1	50	M	1.0, 0.6	ND	−	120, ND	IVS6-8_c.810del/insGC
2	52	F	0.6, 1.0	ring scotoma	−	ND, 27	IVS6-8_c.810del/insGC
3	46	F	0.8, 0.9	paracentral scotoma	+	207, ND	IVS6-8_c.810del/insGC
4	38	F	0.2, 1.5	central/paracentral scotoma	−	274, 296	IVS6-8_c.810del/insGC
5	54	F	1.0, 1.2	central scotoma	+	296, ND	c.518T>G
6	52	M	0.9, 0.7	paracentral scotoma	+	385, ND	IVS6-8_c.810del/insGC
7	54	M	0.5, 0.5	central scotoma	+	311, ND	IVS6-8_c.810del/insGC
8	47	F	1.2, 1.2	ring scotoma	+	312, ND	IVS6-8_c.810del/insGC

*B-wave amplitudes from single white flash ERGs extracting mixed rod and cone responses. (normal range ≥314 µV).

ND, not determined.

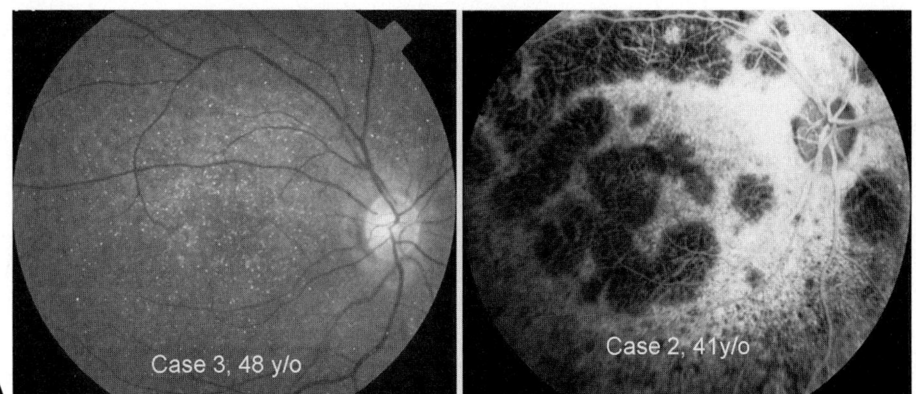

Figure 8.1. Fundus photographs and a fluorescein angiograms of patients with Bietti crystalline corneoretinal dystrophy associated with a mutation of the *CYP4V5* gene (IVS6-8_c.810del/insGC). (**A**) right eye, case 3: numerous crystalline deposits scattered throughout the fundus and diffuse atrophy of the retinal pigment epithelium are seen. (**B**) right eye, case 5: atrophy of the retinal pigment epithelium and choriocapillaris loss at the posterior. The case number and the age (years) are indicated in each photograph.

Slit-lamp examination revealed crystalline deposits in the peripheral cornea in 5 patients. Each patient had a central, a paracentral, or a ring scotoma that was detected by Goldmann kinetic visual perimetry. The scotomas enlarged with age.

All patients characteristically had numerous, small retinal crystalline deposits concentrated in the posterior pole (Figure 8.1A). Fluorescein angiography showed varying degrees of RPE atrophy and choriocapillaris loss at the posterior pole and sometimes extending to the midperiphery (Figure 8.1B). The area of choriocapillaris loss enlarged with age.

Full-field ERGs showed different degrees of rod and cone dysfunction ranging from normal to severe reduction. The large variability in the amplitudes was noted even among patients carrying the same mutation and at a similar age (Figure 8.2). In patients with reduced ERG responses, both rod and cone ERG responses were reduced, and neither was

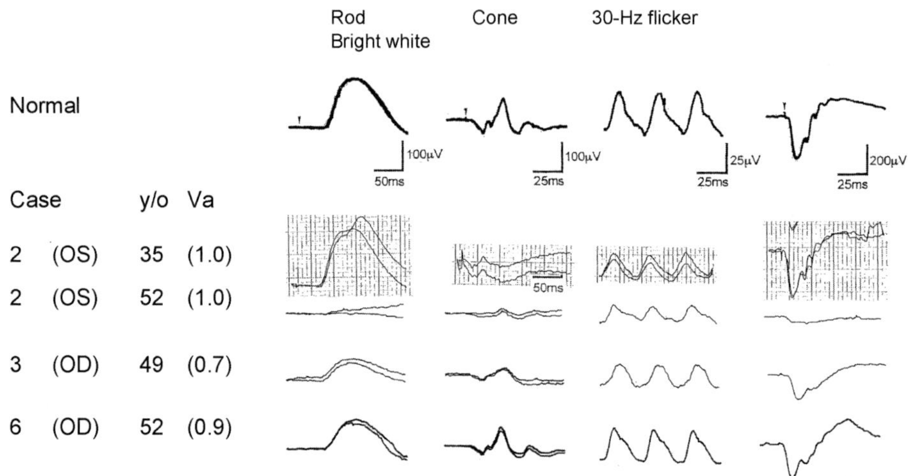

Figure 8.2. Full-field electroretinograms recorded from a normal subject and a patient with Bietti crystalline corneoretinal dystrophy associated with a mutation of the *CYP4V5* gene. The arrows indicate the stimulus onset. The case number and the age (years) of each patient is noted under the case number in the left column.

predominant. There seemed to be no direct correlation between the degrees of reduction of the visual acuity and the degrees of reductions of the full-field ERG amplitudes. In a patient (case 2), the ERG responses recorded after a 17-year follow-up had decreased significantly (Figure 8.2).

5. DISCUSSION

We analyzed the *CYP4V5* gene in 8 unrelated Japanese patients with BCD and identified mutations in the gene in all. Combining these findings with previous results that mutations in the *CYP4V5* gene were found in 23 of 25 unrelated patients with BCD,[11] we conclude that mutations in the *CYP4V5* gene are the major cause of BCD.[13]

Seven of our 8 patients were found to have the IVS6-8_c.810del/insGC mutation in a homozygous state.[13] In the previous study the same mutation was found in 7 of 8 unrelated Japanese BCD families as well as in 7 of 10 unrelated Chinese families.[11] Thus, this mutation is considered to have a high incidence in the Japanese and Chinese populations. A founder effect rather than a mutational hot spot was considered for this mutation, because the mutation has never been identified in other populations including Caucasian.[11,13] The founder of the mutation was likely to be a very old ancestor, because the region of the conserved linked markers extended less than 17.1 kb.[13]

All patients with the *CYP4V5* gene mutations shared characteristic clinical features of BCD such as retinal crystalline deposits, RPE atrophy, and choriocapillaris loss. However, full-field ERGs showed remarkable variability in amplitudes even among patients with the same mutations and at a similar age. These observations would indicate the possibility that other genetic or environmental factors influenced the course of the disease.[13]

6. ACKNOWLEDGMENTS

This study was supported in part by Grant-in Aid for Scientific Research from the Ministry of Education, Culture, Sports, Science, and Technology of Japan (Dr Nakamura, C16591746), and Grant-in Aid from the Ministry of Health, Labor, and Welfare of Japan, Tokyo, Japan (Dr. Miyake).

7. REFERENCES

1. Bietti GB. Ueber familiaeres vorkommen von "retinitis punctata albescens" (verbunden mit "dystrophia marginalis cristallinea corneae"), glitzern des glaskoerpers und anderen degenerativen augenveraenderungen. *Klin Mbl Augenheilk.* 1937;**99**:737-57.
2. Bagolini B, Ioli-Spada G. Bietti's tapetoretinal degeneration with marginal corneal dystrophy. *Am J Ophthalmol.* 1968 Jan;**65**(1):53-60.
3. Welch RB. Bietti's tapetoretinal degeneration with marginal corneal dystrophy crystalline retinopathy. *Trans Am Ophthalmol Soc.* 1977;**75**:164-79.
4. Yagasaki K, Miyake Y. [Crystalline retinopathy]. *Nippon Ganka Gakkai Zasshi.* 1986 May;**90**(5):711-9.
5. Wilson DJ, Weleber RG, Klein ML, Welch RB, Green WR. Bietti's crystalline dystrophy. A clinicopathologic correlative study. *Arch Ophthalmol.* 1989 Feb;**107**(2):213-21.
6. Bernauer W, Daicker B. Bietti's corneal-retinal dystrophy. A 16-year progression. *Retina.* 1992;**12**(1):18-20.
7. Takikawa C, Miyake Y, Yamamoto S. Re-evaluation of crystalline retinopathy based on corneal findings. *Folia Ophthalmol Jpn.* 1992;**43**:969-78.
8. Kaiser-Kupfer MI, Chan CC, Markello TC, Crawford MA, Caruso RC, Csaky KG, Guo J, Gahl WA. Clinical biochemical and pathologic correlations in Bietti's crystalline dystrophy. *Am J Ophthalmol.* 1994 Nov 15;**118**(5):569-82.
9. Yanagi Y, Tamaki Y, Takahashi H, Sekine H, Mori M, Hirato T, Okajima O. Clinical and functional findings in crystalline retinopathy. *Retina.* 2004 Apr;**24**(2):267-74.
10. Hu DN. Ophthalmic genetics in China. *Ophthal Paediat Genet.* 1983;**2**:39-45.
11. Li A, Jiao X, Munier FL, Schorderet DF, Yao W, Iwata F, Hayakawa M, Kanai A, Shy CM, Alan LR, Heckenlively J, Weleber RG, Traboulsi EI, Zhang Q, Xiao X, Kaiser-Kupfer M, Sergeev YV, Hejtmancik JF. Bietti crystalline corneoretinal dystrophy is caused by mutations in the novel gene CYP4V2. *Am J Hum Genet.* 2004 May;**74**(5):817-26.
12. Lee J, Jiao X, Hejtmancik JF, Kaiser-Kupfer M, Chader GJ. Identification, isolation, and characterization of a 32-kDa fatty acid-binding protein missing from lymphocytes in humans with Bietti crystalline dystrophy (BCD). *Mol Genet Metab.* 1998 Oct;**65**(2):143-54.
13. Lin J, Nishiguchi KM, Nakamura M, Dryja TP, Berson EL, Miyake Y. Recessive mutations in the *CYP4V2* gene in East Asian and Middle Eastern patients with Bietti crytalline corneoretinal dystrophy. *J Med Genet.* (in press)

DIAGNOSTIC, CLINICAL, CYTOPATHOLOGICAL AND PHYSIOLOGIC ASPECTS OF RETINAL DEGENERATION

FUNDUS APPEARANCE OF CHOROIDEREMIA USING OPTICAL COHERENCE TOMOGRAPY

Bradley J. Katz[1], Zhenglin Yang[1,2], Marielle Payne[1,2], Yin Lin[1,2,3], Yu Zhao[1,2], Erik Pearson[1,2], Shan Duan[1,2], Shin Kamaya[1,2], Goutam Karan[1,2], and Kang Zhang[1,2]

1. INTRODUCTION

Choroideremia is an X-linked recessive disorder characterized by progressive degeneration of the choroid, RPE and retina Goedblood et al., 1942). Patients initially present with night blindness in the first or second decade that progresses to severe constriction of the visual field. Central acuity is lost late in life (Kril et al.).

The initial fundus change is pigment stippling and focal atrophy of the RPE. With time, areas of choroidal atrophy become apparent with exposure of the underlying choroidal vessels. Eventually, only islands of intact retina and choroid remain in the macula and periphery (Kril et al., 1971). Histopathologically there is loss of the outer retinal layers and RPE. Bruch's membrane may remain, as well as a remnant of the inner retinal layers. Fibrosis of the choroid may also be observed (Rafuse et al., 1968). The pathophysiology of choroideremia is not understood. However, linkage analysis has localized the gene defect to Xq13-q22 (Lewis et al., 1985) and the gene has subsequently been cloned (Bokhoven et al., 1994. The protein product of this gene is a Rab escort protein (REP-1) functioning as a RAB geranylgeranyl transferase involved in intracellular vesicular transport (Seabra et al., 1993).

2. PURPOSE

To characterize the appearance of the fundus of a patient with choroideremia using optical coherence tomography (OCT) and identify the underlying gene defect.

---

[1] Moran Eye Center, Department of Ophthalmology and Visual Science, [2] Program in Human Molecular Biology & Genetics, Eccles Institute of Human Genetics, University of Utah, Salt Lake City, UT and [3] Molecular Biology and Genetics, Sichuan Provincial Medical Academy and Sichuan Provincial People's Hospital, P. R. China.

3. CASE REPORT

A 24-year-old male presented for evaluation of decreased vision. Acuity was 20/25 and visual fields were constricted in each eye. Anterior segments were clear and quiet and intraocular pressures were normal. On fundus examination of both eyes, the optic nerves appeared normal. The retinal arterioles were attenuated. There was an island of intact retina in the center of each macula. The remainder of the fundus was characterized by extensive atrophy of the retina and choriocapillaris with exposure of the underlying large choroidal vasculature (Figures 9.1 and 9.2). A few pigment clumps were seen throughout the fundi.

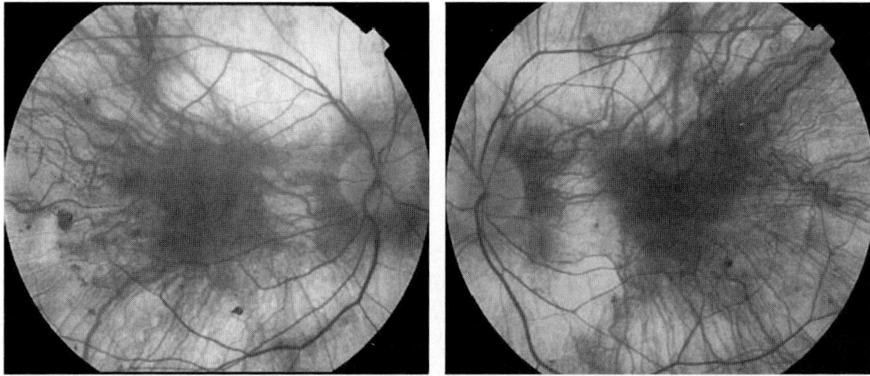

Figure 9.1. Fundus photographs of the posterior pole, OD (above left) and OS (above right). The retinal arterioles are attenuated. Within the macula, there is an island of normal retina. Surrounding this island there is extensive atrophy with exposure of the underlying choroidal vessels. A few areas of pigment clumping are evident.

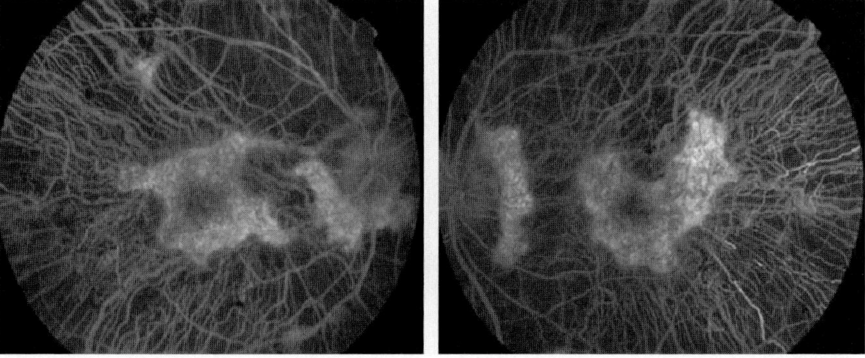

Figure 9.2. Fluorescein angiogram of he posterior pole, OD (above left) and OS (above right) at a late phase of angiogram demonstrating extensive atrophy of choriocapillaris and RPE except small islands of remaining retina in the macula and peripapillary regions.

4. RESULTS

Optical coherence tomography through the patient's optic nerve and macula revealed an abrupt demarcation line between the island of remaining retina within the macula and the surrounding area of chorioretinal atrophy (Figure 9.3). The outer retina and RPE corresponding to the atrophic areas identified by fundus photograghs and fluorescein angiogram were missing. The OCT appearance of the area of chorioretinal atrophy is consistent with previous reports of the histopathology of choroideremia (Figure 9.4).

5. DISCUSSION

We describe the fundus characteristics of choroideremia using OCT. The appearance is consistent with the findings of previous histologic studies. To the best of our knowledge, this is the first report of fundus appearance of choroideremia using OCT. The gene for REP-1 has been successfully re-introduced *in vitro* into deficient lymphocytes and fibroblasts with a recombinant adenovirus (Anand et al., 2003) holding open the possibility that treatment of patients with choroideremia may become available in the future. With the use of high resolution OCT, the sub-retina structures can be visualized in a fine detail. OCT will become a valuable tool in monitoring the effect of retina preservation of patients with choroideremia who undergo treatments with gene or drug based therapies.

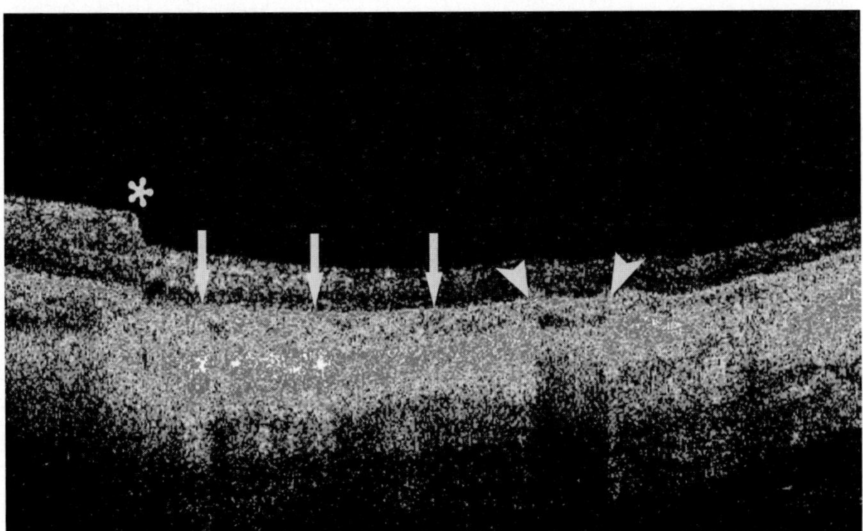

Figure 9.3. Ocular coherence tomography (OCT) through the optic nerve (arrowheads) and macula of the right eye shows an abrupt demarcation (asterisk) between the island of intact retina and the area of atrophy. The outer retina and RPE are absent, but a thin layer of inner retina and Bruch's membrane (arrows) are intact within the area of atrophy, consistent with histopathological samples from other patients with choroideremia (Compare to Figure 9.4). Increased signal from the choroid underlying the atrophic area could be consistent with fibrosis of the choroid. See also color insert.

Figure 9.4. Histopathology of choroideremia in a specimen taken from another patient. There is fibrosis of the choroid and only a single choroidal artery remains. The retinal pigment epithelium and outer nuclear layers are absent. The inner nuclear layer rests against Bruch's membrane. H&E ×330. (Reprinted from: Spencer WH, *Ophthalmic Pathology, An Atlas and Textbook*, 4/e, Figure 9-723, 1997, with permission from Elsevier). See also color insert.

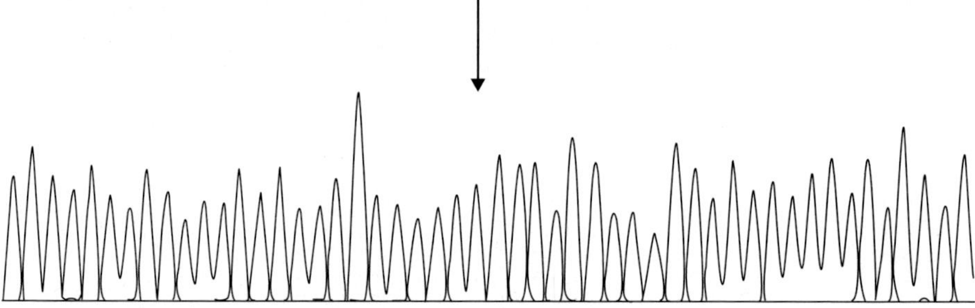

Figure 9.5. DNA tracing of above patient showing a change at nucleotide position 757 from C to T resulting in the placement of a premature stop codon at amino acid 253.

6. ACKNOWLEDGEMENT

This research was supported by National Institutes of Health Grants R01EY14428, R01EY14448 and GCRC M01-RR00064, the Ruth and Milton Steinbach Fund, Ronald McDonald House Charities, the Macular Vision Research Foundation, the Research to Prevent Blindness, Inc., Knights Templar Eye Research Foundation, American Health Assistance Foundation, the Karl Kirchgessner Foundation, Val and Edith Green Foundation, and the Simmons Foundation, an unrestricted grant to the Department of Ophthalmology and Visual Sciences from Research to Prevent Blindness, Inc., New York, NY.

7. REFERENCES

Anand, V., Barral, D. C., Zeng, Y., *et al.*, 2003, Gene therapy for choroideremia: in vitro rescue mediated by recombinant adenovirus, *Vision Res* **43**:919-926.

Goedblood, J., 1942, Mode of Inheritance in Choroideremia, *International Journal of Ophthalmology* **104**:309-315.

Krill A. E., Archer D., 1971, Classification of the choroidal atrophies, *Am J Ophthalmol* **72**(3):562-585.

Lewis, R. A., Nussbaum, R. L., Ferrell, R., 1985, Mapping X-linked ophthalmic diseases. Provisional assignment of the locus for choroideremia to Xq13-q24, *Ophthalmology* **92**:800-806.

Rafuse E. V., McCulloch, C., 1968, Choroideremia: A pathological report, *Can J Ophthalmol* **3**:347-352.

Seabra, M C., Brown, M S., Goldstein, J L., 1993, Retinal degeneration in choroideremia: deficiency of rab geranylgeranyl transferase, Science, **259**:377-381

van Bokhoven, H., van den Hurk, J. A., Bogerd, L., *et al.*, 1994, Cloning and characterization of the human choroideremia gene, *Hum Mol Genet* **3**:1041-1046.

CHAPTER 10

A2E, A FLUOROPHORE OF RPE LIPOFUSCIN, CAN DESTABILIZE MEMBRANE

Janet R. Sparrow, Bolin Cai, Young Pyo Jang, Jilin Zhou, and Koji Nakanishi*

1. INTRODUCTION

Studies of Stargardt disease suggest a role for RPE lipofuscin in the RPE cell atrophy that characterizes macular degeneration. The best known constituent of RPE lipofuscin is the pyridinium bisretinoid, A2E (Eldred and Lasky, 1993; Parish et al., 1998). Amongst the properties of A2E that may be damaging to the RPE cell is its ability to destabilize cell membranes (Sparrow et al., 1999). A hydrophilic head group combined with a pair of hydrophobic side-arms are the structural correlates of this behavior. This amphiphilic structure accounts for the tendency of A2E to aggregate (Sakai et al., 1996; De and Sakmar, 2002), a behavior first recognized in deuterated chloroform ($CDCl_3$), the broadening of the ^1H NMR signal indicating that the protonated pyridinium moieties of A2E were closely packed within the interior of micelles while the hydrophobic chains contacted the solvent. Further evidence of the detergent-like behaviour of A2E has been revealed in experiments demonstrating the ability of A2E to induce concentration-dependent membrane leakage (Sparrow et al., 1999). In studies employing unilamellar vesicles, it has also been shown that A2E, at critical micellar concentrations, can solubilize membranes (De and Sakmar, 2002).

In an effort to further our understanding of the ability of A2E to alter membrane integrity, we developed an experimental paradigm for the detection of membrane blebbing using ARPE-19 cells transduced to express wild-type green fluorescent protein. In addition, we compared the behavior of A2E to the mono-retinoid, A1E, a single side- arm counterpart to A2E. Here we report the results of these studies.

2. METHODS

ARPE-19 cells, A1E and A2E. A human RPE cell line (ARPE-19) was grown as formerly described. A2E and A1E were synthesized using published methods (Parish et al.,

---

*Departments of Ophthalmology and Chemistry, Columbia University, New York, NY 10032.

1998; Jockusch et al., 2004) and were incubated with cells as previously reported (Sparrow et al., 2000). For the imaging of A2E and A1E accumulation by epifluorescence microscopy, the filters and dichroic mirror utilized permitted 330 ± 80 nm excitation and >400 nm emission. For laser scanning confocal microscopy, cell borders were defined by immunolabeling with rabbit antibody to human ZO-1 (Zymed Laboratories, South San Francisco CA) and TRITC-labeled donkey anti-rabbit IgG (Jackson ImmunoResearch, West Grove, PA). Nuclei were stained with propidium iodide (Molecular Probes, Eugene OR) and A1E was visualized with fluorescein-appropriate filters so that its fluorescence appeared green, in contrast with the cell borders and nuclei.

GFP-expressing ARPE-19 cells. A lentivirus-based vector (CMV promotor and VSV-G envelope protein) was used to transfer the gene for wild-type green fluorescent protein (GFP) (Lai et al., 1999) to ARPE-19 cells. For viral transduction the cells were grown to 70-80% confluence and were then incubated with GFP-expressing lentivirus (10^5-10^7 transducing units) in serum free medium. After 24 hours the infection medium was replaced with normal growth medium and maintained for 4 days. The transfected cells were then replated in eight-well plastic chamber slides (LabTek; Nunc, Naperville, IL) and at subconfluent cell densities, synthesized A2E was added to the cells at concentrations of 20 and 100 μM.

Assays of cell permeability. To assay membrane integrity, cultures were incubated with the membrane impermeable dye Dead Red (Molecular Probe, Eugene OR) (Sparrow et al., 1999) and postfixation, were stained with DAPI (6-diamidino-2-phenylindole). Release of the cytoplasmic enzyme lactate dehydrogenase (LDH) into culture media was measured as previously described (Sparrow et al., 1999).

3. RESULTS

A2E-induced membrane blebbing is indicative of the ability of this fluorophore to perturb membrane. To test for evidence that A2E can induce membrane blebbing, we employed the plasma membrane of ARPE-19 cells as a model membrane bilayer and labeled the cytosol by transducing the cells to express wild-type GFP. The cells were then exposed to exogenous A2E at concentrations of 20 μM and 100 μM for 3-4 hours. As shown in Fig. 10.1, GFP-filled membrane blebs formed on A2E-treated but not control cells. The effect was also concentration-dependent with blebbing occurring at 100 μM but not 20 μM A2E. The DAPI-stained nuclei of many of the cells exhibiting membrane blebs were not co-labeled with a membrane impermeable dye (Dead Red; Molecular Probes). Exclusion of the dye indicates that, at least during the early stages of membrane blebbing, the cells retained normal membrane impermeability.

The wedge-shaped structure of A2E influences its ability to penetrate and perturb cell membranes. To begin to understand how the structure of A2E determines the manner in which it interacts with membranes, we designed and synthesized A1E [molecular weight (mw) 352.7; UV λ_{max} 411 and 248 nm], a mono-retinoid single side arm counterpart to A2E that is not naturally occurring (Fig. 10.2). Like A2E and iso-A2E (mw 592; UV λ_{max} 430 and 335 nm), this compound presents with a positively charged pyridine ring and retains both hydrophobic and hydrophilic elements. Just as A2E is amassed by cells in culture (Sparrow et al., 1999), it was evident by confocal microscopy that A1E accumulates in cells (Fig. 10.3). However, A1E accumulation occurred more rapidly. For instance, when we incubated cells with either A2E or A1E at 20 μM for 2 days, A1E fluorescence was readily appar-

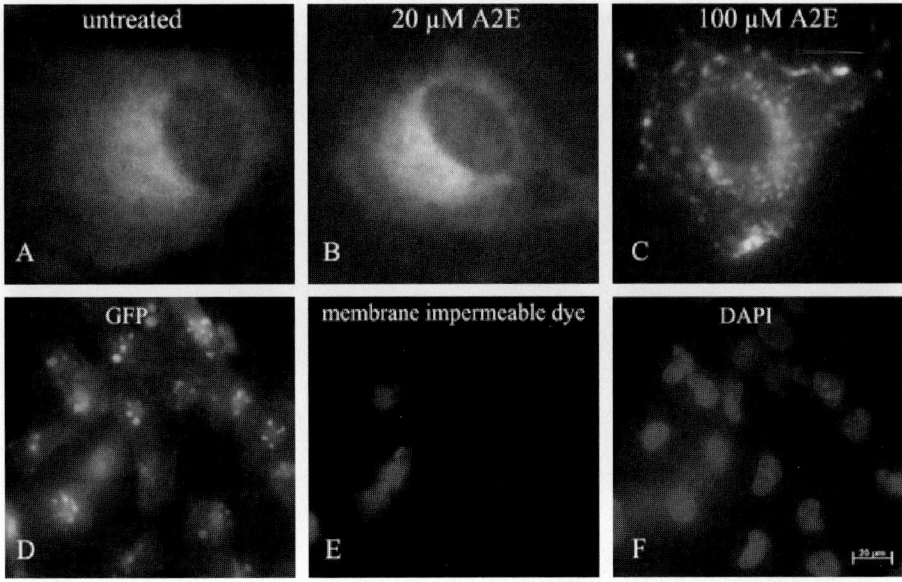

Figure 10.1. A2E induces membrane blebbing. Concentration dependent-bleb formation is visualized in non-confluent ARPE-19 cells transduced to express cytoplasmically-located wild-type green fluorescent protein (GFP) (A–C). At the time of bleb formation (C,D), many of the cells maintain membrane impermeability as evidenced by the absence of nuclear labeling by a membrane impermeable dye (E). The nuclei of the cells are visualized with DAPI (F).

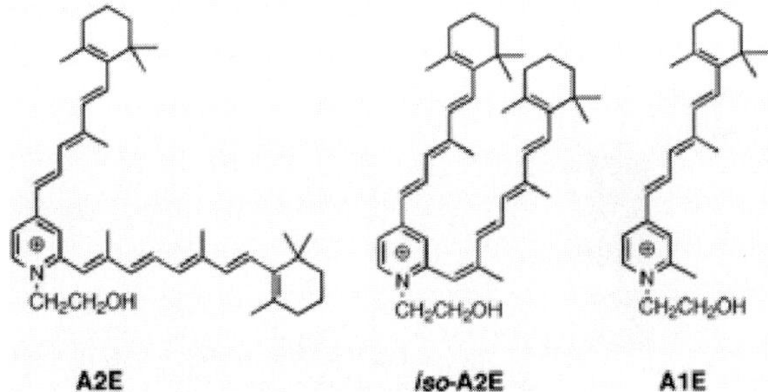

Figure 10.2. Structures of A2E, the photoisomer iso-A2E and the non-biological compound A1E.

ent in the cells, but no A2E fluorescence was yet visible (Fig. 10.4A). Within this 2 day interval, 20 µM A1E, but not A2E, also induced membrane permeabilization (Fig. 10.4B).

Rapid membrane permeabilization induced by A1E was also demonstrated by assaying for the release of cytoplasmic LDH into the culture medium. Thus when cells were incubated with A1E for 2 hours, LDH levels in culture supernatants increased in a concentration-dependent manner. However, although A2E is well known to cause membrane permeabilization (Sparrow et al., 1999), a 2 hour incubation in A2E was not of sufficient duration to elicit a detectable increase in LDH-associated absorbance in the media.

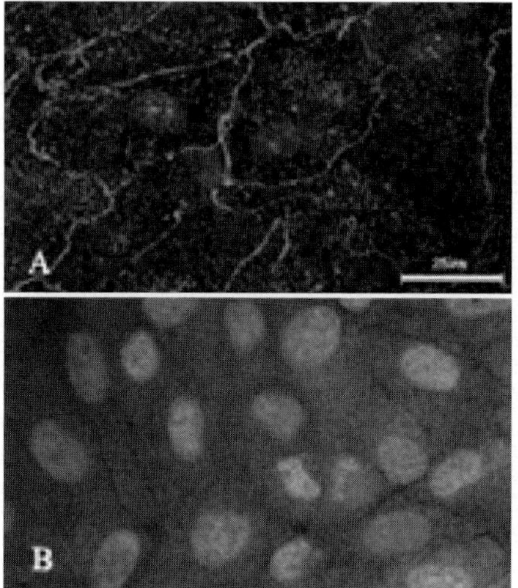

Figure 10.3. A1E is internalized by ARPE-19 cells. Detection by fluorescence confocal microscopy (1 µM optical section). Cell borders were labeled by immunostaining with antibody to ZO-1 (A,B) and nuclei were stained with propidium iodide (B). Internalized A1E presents as punctate-filling of the cells.

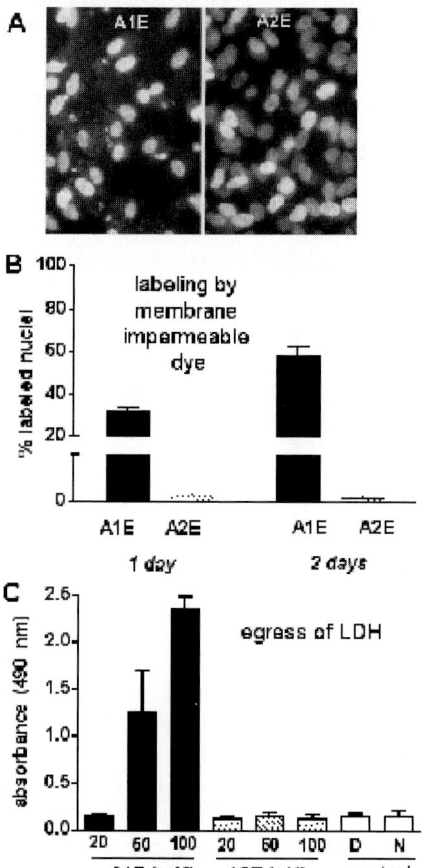

Figure 10.4. Comparison of effects of A1E and A2E on membrane integrity. A. Epifluorescence detection of A1E and A2E in ARPE-19 cells after 2 days of accumulation from 20 µM concentrations in media. Nuclei stained with DAPI. A1E visible in cytoplasmic region. A2E not accumulated to detectable levels. Note reduced nuclear density in A1E cultures due to cell loss. **B**. Percent permeabilized cells was determined by labeling with membrane impermeable dye. All nuclei labeled with DAPI. Percent of labeled cells in A2E-accumulating cultures is similar to that in control untreated cultures (data not shown). Mean ± SEM, 3-5 fields/well; 3 wells/condition. **C**. Loss of membrane integrity as evidenced by egress of LDH into culture medium, assayed colorimetrically after incubating with either A1E or A2E at various concentrations for 2 hours. D, DMSO; N, non-treated. Mean ± SEM.

4. CONCLUSIONS

The properties of A2E that are likely to determine its behavior in a phospholipid bilayer are its amphiphilic structure, size, shape and cationic nature. Since A2E and A1E are both amphiphilic molecules, it is to be expected that both compounds would exhibit detergent-like properties. However the linear configuration of the non-biological compound A1E is more typical of a detergent and probably because of this stream-lined structure, A1E was able to penetrate the membrane more rapidly than A2E. In the current assays in which membrane permeabilization was assessed over a short period of time, A1E also induced a faster rate of loss in membrane integrity. The two widely displaced retinal-derived chains of A2E confer a bulky structure that likely displaces a relatively large area of membrane and may impede passage as it penetrates. Electrostatic attractions between amphiphiles with a cationic head-group and acidic phospholipids, such as phosphatidylserine can also influence the movement of the compound through the membrane and fluorescence anisotropy studies (De and Sakmar, 2002) suggest that this may be the case for A2E, despite the presence of its counterion.

As further evidence of the ability of A2E to perturb membrane integrity, we have shown here using the plasma membrane as a model phospholipid bilayer, that A2E can provoke membrane blebbing. These membrane blisters were observed in cells that had not yet undergone a change in permeability. While the mechanism by which A2E induces membrane blebbing has not been demonstrated, it is possible to speculate as to some of the events. For instance, A2E, because of its cationic head group may distribute preferentially in the inner leaflet of the membrane due to an attraction to negatively charged phosphatidylserine that is concentrated on the cytoplasmic side of the bilayer (Sheetz and Singer, 1974). Within the cytoplasmic leaflet of the membrane, wedge-shaped A2E would become oriented with its hydrophilic pyridinium portion interacting with the polar heads of the phospholipids and its broadly-spaced hydrophobic side-arms intermingling with the phospholipid hydrocarbon tails (Sparrow et al., 1999). Accordingly, the bulky side arms of A2E could force a large separation of the lipid acyl chains and expand the inner leaflet relative to the outer, the negative curvature of the inner leaflet producing a surface protrusion or bleb. Since A2E becomes sequestered within the lysosomal compartment of the cell, at sufficient concentration, A2E may exert similar effects on the lysosomal membrane.

5. ACKNOWLEDGEMENTS

The work was supported by National Institutes of Health Grants EY12951 and GM 34509, Macula Vision Research Foundation and American Health Assistance Foundation. JRS is the recipient of an Alcon Research Institute Award.

6. REFERENCES

De S, Sakmar TP. 2002, Interaction of A2E with model membranes. Implications to the pathogenesis of age-related macular degeneration. *J Gen Physiol* **120**(2):147-157.

Eldred GE, Lasky MR. 1993, Retinal age pigments generated by self-assembling lysosomotropic detergents. *Nature* **361**(6414):724-726.

Jockusch S, Ren RX, Jang YP, Itagaki Y, Vollmer-Snarr HR, Sparrow JR, Nakanishi K, Turro NJ. 2004, Photochemistry of A1E, a retinoid with a conjugated pyridinium moiety: competition between pericyclic photooxygenation and pericyclization. *J Am Chem Soc* **126**(14):4646-4652.

Lai C, Gouras P, Doi K, Lu F, Kjeldbye H, Goff SP, Pawliuk R, Leboulch P, Tsang SH. 1999, Tracking RPE transplants labeled by retroviral gene transfer with green fluorescent protein. *Invest Ophthalmol Vis Sci* **40**(9):2141-2146.

Parish CA, Hashimoto M, Nakanishi K, Dillon J, Sparrow JR. 1998, Isolation and one-step preparation of A2E and iso-A2E, fluorophores from human retinal pigment epithelium. *Proc Natl Acad Sci U S A* **95**(25):14609-14613.

Sakai N, Decatur J, Nakanishi K, Eldred GE. 1996, Ocular age pigment "A2E": An unprecedented pyridinium bis-retinoid. *J Am Chem Soc* **118**:1559-1560.

Sheetz MP, Singer SJ. 1974, Biological membranes as bilayer couples. A molecular mechanism of drug-erythrocyte interactions. *Proc Natl Acad Sci U S A* **71**(11):4457-4461.

Sparrow JR, Nakanishi K, Parish CA. 2000, The lipofuscin fluorophore A2E mediates blue light-induced damage to retinal pigmented epithelial cells. *Invest Ophthalmol Vis Sci* **41**(7):1981-1989.

Sparrow JR, Parish CA, Hashimoto M, Nakanishi K. 1999, A2E, a lipofuscin fluorophore, in human retinal pigmented epithelial cells in culture. *Invest Ophthalmol Vis Sci* **40**(12):2988-2995.

CHAPTER 11

AMINO-RETINOID COMPOUNDS IN THE HUMAN RETINAL PIGMENT EPITHELIUM

Heidi R. Vollmer-Snarr,* McKenzie R. Pew, Mary L. Alvarez,
D. Joshua Cameron, Zhibing Chen, Glenn L. Walker, Josh L. Price,
and Jeffrey L. Swallow

1. INTRODUCTION

For many years, blue light damage associated with retinal pigment epithelial (RPE) cell lipofuscin (LF) has been implicated in the cause of age-related macular degeneration (AMD) (Young, 1988; Winkler et al., 1999; Bressler et al., 2000) and other retinal degenerative diseases, such as Stargardt's disease, (Lopez et al., 1990; Birnbach et al., 1994), Best's macular dystrophy (Weingeist et al., 1982) and cone-rod dystrophy (Rabb et al., 1986). A2E and isomers, pyridinium bis-retinoid compounds and only major blue-light absorbing fluorophores isolated from human LF, (Eldred and Katz, 1988; Eldred and Lasky, 1993; Sakai et al., 1996; Parish et al., 1998;) account for a small fraction of the composition of LF. Proteins account for 30-70% (Schutt et al., 2002; Haralampus-Grynaviski et al., 2003). The percentage of A2E oxidation products, retinol and retinyl palmitate has not been quantified, yet these compounds, too, make up only a fraction of the composition of LF. Evidence for A2E and especially A2E photo-oxidation products' involvement in AMD is strong (Sparrow et al., 1999; Sparrow et al., 2000; Sparrow and Cai, 2001; Ben-Shabat et al., 2002a; Finnemann et al., 2002; Sparrow et al., 2002; Sparrow et al., 2003); however, their involvement does not rule out the possibility of AMD's etiology being multifactorial. There are many compounds in LF that have not been characterized, some of which may also play a role in the pathogenesis of AMD and other retinal degenerative diseases.

The biosynthetic pathway of A2E begins with the formation of A2PE in the rod outer segments (ROS) of the eye (Parish et al., 1998; Liu et al., 2000; Mata et al., 2000; Ben-Shabat et al., 2002b). In the formation of A2PE, one molecule of phosphatidyl ethanolamine reacts with two molecules of all-*trans*-retinal. A2PE's phosphatidyl group is then cleaved to form A2E. A2E is deposited into RPE LF during phagocytosis of ROS. In

---

*Department of Chemistry & Biochemistry, Brigham Young University, Provo, UT 84602.

common with phosphatidyl ethanolamine, other biogenic amines found in the retina (Makino-Tasaka et al., 1985; Drujan et al., 1989; Djamgoz and Wagner, 1992; Gulcan et al., 1993; Taibi and Schiavo, 1993; Witkovsky et al., 1993) and specifically in ROS (Crain et al., 1978; Macaione and Calatroni, 1978; Aveldano and Bazan, 1983; Lentile et al., 1986; Taibi et al., 1995) also have a free amine functionality necessary for reaction with all-*trans*-retinal. Serotonin, tryptamine, norepinephrine, putrescine, spermidine, tyramine, spermine, and dopamine represent a small selection of biogenic amines with known abundances in the retina. All of these amines react with all-*trans*–retinal to form amino-retinoid compounds. These compounds may account for some of the uncharacterized compounds within RPE LF.

2. SYNTHETIC AMINO-RETINOID COMPOUNDS

Serotonin, tryptamine, norepinephrine, putrescine, spermidine, tyramine, spermine, and dopamine were reacted with all-*trans*-retinal in order to form amino-retinoid standard compounds for use in their detection in the RPE (Scheme 11.1). Fast atom bombardment (FAB+) mass spectrometry (MS), high performance liquid chromatography (HPLC) and nuclear magnetic resonance spectroscopy (NMR) data were used to characterize the structures of the products formed in these reactions (figs. 11.1 and 11.2). A mixture of pyridinium, "A2E-like," products and single retinoid side-arm products resulting from Pictet-Spengler and modified Pictet-Spengler type mechanisms were formed in these reactions. Mono- and bis-retinoid compounds were observed in reactions of serotonin, tryptamine and norepinephrine with all-*trans*-retinal; bis- and tetra-retinoid compounds were observed in reactions with putrescine, spermidine and spermine with all-*trans*-retinal; and only bis-retinoid products were observed in the reaction of tyramine with all-*trans*-retinal (fig. 11.1). In the reaction between dopamine and all-*trans*-retinal, the same Pictet-Spengler or mono-retinoid products were observed as reported by Pezzella and Prota (2002); bis-retinoid compounds were also observed (fig. 11.2).

Because the tetra-retinoid compounds observed in the reactions reported above have two bis-retinoid pyridinium ring moieties similar to the single bis-retinoid pyridinium structure of A2E, the bioactivity of these compounds may be similar to A2E. However, because A2E is most damaging to cells through its oxidation products, (Ben-Shabat et al., 2002a; Sparrow et al., 2002; Sparrow et al., 2003) products resulting from reactions with putrescine, spermidine and spermine (tetra-retinoid compounds) may form double the number of oxidation products of A2E. Therefore, their oxidation products may prove to be even more damaging to RPE cells than photo-oxidized A2E. The observed Pictet-Spengler or mono-retinoid reaction products may also exhibit different cytotoxic mechanisms to those observed by A2E. Furthermore, the new bis-retinoid compounds formed may also exhibit alternative mechanisms of RPE cytotoxicity to that of A2E. It is therefore important to determine whether any of these compounds exist in RPE LF.

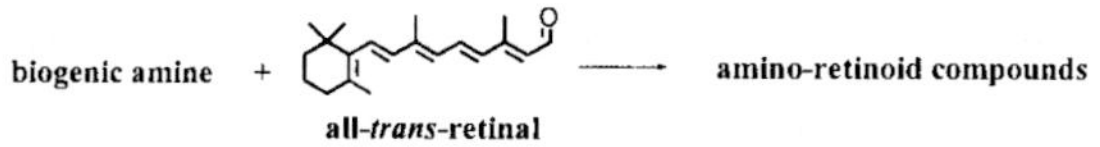

Scheme 11.1. General reaction between all-*trans*-retinal – **1** and any given biogenic amine found in the retina.

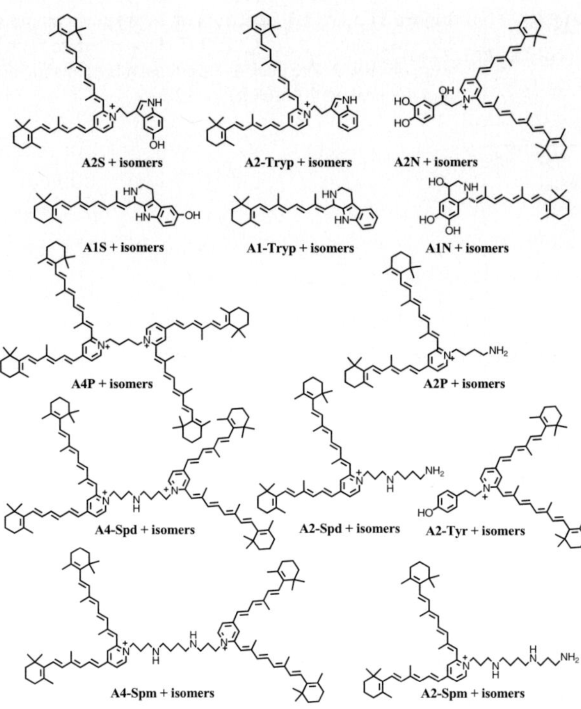

Figure 11.1. Reaction products observed for reactions of serotonin (A2S & A1S), tryptamine (A2- & A1-Tryp), norepinephrine (A2N & A1N), putrescine (A4P & A2P), spermidine (A2- & A4-Spd), tyramine (A2-Tyr) and spermine (A4- and A2-Spm) with all-trans-retinal.

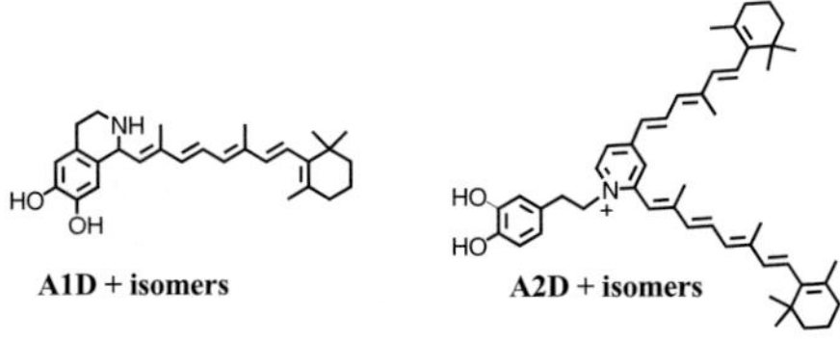

Figure 11.2. Reaction products observed in reaction of dopamine with all-trans-retinal (A1D & A2D).

3. EVIDENCE OF AMINO-RETINOID COMPOUNDS IN THE HUMAN RPE

It has been suggested that extraction of RPE cells provide the same fluorophores as extraction of RPE LF (Eldred and Katz, 1988). Therefore, organic extractions were performed directly on RPE cells. Modifications using dichloromethane were made to extraction procedures described by Eldred and Katz (1988) and Parish *et al.* (1998) which have resulted in the observation of additional peaks on HPLC chromatograms that appear to be amino-retinoid compounds.

During HPLC analysis of the extracted material, a peak was observed with a retention time of eight minutes, which has a very similar UV spectrum to that of A2E (fig. 11.3). This

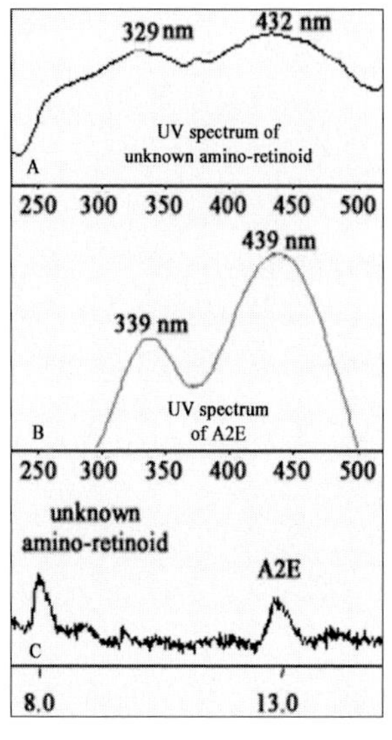

Figure 11.3. A: UV spectrum of an unknown amino-retinoid compound; B: UV spectrum of A2E; C: HPLC chromatogram from which the UV spectra were observed. The peak at 8 minutes represents the unknown amino-retinoid and the peak at 13 minutes is A2E.

data suggests that the compound with the eight minute retention time may be an amino-bis-retinoid. A co-injection of A2E with the extracts confirmed that the new peak was not A2E. A2E was observed with a retention time of 13 minutes. The structure of the unknown amino-retinoid compound is in the process of being elucidated; it is not yet certain if the compound is one of the standard amino-retinoids described above, although the UV matches nicely with several of these compounds.

In addition to the peak observed at eight minutes, retention times and UV spectra associated with A2N and isomers are remarkably similar to those of A2E and isomers, which suggests that bis-retinoid compounds, such as A2N and isomers that are similar to A2E in structure may be masked on HPLC chromatograms by A2E. Co-injections of A2N and A2E have been made and confirm that the 2 sets of peaks are on top of each other. Method sets are being worked out to separate these isomers, and to determine if other bis-retinoid peaks in RPE extracts may be hidden underneath the A2E peaks.

The goals of the research efforts described above were to: 1) chemically synthesize amino-retinoid standard compounds for use in their detection in RPE LF; and 2) perform organic extractions on human RPE cells to determine whether standard compounds or other amino-retinoid compounds may be present. Standard amino-retinoid compounds have been made, which have proved useful in the search for these and related compounds in the RPE. New peaks on HPLC chromatograms of injected human RPE extracts have also been observed, which appear to be yet uncharacterized amino-retinoid compounds. The results of this research are promising, and further research will continue in the synthesis of additional amino-retinoid standard compounds and in the characterization of novel amino-retinoids from the human RPE. A long term goal is to extensively characterize human RPE LF, which has been implicated in the cause of AMD, but has never been completely characterized. As new RPE LF compounds are isolated, a more complete understanding of the role of LF in the RPE will begin to be established. It is hoped that this knowledge will enable new preventative and therapeutic approaches to the retinal degenerative diseases which afflict so many.

4. ACKNOWLEDGEMENTS

The authors graciously thank Drs. Paul Bernstein and Prakash Bhosale for providing human retinal pigment epithelial cells for described experiments and Brigham Young University's Department of Chemistry and Biochemistry for funding.

5. REFERENCES

Aveldano, M. I., and Bazan, N. G., 1983, Molecular species of phosphatidylcholine, -ethanolamine, -serine, and -inositol in microsomal and photoreceptor membranes of bovine retina, *J. Lipid Res.* **24**:620.

Ben-Shabat, S., Itagaki, Y., Jockusch, S., Sparrow, J. R., Turro, N. J., and Nakanishi, K., 2002a, Formation of a nona-oxirane from A2E, a lipofuscin fluorophore related to macular degeneration, and evidence of singlet oxygen involvement. *Angew. Chem. Int. Ed.* **41**:814.

Ben-Shabat, S., Parish, C. A., Vollmer, H. R., Itagaki, Y., Fishkin, N., Nakanishi, K., Sparrow, J. R., 2002b, Biosynthetic studies of A2E, a major fluorophore of RPE lipofuscin. *J. Biol. Chem.* **277**:7183.

Birnbach, C. D., Jarvelainen, M., Possin, D. E., and Milam, A. H., 1994, Histopathology and immunocytochemistry of the neurosensory retina in fundus flavimaculatus, *Ophthalmol.* **101**:1211.

Bressler, S. B., Bressler, N. M., and Gragoudas, E. S., 2000, Age-related macular degeneration: drusen and geographic atrophy, in: *Principles and Practice of Ophthalmology*, Albert, D. M., Jakobiec, F. A., Azar, D. T., and Gragoudas, E. S., eds., vol 3. W. B. Saunders Co, Philadelphia, pp. 1982-1992.

Crain, R. C., Marinetti, G. V., and O'Brien, D. F., 1978, Topology of amino phospholipids in bovine retinal rod outer segment disk membranes, *Biochem.* **17**:4186.

Djamgoz, M. B. A. and Wagner, H. J., 1992, Localization and function of dopamine in the adult vertebrate retina, *Neurochem. Int.* **20**:139.

Drujan, B. D., Jaffe, E. H., Urbina, M., Ayala, C., and Drujan, Y., 1989, Interaction of DA and other biogenic amines in the retina, *Neurol. Neurobiol.* **49**:257.

Eldred, G. E. and Katz, M. L., 1988, Fluorophores of the human retinal pigment epithelium: separation and spectral characterization, *Exp. Eye Res.* **47**:71.

Eldred, G. E. and Lasky, M. R., 1993, Retinal age pigments generated by self-assembling lysosomotropic detergents, *Nature* **361**:724.

Finnemann, S. C., Leung, L. W., and Rodriguez-Boulan, E., 2002, The lipofuscin component A2E selectively inhibits phagolysosomal degradation of photoreceptor phospholipids by the retinal pigment epithelium, *Proc. Natl. Acad. Sci.* USA **99**:3842.

Gulcan, H. G., Alvarez, R. A., Maude, M. B., and Anderson, R. E., 1993, Lipids of human retina, retinal pigment epithelium, and Bruch's membrane/choroid: comparison of macular and peripheral regions, *Invest. Ophthalmol. Visual Sci.* **34**:3187.

Haralampus-Grynaviski, N. M., Lamb, L. E., Clancy, C. M. R., Skumatz, C., Burke, J. M., Sarna, T., and Simon, J. D., 2003, Spectroscopic and morphological studies of human retinal lipofuscin granules, *Proc. Natl. Acad. Sci.* **100**:3179.

Lentile, R., Russo, P., and Macaione, S., 1986, Polyamine localization and biosynthesis in chemically fractionated rat retina, *J. Neurochem.* **47**:1356.

Liu, J., Itagaki, Y., Ben-Shabat, S., Nakanishi, K., and Sparrow, J. R., 2000, The biosynthesis of A2E, a fluorophore of aging retina, involves the formation of the precursor, A2-PE, in the photoreceptor outer segment membrane, *J. Biol. Chem.* **275**:29354.

Lopez, P. F., Maumenee, I. H., de la Cruz, Z., and Green, W. R., 1990, Autosomal-dominant fundus favimaculatus. Clinicopathologic correlation, *Ophthalmol.* **97**:798.

Macaione, S., and Calatroni, A., 1978, Polyamines and ornithine decarboxylase activity in the developing rat retina, *Life Sciences*, **23**:683.

Makino-Tasaka, M., Suzuki, T., Nagai, K., and Miyata, S., 1985, Spatial distribution of visual pigment and dopamine in the bullfrog retina, *Exp. Eye Res.* **40**:767.

Mata, N. L., Weng, J., and Travis, G. H., 2000, Biosynthesis of a major lipofuscin fluorophore in mice and humans with ABCR-mediated retinal and macular degeneration, *Proc. Natl. Acad. Sci. USA*, **97**:7154.

Parish, C. A., Hashimoto, M., Nakanishi, K., Dillon, J., and Sparrow, J. R., 1998, Isolation and one-step preparation of A2E and iso-A2E, fluorophores from human retinal pigment epithelium, *Proc. Natl. Acad. Sci. USA* **95**:14609.

Pezzella, A., and Prota, G., 2002, Formation of novel tetrahydroisoquinoline retinoids by Pictet-Spengler reaction of dopamine and all-trans-retinal under conditions of relevance to biological environments, *Tett. Lett.*, **43**:6719.

Rabb, M. F., Tso, M. O., and Fishman, G. A., 1986, Cone-rod dystrophy. A clinical and histopathologic report, *Ophthalmol.* **93**:1443.

Sakai, N., Decatur, J., Nakanishi, K., and Eldred, G. E., 1996, Ocular age pigment "A2E": An unprecedented pyridinium bisretinoid, *J. Am. Chem. Soc.* **118**:1559.

Schutt, F., Ueberle, B., Schnolzer, M., Holz, F. G., and Kopitz, J., 2002, Proteome analysis of lipofuscin in human retinal pigment epithelial cells, *FEBS Lett.* **528**:217.

Sparrow, J. R., and Cai, B., 2001, Blue light-induced apoptosis of A2E-containing RPE: involvement of caspase-3 and protection by Bcl-2, *Invest. Ophthalmol. Vis. Sci.* **42**:1356.

Sparrow, J. R., Nakanishi, K., and Parish, C. A., 2000, The lipofuscin fluorophore A2E mediates blue light-induced damage to retinal pigmented epithelial cells, *Invest. Ophthalmol. Vis. Sci.* **41**:1981.

Sparrow, J. R., Parish, C. A., Hashimoto, M., and Nakanishi, K., 1999, A2E, a lipofuscin fluorophore, in human retinal pigmented epithelial cells in culture, *Invest. Ophthalmol. Vis. Sci.* **40**:2988.

Sparrow, J. R., Vollmer-Snarr, H. R., Zhou, J., Jang, Y. B., Jockusch, S., Itagaki, Y., and Nakanishi, K., 2003, A2E-epoxides Damage DNA in Retinal Pigment Epithelial cells. Vitamin E and other Antioxidants Inhibit A2E-epoxide formation, *J. Biol. Chem.* **278**:18207.

Sparrow, J. R., Zhou, J., Ben-Shabat, S., Vollmer, H., Itagaki, Y., and Nakanishi, K., 2002, Involvement of oxidative mechanisms in blue light induced damage to A2E-laden RPE, *Invest. Ophthalmol. Vis. Sci.* **43**:1222.

Taibi, G., and Schiavo, M. R., 1993, Simple high-performance liquid chromatographic assay for polyamines and their monoacetyl derivatives, *Journal of Chromatography, Biomedical Applications*, **614**:153.

Taibi, G., Schiavo, M. R., and Nicotra, C., 1995, Polyamines and ripening [maturation] of photoreceptor outer segments in chicken embryos, *International Journal of Developmental Neuroscience*, **13**:759.

Weingeist, T. A., Kobrin, J. L., and Watzke, R. C., 1982, Histopathology of Best's macular dystrophy, *Arch. Ophthalmol.* **100**:1108.

Winkler, B. S., Boulton, M. E., Gottsch, J. D., and Sternberg, P., 1999, Oxidative damage and age-related macular degeneration, *Mol. Vis.* **5**:32.

Witkovsky, P., Nicholson, C., Rice, M. E., Bohmaker, K., and Meller, E., 1993, Extracellular dopamine concentration in the retina of the clawed frog, Xenopus laevis, *Proc. Natl. Acad. Sci.* **90**:5667.

Young, R. W., 1988, Solar radiation and age-related macular degeneration. *Surv Ophthalmol.* **32**:252.

CHAPTER 12

ANNEXINS IN BRUCH'S MEMBRANE AND DRUSEN

Mary E. Rayborn*, Hirokazu Sakaguchi, Karen G. Shadrach,
John W. Crabb, and Joe G. Hollyfield

1. INTRODUCTION

Annexins (also known as lipocortins) are a family of calcium and phospholipid-binding proteins. At least 20 members of this family are known, and they have a wide range of potential functions, such as vesicular transport and trafficking, endocytosis, exocytosis and cell-cell adhesion. Annexins have molecular weights ranging between 30 and 40 kDA (the exception is annexin VI which is 66 kDA) and possess striking structural features. To qualify as an annexin, a protein must have 1) the presence of a conserved 70 amino acid domain repeated either 4 or 8 times in the overall structure (annexin VI has an 8 repeating amino acid domain; whereas the rest have 4), 2) the ability to bind phosopholipids in the presence of calcium. Annexins are exported from the cytosol to the exterior of cells across the plasma membrane by an unknown mehanism. When located extracellular, some annexins have been shown to function as receptors for other extracellular proteins: annexin II binds to tenascin and tissue plasminogen activator, while annexin V binds to collagen (Kojima, 1997).

Several annexins were identified in a recent proteomic study of drusen (Crabb et al, 2002). Specifically, peptides from annexins I, II, IV, and VI were found by LC MS/MS Q-Tof analysis of trypsin digested drusen proteins. To define the precise the distribution of these annexins in drusen and Bruch's membrane/choroid interface, we conducted immunocytochemical studies using a series of commercially available annexin antibodies.

2. METHODS

Bruch's membrane/choroid complexes from 70–90 year old donor eyes were isolated as described previously (Crabb et al, 2002). After fixation in 4% paraformaldehyde, the tissues were embedded in paraffin. Three μm thick sections were cut, placed on microscope slides and deparaffinized. After washing in PBS for 15 min, sections were incubated in 0.3%

* Mary E. Rayborn, et al., Department of Ophthalmic Research, Cole Eye Institute, Cleveland Clinic Foundation, 9500 Euclid Avenue, Cleveland, Ohio 44195.

Table 12.1 Annexin antibodies used in this study.

Antibody	Annexin I	Annexin II	Annexin IV	Annexin VI
Company	Zymed Lab	BD transduction	BD transduction	Zymed Lab
Concentration used	1 μg/ml	5 μg/ml	5 μg/ml	5 μg/ml
Type	Rabbit serum	Mouse IgG1	Mouse IgG1	Mouse IgG1

H_2O_2 in PBS for 20 min at room temperature. Sections were washed in PBS for 10 min and then incubated in 5% BSA and 0.3% triton X-100 in PBS for 2 hours. The sections were then probed with antibodies or anti-serum (below) in 1% BSA in PBS overnight at 4°C. For the control incubations, mouse IgG1 or rabbit normal serum at a concentration corresponding to that of the primary antibody or serum was used. After washing in 0.3% triton X-100 in PBS and incubation in biotin conjugated anti-mouse or- rabbit IgG antibody, the sections were washed in PBS and incubated in avidin and biotin in PBS for 30 min, the sections were washed in PBS and then developed with DAB for 2 minutes. After dehydration, the sections were mounted with mounting medium. The sections were examined by light microscopy and the images were digitized using a Hamamatsu CCD camera and processed with Adobe Photoshop 4.0 software on a Power Macintosh computer.

3. DISCUSSION AND RESULTS

Annexin I immunoreactivity was intense in drusen, both on the surface and in the interior. Annexin I antibody also intensely labeled Bruch's membrane and the choroid. Annexin II immunoreactivity was present on the surface of drusen associated with remaining basal lamina of the RPE that had not been completely removed when the Bruch's membrane/choroid preparations had been initially isolated. No immunoreactivity was present in the interior of drusen, but punctate labeling was evident in the choriocapillaris. Annexin IV showed no immunoreactivity in drusen or in Bruch's membrane, but the choroid appeared to be lightly labeled. Annexin VI showed heavy labeling of drusen and Bruch's membrane. Additionally, this antibody stained the surface of drusen more intensely than any of the other annexin antibodies employed. Examples of this staining pattern are shown in the accompanying figures.

Of the four antibodies used, only annexin I and annexin VI clearly indicate that these two proteins are present in the interior of drusen. Annexin II antibody labeled the basal lamina of the RPE that remained associated with the drusen surface. It was not seen in the interior of drusen, and therefore this protein that was identified in our previous proteomic analysis of drusen must be considered as a part of closely associated cellular contaminants in the drusen preparation utilized (Crabb et al, 2002).

Only annexin IV antibody failed to label any component of the Bruch's membrane-choroid sample. We did observe annexin IV immunoreactivity at the level of the inner limiting membrane of the retina (not shown). Either annexin IV is only occasionally found in drusen and the three samples we have studied did not contain this annexin type, or the antibody employed does not function well with the fixation techniques employed. Since this antibody stains the outer limiting membrane of the retina, this latter interpretation is unlikely.

Annexins are a family of proteins that bind calcium ions and acidic phospholipids. Possible functions include being anti-inflammatory via the inhibition of phospholipase A2

 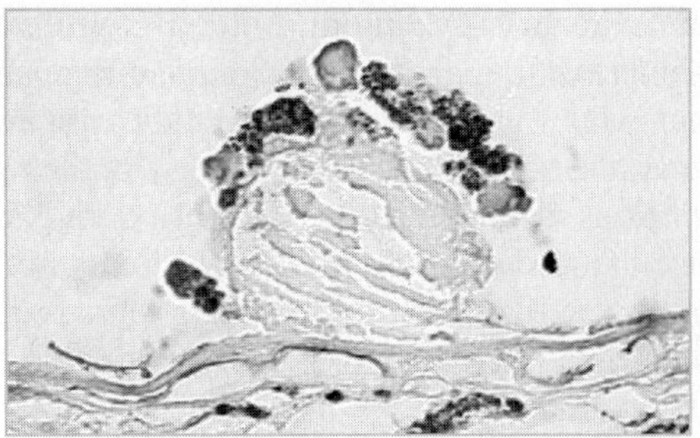

Annexin I. Immunocytochemistry. Serial sections from a 75 year old donor eye. Tissue in the left panel that was treated with the antibody shows intense immunoreactivity over the druse interior, as well as Bruch's membrane and the choroid. The right control panel that was treated with non-immune rabbit serum shows no label. Some remaining melanin from the RPE decorates the upper surface of the druse in both sections.

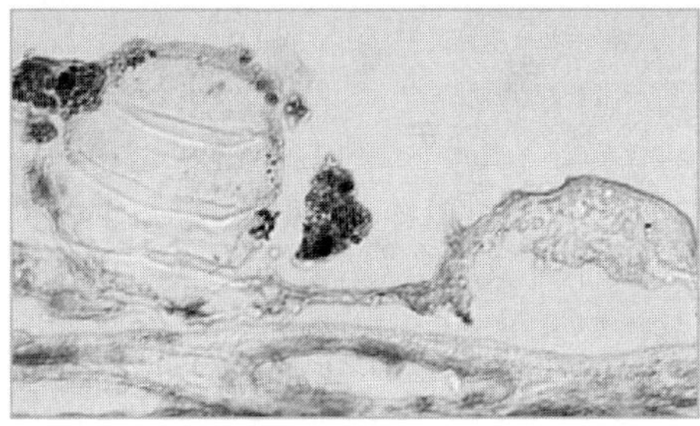 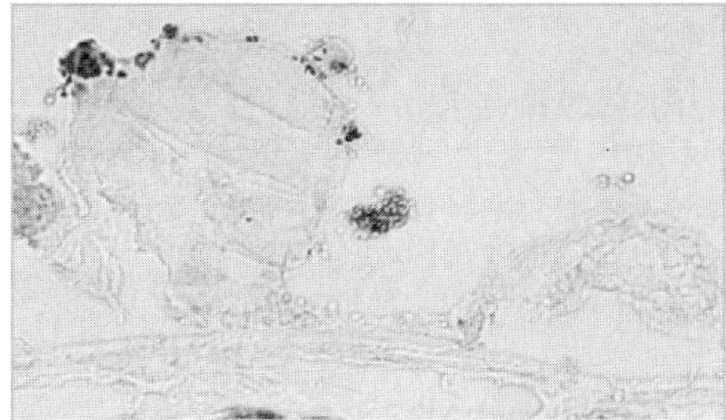

Annexin II. Immunocytochemistry. The left panel treated with the antibody shows no immunoreactivity over the drusen interior and only light labeling of the drusen surface (arrow). Staining in this area represents Annexin II associated with the remaining basal lamina of the RPE. Bruch's membrane is not labeled but choroidal vessels (**) are labeled. Right control panel treated with non-immune mouse IgG shows no labeling. Remaining melanin from the RPE decorates the druse surface in both panels.

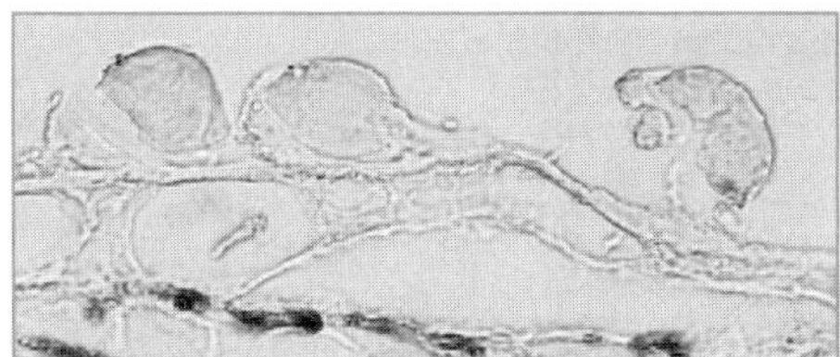 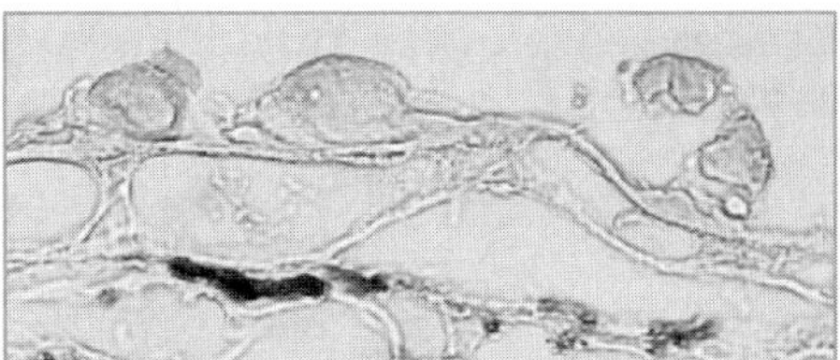

Annexin IV. Immunocytochemistry. No immunoreactivity was evident using this antibody in the panel on the left. It is virtually identical to the control in the panel on the right that employed non-immune mouse IgG.

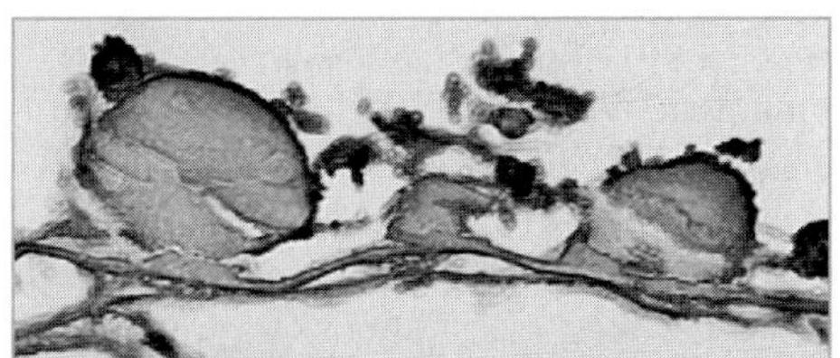 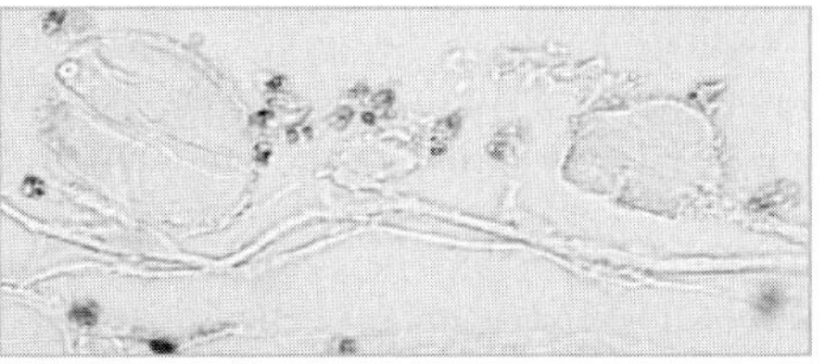

Annexin VI. Immunocytochemistry. Annexin-VI immunoreactivity was evident on the surface and interior of drusen. Labeling of Bruch's membrane is also observed. Intense labeling in remaining RPE debris is present. No labeling is evident in the control section on the right that was treated with non-immune mouse IgG.

(annexins I-VI), inhibiting blood coagulation (annexins I-VI), mediating exocytosis via membrane to membrane fusion, and regulating cytoskeletal organization via actin binding (annexins 1 and 2). Numerous studies are available on identification of annexins in the vascular and skeletal systems, but few on annexins in the visual system. There are reports of annexins identified in the rat cornea during inflammation (Matsuda et al, 1999) and in galactose-induced cataracts in rats (Koshimoto et al, 1992). Zernii et al (2003) recently detected and isolated annexin IV in bovine retinal rod outer segments. Further studies are needed to determine the role of the annexins in the retina and posterior layers of the eye.

4. ACKNOWLEDGEMENTS

Supported by NIH grants EY14240, EY15638 and a Research Center grant from The Foundation Fighting Blindness.

5. REFERENCES

Crabb, J. W., Miyagi, M., Gu, X., Shadrach, K., West, K., Sakaguchi, H., Kamei, M., Hasan, A., Yan, L., Rayborn, M. E., Salomon, R., and Hollyfield, J. G., 2002, Drusen proteome analysis: an approach to the etiology of age-related macular degeneration. Proc Natl Acad Sci. (USA) **99**:14682-14687.

Kojima, K., Yamamoto, K., Irimura, T., Osawa, T., Ogawa, H., and Matsumoto. I., 1996, Characterization of carbohydrate-binding protein p33/41 relation with annexin IV, molecular basis of the doublet forms (p33 and p41), and modulation of the carbohydrate binding activity by phospholipids. J Biol Chem. **271**:7679-7685.

Koshimoto, H. N., Tokuda, M., Matsui, H., Itano, T., Hasegawa, E., and Hatase, O., 1992, Involvement of lipocortin I in development of galactose induced cataracts in rat. Exp Eye Res. **54**:817-820.

Matsuda, A., Tagawa, Y., Yamamoto, K., Matsuda, H., and Kusakabe, M., 1999, Identification and immunohistocemical localization of annexin II in rat cornea. Curr Eye Res. **19**:368-375.

Mollenhauer, J., 1997, Annexins. Cell Mol Life Sci. **53**:506-507.

Zernii, E. Yu., Tikhimirova, P. P., and Senin, I. I., 2003, Detection of annexin IV in bovine retinal rods. Biochemistry (Mosc). **68**:129-160.

PART III

ANIMAL MODELS OF RETINAL DEGENERATION

MOLECULAR MECHANISMS OF PHOTORECEPTOR DEGENERATION IN RP CAUSED BY IMPDH1 MUTATIONS

Aileen Aherne[1], Avril Kennan[1], Paul F. Kenna[1], Niamh McNally[1], G. Jane Farrar[1], and Pete Humphries[1]

1. INTRODUCTION

Mutations within the gene encoding inosine monophosphate dehydrogenase 1 (*IMPDH1*), the rate-limiting enzyme of the *de novo* pathway of guanine nucleotide biosynthesis, have been shown to cause the RP10 form of autosomal dominant retinitis pigmentosa (RP). This form of RP is generally early-onset and severe in those patients that have been identified to date. The two mutations originally identified in large RP10 families in 2002 were Arg224Pro and Asp226Asn substitutions, and since then several additional mutations have been identified in RP families and individual patients (Kennan *et al.*, 2002; Bowne *et al.*, 2002; Daiger *et al.*, 2003; Grover *et al.*, 2004).

IMPDH1 is specifically responsible for the conversion of inosine monophosphate (IMP) to xanthosine monophosphate (XMP), which is a precursor for guanine nucleotides including GMP and GTP. The IMPDH1 protein is expressed in many tissues of the body but it remains unknown why mutations in this gene manifest in clinical symptoms only in retinal tissues. We have previously shown that *IMPDH1* transcript is expressed at high levels specifically in the photoreceptor cells in the outer nuclear layer of the retina (Kennan *et al.*, 2003). It is likely that high levels of IMPDH1 are required here for the generation of GTP and cGMP which are essential at several stages during phototransduction processes, and also for the provision of guanine nucleotides as RNA precursors for rhodopsin transcript, also required in large quantities in photoreceptor cells. Almost all the confirmed RP mutations identified to date, however, are located in a region of the protein referred to as the CBS domain, which is not close to the active site of the IMPDH1 enzyme. We have in fact shown previously that two of these substitutions (Arg224Pro and Asp226Asn), when introduced

---

[1] The Ocular Genetics Unit, Department of Genetics, Trinity College, Dublin 2, Ireland.

into the protein sequence, do not have an adverse effect on the enzyme activity of IMPDH1 (Aherne *et al.*, 2004). It appears therefore that lack of enzyme activity alone is not the cause of disease symptoms.

A mouse deficient in the *Impdh type 1* gene was generated using standard gene targeting techniques by Beverly Mitchell and colleagues, University of North Carolina, USA, and was kindly made available for this study. This group reported that the *Impdh1$^{-/-}$* mouse appears to have a normal phenotype and concluded that IMPDHI enzymatic activity is not essential for normal mouse development or fertility (Gu *et al.*, 2003). In the present study we show following electroretinographic (ERG) and histological analysis, that *Impdh1$^{-/-}$* knockout mice display symptoms of a mild retinopathy. In addition, both wild-type and mutant IMPDH1 proteins have been expressed in cell culture systems in an effort to understand the molecular mechanisms causing photoreceptor degeneration in this form of RP. Results from these studies point to a probable dominant negative effect of mutant IMPDH1 protein causing the RP10 form of RP, and indicate that protein misfolding and aggregation, rather than reduced IMPDH1 enzyme activity or simply deficiency of IMPDH1, is the likely cause of photoreceptor degeneration causing the severe RP phenotype experienced by human subjects.

2. MATERIALS AND METHODS

2.1. Retinal Histology and Electroretinography

Eyes were perfused overnight with a mixture of 2.5% glutaraldehyde and 2% paraformaldehyde on 0.1 M phosphate buffer, pH 7.4, and processed in Epon. Sections 1 μm thick were cut through the optic nerve head, along the vertical meridian of the eye, and were stained with toludine blue for light microscopy. Mouse ERGs were recorded at 6 weeks of age (2 animals), 5 months (6 animals), 8 months (4 animals) 11 months (5 animals) and 13 months (3 animals). Animals were dark-adapted overnight and prepared for electroretinography under dim red light. ERGs were carried out as described fully in Aherne *et al.* 2004. Rod isolated responses were recorded using a dim white flash (−25 dB maximal intensity) presented in the dark-adapted state. The maximal combined rod/cone response to the maximal intensity flash was then recorded. Following light adaptation for 10 minutes to a background illumination of 30 candelas per m^2 presented in Ganzfeld bowl the cone-isolated responses were recorded to the maximal intensity flash (3 candelas per m^2 per second) presented as a single flash and 10 Hz flickers.

2.2. Expression of IMPDH1 Proteins in Mammalian Cells

The Arg224Pro and Asp226Asn amino acid substitutions were introduced into the WT IMPDH1 cDNA sequence using the Quikchange site-directed mutagenesis kit (Stratagene). The WT and two mutant *IMPDH1* cDNA sequences were sub-cloned into the pcDNA3.1/HisC (Invitrogen) vector, which incorporates an N-terminus His-tag in the resulting protein. Human embryonic kidney 293 (HEK 293T) cells were cultured in DMEM with 10% fetal bovine serum at 37°C and an atmosphere of 5% CO_2. Cells were seeded at a density of 10^6 cells per 10 cm dish overnight and were transiently transfected using 20 μg of DNA and the calcium phosphate method. Cells were harvested by trypsinisation after

48 hours, washed twice with PBS and centrifuged gently to pellet. A protein isolation protocol for extraction of cell fractions involved a series of lysis and centrifugation steps which was adapted from Deery *et al.* 2002. Aliquots containing equal amounts of total protein (30 μg) were loaded onto 10% SDS-PAGE gels in duplicate. One gel was stained with Coomassie blue and the proteins on the second gel were transferred by electroblotting onto nitrocellulose membrane. Blots were probed with an Anti His-tag monoclonal antibody (Novagen 200 ng/ml) and HRP-conjugated anti-mouse IgG secondary antibody (Sigma).

3. RESULTS

ERG analysis of retinal function was carried out on *Impdh1⁻ᐟ⁻* mice at a number at a number of different ages. Rod-isolated responses (Fig. 13.1a), maximal dark-adapted, combined rod and cone responses (Fig. 13.1b) and light-adapted cone responses to a single flash (Fig. 13.1c) were recorded at 6 weeks, 5 months, 8 months, 11 months and 13 months of age. By 11 months, all responses showed a significant reduction in a- and b-wave amplitudes and, in the case of the cone responses, significant delay in b-wave timing (Fig. 13.1c). In the maximal, dark-adapted, combined rod and cone responses, progressive reduction in wave amplitudes was noted from 5 months of age indicating disturbance of phototransduction. It appears that the reduction in ERG amplitudes observed in *Impdh1⁻ᐟ⁻* mice are much milder than those recorded in humans with dominant *IMPDH1* mutations, where generally

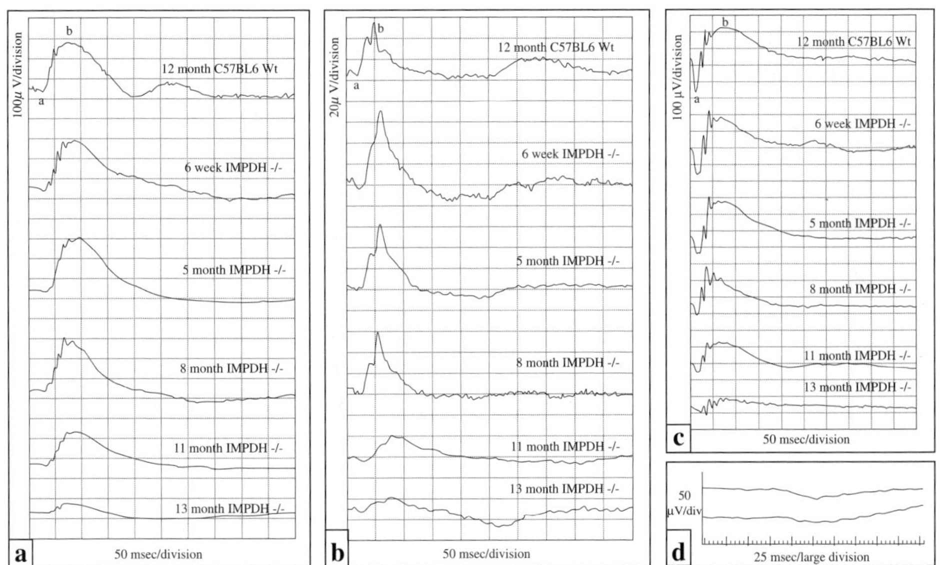

Figure 13.1. ERG analysis of *Impdh1⁻ᐟ⁻* mice. **(a)** Rod-isolated responses, **(b)** light-adapted cone responses to single flash and **(c)** maximal, dark-adapted, combined rod and cone responses were recorded in a 12-month C57 wild-type animal, and *Impdh1⁻ᐟ⁻* animals at 6 weeks of age, 5 months, 8 months, 11 months and 13 months. **(d)** ERG analysis of combined rod and cone response in a 17-year old affected patient harbouring an Arg224Pro IMPDH1 mutation.

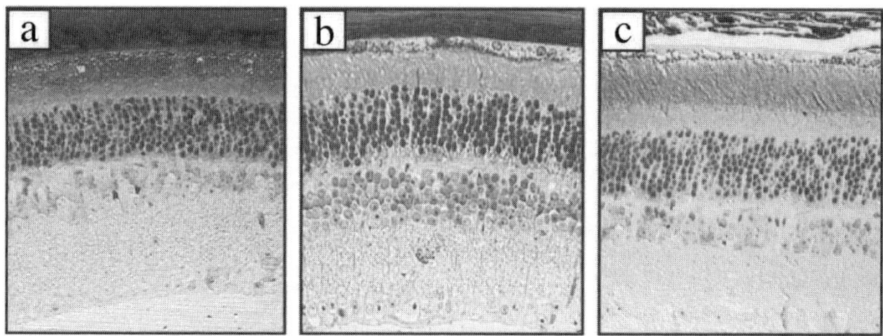

Figure 13.2. Light micrographs of retinal sections from *Impdh1*[-/-] mice. Sections shown are **(a)** 6 month *Impdh1*[-/-], **(b)** 10 month *Impdh1*[-/-], **(c)** 10 month wild-type C57/BL6. The outer nuclear layer thickness of *Impdh1*[-/-] retina is virtually indistinguishable to that of wild-type up to 10 months of age. (Figures adapted from Aherne et al., Human Molecular Genetics: On the molecular pathology of neurodegeneration in IMPH1-based retinitis pigmentosa, 2004, 13:641–50, by permission of Oxford University Press).

all ERG responses are unrecordable by the end of the second decade. The *Impdh1*[-/-] mouse model at 4 months of age, which we estimate to be approximately equivalent to a teenage human subject in terms of relative lifespan, displays no detectable structural or functional degeneration of the retina. In contrast the response in an affected 17-year old male from a large RP10 family to the maximal intensity flash presented in the dark adapted state indicates that there is no recordable photoreceptor activity (Fig. 13.1d) (Jordan *et al.*, 1993; Kennan *et al.*, 2002). Photoreceptor dysfunction in the *Impdh1*[-/-] mouse does not appear to be the result of degeneration of photoreceptor cells, since the outer nuclear layer thickness of the retinas in these mice is similar to wild type retinas up to 10 months of age, and even by 13 months of age we have found that there is only marginal loss of nuclei from the outer nuclear layer of the retina (Fig. 13.2). There is evidence of some degree of disorganisation of photoreceptor outer segments in older animals, but the actual number of ONL nuclei does not appear to be affected. Thus, IMPDH1 does not appear to be essential for normal retinal development or early visual function, and photoreceptor cells appear to survive in it's absence.

Expression studies were carried out on wild-type and mutant IMPDH1 proteins in mammalian cells and involved transfection of constructs expressing IMPDH1 sequences into cells and subsequent extraction of cytosolic and nuclear protein fractions. This analysis indicated that a significant difference in protein solubility was detectable between the wild-type and two mutant forms of the IMPDH1 protein (Fig. 13.3). This suggests that the mutant forms of IMPDH1 protein may be misfolded, causing them to form insoluble aggregates when expressed at high levels in cells.

4. DISCUSSION

Mutations within the IMPDH1 gene are thought to account for 3-5% of autosomal dominant RP (Daiger *et al.*, 2003). Previous studies carried out in this laboratory have shown that two of the confirmed IMPDH1 mutations studied to date (Arg224Pro and Asp226Asn) do not appear to affect the enzymatic activity of IMPDH1 (Aherne *et al.*, 2004). Expres-

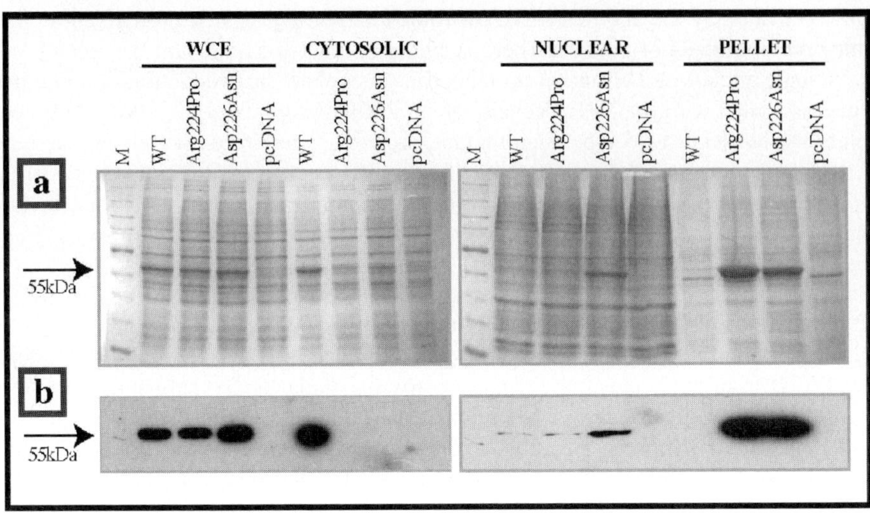

Figure 13.3. Wild-type and mutant IMPDH1 proteins expressed in HEK293T cells. **(a)** SDS-PAGE with Coomassie Blue staining of protein extracts from cells transfected with pcDNA3.1/His wild-type (WT), Arg224Pro and Asp226Asn mutant IMPDH1 sequences, and pcDNA vector alone (negative control). Whole cell extracts (WCE), soluble cytosolic, nuclear and final pelleted protein fractions were extracted, **(b)** Western blot probed with Anti-His antibody (Figure adapted from Aherne et al., Human Molecular Genetics: On the molecular pathology of neurodegeneration in IMPH1-based retinitis pigmentosa, 2004, 13:641–50, by permission of Oxford University Press).

sion of IMPDH1 proteins in mammalian cells has shown that a substantial difference in solubility exists between the wild-type and mutant forms of the protein, with the mutant proteins being less soluble. This is most likely caused by a detrimental effect on normal protein folding brought about by the presence of pathogenic mutations. This is believed to be the situation in the case of the Arg224Pro mutation and has also been predicted by computer molecular modelling studies on wild-type and mutant IMPDH1 proteins (Aherne *et al.*, 2004). Analysis of retinal function and structure in the *Impdh1$^{-/-}$* mouse has shown that this animal model displays symptoms of a mild retinopathy. The retina has the highest metabolic rate of any tissue of the body, with photoreceptors having a particularly high requirement for GTP in visual transduction processes (La Cour, 2002). Therefore, we suggest that depletion of the GTP pool over time is the most likely explanation for the reduction in ERG amplitudes seen in the retinas of older *Impdh1$^{-/-}$* mice.

The *Impdh1$^{-/-}$* mouse model displays relatively normal ERG amplitudes and histological features up to the age of 5 months, which when compared to other mouse models of RP that have been generated, would not be considered a severe degenerative retinopathy. Therefore, it appears unlikely that the autosomal dominant segregation pattern which is characteristic of the RP10 form of retinitis pigmentosa, is caused as a result of 'haploinsufficiency' for the normal *IMPDH1* gene product. Rather, disease pathology is probably caused by a gain-of-function or dominant-negative phenotypic effect exerted by mutant protein. Protein misfolding now appears to be a common theme in discussions of the molecular mechanisms of degeneration in adRP. A number of groups have reported that misfolded RP-associated proteins can initiate the 'unfolded protein response' (UPR) in cells, which is a stress response activated by the appearance of misfolded and/or aggregated proteins in the endoplasmic reticulum (ER) of the cell. There is evidence that rhodopsin and the recently identified carbonic anhydrase IV mutant proteins can upregulate the production of a number of proteins associated with the UPR (Rebello *et al.*, 2004; Frederick *et al.*, 2001). Differences in solubility between wild-type and mutant proteins, similar to those seen here with IMPDH1, have previously been noted with the RP-associated splicing factor protein, PRPF31 (Deery *et al.*, 2002). In addition, protein folding defects associated with rhodopsin mutations have been demonstrated to cause the formation of insoluble aggresomes in cells (Illing *et al.*, 2002; Saliba *et al.*, 2002).

Interestingly, the results shown here may present a new option for therapy in this form of RP. It is possible that the suppression of both wild-type and mutant IMPDH1 alleles may, in principle, be sufficient to remove the major dominant negative pathological effect of mutant IMPDH1 protein. This may be particularly beneficial given that using current gene suppression technologies such as RNA interference, it has proven difficult to target single base pair mutations (Miller *et al.*, 2003). Targeting both wild-type and mutant *IMPDH1* alleles may result in a situation in which photoreceptor neurons continue to survive and remain sufficiently functional such as to provide useful vision, possibly well into adult life. In addition, supplementation with XMP or GTP molecules may be sufficient to ameliorate disease symptoms, without the need for gene replacement. The current focus of our research is on the generation of dominant animal models of RP caused by mutations within the IMPDH1 gene and on the assessment of novel therapies for this form of RP.

5. ACKNOWLEDGEMENTS

The Ocular Genetics Unit of TCD is supported by grants from the from the Wellcome Trust, the Health Research Board of Ireland, Science Foundation Ireland, the British RP Society, the Foundation Fighting Blindness (USA), Fighting Blindness Ireland, and the European Union 5th Framework Programme. The Ocular Genetics Unit is a member of the HEA-Ireland-sponsored Biopharmaceutical Science Network.

6. REFERENCES

Aherne, A., Kennan, A., Kenna P.F., McNally N., Lloyd D.G., Alberts I.L., Kiang A.S., Humphries M.M., Ayuso C., Engel P.C., Gu J.J., Mitchell B.S., Farrar G.J., Humphries P., 2004, On the molecular pathology of neurodegeneration in IMPDH1-based retinitis pigmentosa. *Hum Mol Genet.* **13**:641-650.

Bowne, S.J., Sullivan, L.S., Blanton, S.H., Cepko, C.L., Blackshaw, S., Birch, D.G., Hughbanks-Wheaton, D., Heckenlively, J.R., and Daiger, S.P., 2002, Mutations in the inosine monophosphate dehydrogenase 1 gene (IMPDH1) cause the RP10 form of autosomal dominant retinitis pigmentosa. *Hum Mol Genet.* **11**:559.

Daiger, S.P., Sullivan, L.S., Bowne, S.J., Kennan, A., Humphries, P., Birch, D.G., Heckenlively, J.R.; RP1 Consortium, 2003, Identification of the RP1 and RP10 (IMPDH1) genes causing autosomal dominant RP. *Adv Exp Med Biol.* **533**:1-11.

Deery, E.C., Vithana, E.N., Newbold, R.J., Gallon, V.A., Bhattacharya, S.S., Warren, M.J., Hunt, D.M. and Wilkie, S.E., 2002, Disease mechanism for retinitis pigmentosa (RP11) caused by mutations in the splicing factor gene PRPF31. *Hum Mol Genet.* **11**:3209-3219.

Frederick, J.M., Krasnoperova, N.V., Hoffmann, K., Church-Kopish, J., Ruther, K., Howes, K., Lem,, J. and Baehr, W., 2001, Mutant rhodopsin transgene expression on a null background. *Invest Ophthalmol Vis Sci.* **42**:826-833.

Grover, S., Fishman, G.A. and Stone, E.M., 2004, A novel IMPDH1 mutation (Arg231Pro) in a family with a severe form of autosomal dominant retinitis pigmentosa. *Ophthalmology.* **111**:1910-1916.

Gu, J.J., Tolin, A.K., Jain, J., Huang, H., Santiago, L. and Mitchell, B.S., 2003, Targeted Disruption of the Inosine 5'-Monophosphate Dehydrogenase Type I Gene in Mice. *Mol Cell Biol.* **23**:6702-6712.

Illing, M.E., Rajan, R.S., Bence, N.F. and Kopito, R.R., 2002, A rhodopsin mutant linked to autosomal dominant retinitis pigmentosa is prone to aggregate and interacts with the ubiquitin proteasome system. *J Biol Chem.* **277**:34150-34160.

Jordan, S.A., Farrar, G.J., Kenna, P., Humphries, M.M., Sheils, D.M., Kumar-Singh, R., Sharp, E.M., Ayuso, C., Benitez, J., and Humphries, P., 1993, Localization of an autosomal dominant retinitis pigmentosa gene to chromosome 7q. *Nat Genet.* **4**:54.

Kennan, A., Aherne, A., Palfi, A., Humphries, M., McKee, A., Stitt, A., Simpson, D.A., Demtroder, K., Orntoft, T., Ayuso, C., Kenna, P.F., Farrar, G.J., and Humphries, P., 2002, Identification of an IMPDH1 mutation in autosomal dominant retinitis pigmentosa (RP10) revealed following comparative microarray analysis of transcripts derived from retinas of wild-type and Rho (-/-) mice. *Hum Mol Genet.* **11**:547.

Kennan, A., Aherne, A., Bowne, S.J., Daiger, S.P., Farrar, G.J., Kenna, P.F., Humphries, P., 2003, On the role of IMPDH1 in retinal degeneration. *Adv Exp Med Biol* **533**:13-18.

Kopito, R.R., 2000, Aggresomes, inclusion bodies and protein aggregation. *Trends Cell Biol.* **10**:524-530.

La Cour, M., 2002, The retinal pigment epithelium. In Kaufman, P.L., Alm, A. (eds), *Adler's Physiology of the Eye.* 10th Ed. Mosby, St. Louis, pp. 348-357.

Miller, V.M., Xia, H., Marrs, G.L., Gouvion, C.M., Lee, G., Davidson, B.L. and Paulson, H.L., 2003, Allele-specific silencing of dominant disease genes. *Proc. Natl. Acad. Sci. U S A,* **100**:7195-7200.

Rebello, G., Ramesar, R., Vorster, A., Roberts, L., Ehrenreich, L., Oppon, E., Gama, D., Bardien, S., Greenberg, J., Bonapace, G., Waheed, A., Shah, G.N. and Sly, W.S., 2004, Apoptosis-inducing signal sequence mutation in carbonic anhydrase IV identified in patients with the RP17 form of retinitis pigmentosa. *Proc. Natl. Acad. Sci.* **27**:6617-6622.

Saliba, R.S., Munro, P.M., Luthert, P.J. and Cheetham, M.E., 2002, The cellular fate of mutant rhodopsin: quality control, degradation and aggresome formation. *J Cell Sci.* **115**:2907-2918.

CHAPTER 14

BIOCHEMICAL FUNCTION OF THE LCA LINKED PROTIEN, ARYL HYDROCARBON RECEPTOR INTERACTING PROTEIN LIKE–1 (AIPL1)

Role of AIPL1 in retina

Matthew L. Schwartz, James B. Hurley, and Visvanathan Ramamurthy*

1. INTRODUCTION

Leber congenital amaurosis (LCA) is a clinically and genetically heterogeneous form of early-onset retinal dystrophy that is usually recessively inherited. LCA is the most rapid and severe form of congenital blindness, and it represents approximately 5% of all inherited retinopathies.[1] Clinically, LCA is characterized by severely impaired vision and a weak or absent electroretinogram evident within the first year of life. To date, seven genes have been independently linked to LCA.[2] The majority of mutations implicated in the causation of LCA are genetically consistent with recessively inherited loss-of-function pathogenesis mechanisms.[2]

The gene *AIPL1* was originally identified by genetic analysis of patients with LCA.[3] The gene was given the name *AIPL1* (*a*ryl hydrocarbon receptor-*i*nteracting *p*rotein like-1) because it encodes a protein (AIPL1) with sequence homology (49% identity, 69% similarity) to the protein AIP (*a*ryl hydrocarbon receptor-interacting *p*rotein).[3] Human *AIPL1* contains 6 exons encoding a 384 amino acid protein that contains 3 tetratricopeptide repeat (TPR) domains and a C-terminal proline-rich region.[3] The TPR domains are highly conserved amongst mammals, whereas the proline-rich region is thought to be present only in primates and shows considerable sequence variation amongst primates.[4] AIPLI is expressed exclusively in the retina and the pineal gland.[3, 5, 6] During photoreceptor development in humans, AIPL1 is expressed in both rod and cone photoreceptors, but it's expression is restricted to rods in the adult.[7]

AIPL1 mutations result in the most clinically severe forms of LCA, and it is estimated that *AIPL1* mutations are responsible for approximately 7% of all LCA.[8] LCA-linked muta-

---

*Department of Biochemistry, University of Washington, Seattle, WA 98195, U.S.A.

89

tions include missense mutations, nonsense mutations, and short deletions.[8] *AIPL1* mutations linked to LCA are present in either the N-terminal immunophilin like domain (class I) or in the TPR domain (class II). Mutations in the C-terminal proline rich domain (class III) have been linked to dominant cone-rod dystrophy or juvenile retinitis pigmentosa and LCA.[9]

2. ROLE OF AIPL1 IN RETINA- FINDINGS FROM AIPL1 DEFICIENT MICE

To elucidate the role of AIPL1 in the retina and to develop an animal model to study LCA caused by AIPL1 deficiency, we created an *AIPL1* knock out mouse. *AIPL1*[-/-] mice demonstrate a phenotype consistent with LCA. Specifically, AIPL1[-/-] mice exhibit no measurable electroretinogram (ERG) response at any age and are completely blind at birth[10] (Fig. 14.1). Ultra-structural details of the retina analyzed by electron microscopy show no obvious difference between wild type and knock out mice at post natal day 8 (P8) (Fig. 14.1).[10] At P8, AIPL1 deficient mice show a normal complement of rod and cone photoreceptor cells with morphologically normal rod and cone outer segments. This suggests that AIPL1 does not play an essential role in the initial formation of rods and cone photoreceptor cells. At P11, the photoreceptor layer of the retina in the knock out mouse is morphologically indistinguishable form that of wild type by light microscopy, but ultra structural details observed by electron microscopy show disorganized and fragmented outer segments compared to wild type.[10] This suggests that AIPL1 plays an essential role in either maintaining the outer segment and/or further photoreceptor cell differentiation after P9 in mice. The photoreceptor nuclear layer is reduced to half at P14 and at P18 the photoreceptor nuclear layer is only 1 cell thick in mice lacking AIPL1.[10] By four weeks after birth, the degeneration is complete (Fig. 14.1). Both rod and cone photoreceptor cells degenerate at a similarly rapid rate.[10]

At the molecular level, differences between wild type and AIPL1[-/-] retinas appear earlier than the morphological differences. At P8, prior to the onset of retinal degeneration, AIPL1 deficient mice show reduced levels of cGMP phosphodiesterase (PDE) protein.[10] The reduction in level of PDE is specific, as the levels of other photoreceptor-specific proteins such as Rhodopsin (Rho), Guanylyl cyclase (GC-E) are normal.[10] All three subunits of PDE $\alpha\beta\gamma$ are reduced by 90% despite normal levels of the mRNAs.[10] Additionally, no cGMP-dependent PDE activity can be detected in the knockout retinas, implying that the PDE that is present in AIPL1[-/-] is dysfunctional. Consistent with the loss of PDE activity, cGMP levels are high starting at P8.[10] Destabilization of rod cGMP PDE as a pathogenic mechanism has precedent in the mouse. The well-characterized *rd* (retinal dystrophic) mouse results from a truncation mutation and/or a viral insertion in the rod PDEβ gene that causes a reduction in PDEβ mRNA and loss of rod cGMP PDE.[11,12] In *rd/rd* mice, there is a pattern of rapid photoreceptor cell degeneration similar to the degeneration that occurs in AIPL1[-/-] mice.[12] However, there are significant differences between *rd/rd* and AIPL1[-/-] mice.

In AIPL1[-/-] mice, both rods and cone photoreceptor cells degenerate at a similar rapid rate, whereas in *rd/rd* mice, rods degenerate faster than cones.[10,12] In AIPL1[-/-] mice there is no recordable ERG at any age, whereas in *rd/rd* mice there are some cone responses at postnatal day 12.[10,12] This is consistent with the fact that in humans, mutations in AIPL1 cause severe blindness that affects both rods and cones whereas deficiencies in PDE cause retinitis pigmentosa, primarily a rod disease.[13]

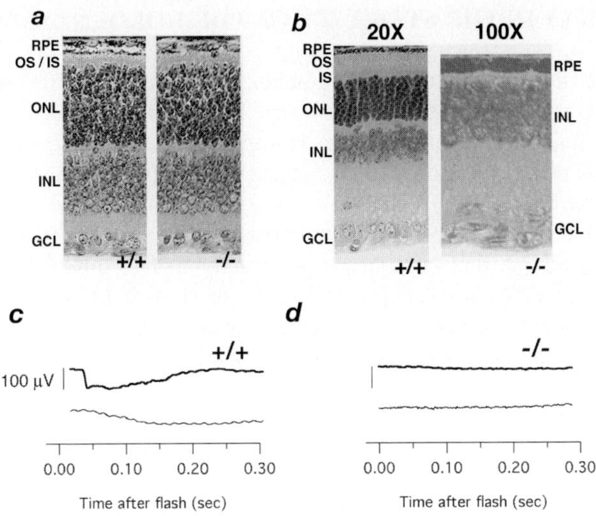

Figure 14.1. Rapid degeneration (*a-b*) and electrophysiological responses (*c-d*) in mice lacking AIPL1. Paraffin sections (5 µm) from whole eye stained with hematoxylin and eosin. RPE, retinal pigment epithelium; OS, outer segments; IS, inner segments; ONL, outer nuclear layer; INL, inner nuclear layer; GCL, ganglion cell layer *a)* At P8, there is no difference in the thickness of ONL between wild-type and *Aipl1$^{-/-}$* littermates *b)* By P30, the degeneration of photoreceptor is complete with no ONL. *Aipl1$^{-/-}$* section is shown at 100X magnification compared to wild-type, which is shown at 20X magnification. *c and d)* ERG responses to moderate (300 nJ/cm^2, lighter traces) and bright (300 µJ/cm^2, darker traces) flashes recorded from P12 wild-type (*c*) and P12 *Aipl1$^{-/-}$* retinas. No ERG responses could be elicited from any of the *Aipl1$^{-/-}$* mice examined. (Figure reproduced from Ramamurthy *et al.*, PNAS (2004), **101**, 38, 13897-902).

In addition to the knockout mouse, an AIPL1 knockdown mouse was created in which AIPL1 expression was reduced to 20-25% of wild-type levels.[14] The knockdown mouse exhibits normal development and normal retinal morphology up to three months of age. At three months, the rod photoreceptor outer segments become disorganized, and by 8 months, more than half of the photoreceptors are lost. Similar to the AIPL1 knock out mouse, PDE protein levels were drastically reduced despite normal mRNA message levels prior to the onset of retinal degeneration.[14] The knockdown mice exhibit ERG responses with lower gain and a longer response delay, consistent with a reduced level of rod PDE.[14] Surprisingly, cGMP levels were found to be lower in the AIPL1 knockdown rods. This is in contrast to the AIPL1 knockout mice, where reduced levels of PDE result in higher levels of cGMP.[14]

The phenotype of the AIPL1 knockdown mice further supports the hypothesis that the LCA-related defect in AIPL1 is caused by a defect in the maintenance of rod photoreceptors, as the knockdown mice show normal development and persist with functional photoreceptors for several months prior to onset of the degeneration. The AIPL1 deficient mice suggest that the essential role of AIPL1 in retina is to stabilize the active rod cGMP phosphodiesterase holoenzyme.

3. ROLE OF AIPL1 IN THE STABILITY OF PDE HOLOENZYME

At P8, before the degeneration, AIPL1 deficient mice express 10% or less of all three PDE subunits (α,β and γ) despite normal message levels.[10] The link between AIPL1 and the stability of PDE holoenzyme is not clear. However, a previous yeast-two hybrid screen together with HEK cell expression studies suggest the involvement of the post-translational modification farnesylation.[6]

Farnesylation, a type of prenylation, is a post-translational modification that occurs at the C-termini of proteins that contain a C-terminal "CaaX" box signal sequence (Cys-aliphatic-aliphatic-specific amino acid). Farnesylation is a multi-step process. In the first step, farnesyl transferase (FTase) catalyzes the covalent attachment of a farnesyl (C-15) group to the conserved cysteine of the CaaX box. The farnesylated protein is then targeted to the endoplasmic reticulum where the C-terminal three amino acid residues (-aaX) are removed, and the exposed farnesyl cysteine is carboxymethylated.[15] Geranylgeranylation is a similar modification that adds a geranylgeranyl (C20) group instead of the farnesyl group to the C-terminus of the protein. PDE-α and PDE-β are known to be farnesylated and ger-anylgeranylated, respectively.[16] The prenylation of PDE-α and β subunits is essential for stability and membrane interactions.[17] PDE-α and β subunit mutants, in which conserved CaaX box cysteine is replaced by serine to prevent prenylation, are unstable and degrade rapidly when expressed in insect cells.[17] It is not clear whether the methylation or the prenylation modification is essential for the stability of PDE. Inhibition of prenylation in adult rat retina causes the whole retinal cytoarchitecture and photoreceptor structure to fall apart rapidly, suggesting that prenylation is required for photoreceptor structure maintenance.[18]

The three TPR domains of AIPL1 suggest that the protein is involved in multi-protein complexes with chaperone, transcriptional, cell-cycle regulation, or protein transport activities.[19] TPR domains participate in protein-protein interactions by interacting with the C-termini of proteins.[19] In agreement with this, AIPL1 interacts with proteins that have a conserved farnesylation signal at their C-termini.[6] Furthermore, in cultured human embryonic kidney cells (HEK-293), AIPL1 enhances the farnesylation of proteins.[6] However, the mechanism of this enhancement is not presently known. In AIPL1[-/-] mice, photoreceptor degeneration seems to be primarily due to loss of PDE subunits. AIPL1 specifically enhances the stability of all three PDE subunits. Surprisingly, the stability of other known farnesylated retinal proteins, such as rhodopsin kinase and transducin gamma subunit are not altered in the absence of AIPL1.[10,14] This suggests that the role of AIPL1 is complex and that it may not play a general role in enhancing the farnesylation of retinal proteins. Alternatively, this could reflect the importance of prenylation specifically for stability of PDE.

AIPL1 also interacts with Nedd8 ultimate buster (NUB1), a protein ubiquitously expressed and thought to be involved in proteolysis.[20] AIPL1-NUB1 interaction could play a role in the stability of PDE. The presence of an immunophilin like domain in AIPL1 suggests that AIPL1 may play a role as a chaperone of PDE subunits. AIPL1 could either stabilize them or aid in the assembly of the three PDE subunits. Further studies are warranted to understand the requirement of AIPL1 for the stability of rod PDE subunits. At present it is not known if AIPL1 also plays a role in the stability of cone PDE subunits. It seems unlikely, as AIPL1 is not expressed in mature human cones.

It has been suggested that AIPL1 like other immunophilin and TPR containing proteins, could be involved in protein transport. However, our experiments so far do not support a role for AIPL1 in the transport of retinal proteins frorm inner to outer segments.[10] Most of the retinal proteins including the residual PDE are localized normally in the absence of AIPL1.

4. AIPL1 AND RETINAL DEVELOPMENT

The phenotype of AIPL1 deficient mice shows that AIPL1 is necessary for either photoreceptor differentiation and/or maintenance.[10] It has been suggested that AIPL1 plays a critical role in the early development of both rod and cone photoreceptors. *AIPL1* is expressed in both rod and cone photoreceptors in early human retinal development, and mutations in the *AIPL1* gene are associated with severe blinding disease.[7,9] NUB1, a protein that interacts with AIPL1, is thought to be involved in cell-cycle progression.[20] This has lead to the hypothesis that AIPL1 plays an early role in regulating retinal cell fate decision and or retinal progenitor cell proliferation.[20] However, we have not seen much evidence so far to support a developmental role for AIPL1.[10] Ultrastructural details of retina analyzed at post-natal day 8 do not show any obvious alterations in mice lacking AIPL1.[10] A recent *AIPL1* knock out mouse model with a rapid retinal degeneration similar to our *AIPL1* knockout, shows no significant defects in number of rods, cones or any second order neuronal cells, such as bipolar, amacrine or ganglion cells.[21] However, unlike *rd/rd* mice, which also have a defective PDE, AIPL1[-/-] never exhibit any electrical response to light at any age tested, reflecting the rapid cone degeneration seen in AIPL1[-/-] mice in comparison to *rd/rd* mice.[10,21] It is possible that the rapid cone degeneration seen in AIPL1[-/-] is an indirect effect due to the early and severe rod photoreceptor cell degeneration. In agreement with this, recent studies show that rod photoreceptor cells produce viability factors (RdCVF) that are essential for survival of cones.[22] Alternatively, AIPL1 may play an important role in early cones, which is consistent with its early expression in cones.[7] In mature human retina, AIPL1 is expressed exclusively in rod photoreceptor cells suggesting that AIPL1 is necessary for maintenance of rod photoreceptors.[7] Whether the shift in the expression from developing cone and rod photoreceptor cells to only mature rods reflects a shift in the function of AIPL1 is presently not known. More studies are needed to address the role of AIPL1 in developing cones, as these will be crucial in the design of suitable therapies for treating patients with AIPL1 associated LCA.

5. CONCLUSIONS

The AIPL1 knockout mouse replicates the human LCA caused by AIPL1 mutations and is a suitable animal model to test novel therapies. Similar to human patients with LCA, mice lacking AIPL1 do not exhibit any electrical responses. Both rod and cone photoreceptor cells degenerate rapidly. The role of AIPL1 in cones is not clear. However, the rapid cone degeneration in *AIPL1[-/-]* mice suggests that AIPL1 is necessary for the survival of the cones. Our study shows that the rapid rod photoreceptor cell degeneration is due to lack of stable PDE trimeric holenzyme. The mechanism by which AIPL1 contributes to the stability of PDE is unknown. Further study of the biochemical function of AIPL1 is required to understand the significance of AIPL1 to photoreceptor differentiation and or maintenance. In addition to the direct application to therapies for LCA patients, such information may enhance our understanding of photoreceptor development and long-term photoreceptor survival.

6. REFERENCES

1. Cremers, F.P., van den Hurk, J.A. & den Hollander, A.I. Molecular genetics of Leber congenital amaurosis. *Hum Mol Genet* **11**:1169-76 (2002).
2. Hanein, S. et al. Leber congenital amaurosis: comprehensive survey of the genetic heterogeneity, refinement of the clinical definition, and genotype-phenotype correlations as a strategy for molecular diagnosis. *Hum Mutat* **23**:306-17 (2004).
3. Sohocki, M.M. et al. Mutations in a new photoreceptor-pineal gene on 17p cause Leber congenital amaurosis. *Nat Genet* **24**:79-83 (2000).
4. Sohocki, M.M., Sullivan, L.S., Tirpak, D.L. & Daiger, S.P. Comparative analysis of aryl-hydrocarbon receptor interacting protein-like 1 (Aipl1), a gene associated with inherited retinal disease in humans. *Mamm Genome* **12**:566-8 (2001).
5. van der Spuy, J. et al. The Leber congenital amaurosis gene product AIPL1 is localized exclusively in rod photoreceptors of the adult human retina. *Hum Mol Genet* **11**:823-31 (2002).
6. Ramamurthy, V. et al. AIPL1, a protein implicated in Leber's congenital amaurosis, interacts with and aids in processing of farnesylated proteins. *Proc Natl Acad Sci U S A* **100**:12630-5 (2003).
7. van der Spuy, J. et al. The expression of the Leber congenital amaurosis protein AIPL1 coincides with rod and cone photoreceptor development. *Invest Ophthalmol Vis Sci* **44**:5396-403 (2003).
8. Dharmaraj, S. et al. The phenotype of Leber congenital amaurosis in patients with AIPL1 mutations. *Arch Ophthalmol* **122**:1029-37 (2004).
9. Sohocki, M.M. et al. Prevalence of AIPL1 mutations in inherited retinal degenerative disease. *Mol Genet Metab* **70**:142-50 (2000).
10. Ramamurthy, V., Niemi, G.A., Reh, T.A. & Hurley, J.B. Leber congenital amaurosis linked to AIPL1: a mouse model reveals destabilization of cGMP phosphodiesterase. *Proc Natl Acad Sci U S A* **101**:13897-902 (2004).
11. Pittler, S.J. & Baehr, W. Identification of a nonsense mutation in the rod photoreceptor cGMP phosphodiesterase beta-subunit gene of the rd mouse. *Proc Natl Acad Sci U S A* **88**:8322-6 (1991).
12. Farber, D.B., Flannery, J.G. & Bowes-Rickman, C. The *rd* Mouse Story: Seventy Years of Research on an Animal Model of Inherited Retinal Degeneration. *Prog in Retinal and Eyes Res* **13**:31-65 (1994).
13. Huang, S.H. et al. Autosomal recessive retinitis pigmentosa caused by mutations in the alpha subunit of rod cGMP phosphodiesterase. *Nat Genet* **11**:468-71 (1995).
14. Liu, X. et al. AIPL1, the protein that is defective in Leber congenital amaurosis, is essential for the biosynthesis of retinal rod cGMP phosphodiesterase. *Proc Natl Acad Sci U S A* **101**:13903-8 (2004).
15. Choy, E. et al. Endomembrane trafficking of ras: the CAAX motif targets proteins to the ER and Golgi. *Cell* **98**:69-80 (1999).
16. Anant, J.S. et al. In vivo differential prenylation of retinal cyclic GMP phosphodiesterase catalytic subunits. *J Biol Chem* **267**:687-90 (1992).
17. Qin, N. & Baehr, W. Expression and mutagenesis of mouse rod photoreceptor cGMP phosphodiesterase. *J Biol Chem* **269**:3265-71 (1994).
18. Pittler, S.J., Fliesler, S.J., Fisher, P.L., Keller, P.K. & Rapp, L.M. In vivo requirement of protein prenylation for maintenance of retinal cytoarchitecture and photoreceptor structure. *J Cell Biol* **130**:431-9 (1995).
19. Blatch, G.L. & Lassle, M. The tetratricopeptide repeat: a structural motif mediating protein-protein interactions. *Bioessays* **21**:932-9 (1999).
20. Akey, D.T. et al. The inherited blindness associated protein AIPL1 interacts with the cell cycle regulator protein NUB1. *Hum Mol Genet* **11**:2723-33 (2002).
21. Dyer, M.A. et al. Retinal degeneration in Aipl1-deficient mice: a new genetic model of Leber congenital amaurosis. *Brain Res Mol Brain Res* **132**:208-20 (2004).
22. Leveillard, T. et al. Identification and characterization of rod-derived cone viability factor. *Nat Genet* **36**:55-9 (2004).

CHAPTER 15

CHARACTERIZATION OF MOUSE MUTANTS WITH ABNORMAL RPE CELLS

Chun-hong Xia[1], Haiquan Liu[1], Meng Wang[1], Debra Cheung[1], Alex Park[1], Yang Yang[1], Xin Du[2], Bo Chang[3], Bruce Beutler[2], and Xiaohua Gong

1. INTRODUCTION

Retinal pigment epithelium (RPE) is essential for the function and survival of photoreceptor cells by playing supporting roles including shedding the outer segments of the photoreceptor cells, removing metabolic wastes, transporting nutrients and maintaining visual cycle. RPE defects have been found in various human retinal disorders, such as age-related macular degeneration (Zarbin, 1998), Best disease (Petrukhin et al., 1998; Marmorstein et al., 2000), Sorsby fundus dystrophy (Weber et al., 1994; Ruiz et al., 1996; Della et al., 1996), and childhood-onset severe retinal dystrophy (Gu et al., 1997). Animal models with RPE defects have been used to study the molecular basis for the function of the RPE cells. The Royal College of Surgeons (RCS) rat, a model for recessive inherited retinal degeneration, is characterized by the dysfunction of RPE due to a null mutation of the receptor tyrosine kinase *Mertk* gene (D'Cruz et al., 2000). RPE cells fail to shed the outer segments of the photoreceptor cells in the RCS rat (Mullen et al., 1996). Recently, mice with a targeted disruption of the *Mertk* gene manifest retinal dystrophy similar to RCS rats (Duncan et al., 2003). In addition, mutated *Mertk* gene has been identified in patients with retinitis pigmentosa (Gal et al., 2002).

In order to identify new gene mutations that cause eye diseases and to establish important animal models for human eye diseases, we have carried out a forward genetic study by using clinical methods to screen a mouse germline mutagenesis program (Hoebe et al., 2003). Mice of C57BL/6J background mutated by the alkylating agent N-ethyl-N-nitrosourea (ENU) have been screened for eye defects based on their fundus abnormality. We have identified and determined that mouse lines, BEMr15 (r15) and BEMr18 (r18), are two recessive mutations. Furthermore, histopathological data show abnormal RPE cells, dis-

[1] School of Optometry and Vision Science Program, University of California, Berkeley, Berkeley, CA 94720-2020, USA; [2] Department of Immunology, The Scripps Research Institute, La Jolla, CA; [3] The Jackson Laboratory, Bar Harbor, ME. Corresponding author: X. Gong, E-mail: xgong@berkeley.edu.

organized outer segments of photoreceptor cells and a progressive reduction of outer nuclear layers in both mutant lines. Here, we present the preliminary morphological characterization of RPE and photoreceptor cells in both r15 and r18 mutant mice. Since some unique features in these mutants have not been observed in any previous animal models, we hypothesize that both mouse lines are new mutations in genes that play important roles in the RPE cells. These mutations could be useful new animal models for studying the functions of RPE cells and the degeneration of photoreceptor cells.

2. BEMR15 SHOWS DISORGANIZED RPE CELLS AND PROGRESSIVE RETINAL DEGENERATION

The wild type male mice of C57BL/6J strain were intraperitoneally injected with ENU to become the F_0 generation of ENU-induced mutant mice (Du et al., 2004). Each F_0 male was bred with normal C57BL/6J female to generate the F_1 male animals, which mate with wild type C57BL/6J female to generate F_2 female, and then the backcross of F_2 female mice to the F_1 male mice produced the F_3 mice. Both F_1 (for dominant trait) and F_3 (for recessive trait) mice at the age of 2 to 3 months were examined for their retinal fundus by indirect ophthalmoscope.

The founder of the BEMr15 line is a F_3 mouse and additional genetic test has verified its recessive inheritance. The r15 homozygous mice are viable and develop retinal phenotype showing depigmented patches in the fundus photo (Fig. 15.1). These depigmented patches could be observed in 3-week-old mice whose retinal vessels appear normal. Like many other mouse mutants that develop retinal degeneration, retinal vessel attenuation is also observed in the r15 mutant as mice grow old (Fig. 15.1).

Histology analysis, using plastic sections stained with toluidine blue, has been performed to examine the retinal morphology at different age. As shown in Fig. 15.2, r15 homozygous mutant has disorganized outer segments of photoreceptor cells and obvious disrupted RPE cell layer. Abnormal changes that include attenuated RPE layer, vacuoles and hypertrophy of RPE cells are observed. Most strikingly, abnormal pigmented structures appear in the subretinal space as well as in the deep regions of the outer segments (arrows in Fig. 15.2B, 15.2D, and 15.2F). The loss of photoreceptor is less obvious in r15 homozygous mutant at the age of 2 months. However, homozygous mutant shows a loss of 4-5 layers of photoreceptor cells at the age of 6.5-months (Fig. 15.2B) and contains only a few layers

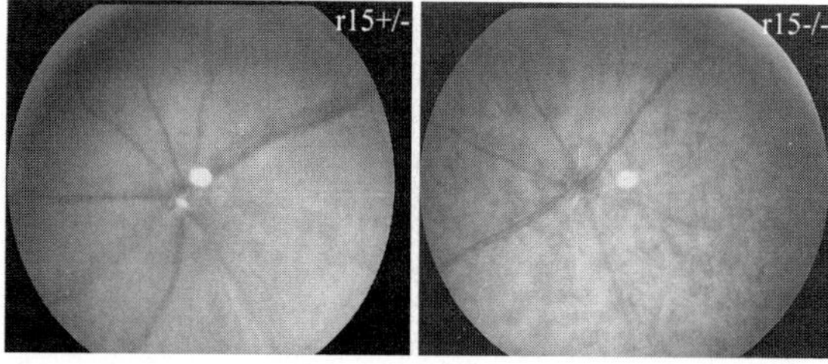

Figure 15.1. Fundus photos of the eyes of 12-months old littermates. The r15 homozygous mutant shows depigmented patches with retinal vessel attenuation (right panel) while the heterozygous eye seems normal.

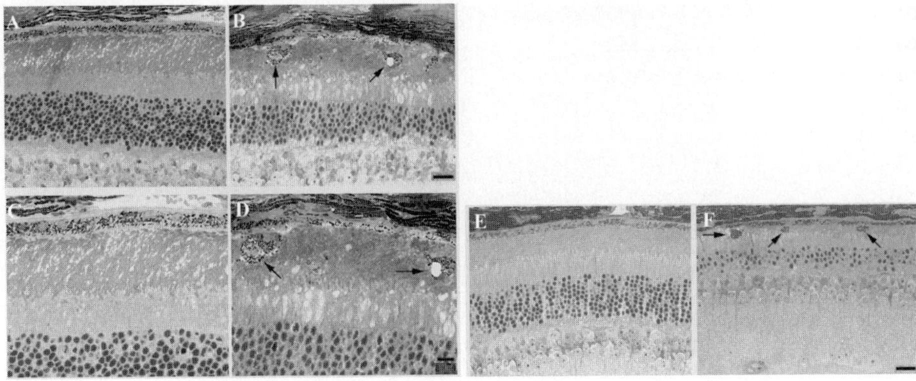

Figure 15.2. Light micrographs of the toluidine blue stained retinal sections show abnormal RPE cells and a loss of photoreceptor cells in the homozygous r15 mice but not in the heterozygous mice. (A) and (C) show relative normal retina of 6.5-month-old r15 heterozygous mouse; (B) and (D) show abnormal RPE cell layer and displaced RPE cells in the subretinal space of 6.5-month-old r15 homozygous mutant retina; (E) is a section of 17.5-month-old r15 heterozygous control; (F) is a section of 17.5-month-old r15 homozygous mutant. Note the reduced photoreceptor layers in the homozygous sections. Arrows (in B, D and F) indicate the abnormal pigmented structures. Scale bars: 20 μm in (A), (B), (E), and (F); 10 μm in (C) and (D).

of photoreceptor nuclear at the age of 17.5-months (Fig. 15.2F). Therefore, the r15 mutation develops a slow degeneration of photoreceptor cells.

Transmission electron microscopic analysis verifies that the RPE cells of homozygous r15 mutant lack apical microvilli and form membranous whorls in the subretinal space (Fig. 15.3B). It suggests that mutant RPE cells may fail to shed the outer segments of photoreceptor cells, and the unphagocytosed outer segments could form membranous whorls at the subretinal space. Mutant RPE cells contain intracellular vacuoles (Fig. 15.3D) and degenerative RPE layer gives rise to the aberrantly displaced structures in the subretinal space (data not shown). Therefore, the r15 mutation recapitulates some of the defects observed in mouse *Mertk* mutation, this suggests that the causative gene for r15 mutation is essential for the phagocytosis of RPE cells. Currently, it is not clear whether r15 mutation is mechanistically related to the *Mertk* mutation. We continue our efforts to map the chromosome location of the r15 mutation and to identify the causative gene. We believe that the r15 mutant line provides us an alternative model to study the molecular basis for the regulation and function of the RPE phagocytosis.

2.1. BEMR18 is Another ENU-Induced Mouse Mutation with RPE Abnormality

BEMr18 is also an ENU-mutagenized recessive mutation that shows retinal vessel attenuation. Histology analysis reveals that RPE cell hypertrophy can be observed in the r18 homozygous mice at as early as 4-weeks old (data not shown). RPE cell hypertrophy and RPE layer disorganization become more obvious as the mice grow old (Fig. 15.4D and 4E). Shortening of outer segments and loss of 3-4 layers of photoreceptor cells have been observed in the 1-year-old r18 homozygous mutant, but not in the heterozygous control littermate (Fig. 15.4).

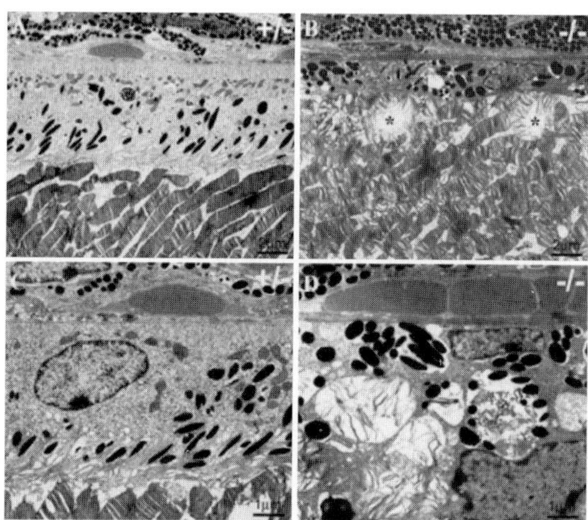

Figure 15.3. Electron micrographs show relative normal RPE and outer segments in the 6.5-month-old r15 heterozygous control (A and C), and disorganized photoreceptor outer segments with membranous whorls (asterisks in B) and vacuoles in the RPE cells in the homozygous littermate (B and D). Scale bars: 2 μm in (A) and (B); 1 μm in (C) and (D).

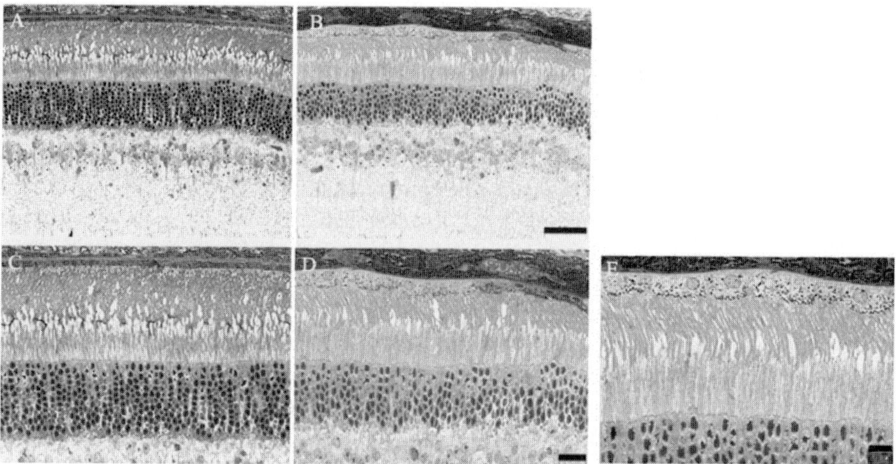

Figure 15.4. Histology analysis of 1-year-old heterozygous (A and C) and homozygous (B, D, and E) r18 littermates. RPE hypertrophy and loss of outer nuclear layers are seen in homozygous r18 retina compared to the heterozygous retina. Scale bars: 50 μm in (A) and (B); 20 μm in (C) and (D); 10 μm in (E).

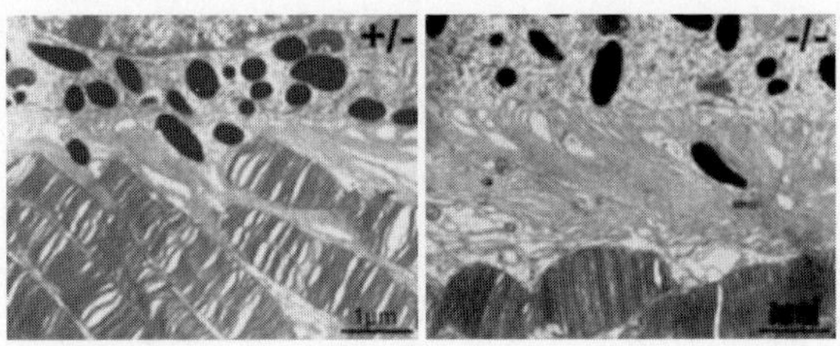

Figure 15.5. Electron micrographs show the apical sides of RPE cells of 2-month-old r18 heterozygous and homozygous littermates. Scale bars: 1 μm.

EM study of the r18 mutant retina reveals that substantial amount of microvilli from the apical side of RPE cells accumulate in the subretinal space in the homozygous mutation but not in the heterozygous control (Fig. 15.5).

Thus, in comparison to homozygous r15 mutation, r18 homozygous mutation represents a different phenotype in the RPE cells with a much slower degeneration of photoreceptor cells. We hypothesize that both r15 and r18 gene mutations either directly or indirectly modulate the apical microvilli to cause the degeneration of photoreceptor cells. We believe that further investigation of both mutations will lead to new knowledge for genes that play important roles in the functions of RPE cells.

3. ACKNOWLEDGEMENTS

We thank the National Eye Institute for the travel award provided to CX to attend this meeting. XG is supported by NIH grants RO1 EY12808 and RO1 EY13849.

4. REFERENCES

Della, N. G., Campochiaro, P. A., Zack, D. J., 1996, Localization of TIMP-3 mRNA expression to the retinal pigment epithelium. *Invest. Ophthalmol. Vis. Sci.* **37**:1921-1924.

D'Cruz, P. M., Yasumura, D., Weir, J., Matthes, M. T., Abderrahim, H., LaVail, M. M., Vollrath, D., 2000, Mutation of the receptor tyrosine kinase gene Mertk in the retinal dystrophic RCS rat. *Hum. Mol. Genet.* **9**:645-651.

Du, X., Tabeta, K., Hoebe, K., Liu, H., Mann, N., Mudd, S., Crozat, K., Sovath, S., Gong, X., Beutler, B., 2004, Velvet, a dominant Egfr mutation that causes wavy hair and defective eyelid development in mice. *Genetics.* **166**:331-340.

Duncan, J. L., LaVail, M. M., Yasumura, D., Matthes, M. T., Yang, H., Trautmann, N., Chappelow, A. V., Feng, W., Earp, H. S., Matsushima, G. K., Vollrath, D., 2003, An RCS-like retinal dystrophy phenotype in mer knockout mice. *Invest. Ophthalmol. Vis. Sci.* **44**:826-838.

Gal, A., Li, Y., Thompson, D. A., Weir, J., Orth, U., Jacobson, S. G., Apfelstedt-Sylla, E., Vollrath, D., 2000, Mutations in MERTK, the human orthologue of the RCS rat retinal dystrophy gene, cause retinitis pigmentosa. *Nat. Genet.* **26**:270-271.

Gu, S. M., Thompson, D. A., Srikumari, C. R., Lorenz, B., Finckh, U., Nicoletti, A., Murthy, K. R., Rathmann, M., Kumaramanickavel, G., Denton, M. J., Gal, A., 1997, Mutations in RPE65 cause autosomal recessive childhood-onset severe retinal dystrophy. *Nat. Genet.* **17**:194-197.

Hoebe, K., Du, X., Goode, J., Mann, N., Beutler, B., 2003, Lps2: a new locus required for responses to lipopolysaccharide, revealed by germline mutagenesis and phenotypic screening. *J. Endotoxin. Res.* **9**:250-255.

Marmorstein, A. D., Marmorstein, L. Y., Rayborn, M., Wang, X., Hollyfield, J. G., Petrukhin, K., 2000, Bestrophin, the product of the Best vitelliform macular dystrophy gene (VMD2), localizes to the basolateral plasma membrane of the retinal pigment epithelium. *Proc. Natl. Acad. Sci. U S A* **97**:12758-12763.

Mullen, R. J., LaVail, M. M., 1996, Inherited retinal dystrophy: primary defect in pigment epithelium determined with experimental rat chimeras. *Science* **192**:799-801.

Petrukhin, K., Koisti, M. J., Bakall, B., Li, W., Xie, G., Marknell, T., Sandgren, O., Forsman, K., Holmgren, G., Andreasson, S., Vujic, M., Bergen, A. A., McGarty-Dugan, V., Figueroa, D., Austin, C. P., Metzker, M. L., Caskey, C.T., Wadelius, C., 1998, Identification of the gene responsible for Best macular dystrophy. *Nat. Genet.* **19**:241-7.

Ruiz, A., Brett, P., Bok, D., 1996, TIMP-3 is expressed in the human retinal pigment epithelium. *Biochem. Biophys. Res. Commun.* **226**:467-474.

Weber, B. H., Vogt, G., Pruett, R. C., Stohr, H., Felbor, U., 1994, Mutations in the tissue inhibitor of metalloproteinases-3 (TIMP3) in patients with Sorsby's fundus dystrophy. *Nat. Genet.* **8**:352-356.

Zarbin, M. A., 1998, Age-related macular degeneration: review of pathogenesis. *Eur. J. Ophthalmol.* **8**:199-206.

CHAPTER 16

ROD AND CONE PIGMENT REGENERATION IN *RPE65*[-/-] MICE

Baerbel Rohrer[1,2]* and Rosalie Crouch[1]

1. INTRODUCTION

RPE65 is a major protein in the retinal pigment epithelium (RPE) (Hamel et al., 1993), where it is required for the regeneration of 11-*cis* retinal, the native ligand of rod and cone opsins, in the dark (Redmond et al., 1998). Therefore, the retina of the *Rpe65*[-/-] mouse is almost completely depleted of 11-*cis* retinal, resulting in a minimal level of photosensitivity. This observation poses several questions, of which we will only address three: first, which cell type (rods and/or cones) is responsible for the remaining photosensitivity; second, if only one cell type remains photosensitive, what happens to the other one; and third, what is the chromophore that enables the formation of light-sensitive pigment? Early reports have disagreed whether the remaining photosensitivity can be attributed to rod (Seeliger et al., 2001) or cone function (Redmond et al., 1998; Ekesten et al., 2001). Double knockout experiments, crossing the *Rpe65*[-/-] with either the rhodopsin or the cone cGMP-gated channel knockout, respectively, revealed that the remaining photosensitivity in the young adult and old *Rpe65*[-/-] mouse retina (>6 weeks-of-age) can be attributed to rod sensitivity (Seeliger et al., 2001). This leaves open the question as to the possible fate of the cone photoreceptors in the absence of RPE65. Likewise, early reports demonstrated a slow degeneration in particular of the rod photoreceptors (Redmond et al., 1998), suggesting that the accumulation of the retinyl ester in the RPE might contribute to the demise of the photoreceptors. However, in the *Rpe65*[-/-]*::Gnat1*[-/-] mouse, in which similar elevated amounts of retinyl ester have been reported to accumulate in the RPE, no rod degeneration occurs (Woodruff et al., 2003). And finally, with respect to the available chromophore; several groups have tried to obtain a spectrum of the pigment from pooled tissue and have failed to do so (Ablonzcy et al., 2001; C. H. Remé, personal communication). Thus, here we would like

[1]Departments of Ophthalmology[1] and Physiology and Neuroscience[2], Medical University of South Carolina, Charleston, SC. *Department of Ophthalmology, Medical University of South Carolina, 167 Ashley Avenue, SEI 511, Charleston, South Carolina, 29425; phone: (843) 792-5086; fax (843) 792-1723; e-mail: rohrer@musc.edu.

to further address these three key issues and how our laboratories have investigated them over the past 5 years.

2. RESULTS

2.1. Rod Responses in *Rpe65*[-/-] Mice

Patients, in which the function of RPE65 is eliminated, have impaired light sensitivity, with the main loss in sensitivity under dark-adapted conditions (e.g., Van Hooser et al., 2000). This has led to the classification of this condition in Rpe65 patients as a rod-cone dystrophy. However, Seeliger and coworkers (2001) have demonstrated convincingly that the remaining light sensitivity in the young adult *Rpe65*[-/-] mouse retina (>6 weeks-of-age) results from the activity of rod photoreceptors. Interestingly, this activity, and thus the mechanism for chromophore generation, persists even in old animals (18 months-of-age) (Rohrer et al., 2003). However, what was puzzling was that these apparent rod responses from the *Rpe65*[-/-] mouse retina appeared to have different kinetics than those recordable from the age-matched wild type mice. This was easiest seen in the scotopic flicker ERG (see Figure 16.1). Responses with faster kinetics are indicative of light-adapted rod responses. Light-adapted responses might also in part explain the reduced light-sensitivity (Cornwall and Fain, 1994; Cornwall et al., 1990). This hypothesis was tested by Woodruff and colleagues (2003), who used single cell recordings in isolated *Rpe65*[-/-] mouse rods. In this preparation they were able to demonstrate that *Rpe65*[-/-] mouse rods have significantly smaller circulating current, reduced light sensitivity, and accelerated response kinetics. In accordance with the light-adaptation hypothesis, these rods were also found to exhibit lower intradiscal calcium concentrations, due to the partial closure of the cGMP-gated cation channels in the photoreceptor outer membrane.

The lack of RPE65 has been shown to result in slow rod photoreceptor degeneration (Redmond et al., 1998; Katz and Redmond, 2001; Rohrer et al., 2003). Opsin in the absence of ligand has been shown to activate the photoreceptor signal transduction cascade, albeit at a much lower rate than activated rhodopsin (Cornwall and Fain, 1994; Cornwall et al.,

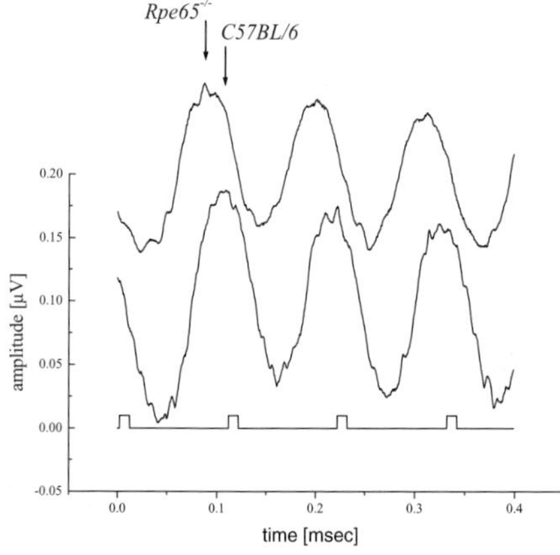

Figure 16.1. Flicker ERGs in young *Rpe65*[-/-] and *C57BL/6* mice. 10 Hz flicker ERGs were recorded under scotopic conditions. Light intensities were adjusted to elicit flicker ERGs of approximately equal amplitudes (*Rpe65*[-/-]: 1.9×10^{12} photons/mm²; *C57BL/6*: 3.1×10^{7} photons/mm²). Flicker ERG responses peaked earlier in the mutant than in the wild type responses. (Redrawn from Ablonczy et al., 2001; © Journal of Biological Chemistry.)

1995). Interestingly, this constant background activity has been proposed to lead to photoreceptor degeneration (Fain and Lisman, 1993). However, rod photoreceptors also possess a protective mechanism to prevent the activity of bleached, unliganded opsin, which is opsin phosphorylation and arrestin binding. We have shown that in the *Rpe65*[-/-] mouse retina, opsin is constitutively phosphorylated at an elevated level of ~25%. This level of phosphorylation is approximately half of the maximal light-inducible (transient) amount of phosphorylation in the wild type retina (Ablonczy et al., 2001). In addition, this constitutively phosphorylated opsin was found to bind arrestin at similar levels as bleached wild type opsin (Crouch et al., 2003). Accordingly, it was not surprising that arrestin distribution in the *Rpe65*[-/-] retina has been shown to mimic that of a light-adapted wild type retina (Mendez et al., 2003). We therefore argue that the kinetics of rod degeneration is controlled by both constitutive opsin activity and constitutive opsin phosphorylation.

2.2. Cone Responses in *Rpe65*[-/-] Mice

Rod photoreceptors appear to be more susceptible to degeneration either during aging (e.g., Gresh et al., 2003; Curcio, 2001) or as a consequence of insults (light damage; Cicerone, 1976). Thus, we chose to revisit the question of cone function and survival in the *Rpe65*[-/-] mouse. Classically, cone responses are recorded in single flash experiments in the presence of a rod-adapting background, or in flicker ERGs. However, due to the reduced light sensitivity in the *Rpe65*[-/-] rods, it is impossible to bleach them sufficiently to be certain that they no longer contribute to the elicited light response. To be able to record unequivocally from *Rpe65*[-/-] cones, we obtained the *Rpe65*[-/-]*::Rho*[-/-] cross, in which rod responses are eliminated genetically (Seeliger et al., 2001). Elimination of rhodopsin causes a delayed rod degeneration (onset ~P60), thus all experiments were performed in animals <1 month-of-age. Interestingly, if photopic ERGs were recorded using normal averaging conditions (averaging 3-5 traces) using a maximal white flash of $~2.2 \times 10^{13}$ photons/mm^2, no ERGs could be recorded; however, if averaging was increased to 50 traces, a small but reliable cone ERG response ($13.7 \pm 1.87 \mu V$; n = 9) could be recorded from these mice.

If constitutive cone opsin activity were to contribute to the reduced sensitivity of the cones, one would predict that cone degeneration would likewise occur. However, as the amplification in the signal transduction cascade is lower in cones then in rods, one might predict cone degeneration to be slower than rod degeneration. However, the small cone ERG response was virtually eliminated by 4 months-of-age (unpublished results; BR 2004), suggesting that a different mechanism is responsible for the demise of the cones. Interpretation of these data is complicated by the finding that RPE65 is present in all mouse cones (Znoiko et al., 2002). The role of this protein in cones is unknown.

2.3. Rod and Cone Opsin Distribution in the *Rpe65*[-/-] Mouse Retina

Recently, Chapple and coworkers (2001) have speculated that the chromophore 11-*cis* retinal might act as a pharmacological chaperone that facilitates opsin folding into an appropriate conformational state that allows for its proper transport through the trans-Golgi network and integration into the photoreceptor outer membrane. Likewise, it would allow for appropriate posttranslational modifications. Wild type rhodopsin is a relatively stable protein that appears to be transported appropriately in the absence of 11-*cis* retinal (Figure 16.2A). However, the P23H mutation, which does not fold properly, can only be targeted

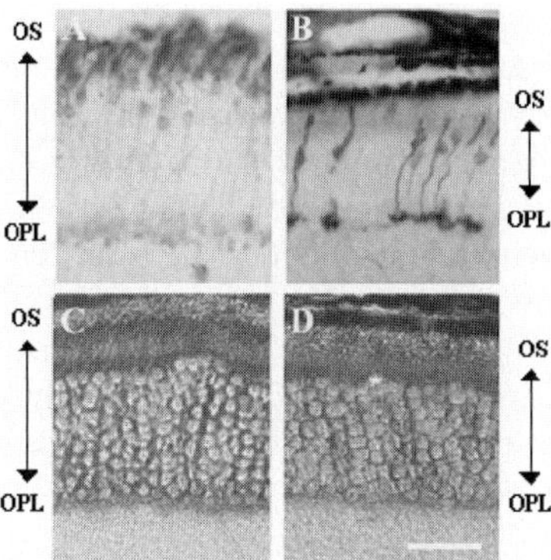

Figure 16.2. UV-cone and rod opsin distribution in C57BL/6 (*wt*) and *Rpe65[-/-]* mice. (Top row) UV cone opsin localization at P25 (using a polyclonal antibody against mouse UV cone opsin, generously provided by J. Chen, University of Southern California, Los Angels, CA, USA). **(A)** In the *wt* retina, cone opsin was localized predominantly to the outer segment (OS). **(B)** UV cone apoprotein was found to be distributed throughout the entire cone in *Rpe65[-/-]::Rho[-/-]* mice. [Green cone opsin (localized with a polyclonal antibody against human MWL cone opsin, generously provided by J. Nathans, Johns Hopkins University School of Medicine, Baltimore, MD, USA) showed a similar distribution profile, but is not depicted here.] **(Bottom row)** Rod opsin localization at P30 (using monocolonal antibody against bovine rhodopsin, generously provided by R. Molday; University of British Columbia, Vancouver, BC, Canada). Rhodopsin **(C)** and apoprotein **(D)** localization was indistinguishable between the *wt* and *Rpe65[-/-]* mouse retina, respectively. Scalebar: 15 μm.

appropriately in the presence of ligand (either 9-*cis* or 11-*cis* retinal; Noorwez et al., 2003). The accumulation of misfolded P23H rhodopsin apparently leads to photoreceptor degeneration (e.g., Olsson et al., 1992). Cone opsins appear to be more thermally unstable (Matsumoto et al., 1975), suggesting that they may be more dependent on a pharmacological chaperone for stabilization. We therefore tested the hypothesis that the cone opsin apoproteins are mislocalized in the cones of the *Rpe65[-/-]::Rho[-/-]* mice. Cone opsin in the absence of 11-*cis* retinal was found to be localized throughout the cones. Here we show the distribution of UV cone opsin (Figure 16.2B), a distribution profile that is recapitulated by green cone opsin (data not shown). Whether cone opsin is misfolded in the *Rpe65[-/-]::Rho[-/-]* retinas will be difficult to determine. One of the presumed results of misfolding is the lack of appropriate glycosylation. However, these retinas contain such small amounts of cone opsin that Western blotting to determine a difference in molecular weight will be a challenge.

2.4. Isolation of the Chromophore for Rhodopsin

Rpe65[-/-] mouse photoreceptors have been found to exhibit minimal levels of light sensitivity. However, even after careful pooling of retinas isolated from *Rpe65[-/-]* mice, no pigment could be measured. We recently found, however, that prolonged dark-adaptation resulted in slow accumulation of chromophore (~0.4 pmol/day), demonstrating that the process is very inefficient (Fan et al., 2003). Young adult *Rpe65[-/-]* mice were transferred into complete darkness, and retinas were collected at weekly intervals. The two retinas from each animal were pooled for pigment measurements in 1% dodecylmaltoside, determining the

Figure 16.3. Isolation of isorhodopsin from *Rpe65^{-/-}* mouse retinas. Absorption spectra of endogenous pigments isolated after 2 months of dark adaptation from a wild type (black trace) and a *Rpe65^{-/-}* mouse (grey trace). Nomograms were fitted to the difference spectra to determine a peak of the absorbance. The wild type spectrum could be fitted with a nomogram with a peak absorbance of 500 nm, which corresponds to rhodopsin, whereas the pigment isolated from the *Rpe65^{-/-}* mouse had a blue-shifted spectrum, indicative of isorhodopsin, which is formed in the presence of 9-*cis* retinal. (Adapted from Fan et al., 2003; © PNAS.)

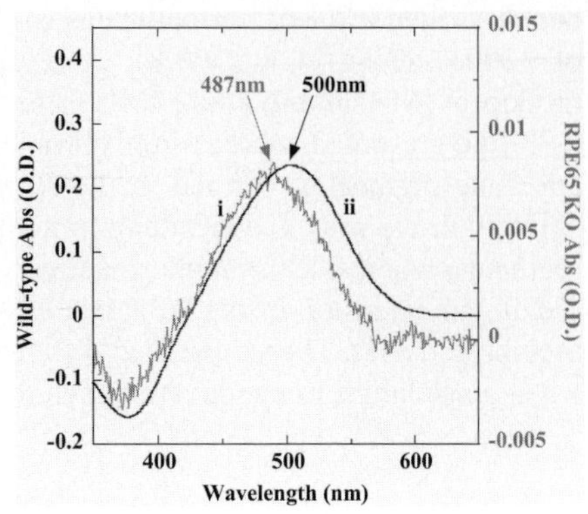

difference spectra from measurements before and after bleaching. Interestingly, the pigment isolated from the *Rpe65^{-/-}* mouse retinas was found to be isorhodopsin, which is the rod pigment regenerated with 9-*cis* rather then 11-*cis* retinal (Figure 16.3). The presence of 9-*cis* retinal in these retinas and the RPE was confirmed by higher performance liquid chromatography (Fan et al., 2003). It is unclear at this point as to how 9-*cis* retinal is formed; the only definitive statements that can be made are that it is independent of light (see Figure 16.3), the light-dependent, RGR-mediated mechanism (unpublished results; J. Fan, 2003) and RPE65 (Fan et al., 2003). In addition, it is unclear whether 9-*cis* retinal is also the chromophore for the cones, as the amount of cone pigment generated in these retinas is too minute to be measured. Single cell spectrophotometry or electrophysiology may shed some light on this question. However, our experiments do provide indirect evidence that the proposed Müller-cell-mediated mechanisms for chromophore regeneration in cone-dominant retinas, which is independent of RPE65 (Mata et al., 2003), does not exist in the rod-dominant mouse retina.

3. DISCUSSION

Here we have discussed the differences in behavior between rods and cones in the absence of RPE65 and thus endogenous chromophore. We have presented evidence that both rod and cone responses can be recorded in the young *Rpe65^{-/-}* mice. Rods degenerate very slowly and those kinetics are controlled by the constitutive activity of the free opsin and opsin phosphorylation. In the absence of chromophore, the apoprotein appears to be folding correctly and thus proper targeting to the rod outer segments occurs. In contrast, cones degenerate very quickly in the absence of chromophore. Cone opsin is not targeted properly to the outer segments, but is distributed throughout the entire cell membrane. We are suggesting that folding and thus targeting of the cone opsin to the cone outer segment requires a chromophore as a pharmacological chaperone.

In addition, we have demonstrated that the chromophore available for the formation of pigment in the *Rpe65^{-/-}* mouse retina is 9-*cis* retinal. For the sake of argument, let's assume that the cones in the *Rpe65^{-/-}* mouse retina use the same chromophore. Based on the amount of chromophore generated (~0.4 pmol/day; Fan et al., 2003), the *Rpe65^{-/-}* mouse retina contains less then 0.2% of normal levels of chromophore. As rods outnumber cones by a factor of ~100:1 (Jeon et al., 1998), cone opsin has a lower affinity for chromophore than rhodopsin (Matsumoto et al., 1975) and the cone outer segments are further away from the RPE (the presumed source of the chromophore); it is, therefore, difficult to understand how any cone pigment is formed in the *Rpe65[-/-]* mouse retina. We suggest that chromophore delivery in the rods is dependent upon diffusion of IRBP-bound chromophore to the outer segments, whereas chromophore delivery to the cones is aided by the presence of the cone sheath, an extension of the RPE that ensheathes the entire cone outer segment (Fisher and Steinberg, 1982). These observed differences between rods and cones need to be addressed when considering treatment strategies for LCA patients.

4. ACKNOWLEDGEMENTS

Funding was provided by NIH grants EY-13520, EY-04939, EY-14793; Foundation Fighting Blindness; and an unrestricted grant to MUSC from Research to Prevent Blindness, Inc., New York, NY. RKC is a RPB Senior Scientific Investigator. The authors thank Jie Fan, Jian-xing Ma, Sergey Znoiko and Patrice Goletz for contributing experiments for this review; Mathias Seeliger, Michael Redmond and Peter Humphries for contributing mice, Jeannie Chen, Jeremy Nathans and Robert Molday for providing antibodies and Luanna Bartholomew for editorial assistance.

5. REFERENCES

Ablonczy, Z., Kono, M., Crouch, R. K., and Knapp, D. R., 2001, Mass spectrometric analysis of integral membrane proteins at the subnanomolar level: Application to recombinant photopigments. *Anal Chem* **73**:4774-4779.

Chapple, J. P., Grayson, C., Hardcastle, A. J., Saliba, R. S., van der Spuy, J. and Cheetham, M. E., 2001, Unfolding retinal dystrophies: A role for molecular chaperones? *Trends Mol Med* **7**:414-421.

Cicerone, C. M., 1976, Cones survive rods in the light-damaged eye of the albino rat. *Science* **194**:1183-1185.

Cornwall, M. C., and Fain, G. L., 1994, Bleached pigment activates transduction in isolated rods of the salamander retina. *J Physiol* **480**:261-279.

Cornwall, M. C., Fein, A., and MacNichol Jr., E. F., 1990, Cellular mechanisms that underlie bleaching and background adaptation. *J Gen Physiol* **96**:345-372.

Cornwall, M. C., Matthews, H. R., Crouch, R. K., and Fain, G. L., 1995, Bleached pigment activates transduction in salamander cones. *J Gen Physiol* **106**:543-557.

Crouch, R. K., Znoiko, S., Kono, M., Rohrer, B., Goletz, P. W., Gresh, J., Redmond, T. M., and Ma, J. X., 2003, Can delivery of 11-*cis* retinal to the RPE65 KO mouse restore normal rod and cone function? *Invest Ophthalmol Vis Sci* **44-CD**:PR# 44.

Curcio, C. A., 2001, Photoreceptor topography in ageing and age-related maculopathy. *Eye* **15**:376-383.

Ekesten, B., Gouras, P., and Salchow, D. J., 2001, Ultraviolet and middle wavelength sensitive cone responses in the electroretinogram (ERG) of normal and Rpe65[-/-] mice. *Vision Res* **41**:2425-2433.

Fain, G. L., and Lisman, J. E., 1993, Photoreceptor degeneration in vitamin A deprivation and retinitis pigmentosa: the equivalent light hypothesis. *Exp Eye Res* **57**:335-340.

Fan, J., Rohrer, B., Moiseyev, G., Ma, J. X., and Crouch, R. K., 2003, Isorhodopsin rather than rhodopsin mediates rod function in RPE65 knock-out mice. *Proc Natl Acad Sci U S A* **100**:13662-13667.

Fisher, S. K., and Steinberg, R. H., 1982, Origin and organization of pigment epithelial apical projections to cones in cat retina. *J Comp Neurol* **206**:131-145.

Gresh, J., Goletz, P. W., Crouch, R. K., and Rohrer, B., 2003, Structure-function analysis of rods and cones in juvenile, adult, and aged C57bl/6 and Balb/c mice. *Vis Neurosci* **20**:211-220.

Hamel, C. P., Tsilou, E., Harris, E., Pfeffer, B. A., Hooks, J. J., Detrick, B., and Redmond, T. M., 1993, A developmentally regulated microsomal protein specific for the pigment epithelium of the vertebrate retina. *J Neurosci Res* **34**:414-425.

Jeon, C. J., Strettoi, E., and Masland, R. H., 1998, The major cell populations of the mouse retina. *J Neurosci* **18**:8936-8946.

Katz, M. L., and Redmond, T. M., 2001, Effect of Rpe65 knockout on accumulation of lipofuscin fluorophores in the retinal pigment epithelium. *Invest Ophthalmol Vis Sci* **42**:3023-3030.

Mata, N. L., Radu, R. A., Clemmons, R. C., and Travis, G. H., 2003, Isomerization and oxidation of vitamin A in cone-dominant retinas: A novel pathway for visual-pigment regeneration in daylight. *Neuron* **36**:69-80.

Matsumoto, H., Tokunaga, F., and Yoshizawa, T., 1975, Accessibility of the iodopsin chromophore. *Biochim Biophys Acta* **404**:300-308.

Mendez, A., Lem, J., Simon, M., and Chen, J., 2003, Light-dependent translocation of arrestin in the absence of rhodopsin phosphorylation and transducin signaling. *J Neurosci* **23**:3124-3129.

Noorwez, S. M., Kuksa, V., Imanishi, Y., Zhu, L., Filipek, S., Palczewski, K., and Kaushal, S., 2003, Pharmacological chaperone-mediated in vivo folding and stabilization of the P23H-opsin mutant associated with autosomal dominant retinitis pigmentosa. *J Biol Chem* **278**:14442-14450.

Olsson, J. E., Gordon, J. W., Pawlyk, B. S., Roof, D., Hayes, A., Molday, R. S., Mukai, S., Cowley, G. S., Berson, E. L., and Dryja, T. P., 1992, Transgenic mice with a rhodopsin mutation (Pro23His): A mouse model of autosomal dominant retinitis pigmentosa. *Neuron* **9**:815-830.

Redmond, T. M., Yu, S., Lee, E., Bok, D., Hamasaki, D., Chen, N., Goletz, P., Ma, J. X., Crouch, R. K. and Pfeifer, K., 1998, Rpe65 is necessary for production of 11-*cis*-vitamin A in the retinal visual cycle. *Nat Genet* **20**:344-351.

Rohrer, B., Goletz, P., Znoiko, S., Ablonczy, Z., Ma, J. X., Redmond, T. M., and Crouch, R. K., 2003, Correlation of regenerable opsin with rod ERG signal in Rpe65-/- mice during development and aging. *Invest Ophthalmol Vis Sci* **44**:310-315.

Seeliger, M. W., Grimm, C., Stahlberg, F., Friedburg, C., Jaissle, G., Zrenner, E., Guo, H., Reme, C. E., Humphries, P., Hofmann, F., Biel, M., Fariss, R. N., Redmond, T. M., and Wenzel, A., 2001, New views on RPE65 deficiency: The rod system is the source of vision in a mouse model of Leber congenital amaurosis. *Nat Genet* **29**:70-74.

Van Hooser, J. P., Aleman, T. S., He, Y. G., Cideciyan, A. V., Kuksa, V., Pittler, S.J., Stone, E. M., Jacobson, S. G., and Palczewski, K., 2000, Rapid restoration of visual pigment and function with oral retinoid in a mouse model of childhood blindness. *Proc Natl Acad Sci USA* **97**:8623-8628.

Woodruff, M. L., Wang, Z., Chung, H. Y., Redmond, T. M., Fain, G. L., and Lem, J., 2003, Spontaneous activity of opsin apoprotein is a cause of Leber congenital amaurosis. *Nat Genet* **35**:158-164.

Znoiko, S. L., Crouch, R. K., Moiseyev, G., and Ma, J. X., 2002, Identification of the RPE65 protein in mammalian cone photoreceptors. *Invest Ophthalmol Vis Sci* **43**:1604-1609.

CHAPTER 17

INITIAL OBSERVATIONS OF KEY FEATURES OF AGE-RELATED MACULAR DEGENERATION IN *APOE* TARGETED REPLACEMENT MICE

For contributed volumes

Goldis Malek[1A], Brian Mace[1B], Peter Saloupis[1A], Donald Schmechel[1B,C], Dennis Rickman[1A,B], Patrick Sullivan[1B], and Catherine Bowes Rickman[1A,D]

1. INTRODUCTION

Age-related macular degeneration (AMD) is a late-onset, neurodegenerative disease of the retina and is the leading cause of catastrophic vision loss in the elderly. AMD usually occurs in people over the age of 65 years (Javitt et al., 2003) and accounts for approximately 50% of registered blindness in Western Europe and North America (Mitchell et al., 1998; Vingerling et al., 1995a). It develops as either dry AMD, geographic atrophy, or wet AMD (exudative) (Bird et al., 1995; Green, 1999). Dry AMD is characterized by the presence of sub-retinal pigment epithelium (RPE) deposits including drusen, basal linear deposits and basal laminar deposits (Curcio and Millican, 1999; Sarks, 1976). Geographic atrophy is characterized by RPE atrophy and wet or exudative AMD is characterized by choroidal neo-vascularization (CNV) (1991) and more recently, retinal angiomatous proliferation (Yannuzzi et al., 2001). In AMD with CNV, tufts of newly formed, functionally incompetent, blood vessels proliferate from the choroid, break through a thickened and fragile Bruch's membrane (Grossniklaus and Green, 2004), and extend laterally into the sub-RPE (Sarks et al., 1980; Tobe et al., 1998a). These vessels in turn, may erode through the RPE, infiltrate the neural sensory retina, and communicate with the retinal circulation in what has been referred to as a retinal-choroidal anastomosis. This is common in the end stage of disciform disease (Yannuzzi et al., 2001).

AMD has been difficult to study due to its late onset, complex genetics and strong environmental components that play a role in the transition from normal aging to disease. To

---

[1A]Department of Ophthalmology, [1B]Department of Neurobiology, [1C]Joseph and Kathleen Bryan Alzheimer's Disease Research Center, [1D]Department of Cell Biology, Duke University Medical Center, Durham, North Carolina 27710, USA.

further understand the pathogenesis of AMD and test potential treatments for the disease, the search for an animal model of AMD has been extensive and so far unsuccessful. This is not surprising as it is likely that many different combinations of environmental and genetic risk factors may lead to development of this multifactorial disease. Here we present a murine model of AMD developed using factors that are each, individually, associated with an increased risk for the disease.

2. AMD RISK FACTORS

Although the etiology of AMD remains largely unknown, numerous studies have implicated both genetic and environmental influences, which are reflected in the number of risk factors identified to date (Vingerling et al., 1995b). Known risk factors for AMD include: advanced age, environmental factors including cigarette smoking, nutritional factors such as diets high in fat and cholesterol, systemic diseases including hypercholesterolemia and atherosclerosis, gender, ethnicity, and family history (2000; Cho et al., 2001; Christen et al., 1996; Delcourt et al., 2001). There is strong evidence that genetics play an important role in the pathogenesis of AMD (Heiba et al., 1994; Klaver et al., 1998b; Seddon et al., 1997), such that allelic variations in apolipoprotein E (apoE = protein; *APOE* = gene), fibulin-5 (Stone et al., 2004), and ABCA4 (Allikmets, 2000), have each, to some degree, been associated with increased risk for AMD.

2.1. ApoE and Neurodegenerative Disease

The *APOE* gene has been identified as a risk factor for AMD in numerous epidemiological studies (Klaver et al., 1998a; Schmidt et al., 2000; Schultz et al., 2003; Simonelli et al., 2001; Souied et al., 1998), as well as studies using *APOE* knockout mice (Dithmar et al., 2000) and *APOE* (*)E3-Leiden mice that carry a dysfunctional form of human *APOE*-E3 (Kliffen et al., 2000). ApoE forms a lipid protein complex and is an important regulator of cholesterol and lipid clearance, transport and distribution (Mahley, 1988; Mahley and Rall, 2000). It is expressed at high levels in the retina, and is involved in processes that have been implicated in the pathological formation of sub-RPE deposits including drusen (Anderson et al., 2001; Klaver et al., 1998a). ApoE protein has been localized in the eye to RPE, Müller cells, Bruch's membrane and drusen (Hageman et al., 1999; Klaver et al., 1998a) and *APOE* mRNA is present in the neural retina and RPE/choroid (Hageman et al., 1999). In humans, the *APOE* gene is polymorphic and encodes three isoforms: E2, E3, and E4, with frequencies of 7%, 77% and 15%, respectively, in the general population (Davignon et al., 1988). Epidemiological studies of the association of *APOE* allele with AMD have yielded conflicting results regarding the *APOE* isoform-associated risk, with many studies describing a protective effect in E4 carriers and detrimental effect in E2 carriers (Schmidt et al., 2002; Schmidt et al., 2000) while others have not been able to find this association (Smith et al., 2001). This is in contrast to its role in other neurodegenerative and systemic diseases including Alzheimer's, atherosclerosis, amyotrophic lateral sclerosis, and stroke in which there is a very strong negative association of the E4 allele with disease (Kalaria, 1997; Lacomblez et al., 2002; Weller and Nicoll, 2003). In Alzheimer's disease, for example, there is a 90% risk for the disease associated with the expression of the E4 allele (Higgins et al., 1997).

2.2. Animal Models of AMD

To date there is no generally accepted experimental animal model for dry or exudative AMD. Reproducibility of existing models is limited by technical artifacts or because, as in exudative AMD in humans, with the induction of CNV, there is also a concomitant non-specific, local inflammatory reaction (Mori et al., 2001; Spilsbury et al., 2000). Though previous studies attempting to simulate AMD using transgenic mice have been very useful, many of the resulting animal models were produced by manipulating genes that are not known AMD risk factors. For example, animals expressing mutant forms of a short chain collagen gene, *CTRP5* (Hayward et al., 2003), cathepsin D (Rakoczy et al., 2002), or the very low density lipoprotein receptor gene (Heckenlively et al., 2003) develop various aspects of AMD. Transgenic knockouts of genes expressing proteins that are normally present in the retina at low levels [monocyte chemoattractant protein-1 (MCP-1 or Ccl-2) and chemokine receptor-2 (Ccr-2)(Ambati et al., 2003)] have also produced models of AMD. AMD has been modeled using phototoxicity (Cousins et al., 2002), senescence acceleration (Majji et al., 2000), high fat diets (Cousins et al., 2002; Dithmar et al., 2000; Fliesler et al., 2000; Kliffen et al., 2000; Miceli et al., 2000; Ong et al., 2001), *in vitro* procedures such as laser photocoagulation of Bruch's membrane in primates (Miller et al., 1986), rats (Dobi et al., 1989; Frank et al., 1989; Hikichi et al., 2002) and mice (Tobe et al., 1998a; Tobe et al., 1998b), elevated levels of bFGF in minipigs (Soubrane et al., 1994) and rabbits (Kimura et al., 1995), implantation of tumors on or in the eye to stimulate angiogenesis (Ambati et al., 2002), and exposure of newborn animals to high oxygen (McLeod et al., 2002). Collectively, none of these models have successfully correlated their histological, morphological and biochemical features to the key features of AMD.

3. DESIGN OF A NEW ANIMAL MODEL

As described above, epidemiological studies have demonstrated an *APOE* isoform-associated risk for AMD, supporting a role for *APOE* isoform-specific effects in the pathogenesis of AMD. Therefore, we are studying transgenic mice expressing each of the three major human *APOE* isoforms in order to rigorously analyze the role of *APOE* in the development of AMD.

3.1. *APOE* Targeted-Replacement Mice

To date, transgenic animals made by pronuclear injection of human DNA have been used to study the effect of *APOE* expression, but this method produces mice with varying levels of transgene expression due to differences in insertion of the transgene constructs within the host genome, and transgene copy number. This, as well as expression of the endogenous mouse *ApoE*, complicates interpretation of the effect of the foreign *APOE* (Sullivan et al., 1997). To overcome these two complications we used targeted gene replacement (TR) mice generated by homologous recombination of human *APOE* coding sequences with the endogenous murine *ApoE* gene, made by our collaborator Dr. Patrick Sullivan (Sullivan et al., 1997). These mice express one of the three common human *APOE* isoforms (E2, E3 or E4) at levels that are in the physiological range. Briefly, coding sequences for the mouse *ApoE* gene were replaced with coding sequences for human *APOE*2, E3 or E4, without disturbing any of the known 5′ or 3′ murine regulatory sequences. This gene replacement strategy results in animals that express human *APOE*2, E3 or E4 mRNAs, identical in tissue distribution and levels to that of mouse *APOE* mRNA in wild type animals (Sullivan et al., 2004; Sullivan et al., 1997). It should be pointed out that in humans the most common *APOE* allele is E3 which is equivalent to the murine wild type *ApoE*. On the other hand, 5-10% of *APOE*2 humans will develop Type III hyperlipoproteinemia, whereas 100% of *APOE*2 TR mice on a high fat diet develop Type III hyperlipoproteinemia (Sullivan et al., 1998).

3.2. Experimental Protocol and Diet Paradigm

APOE expression alone is not sufficient to cause neurodegenerative disease in humans, but is instead a *risk* factor that when combined with other insults, collectively, leads to development of Alzheimer's disease, stroke and putatively AMD. Therefore, we used two other known *risk* factors in our animal model to mimic a multifactorial combination that would produce characteristic features of AMD. The factors were: advanced age and a high fat/high cholesterol diet.

Aged male and female (aged 65-127 weeks) *APOE* TR mice of each isoform (E2, E3, E4) were bred and housed conventionally and fed a high fat/high cholesterol diet rich in cholate (HF-C) and water ad libitum for 8 weeks. Age-matched controls were fed standard rodent chow and water ad libitum. At the end of the eight week diet exposure, mice were euthanized with an overdose of anesthesia, and the eyes were collected for analysis. Histologically, 10 μm thick cryosections and 1 μm plastic sections were examined at the light microscopic level and ultra thin sections were examined using a transmission electron microscope (EM).

4. HISTOLOGICAL CHARACTERIZATION OF AGED *APOE* TR MICE

4.1. Light Microscopic Evaluation

There were no remarkable abnormalities by light microscopy in the retina, RPE, Bruch's membrane and choroid of aged *APOE* TR mice maintained on normal mouse chow diet (Fig. 17.1A). In aged *APOE*3 TR mice fed the HF-C (TR-ch), there were no detectable retinal changes and only minor RPE changes including RPE vacuolization. This is not surprising since as mentioned before, in humans the *APOE*3 allele is equivalent to the murine wild type *APOE*. *APOE*2 TR-ch mice showed more RPE vacuolization and were further compromised by RPE blebbing and Bruch's membrane thickening (Fig. 17.1B, arrow), and basal deposit formation, some RPE and photoreceptor degeneration. These changes were also documented, in the *APOE*4 TR-ch mouse eyes, which showed the most extensive AMD-related changes overall. Some *APOE*4 TR-ch eyes also showed 'drusen-like' basal deposits (Fig. 17.1C, arrowhead), and neovascularization (NV) of varying severities (Fig. 17.1D). The NV ranged from mild, confined to an area adjacent to the RPE and Bruch's membrane without any disruption of the neural retina, to more extensive, proliferating through the RPE, through the sub-retinal space and into the neural retina. Though it appears the NV origi-

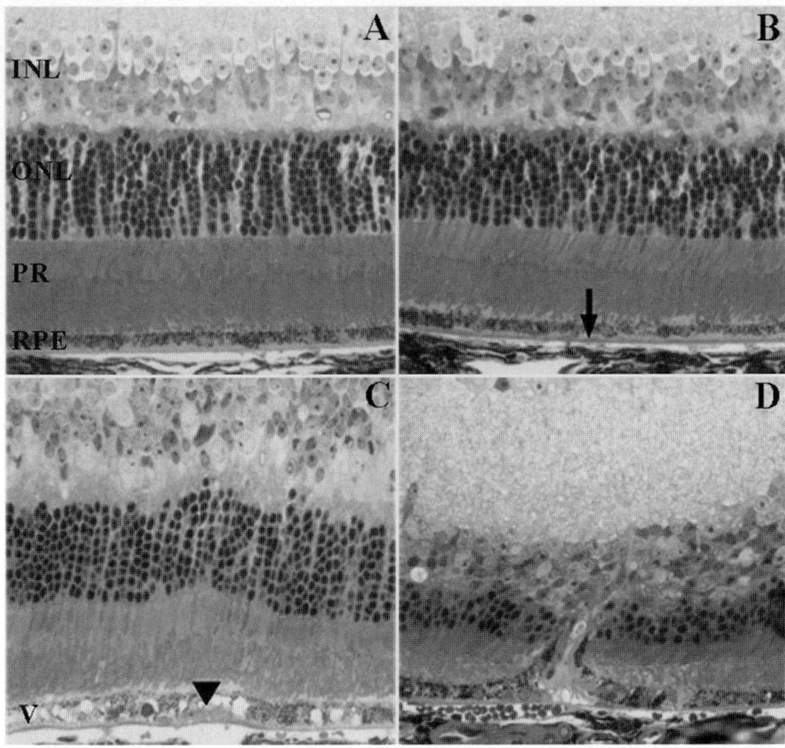

Figure 17.1. Light micrographs from plastic sections of posterior retinas of *APOE* TR mice stained with toluidine blue. A. *APOE*3 TR mouse fed normal mouse chow. B. *APOE*2 TR mouse on cholate diet with thickening of Bruch's membrane (arrow). C. *APOE*4 TR mouse on cholate diet with 'drusen-like' basal deposit (arrowhead) and vacuoles (V). D. *APOE*4 TR mouse on cholate diet with neovascularization spanning the RPE and the overlying neural retinal layers. Magnification 20X. INL = inner nuclear layer, ONL = outer nuclear layer, PR = photoreceptors, RPE = retinal pigment epithelium.

nates from the choroid, the source remains to be conclusively determined using fluorescein angiography and/or further histological analysis of sequential sections through the NV.

4.2. Electron Microscopic Evaluation

EM was used to further evaluate the histological abnormalities documented at the light microscopic level. In all aged *APOE* TR mice maintained on the normal diet the retina, RPE, and Bruch's membrane remained essentially normal. EM analysis of RPE vacuoles initially observed by light microscopy, in *APOE*4 TR-ch mice, revealed the presence of membraneous material (Fig. 17.2A). In the retinas of *APOE*2 and E4 TR-ch mice RPE basal infoldings were disorganized or absent (Fig. 17.2B). In *APOE*4 TR-ch mouse eyes there was Bruch's membrane thickening (Fig. 17.2C), as well as thick basal deposits between RPE and Bruch's membrane (Fig. 17.2D). In *APOE*4 TR-ch mice with neovascularization, thin or absent Bruch's membrane was observed with vessels adjacent to Bruch's membrane in both the sub-RPE and choroidal regions (not shown).

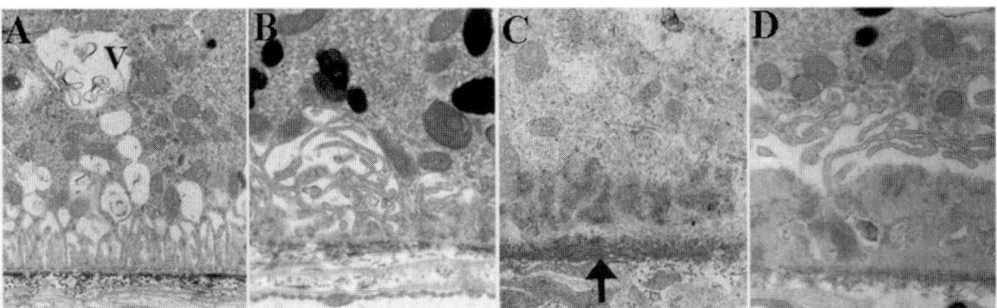

Figure 17.2. Electron micrographs of the basal RPE/Bruch's membrane interface at 5000 X magnification of *APOE*4 TR mice on high fat/high cholesterol cholate diet. A. Normal basal infoldings with vacuoles that contain whorls of membranous material. B. Disorganized RPE basal infoldings. C. Thickened Bruch's membrane (arrow). D. Thick basal laminar deposit.

5. CONCLUSIONS

Our findings suggest that specific *APOE* isoform expression alone results in only mild retinal changes with aging. However, in the *APOE*2 and especially *APOE*4 expressing animals, the combination of increased age (65-127 weeks) and a high fat diet rich in cholate (which enhances cholesterol absorption), resulted in more extensive degenerative changes in the retina/choroid reminiscent of both dry and wet AMD changes in the human eye. Hallmarks of AMD including thick diffuse sub-RPE deposits (Figs. 17.1B and 17.2D) and drusen like deposits (Fig. 17.1C), thickening of BrM (Fig. 17.2C), patchy regions of RPE atrophy overlying photoreceptor degeneration, and neovascularization (Fig. 17.1D) were detected in these animals. This is the first report of spontaneously occurring neovascularization and retinal/RPE changes in an animal model developed by combining the following known risks associated with AMD: advanced age, environmental risk (high fat, high cholate diet), and genetic risk [*APOE* isoform expression, using *APOE* targeted replacement mice (TR)]. We are currently continuing to examine the features of this model more closely and correlate similarities between this model and the human form of AMD, through imaging of the neovascularization using Fluorescein angiography, immunohistochemical localization of hallmark growth factors and proteins found in human CNV and basal deposits/drusen, in the mouse NV and basal deposits.

6. ACKNOWLEDGEMENTS

The authors gratefully acknowledge Drs. Cynthia Toth and Lincoln Johnson for scientific discussions, and valuable input, and the following funding agencies: NEI RO1 EY11286 (CBR), NEI P30 EY05722 (Core), NIA P50 AG05128-20 (DS, GM), Research to Prevent Blindness Career Development Award (CBR), AHAF MDR 2004 (CBR). We also sincerely thank the National Eye Institute and Foundation for Fighting Blindness for the travel award provided to GM to attend this meeting.

7. REFERENCES

No authors listed, 1991. Subfoveal neovascular lesions in age-related macular degeneration. Guidelines for evaluation and treatment in the macular photocoagulation study. Macular Photocoagulation Study Group. *Arch Ophthalmol* **109**:1242-1257.

No authors listed, 2000. Risk factors associated with age-related macular degeneration. A case-control study in the age-related eye disease study: age-related eye disease study report number 3. Age-Related Eye Disease Study Research Group. *Ophthalmology* **107**:2224-2232.

Allikmets, R., 2000. Further evidence for an association of ABCR alleles with age-related macular degeneration. The International ABCR Screening Consortium. *Am J Hum Genet* **67**:487-491.

Ambati, B. K., Joussen, A. M., Ambati, J., Moromizato, Y., Guha, C., Javaherian, K., Gillies, S., O'Reilly, M. S., and Adamis, A. P. 2002. Angiostatin inhibits and regresses corneal neovascularization. *Arch Ophthalmol* **120**:1063-1068.

Ambati, J., Anand, A., Fernandez, S., Sakurai, E., Lynn, B. C., Kuziel, W. A., Rollins, B. J., and Ambati, B. K. 2003. An animal model of age-related macular degeneration in senescent Ccl-2- or Ccr-2-deficient mice. *Nat Med* **9**:1390-1397.

Anderson, D. H., Ozaki, S., Nealon, M., Neitz, J., Mullins, R. F., Hageman, G. S., and Johnson, L. V. 2001. Local cellular sources of apolipoprotein E in the human retina and retinal pigmented epithelium: implications for the process of drusen formation. *Am J Ophthalmol* **131**:767-781.

Bird, A. C., Bressler, N. M., Bressler, S. B., Chisholm, I. H., Coscas, G., Davis, M. D., de Jong, P. T., Klaver, C. C., Klein, B. E., Klein, R., et al. 1995. An international classification and grading system for age-related maculopathy and age-related macular degeneration. The International ARM Epidemiological Study Group. *Surv Ophthalmol* **39**:367-374.

Cho, E., Hung, S., Willett, W. C., Spiegelman, D., Rimm, E. B., Seddon, J. M., Colditz, G. A., and Hankinson, S. E. 2001. Prospective study of dietary fat and the risk of age-related macular degeneration. *Am J Clin Nutr* **73**:209-218.

Christen, W. G., Glynn, R. J., Manson, J. E., Ajani, U. A., and Buring, J. E. 1996. A prospective study of cigarette smoking and risk of age-related macular degeneration in men. *Jama* **276**:1147-1151.

Cousins, S. W., Espinosa-Heidmann, D. G., Alexandridou, A., Sall, J., Dubovy, S., and Csaky, K. 2002. The role of aging, high fat diet and blue light exposure in an experimental mouse model for basal laminar deposit formation. *Exp Eye Res* **75**:543-553.

Curcio, C. A., and Millican, C. L. 1999. Basal linear deposit and large drusen are specific for early age-related maculopathy. *Arch Ophthalmol* **117**:329-339.

Davignon, J., Gregg, R. E., and Sing, C. F. 1988. Apolipoprotein E polymorphism and atherosclerosis. *Arteriosclerosis* **8**:1-21.

Delcourt, C., Michel, F., Colvez, A., Lacroux, A., Delage, M., and Vernet, M. H. 2001. Associations of cardiovascular disease and its risk factors with age-related macular degeneration: the POLA study. *Ophthalmic Epidemiol* **8**:237-249.

Dithmar, S., Curcio, C. A., Le, N. A., Brown, S., and Grossniklaus, H. E. 2000. Ultrastructural changes in Bruch's membrane of apolipoprotein E-deficient mice. *Invest Ophthalmol Vis Sci* **41**:2035-2042.

Dobi, E. T., Puliafito, C. A., and Destro, M. 1989. A new model of experimental choroidal neovascularization in the rat. *Arch Ophthalmol* **107**:264-269.

Fliesler, S. J., Richards, M. J., Miller, C., Peachey, N. S., and Cenedella, R. J. 2000. Retinal structure and function in an animal model that replicates the biochemical hallmarks of desmosterolosis. *Neurochem Res* **25**:685-694.

Frank, R. N., Das, A., and Weber, M. L. 1989. A model of subretinal neovascularization in the pigmented rat. *Curr Eye Res* **8**:239-247.

Green, W. R. 1999. Histopathology of age-related macular degeneration. *Mol Vis* **5**:27.

Grossniklaus, H. E., and Green, W. R. 2004. Choroidal neovascularization. *Am J Ophthalmol* **137**:496-503.

Hageman, G. S., Mullins, R. F., Russell, S. R., Johnson, L. V., and Anderson, D. H. 1999. Vitronectin is a constituent of ocular drusen and the vitronectin gene is expressed in human retinal pigmented epithelial cells. *Faseb J* **13**:477-484.

Hayward, C., Shu, X., Cideciyan, A. V., Lennon, A., Barran, P., Zareparsi, S., Sawyer, L., Hendry, G., Dhillon, B., Milam, A. H., *et al.* 2003. Mutation in a short-chain collagen gene, CTRP5, results in extracellular deposit formation in late-onset retinal degeneration: a genetic model for age-related macular degeneration. *Hum Mol Genet* **12**:2657-2667.

Heckenlively, J. R., Hawes, N. L., Friedlander, M., Nusinowitz, S., Hurd, R., Davisson, M., and Chang, B. 2003. Mouse model of subretinal neovascularization with choroidal anastomosis. *Retina* **23**:518-522.

Heiba, I. M., Elston, R. C., Klein, B. E., and Klein, R. 1994. Sibling correlations and segregation analysis of age-related maculopathy: the Beaver Dam Eye Study. *Genet Epidemiol* **11**:51-67.

Higgins, G. A., Large, C. H., Rupniak, H. T., and Barnes, J. C. 1997. Apolipoprotein E and Alzheimer's disease: a review of recent studies. *Pharmacol Biochem Behav* **56**:675-685.

Hikichi, T., Mori, F., Sasaki, M., Takamiya, A., Nakamura, M., Shishido, N., Takeda, M., Horikawa, Y., Matsuoka, H., and Yoshida, A. 2002. Inhibitory effect of bucillamine on laser-induced choroidal neovascularization in rats. *Curr Eye Res* **24**:1-5.

Javitt, J. C., Zhou, Z., Maguire, M. G., Fine, S. L., and Willke, R. J. 2003. Incidence of exudative age-related macular degeneration among elderly Americans. *Ophthalmology* **110**:1534-1539.

Kalaria, R. N. 1997. Arteriosclerosis, apolipoprotein E, and Alzheimer's disease. *Lancet* **349**:1174.

Kimura, H., Sakamoto, T., Hinton, D. R., Spee, C., Ogura, Y., Tabata, Y., Ikada, Y., and Ryan, S. J. 1995. A new model of subretinal neovascularization in the rabbit. *Invest Ophthalmol Vis Sci* **36**:2110-2119.

Klaver, C. C., Kliffen, M., van Duijn, C. M., Hofman, A., Cruts, M., Grobbee, D. E., van Broeckhoven, C., and de Jong, P. T. 1998a. Genetic association of apolipoprotein E with age-related macular degeneration. *Am J Hum Genet* **63**:200-206.

Klaver, C. C., Wolfs, R. C., Assink, J. J., van Duijn, C. M., Hofman, A., and de Jong, P. T. 1998b. Genetic risk of age-related maculopathy. Population-based familial aggregation study. *Arch Ophthalmol* **116**:1646-1651.

Kliffen, M., Lutgens, E., Daemen, M. J., de Muinck, E. D., Mooy, C. M., and de Jong, P. T. 2000. The (APO*)E3-Leiden mouse as an animal model for basal laminar deposit. *Br J Ophthalmol* **84**, 1415-1419.

Lacomblez, L., Doppler, V., Beucler, I., Costes, G., Salachas, F., Raisonnier, A., Le Forestier, N., Pradat, P. F., Bruckert, E., and Meininger, V. 2002. APOE: a potential marker of disease progression in ALS. *Neurology* **58**:1112-1114.

Mahley, R. W. 1988. Apolipoprotein E: cholesterol transport protein with expanding role in cell biology. *Science* **240**:622-630.

Mahley, R. W., and Rall, S. C., Jr. 2000. Apolipoprotein E: far more than a lipid transport protein. *Annu Rev Genomics Hum Genet* **1**:507-537.

Majji, A. B., Cao, J., Chang, K. Y., Hayashi, A., Aggarwal, S., Grebe, R. R., and De Juan, E., Jr. 2000. Age-related retinal pigment epithelium and Bruch's membrane degeneration in senescence-accelerated mouse. *Invest Ophthalmol Vis Sci* **41**:3936-3942.

McLeod, D. S., Taomoto, M., Cao, J., Zhu, Z., Witte, L., and Lutty, G. A. 2002. Localization of VEGF receptor-2 (KDR/Flk-1) and effects of blocking it in oxygen-induced retinopathy. *Invest Ophthalmol Vis Sci* **43**:474-482.

Miceli, M. V., Newsome, D. A., Tate, D. J., Jr., and Sarphie, T. G. 2000. Pathologic changes in the retinal pigment epithelium and Bruch's membrane of fat-fed atherogenic mice. *Curr Eye Res* **20**:8-16.

Miller, H., Miller, B., and Ryan, S. J. 1986. The role of retinal pigment epithelium in the involution of subretinal neovascularization. *Invest Ophthalmol Vis Sci* **27**:1644-1652.

Mitchell, P., Smith, W., and Wang, J. J. 1998. Iris color, skin sun sensitivity, and age-related maculopathy. The Blue Mountains Eye Study. *Ophthalmology* **105**:1359-1363.

Mori, K., Ando, A., Gehlbach, P., Nesbitt, D., Takahashi, K., Goldsteen, D., Penn, M., Chen, C. T., Melia, M., Phipps, S., *et al.* 2001. Inhibition of choroidal neovascularization by intravenous injection of adenoviral vectors expressing secretable endostatin. *Am J Pathol* **159**:313-320.

Ong, J. M., Zorapapel, N. C., Rich, K. A., Wagstaff, R. E., Lambert, R. W., Rosenberg, S. E., Moghaddas, F., Pirouzmanesh, A., Aoki, A. M., and Kenney, M. C. 2001. Effects of cholesterol and apolipoprotein E on retinal abnormalities in ApoE-deficient mice. *Invest Ophthalmol Vis Sci* **42**:1891-1900.

Rakoczy, P. E., Zhang, D., Robertson, T., Barnett, N. L., Papadimitriou, J., Constable, I. J., and Lai, C. M. 2002. Progressive age-related changes similar to age-related macular degeneration in a transgenic mouse model. *Am J Pathol* **161**:1515-1524.

Sarks, S. H. 1976. Ageing and degeneration in the macular region: a clinico-pathological study. *Br J Ophthalmol* **60**:324-341.

Sarks, S. H., Van Driel, D., Maxwell, L., and Killingsworth, M. 1980. Softening of drusen and subretinal neovascularization. *Trans Ophthalmol Soc U K* **100**:414-422.

Schmidt, S., Klaver, C., Saunders, A., Postel, E., De La Paz, M., Agarwal, A., Small, K., Udar, N., Ong, J., Chalukya, M., *et al.* 2002. A pooled case-control study of the apolipoprotein E APOE gene in age-related maculopathy. *Ophthalmic Genet* **23**:209-223.

Schmidt, S., Saunders, A. M., De La Paz, M. A., Postel, E. A., Heinis, R. M., Agarwal, A., Scott, W. K., Gilbert, J. R., McDowell, J. G., Bazyk, A., *et al.* 2000. Association of the apolipoprotein E gene with age-related macular degeneration: possible effect modification by family history, age, and gender. *Mol Vis* **6**:287-293.

Schultz, D. W., Klein, M. L., Humpert, A., Majewski, J., Schain, M., Weleber, R. G., Ott, J., and Acott, T. S. 2003. Lack of an association of apolipoprotein E gene polymorphisms with familial age-related macular degeneration. *Arch Ophthalmol* **121**:679-683.

Seddon, J. M., Ajani, U. A., and Mitchell, B. D. 1997. Familial aggregation of age-related maculopathy. *Am J Ophthalmol* **123**:199-206.

Simonelli, F., Margaglione, M., Testa, F., Cappucci, G., Manitto, M. P., Brancato, R., and Rinaldi, E. 2001. Apolipoprotein E polymorphisms in age-related macular degeneration in an Italian population. *Ophthalmic Res* **33**:325-328.

Smith, W., Assink, J., Klein, R., Mitchell, P., Klaver, C. C., Klein, B. E., Hofman, A., Jensen, S., Wang, J. J., and de Jong, P. T. 2001. Risk factors for age-related macular degeneration: Pooled findings from three continents. *Ophthalmology* **108**:697-704.

Soubrane, G., Cohen, S. Y., Delayre, T., Tassin, J., Hartmann, M. P., Coscas, G. J., Courtois, Y., and Jeanny, J. C. 1994. Basic fibroblast growth factor experimentally induced choroidal angiogenesis in the minipig. *Curr Eye Res* **13**:183-195.

Souied, E. H., Benlian, P., Amouyel, P., Feingold, J., Lagarde, J. P., Munnich, A., Kaplan, J., Coscas, G., and Soubrane, G. 1998. The epsilon4 allele of the apolipoprotein E gene as a potential protective factor for exudative age-related macular degeneration. *Am J Ophthalmol* **125**:353-359.

Spilsbury, K., Garrett, K. L., Shen, W. Y., Constable, I. J., and Rakoczy, P. E. 2000. Overexpression of vascular endothelial growth factor (VEGF) in the retinal pigment epithelium leads to the development of choroidal neovascularization. *Am J Pathol* **157**:135-144.

Stone, E. M., Braun, T. A., Russell, S. R., Kuehn, M. H., Lotery, A. J., Moore, P. A., Eastman, C. G., Casavant, T. L., and Sheffield, V. C. 2004. Missense variations in the fibulin 5 gene and age-related macular degeneration. *N Engl J Med* **351**:346-353.

Sullivan, P. M., Mace, B. E., Maeda, N., and Schmechel, D. E. 2004. Marked regional differences of brain human apolipoprotein E expression in targeted replacement mice. *Neuroscience* **124**:725-733.

Sullivan, P. M., Mezdour, H., Aratani, Y., Knouff, C., Najib, J., Reddick, R. L., Quarfordt, S. H., and Maeda, N. 1997. Targeted replacement of the mouse apolipoprotein E gene with the common human APOE3 allele enhances diet-induced hypercholesterolemia and atherosclerosis. *J Biol Chem* **272**:17972-17980.

Sullivan, P. M., Mezdour, H., Quarfordt, S. H., and Maeda, N. 1998. Type III hyperlipoproteinemia and spontaneous atherosclerosis in mice resulting from gene replacement of mouse Apoe with human Apoe*2. *J Clin Invest* **102**:130-135.

Tobe, T., Okamoto, N., Vinores, M. A., Derevjanik, N. L., Vinores, S. A., Zack, D. J., and Campochiaro, P. A. 1998a. Evolution of neovascularization in mice with overexpression of vascular endothelial growth factor in photoreceptors. *Invest Ophthalmol Vis Sci* **39**:180-188.

Tobe, T., Ortega, S., Luna, J. D., Ozaki, H., Okamoto, N., Derevjanik, N. L., Vinores, S. A., Basilico, C., and Campochiaro, P. A. 1998b. Targeted disruption of the FGF2 gene does not prevent choroidal neovascularization in a murine model. *Am J Pathol* **153**:1641-1646.

Vingerling, J. R., Dielemans, I., Hofman, A., Grobbee, D. E., Hijmering, M., Kramer, C. F., and de Jong, P. T. 1995a. The prevalence of age-related maculopathy in the Rotterdam Study. *Ophthalmology* **102**:205-210.

Vingerling, J. R., Klaver, C. C., Hofman, A., and de Jong, P. T. 1995b. Epidemiology of age-related maculopathy. *Epidemiol Rev* **17**:347-360.

Weller, R. O., and Nicoll, J. A. 2003. Cerebral amyloid angiopathy: pathogenesis and effects on the ageing and Alzheimer brain. *Neurol Res* **25**:611-616.

Yannuzzi, L. A., Negrao, S., Iida, T., Carvalho, C., Rodriguez-Coleman, H., Slakter, J., Freund, K. B., Sorenson, J., Orlock, D., and Borodoker, N. 2001. Retinal angiomatous proliferation in age-related macular degeneration. *Retina* **21**:416-434.

CHAPTER 18

ALTERED RHYTHM OF PHOTORECEPTOR OUTER SEGMENT PHAGOCYTOSIS IN β5 INTEGRIN KNOCKOUT MICE

Emeline F. Nandrot[1] and Silvia C. Finnemann[1,2]*

1. INTRODUCTION

In the retina photoreceptor and retinal pigment epithelial cells (RPE cells) are in close contact. The outer segment portions of photoreceptors consist of stacked membranous disks containing the phototransduction machinery. These disks are permanently produced thereby lengthening outer segments. To maintain constant outer segment length photoreceptors eliminate their most aged tips by daily shedding (Young, 1967).

RPE cells form a polarized monolayer. They extend apical microvilli that ensheath photoreceptor outer segments. Outer segment shedding by photoreceptors precedes a burst of phagocytosis by the RPE that efficiently clears photoreceptor outer segment fragments (POS) from the subretinal space and recycles their components (Young and Bok, 1969). POS shedding and subsequent phagocytosis by RPE cells are crucial for photoreceptor cell function and survival. A single gene defect in the *mer* gene encoding the receptor tyrosine kinase Mer abolishes efficient POS phagocytosis by RPE cells in the Royal College of Surgeons (RCS) rat strain (D'Cruz et al., 2000; Nandrot et al., 2000). Failure of RCS RPE to ingest POS causes debris accumulation and rapid photoreceptor degeneration illustrating the importance of RPE phagocytosis (Mullen and LaVail, 1976).

RPE cells do not usually divide in the adult retina. Each RPE cell faces around 30 outer segments depending on the area of the retina. As photoreceptor shed ~7% of their outer segment mass each day, every RPE cell digests 25,000 to 30,000 disks each and every day of life (Besharse and Defoe, 1998). Therefore, RPE cells are the most active phagocytes in the body. Synchronized POS clearance is tightly regulated and any delay in completing the shedding or digestion process can cause accumulation of undigested material. Indeed, auto-fluorescent inclusion bodies commonly accumulate in human RPE cells of age. These lipo-

*Margaret M. Dyson Vision Research Institute, [1]Department of Ophthalmology and [2]Department of Cell and Developmental Biology, Weill Medical College of Cornell University, Box 233, 1300 York Avenue, New York, NY10021, U.S.A.

fuscin storage bodies contain a complex mix of proteins and lipids and likely results from incomplete turnover of POS material (Feeney, 1978). *In vitro* studies have recently shown that lipofuscin components may directly impair RPE function and viability (Holz et al., 1999; Finnemann et al., 2002). These data suggest that defective digestion of POS by RPE cells may contribute to development or progression of age-related retinal diseases such as age-related macular degeneration.

Outer segment renewal in higher vertebrates is synchronized by circadian rhythms influenced by the daily dark-light cycle (Goldman et al., 1980). Animal studies in rod- or cone-dominant species revealed that rods mainly shed their POS ~2 hours after onset of light and cones shed ~2 hours after dusk (LaVail, 1976; Young, 1977). The increase in the number of phagosomes present in RPE cells at these two time points suggests a peak in phagocytic activity every 12 or 24 hours for RPE cells depending on whether they serve rods, cones or both. No untimely phagocytosis has been observed so far, suggesting that RPE cells may downregulate their phagocytic activity if not "on duty". Currently, mechanisms that regulate RPE phagocytosis are only poorly understood.

2. THE PHAGOCYTIC MACHINERY OF THE RPE

Three plasma membrane receptors have been shown to fullfill distinct roles in RPE phagocytosis. These are the Mer tyrosine kinase receptor MerTK (D'Cruz et al., 2000; Nandrot et al., 2000), the scavenger receptor CD36 (Ryeom et al., 1996), and the adhesion receptor $\alpha v \beta 5$ integrin (Finnemann et al., 1997; Miceli et al., 1997; Lin and Clegg, 1998). MerTK is involved in the internalization step of phagocytosis (Edwards and Szamier, 1977). MerTK deficient macrophages have a reduced capacity to phagocytose apoptotic cells confirming that OS phagocytosis by the RPE is a mechanism similar to the macrophage clearance mechanism for apoptotic cells (Finnemann and Rodriguez-Boulan, 1999; Scott et al., 2001). *In vitro* experiments have shown that CD36 ligation regulates the rate of POS internalization suggesting a role for CD36 in phagocytic signaling (Finnemann and Silverstein, 2001).

$\alpha v \beta 5$ expression at the apical surface of rodent RPE *in vivo* coincides with postnatal establishment of mature interactions between photoreceptors and RPE including the onset of POS renewal (Ratto et al., 1991; Finnemann et al., 1997). Stable binding of POS to the cell surface of RPE cells in culture is largely dependent on $\alpha v \beta 5$ integrin receptors (Finnemann et al., 1997). In addition to POS recognition or tethering RPE cells also use $\alpha v \beta 5$ integrin receptors to activate signaling pathways through focal adhesion kinase that are necessary to activate MerTK (Finnemann, 2003; Nandrot et al., 2004; and see chapter by Finnemann and Nandrot in this volume). As $\beta 5$ integrin only dimerizes with αv integrin subunits, $\beta 5$ integrin knockout mice provide the opportunity to study RPE cells that permanently and exclusively lack $\alpha v \beta 5$ receptors.

3. IMPACT OF LACK OF $\alpha v \beta 5$ INTEGRIN ON RPE FUNCTION

$\beta 5$ knockout mice are viable and fertile, and have normal lifespan (Huang et al., 2000). When we first examined their retinal tissues we did not find gross anatomical abnormalities in $\beta 5$ null mice regardless of age (Nandrot et al., 2004). We then set out to compare wild-

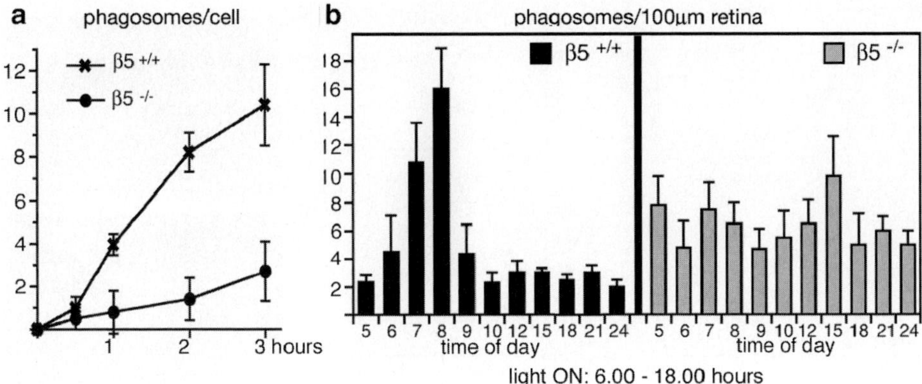

Figure 18.1. β5 integrin deficient RPE cells display defects in POS phagocytosis. **(a)** After 1 to 3 hours of POS phagocytic challenge *in vitro*, β5–/– cells in primary culture phagocytosed less OS than β5+/+ cells. **(b)** Phagosomes were counted on cross sections of β5+/+ and β5–/– eyecups at different time-points of the 24-hour phagocytic cycle. The phagocytic peak detected in β5+/+ mice 2 hours after light onset is absent in β5–/– mice. Modified from Nandrot et al. (2004), J. Exp. Med. 200:1539-1545 by copyright permission of The Rockefeller University Press.

type and β5 knockout RPE phagocytic function both *in vitro* and *in vivo*. Finally, we investigated vision and RPE cell ultrastructure as a function of age in β5 knockout and wild-type control mice.

To test POS phagocytosis *in vitro* we isolated and maintained in primary culture RPE cells from β5 null and wild-type control animals. Strikingly, β5 null RPE cells failed to efficiently take up OS. This decrease was not due to slower uptake kinetics as uptake by β5 null cells remained low at all times of phagocytic stimulation (Figure 18.1a).

To check POS phagocytosis *in vivo* we counted phagosomes present in RPE cells of 4 week old wild-type and β5 null mice at different time points of the 24-hour period. As expected, we detected a peak in the number of phagosomes in wild-type retina ~2 hours after the light onset occuring at 6 AM in our facility (Figure 18.1b) (LaVail, 1976). This peak was lost in β5 null retina in which we counted equal numbers of phagosomes at all time points tested. However, phagosome counts in β5 null retina were similar or higher than phagosome counts in wild-type retina outside of the phagocytosis peak. Thus, phagocytosis in β5 knockout retina is not abolished but loses its temporal regulation. Absence of αvβ5 integrin eliminates the burst of phagocytic activity required for synchronized POS engulfment.

These results show that POS phagocytosis by RPE cells is impaired in 1-month-old β5 null mice. However, β5 null mice develop impaired vision only at a much older age. Scotopic electroretinograms (ERGs) recorded in 1-year-old animals showed a strongly attenuated response in β5 null compared to wild-type mice (Figure 18.2a). The reduction of retinal responses to light stimuli was progressive from 4 months of age in β5 null animals whereas responses did not vary greatly for wild-type animals (Figure 18.2b).

Finally, we detected excessive autofluorescent lipofuscin deposits in RPE of 1-year-old β5 null mice by wide-field fluorescence microscopy (Nandrot et al., 2004). These lipofuscin granules appeared as dense opaque bodies on ultrathin sections examined by electron

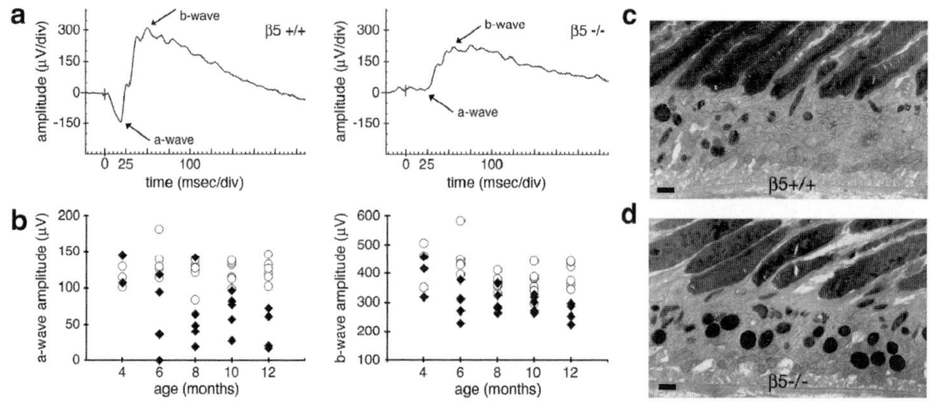

Figure 18.2. Age-related retinal changes in β5–/– mice. **(a)** At 1 year of age, scotopic ERG responses are greatly reduced in β5–/– mice. **(b)** ERGs were recorded for wild-type (♦) and β5–/– (○) mice between the ages of 4 and 12 months. Both a- and b-waves declined in β5–/– mice older than 4 months. RPE cells of 1-year-old wild-type **(c)** and β5–/– **(d)** mice were examined by electron microscopy on ultrathin sections. Electron dense inclusion bodies were detected in β5–/– RPE in larger numbers than in wild-type RPE. Modified from Nandrot et al. (2004), J. Exp. Med. 200:1539-1545 by copyright permission of The Rockefeller University Press.

microscopy. We observed few of these inclusion bodies in wild-type RPE (Figure 18.2c), whereas they were present in large numbers in β5 null RPE (Figure 18.2d).

4. PERSPECTIVE

Our results show that lack of αvβ5 integrin receptors eliminates the rhythm of POS phagocytosis in the retina. With age, mice lacking αvβ5 integrin progressively lose vision and their RPE accumulates lipofuscin granules. Strikingly, even with an early phagocytosis defect β5 null mice only develop late onset pathology. Our findings suggest that constant instead of rhythmic POS phagocytosis by RPE cells may impair POS digestion causing gradual accumulation of autofluorescent compounds as lipofuscin.

Taken together these data emphasize the importance of timely regulation of RPE phagocytosis. αvβ5 integrin activity synchronizes this daily process that is essential for vision (Nandrot et al., 2004). The age-related changes we observed in retinas of β5 knockout mice share some characteristics of retinas of humans with age-related macular degeneration. The β5 knockout mouse strain may thus provide a useful animal model to study lipofuscin buildup with age and gene therapies to delay or reverse its accumulation.

5. ACKNOWLEDGMENTS

This work was supported by NIH grants EY13295 and EY14184, by a Karl Kirchgessner research grant, and by the Irma T. Hirschl/Monique Weill-Caulier Trust.

6. REFERENCES

Besharse, J., and Defoe, D., 1998, The role of the retinal pigment epithelium in photoreceptor membrane turnover, in: *The retinal pigment epithelium*, M. F. Marmor and T. J. Wolfensberger, ed., Oxford University Press, New York, Oxford, pp. 152-172.

D'Cruz, P. M., Yasumura, D., Weir, J., Matthes, M. T., Abderrahim, H., LaVail, M. M., and Vollrath, D., 2000, Mutation of the receptor tyrosine kinase gene Mertk in the retinal dystrophic RCS rat, *Hum. Mol. Genet.* **9**:645-651.

Edwards, R. B., and Szamier, R. B., 1977, Defective phagocytosis of isolated rod outer segments by RCS rat retinal pigment epithelium in culture. *Science.* **197**:1001-1003.

Feeney, L., 1978, Lipofuscin and melanin of human retinal pigment epithelium. Fluorescence, enzyme cytochemical, and ultrastructural studies, *Invest. Ophthalmol. Vis. Sci.* **17**:583-600.

Finnemann, S. C., Bonilha, V. L., Marmorstein, A. D., and Rodriguez-Boulan, E., 1997, Phagocytosis of rod outer segments by retinal pigment epithelial cells requires αvβ5 integrin for binding but not for internalization, *Proc. Natl. Acad. Sci. U. S. A.* **94**:12932-12937.

Finnemann, S. C., and Rodriguez-Boulan, E., 1999, Macrophage and retinal pigment epithelium phagocytosis: apoptotic cells and photoreceptors compete for αvβ3 and αvβ5 integrins, and protein kinase C regulates αvβ5 binding and cytoskeletal linkage, *J. Exp. Med.* **190**:861-874.

Finnemann, S. C., and Silverstein, R. L., 2001, Differential roles of CD36 and αvβ5 integrin in photoreceptor phagocytosis by the retinal pigment epithelium, *J. Exp. Med.* **194**:1289-1298.

Finnemann, S. C., Leung, L. W., and Rodriguez-Boulan, E., 2002, The lipofuscin component A2E selectively inhibits phagolysosomal degradation of photoreceptor phospholipid by the retinal pigment epithelium, *Proc. Natl. Acad. Sci. U. S. A.* **99**:3842-3847.

Finnemann, S. C., 2003, Focal adhesion kinase signaling promotes phagocytosis of integrin-bound photoreceptors, *EMBO J.* **22**:4143-4154.

Goldman, A. I., Teirstein, P. S., and O'Brien, P. J., 1980, The role of ambient lighting in circadian disc shedding in the rod outer segment of the rat retina, *Invest. Ophthalmol. Vis. Sci.* **19**:1257-1267.

Holz, F. G., Schutt, F., Kopitz, J., Eldred, G. E., Kruse, F. E., Volcker, H. E., and Cantz, M., 1999, Inhibition of lysosomal degradative functions in RPE cells by a retinoid component of lipofuscin, *Invest. Ophthalmol. Vis. Sci.* **40**:737-743.

Huang, X., Griffiths, M., Wu, J., Farese, R. V., Jr., and Sheppard, D., 2000, Normal development, wound healing, and adenovirus susceptibility in β5-deficient mice, *Mol. Cell. Biol.* **20**:755-759.

LaVail, M. M., 1976, Rod outer segment disk shedding in rat retina: relationship to cyclic lighting, *Science.* **194**:1071-1074.

Lin, H., and Clegg, D. O., 1998, Integrin αvβ5 participates in the binding of photoreceptor rod outer segments during phagocytosis by cultured human retinal pigment epithelium, *Invest. Ophthalmol. Vis. Sci.* **39**:1703-1712.

Miceli, M. V., Newsome, D. A., and Tate, D. J., Jr., 1997, Vitronectin is responsible for serum-stimulated uptake of rod outer segments by cultured retinal pigment epithelial cells, *Invest. Ophthalmol. Vis. Sci.* **38**:1588-1597.

Mullen, R. J., and LaVail, M. M., 1976, Inherited retinal dystrophy: primary defect in pigment epithelium determined with experimental rat chimeras, *Science.* **192**:799-801.

Nandrot, E., Dufour, E. M., Provost, A. C., Pequignot, M. O., Bonnel, S., Gogat, K., Marchant, D., Rouillac, C., Sepulchre de Conde, B., Bihoreau, M. T., Shaver, C., Dufier, J. L., Marsac, C., Lathrop, M., Menasche, M., and Abitbol, M. M., 2000, Homozygous deletion in the coding sequence of the c-mer gene in RCS rats unravels general mechanisms of physiological cell adhesion and apoptosis, *Neurobiol. Dis.* **7**:586-599.

Nandrot E. F., Kim, Y., Brodie, S. E., Huang, X., Sheppard, D., and Finnemann, S. C., 2004, Loss of synchronized retinal phagocytosis and age-related blindness in mice lacking αvβ5 integrin, *J. Exp. Med.* **200**:1539-1545.

Ryeom, S. W., Sparrow, J. R., and Silverstein, R. L., 1996, CD36 participates in the phagocytosis of rod outer segments by retinal pigment epithelium, *J. Cell Sci.* **109**:387-395.

Scott, R. S., McMahon, E. J., Pop, S. M., Reap, E. A., Caricchio, R., Cohen, P. L., Earp, H. S., and Matsushima, G. K., 2001, Phagocytosis and clearance of apoptotic cells is mediated by MER, *Nature.* **411**:207-211.

Young, R.W., 1967, The renewal of photoreceptor cell outer segments, *J. Cell Biol.* **33**:61-72.

Young, R. W., 1977, The daily rhythm of shedding and degradation of cone outer segment membranes in the lizard retina, *J. Ultrastruct. Res.* **61**:172-185.

Young, R. W., and Bok, D., 1969, Participation of the retinal pigment epithelium in the rod outer segment renewal process, *J. Cell Biol.* **42**:392-403.

LIGHT/DARK TRANSLOCATION OF ALPHATRANSDUCIN IN MOUSE PHOTORECEPTOR CELLS EXPRESSING G90D MUTANT OPSIN

Zack A. Nash[1] and Muna I. Naash[1]

1. INTRODUCTION

Genomic analysis, from patients with retinal diseases such as retinitis pigmentosa (RP) and congenital night blindness (CNB), has provided convincing evidence that various sub-types of retinal diseases can result from mutations in the gene encoding rod opsin, a protein that binds 11-*cis* retinal to form the visual pigment, rhodopsin. Over 100 different mutations in this gene have been documented, and shown to co-segregate with autosomal dominant RP (Berson et al., 2002; Farrar et al., 2002; Nour and Naash, 2003) as well as a few other mutations with CNB (Dryja et al., 1993; Sieving et al., 1995; al Jandal et al., 1999). Interestingly when these CNB causing mutations are expressed in COS cells, they constitutively activated transducin in the dark while in the absence of 11-*cis* retinal (Dryja et al., 1993; Rao et al., 1994; Gross et al., 2003a; Gross et al., 2003b). Thus, the *in vitro* data suggests that the CNB mutations generate a persistent 'dark light' that saturates the rod photocurrent and severely depresses rod sensitivity in a manner similar to that produced by light. One of these CNB mutations resulted from the replacement of glycine at position 90 by aspartic acid (G90D), which has been shown to associate with an unusual trait of congenital stationary night blindness (CSNB) (Sieving et al., 1995).

Expression of the G90D mutation in transgenic mice has recently been described (Sieving et al., 2001; Naash et al., 2004) and shown to have a persistent state of light adaptation (Sieving et al., 2001). Comprehensive studies of the functional abnormalities and the structural changes at the light- and electron-microscopic levels induced by the G90D mutation have recently been described with three transgenic lines that express the transgene at different levels (Naash et al., 2004). A direct relation between level of transgene expression and extent of photoreceptor degeneration has been observed. Higher levels of G90D opsin

[1] Cell Biology, University of Oklahoma Health Sciences Center, 940 Stanton L. Young Blvd., Oklahoma City, OK 73104.

expression produced earlier signs of retinal degeneration and more severe disruption of photoreceptor morphology (Naash et al., 2004). Furthermore, the rod a-wave amplitudes and the amounts of photosensitive pigment were substantially reduced at early ages. Thus, these findings suggest a failure of G90D opsin to efficiently complex with the chromophore to form a light-sensitive rhodopsin in the rod outer segments (ROS). Due to the significant loss of rod sensitivity observed in these G90D retinas; this study addresses the effect of such a desensitized rod environment on the light-dependent translocation of transducin alpha subunit (Tα). Studies have shown that upon illumination, a substantial proportion of Tα found to rapidly move out of the ROS and return only in the dark (Whelan and McGinnis, 1988). The process of Tα movement out of the ROS seems to be mediated by the activation of transducin by photo-isomerized rhodopsin, while the reverse movement depends on the deactivation of the transduction cascade (Hardie, 2003).

2. MATERIALS AND METHODS

2.1. Generation of G90D Transgenic Mice

A 15 kb mouse opsin genomic fragment was used to generate the G90D line. This fragment contains 6.0 kb of the promoter region, all the introns and exons, and 3.5 kb of the 3′ untranslated region containing all of the multiple polyadenylation signals. Mutations were introduced by site directed mutagenesis in PCR reactions as described in previous work (Naash et al., 2004). Ten potential founders were identified and eight of them passed the transgene to their offspring. Transgenic mice were mated to rhodopsin knockout mice (–/–) (Lem et al., 1999) to express the transgene in different backgrounds of wild-type (wt) opsin (i.e., +/+, +/–, or –/–). All mice studied here were maintained in the breeding colony under cyclic light (12L:12D) conditions; cage illumination was ~7 foot-candles during the light cycle. Light-adapted animals were taken after 4 hours of being in their light cycle while dark-adapted animals were kept in the dark for at least 4-6 hours. All procedures were approved by the local Institutional Animal Care and Use Committees and adhered to the ARVO Statement for the Use of Animals in Ophthalmic and Vision Research.

2.2. Histology and Immunohistochemistry

For histology using light microscopy, animals were sacrificed, and eyes were fixed in Davidson's fixative (95% ethanol, 10% neutral buffered formalin, glacial acetic acid, and distilled water) overnight at 4°C. The eyes were then washed 3 times in PBS and stored in 70% ethanol prior to serial dehydration and embedment in paraffin. Eyes were sectioned (6- to 8-μm thickness) on a standard microtome and stained with hematoxylin/eosin.

For immunohistochemistry, animals were sacrificed and eyes were fixed in 0.1 M phosphate buffer (pH 7.4) containing 4% formaldehyde. Tissues were incubated overnight with 10% sucrose in 0.1 M phosphate buffer (pH 7.4) and then stored in 30% sucrose phosphate buffer. The eyes were embedded and processed as described previously (Ding et al., 2004; Nour et al., 2004). The sections were then incubated overnight at 4°C with either 1D4 (a generous gift of Dr. Robert Molday) or with a polyclonal antibody to Tα (purchased from Santa Cruz Biotechnology) at 1 : 100 dilution. Three retinal sections per slide were incubated with 1% BSA in PBS without the primary antibody as a negative control. Sections were treated with goat-anti-rabbit FITC (diluted 1:100) for 1 hr at room temperature and photographed with a Zeiss universal microscope equipped for epifluorescence. Digitally captured images were saved as TIFF files on a computer and imported into Adobe PHOTOSHOP™ 8.0, where color pictures were generated.

3. RESULTS

3.1. Histological Analyses of the G90D Retinas

The G90D mutation is located in the middle of the second transmembrane domain of the molecule (Fig. 19.1). In a recently published study, a direct relation between the level of transgene expression and the extent of photoreceptor degeneration in three G90D lines was demonstrated. In this study, the histological abnormalities at one month of age in one line that expresses the transgene at a level equivalent to one wt allele was evaluated and compared to non-transgenic littermates in all rhodopsin genetic backgrounds (Fig. 19.2A (A-F)). Transgenic retinal sections on the rhodopsin wt (+/+) and hemi (+/−) backgrounds showed no obvious signs of abnormalities when compared to their cognate controls (Fig. 19.2A (A-D)). The length and organization of photoreceptor inner and outer segments as well as the thickness of the outer nuclear layer (ONL) were also similar to controls. Retinal section in Fig. 19.2A (F) shows how the incorporation of one copy of the transgene in the rhodopsin null background (−/−) results in the development of ROS and survival of the photoreceptors. The retinal structure of rhodopsin null (−/−) animals have a normal complement of rod nuclei at early postnatal stages, but the cells fail to form ROS (Fig. 19.2B (E)).

Figure 19.2B shows the localization of the wt and mutant proteins in the G90D retinas in all rhodopsin genetic backgrounds. Using polyclonal anti-opsin antibody (1D4), which recognizes both wt and mutant opsins, on all sections from the three genetic backgrounds showed proper translocation of wt opsin as well as for the G90D protein from the site of synthesis in the rod inner segments to the disc membranes of the ROS. No labeling was seen either in the inner nuclear layer or in the cellular debris at the photoreceptor/RPE interface of rhodopsin null retina (Fig. 19.2B (F)). However, strong immunoreactivity was detected in the ROS of the G90D retinas in the rhodopsin null background (Fig. 19.2B (D))

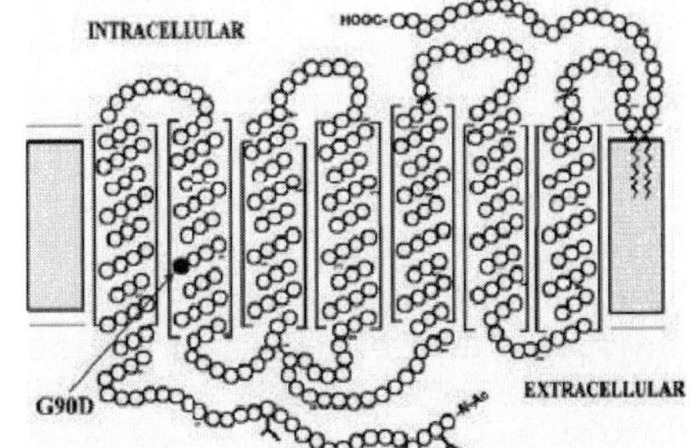

Figure 19.1. Diagram of the mouse opsin molecule showing the location of the G90D mutation in the second transmembrane domain.

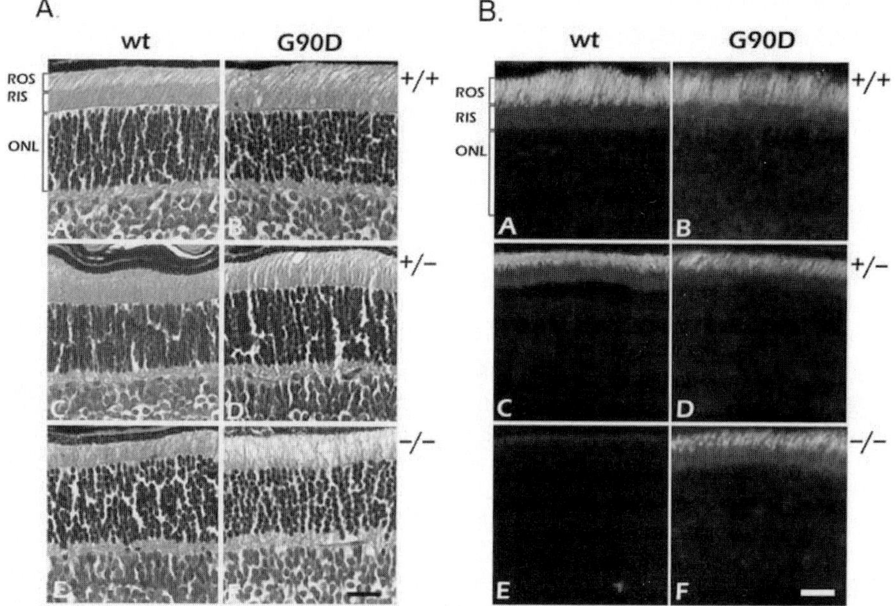

Figure 19.2. (A). Structural analysis at the light level of one month old wt and G90D retinas on all rhodopsin genetic backgrounds. (B). Immunofluorescence of frozen sections of the same animals used in panel A labeled with 1D4. Immunofluorescence was confined almost exclusively to the ROS and rod inner segments (RIS). The fluorescent intensity appeared to be greatest in the ROS. No labeling was detected in the outer nuclear layer (ONL). Scale bar: 25 μm.

which indicates that G90D opsin has been properly localized and integrated into the ROS disc membranes.

3.2. Light-Dependent Translocation of Transducin Alpha in the G90D Retinas

In wt mouse retinas, light triggers the translocation of transducin subunits from the outer segment in the dark, to the inner segment in the light. This process is dependent on photoisomerization of rhodopsin (Hardie, 2003). This experiment, tested whether the desensitization state of the G90D retinas interferes with the translocation of Tα in rod cells. Comparisons of the Tα antibody staining patterns of dark- and light-adapted wt and G90D retinas revealed that the majority of Tα resides in the ROS of dark-adapted G90D retinas from all genetic backgrounds, whereas in the light-adapted state Tα is found throughout the G90D photoreceptor layer, including the RIS and the synapse, similar to the distribution pattern of Tα in non-transgenic controls. These results suggest that Tα redistribution is not effected by the desensitization state imposed by the G90D mutation. Tα translocation was also obvious in the G90D retinas on the rhodopsin null background (Fig. 19.3K&L). Since Tα level in the rhodopsin null retinas is significantly low, its light-dependent translocation in the absence of rhodopsin was not apparent (Fig. 19.3I&J).

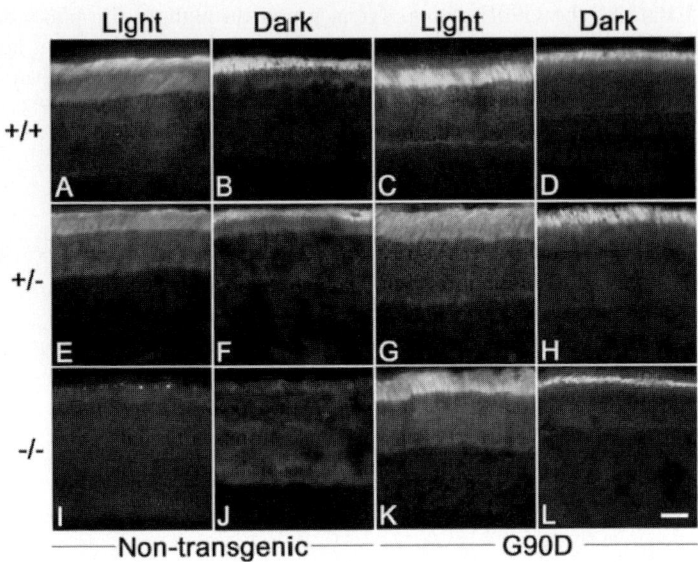

Figure 19.3. Sub cellular distribution of Tα in rod cells in dark- and light-adapted wt and G90D retina from all rhodopsin genetic backgrounds (+/+, +/−, and −/−). The patterns of immuno-staining in sections taken from dark- and light-adapted wt and G90D retinas are similar suggesting that the desensitization state of the G90D rods has no effect on the translocation of transducin. Scale bars: 25 μm; (insets).

4. DISCUSSION

Light-dependent translocation of phototransduction proteins between the inner and outer segments of photoreceptor cells has been thought to play an important role in the adaptation of these cells to light (Whelan and McGinnis, 1988; Zhang et al., 2003; Elias et al., 2004; Lee and Montell, 2004). However, the mechanism by which simultaneous translocation of these proteins remains unknown. Studies have demonstrated that some of these movements occur within minutes following the onset of light (Elias et al., 2004). It has been proposed that the major function of these translocations is to provide protection of photoreceptors in bright light rather than increased vision in dim light (Elias et al., 2004).

The present study tested the light-dependent trafficking of Tα in transgenic retinas carrying the G90D mutation in the opsin gene, a mutation that has been shown to cause a form of CSNB in humans (Sieving et al., 1995), constitutive activation of opsin in COS cells (Rao et al., 1994), and a persistent state of light adaptation ("dark light") in transgenic mice (Sieving et al., 2001; Naash et al., 2004). Since photoreceptor degeneration is directly related to the levels of transgene expression due to opsin over-expression (Naash et al., 2004), a transgenic line was used in this study that expresses G90D opsin at a level equivalent to that of wt. Although no signs of photoreceptor degeneration was seen in this line at early ages (Fig. 19.2A), retinal morphological changes were evident in older animals on different rhodopsin backgrounds (Naash et al., 2004). These changes are attributed to the presence of the G90D opsin rather than opsinoverexpression (Tan et al., 2001). As shown in Figure 19.3, the majority of Tα resides in ROS in dark-adapted wt and G90D retinas. In the light-adapted state of the G90D retina, Tα is found throughout the photoreceptor layer, including the RIS and the synapse, similar to the distribution of Tα in wt mouse retinas. These results suggest that Tα redistribution is independent of the activation status of rhodopsin, an observation that was previously reported (Mendez et al., 2003).

5. ACKNOWLEDGMENT

This research was funded by National Eye Institute (NEI) grant EY-10609 (MIN), Department of Ophthalmology NEI Core grant (EY-01792), the Foundation Fighting Blindness (MIN). Dr. Naash is a recipient of the Research to Prevent Blindness James S. Adams Scholar award.

6. REFERENCES

al Jandal, N., Farrar, G.J., Kiang, A.S., Humphries, M.M., Bannon, N., Findlay, J.B. et al., 1999, A novel mutation within the rhodopsin gene (Thr-94-Ile) causing autosomal dominant congenital stationary night blindness. *Hum. Mutat.* **13**:75-81.

Berson, E.L., Rosner, B., Weigel-DiFranco, C., Dryja, T.P., and Sandberg, M.A., 2002, Disease progression in patients with dominant retinitis pigmentosa and rhodopsin mutations. *Invest Ophthalmol. Vis. Sci.* **43**:3027-3036.

Ding, X.Q., Nour, M., Ritter, L.M., Goldberg, A.F., Fliesler, S.J., and Naash, M.I., 2004, The R172W mutation in peripherin/rds causes a cone-rod dystrophy in transgenic mice. *Hum. Mol. Genet.* **13**:2075-2087.

Dryja, T.P., Berson, E.L., Rao, V.R., and Oprian, D.D., 1993, Heterozygous missense mutation in the rhodopsin gene as a cause of congenital stationary night blindness. *Nat. Genet.* **4**:280-283.

Elias, R.V., Sezate, S.S., Cao, W., and McGinnis, J.F., 2004, Temporal kinetics of the light/dark translocation and compartmentation of arrestin and alpha-transducin in mouse photoreceptor cells. *Mol. Vis* **10**:672-681.

Farrar, G.J., Kenna, P.F., and Humphries, P., 2002, On the genetics of retinitis pigmentosa and on mutation-independent approaches to therapeutic intervention. *EMBO J.* **21**:857-864.

Gross, A.K., Rao, V.R., and Oprian, D.D., 2003a, Characterization of rhodopsin congenital night blindness mutant T94I. *Biochemistry* **42**:2009-2015.

Gross, A.K., Xie, G., and Oprian, D.D., 2003b, Slow binding of retinal to rhodopsin mutants G90D and T94D. *Biochemistry* **42**:2002-2008.

Hardie, R.C., 2003, Phototransduction: shedding light on translocation. *Curr. Biol.* **13**:R775-R777.

Lee, S.J. and Montell, C., 2004, Light-dependent translocation of visual arrestin regulated by the NINAC myosin III. *Neuron* **43**:95-103.

Lem, J., Krasnoperova, N.V., Calvert, P.D., Kosaras, B., Cameron, D.A., Nicolo, M. et al., 1999, Morphological, physiological, and biochemical changes in rhodopsin knockout mice. *Proc. Natl. Acad. Sci. U. S. A.* **96**:736-741.

Mendez, A., Lem, J., Simon, M., and Chen, J., 2003, Light-dependent translocation of arrestin in the absence of rhodopsin phosphorylation and transducin signaling. *J. Neurosci.* **23**:3124-3129.

Naash, M.I., Wu, T.H., Chakraborty, D., Fliesler, S.J., Ding, X.Q., Nour, M. et al., 2004, Retinal abnormalities associated with the G90D mutation in opsin. *J. Comp Neurol.* **478**:149-163.

Nour, M., Ding, X.Q., Stricker, H., Fliesler, S.J., and Naash, M.I., 2004, Modulating expression of peripherin/rds in transgenic mice: critical levels and the effect of overexpression. *Invest Ophthalmol Vis Sci* **45**:2514-2521.

Nour, M. and Naash, M.I., 2003, Mouse models of human retinal disease caused by expression of mutant rhodopsin. A valuable tool for the assessment of novel gene therapies. *Adv. Exp. Med. Biol.* **533**:173-179.

Rao, V.R., Cohen, G.B., and Oprian, D.D., 1994, Rhodopsin mutation G90D and a molecular mechanism for congenital night blindness. *Nature* **367**:639-642.

Sieving, P.A., Fowler, M.L., Bush, R.A., Machida, S., Calvert, P.D., Green, D.G. et al., 2001, Constitutive "light" adaptation in rods from G90D rhodopsin: a mechanism for human congenital nightblindness without rod cell loss. *J. Neurosci.* **21**:5449-5460.

Sieving, P.A., Richards, J.E., Naarendorp, F., Bingham, E.L., Scott, K., and Alpern, M., 1995, Dark-light: model for nightblindness from the human rhodopsin Gly-90–>Asp mutation. *Proc. Natl. Acad. Sci. U. S. A.* **92**:880-884.

Tan, E., Wang, Q., Quiambao, A.B., Xu, X., Qtaishat, N.M., Peachey, N.S. et al., 2001, The relationship between opsin overexpression and photoreceptor degeneration. *Invest Ophthalmol. Vis. Sci.* **42**:589-600.

Whelan, J.P. and McGinnis, J.F. (1988) Light-dependent subcellular movement of photoreceptor proteins. *J. Neurosci. Res.* **20**:263-270.

Zhang, H., Huang, W., Zhang, H., Zhu, X., Craft, C.M., Baehr, W., and Chen, C.K., 2003, Light-dependent redistribution of visual arrestins and transducin subunits in mice with defective phototransduction. *Mol. Vis.* **9**:231-237.

CHAPTER 20

SLOWED PHOTORESPONSE RECOVERY AND AGE-RELATED DEGENERATION IN CONES LACKING G PROTEIN-COUPLED RECEPTOR KINASE 1

Xuemei Zhu, Bruce Brown, Lawrence Rife, and Cheryl M. Craft*

1. INTRODUCTION

The vertebrate retina is a specialized neural network that contains very sensitive signal transducers—the rod and cone photoreceptors. Rods function in near darkness (scotopic) and are responsible for dim light vision, while cones operate in bright light (photopic) and provide daytime, high acuity color vision. In the human retina, rods are the dominant photoreceptor cell type and comprise about 95% of all photoreceptor cells, while cones account for only 5% of the cells. Yet in a bright light environment, normal cone function is essential for visual perception since rods become saturated and are rendered nonfunctional.

Retinitis pigmentosa (RP) and age-related macular degeneration (AMD) are two major causes of visual loss due to photoreceptor degeneration. In RP, rod degeneration results in night blindness and loss of peripheral vision. Inevitably, cones are also lost following the disappearance of rods through currently unknown mechanisms, resulting in blindness.[1] AMD is the leading cause of visual impairment and legal blindness in elderly people in the Western world.[2] In AMD, central vision is compromised initially, and although rod cell death occurs prior to cone loss, it is the subsequent cone cell death that eventually leads to complete visual loss.[3-5] During the last decade, significant advances have been made in understanding the mechanisms leading to rod cell death; however, those underlying cone loss are still poorly delineated. This is due to the paucity of cones in the mammalian retina that makes the study of cone function and disease-related processes difficult.

The neural retinal leucine zipper (*Nrl*) knockout (KO, $^{-/-}$) mouse has a pure-cone retina due to a cell fate switch from rod to S cone during retinal development.[6] ERG analysis of $Nrl^{-/-}$ mice reveals that the amplitude of light-adapted ERG responses elicited by maximum stimulus does not change significantly up to 31 weeks of age, suggesting the cones in these

* Xuemei Zhu, Bruce Brown, Lawrence Rife and Cheryl M. Craft, the Mary D. Allen Laboratory for Vision Research, Doheny Eye Institute, Department of Ophthalmology and Cell & Neurobiology, Keck School of Medicine of the University of Southern California, Los Angeles, California, 90033-9224.

mouse retinas survive without rod function.[6] In this study, we describe an age-dependent retinal degeneration in the pure-cone *Nrl*[-/-] mice lacking G protein-coupled receptor kinase 1 (GRK1). The *Nrl*[-/-]*Grk1*[-/-] mice may provide a useful model for studying the molecular mechanisms of cone cell death in AMD and other retinal diseases.

2. METHODS

2.1. Animals

Nrl[-/-]*Grk1*[-/-] mice were generated by crossing the *Nrl*[-/-] mice[6] with the *Grk1*[-/-] mice[7] as described previously.[8] The *Nrl*[-/-] were generously provided by Anand Swaroop and Alan Mears (University of Michigan), and the *Grk1*[-/-] mice were provided by Jason Chen (University of Utah). Both the *Nrl*[-/-] and the *Nrl*[-/-]*Grk1*[-/-] mice were born and maintained in total darkness. All animals were treated according to the guidelines established by the Institute for Laboratory Animal Research.

2.2. Immunoblot Analysis

Three mice from each genotype and each age group were killed at mid-day under room light. From each animal, the retina was dissected from one eye, and the other eye was fixed for histological and immunohistochemistry analysis (data not shown). The retinas were flash frozen on dry ice and kept at $-80°C$ until use. Retinas were homogenized, and an equal amount of proteins from each retina was resolved on replicate 11.5% SDS-PAGE and transferred to PVDF membranes (Millipore Corp., Bedford, MA). The blots were incubated with rabbit polyclonal antibodies against mouse cone arrestin (mCAR),[9] S or M opsin[8] followed by a horseradish peroxidase (HRP) conjugated goat anti-rabbit secondary antibody and visualized by an Enhanced Chemiluminescence (ECL) Kit (Amersham, Arlington Heights, IL).[10]

2.3. Electroretinography (ERG)

ERGs were recorded as previously described.[11] Mice were dark-adapted overnight, and their eyes were dilated with topical administration of phenylephrine (2.5%) and tropicamide (0.5%). Mice were anesthetized via an intraperitoneal injection of ketamine HCl (100 mg/kg) and xylazine HCl (10 mg/kg), and the cornea anesthetized with 0.5% tropical tetracaine. The mouse was placed in an aluminum foil-lined Faraday cage and a DLT fiber electrode placed on the right cornea. A platinum reference electrode was placed on the lower eyelid and another ground electrode on the epsilateral ear. Photic stimuli of 10-µs duration were delivered through one arm of a coaxial cable using a Grass PS22 xenon flash. The cable delivered the flash 5 mm from the surface of the cornea, and flashes were attenuated with neutral density filters held to a window of fixed f-stop. Dark-adapted maximum responses (mesopic ERG) were measured using the non-attenuated light stimulus (10^0). Photopic ERGs were measured using the same stimulus with a 6-foot candle (fc) white background light delivered through the other arm of the coaxial cable. The half-amplitude bandwidth of the system was 0.01-100 Hz. Responses were recorded on a Coopervision/Nicolet Electrovisual Diagnostic System, and amplified potentials were displayed on a storage oscilloscope for viewing and photographic recording.

3. RESULTS

3.1. Delayed Photoresponse Recovery in *Nrl*[-/-]*Grk1*[-/-] Mice

In recently completed work, we have demonstrated that GRK1 is responsible for light-dependent phosphorylation of both S and M cone opsins in the mouse retina.[8] To evaluate the effect of GRK1 deletion on phototransduction in cones, we recorded ERGs from *Nrl*[-/-] and *Nrl*[-/-]*Grk1*[-/-] mice.

As shown in Figure 20.1A, neither *Nrl*[-/-] nor *Nrl*[-/-]*Grk1*[-/-] mice had any ERG response to low intensity light stimulation, which caused a strong rod response in WT mouse, consistent with the pure-cone phenotype of the *Nrl*[-/-] mouse.[6] Strong ERG responses were elicited in both *Nrl*[-/-] and *Nrl*[-/-]*Grk1*[-/-] mice with high intensity light (Figure 20.1B). Interestingly, the amplitude of the b-wave was reduced by approximately 50% with a 6-fc white background light (6-fc bgd) in the *Nrl*[-/-]*Grk1*[-/-] mouse, compared to that of the *Nrl*[-/-] mouse whose b-wave amplitude was not significantly affected by the background light (Figure 20.1B & C).

We also investigated the time course of recovery of dark-adapted ERGs in *Nrl*[-/-] and *Nrl*[-/-]*Grk1*[-/-] mice. Paired-flash ERG responses of dark-adapted mice were recorded using high intensity (10^0) white flashes at inter-stimulus intervals (ISIs) specified in seconds (s) to the left of the traces (Figure 20.2). Complete recovery of the ERG responses in the *Nrl*[-/-] mice was achieved in about 0.625 sec, but for the *Nrl*[-/-]*Grk1*[-/-] mice, the ERG responses were not totally recovered even after 5 sec, suggesting a delayed recovery of cone responses in *Nrl*[-/-]*Grk1*[-/-] mice.

3.2. Age-Dependent Cone Degeneration in the *Nrl*[-/-]*Grk1*[-/-] Mouse Retina

Morphological studies showed that the outer nuclear layer (photoreceptor layer) of the *Nrl*[-/-]*Grk1*[-/-] mice was significantly thinner than that of the age-matched *Nrl*[-/-] mice, and was progressively thinner with advancing age, suggesting an age-dependent photoreceptor degeneration in the *Nrl*[-/-]*Grk1*[-/-] mouse retina (X. Zhu *et al.*, in preparation). ERG analyses of retinal function of mice reared in total darkness revealed that the b-wave amplitude of

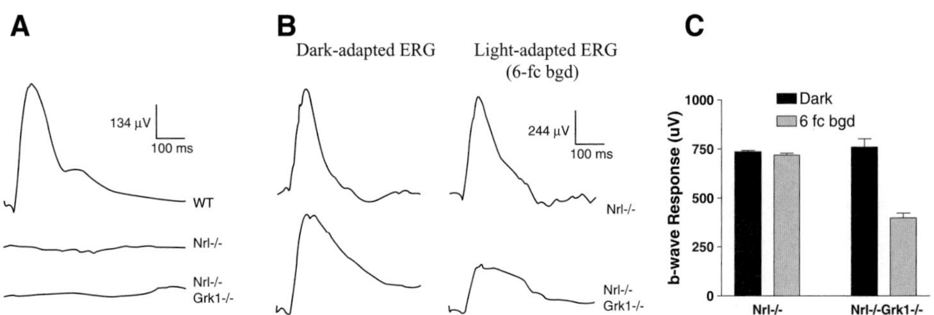

Figure 20.1. Electroretinography. A. Dark-adapted ERG responses with low stimulus intensity (10^{-3}) from WT, *Nrl*[-/-] and *Nrl*[-/-]*Grk*[-/-] mice. B. Dark- and light-adapted ERGs with high stimulus intensity (10^0) from *Nrl*[-/-] and *Nrl*[-/-]*Grk1*[-/-] mice. C. Graph representation of the b-wave amplitude from ERG recordings of *Nrl*[-/-] and *Nrl*[-/-]*Grk1*[-/-] mice. The data represent mean ± Standard Deviation (SD) of six mice of each genotype.

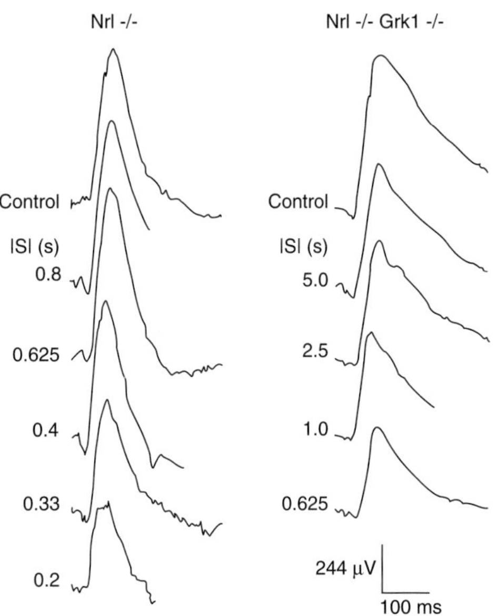

Figure 20.2. Recovery of ERG responses after a conditioning flash in $Nrl^{-/-}$ and $Nrl^{-/-}Grk1^{-/-}$ mice. A high intensity white flash was delivered after a conditioning flash at the same intensity at inter-stimulus intervals (ISIs) specified in seconds (s) to the left of the traces. Responses obtained without the preceding conditioning flashes are marked as controls.

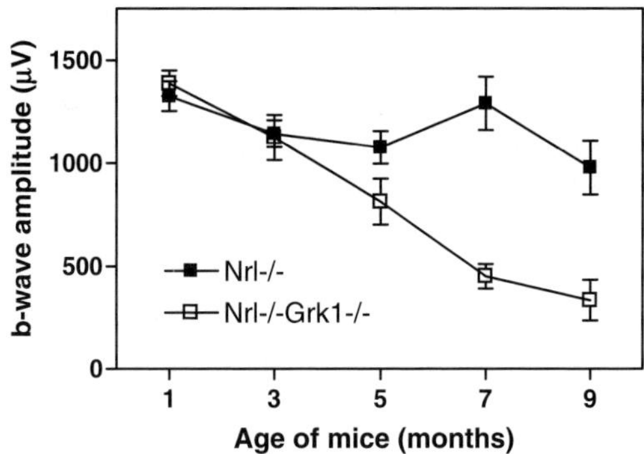

Figure 20.3. Maximum b-wave amplitude of dark-adapted ERG responses in $Nrl^{-/-}$ and $Nrl^{-/-}Grk1^{-/-}$ mice.

mesopic ERGs decreased rapidly with increasing age in the $Nrl^{-/-}Grk1^{-/-}$ mice, while that of the $Nrl^{-/-}$ mice remained unchanged up to 9 months of age (Figure 20.3). Animals reared in either 12:12 hr cyclic light or bright constant light had similar ERG b-wave amplitude to those of the same genotype raised in total darkness (X. Zhu *et al.*, in preparation), suggesting that the functional decrease and degeneration of the $Nrl^{-/-}Grk1^{-/-}$ photoreceptors are dependent on age but independent of light under these conditions, in contrast to the light-induced rod degeneration in the $Grk1^{-/-}$ mouse retina.[7]

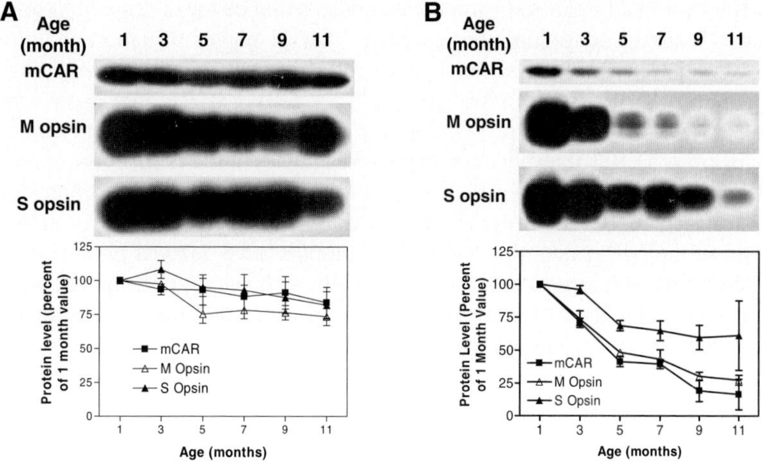

Figure 20.4. Immunoblot analysis of mCAR, S and M Opsins. $Nrl^{-/-}$ (A) and $Nrl^{-/-}Grk1^{-/-}$ mice (B) reared in total darkness were killed, and one retina was dissected and frozen. Frozen retinas were homogenized and equal amounts of proteins were applied to an SDS-PAGE. In each panel, a representative immunoblot and a histogram representing quantitative data (Mean ± SEM) from 3 immunoblots are shown per age group.

3.3. Both S and M Cones Die with Age in the $Nrl^{-/-}Grk1^{-/-}$ Mouse Retina

To determine if both S and M cones degenerate in the $Nrl^{-/-}Grk1^{-/-}$ mouse retina, we determined protein expression levels by immunoblot analyses for S and M opsins, as well as for mCAR, which is expressed in all cones.[8,9] The expression levels of all three photoreceptor-specific proteins decrease dramatically with age in the $Nrl^{-/-}Grk1^{-/-}$ mouse retina but do not change significantly in the $Nrl^{-/-}$ mouse retina up to 11 months of age (Figure 20.4). These results suggest that both S and M cones degenerate with age in the $Nrl^{-/-}Grk1^{-/-}$ mouse retina.

4. DISCUSSION

Morphological and biochemical analysis of the $Nrl^{-/-}$ retina suggest a complete lack of rods and increased number of S cones.[6] Single cell recordings reveal that the $Nrl^{-/-}$ photoreceptors have similar responses to the cones of the rhodopsin$^{-/-}$ mice,[12,13] which are used as a pure cone animal model for studying cone function.[14] Therefore, the photoreceptors of the $Nrl^{-/-}$ mouse retina are functionally and biochemically cones.

In biochemical experiments utilizing these pure-cone retinas, we have found that both S and M opsins are phosphorylated following light exposure, and that cone arrestin preferentially binds to light-activated, phosphorylated cone opsins, suggesting that cones in the $Nrl^{-/-}$ mouse retina may have a similar signal shutoff pathway to that of the wildtype mouse rods.[8]

GRK1 is expressed in all rods and cones in the retina of rod-dominant human, monkey and mouse[8,15-17] and of cone-dominant chicken.[18] GRK1 phosphorylation of light-activated rhodopsin is required for normal inactivation of rhodopsin *in vivo*.[7,19] Although a cone-specific GRK (GRK7) exists in other species,[16,20-22] GRK1 is the only opsin kinase found in the mouse retina[16,22,23] and is responsible for light-dependent phosphorylation of both S and M cone opsins.[8] ERG analysis of cone photoresponses of the *Grk1*[-/-] mouse retina reveals that GRK1 plays a critical role in the inactivation of murine cone phototransduction.[17] In the studies presented here, we demonstrate that when the *Grk1* gene is simultaneously knocked out in the *Nrl*[-/-] mice, the *Nrl*[-/-]*Grk1*[-/-] animals have delayed photoresponse recovery, and their retinal function decreases dramatically with age, indicating that lack of GRK1 function in cones can lead to cone cell degeneration, in addition to the delayed photoreseponse recovery reported previously.[17]

In another study using single cell recordings of the *Nrl*[-/-]*Grk1*[-/-] photoreceptors, Pugh and collaborators have confirmed the critical role of GRK1 in cone phototransduction shutoff and have shown that the shutoff of M opsin is more dramatically delayed than that of S opsin in the *Nrl*[-/-]*Grk1*[-/-] retina.[24] These results suggest a potential alternative shutoff pathway for S opsin in the *Nrl*[-/-]*Grk1*[-/-] mouse retina. Our biochemical experiments show that both S and M opsins decrease with age, but the S opsin decrease appears to be slower than that of M opsin in the *Nrl*[-/-]*Grk1*[-/-] retina (X. Zhu *et al.*, in preparation). Further immuno-histochemistry studies are needed to determine if M cones die earlier or faster than S cones in the *Nrl*[-/-]*Grk1*[-/-] mouse retina. Because the cone cell degeneration in the *Nrl*[-/-]*Grk1*[-/-] mouse retina is light-independent, we postulate that GRK1 may play other roles that are related to regulation of cellular proliferation and/or apoptosis in cone photoreceptors of the mouse retina. Experiments are underway to define the molecular mechanisms leading to cone cell death in the *Nrl*[-/-]*Grk1*[-/-] mouse retina. We believe that these mice will provide a valuable model for studying the molecular pathways of cone photoreceptor degeneration and for testing preventive and therapeutic strategies for rescuing cone function, thus preserving vision, in various retinal degenerative diseases.

5. ACKNOWLEDGEMENTS

We thank Mary D. Allen for over a decade of generous support to our vision research program. CMC is the Mary D. Allen Chair in Vision Research, Doheny Eye Institute (DEI). These studies were also supported, in part, from the NEI Core Vision Research Center grant EY03040 (DEI), the Tony Gray Foundation and Dorie and Fred Miller. The authors also wish to thank Drs. Alan Mears and Anand Swaroop for providing the *Nrl*[-/-] mice and Dr. Ching-Kang Jason Chen for the *Grk1*[-/-] mice.

6. REFERENCES

1. Hicks D, Sahel J. The implications of rod-dependent cone survival for basic and clinical research. *Invest Ophthalmol Vis Sci.* 1999;**40**:3071-3074.
2. Klein R, Klein BE, Linton KL. Prevalence of age-related maculopathy. The Beaver Dam Eye Study. *Ophthalmology.* 1992;**99**:933-943.
3. Curcio CA, Millican CL, Allen KA, Kalina RE. Aging of the human photoreceptor mosaic: evidence for selective vulnerability of rods in central retina. *Invest Ophthalmol Vis Sci.* 1993;**34**:3278-3296.

4. Curcio CA. Photoreceptor topography in ageing and age-related maculopathy. *Eye.* 2001;**15**:376-383.

5. Reme CE, Grimm C, Hafezi F, Iseli HP, Wenzel A. Why study rod cell death in retinal degenerations and how? *Doc Ophthalmol.* 2003;**106**:25-29.

6. Mears AJ, Kondo M, Swain PK, et al. Nrl is required for rod photoreceptor development. *Nat Genet.* 2001;**29**:447-452.

7. Chen CK, Burns ME, Spencer M, et al. Abnormal photoresponses and light-induced apoptosis in rods lacking rhodopsin kinase. *Proc Natl Acad Sci U S A.* 1999;**96**:3718-3722.

8. Zhu X, Brown B, Li A, et al. GRK1-dependent phosphorylation of S and M opsins and their binding to cone arrestin during cone phototransduction in the mouse retina. *J Neurosci.* 2003;**23**:6152-6160.

9. Zhu X, Li A, Brown B, et al. Mouse cone arrestin expression pattern: light induced translocation in cone photoreceptors. *Mol Vis.* 2002;**8**:462-471.

10. Zhu X, Craft CM. Modulation of CRX transactivation activity by phosducin isoforms. *Mol Cell Biol.* 2000;**20**:5216-5226.

11. Chen P, Hao W, Rife L, et al. A photic visual cycle of rhodopsin regeneration is dependent on Rgr. *Nat Genet.* 2001;**28**:256-260.

12. Nikonov SS, Daniele LL, Mears AJ, Swaroop A, Pugh EN. Functional properties of photocurrents of single photoreceptors of the Nrl–/– mouse. *ARVO Abstract.* 2003;

13. Nikonov SS, Daniele L, Zhu X, et al. Photoreceptors of the Nrl knockout mouse: Are they cones. *The eighth Annual Vision Research Conference.* 2004;Abstract:

14. Jaissle GB, May CA, Reinhard J, et al. Evaluation of the rhodopsin knockout mouse as a model of pure cone function. *Invest Ophthalmol Vis Sci.* 2001;**42**:506-513.

15. Zhao X, Huang J, Khani SC, Palczewski K. Molecular forms of human rhodopsin kinase (GRK1). *J Biol Chem.* 1998;**273**:5124-5131.

16. Weiss ER, Ducceschi MH, Horner TJ, et al. Species-specific differences in expression of G-protein-coupled receptor kinase (GRK) 7 and GRK1 in mammalian cone photoreceptor cells: implications for cone cell phototransduction. *J Neurosci.* 2001;**21**:9175-9184.

17. Lyubarsky AL, Chen C, Simon MI, Pugh EN, Jr. Mice lacking G-protein receptor kinase 1 have profoundly slowed recovery of cone-driven retinal responses. *J Neurosci.* 2000;**20**:2209-2217.

18. Zhao X, Yokoyama K, Whitten ME, et al. A novel form of rhodopsin kinase from chicken retina and pineal gland. *FEBS Lett.* 1999;**454**:115-121.

19. Chen J, Makino CL, Peachey NS, Baylor DA, Simon MI. Mechanisms of rhodopsin inactivation in vivo as revealed by a COOH-terminal truncation mutant. *Science.* 1995;**267**:374-377.

20. Hisatomi O, Matsuda S, Satoh T, et al. A novel subtype of G-protein-coupled receptor kinase, GRK7, in teleost cone photoreceptors. *FEBS Lett.* 1998;**424**:159-164.

21. Weiss ER, Raman D, Shirakawa S, et al. The cloning of GRK7, a candidate cone opsin kinase, from cone- and rod-dominant mammalian retinas. *Mol Vis.* 1998;**4**:27.

22. Chen CK, Zhang K, Church-Kopish J, et al. Characterization of human GRK7 as a potential cone opsin kinase. *Mol Vis.* 2001;**7**:305-313.

23. Caenepeel S, Charydczak G, Sudarsanam S, Hunter T, Manning G. The mouse kinome: discovery and comparative genomics of all mouse protein kinases. *Proc Natl Acad Sci U S A.* 2004;**101**:11707-11712.

24. Nikonov SS, Daniele L, Zhu X, et al. Photorecptors of Nrl-/- mice co-express functional S- and M-opsins having distinct inactivation mechanisms. *J Gen Physiol.* 2005; In Press.

CHAPTER 21

TRANSGENIC ANIMAL STUDIES OF HUMAN RETINAL DISEASE CAUSED BY MUTATIONS IN PERIPHERIN/*RDS*

Xi-Qin Ding and Muna I. Naash[1]

1. INTRODUCTION

The photoreceptor disk membrane protein peripherin/*rds* is essential for the outer segment morphogenesis and integrity. Peripherin/*rds* associates with itself and with its homologue Rom-1 to form homo- and hetero-complexes, which are necessary for its structural role (Goldberg et al., 1995; Molday, 1998). More than seventy different pathogenic mutations in the peripherin/*rds* gene have been identified. These mutations are divided primarily into two categories: those associated with classic retinitis pigmentosa (RP), and those associated with various forms of macular dystrophy (MD). In fact, mutations in peripherin/*rds* account for 5-10% of RP causes, and is a major cause for MD (Kohl et al., 1998; Molday, 1998; http://www.sph.uth.tmc.edu/RetNet; http://www.retina-international.org/sci-news/rdsmut.htm). Insights into the functional significance, structural role, and pathogenic effects of this protein have been accumulating considerably since its initial description; this is largely accomplished by the use of laboratory animal models. Use of transgenic or knock-out animals holds great potential for the investigation of retinal disease pathogenesis and the exploration of therapeutic interventions. Table 21.1 summarizes the animal models used to investigate the disease-causing mutations in peripherin/*rds*. In addition to the pathogenesis study, transgenic mouse and Xenopus laevis expressing the wild type peripherin/*rds* or the C-terminus have also been used to explore the structural and functional significance of the protein (Loewen et al., 2003; Ritter et al., 2004).

[1] Cell Biology, University of Oklahoma Health Sciences Center, 940 Stanton L. Young Blvd, Oklahoma City, OK 73104, U.S.A. Corresponding author Muna I. Naash, E-mail muna-naash@ouhsc.edu.

Table 21.1. Transgenic animal models of retinal diseases caused by mutations in peripherin/*rds*.

Mutations and Diseases	Animal Models	Phenotype in the Animals	Pathogenesis of Retinal Degeneration	References
L185P/Rom-1 null (RP)	Transgenic mice	Rod degeneration	Haploinsufficiency	Kedzierski et al., 2001
	Transgenic Xenopus	Aggregated in the inner segment		Loewen et al., 2003
C214S (RP)	Transgenic mice	Rod degeneration	Haploinsufficiency	Stricker et al., 2003
	Transgenic Xenopus	Aggregated in the inner segment		Loewen et al., 2003
P216L (RP)	Transgenic mice	Rod degeneration	Dominant negative effect	Kedzierski et al., 1997
	Transgenic Xenopus	Rod degeneration	Haploinsufficiency	Loewen et al., 2003
307del (RP)	Target-deleting mice	Rod degeneration	Dominant negative effect Haploinsufficiency	McNally et al., 2002
R172W (MD)	Transgenic mice	Cone-dominant degeneration	Dominant negative effect	Li et al., 1999; Ding et al., 2004

2. TRANSGENIC ANIMAL MODELS OF RP CAUSED BY MUTATION IN PERIPHERIN/*RDS*

The majority of the RP-causing mutations in peripherin/*rds* exerts a dominant effect such as the P216L (Kajiwara et al., 1991) and C214S mutations (Saga et al., 1993); a few fall into the digenic group for example the double heterozygous for a L185P mutation in peripherin/*rds* and a second null mutation in Rom-1 (Dryja et al., 1997). In the first transgenic mouse model for peripherin/*rds* mutation-linked autosomal dominant RP, Kedzierski et al., (1997) described an expression-level-dependent photoreceptor degeneration and outer segment shortening in the P216L transgenic mice. Expression of the P216L transgene on the *rds*$^{+/-}$ and *rds*$^{-/-}$ background resulted in a faster rate of photoreceptor degeneration and outer segment dysplasia than that seen in the non-transgenic controls. Thus, the phenotype seen in P216L retina is caused by both direct dominant effect of the mutant protein and a consequence of haploinsufficiency. In the study by Stricker et al., (2003), the pathogenesis of the C214S mutation was examined in several transgenic lines with different expression levels of the transgene. Although, comparable amount of transgene message was formed in the transgenic retinas, only a very small amount of the C214S protein was detected. Moreover, ectopic expression of the C214S mutant protein was observed in the inner retinal cells of transgenic mice (Stricker et al., 2003) . The phenotype of photoreceptor degeneration seen in these transgenic mice resembles the symptom in patients with the same mutation. Thus, the haploinsufficiency resulted from the fatal mutation contributes to the retinopathy caused by the C214S mutation.

In a digenic RP mouse model (Kedzierski et al., 2001), in which both the L185P mutation and levels of peripherin/*rds* and Rom-1 closely matched those predicted for the corresponding human diseases, photoreceptor degeneration in these mice was shown to be faster than that in the monogenic controls. From this model, it was proposed that deficiency of peripherin/*rds* and Rom-1 might be the main cause of photoreceptor degeneration and that the threshold level for the combined abundance of peripherin/*rds* and Rom-1 is approximately 60% of the wild type. Below this level, the extent of outer segment disorganization may result in clinically significant photoreceptor degeneration.

Transgenic Xenopus, which express wild type peripherin/*rds* and the autosomal dominant RP-linked mutants as GFP-fusion proteins in rod photoreceptors, was recently established to identify the determinants required for peripherin/*rds* targeting to disk membranes and to elucidate the mutation pathogenesis (Loewen et al., 2003). The wild type and the P216L mutant were properly assembled as tetramers and targeted to disk membranes as visualized by confocal and electron microscopy. In contrast, the C214S and L185P mutants, which form homodimers but not tetramers, were retained in the inner segments. The finding of mislocalization of the C214S mutant was consistent with that observed in transgenic mice (Stricker et al., 2003). From these studies, it was proposed that tetramerization is required for peripherin/*rds* targeting and incorporation into disk membranes and that a checkpoint between the photoreceptor inner and outer segments allows only correctly assembled peripherin/*rds* tetramers to be incorporated into nascent disk membranes. Thus, the tetramerization-defective mutants (C214S and L185P) cause RP through a deficiency in wild type peripherin/*rds*, whereas tetramerization-proficient P216L peripherin/*rds* causes RP through a dominant negative effect. Further studies on this model indicated that the introduction of a new N-linked oligosaccharide chain might contribute to the defect of the P216L mutant protein (Loewen et al., 2003).

McNally et al., (2002) introduced a targeted single-base deletion at codon 307 of the peripherin/*rds* gene in mice, similar to a human mutation reported by Apfelstedt-Sylla et al., (1995) in which patients suffered from a slowly progressive form of autosomal dominant RP. The mutation in the human gene causes a frameshift which results in a stop codon after a further 16 triplets, and the expected protein is 26 amino acids shorter than the wild-type peripherin/*rds*. The frameshift in the mouse gene is predicted to result in alteration of the last 40 amino acids of the C-terminus of the protein and the addition of an extra 11 amino acids. In heterozygous and homozygous peripherin/*rds*-307del mice, the induced retinopathy, as evaluated by histopathologic and electroretinographic analysis, appeared to be more rapid when compared with $rds^{+/-}$ or $rds^{-/-}$ mice. Thus, the pathogenesis of this mutation in patients may involve both the dominant-negative effect and the haploinsufficiency.

3. TRANSGENIC ANIMAL MODEL OF MD CAUSED BY MUTATIONS IN PERIPHERIN/*RDS*

The second category of the disease-causing mutations contains those mutations affecting the central macular regions of the retina. To date, more than thirty different mutations in peripherin/*rds* have been identified in patients diagnosed with MD or different forms of cone-rod dystrophy. Wells et al., (1993) first described a substitution of arginine with tryptophan in codon 172 (R172W mutation) in patients with MD. Later, the same mutation was reported in patients of English, Japanese, Swiss, and Spanish origins, with similar patterns of retinopathy (Wroblewski et al., 1994; Nakazawa et al., 1995; Jacobson et al., 1996; Milla et al., 1998; Payne et al., 1998). In addition, other mutations in this position, including R172G and R172Q, were also found to associate with MD (Nakazawa et al., 1995; Payne et al., 1998). Although symptoms in a majority of the patients suggest a cone-dominant defect, one report has shown a more diffuse and progressive retinal degeneration in patients with this mutation (Ekstrom et al., 1998).

The pathogenesis of MD related to the R172W mutation was explored in the transgenic mouse model (Li et al., 1999; Ding et al., 2004). The phenotype in these mice resembles the clinical symptoms of patients carrying this mutation. Functional, structural and biochemical analyses showed a direct correlation between transgene expression levels and the onset/severity of the phenotype. Transgenic mice from the low expresser line (40% of wild type) showed a mild, late-onset cone dystrophy in which cone functional deficits were associated with reduction in cone density as early as nine months of age (Li et al., 1999). However, the high expresser line (75% of the wild type) revealed an early onset, autosomal dominant cone-rod dystrophy (Ding et al., 2004). The cone-dominant degeneration induced by the R172W mutation was well documented in the transgenic mice with mutant protein expressed on the different *rds* background. When expressed on the wild type background, both cone and rod structure and function were significantly diminished with more severe cone defect. The phenotype seen in the transgenic retina on the wild type background is not an effect of peripherin/*rds* over-expression. This has been demonstrated in a study by Nour et al., (2004) in which over-expression of wild type peripherin/*rds* did not alter retinal function and structure. When the R172W transgene expressed on the *rds*$^{+/-}$ background, cone ERG responses were diminished to 41% of the wild type level while rod function and outer segment structure were improved. Expression of the R172W mutant in *rds*$^{-/-}$ mice rescued the rod function to 30% of the wild type level but no rescue of cone function was observed. The functional and structural characteristics of the transgenic retinas from the high expresser line on different *rds* genetic background are summarized in Table 21.2. Biochemical studies of the mutant protein isolated from transgenic mice on the *rds*$^{-/-}$ background showed no abnormalities in complex formation and association with Rom-1. However, the R172W protein was more sensitive to limited tryptic digestion, suggesting a change in the protein conformation that possibly contributes to the cone-specific phenotype (Ding et al., 2004). As the first animal model for peripherin/*rds*-associated cone-rod dystrophy, the R172W mice provides a valuable tool for studying the pathophysiology of human macular dystrophies and for development of the therapeutic interventions.

Table 21.2. Functional and structural characteristics of the R172W transgenic mice. Retinal characteristics of R172W mice were compared with the age-matched, non-transgenics on the same *rds* genetic background. WT, wild type; OS, outer segment; IS, inner segment. (Adapted from Ding et al., Human Molecular Genetics, 2004, V.13, Issue 18, 2075–2087, by permission of Oxford University Press).

Transgenic Mice	Rod ERG Response	Cone ERG Response	Light-microscopic Appearance	Electron-microscopic Appearance
R172W$^{+/+}$/*rds*$^{+/+}$	40% reduction of the WT	75% reduction of the WT	Disorganization and shortening in OS length	Disruption of OS structure
R172W$^{+/-}$/*rds*$^{+/-}$	81% increase of the *rds*$^{+-}$	60% reduction of the *rds*$^{+-}$	Improvement on OS structure	Improvement on OS structure
R172W$^{+/+}$/*rds*$^{-/-}$	30% rescue of the WT	No rescue	Partial rescue and OS formation	Restoration in OS and IS structure

4. TRANSGENIC ANIMAL MODELS USED TO STUDY THE STRUCTURE AND FUNCTION RELATIONSHIP OF PERIPHERIN/*RDS*

Transgenic animal models have been applied to the study of the structural and functional role of peripherin/*rds*. Using transgenic mouse line expressing the wild type peripherin/*rds*, Nour et al., (2004) has demonstrated *in vivo* the critical level of peripherin/*rds* needed to maintain photoreceptor structure and ERG function. Total peripherin/*rds* levels in the retina were modulated by crossing the wild type transgenic mice into different *rds* genetic backgrounds and the consequences of peripherin/*rds* over-expression in both rods and cones were assessed morphologically and functionally. A positive correlation was observed between peripherin/*rds* expression levels and the structural and functional integrity of photoreceptor outer segments. Over-expression of peripherin/*rds* caused no detectable adverse effects on the structure and function of rods or cones (Nour et al., 2004). In the study by Kedzierski et al., (1999), the functional role of the intradiscal D2 loop of peripherin/*rds* was examined in transgenic mice expressing a chimeric protein containing the D2 loop of peripherin/*rds* in the context of Rom-1. The chimeric protein was able to form covalent homodimers and to interact non-covalently with itself, wild type peripherin/*rds*, and Rom-1, and displayed a more stable interaction with peripherin/*rds* when compared to the authentic Rom-1. This study proposed that peripherin/*rds* is about 2.5-fold more abundant than Rom-1 and the complexes formed may extend the entire circumference of the disc (Kedzierski et al., 1999).

Through use of transgenic Xenopus expressing the GFP-C-terminus of peripherin/*rds*, the participation of the C-terminus in rod outer segment targeting and alignment of disk incisures was studied recently (Ritter et al., 2004; Tam et al., 2004). The C-terminus peripherin/*rds* fusion protein localized uniformly to disk membranes while the Rom-1 C-terminus did not promote rod outer segment localization. The GFP-fusion proteins did not immunoprecipitate with peripherin/*rds* or Rom-1, suggesting that this region does not form intermolecular interactions and is not involved in subunit assembly. Interestingly, presence of GFP-peripherin/*rds* fusions correlated with disrupted outer segments structure which may reflect competition of the fusion proteins for other proteins that interact with peripherin/*rds*.

5. ACKNOWLEDGEMENTS

This work was supported by the National Eye Institute EY-10609 (MIN), Fight For Sight (XQD), and the Knights Templar Eye Foundation, Inc. (XQD). We thank the National Eye Institute for the Travel Award provided to XQD to attend this meeting.

6. REFERENCES

Apfelstedt-Sylla, E., Theischen, M., Ruther, K., Wedemann, H., Gal, A., and Zrenner, E., 1995, Extensive intrafamilial and interfamilial phenotypic variation among patients with autosomal dominant retinal dystrophy and mutations in the human RDS/peripherin gene, *Br J Ophthalmol* **79**:28-34.

Ding, X.Q., Nour, M., Ritter, L.M., Goldberg, A.F., Fliesler, S.J., and Naash, M.I., 2004, The R172W mutation in peripherin/rds causes a cone-rod dystrophy in transgenic mice, *Hum Mol Genet* **13**:2075-2087.

Dryja, T.P., Hahn, L.B., Kajiwara, K., and Berson, E.L., 1997, Dominant and digenic mutations in the peripherin/RDS and ROM1 genes in retinitis pigmentosa, *Invest Ophthalmol Vis Sci* **38**:1972-1982.

Ekstrom, U., Andreasson, S., Ponjavic, V., Abrahamson, M., Sandgren, O., Nilsson-Ehle, P., and Ehinger, B., 1998, A Swedish family with a mutation in the peripherin/RDS gene (Arg-172-Trp) associated with a progressive retinal degeneration, *Ophthalmic Genet* **19**:149-156.

Goldberg, A.F., Moritz, O.L., and Molday, R.S., 1995, Heterologous expression of photoreceptor peripherin/rds and Rom-1 in COS-1 cells: assembly, interactions, and localization of multisubunit complexes, *Biochemistry* **34**:14213-14219.

Jacobson, S.G., Cideciyan, A.V., Maguire, A.M., Bennett, J., Sheffield, V.C., and Stone, E.M., 1996, Preferential rod and cone photoreceptor abnormalities in heterozygotes with point mutations in the RDS gene, *Exp Eye Res* **63**:603-608.

Kajiwara, K., Hahn, L.B., Mukai, S., Travis, G.H., Berson, E.L., and Dryja, T.P., 1991, Mutations in the human retinal degeneration slow gene in autosomal dominant retinitis pigmentosa, *Nature* **354**:480-483.

Kedzierski, W., Lloyd, M., Birch, D.G., Bok, D., and Travis, G.H., 1997, Generation and analysis of transgenic mice expressing P216L-substituted rds/peripherin in rod photoreceptors, *Invest Ophthalmol Vis Sci* **38**:498-509.

Kedzierski, W., Nusinowitz, S., Birch, D., Clarke, G., McInnes, R.R., Bok, D., and Travis, G.H., 2001, Deficiency of rds/peripherin causes photoreceptor death in mouse models of digenic and dominant retinitis pigmentosa, *Proc Natl Acad Sci U S A* **98**:7718-7723.

Kedzierski, W., Weng, J., and Travis, G.H., 1999, Analysis of the rds/peripherin.rom1 complex in transgenic photoreceptors that express a chimeric protein, *J Biol Chem* **274**:29181-29187.

Kohl, S., Giddings, I., Besch, D., Apfelstedt-Sylla, E., Zrenner, E., and Wissinger, B., 1998, The role of the peripherin/RDS gene in retinal dystrophies, *Acta Anat (Basel)* **162**:75-84.

Li, C., Peachey, N.S., and Naash, M.I., 1999, Animal models with mutations in the peripherin/rds gene, *Invest Ophthalmol Vis Sci* **40**:S202.

Loewen, C.J., Moritz, O.L., Tam, B.M., Papermaster, D.S., and Molday, R.S., 2003, The role of subunit assembly in peripherin-2 targeting to rod photoreceptor disk membranes and retinitis pigmentosa, *Mol Biol Cell* **14**:3400-3413.

McNally, N., Kenna, P.F., Rancourt, D., Ahmed, T., Stitt, A., Colledge, W.H., Lloyd, D.G., Palfi, A., O'Neill, B., Humphries, M.M., et al., 2002, Murine model of autosomal dominant retinitis pigmentosa generated by targeted deletion at codon 307 of the rds-peripherin gene, *Hum Mol Genet* **11**:1005-1016.

Milla, E., Heon, E., Piguet, B., Ducrey, N., Butler, N., Stone, E., Schorderet, D.F., and Munier, F., 1998, Mutational screening of peripherin/RDS genes, rhodopsin and ROM-1 in 69 index cases with retinitis pigmentosa and other retinal dystrophies, *Klin Monatsbl Augenheilkd* **212**:305-308.

Molday, R.S., 1998, Photoreceptor membrane proteins, phototransduction, and retinal degenerative diseases. The Friedenwald Lecture, *Invest Ophthalmol Vis Sci* **39**:2491-2513.

Nakazawa, M., Wada, Y., and Tamai, M., 1995, Macular dystrophy associated with monogenic Arg172Trp mutation of the peripherin/RDS gene in a Japanese family, *Retina* **15**:518-523.

Nour, M., Ding, X.-Q., Stricker, H., Fliesler, S.J., and Naash, M.I., 2004, Modulating expression of peripherin/rds in transgenic mice: Critical levels and the effect of over-expression, *Invest Ophthalmol Vis Sci* **45**:2514-2521.

Payne, A.M., Downes, S.M., Bessant, D.A., Bird, A.C., and Bhattacharya, S.S., 1998, Founder effect, seen in the British population, of the 172 peripherin/RDS mutation-and further refinement of genetic positioning of the peripherin/RDS gene, *Am J Hum Genet* **62**:192-195.

Ritter, L.M., Boesze-Battaglia, K., Tam, B.M., Moritz, O.L., Khattree, N., Chen, S.C., and Goldberg, A.F., 2004, Uncoupling of photoreceptor peripherin/rds fusogenic activity from biosynthesis, subunit assembly, and targeting: a potential mechanism for pathogenic effects, *J Biol Chem* **279**:39958-39967.

Saga, M., Mashima, Y., Akeo, K., Oguchi, Y., Kudoh, J., and Shimizu, N., 1993, A novel Cys-214-Ser mutation in the peripherin/RDS gene in a Japanese family with autosomal dominant retinitis pigmentosa, *Hum Genet* **92**:519-521.

Stricker, H.M., Ding, X.-Q., Quiambao, A.B., and Naash, M.I., 2003, The C214S mutation in peripherin/rds confers protein instability, *Protein Science* **12**(suppl.):111.

Tam, B.M., Moritz, O.L., and Papermaster, D.S., 2004, The C terminus of peripherin/rds participates in rod outer segment targeting and alignment of disk incisures, *Mol Biol Cell* **15**:2027-2037.

Wells, J., Wroblewski, J., Keen, J., Inglehearn, C., Jubb, C., Eckstein, A., Jay, M., Arden, G., Bhattacharya, S., Fitzke, F., et al., 1993, Mutations in the human retinal degeneration slow (RDS) gene can cause either retinitis pigmentosa or macular dystrophy, *Nat Genet* **3**:213-218.

Wroblewski, J.J., Wells, J.A., 3rd, Eckstein, A., Fitzke, F., Jubb, C., Keen, T.J., Inglehearn, C., Bhattacharya, S., Arden, G.B., Jay, M., et al., 1994, Macular dystrophy associated with mutations at codon 172 in the human retinal degeneration slow gene, *Ophthalmology* **101**:12-22.

CHAPTER 22

TRANSGENIC EXPRESSION OF LEUKEMIA INHIBITORY FACTOR INHIBITS BOTH ROD AND CONE GENE EXPRESSION

Gp130 regulates cone gene expression

John D. Ash[1] and Dianca R. Graham[1]

1. INTRODUCTION

Leukemia inhibitory factor (LIF) is a member of the interleukin 6 (IL-6) family of cytokines, which also includes oncostatin-M, ciliary neurotrophic factor (CNTF), inter-leukin-11, and cardiotrophin-1. Members of this family are grouped together based on acti-vation of a common tyrosine kinase receptor, gp130. (Ip et al, 1992) The expression of activating ligands of gp130, including LIF and CNTF, have been localized to Muller glial cells and microglial cells in the retina. (Harada et al, 2002; Kirsch et al, 1997; Neophytou et al, 1997; Walsh et al, 2001) While little is known about the regulated expression of LIF in the eye, CNTF has been shown to be up-regulated in the retina following injury or stress. (Cao et al, 1997; Wen et al, 1995; Wen et al, 1998) This up regulation has led to the hypo-thesis that activation of gp130 is an endogenous mechanism for neuroprotection in the retina. In support of this hypothesis, it has been shown that activating gp130 either by direct injections of LIF or CNTF into the eye or by their expression from transduced cells is effec-tive at protecting retinal neurons from cell death. This includes cell death caused by pro-longed exposure to constant light and inherited retinal degenerative mutations (Bok et al, 2002; Cayouette and Gravel, 1997; LaVail et al, 1998; LaVail et al, 1992). CNTF and LIF are both currently in clinical trials for the treatment of neurological degenerative disorders including amyotrophic lateral sclerosis (Festoff, 1996) and retinitis pigmentosa (www.clin-icaltrials.gov). While CNTF and LIF are effective at preventing cell death, the mechanism has not yet been identified.

[1] Departments of Ophthalmology and Cell Biology, University of Oklahoma Health Science Center, 908 Stanton L. Young Blvd, Oklahoma City, OK 73104. Corresponding author John D. Ash, Tel: (405) 271-3642; Fax: (405) 271-3721; E-mail: john-ash@ouhsa.edu.

In order to study the effects of gp130 activation in the retina we have generated transgenic mice that express human LIF in the eye. We have utilized the αA-crystallin promoter to drive LIF expression specifically in the lens. We have named these mice αA-LIF. In these mice LIF is expressed specifically in the lens fiber cells beginning around embryonic day 11. The lenses in these mice continue to express LIF throughout the life of the mouse. In a previous study we have used these mice to show that activation of gp130 in the retina inhibits photoreceptor differentiation. (Ash, 2001) In search of the mechanism, we have previously reported that LIF expression in development does not prevent the differentiation of CRX expressing photoreceptor cells. (Ash, 2001) However, rod photoreceptors do not expression of NRL. Without this essential transcription factor rod photoreceptors cannot induce expression of rod photoreceptor transduction genes including opsin. In the current study we analyzed the effects of gp130 activation on cone differentiation.

2. MATERIALS AND METHODS

2.1. Transgenic Mice

All procedures were in accordance with the ARVO Statement for the Use of Animals in Ophthalmic and Vision Research. The generation of the αA-LIF transgenic mice were described previously. (Ash, 2001) Founder mice were mated to FVB/N mice in order to establish transgenic lines. The FVB mouse strain is homozygous for the retinal degeneration mutation (pdeb^{rd1}, formally known as rd1), which is caused by a mutation in the gene encoding rod-specific cGMP phosphodiesterase. (Bowes et al, 1990) This autosomal recessive defect results in rapid rod cell death followed by a slow loss of cones. To analyze the effects of LIF on normal retinas, LIF transgenic mice were mated to C57BL/6$^{pdeb+/+}$ mice. The F1 progeny of this mating are heterozygous for pdeb^{rd1} (*rd1$^{+/-}$*) and do not have photoreceptor degeneration. Therefore, the changes in retinal phenotype and function described for mice in the C57BL/6 background are caused by the expression of LIF in the lens and not a complication of the pdeb^{rd1} mutation.

2.2. Immunofluorescence

Eyes were enucleated, fixed in 10% neutral buffered formalin for 24 hours, dehydrated, and embedded in paraffin. To detect M- and S-opsin expression, paraffin embedded tissue sections on glass slides were rehydrated in an ethanol series then blocked in 10% horse serum in PBS for 30 minutes at room temperature prior to incubation with primary antibody. The rabbit anti-M-opsin and anti-S-opsin antibodies (Chemicon, Temecula, CA) were diluted 1:200 in 10% horse serum/PBS. Sections were incubated overnight at 4°C, washed three times in PBS, then incubated for one hour with Alexa-595, anti-rabbit secondary antibody (Molecular probes) diluted 1:500 in 10% horse serum/PBS. Slides were coverslipped with fluorescent mounting media containing DAPI (Vector labs). Fluorescent complexes were detected using a Nikon Eclipse fluorescent microscope.

2.3. Semi-Quantitative RT-PCR Analysis

In these experiments 1 μg of RNA was reverse transcribed and diluted to 200 ml. 5 μl and 10 ul of cDNA was used for each TrBβ2 and Nr2E3 PCR reaction, respectively. PCR amplification was performed according to the temperature profile: 95°C for 6 minutes, 24 or 35 cycles of 95°C for 30 seconds, 60°C for 30 seconds followed by extension at 72°C for 1 minute. Primers used were: Trβ2 primers (5′-GCTAGCCAAGCGGAAGCTT-ATAGA-3′) and (5′-TGGGCGATCTGAAGACATTAGCAG-3′), Nr2e3 primers (5′-CAGTGGCTTCTTCAAGAGAGTGT-3′) and (5′-CCACTGTATGGCTCCAAGAAG-GAA-3′), and beta-actin primers (5′-TCTACGAGGGCTATGCTCTCC-3′) and (5′-TCTTTGATGTCACGCACGATTTC-3′). Following PCR reactions, samples were then loaded into a 1% agarose gel.

3. RESULTS

3.1. Activation of gp130 Blocks Cone Opsin Expression

To determine whether or not cones were blocked in differentiation we analyzed the expression of cone opsins by immunofluorescence. In P14 non-transgenic mice we were able to detect both S- and M-opsin in our retinal sections (fig. 22.1A, and B). However, we

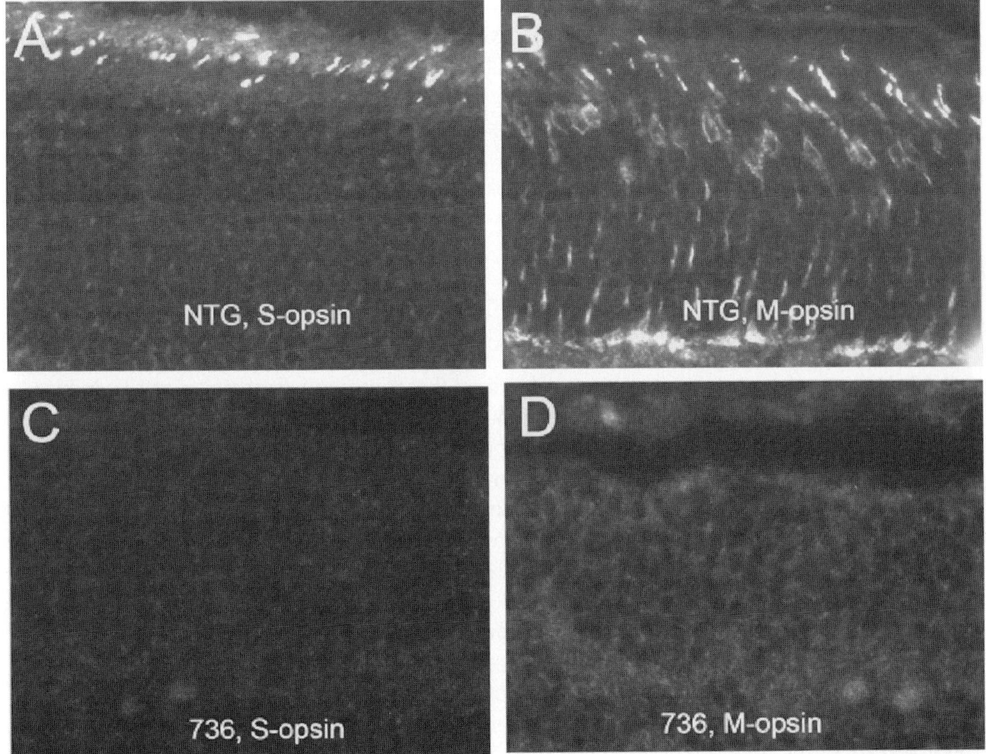

Figure 22.1. Immunofluorescent images of retinas stained with antibodies to S-opsin (A and C), or M-opsin (B and D), from non-transgenic (A and B), and LIF expressing transgenic mice (C and D). Photoreceptor outer segments are oriented to the top of each panel.

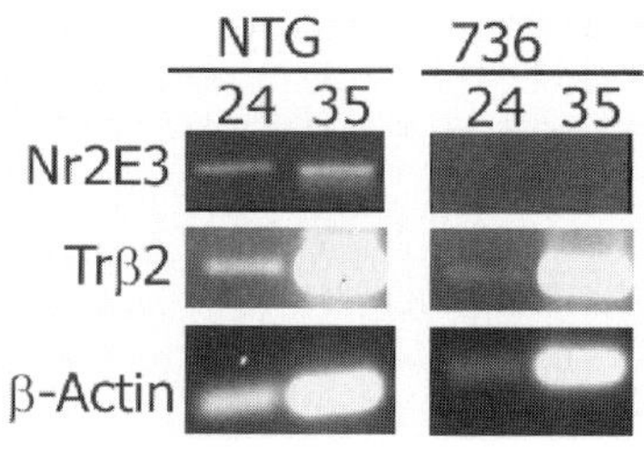

Figure 22.2. Semi-quantitative RT-PCR analysis of Nr2E3, and Trβ2 expression from non-transgenic mice (NTG) and LIF expressing transgenic mice (736). Samples were run for either 24 or 35 cycles.

were unable to detect either cone opsin in retinas from αA-LIF transgenic mice (fig. 22.1C and D). We have confirmed that cone opsin expression was undetectable in the 736 αA-IF transgenic mice by semi-quantitative RT-PCR and Western blots (not shown).

3.2. Activation of gp130 Does Not Inhibit Trβ2 Expression.

Cone gene expression has been shown to require the transcription factor CRX. (Furukawa et al, 1999) We have previously shown that photoreceptors in the αA-LIF transgenic mice have detectable levels of CRX. (Ash, 2001) Therefore the absence of cone opsins cannot be explained by the absence of CRX. M-opsin has been shown to require the expression of the thyroid hormone receptor beta 2 (TRβ2). (Ng et al, 2001) To determine if Trβ2 expression was inhibited by gp130 activation, we used semi-quantitative rt-PCR to measure its expression. As evident in Figure 22.2, the expression of the rod nuclear receptor Nr2E3 is inhibited by gp130 activation. Since Nr2E3 expression is dependent on NRL, the absence of Nr2E3 is the expected result from the lack of NRL expression. (Mears et al, 2001) In contrast to Nr2E3, Trβ2 was expressed in αA-LIF transgenic retinas (fig. 22.2).

4. DISCUSSION

Relevance to retinal degenerations. We found that transgenic expression of LIF induces photoreceptor survival in *pdeb^rd1* homozygous mice (Ash, 2001), but at the same time reduces the expression of phototransduction genes and reduces light-mediated responses in both rods and cones. Recent studies have shown that viral delivery of CNTF in vivo is also effective at preventing photoreceptor cell death from the P216L and Rd2 mutations in rds/peripherin or the P23H mutation in rhodopsin. (Bok et al, 2002; Liang et al, 2001) These studies also have shown that while CNTF prevents cells from dying, it reduces their ability to respond to light. Collectively, the data show that activation of gp130 either by LIF or CNTF can protect photoreceptors from cell death but at the cost of reduced photoreceptor function. The observed changes in gene expression and photoreceptor function suggests the possibility that neuroprotection is the result of reduced phototransduction gene expression in both rods and cones. For example, we have shown that LIF transgenic mice do not express phototransduction genes including pdeb (data not shown). In the absence of protein expression the pdeb^rd1 mutation is not able to cause photoreceptor cell death. For the same reason, mutations in rhodopsin or peripherin that normally cause retinal degeneration can not cause photoreceptor cell death when the expression of mutant proteins are dramatically reduced. Alternatively, neuroprotection and inhibition of phototransduction gene expression may be the result of gp130 signaling through two independent pathways. Activated gp130 signals through the JAK/STAT (Heinrich et al, 2003; Stahl et al, 1994), mitogen activated protein kinase pathway (Erk1/2), and the phosphatidylinositol 3 kinase (PI3K) pathway. (Boulton et al, 1994; Oh et al, 1998) Multiple pathways have been shown to mediate cell survival, and more recently the JAK/STAT pathway has been shown to be important for inhibiting phototransduction gene expression. (Alonzi et al, 2001; Ozawa et al, 2004; Zhang et al, 2004) More work needs to be done to determine the mechanism by which each pathway regulates neuroprotection and regulates phototransduction gene expression. In pursuit of the mechanisms by which gp130 regulates phototransduction gene expression we have been analyzing the expression of transcription factors that are necessary for expression of phototransduction genes.

Relevance to transcriptional regulation in cone photoreceptors. As retinal progenitors differentiate into rods one of the earliest genes expressed is the cone-rod homeobox transcription factor CRX. (Furukawa et al, 1997) In the αA-LIF mice, CRX expression is not inhibited by gp130 activation, demonstrating that early photoreceptor fate decisions are not inhibited. (Ash, 2001) As rods continue to differentiate they begin to express NRL and then Nr2e3. (Mears et al, 2001) In combination with CRX, NRL and Nr2e3 are required for high-level opsin expression (Cheng et al, 2004; Mitton et al, 2000). In rods we have shown that gp130 activation does not inhibit CRX expression but does inhibit NRL and Nr2e3. Mice lacking NRL or Nr2e3 have a condition known as enhanced S-cone syndrome. (Mears et al, 2001) Without NRL or Nr2e3, photoreceptors differentiate into S-cones rather than into rods. Photoreceptors from the αA-LIF mice did not express S-opsin, despite the expression of CRX. This demonstrates that while S-cone gene expression may be the default pathway, the differentiation of S-cones requires more than CRX expression. M-opsin expression is known to require the expression of CRX and Trβ2. (Ng et al, 2001) Our data show that while CRX and Trβ2 are both expressed following gp130 activation, M-opsin is not. Our results demonstrate that cone differentiation is blocked by gp130 activation, and that CRX and Trβ2 are not sufficient to drive their differentiation.

5. REFERENCES

Alonzi, T., Middleton, G., Wyatt, S., Buchman, V., Betz, U. A., Muller, W., Musiani, P., Poli, V., and Davies, A. M., 2001, Role of STAT3 and PI 3-kinase/Akt in mediating the survival actions of cytokines on sensory neurons, *Mol.Cell Neurosci.*, **18**:270-282

Ash, J. D., 2001, Leukemia inhibitory factor prevents photoreceptor cell death in rd-/- mice by blocking functional differentiation., in:*New insights into retinal degenerative diseases and experimental therapy*Anderson, R. E., LaVail, M. M., and Hollfield, J. G. ed.Kluwer Academic / Plenum Publishers, New York, pp. 135-144

Bok, D., Yasumura, D., Matthes, M. T., Ruiz, A., Duncan, J. L., Chappelow, A. V., Zolutukhin, S., Hauswirth, W., and LaVail, M. M., 2002, Effects of adeno-associated virus-vectored ciliary neurotrophic factor on retinal structure and function in mice with a P216L rds/peripherin mutation, *Exp.Eye Res.*, **74**:719-735

Boulton, T. G., Stahl, N., and Yancopoulos, G. D., 4-15-1994, Ciliary neurotrophic factor/leukemia inhibitory factor/interleukin 6/oncostatin M family of cytokines induces tyrosine phosphorylation of a common set of proteins overlapping those induced by other cytokines and growth factors, *J.Biol.Chem.*, **269**:11648-11655

Bowes, C., Li, T., Danciger, M., Baxter, L. C., Applebury, M. L., and Farber, D. B., 10-18-1990, Retinal degeneration in the rd mouse is caused by a defect in the beta subunit of rod cGMP-phosphodiesterase, *Nature*, **347**:677-680

Cao, W., Wen, R., Li, F., LaVail, M. M., and Steinberg, R. H., 1997, Mechanical injury increases bFGF and CNTF mRNA expression in the mouse retina, *Exp.Eye Res.*, **65**:241-248

Cayouette, M. and Gravel, C., 3-1-1997, Adenovirus-mediated gene transfer of ciliary neurotrophic factor can prevent photoreceptor degeneration in the retinal degeneration (rd) mouse, *Hum.Gene Ther.*, **8**:423-430

Cheng, H., Khanna, H., Oh, E. C., Hicks, D., Mitton, K. P., and Swaroop, A., 8-1-2004, Photoreceptor-specific nuclear receptor NR2E3 functions as a transcriptional activator in rod photoreceptors, *Hum.Mol.Genet.*, **13**:1563-1575

Festoff, B. W., 1996, Amyotrophic lateral sclerosis: current and future treatment strategies, *Drugs*, **51**:28-44

Furukawa, T., Morrow, E. M., and Cepko, C. L., 11-14-1997, Crx, a novel otx-like homeobox gene, shows photoreceptor-specific expression and regulates photoreceptor differentiation, *Cell*, **91**:531-541

Furukawa, T., Morrow, E. M., Li, T., Davis, F. C., and Cepko, C. L., 1999, Retinopathy and attenuated circadian entrainment in Crx-deficient mice, *Nat.Genet.*, **23**:466-470

Harada, T., Harada, C., Kohsaka, S., Wada, E., Yoshida, K., Ohno, S., Mamada, H., Tanaka, K., Parada, L. F., and Wada, K., 11-1-2002, Microglia-Muller glia cell interactions control neurotrophic factor production during light-induced retinal degeneration, *J.Neurosci.*, **22**:9228-9236

Heinrich, P. C., Behrmann, I., Haan, S., Hermanns, H. M., Muller-Newen, G., and Schaper, F., 5-29-2003, Principles of IL-6-type cytokine signalling and its regulation, *Biochem.J.*, **374**(Pt1):1-20

Ip, N. Y., Nye, S. H., Boulton, T. G., Davis, S., Taga, T., Li, Y., Birren, S. J., Yasukawa, K., Kishimoto, T., Anderson, D. J., et al., 6-26-1992, CNTF and LIF act on neuronal cells via shared signaling pathways that involve the IL-6 signal transducing receptor component gp130, *Cell*, **69**:1121-1132

Kirsch, M., Lee, M. Y., Meyer, V., Wiese, A., and Hofmann, H. D., 1997, Evidence for multiple, local functions of ciliary neurotrophic factor (CNTF) in retinal development: expression of CNTF and its receptors and in vitro effects on target cells, *J.Neurochem.*, **68**:979-990

LaVail, M. M., Unoki, K., Yasumura, D., Matthes, M. T., Yancopoulos, G. D., and Steinberg, R. H., 12-1-1992, Multiple growth factors, cytokines, and neurotrophins rescue photoreceptors from the damaging effects of constant light, *Proc.Natl.Acad.Sci.U.S.A*, **89**:11249-11253

LaVail, M. M., Yasumura, D., Matthes, M. T., Lau-Villacorta, C., Unoki, K., Sung, C. H., and Steinberg, R. H., 1998, Protection of mouse photoreceptors by survival factors in retinal degenerations, *Invest Ophthalmol.Vis.Sci.*, **39**:592-602

Liang, F. Q., Aleman, T. S., Dejneka, N. S., Dudus, L., Fisher, K. J., Maguire, A. M., Jacobson, S. G., and Bennett, J., 2001, Long-term protection of retinal structure but not function using RAAV.CNTF in animal models of retinitis pigmentosa, *Mol.Ther.*, **4**:461-472

Mears, A. J., Kondo, M., Swain, P. K., Takada, Y., Bush, R. A., Saunders, T. L., Sieving, P. A., and Swaroop, A., 2001, Nrl is required for rod photoreceptor development, *Nat.Genet.*, **29**:447-452

Mitton, K. P., Swain, P. K., Chen, S., Xu, S., Zack, D. J., and Swaroop, A., 9-22-2000, The leucine zipper of NRL interacts with the CRX homeodomain. A possible mechanism of transcriptional synergy in rhodopsin regulation, *J.Biol.Chem.*, **275**:29794-29799

Neophytou, C., Vernallis, A. B., Smith, A., and Raff, M. C., 1997, Muller-cell-derived leukaemia inhibitory factor arrests rod photoreceptor differentiation at a postmitotic pre-rod stage of development, *Development*, **124**:2345-2354

Ng, L., Hurley, J. B., Dierks, B., Srinivas, M., Salto, C., Vennstrom, B., Reh, T. A., and Forrest, D., 2001, A thyroid hormone receptor that is required for the development of green cone photoreceptors, *Nat.Genet.*, **27**:94-98

Oh, H., Fujio, Y., Kunisada, K., Hirota, H., Matsui, H., Kishimoto, T., and Yamauchi-Takihara, K., 4-17-1998, Activation of phosphatidylinositol 3-kinase through glycoprotein 130 induces protein kinase B and p70 S6 kinase phosphorylation in cardiac myocytes, *J.Biol.Chem.*, **273**:9703-9710

Ozawa, Y., Nakao, K., Shimazaki, T., Takeda, J., Akira, S., Ishihara, K., Hirano, T., Oguchi, Y., and Okano, H., 2004, Downregulation of STAT3 activation is required for presumptive rod photoreceptor cells to differentiate in the postnatal retina, *Mol.Cell Neurosci.*, **26**:258-270

Stahl, N., Boulton, T. G., Farruggella, T., Ip, N. Y., Davis, S., Witthuhn, B. A., Quelle, F. W., Silvennoinen, O., Barbieri, G., Pellegrini, S., et al., 1-7-1994, Association and activation of Jak-Tyk kinases by CNTF-LIF-OSM-IL-6 beta receptor components, *Science*, **263**:92-95

Walsh, N., Valter, K., and Stone, J., 2001, Cellular and subcellular patterns of expression of bFGF and CNTF in the normal and light stressed adult rat retina, *Exp.Eye Res.*, **72**:495-501

Wen, R., Cheng, T., Song, Y., Matthes, M. T., Yasumura, D., LaVail, M. M., and Steinberg, R. H., 1998, Continuous exposure to bright light upregulates bFGF and CNTF expression in the rat retina, *Curr.Eye Res.*, **17**:494-500

Wen, R., Song, Y., Cheng, T., Matthes, M. T., Yasumura, D., LaVail, M. M., and Steinberg, R. H., 1995, Injury-induced upregulation of bFGF and CNTF mRNAS in the rat retina, *J.Neurosci.*, **15**:7377-7385

Zhang, S. S., Wei, J., Qin, H., Zhang, L., Xie, B., Hui, P., Deisseroth, A., Barnstable, C. J., and Fu, X. Y., 2004, STAT3-mediated signaling in the determination of rod photoreceptor cell fate in mouse retina, *Invest Ophthalmol.Vis.Sci.*, **45**:2407-2412

A ROLE FOR BHLH TRANSCRIPTION FACTORS IN RETINAL DEGENERATION AND DYSFUNCTION

Mark E. Pennesi, Debra E. Bramblett, Jang-Hyeon Cho, Ming-Jer Tsai, and Samuel M. Wu*

1. INTRODUCTION

The basic helix loop helix (bHLH) transcription factors collectively mediate cellular differentiation in almost every type of tissue including the retina (Murre *et al.* 1989; Jan and Jan 1993; Cepko 1999). Class A factors are ubiquitously expressed throughout mammalian tissue, while the expression of class B factors are cell type specific. These factors have both a DNA binding domain and helix loop helix domain (HLH) protein dimerization domain. Class B factors usually heterodimerize with the ubiquitously expressed, bHLH factors, such as E12/E47. Because of their importance during photoreceptor development, bHLH factors are candidate genes for photoreceptor degeneration. We have examined the roles of two bHLH factors, both which are expressed during retinal development, but also share the property of continued expression in the adult retina.

Beta2, a bHLH transcription factor, was cloned as a regulator for insulin gene expression (Naya *et al.* 1995). It was also isolated from embryos and referred to as *NeuroD* because it could convert epidermal cell fate into neuronal (Lee *et al.* 1995). *Beta2/NeuroD* is widely expressed throughout the nervous system and thought to act as a neuronal differentiator (Lee 1997; Cho and Tsai 2004). Mice homozygous null for this gene have decreased insulin production and defects in the limbic, vestibular, and auditory systems (Naya *et al.* 1997; Liu *et al.* 2000a; Liu *et al.* 2000b; Kim *et al.* 2001). Studies in rodent retinal explants demonstrated multiple roles for this gene in retinal development and predicted its importance in photoreceptor survival (Morrow *et al.* 1999). Expression in the mouse retina begins at E10.5 in the outer neuroblastic layer and encompasses all three retinal layers by E18.5 (Brown *et al.* 1998; Morrow *et al.* 1999). Additionally, robust expression of *Beta2/NeuroD* in the

---

*Mark E. Pennesi, Samuel M. Wu, Department of Ophthalmology, Debra E. Bramblett. Jang Cho, Ming-Jer Tsai, Department of Molecular and Cellular Biology, Baylor College of Medicine, Houston, Texas, 77030. Current Address for Debra Bramblett: Department of Biology, University of St. Thomas, Houston TX 77006.

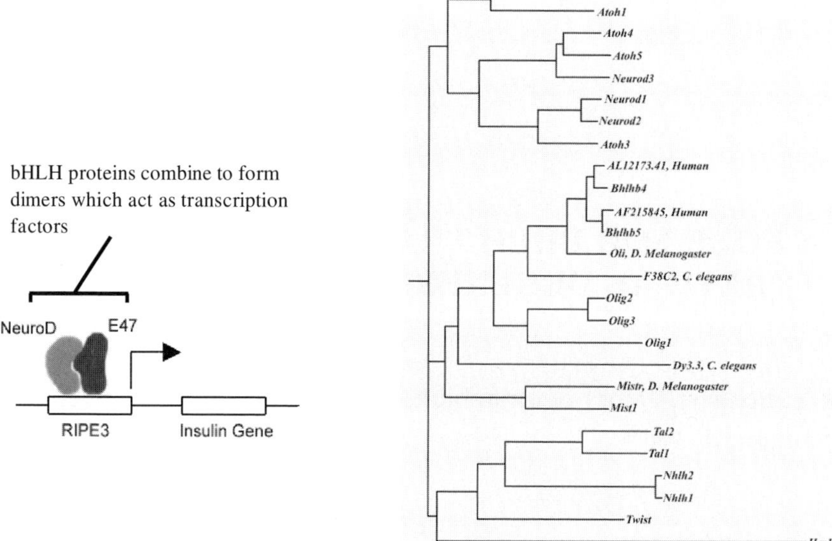

Figure 23.1. Schematic showing how bHLH proteins combine to form transcription factors. Family tree showing the relationship of *Beta2/NeuroD* and *Bhlhb4* (Bramblett *et al.* 2002).

outer nuclear layer continues in the adult mouse, suggesting a possible role for this gene in not only the development of these cells, but also in their survival (Pennesi *et al.* 2003).

Bhlhb4 is a previously uncharacterized bHLH transcription factor related to *Beta2/NeuroD* which is also highly expressed in the retina (Bramblett *et al.* 2002). Figure 23.1 shows the relationship of *Beta2/NeuroD* and *Bhlhb4* within the family of class B of bHLH genes. Both the temporal and spatial expression patterns of *Bhlhb4* differ from that of *Beta2/NeuroD*. Its expression was first detected at P5 in a restricted population of cells in the INL, which morphologically and immunohistochemically resemble rod bipolar cells. Expression of this gene transiently drops between P9 and P12, but returns at P14 and is maintained in the adult retina. Thus, it appeared that *Bhlhb4* may have both a developmental role and as well as a functional role in the adult retina.

To explore the roles of *Beta2/NeuroD* and *Bhlhb4* in retinal development and to help elucidate their role in the adult retina, we developed mice lacking these genes and studied their retinal histology and function.

2. METHODS

Detailed methods are described elsewhere (Pennesi *et al.* 2003; Bramblett *et al.* 2004). *Beta2/NeuroD* null mice were generated on the 129/SvJ background as previously described (Liu *et al.* 2000b). *Bhlhb4* null mice were generated on the 129/SvEv background and backcrossed to C57Bl6 mice (Bramblett *et al.* 2004). For histology, eyes were placed in 4%

paraformaldehyde containing PBS. Eyecups were dehydrated, orientated, and embedded in JB-4 for cutting. Sections were stained with hematoxylin and eosin. For ERGs, mice were anesthetized under dim red light with a solution of ketamine and xylazine. Methylcellulose gel and a platinum electrode were applied to the cornea. Flashes for scotopic and a-wave measurements were generated by a Grass PS-33+ photostimulator and a 1500-watt Novatron xenon flash, respectively.

3. RETINAL HISTOLOGY FROM KNOCKOUT MICE

Figure 23.2 shows light micrographs of retinas taken from adult *BETA2/NeuroD* and *Bhlhb4* mice. In *BETA2/NeuroD* null mice the cell loss was most prominent in the outer nuclear layer (ONL). The ONL, which contains the photoreceptors, normally measures 10-12 cells thick (Carter-Dawson and LaVail 1979). In *BETA2/NeuroD* null mice at 2 months of age, the thickness of the ONL was reduced to 5-6 cells. In addition, there also appeared to be a thinning of the outer plexiform layer (OPL) and a shortening of the outer and inner segments when analyzed by electron microscopy (Pennesi *et al.* 2003). There was no change in the thickness or appearance of the INL, inner plexiform layer (IPL), or ganglion cell layer (GCL) in null mice. Light microscopy from 18-month-old –/– mice revealed that the ONL was completely devoid of photoreceptors.

Bhlhb4 null mice displayed a markedly different phenotype. Unlike *Beta2/NeuroD* null retinas, which exhibit a degeneration of photoreceptors, the ONL of *Bhlhb4* null retinas were normal in thickness and cell count. However, there was a notable difference in thickness of the INL. Detailed cell counts revealed that the average ratio of the number of INL cells to the number of ONL cells in the *Bhlhb4* per section was, 21 percent less than wild type. Immunohistochemical studies showed that the identity of the missing cells was the rod bipolar cell, and that the death of these cells occurred during the postnatal development of the retina (Bramblett *et al.* 2004).

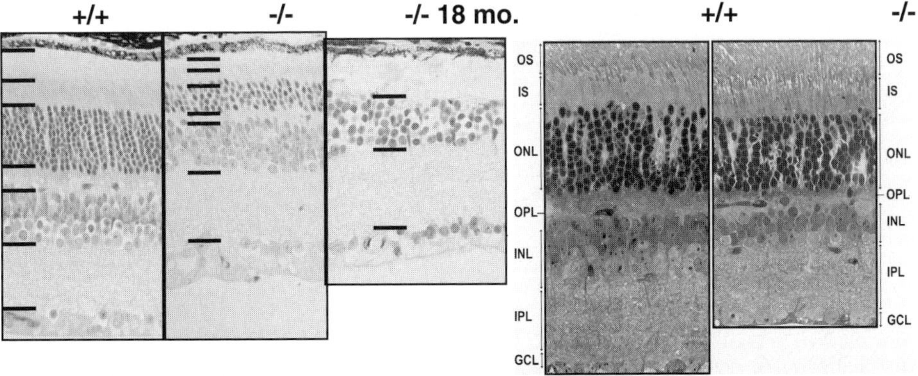

Figure 23.2. On the left are retinal slices from two month old *Beta2/NeuroD* +/+ mice, two month old –/– mice, and, 18 month old –/– mice. On the right are retinal slices from two month old *Bhlhb4* +/+ and –/– mice.

4. ELECTRORETINOGRAMS FROM KNOCKOUT MICE

Under scotopic conditions, the electroretinogram (ERG) detects responses from rod-driven circuitry, while under photopic conditions it detects responses from cone-driven circuitry. Figure 23.3 shows scotopic ERG recordings from the *BETA2/NeuroD* and *Bhlhb4* lines of mice. In *BETA2/NeuroD* wild-type mice, the maximum scotopic b-wave (b_{max}) measured 650 μV. Heterozygous mice were indistinguishable from control mice for this and other tests of visual function. Homozygous null mice had a severe decrease in the scotopic b-wave with b_{max} measuring 300 μV. To establish if the decrease in the ERG worsened with age, we tested several mice older than 9 months. Neither rod-driven nor cone-driven responses were detectable from null mice at these ages. ERGs from corresponding control mice were only slightly decreased, consistent with aging (data not shown). In *Bhlhb4* wild-type mice, b_{max} measured 640 μV. Heterozygous mice demonstrated a small, but significant decrease in the scotopic b-waves with b_{max} averaging 485 μV. Scotopic b-wave recordings

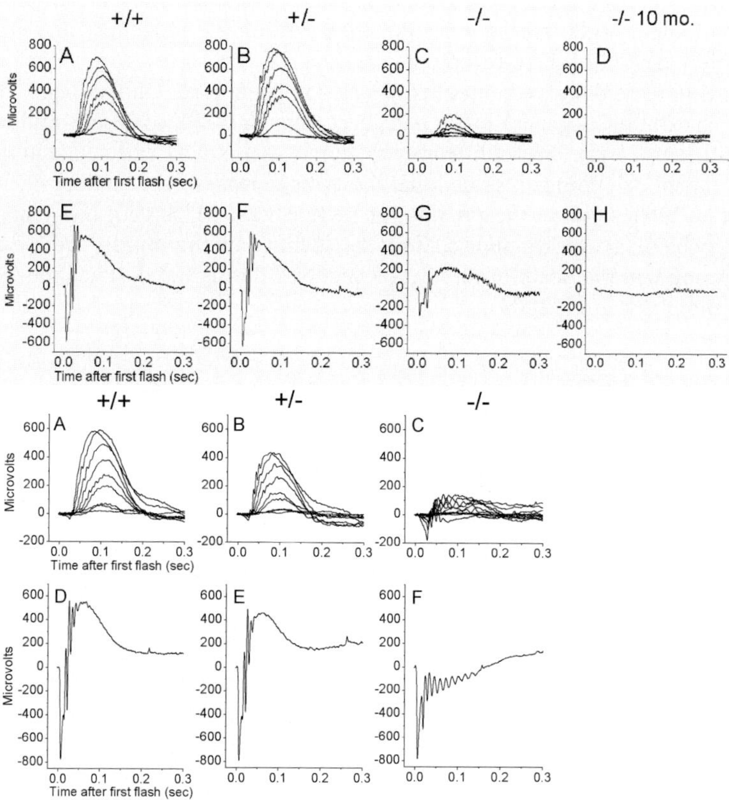

Figure 23.3. The top row shows scotopic ERGs recorded from two month old *Beta2/NeuroD* mice, while the second row shows the response to a saturating flash. The bottom two rows show the responses to similar stimuli in *Bhlhb4* mice.

from *Bhlhb4* null mice were profoundly decreased with b_{max} measuring only 165 µV. To directly characterize rod photoreceptor function, we used intense flashes to measure the scotopic a-wave. In *Beta2/NeuroD* wild-type mice this stimulus elicited a saturated a-wave, or a_{max}, measuring 600 µV. In null mice, the saturated a-wave was reduced by almost half to 300 µV. In *Bhlhb4* wild-type mice, a_{max} measured about 825 µV. While, in heterozygous and null mice a_{max}, measured 700 µV and 825 µV, respectively. In contrast to the *Beta2/NeuroD* mice, there was no significant difference between wild-type, heterozygous, or *Bhlhb4* null mice.

5. DISCUSSION

While both *Beta2/NeuroD* and *Bhlhb4* null mice showed decreased scotopic b-waves, the origin of the deficit in each was different. The scotopic b-wave is the extracellular field potential that primarily arises from rod bipolar cells in response to dim flashes of light (Pugh *et al.* 1998). The maximum amplitude of the scotopic b-wave is dependent on the number and activity of bipolar cells, the integrity of the photoreceptor bipolar synapse, and the number and activity of rod photoreceptor cells. In *Beta2/NeuroD* null mice, there was a 50% reduction in both the scotopic b-wave and rod a-wave. There was no change in the number of rod bipolar cells and although we cannot rule out synaptic defects, the most likely cause of the diminished b-wave is decreased input to the bipolar cell due to the loss of photoreceptors and functional impairment of remaining photoreceptors due to shortened outer segments. In *Bhlhb4* null mice, ERG recordings displayed a dramatic reduction in the scotopic b-wave with a preserved a-wave. With normal rod morphology, the disruption of b-wave in these mice is certainly due to the death of the rod bipolar cells. The residual rod b-wave at observed higher intensities likely represents contribution from cone bipolar cells, since these stimuli are above the cone threshold.

The cause of death of photoreceptors in *Beta2/NeuroD* null mice and bipolar cells in *Bhlhb4* null mice is not clear. The loss of *Beta2/NeuroD* has been linked to the down regulation of developmental markers and defects in differentiation in the pancreas, dentate gyrus, and auditory system (Naya *et al.* 1997; Mutoh *et al.* 1997; Liu *et al.* 2000b; Kim *et al.* 2001). For example, *Beta2/NeuroD* is necessary for the expression of insulin in β-cells of the pancreas and proper formation of the islets. *Beta2/NeuroD* may play similar role in the retina by inducing and maintaining the expression of photoreceptor specific genes. *Bhlhb4* likely plays a similar role in the terminal differentiation and survival of rod bipolar cells, although no specific gene targets have been elucidated.

While the importance of bHLH genes in retinal development has been known for some time, the idea that continued expression of these genes may be necessary for survival of retinal cells is new and implies that mutations in these genes may result in degenerative diseases of the retina. No visual disease in humans has been linked to either the *Beta2/NeuroD* or *Bhlhb4* loci. Heterozygous mutations in *BETA2/NeuroD* are associated with the development of both type 1 and type 2 diabetes mellitus in humans, but have not been implicated in retinal degeneration (Malecki *et al.* 1999; Iwata *et al.* 1999). The loss of *Bhlhb4* leads to a ERG phenotype that is commonly referred to as "negative ERG". In humans, "negative ERG" is often associated with congenital stationary night blindness (CSNB) (Dryja 2000). *Bhlhb4* is excluded as the determinate of X-linked CSNB because this has been mapped to the distal end of human chromosome 20 (Bramblett *et al.* 2002). However, variations of

CSNB exist, including autosomal dominant and recessive forms (Dryja 2000; Fitzgerald *et al.* 2001) and perhaps *Bhlhb4* plays a role in these disorders. Our results indicate that the loss of bHLH factors could play a role in retinal disease and should be screened in the future for mutations.

6. REFERENCES

Bramblett, D. E., Copeland, N. G., Jenkins, N. A., and Tsai, M. J. (2002) BHLHB4 is a bHLH transcriptional regulator in pancreas and brain that marks the dimesencephalic boundary. *Genomics* **79:**402-412.

Bramblett, D. E., Pennesi, M. E., Wu, S. M., and Tsai, M. J. (2004) The transcription factor Bhlhb4 is required for rod bipolar cell maturation. *Neuron* **43:**779-793.

Brown, N. L., Kanekar, S., Vetter, M. L., Tucker, P. K., Gemza, D. L., and Glaser, T. (1998) Math5 encodes a murine basic helix-loop-helix transcription factor expressed during early stages of retinal neurogenesis. *Development* **125:**4821-4833.

Carter-Dawson, L. D. and LaVail, M. M. (1979) Rods and cones in the mouse retina. I. Structural analysis using light and electron microscopy. *J Comp Neurol* **188:**245-262.

Cepko, C. L. (1999) The roles of intrinsic and extrinsic cues and bHLH genes in the determination of retinal cell fates. *Curr Opin Neurobiol* **9:**37-46.

Cho, J. H. and Tsai, M. J. (2004) The role of BETA2/NeuroD1 in the development of the nervous system. *Mol Neurobiol* **30:**35-47.

Dryja, T. P. (2000) Molecular genetics of Oguchi disease, fundus albipunctatus, and other forms of stationary night blindness: LVII Edward Jackson Memorial Lecture. *Am J Ophthalmol* **130:**547-563.

Fitzgerald, K. M., Hashimoto, T., Hug, T. E., Cibis, G. W., and Harris, D. J. (2001) Autosomal dominant inheritance of a negative electroretinogram phenotype in three generations. *Am J Ophthalmol* **131:**495-502.

Iwata, I., Nagafuchi, S., Nakashima, H., Kondo, S., Koga, T., Yokogawa, Y., Akashi, T., Shibuya, T., Umeno, Y., Okeda, T., Shibata, S., Kono, S., Yasunami, M., Ohkubo, H., and Niho, Y. (1999) Association of polymorphism in the NeuroD/BETA2 gene with type 1 diabetes in the Japanese. *Diabetes* **48:**416-419.

Jan, Y. N. and Jan, L. Y. (1993) HLH proteins, fly neurogenesis, and vertebrate myogenesis. *Cell* **75,** 827-830.

Kim, W. Y., Fritzsch, B., Serls, A., Bakel, L. A., Huang, E. J., Reichardt, L. F., Barth, D. S., and Lee, J. E. (2001) NeuroD-null mice are deaf due to a severe loss of the inner ear sensory neurons during development. *Development* **128:**417-426.

Lee, J. E. (1997) NeuroD and neurogenesis. *Dev Neurosci* **19:**27-32.

Lee, J. E., Hollenberg, S. M., Snider, L., Turner, D. L., Lipnick, N., and Weintraub, H. (1995) Conversion of Xenopus ectoderm into neurons by NeuroD, a basic helix-loop-helix protein. *Science* **268:**836-844.

Liu, M., Pereira, F. A., Price, S. D., Chu, M. J., Shope, C., Himes, D., Eatock, R. A., Brownell, W. E., Lysakowski, A., and Tsai, M. J. (2000a) Essential role of BETA2/NeuroD1 in development of the vestibular and auditory systems. *Genes Dev* **14:**2839-2854.

Liu, M., Pleasure, S. J., Collins, A. E., Noebels, J. L., Naya, F. J., Tsai, M. J., and Lowenstein, D. H. (2000b) Loss of BETA2/NeuroD leads to malformation of the dentate gyrus and epilepsy. *Proc Natl Acad Sci U S A* **97:**865-870.

Malecki, M. T., Jhala, U. S., Antonellis, A., Fields, L., Doria, A., Orban, T., Saad, M., Warram, J. H., Montminy, M., and Krolewski, A. S. (1999) Mutations in NEUROD1 are associated with the development of type 2 diabetes mellitus. *Nat Genet* **23:**323-328.

Morrow, E. M., Furukawa, T., Lee, J. E., and Cepko, C. L. (1999) NeuroD regulates multiple functions in the developing neural retina in rodent. *Development* **126:**23-36.

Murre, C., McCaw, P. S., and Baltimore, D. (1989) A new DNA binding and dimerization motif in immunoglobulin enhancer binding, daughterless, MyoD, and myc proteins. *Cell* **56:**777-783.

Mutoh, H., Fung, B. P., Naya, F. J., Tsai, M. J., Nishitani, J., and Leiter, A. B. (1997) The basic helix-loop-helix transcription factor BETA2/NeuroD is expressed in mammalian enteroendocrine cells and activates secretin gene expression. *Proc Natl Acad Sci U S A* **94:**3560-3564.

Naya, F. J., Huang, H. P., Qiu, Y., Mutoh, H., DeMayo, F. J., Leiter, A. B., and Tsai, M. J. (1997) Diabetes, defective pancreatic morphogenesis, and abnormal enteroendocrine differentiation in BETA2/neuroD-deficient mice. *Genes Dev* **11:**2323-2334.

Naya, F. J., Stellrecht, C. M., and Tsai, M. J. (1995) Tissue-specific regulation of the insulin gene by a novel basic helix-loop-helix transcription factor. *Genes Dev* **9**:1009-1019.

Pennesi, M. E., Cho, J. H., Yang, Z., Wu, S. H., Zhang, J., Wu, S. M., and Tsai, M. J. (2003) BETA2/NeuroD1 null mice: a new model for transcription factor-dependent photoreceptor degeneration. *J Neurosci* **23**:453-461.

Pugh, E. N., Falsini, B., and Lyubarsky, A. L. (1998) The origin of the major rod- and cone-driven components of the rodent electrogram and the effect of light rearing history on the magnitude of these components. In: Photostasis and related phenomena, pp. 93-128. Eds T. P. Williams, A. B. Thisitle. Plenum Press: New York.

CHAPTER 24

CHARACTERISATION OF A MODEL FOR RETINAL NEOVASCULARISATION

VEGF model characterisation

Pauline E. van Eeden[1], Lisa Tee[1], Wei-Yong Shen[2,3], Sherralee Lukehurst[1], Chooi-May Lai[3], P. Elizabeth Rakoczy[3], Lyn D. Beazley[1,4], and Sarah A. Dunlop[1,4]

1. INTRODUCTION

Retinal neovascularisation is a major clinical complication of diabetic retinopathy that takes place late in the disease process and constitutes the most damaging phase resulting in loss of vision (Klein et al., 1984). Neovascularisation is defined as the growth of new blood vessels which, in a disease process such as diabetic retinopathy, occurs in abnormal retinal locations. Long term consequences of retinal neovascularisation include the formation of epiretinal membranes and retinal detachment (Smith et al., 1999). In addition, new blood vessels lack a patent blood retinal barrier and exhibit leukostasis presumably resulting in cytotoxic damage (Ishida et al., 2003; Qaum et al., 2001).

The precise mechanisms underlying retinal neovascularisation have not been fully elucidated but have been linked to hypoxia and are thought to be mediated by various growth factors including vascular endothelial growth factor (VEGF) (Witmer et al., 2003). VEGF has been shown to induce pathological changes similar to those seen in diabetic retinopathy (Lu et al., 1999; Tolentino et al., 2002; Tolentino et al., 1996). Furthermore, VEGF is upregulated in both animal models of diabetic retinopathy and in human sufferers (Aiello et al., 1994; Amin et al., 1997; Sone et al., 1997). To gain further insights eye-specific VEGF transgenic mouse models have been developed. However, in these models ocular neovascularisation is both extensive and rapid, resulting in severe retinal damage (Ohno-Matsui et al., 2002; Okamoto et al., 1997).

[1] School of Animal Biology, [2] Department of Molecular Ophthalmology, Lions Eye Institute, [3] Centre for Ophthalmology and Visual Science and [4] The Western Australian Institute for Medical Research, The University of Western Australia, Nedlands, Western Australia, 6009. [1,4] Correspondence to Professor Sarah Dunlap, Tel: 618 6488 1403; Fax: 618 6488 1029; E-mail: sarah@cyllene.uwa.edu.au.

Our research team has recently generated six transgenic mouse lines via microinjection of a DNA construct containing the human $VEGF_{165}$ (hVEGF) gene driven by a truncated mouse rhodopsin promoter (Lai et al., 2004). Of these, one (Line 029) displayed slow, progressive retinal neovascularisation allowing us to chart the spatio-temporal changes in retinal neovascularisation at early stages prior to retinal damage and provide a useful model in which to test therapies for diabetic retinopathy.

2. MATERIALS AND METHODS

Animal care and anaesthesia followed the ARVO Statement for the Use of Animals in Ophthalmic and Vision Research and with approval from the Animal Ethics Committee at The University of Western Australia, Australia. Mice were housed in cages at a constant temperature of 22°C, with a $12:12$ hour light/dark cycle and food and water were available *ad libitum*. Wildtype and litter-mate transgenic fourth generation animals were examined at 1 and 4 weeks postnatal. Genotyping was performed using multiplex PCR amplfication for the $VEGF_{165}$ transgene and GAPDH. Retinal vascular beds were visualised in retinal wholemounts stained with the iso-lectin *Griffonia simplicifolia* IB4 and viewed in the confocal microscope (MRC 1000, BioRad, Hercules, CA, Kalman filtering). Images were collected in z-series throughout the depth of the retina noting the levels at which capillary beds had formed. In addition, the presence of microaneurysms and capillary drop-out, clinical features common in diabetic retinopathy, were noted. Paraffin embedded eyes were sectioned at 6μm so as to reveal the naso-temporal retinal axis, and then stained with haemotoxylin and eosin. Retinal thickness, namely the distance between the vitread aspect of the nerve fibre layer (NFL) and the sclerad aspect of the photoreceptor outer segments, was measured using ImagePro Plus. Measurements (mean ± standard deviation) were taken from central retina, the most developmentally advanced region, selecting locations 200μm on dorsal and ventral sides of the optic disk.

3. RESULTS

In wildtype animals at 1 week, retinal wholemounts revealed that blood vessels had grown from the optic disk to almost reach the retinal periphery establishing a capillary bed within the NFL (Fig. 24.1A); some capillaries had extended from the central NFL to the vitread aspect of the inner nuclear layer (INL; Fig. 24.1B). By 4 weeks, the retinal vasculature was mature, forming 3 regularly arrayed pan-retinal capillary beds within the NFL as well as in both the vitread and sclerad aspects of the INL (Figs. 24.1C-E).

By contrast, in transgenics, retinal vascular growth was accelerated and appeared abnormal. By 1 week, as in wildtypes, vessels occupied the NFL and vitread aspect of the INL pan-retinally but were abnormally sparse with few branch points and vessel diameter appeared larger, indicative of capillary dropout (Fig. 24.1F). In addition, abnormal vessels resembling microaneurysms extended into the sclerad aspect of the INL (Fig. 24.1G). Furthermore, we noted some individual variability in the progression of neovascularisation; some animals at 1 week appeared to be more advanced than others having microaneurysms that had penetrated the outer plexiform layer. At 4 weeks, a more pronounced pattern was seen: capillary beds in the NFL and the vitread and sclerad aspect of the INL were sparse

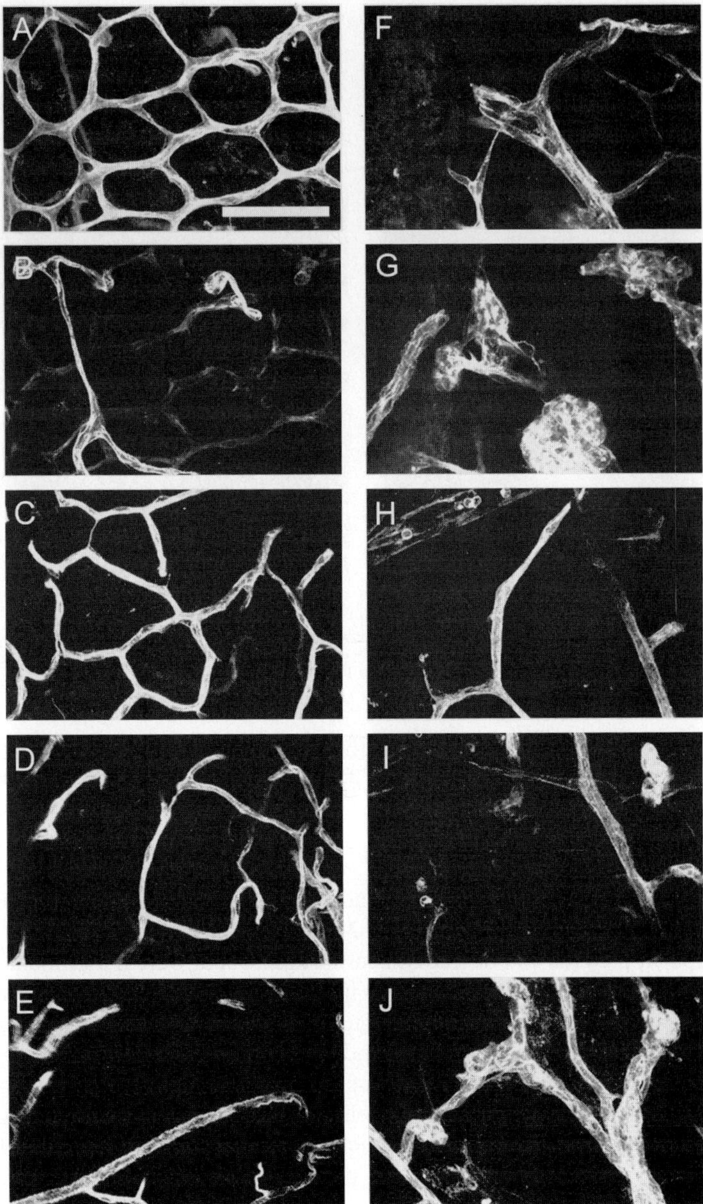

Figure 24.1A-J. Confocal images from retinal wholemounts at 1 (A,B,F&G) and 4 (C-E) and (H-J) weeks postnatal in wildtype (A-E) and transgenic (F-J). Images represent stacks at 2 μm intervals throughout the depth of the retina with the retinal ganglion cell layer uppermost. At 1 week in wildtype animals, a vascular bed had formed in the NFL (A) and vessels have reached the vitread aspect of the INL (B). By 4 weeks, 3 regularly arrayed capillary beds were seen in the NFL (C) and the vitread (D) and sclerad (E) aspects of the INL. In transgenics at 1 week, the vascular bed in the NFL was abnormal being sparse with few branch points (F) and with microaneurysms (G). At 4 weeks, all 3 capillary beds were sparsely (H,I&J) and microaneurysms were seen (J). Scale bar: 100 μm.

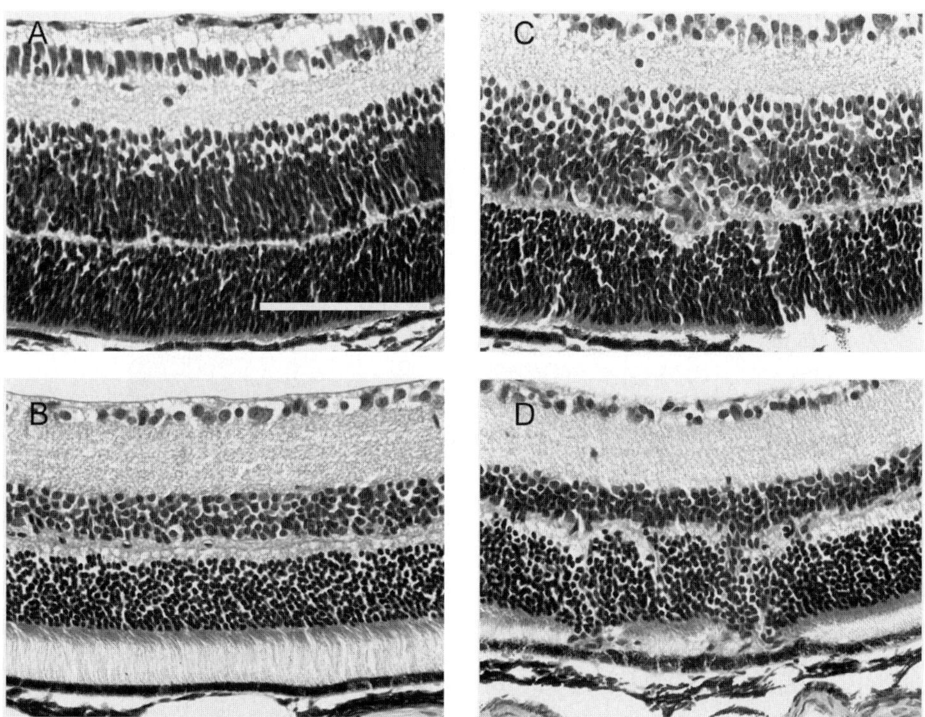

Figure 24.2A-D. Sectioned retinae from wildtype (A,B) and transgenic (C,D) animals at 1 (A,C) and 4 (B,D) weeks. Microaneurysms were readily distinguishable (C,D) and were associated with disruptions to retinal layering. Scale bar: 100 µm. NFL uppermost.

and vessel diameter abnormally large, again indicative of capillary dropout (Figs. 24.1H,I,J). In addition, microaneurysms were seen (Fig. 24.1J). In contrast to 1 week, by 4 weeks variation in the spatial patterns and degree of neovascularisation, were not as marked, although by this later stage some animals appeared to have a more severe phenotype than others.

Sectioned material confirmed the spatio-temporal pattern of neovascularisation seen in the retinal wholemounts of both wildtypes and transgenics. Compared to wildtypes at both 1 and 4 weeks (Fig. 24.2A,B), transgenics at both ages showed disruptions to the retinal layering that were associated with microaneurysms (Figs. 24.2C,D). In addition, overall retinal thickness was reduced in the transgenics compared to wildtypes at both 1 (wildtype: 190 ± 27.9 µm; transgenic: 169.6 + 9.9 µm; p < 0.05) and 4 (wild type: 180.5 + 9.4 µm; transgenic: 168.1 + 18.1 µm; p < 0.1) weeks.

4. CONCLUSION

The transgenic model (Line 029) displays a number of features, such as neovascularisation, microaneurysms and capillary dropout, characteristics that are the hallmark of early stages of human proliferative diabetic retinopathy (Table 24.1). In addition, compared to the

Table 24.1. Comparison of retinal and vascular changes observed in the eyes of patients with diabetic retinopathy and the VEGF transgenic mouse model.

	Retinal and vascular changes	Human	Transgenic mouse
Background retinopathy	Microaneurysm formation, tiny haemorrhages and dilated blood vessels	✓	✓
	Capillary occlusion and capillary drop out	✓	✓
	Basement membrane thickening and pericyte loss	✓	✓
			Pericyte loss undetermined
Proliferative retinopathy	Neovascularisation	✓	✓
	Haemorrhage	✓	✓
	Traction/retinal detachment	✓	✓

rapid onset of neovascularisation observed in other VEGF transgenic models (Ohno-Matsui et al., 2002; Okamoto et al., 1997), these features develop relatively slowly in the mouse model described here. Our transgenic model will thus allow better chronological resolution of pathological changes associated with abnormal blood vessel growth.

5. ACKNOWLEDGEMENTS

We acknowledge the financial support provided by the National Health and Medical Research Council (Australia), the Juvenile Diabetes Research Foundation (USA) and WestPac Australia. This work is part of the research effort of the Diabetic Retinopathy Consortium, Perth, Western Australia.

The authors would like to thank Carole Bartlett for her technical support, the Centre for Microscopy and Microanalysis, University of Western Australia for the use of the Laser confocal microscope and the Animal Resource Centre, Western Australia for the animal husbandry.

6. REFERENCES

Aiello, L. P., Avery, R. L., Arrigg, P. G., Keyt, B. A., Jampel, H. D., Shah, S. T., Pasquale, L. R., Thieme, H., Iwamoto, M. A., Park, J. E., and et al., 1994, Vascular endothelial growth factor in ocular fluid of patients with diabetic retinopathy and other retinal disorders, *N Engl J Med* **331**(22):1480-7.

Amin, R. H., Frank, R. N., Kennedy, A., Eliott, D., Puklin, J. E., and Abrams, G. W., 1997, Vascular endothelial growth factor is present in glial cells of the retina and optic nerve of human subjects with nonproliferative diabetic retinopathy, *Invest Ophthalmol Vis Sci* **38**(1):36-47.

Ishida, S., Usui, T., Yamashiro, K., Kaji, Y., Ahmed, E., Carrasquillo, K. G., Amano, S., Hida, T., Oguchi, Y., and Adamis, A. P., 2003, VEGF164 is proinflammatory in the diabetic retina, *Invest Ophthalmol Vis Sci* **44**(5):2155-62.

Klein, R., Klein, B. E., Moss, S. E., Davis, M. D., and DeMets, D. L., 1984, The Wisconsin epidemiologic study of diabetic retinopathy. III. Prevalence and risk of diabetic retinopathy when age at diagnosis is 30 or more years, *Arch Ophthalmol* **102**(4):527-32.

Lai, C-M., Dunlop, S. A., May, L. A., Gorbatov, M., Brankov, M., Shen, W-Y., Binz, N., Graham, C. E., Barry, C. J., Constable, I. J., Beazley, L. D., and Rakoczy, P. E., 2004, Generation of transgenic mice with mild and severe retinal neovascularisation. *Brit J Ophthalmol* (in press).

Lu, M., Perez, V. L., Ma, N., Miyamoto, K., Peng, H. B., Liao, J. K., and Adamis, A. P., 1999, VEGF increases retinal vascular ICAM-1 expression in vivo, *Invest Ophthalmol Vis Sci* **40**(8):1808-12.

Ohno-Matsui, K., Hirose, A., Yamamoto, S., Saikia, J., Okamoto, N., Gehlbach, P., Duh, E. J., Hackett, S., Chang, M., Bok, D., Zack, D. J., and Campochiaro, P. A., 2002, Inducible expression of vascular endothelial growth factor in adult mice causes severe proliferative retinopathy and retinal detachment, *Am J Pathol* **160**(2):711-9.

Okamoto, N., Tobe, T., Hackett, S. F., Ozaki, H., Vinores, M. A., LaRochelle, W., Zack, D. J., and Campochiaro, P. A., 1997, Transgenic mice with increased expression of vascular endothelial growth factor in the retina: a new model of intraretinal and subretinal neovascularization, *Am J Pathol* **151**(1):281-91.

Qaum, T., Xu, Q., Joussen, A. M., Clemens, M. W., Qin, W., Miyamoto, K., Hassessian, H., Wiegand, S. J., Rudge, J., Yancopoulos, G. D., and Adamis, A. P., 2001, VEGF-initiated blood-retinal barrier breakdown in early diabetes, *Invest Ophthalmol Vis Sci* **42**(10):2408-13.

Smith, G., McLeod, D., Foreman, D., and Boulton, M., 1999, Immunolocalisation of the VEGF receptors FLT-1, KDR, and FLT-4 in diabetic retinopathy, *Br J Ophthalmol* **83**(4):486-94.

Sone, H., Kawakami, Y., Okuda, Y., Sekine, Y., Honmura, S., Matsuo, K., Segawa, T., Suzuki, H., and Yamashita, K., 1997, Ocular vascular endothelial growth factor levels in diabetic rats are elevated before observable retinal proliferative changes, *Diabetologia* **40**(6):726-30.

Tolentino, M. J., McLeod, D. S., Taomoto, M., Otsuji, T., Adamis, A. P., and Lutty, G. A., 2002, Pathologic features of vascular endothelial growth factor-induced retinopathy in the nonhuman primate, *Am J Ophthalmol* **133**(3):373-85.

Tolentino, M. J., Miller, J. W., Gragoudas, E. S., Jakobiec, F. A., Flynn, E., Chatzistefanou, K., Ferrara, N., and Adamis, A. P., 1996, Intravitreous injections of vascular endothelial growth factor produce retinal ischemia and microangiopathy in an adult primate, *Ophthalmology* **103**(11):1820-8.

Witmer, A. N., Vrensen, G. F., Van Noorden, C. J., and Schlingemann, R. O., 2003, Vascular endothelial growth factors and angiogenesis in eye disease, *Prog Retin Eye Res* **22**(1):1-29.

CHAPTER 25

A TWO-ALTERNATIVE, FORCED CHOICE METHOD FOR ASSESSING MOUSE VISION

Yumiko Umino, Bridget Frio, Maryam Abbasi, and Robert Barlow

1. INTRODUCTION

The retina is an important model for studies of neurodegeneration. More than 100 gene mutations are known to cause retinal degeneration (Chader, 2002; Pacione et al., 2003; Rattner et al., 1999). Studies of retinal degeneration in mammals have focused on mouse because of the knowledge of its genome together with the ease of its breeding and husbandry (Naash et al., 2004; Olsson et al., 1992; Pinto et al., 2004). Assessment of the progression of retinal degeneration generally involves measurements of retinal sensitivity using the electroretinogram (ERG) and analysis of retinal anatomy using histological and histochemical techniques. In addition to knowledge of the anatomical and physiological consequences of retinal degeneration, it is important to assess its affect on visual function, that is, to answer the critical question: "How well can a mouse see?"

Mouse vision has been studied with discrimination tasks that reinforce behavior (Prusky et al., 2000; Gianfranceschi et al., 1999). These tasks require extensive testing and are generally not effective with younger mice. Optomotor responses, however, are reflexive and evoked by mice of all ages. Taking advantage of these attributes, Prusky et al. (2004) recently developed a computer-controlled technique that rapidly measures mouse vision by observing optomotor responses. This behavioral test incorporates an "observational" (OB) technique in which the investigator observes the animal respond to a rotating pattern that is visible to both the mouse and the investigator. To avoid observer bias we have developed an objective method that incorporates a two-alternative forced choice (TAFC) protocol in which the investigator chooses the direction of pattern rotation based only on the animal's behavior. Here we describe the TAFC protocol and compare its results with those measured by the OB protocol.

Center for Vision Research, Department of Ophthalmology, Upstate Medical University, Syracuse, NY. *Center for Vision Research, 3258 Weiskotten Hall, Upstate Medical University, 750 East Adams Street, Syracuse, NY 13210; E-mail: barlowr@upstate.edu.

2. BEHAVIORAL TESTING

2.1. Setup and Animals

We measured visual acuity and contrast sensitivity of mice by observing their opto-motor behavior with a computer-controlled threshold measuring system (Prusky et al., 2004). The program (OptoMotry; CerebralMechanics, Lethbridge, Alberta, Canada) runs on a dual processor (G5 Power Mac, Apple Computer) controlling four monitors. We tested C57/BL6J mice that had a single copy of the glucagon receptor gene, *Gcgr* (Gelling et al., 2003). The transgenic Gcgr+/− mice, ranging in age from 1 to 6 months, served as controls for a study of the influence of chronic hypoglycemia on retinal degeneration and visual function (Barlow et al., 2004). Briefly, a mouse is placed on a pedestal situated in the center of a square array of monitors that display a rotating sinusoidal grating. The OptoMotry program controls the speed, direction of rotation, spatial frequency and contrast of the vertical sinusoidal gratings. The luminance of the monitors was $0.4\,cd/m^2$ at minimum (black) level and $155\,cd/m^2$ at maximum (white) mean level. The observer monitors the mouse via an overhead closed-circuit TV camera inside the testing chamber. If it can detect the rotating patterns, the mouse generally exhibits optomotor responses which are reflexive head movements in the direction of pattern rotation. When the mouse has become accustomed to the pedestal, the observer initiates a 5-second trial after which the monitors return to a homogenous gray. Mice were tested during the first four hours of their daytime light cycle (14h light and 10h dark). Observers were unaware of the genotype, sex, and age of the mice as well as their previously recorded thresholds. We measured visual acuity and contrast sensitivity using two techniques.

2.2. Observational Technique

The first protocol termed "observational" (OB) technique was developed by Prusky et al. (2004). Following their protocol, we measured visual acuity with a staircase procedure that systematically increased the spatial frequency of the grating (100% contrast) until the animal no longer exhibited detectable responses. The highest spatial frequency to which the animal responded was defined as their acuity. Next we measured contrast sensitivity in response to moving gratings with a spatial frequency of one half that of their visual acuity. Contrast of the pattern was decreased systematically in a staircase manner until the animal no longer appeared to respond. The lowest contrast (in percent) to which the mouse responded is the contrast threshold. In all tests using this protocol both the animal and the observer see the rotating grating.

2.3. Two-Alternative Forced-Choice Technique

To avoid observer bias we developed a TAFC technique (Bilotta and Powers, 1991; Solessio et al., 2004) with the programming aid of G. T. Prusky and R. M. Douglas. The physical setup was the same as used in the OB technique with the important exception that the rotating grating was masked from the observer's view. The observer viewed the mouse on the pedestal via the closed-circuit TV but was "blind" to the stimulus. The task of the observer was to choose the direction of pattern rotation based on the animal's behavior, specifically, its reflexive optomotor head movements. In each trial the computer-controlled protocol randomly selected the direction of rotation of the grating, and the experimenter assessed the mouse's behavior for a five-second period. At the end of each 5-s trial the observer answered whether the pattern rotated clockwise or counterclockwise and received an auditory feedback indicating correct or incorrect response. The computer changed spatial frequency and/or contrast using a staircase paradigm and converged on a threshold based on the observers responses. Threshold is defined as 70% correct responses by the observer.

3. RESULTS OF THE TWO BEHAVIORAL TECHNIQUES

Figure 25.1 compares the visual acuities and contrast sensitivities measured for 14 mice using the OB and TAFC techniques. We pooled the data because the results for both acuity and contrast did not differ significantly with the age of the mice. We found that both techniques yielded the same visual acuity: 0.39+/−0.02 c/d and 0.39+/−0.01 c/d, respectively, for the OB and TAFC techniques.

The two techniques did not yield the same thresholds for contrast sensitivity. Expressed in terms of percent contrast, the threshold of 7.1+/−0.5% measured with the TAFC technique is significantly lower than that of 9.1+/−0.4% measured with the OB technique.

4. DISCUSSION

The visual acuity we measured with both the OB and TAFC techniques (0.39 c/d) is equal to that (0.4 c/d) reported by Prusky et al. (2004) using the OB technique with C57/BL6 mice. The contrast sensitivity we measured with the OB technique of 9.1% contrast at 0.2 c/d also agrees with that of ~10% contrast at the same spatial frequency reported by

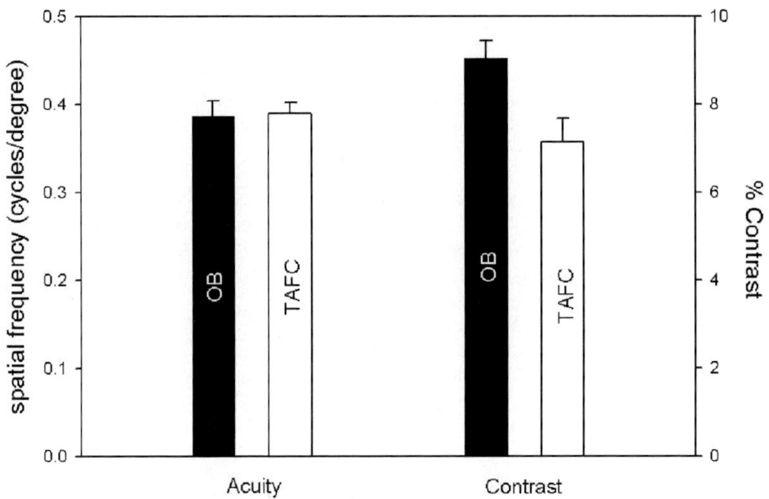

Figure 25.1. Visual acuity and contrast sensitivity measured with two techniques: Observational (OB) and two-alternative forced choice (TAFC). Bar give mean data for fourteen mice +/− standard error.

Prusky et al. (2004) using the same technique. However, for unknown reasons we measured a significantly lower contrast threshold (7.1% contrast) with the TAFC technique.

We developed the TAFC technique to avoid observer bias. This objective psychophysical technique not only produced reliable, reproducible results but also was well received by those testing the mice. Specifically, each of the three observers in this study (M.A., P.A. and B.F.) reported that they felt more comfortable using the TAFC technique for three reasons: first, they were "blind" to the optomotor stimulus and thus could not be influenced by it; second, their duty was only to detect the direction of pattern movement based on the animal's behavior; and third, the TAFC technique is easy to learn. They also reported that they felt more confident with the results. In view of the growing role of mice in retinal degeneration studies such an objective technique for assessing visual function using optomotor responses should prove useful.

5. ACKNOWLEDGMENTS

We thank Glen T. Prusky of the University of Lethbridge (Alberta, Canada) and Robert M. Douglas of the University of British Colombia (Vancouver, Canada) for their modification of the computer program, OptoMotry, to accommodate the TAFC technique. Supported the National Eye Institute (EY00667 and EY12286), Research to Prevent Blindness, and the Lions of Central NY.

6. REFERENCES

Barlow, R., Umino, Y., Loi, T. et al., 2004, Retinal degeneration in a hypoglycemic mouse, *Invest Ophthalmol Vis Sci.* **44**:ARVO E-Abstract

Bilotta, J., Powers, M.K., 1991, Spatial contrast sensitivity of goldfish: mean luminance, temporal frequency and a new psychophysical technique, *Vision Res.* **31**(3):577-585.

Chader, G.J., 2002, Animal models in research on retinal degenerations: past progress and future hope, *Vision Res.* **42**(4):393-399.

Gelling, R.W., Du, X.Q., Dichmann, D.S. et al., 2003, Lower blood glucose, hyperglucagonemia, and pancreatic α cell hyperplasia in glucagon receptor knockout mice, *Proc Natl Acad Sci* USA. **100**(3):1438-1443.

Gianfranceschi, L., Fiorentini, A., Maffei, L., 1999, Behavioral visual acuity of wild type and *bcl2* transgenic mouse. *Vision Res.* **39**(3):569-574

Pacione, L.R., Szego, M.J., Ikeda, S. et al., 2003, Progress toward understanding the genetic and biochemical mechanisms of inherited photoreceptor degenerations, *Annu Res Neurosci.* **26**:657-700.

Pinto, L.H., Vitaterna, M.H., Siepka, S.M. et al., 2004, Results from screening over 9000 mutation-bearing mice for defects in the electroretinogram and appearance of the fundus, *Vision Res.* **44**(28):3335-3345.

Prusky, G.T., West, P.W.R., and Douglas, R.M., 2000, Behavioral assessment of visual acuity in mice and rats, *Vision Res.* **40**(16):2201-2209.

Prusky, G.T., Alam, N.M., Beekman, S. et al., 2004, Rapid quantification of adult and developing mouse spatial vision using a virtual optomotor system, *Invest Ophthalmol Vis Sci.* **45**(12):4611-4616.

Rattner, A., Sun, H., and Nathans, J., 1999, Molecular genetics of human retinal disease, *Annu Rev Genet.* **33**:89-131.

Solessio, E., Scheraga, D., Engbretson, G.A. et al., 2004, Circadian modulation of temporal properties of the rod pathway in larval *Xenopus*, *J Neurophysiol.* **92**:2672-2684.

CHAPTER 26

CONDITIONAL GENE KNOCKOUT SYSTEM IN CONE PHOTORECEPTORS

Yun-Zheng Le[1,4], John D. Ash[3,4], Muayyad R. Al-Ubaidi[1], Ying Chen[2], Jian-Xing Ma[1,2], and Robert E. Anderson[1,3,4]

1. INTRODUCTION

The generation of gene knockout mice by injection of homologous gene targeted embryonic stem (ES) cells into blastocyst has rapidly advanced our knowledge of gene function in mammals. However, knockout of essential genes often causes embryonic or neonatal lethality and thus obscures the particular role of genes in a target tissue or in adults. To circumvent this problem and disrupt genes in a temporal and spatial fashion, the Cre/*lox* system based gene targeting strategy has become a method of choice (for review, see Le and Sauer (2000)). This method can also be used in dissection of the roles of multifunctional genes in a particular tissue/cell-type. To study the function of widely expressed essential genes in the retinas, we are systematically establishing Cre/*lox* conditional knockout systems in retinal cells. This report is a review of our recent work on the generation and characterization of the cone-specific *cre* mice (Le et al., 2004).

2. MATERIALS AND METHODS

2.1. Generation of Transgenic Mice

The transgene vector (pLE109), carrying a 6.3-kb human red/green pigment (HRGP) gene promoter, a translationally optimized *cre* and an intron-containing mouse metallothionein polyA, was derived from plasmids pBS185 and pJHN60 (Lakso et al., 1992; Wang et al., 1992). The gel-purified transgene DNA, a restriction enzyme NotI digested fragment, was used in injection of FVB/N background zygotes to generate transgenic mice. Geno-

[1] Departments of Cell Biology, [2] Medicine, and [3] Ophthalmology, University of Oklahoma Health Sciences Center, and [4] Dean A. McGee Eye Institute, Oklahoma City, OK 73104.

typing of the *cre* transgene and the Cre-activatable *lacZ* reporter gene with PCR was performed using mouse tail DNA according to the conditions described previously (Le et al., 2003; Le and Sauer, 2000).

2.2. β-Galactosidase Staining and Immunohistochemistry

Cre function in transgenic mice was analyzed with β-galactosidase staining assays using the retinas of double transgenic F1 of HRGP-*cre* and Cre-activatable *lacZ* reporter (R26R) mice (Soriano, 1999). The eyes were removed and fixed with fixation buffer (2% formaldehyde in PBS) for 10 minutes. Following the removal of the lens and vitreous, the eyecup was fixed in fixation buffer for 10 more minutes, washed with PBS, and incubated overnight at RT in X-gal staining solution. After staining, the retinas of one eye were dissected to remove sclera and the β-galactosidase staining in the flat mount retinas was observed under a dissecting microscope. The remaining eyecup was used in paraffin embedding, sectioning (5 μm thick), and microscopy.

To localize Cre expression, cryosections (10-μm thick) were blocked with 10% horse serum in PBS, incubated with anti-Cre polyclonal antibody (Novagen, San Diego CA), anti-M-opsin polyclonal antibody (Chemicon, Temecula, CA), and/or peanut agglutinin (Vector Laboratories, Burlingame, CA) at 4°C overnight, and incubated with corresponding fluorescent chromophore (Alexa 488 or Alexa 568) conjugated secondary antibody (Molecular Probe, Eugene, OR). To analyze cone distribution, the dissected anterior part of the eye containing the whole retina with lens and cornea was fixed with fixation buffer, washed with 0.5% NaBH$_4$ for 5 min at room temperature to diminish the autofluorescence, blocked overnight with blocking buffer (3% IgG free BSA, 5% normal goat serum in washing buffer) at 4°C with gently shaking, and incubated with TRITC-labeled lectin (Sigma, St. Louis, MO).

2.3. Electroretinography (ERG)

For assessment of retinal function of transgenic mice, both scotopic and photopic ERG were measured with a UTAS-E 3000 ERG system (LKC technologies, Inc., Gaithersburg, MD) according to established conditions (Xu et al., 2000).

3. RESULTS AND DISCUSSION

3.1. Generation of Cone-Specific Transgenic Cre Transgenic Mice

PCR screening of potential transgenic founders with genomic DNA of the 143 newborn mice obtained from zygote injection identified 31 *cre*-positive mice. All the *cre*-positive mice appeared to be normal in size, morphology, or behavior. Semi-quantitative RT-PCR analysis (data not shown) of Cre expression in retinas from 10 to 15-day-old transgenic mice suggested that 12 lines expressed *cre* mRNA in their retinas. All of them were capable of transmitting the *cre* through germline in a Mendelian fashion and were characterized further.

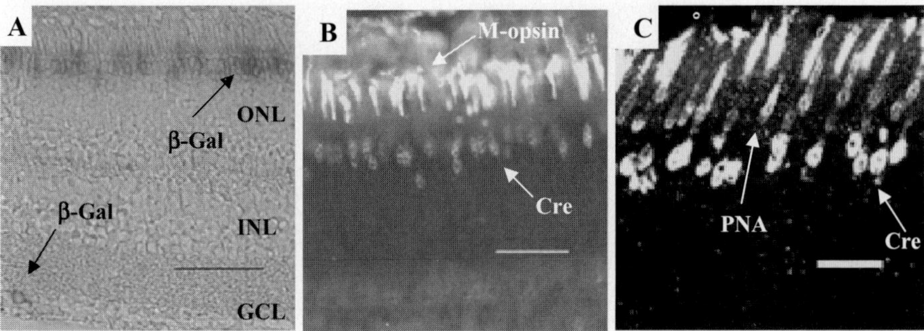

Figure 26.1. Localization and functional analysis of Cre expression (adopted from Le et al., 2004). A: β-Galactosidase staining of retinal section of double transgenic HRGP-*cre*/R26R mice. ONL: outer nuclear layer; INL: inner nuclear layer; GCL: ganglion cell layer. B: Fluorescent microscopy of retinal section stained for Cre (cone nuclei) and M-Opsin (cone outer segment). C: Confocal microscopy of retinal section stained for Cre, M-opsin and peanut agglutinin. The scale bar equals to 50 μm in A and B, and 20 μm in C.

3.2. Analysis of Cre Expression

To localize Cre expression and perform Cre functional assays, we decided to use a Cre-activatable *lacZ* reporter mouse line R26R (Soriano, 1999). This mouse strain carries a *loxP* flanked STOP (un-functional) sequence that prevents the expression of *lacZ* reporter gene (Soriano, 1999). Upon a Cre-mediated recombination that removes the STOP sequence, the *lacZ* reporter gene is expressed under the control of a generalized promoter ROSA26; therefore, the cells expressing Cre recombinase are blue after β-galactosidase staining. β-Galactosidase assays using 6 to 8 week-old double transgenic HRGP-*cre*/R26R mice (F1 of FVB/N and C57B6 genetic background) showed that only two Cre-expressing candidate lines had efficient Cre expression in the retina, judging by X-gal staining in retinal flat-mount (Le et al., 2004) and sections (Figure 26.1A). X-gal staining was strong near the center of the retina (Le et al., 2004). This expression pattern is consistent with mouse cone distribution (Jeon et al., 1998). X-gal staining of retinal sections showed that the β-galactosidase activity was localized to the presumptive cone photoreceptor cells (Figure 26.1A). A small number of presumptive ganglion cells were positive with X-gal staining (Figure 26.1A). The remaining candidate transgenic mouse lines had only limited amount of Cre activities and the β-galactosidase staining was punctuated. These strains were not characterized further.

To further confirm that Cre was expressed in cone photoreceptors, retinal sections were stained with anti-Cre antibody and Cre was exclusively localized to the nucleus of cone photoreceptors (data not shown). This result was consistent with our earlier observation that Cre carried a nuclear localization signal (Gagneten et al., 1997; Le et al., 1999). Double labeling of retinal sections with anti-Cre and anti-M-opsin antibodies showed that cells expressing M-opsin (cone outer segment) also had Cre expression (nuclei) (Figure 26.1B). This result suggested that almost all M-opsin-expressing cells expressed Cre recombinase. Since most mouse cone cells express both M- and S-opsins and only a small percentage of cone cells express a single type of pigment (Applebury et al., 2000), retinal sections were

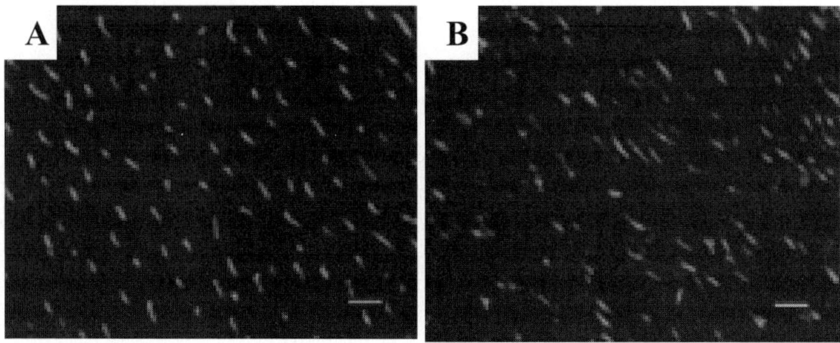

Figure 26.2. Cone distribution in double transgenic HRGP-*cre*/R26R mice (adopted from Le et al., 2004). Lectin stained cone photoreceptors in 6 month old transgenic (A) and wild-type littermate (B). The scale bars equal to 16 µm. Reprinted with permission from Molecular Vision.

stained for Cre and peanut agglutinin (PNA). The results suggested that almost all cone photoreceptors were Cre-positive in our transgenic mice (Figure 26.1C).

3.3. Distribution and Function of Cone Photoreceptors

Cre is a DNA recombinase and there is evidence suggesting that over-expression of Cre may cause rod photoreceptor degeneration (Le and Anderson, unpublished observation; Fenier L et al., *IVOS*: 2004 ARVO E-Abstract 3596; Chen, C et al., *IVOS*: 2004 ARVO E-Abstract 3597), presumably due to undesired recombination that causes chromosomal rearrangement (Loonstra et al., 2001; Schmidt et al., 2000). Since the goal of generating cone-specific Cre transgenic mice is for gene function studies, it is necessary to determine if the cones are normal in the mice. Therefore, the distribution of cones in six-month-old HRGP-*cre*/R26R mice was analyzed. Fluorescent microscopic analysis of lectin-labeled retinas indicated that there were no significant differences in cone density and distribution between the transgenic mice and wild-type littermates (Figure 26.2). This result suggested that there was no cone photoreceptor degeneration.

To further confirm that cone photoreceptors were normal, six-month-old F1 HRGP-*cre*/R26R mice were used in ERG analysis. The photopic ERG suggested that there were no significant differences between the transgenic mice and the wild-type littermates, as shown in Figure 26.3. In addition, the scotopic ERG data showed that rod function was normal in HRGP-*cre*/R26R mice (data not shown). This result indicated that the insertion of the transgene did not cause any changes in retinal function.

4. SUMMARY

To study function of widely expressed essential genes, we established a conditional knockout system for cone photoreceptor cells. Our goal is to generate a useful genetic system that can be utilized to disrupt gene function efficiently in cone photoreceptor cells. Functional assay using a Cre-activatable *lacZ* reporter gene suggested that HRGP-*cre* mice

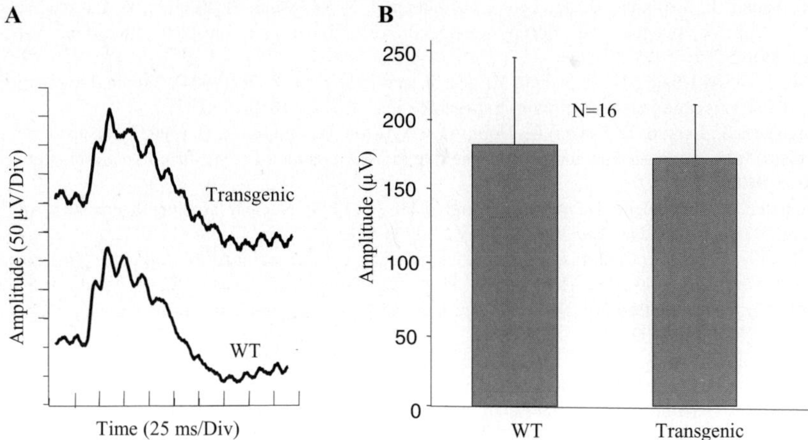

Figure 26.3. Cone function in 6-month old HRGP-*cre*/R26R mice (adopted from Le et al., 2004). A. Representative photopic ERG of transgenic mouse and wild-type littermate. B. Average b-wave amplitude from 16 eyes of the transgenic mice and wild-type littermates.

had widely expressed functional Cre in cone photoreceptors. Since nearly all cone photoreceptor cells express Cre in our transgenic mice, they will be efficient in carrying out Cre-mediated gene activation and inactivation in both M- and S-cone photoreceptors. In addition, we have not observed any apparent ectopic expression in HRGP-*cre* mice (Le et al., 2004); thus, conditional knockout of essential genes with these mice is not likely to cause any detrimental effect in non-ocular tissues.

5. ACKNOWLEDGEMENTS

The authors thank C. Kontas, L. Cowley, P. Pierce, J. Woods, M. Agbaga, D. Vuong, and M. Dittmar for technical assistance; M. Zhu for technical advice; Drs. J. Nathans and B. Sauer for providing the plasmids used in this study; and the Transgenic Core Facility at the Oklahoma Medical Research Foundation for generating the transgenic mice. This study was supported by OCAST contract HR01-083; NIH grants EY00871, EY12190, EY04149, EY015299, and RR17703; Research to Prevent Blindness, Inc.; and the Foundation Fighting Blindness, Inc.

6. REFERENCES

Applebury, M. L., Antoch, M. P., Baxter, L. C., Chun, L. L., Falk, J. D., Farhangfar, F., Kage, K., Krzystolik, M. G., Lyass, L. A., and Robbins, J. T. (2000). The murine cone photoreceptor: a single cone type expresses both S and M opsins with retinal spatial patterning. *Neuron* **27:**513-23.

Gagneten, S., Le, Y., Miller, J., and Sauer, B. (1997). Brief expression of a GFP cre fusion gene in embryonic stem cells allows rapid retrieval of site-specific genomic deletions. *Nucleic Acids Res* **25:**3326-31.

Jeon, C. J., Strettoi, E., and Masland, R. H. (1998). The major cell populations of the mouse retina. *J Neurosci* **18:**8936-46.

Lakso, M., Sauer, B., Mosinger, B., Jr., Lee, E. J., Manning, R. W., Yu, S. H., Mulder, K. L., and Westphal, H. (1992). Targeted oncogene activation by site-specific recombination in transgenic mice. *Proc Natl Acad Sci U S A* **89:**6232-6.

Le, Y., Ash, J. D., Al-Ubaidi, M. R., Chen, Y., Ma, J., and Anderson, R. E. (2004). Targeted expression of Cre recombinase to cone photoreceptors in transgenic mice. *Mol Vis* **10:**1011-1018.

Le, Y., Gagneten, S., Larson, T., Santha, E., Dobi, A., v Agoston, D., and Sauer, B. (2003). Far-upstream elements are dispensable for tissue-specific proenkephalin expression using a Cre-mediated knock-in strategy. *J Neurochem* **84:**689-97.

Le, Y., Gagneten, S., Tombaccini, D., Bethke, B., and Sauer, B. (1999). Nuclear targeting determinants of the phage P1 cre DNA recombinase. *Nucleic Acids Res* **27:**4703-9.

Le, Y., and Sauer, B. (2000). Conditional gene knockout using cre recombinase. *Methods Mol Biol* **136:**477-85.

Loonstra, A., Vooijs, M., Beverloo, H. B., Allak, B. A., van Drunen, E., Kanaar, R., Berns, A., and Jonkers, J. (2001). Growth inhibition and DNA damage induced by Cre recombinase in mammalian cells. *Proc Natl Acad Sci U S A* **98:**9209-14.

Schmidt, E. E., Taylor, D. S., Prigge, J. R., Barnett, S., and Capecchi, M. R. (2000). Illegitimate Cre-dependent chromosome rearrangements in transgenic mouse spermatids. *Proc Natl Acad Sci U S A* **97:**13702-7.

Soriano, P. (1999). Generalized lacZ expression with the ROSA26 Cre reporter strain. *Nat Genet* **21:**70-1.

Wang, Y., Macke, J. P., Merbs, S. L., Zack, D. J., Klaunberg, B., Bennett, J., Gearhart, J., and Nathans, J. (1992). A locus control region adjacent to the human red and green visual pigment genes. *Neuron* **9:**429-40.

Xu, X., Quiambao, A. B., Roveri, L., Pardue, M. T., Marx, J. L., Rohlich, P., Peachey, N. S., and Al-Ubaidi, M. R. (2000). Degeneration of cone photoreceptors induced by expression of the Mas1 protooncogene. *Exp Neurol* **163:**207-19.

CHAPTER 27

REGULATION OF TIGHT JUNCTION PROTEINS IN CULTURED RETINAL PIGMENT EPITHELIAL CELLS AND IN VEGF OVEREXPRESSING TRANSGENIC MOUSE RETINAS

Reza Ghassemifar[1,2,]*, Chooi-May Lai[2], and P. Elizabeth Rakoczy[2]

1. INTRODUCTION

Tight junctions (TJ) are specialized multiprotein complexes which act to seal the intercellular space and thereby generate a permeability barrier required for transport processes (Matter and Balda, 1999). In addition, by regulation of the TJs, the paracellular pathway may be opened for selective transport of molecules, ions (Madara et al., 1992) and neutrophils (Huber et al., 2000). Thus far several interacting constituents comprising of transmembrane proteins including occludins (OCLN-TM4$^+$/TM4$^-$ and 1B), claudins with >20 members and a series of cytoplasmic plaque proteins including ZO-1, -2, or -3, cingulin, 7H6 and symplekin have been identified in several tissue types of different species (Citi and Cordenonsi, 1998; Furuse et al., 1993; Ghassemifar et al., 2002; Matter and Balda, 1999; Mitic and Anderson, 1998). While the inner blood–retinal barrier, the retinal vascular endothelial cells, maintain a restricted and regulated trans-/paracellular transport from the blood to the surrounding tissue, the outer blood-retinal barrier, the retinal pigment epithelium (RPE), provides a permeability barrier between the retina and the choroid allowing vectorial exchange of solutes between these layers (Thumann, 2001). The RPE plays an important role in the proper function and maintenance of the photoreceptors by releasing differentiation and survival promoting factors. Despite the identification of TJ proteins in retinal tissue, little is known about the effect of pathological insult, such as hypoxia, on the expression of TJ proteins (Mark and Davis, 2002). It has also been reported that factors present in embryonic or whole eye-extract can influence the neurons of chick sympathetic ganglia to express choline acetyltransferase (Iacovitti et al., 1987) and also support full sur-

---

[1] Department of Molecular Ophthalmology, Lions Eye Institute, and The University of Western Australia, 2 Verdun Street, Nedlands, Western Australia 6009, Australia and [2] Centre for Ophthalmology and Visual Sciences, The University of Western Australia, Australia. *Corresponding author.

vival ganglion neurons for up to 6 days in culture (Margiotta and Howard, 1994). Retinal hypoxia has been reported to increase production of vascular endothelial growth factor (VEGF), a known stimulant of retinal neovascularization (Okamoto et al., 1997). The aim of this study was to investigate the regulation of tight junction (TJ) gene expression in cultured retinal pigment epithelial (RPE) cells and in the retina of normal and vascular endothelial growth factor (VEGF) transgenic mice.

2. MATERIALS AND METHODS

2.1. Generation of Transgenic Mice

The human $VEGF_{165}$ isoform (Chavand et al., 2001) and the truncated murine opsin promoter containing 1.4 kb of a 5′ upstream regulatory sequence of the opsin gene (May et al., 2003) were used to generate the pcDNA.CMV.VEGF construct for production of transgenic mouse (Lai et al., 2004 BJO-in press). All mice experimentation was performed in accord with the guidelines of the Association for Research in Vision and Ophthalmology on the use of animals in research and following the guidelines of the Animal Ethics Committee at The University of Western Australia, Australia.

2.2. Effects of Eye Extract and Hypoxia on RPE Cell Culture

Human RPE cells were isolated from the retina of a 51-year-old donor as previously described (Kennedy et al., 1996) and were maintained in Dulbecco's modified Eagle's medium (DMEM) supplemented with 10% fetal bovine serum. For eye-extract experiments, immediately following enucleation of the eyes of four wild-type female C57/B1 mice, the anterior chamber and lenses were removed and the remaining eyecups were homogenized in culture medium using sterile pestle. The homogenate was centrifuged at $100\,g$ for one minute to remove debris. The supernatant was diluted to final volume of 14 mL with DMEM (Sigma) containing 10% fetal bovine serum. RPE cells were seeded onto 6-well culture plates with or without 13 mm coverslips at 5×10^5 cells/well and cultured in a humidified incubator. Twenty-four hours later plates with or without coverslips received 1 and 2.5 mL of diluted eye extract respectively, and control plates received the same volume of culture medium only. All plates were incubated for an additional 24 hours before being analysed. For hypoxia experiments, RPE cells were seeded onto 6-well culture plates (Australian Biosearch) with or without 13 mm coverslips (Biolab) at 5×10^5 cells/well and cultured under normal (37°C, 5% CO_2 and 20% O_2) condition. Twenty-four hours later cells were either incubated under normoxic (20% O_2) or hypoxic (2% O_2) conditions for additional 24 hours prior to being analysed.

2.3. Antibodies and Immunoconfocal Microscopy

For immunoconfocal the eyes of mice were enucleated and one eye per animal was embedded in optimum cutting temperature compound (OCT) (Tissue Tek) and snap frozen using liquid nitrogen. Immunostaining was performed as described previously (Ghassemifar et al., 2002) with minor modifications. Briefly, fixed in ice-cold methanol, cells on coverslips and or cryosections on slides were rehydrated in PBS/0.5% BSA for 15 minutes prior to a 1 hour incubation at room temperature (rt) with either polyclonal anti-ZO-1 (Zymed) (2 µg/mL in PBS/0.5% BSA) or monoclonal anti-occludin (Zymed) (3 µg/mL in PBS/0.5%BSA). A further 3 washes were followed by a 1 hour incubation at rt with FITC-conjugated goat anti-mouse antibodies (1 : 500 in PBS/0.5% BSA) and/or TRITC-conjugated goat anti-rabbit antibody (1 : 500 in PBS/0.5% BSA). Specimens were viewed on a Bio-Rad MRC 1024 UV Laser Scanning Confocal Microscope system.

2.4. RT-PCR

Total RNA was extracted from the collected cells and remaining eyes using QIAzol lysis reagent (QIAGEN), according to manufacturer's instructions. Two-hundred nanograms of total RNAs were reverse transcribed in 50 µl RT reactions into cDNA. Directly from this reaction, 2.5 µl (5%) of each cDNA were used as template for PCRs containing 0.5 µM of each gene-specific primers for detection of OCLN-TM4$^+$/TM4$^-$ 966F (5′-tagtgagtgctatcctgggcat-3′) and 1585R (5′-tgcaggtgctcttttgaaggt-3′), ZO-1α$^+$/α$^-$ 3120F (5′-gagaggactcctctggaatg-3′) and 3818R (5′-caaggtcttgagagtgctga-3′), VEGF 1147F (5′-catcacgaagtggtgaagtt-3′) and 1533R (5′-aacgctccaggacttatacc-3′), and β-actin 188F (5′-aggcaccagggcgtgat-3′) and 711R (5′-ttaatgtcacgcacgatttc-3′) (Proligo).

3. RESULTS

3.1. Expression and Immunolocalization of TJs in RPE Cells

The ZO-1α$^+$ transcripts were differentially upregulated in a dose dependant manner (Figure 27.1A). However, the transcription levels of ZO-1α$^-$ isoform remained unchanged in all three groups. The transcription levels of OCLN-TM4$^-$ isoform was also differentially upregulated in RPE cells incubated with eye extract compared to controls. Moreover, the results showed a downregulation of VEGF transcripts at the highest eye extract concentration compared to the increased transcription levels with the lower eye extract concentration and the control. Moreover, the results showed that within hypoxic group and/or when compared to normoxic controls, there were a noticeable upregulation of ZO-1α$^-$ compared to α$^+$ transcript (Figure 27.1B). Moreover, RPE cells under hypoxia upregulated OCLN-TM4$^+$ more than TM4$^-$ transcripts both within and/or when compared to normoxic controls. The

A   B

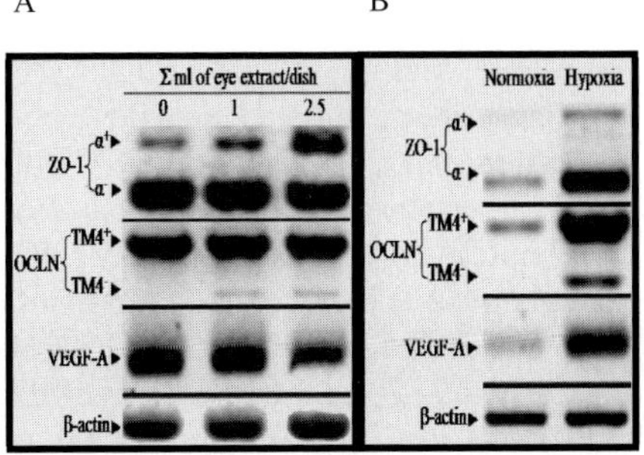

Figure 27.1. Composite images of RT-PCR analysis of ZO-1α+/α−, OCLN-TM4+/TM4−, VEGF and β-actin transcripts during (A) addition of 0 mL, 1 mL and 2.5 mL of eye extract and (B) normoxia and hypoxia, in cultured RPE cells.

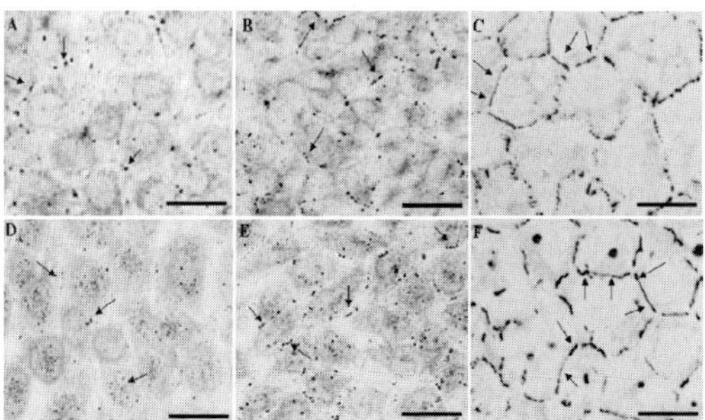

Figure 27.2. (A) Occludin and (D) ZO-1 labelling of control RPE cells showing a few number of junctional staining at cell borders (arrows). (B) Occludin and (E) ZO-1 labelling of RPE cells incubated with 1 mL/dish of mouse eye extract; note significant increase in occludin and ZO-1 junctional staining at cell borders (arrows). (C) Occludin and (F) ZO-1 labelling of confluent HEK-293 cells used as a positive control. Bars, 20 μm.

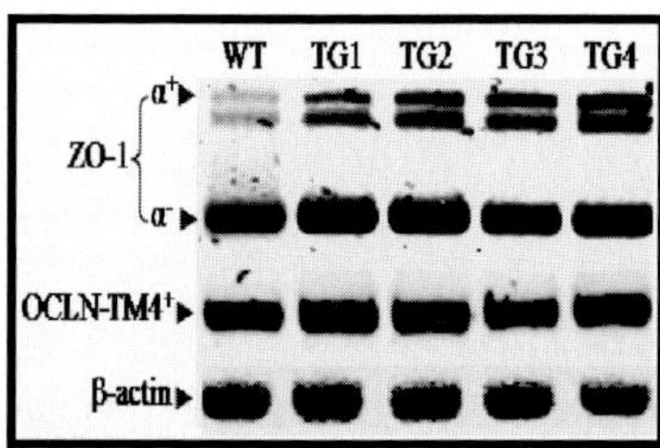

Figure 27.3. Composite images of RT-PCR analysis of ZO-1α+/α−, OCLN-TM4$^+$ and β-actin transcript expression in retinal tissue from transgenic (TG1-TG4) and wild type mice (WT) control.

expression of VEGF transcripts were also significantly higher during hypoxia compared to normoxic controls. RPE cells incubated with eye extract increased the assembly of occludin (Figure 27.2C) and ZO-1 (Figure 27.2D) proteins at the TJ compared to control RPE cells that showed very little if none occludin (Figure 27.2A) and ZO-1 (Figure 27.2B) junctional localization. Panels 2E and 2F are human embryonic kidney cell line (HEK-293) used as positive staining controls for occludin and ZO-1, respectively.

3.2. Expression and Immunolocalization of TJs in Transgenic Mouse Eyes

VEGF over-expressing retinal tissue differentially upregulated the expression of ZO-1α$^+$ transcripts compared to wild type control retinal tissue (Figure 27.3). However, the expression of ZO-1α$^-$ and OCLN-TM4$^+$ transcripts remained uniformly unchanged in all groups. Hematoxylin-eosin stained cryosections showed severe morphological changes in transgenic mice retinas (Figures 27.4F and 27.4K) compared to the retina of wild type control mouse (Figure 27.4A). Immunofluorescence staining revealed an increased TJ localization of occludin (Figures 27.4G-27.4H and 27.4L-27.4N) and ZO-1 (Figures 27.4I-27.4J and 27.4O-27.4P) in the RPE layer of transgenic mice compared to occludin (Figures 27.4B and 27.4C) and ZO-1 (Figures 27.4D and 27.4E) in RPE layer of wild type mouse.

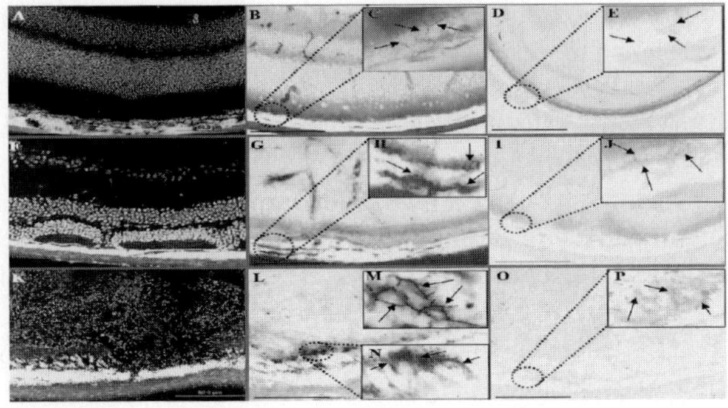

Figure 27.4. Light microscopy and Hematoxylin-eosin stained cryosections show abnormal morphology of transgenic mouse retina (F and K) compared to normal morphology of wild type mouse retina (A). Immunofluorescence staining show occludin [(G with inset H: arrows showing increased junctional staining) and (L with insets M: arrows showing increased junctional and N: arrows showing massive cytoplasmic staining)] and ZO-1 [(I with inset J: arrows showing abnormal punctuated junctional staining) and (O with inset P: arrows showing massive junctional and cytoplasmic staining)] in RPE layer of transgenic mouse retina compared to occludin [(B with inset C: arrows showing normal junctional staining) and ZO-1 (D with inset E: arrows showing normal junctional staining) in the RPE of wild type mouse retina.

4. DISCUSSION

This study demonstrated that human RPE cells cultured in the presence of eye extract differentially upregulated the expression of ZO-1α^+ transcripts in a dose dependant manner while, the transcription levels of the ZO-1α^- isoform remained unchanged in all groups. Previous studies have demonstrated that factors present in embryonic eye-extract can influence the neurons of chick sympathetic ganglia to express choline acetyl-transferase and also whole-eye extract support full survival of E13 dorsal root ganglion neurons for up to 6 days in culture (Fischer et al., 2002). Although the function of the α-domain is still under investigations, the ZO-1α^+ isoform is found in conventional epithelial TJs while ZO-1α^- is present in endothelial junctions and highly specialised epithelial TJs characteristic of Sertoli cells and renal podocytes coinciding with our findings showing high expression of ZO-1α^- in RPE cells suggesting the highly differentiated and specialized character of RPE cells. Moreover, in RPE cells, eye extract caused the upregulation of the OCLN-TM4$^-$ isoform. Nonetheless, it has been suggested that the low level expression of OCLN-TM4$^-$ may contribute to the regulation of occludin function (Ghassemifar et al., 2003). These results were supported by immunoconfocal data showing RPE cells incubated with eye extract increased the assembly of ZO-1 and occludin proteins at the TJ. Moreover, RPE cells downregulated VEGF transcripts at the higher eye extract concentration compared to the controls, which could explain the existence of inbuilt mechanisms within the eye to control the overexpression of VEGF in order to minimise the potential for post developmental neovascularization. During hypoxia, ZO-1α^- transcripts were upregulated compared to the α^+ in RPE cells. Likewise, the OCLN-TM4$^+$ transcripts were upregulated more than TM4$^-$ and when compared to normoxic controls, both OCLN-TM4$^+$/TM4$^-$ transcripts were notably higher during hypoxia. It has been suggested that hypoxia increases the paracellular flux across the cell monolayer via the release of VEGF, which in turn leads to the relocalization, decreased expression, and enhanced phosphorylation of ZO-1 (Fischer et al., 2002). Overexpression of occludin has been shown to increase trans-epithelial electrical resistance (TER) in MDCK cells (Mccarthy et al., 1996) and confers adhesiveness in fibroblasts (Van Itallie and Anderson, 1997). It has been shown that RPE cells in the presence of dimethyl sulfoxide increases assembly of occludin and ZO-1 at cell borders parallel with the increases in TER that occurred with decreases in inulin and dextran permeability (Konari et al., 1995). However, anti-occludin antisense oligonucleotides decrease barrier permeability to solutes in arterial endothelial cells (Kevil et al., 1998). The expression of VEGF transcripts was also significantly higher during hypoxia compared to normoxic controls, coinciding with previous findings that have shown hypoxia upregulates the transcription of VEGF (Carmeliet et al., 1998). In the RPE cells, the upregulation of VEGF during hypoxia is directly corre-lated to the upregulation of TJ proteins in order to create a less permeable RPE layer that could stand against the ingrowth of blood vessels. In summary, we have shown that the eye extract contains constituents, which are to our knowledge directly or indirectly control the regulation of TJ proteins expression and translation. This is the first report to show that in contrast to endothelial cells, hypoxia upregulates and increase the expression and membrane assembly of TJ proteins in RPE cells both *in vitro* and *in vivo*. These results suggest that RPE cells, in the presence of excessive amount of VEGF, probably caused by hypoxia main-tain their transcriptional and post translational processing of TJs to create a less permeable RPE layer to withstand CNV.

5. ACKNOWLEDGMENT

The authors acknowledge the Lions Eye Institute, Lions Save-Sight Foundation and Juvenile Diabetes Research Foundation, USA for the financial support.

6. REFERENCES

Carmeliet, P., Dor, Y., Herbert, J. M., Fukumura, D., Brusselmans, K., Dewerchin, M., Neeman, M., Bono, F., Abramovitch, R., Maxwell, P., Koch, C. J., Ratcliffe, P., Moons, L., Jain, R. K., Collen, D., Keshert, E. and Keshet, E., Role of hif-1alpha in hypoxia-mediated apoptosis, cell proliferation and tumour angiogenesis, *Nature* **394**(6692):485-90 (1998).

Chavand, O., Spilsbury, K. and Rakoczy, P. E., Addition of a c-myc epitope tag within the vegf protein does not affect in vitro biological activity, *Biochem Cell Biol* **79**(1):107-12 (2001).

Citi, S. and Cordenonsi, M., Tight junction proteins, *Biochim Biophys Acta* **1448**(1):1-11 (1998).

Fischer, S., Wobben, M., Marti, H. H., Renz, D. and Schaper, W., Hypoxia-induced hyperpermeability in brain microvessel endothelial cells involves vegf-mediated changes in the expression of zonula occludens-1, *Microvasc Res* **63**(1):70-80 (2002).

Furuse, M., Hirase, T., Itoh, M., Nagafuchi, A., Yonemura, S., Tsukita, S. and Tsukita, S., Occludin: A novel inte-gral membrane protein localizing at tight junctions, *J Cell Biol* **123**(6 Pt 2):1777-88 (1993).

Ghassemifar, M. R., Sheth, B., Papenbrock, T., Leese, H. J., Houghton, F. D. and Fleming, T. P., Occludin tm4(−): An isoform of the tight junction protein present in primates lacking the fourth transmembrane domain, *J Cell Sci* **115**(Pt 15):3171-80 (2002).

Ghassemifar, M. R., Eckert, J. J., Houghton, F. D., Picton, H. M., Leese, H. J. and Fleming, T. P., Gene expression regulating epithelial intercellular junction biogenesis during human blastocyst development in vitro, *Mol Hum Reprod* **9**(5):245-52 (2003).

Huber, D., Balda, M. S. and Matter, K., Occludin modulates transepithelial migration of neutrophils, *J Biol Chem* **275**(8):5773-8 (2000).

Iacovitti, L., Teitelman, G., Joh, T. H. and Reis, D. J., Chick eye extract promotes expression of a cholinergic enzyme in sympathetic ganglia in culture, *Brain Res* **430**(1):59-65 (1987).

Kennedy, C. J., Rakoczy, P. E. and Constable, I. J., A simple flow cytometric technique to quantify rod outer segment phagocytosis in cultured retinal pigment epithelial cells, *Curr Eye Res* **15**(9):998-1003 (1996).

Kevil, C. G., Okayama, N., Trocha, S. D., Kalogeris, T. J., Coe, L. L., Specian, R. D., Davis, C. P. and Alexander, J. S., Expression of zonula occludens and adherens junctional proteins in human venous and arterial endothelial cells: Role of occludin in endothelial solute barriers, *Microcirculation* **5**(2-3):197-210 (1998).

Konari, K., Sawada, N., Zhong, Y., Isomura, H., Nakagawa, T. and Mori, M., Development of the blood-retinal barrier in vitro: Formation of tight junctions as revealed by occludin and zo-1 correlates with the barrier function of chick retinal pigment epithelial cells, *Exp Eye Res* **61**(1):99-108 (1995).

Madara, J. L., Parkos, C., Colgan, S., Nusrat, A., Atisook, K. and Kaoutzani, P., The movement of solutes and cells across tight junctions, *Ann N Y Acad Sci* **664**:47-60 (1992).

Margiotta, J. F. and Howard, M. J., Eye-extract factors promote the expression of acetylcholine sensitivity in chick dorsal root ganglion neurons, *Dev Biol* **163**(1):188-201 (1994).

Mark, K. S. and Davis, T. P., Cerebral microvascular changes in permeability and tight junctions induced by hypoxia-reoxygenation, *Am J Physiol Heart Circ Physiol* **282**(4):H1485-94 (2002).

Matter, K. and Balda, M. S., Occludin and the functions of tight junctions, *Int Rev Cytol* **186**:117-46 (1999).

May, L. A., Lai, C. M. and Rakoczy, P. E., In vitro comparison studies of truncated rhodopsin promoter fragments from various species in human cell lines, *Clin Experiment Ophthalmol* **31**(5):445-50 (2003).

McCarthy, K. M., Skare, I. B., Stankewich, M. C., Furuse, M., Tsukita, S., Rogers, R. A., Lynch, R. D. and Schneeberger, E. E., Occludin is a functional component of the tight junction, *J Cell Sci* **109**(Pt 9):2287-98 (1996).

Mitic, L. L. and Anderson, J. M., Molecular architecture of tight junctions, *Annu Rev Physiol* **60**:121-42 (1998).

Okamoto, N., Tobe, T., Hackett, S. F., Ozaki, H., Vinores, M. A., LaRochelle, W., Zack, D. J. and Campochiaro, P. A., Transgenic mice with increased expression of vascular endothelial growth factor in the retina: A new model of intraretinal and subretinal neovascularization, *Am J Pathol* **151**(1):281-91 (1997).

Thumann, G. a. H., D. R., in: *Retina*, edited by S. J. Ryan (Mosby, St. Louis, 2001) 104-21.

Van Itallie, C. M. and Anderson, J. M., Occludin confers adhesiveness when expressed in fibroblasts, *J Cell Sci* **110**(Pt 9):1113-21 (1997).

CHAPTER 28

PATHOLOGICAL HETEROGENEITY OF VASOPROLIFERATIVE RETINOPATHY IN TRANSGENIC MICE OVEREXPRESSING VASCULAR ENDOTHELIAL GROWTH FACTOR IN PHOTORECEPTORS

Wei-Yong Shen[1,2], Yvonne K.Y. Lai[1,2], Chooi-May Lai[2], Nicolette Binz[1,2], Lyn D. Beazley[3,4], Sarah A. Dunlop[3,4], and P. Elizabeth Rakoczy[2]

1. INTRODUCTION

Retinal neovascularization is a feature shared by many disease processes including diabetic retinopathy, retinopathy of prematurity, branch retinal vein occlusion and central retinal vein occlusion, which are collectively referred to as ischemic retinopathy (Campochiaro, 2000). Retinal neovascularization is the most common cause of blindness in young diabetic patients. Investigations of the pathogenic mechanisms and therapeutic interventions for retinal neovascularization require reproducible and clinically related animal models. Currently, all diabetic models exhibit only early retinal vasculopathy after 1 or 2 years of the disease (Kondo and Kahn, 2004). The lack of retinal neovascularization in diabetic models is probably due to the natural short life span of rodents (2-3 years). In humans, DR is detected only after at least 3 years of diabetes (Dorchy *et al.*, 2002). As angiogenesis is tightly controlled by the relative balance of stimulators and inhibitors, a shift in their balance, such as increased expression of vascular endothelial growth factor (VEGF) or decreased production of pigment epithelium-derived factor, would initiate angiogenesis (Okamoto *et al.*, 1997; Ruberte *et al.*, 2004; Renno *et al.*, 2002). It is clear from literature that ischemia-induced upregulation of VEGF is a potent mediator of retinal neovascularization (Campochiaro, 2000; Miller, 1997). Animal models of retinal neovascularization have been

[1] Department of Molecular Ophthalmology, Lions Eye Institute, affiliated with the University of Western Australia, Nedlands, 6009, Western Australia, Australia. [2] Centre for Ophthalmology and Visual Science, [3] School of Animal Biology and [4] Western Australian Institute for Medical Research, The University of Western Australia. Corresponding author: P.E. Rakoczy, E-mail: rakoczy@cyllene.uwa.edu.au.

established by oxygen-induced retinal ischemia, photodynamically-induced retinal branch vein occlusion and intravitreal implantation of VEGF sustained-release pellets (Campochiaro, 2000; Saito *et al.*, 1997; Ozaki *et al.*, 1997; Tolentino *et al.*, 2002). However, retinal neovascularization in these animal models is either transient or occurs with a delayed onset.

Advanced technology in gene delivery and transgene manipulation has made it possible to generate long-term and reproducible animal models of retinal neovascularization (Wang *et al.*, 2003; Rakoczy *et al.*, 2003; Okamoto *et al.*, 1997; Ruberte *et al.*, 2004). We recently generated a transgenic model of retinal neovascularization by manipulating photoreceptor-specific overexpression of human VEGF (hVEGF) in the eye (Lai *et al.*, in press). A total of four transgenic lines were generated, with one expressing low hVEGF levels and showing correspondingly mild clinical changes such as focal fluorescein leakage, microaneurysms, venous tortuosity, capillary non-perfusion and minor neovascularisation (Lai *et al.*, in press). By contrast, the other three lines expressed high hVEGF levels accompanied by concomitant severe phenotypes. In this study, several generations of two transgenic lines showing mild or severe vasoproliferative retinopathy (lines 029 and 056, Lai *et al.*, in press) were chosen for further characterization.

2. MATERIALS AND METHODS

2.1. Animals

Transgenic mice produced using a pcDNA.opsin.VEGF165 construct driven by the mouse rhodopsin promoter were used in this study. The offspring were screened for transgenic animals by Southern blot analysis showing the presence of a 2.1 kb fragment containing the truncated mouse rhodopsin promoter and human VEGF165 (hVEGF165) fragments, which was then confirmed by polymerase chain reaction amplification of tail DNA. The heterozygote transgenic offspring of transgenic lines 029 and 056 were used for this investigation (Lai *et al.*, in press).

2.2. Fundus Fluorescein Angiography and Retinal Perfusion with Fluorescein-Labeled Dextran

Five generations of line 029 and two generations of line 056 were examined by fundus fluorescein angiography with a modified portable Kowa Genesis camera after intraperitoneal injection of 0.05 ml of 10% sodium fluorescein as described previously (Zaknich *et al.*, 2002). For retinal perfusion, selected mice were deeply anesthesized and initially perfused with 10 ml of PBS through the left ventricle into the aorta to flush out circulating blood, followed by 2 ml fluorescein-labeled dextran (FITC-dextran, 50 mg/ml, molecular weight, 2.0×106; Sigma, St. Louis, MO). The perfused eyes were fixed in 2% paraformaldehyde for 30 min and then flat-mounted for fluorescence microscopy.

2.3. Histology

Eyes were enucleated from transgenic mice showing mild, moderate or severe vasoproliferative retinopathy on angiograms. The enucleated eyes were fixed in 4%

paraformaldehyde for 4 hours, embedded in paraffin, sectioned and stained with hematoxylin and eosin.

3. RESULTS

Increased VEGF production in photoreceptors of transgenic mice led to the progressive development of early, moderate and late stages of diabetic-like retinopathy in line 029 and 056. Line 029 transgenic mice predominantly demonstrated mild and moderate vasoproliferation in the retina (Fig. 28.1A-F, Table 28.1). In eyes with mild retinopathy, fluorescein angiography showed scattered fluorescein leaky spots but with relatively regular arcades of retinal capillaries between leaking lesions (Fig. 28.1A). Perfusion with fluorescein-labeled dextran revealed relatively well-defined retinal capillaries with microaneurysm-like changes under high magnification (Fig. 28.1B and 28.1C). In eyes with moderate vasoproliferation, fluorescein angiography showed more confluent fluorescein leaky spots (Fig. 28.1D). Per-

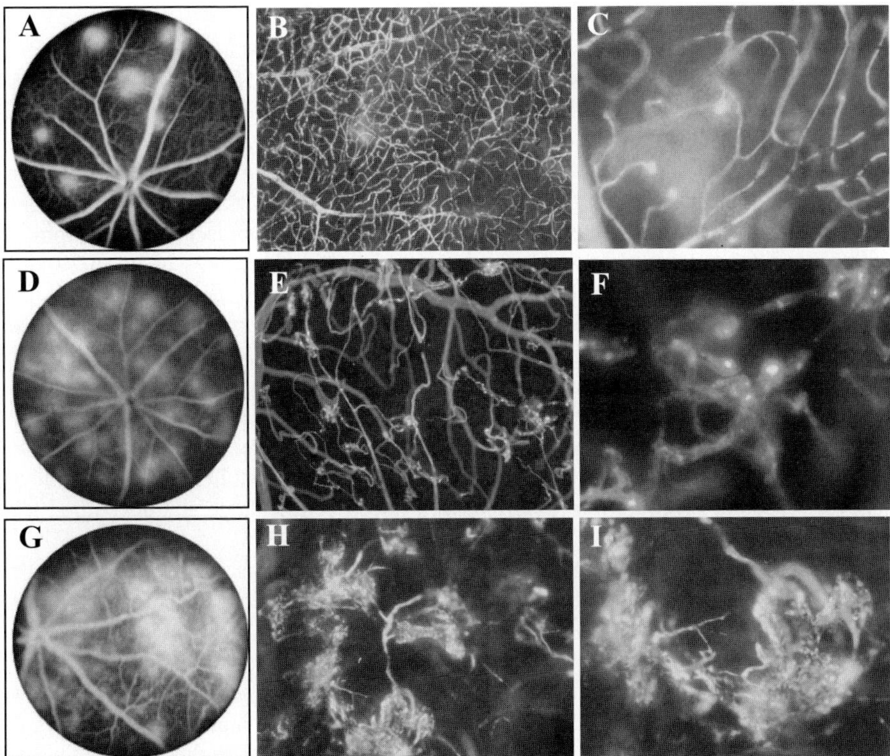

Figure 28.1. Retinal vascular changes of transgenic mice of lines 029 (A-F) and 056 (G-I) showing mild (A-C), moderate (D-F) and severe vasoproliferative retinopathy, respectively. A, D and G, fluorescein angiography; B, C, E, F, H and I, fluorescence microscopy of retinae perfused with fluorescein-labeled dextran. C, F and I are higher magnifications of B, E and H respectively.

Table 28.1. Graded retinopathy in transgenic mice of lines 029 and 056 aged 6-8 weeks.

Generation	n	Graded Retinopathy (%)		
		Mild	Moderate	Severe
Line 029				
A	7	1 (14%)	4 (57%)	2 (29%)
B	4	1 (25%)	3 (75%)	0 (0%)
C	62	19 (30%)	37 (60%)	6 (10%)
D	93	20 (22%)	46 (49%)	27 (29%)
E	18	7 (39%)	9 (50%)	2 (11%)
Subtotal	184	47 (26%)	100 (54%)	37 (20%)
Line 056				
A	11	3 (27%)	1 (9%)	7 (64%)
B	6	0 (0%)	0 (0%)	6 (100%)
Subtotal	17	3 (18%)	1 (6%)	13 (76%)

fusion with fluorescein-labeled dextran revealed obvious neovascular proliferation accompanied by the loss of capillaries in the retina (Fig. 28.1E and 28.1F). In contrast to line 029, line 056 transgenic mice predominantly demonstrated severe vasoproliferative retinopathy (Fig. 28.1G-I, Table 28.1). In these eyes, fluorescein angiography showed heavy fluorescein leakage or pooling (Fig. 28.1G) and perfusion with fluorescein-labeled dextran revealed massive neovascular proliferation in the retina (Fig. 28.1H and 28.1I).

Two hundred and one transgenic mice were examined by fluorescein angiography at 6-8 weeks old. These included 184 transgenic mice from 5 generations of line 029 and 17 transgenic mice from 2 generations of line 056 (Table 28.1). Flourscein angiography showed that 80% of line 029 transgenic mice showed mild/moderate retinopathy while 20% demonstrated severe retinal vascular changes (Table 28.1). In contrast, 76% of the transgenic mice of line 056 showed severe vasoproliferative retinopathy, with only 24% of the transgenic mice developing mild/moderate retinopathy (Table 28.1). Selected animals were examined by fluorescein angiography periodically. The mild and moderate vascular changes of line 029 remained relatively stable for at least 3 months but, in contrast, line 056 transgenic mice with severe vascular proliferation rapidly ceased fluorescein leakage and developed retinal capillary loss with time.

Histologically, eyes with mild retinopathy on angiograms demonstrated scattered vasoproliferation in the subretinal space but without obvious retinal degeneration, although photoreceptors were slightly disturbed (Fig. 28.2A and 28.2B). In eyes showing moderate vasoproliferation, histology revealed a greater number of neovascular lesions. The intraretinal microvascular abnormalities seemed to originate from deep retinal capillaries growing into the subretinal space with an attempt to communicate with choroidal capillaries (Fig. 28.2C and 28.2D). The disturbance to photoreceptors was obvious in these eyes (Fig. 28.2C and 28.2D). In eyes showing heavy fluorescein leakage on angiograms, histology demonstrated massive intraretinal microvascular abnormalities and neovascular proliferation in the subretinal space, accompanied by severe retinal folding and disturbance to photoreceptors (Fig. 28.2E and 28.2F).

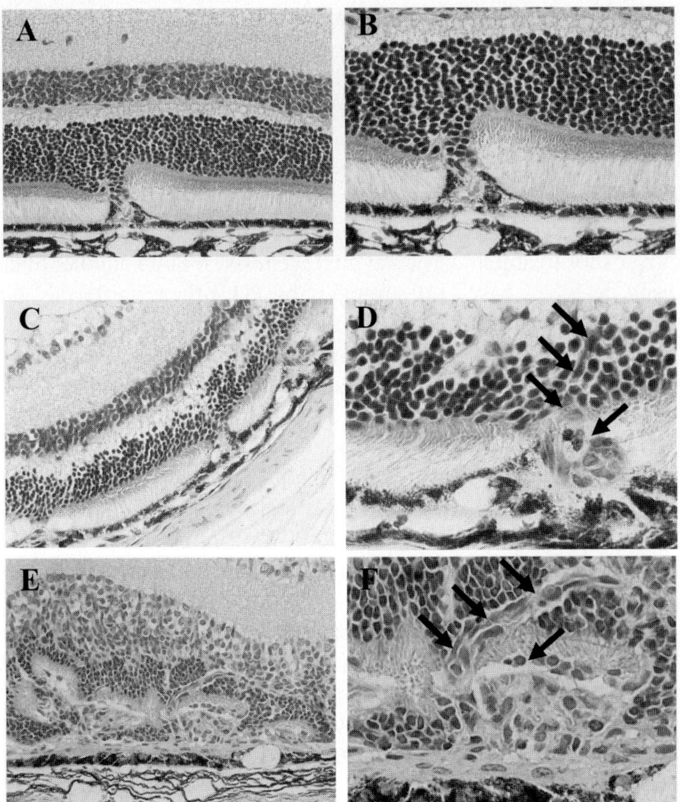

Figure 28.2. Histological changes of transgenic mice of lines 029 (A-D) and 056 (E-F) showing intraretinal microvascular abnormalities and vascular proliferation in the subretinal space. The arrows in D and F indicate the growth of new vessels from the outer nuclear layer into the subretinal space. B, D and F are higher magnifications of A, C and E, respectively.

4. DISCUSSION

In this study, we examined five generations of line 029 and two generations of line 056 transgenic mice with a rhodopsin promoter driven hVEGF overexpression in photoreceptors. We found that mice from transgenic line 029 mainly presented mild or moderate proliferative retinopathy with limited disturbance to the neural retina, while mice from line 056 predominantly developed severe vasoproliferative retinopathy with dramatic destruction to the neural retina. In our previous study, both lines 029 and 056 produced increased hVEGF expression but the levels of hVEGF protein in eyes of line 056 were 10-fold higher than those in eyes of line 029, indicating that the pathological phenotypes in the retinae were closely associated with the overexpression of hVEGF in the retina (Lai *et al.*, in press). In this study, the results obtained from five generations of line 029 and two generations of line 056 suggested that the features presented in both transgenic lines remained relatively steady for as many as five generations. Both transgenic mice lines may prove very useful for testing anti-angiogenic therapies and for investigations of cellular and molecular mechanisms of VEGF-induced retina neovascularization in the eye.

It is very clear that angiogenic factors belonging to the VEGF and angiopoietin protein families play critical roles in retinal neovascularization that occurs in diseases such as diabetic retinopathy, retinopathy of prematurity, retinal branch vein occlusion and retinal central vein occlusion (Campochiaro, 2000; Miller, 1997). Understanding the steps in the angiogenic processes and the angiogenic factors involved in angiogenesis has led to the development of strategies for treatment of ocular angiogenesis. A large number of anti-angiogenic agents have been designed based on strategies by (1) either interfering with VEGF and angiopoietin proteins and their receptors or the downstream signaling or (2) upregulating endogenous angiogenic inhibitors or administration of exogenous inhibitors to counter the angiogenic effect of angiogenic factors (Campochiaro, 2002; Bainbridge *et al.*, 2003). However, evaluation of long-term efficacy of anti-angiogenic therapies has been challenged by transient ocular neovascularization in most animal models. For example, retinal neovascularization in the murine model of ischemia-induced retinopathy occurs as a short-lived response. After post-natal day 21 no further proliferation occurs and the new vessels regress spontaneously (Igarashi *et al.*, 2003; Rota *et al.*, 2004; Bainbridge *et al.*, 2003). In addition, intraocular injection in newly born pups is problematic. In the present study, we have generated a transgenic line (029) with increased expression of VEGF accompanied by mild to moderate vasoproliferation in the retina at an adult age (4 weeks postnatal), and the features of vasoproliferative retinopathy remain stable for at least 8 weeks (12 weeks postnatal). This particular model provides an excellent opportunity to intervene at an early stage in the development of retinal neovascularization in the adult animal.

Although many studies have investigated blood vessel growth in the retina, relatively few studies have addressed the damage to retinal neurons during the processes of retinal neovascularization. In the present model, we also observed disturbance to photoreceptors when new vessels occurred within or adjacent to the outer nuclear layer, and the extent of neural damage was closely associated with the severity of retinal vascular proliferation. There are two basic hypotheses that account for the neural damage accompanied with retinal neovascularization in this model: First, the increased expression of VEGF in photoreceptors result in loss of blood-retinal barrier integrity, which initially manifests as an increase in vascular permeability, causing a failure to control the composition of the extracellular fluid in the retina, which in turn leads to edema and neuronal cell loss. Alternatively, the long-term overexpression of VEGF may initiate apoptosis in the neural retina, leading to gradual loss of neurons. It is not clear which hypothesis will be found to be correct and, in fact, it is likely that vascular permeability and neuronal apoptosis are closely linked components in this model. In a recent study, the retinal neural damage in this model can be remarkably attenuated by recombinant adeno-associated virus mediated transfer of soluble Flt (sFlt) receptor, a heparin-binding protein that complexes VEGF with high affinity, therefore inhibiting the mitogenic response to VEGF by directly sequestering VEGF and in a dominant negative manner via heterodimerizing with the extracellular ligand-binding region spanning FLT-1 and KDR receptors (unpublished data). Since it is now apparent that retinal vascular abnormality and damage to retinal neurons are equally problematic in patients with diabetes (Barber, 1998; Barber *et al.*, 2003), our transgenic models may provide a useful tool for studies on microvascular changes and VEGF-induced retinal neural damage in diabetic retinopathy.

5. ACKNOWLEDGEMENTS

This work was supported by the Juvenile Diabetes Research Foundation International, Australian National Health and Medical Research Council and Westpac (Australia). This work is part of the research effort of the Diabetic Retinopathy Consortium, Perth, Western Australia. We thank the Foundation Fighting Blindness for the Young Investigator Award provided to WYS to attend this meeting.

6. REFERENCES

Bainbridge JW, Mistry AR, Thrasher AJ, Ali RR (2003). Gene therapy for ocular angiogenesis. *Clin Sci* **104**:561-75.

Barber AJ (2003). A new view of diabetic retinopathy: a neurodegenerative disease of the eye. *Prog Neuropsychopharmacol Biol Psychiatry* **27**:283-90.

Barber AJ, Lieth E, Khin SA, Antonetti DA, Buchanan AG, Gardner TW (1998). Neural apoptosis in the retina during experimental and human diabetes. Early onset and effect of insulin. *J Clin Invest* **102**:783-91.

Campochiaro PA (2000). Retinal and choroidal neovascularization. *J Cell Physiol* **184**:301-10.

Campochiaro PA (2002). Gene therapy for retinal and choroidal diseases. *Expert Opin Biol Ther* **2**:537-44.

Dorchy H, Claes C, Verougstraete C (2002). Risk factors of developing proliferative retinopathy in type 1 diabetic patients: role of BMI. *Diabetes Care* **25**:798-9.

Igarashi T, Miyake K, Kato K et al (2003). Lentivirus-mediated expression of angiostatin efficiently inhibits neovascularization in a murine proliferative retinopathy model. *Gene Ther* **10**:219-26.

Kondo T, Kahn CR (2004). Altered insulin signaling in retinal tissue in diabetic states. *J Biol Chem* **279**:37997-8006.

Lai CM, Dunlop SA, May LA et al (2004). Generation of transgenic mice with mild and severe retinal neovascularisation. *Br J Ophthalmol* (in press).

Miller JW (1997). Vascular endothelial growth factor and ocular neovascularization. *Am J Pathol* **151**:13-23.

Okamoto N, Tobe T, Hackett SF et al (1997). Transgenic mice with increased expression of vascular endothelial growth factor in the retina: a new model of intraretinal and subretinal neovascularization. *Am J Pathol* **151**:281-91.

Ozaki H, Hayashi H, Vinores SA, Moromizato Y, Campochiaro PA, Oshima K (1997). Intravitreal sustained release of VEGF causes retinal neovascularization in rabbits and breakdown of the blood-retinal barrier in rabbits and primates. *Exp Eye Res* **64**:505-17.

Rakoczy PE, Brankov M, Fonceca A, Zaknich T, Rae BC, Lai CM (2003). Enhanced recombinant adeno-associated virus-mediated vascular endothelial growth factor expression in the adult mouse retina: a potential model for diabetic retinopathy. *Diabetes* **52**:857-63.

Renno RZ, Youssri AI, Michaud N, Gragoudas ES, Miller JW (2002). Expression of pigment epithelium-derived factor in experimental choroidal neovascularization. *Invest Ophthalmol Vis Sci* **43**:1574-80.

Rota R, Riccioni T, Zaccarini M et al (2004). Marked inhibition of retinal neovascularization in rats following soluble-flt-1 gene transfer. *J Gene Med* **6**:992-1002.

Ruberte J, Ayuso E, Navarro M et al (2004). Increased ocular levels of IGF-1 in transgenic mice lead to diabetes-like eye disease. *J Clin Invest* **113**:1149-57.

Saito Y, Park L, Skolik SA et al (1997). Experimental preretinal neovascularization by laser-induced venous thrombosis in rats. *Curr Eye Res* **16**:26-33.

Tolentino MJ, McLeod DS, Taomoto M, Otsuji T, Adamis AP, Lutty GA (2002). Pathologic features of vascular endothelial growth factor-induced retinopathy in the nonhuman primate. *Am J Ophthalmol* **133**:373-85.

Wang F, Rendahl KG, Manning WC, Quiroz D, Coyne M, Miller SS (2003). AAV-mediated expression of vascular endothelial growth factor induces choroidal neovascularization in rat. *Invest Ophthalmol Vis Sci* **44**:781-90.

Zaknich T, Shen WY, Barry CJ, Brankov M, Rakoczy PE (2002). Modification of clinical cameras for documentation of small laboratory animals. *Ophthalmic Photography* **24**:66-9.

CHAPTER 29

LASER PHOTOCOAGULATION: OCULAR RESEARCH AND THERAPY IN DIABETIC RETINOPATHY

Caroline E. Graham*, Nicolette Binz*, Wei-Yong Shen*, Ian J. Constable*, and Elizabeth P. Rakoczy*

1. INTRODUCTION

Diabetic retinopathy is a severe complication of diabetes leading to some degree of vision impairment in long-term diabetes sufferers. Currently, the most successful treatment available for diabetic retinopathy is laser photocoagulation, a therapy that destroys part of the retina to save central vision. The principal aim of laser photocoagulation in the treatment of diabetic retinopathy is to effect regression of abnormal vessels, reduce oxygen tension and reverse angiogenesis in the retina. Although laser photocoagulation has been employed for more than 30 years, its underlying molecular mechanisms remain unknown. Research is now focused on identifying and understanding these factors, to ultimately develop therapies to protect against the initiation and progression of neovascularisation.

2. DIABETIC RETINOPATHY

Diabetic retinopathy involves changes in the retinal microvasculature and is believed to be a result of long-term hyperglycaemia and possibly hypertension (Porta and Allione, 2004). Early changes are seen in the blood flow of the retina along with thickening of the basement membrane and/or pericyte loss. Non-proliferative diabetic retinopathy progresses with damage to the endothelial cells lining the capillaries, resulting in a breakdown in the blood-retina barrier. This stage is characterised by the presence of microaneurysms, intra-retinal haemorrhages, macular oedema, cotton wool spots and deposits of hard exudates formed from precipitating blood products. Capillary damage causes decreased oxygenation in part of the retina and this localised ischemia provides the stimulus for upregulation of angiogenic factors such as VEGF, thus worsening these microvascular changes.

*Lions Eye Institute and Centre for Ophthalmology and Visual Science, The University of Western Australia, 2 Verdun Street, Nedlands, Australia 6009. Carolineg@lei.org.au.

Proliferative diabetic retinopathy is characterised by the presence of abnormal vessels arising from the optic disk or retina. New vessels develop as fine, leaky structures and bleed easily into the retina and vitreous, leading to the development of haemorrhages that can cause sudden vision loss. This is accompanied with thickening of the extracellular matrix that instigates contraction of the fibrotic component of vessels, resulting in retinal detachment. Central vision can also be affected as these retinal changes move towards the macula. In severe cases, proliferating vessels obscure normal retinal blood flow and can result in the development of neovascular glaucoma, ultimately leading to severe loss of vision or permanent blindness.

3. LASER PHOTOCOAGULATION THERAPY

Laser photocoagulation of the retina is a non-invasive laser treatment and remains the primary therapy for proliferative retinopathies such as diabetic retinopathy. Current photocoagulation treatment is based on the original developments of ruby, argon and krypton lasers first utilised in the late 1960's [Reviewed in Petrovic and Bhistkul (1999)]. Tunable dye and diode lasers were subsequently introduced, making available a range of useful techniques for the treatment of proliferative retinopathies.

Laser photocoagulation acts by focusing laser energy on the retinal pigment epithelium (RPE), damaging the outer retinal layers whilst leaving Bruch's membrane intact. The main site of energy absorption is the melanin within the RPE and choroid as it has an absorption spectrum of between 400-700 nm. The different types of lasers used in photocoagulation include argon (emission at 488 nm, blue/green; 514 nm, green), Nd:YAG (532 nm, green) krypton (647 nm, red) diode (810 nm, infrared) and tuneable dyes such as rhodamine that emit over a selected range of wavelengths. Retinal damage caused by laser photocoagulation can be reduced by decreasing the wavelength, spot size, irradiance and exposure duration (Mainster, 1999). Argon laser is strongly absorbed by melanin and haemoglobin and therefore makes an excellent source for direct targeting of vessels. Krypton laser is less absorbed by melanin and produces deeper and more painful chorioretinal lesions and is therefore less often used in treating retinopathies where induction of choroidal neovascularisation should be avoided.

3.1. Histological Changes

The observed histopathological changes following laser treatment, are a result of the heat transfer out of the absorbing RPE and choroid and subsequent denaturation of the surrounding tissue (Roider *et al.*, 1998). The use of short wavelength lasers such as argon minimises damage to the choroid and therefore is less likely to induce choroidal neovascularisation, detrimental to diabetic retinopathy treatment. The major sites of damage following argon laser photocoagulation are the RPE and outer retinal layers and within the lasered site, the outer segments of the photoreceptors are destroyed (Figure 29.1B). An initial inflammatory response occurs and coagulated cells in the lesion core are removed by phagocytosis. The RPE monolayer can lift to form a gap at the site where photoreceptor outer segments were. This is followed by RPE cell migration and proliferation into multiple layers. Müller cells also proliferate and migrate into areas of damage and interdigitate into RPE cells forming a glial scar (Lewis *et al.*, 1992).

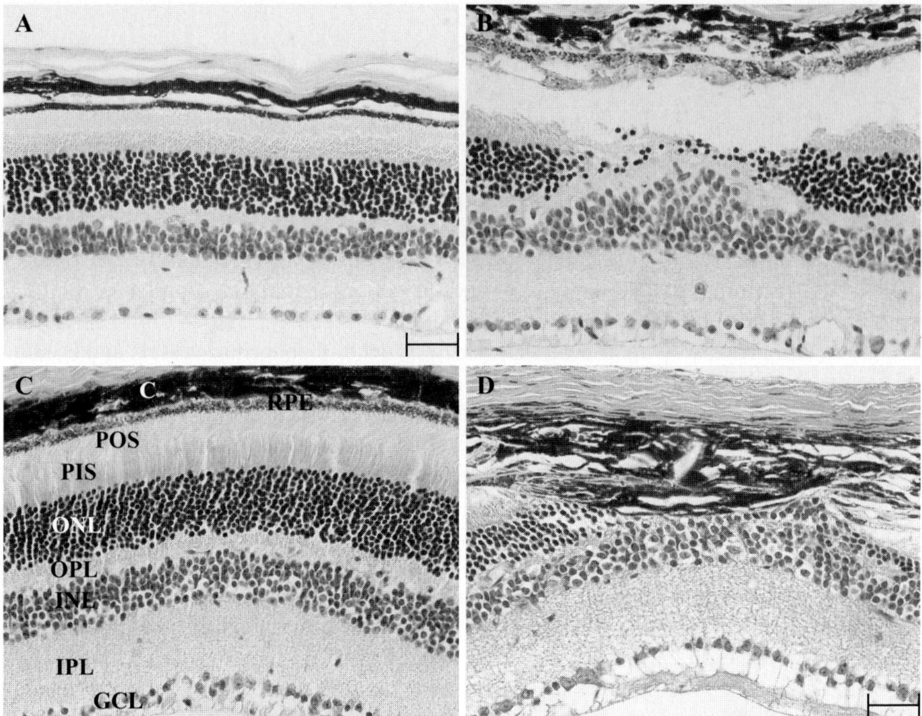

Figure 29.1. Histology of control and lasered mouse retinae at 3 days and 90 days post-argon laser photocoagulation. A, Control retina and B, lasered retina at 3 days post-treatment. C, Control retina and D, lasered retina at 90 days post-treatment. C, choroid; RPE, retinal pigment epithelium; POS, photoreceptor outer segments; PIS, photoreceptor inner segments; ONL, outer nuclear layer; OPL, outer plexiform layer; INL, inner nuclear layer; IPL, inner plexiform layer; GCL, ganglion cell layer. Magnification 40×. Scale bars represent 10 μm.

Late changes in the retina following laser photocoagulation include further extension of glial processes and denser areas of glial scars. The outer nuclear layer reduces in thickness over time (Figure 29.1D) and the number of apoptotic cells present in the laser lesion decreases. Interestingly, photoreceptors adjacent to laser lesions appear to have increased survival compared to photoreceptors in normal retinae as shown by an increase in basic fibroblast growth factor (bFGF)-immuno-reactive cells in this area (Xiao *et al.*, 1998, 1999). The mechanism contributing to the increased survival of these photoreceptors may be due to the increase in bFGF, suppressing apoptosis in these cells.

The benefits of laser photocoagulation therapy are not without risks. Potentially dangerous complications such as haemorrhaging, corneal burns and the development of cataracts are risks in laser therapy. Ultimately, understanding the underlying molecular mechanisms which produce the beneficial effects of laser photocoagulation could lead to the development of non-destructive treatments, thereby circumventing these potential hazards.

3.2. Changes in Gene Expression

3.2.1. Early Changes

High oxygen demand in the diabetic retina is thought to play a role in the initiation and progression of microvascular changes. The development of areas of hypoxia stimulates the up-regulation of angiogenic factors such as vascular endothelial growth factor (VEGF) and angiopoietins, potentiating neovascularisation (Oh *et al.*, 1999, Park *et al.*, 2003, Yancopoulos *et al.*, 2000). Laser therapy decreases the oxygen demand of the tissue by localised destruction of the photoreceptors and the consequential development of glial scars facilitates the diffusion of oxygen through the retina. It is also thought to diminish the stimulation of angiogenesis by photocoagulation of normal and abnormal vessels and/or stimulate the expression of anti-angiogenic factors.

Many studies have been conducted to examine the effects of laser photocoagulation on specific factors in the rat and mouse. These studies clearly demonstrated that laser photocoagulation does not only destroy oxygen-demanding photoreceptor cells within the laser lesions, but does have a very important and significant effect on the expression of genes within the retina. These include bFGF/FGF2, epithelial growth factor (EGF), transforming growth factor alpha and beta (TGFα, TGFβ), insulin growth factor I (IGF-I), glial fibrillary acidic protein (GFAP), platelet derived growth factor (PDGF) and VEGF among others (Humphrey *et al.*, 1997; Xiao *et al.*, 1999). However, these factors are mainly associated with wound healing, an early response to the treatment.

With the advent of array-based gene expression studies, we can now examine the entire gene expression profile of any given tissue, identifying many novel associations and functions for both known and unknown genes. A previous study in our laboratory aimed to identify those genes that were affected by laser photocoagulation (Wilson *et al.*, 2003). This study demonstrated the effect on gene expression of a normal mouse eye three days post-argon laser photocoagulation. Angiogenic factors such as FGF14 and FGF16 were found to be down-regulated whilst angiotensin II type 2 receptor, a potent inhibitor of VEGF and VEGF-induced angiogenesis, was significantly up-regulated. As expected, proteins implicated in tissue remodelling and wound healing were also differentially expressed.

3.2.2. Late Changes

Few studies have followed the expression of factors past the initial wound healing response. Xiao *et al.* (1998) demonstrated a sustained increase (up to 180 days) in bFGF immuno-reactive photoreceptor cells adjacent to the laser lesions. GFAP-positive glial cells were also evident for more than 30 days post-treatment. Zhang *et al.* (1993) reported that RPE cells became aFGF and bFGF-positive while losing their CRALBP-immuno-reactivity for up to 80 days post laser treatment. These studies demonstrated some of the changes that occurred following laser photocoagulation and indicated that cells within these damaged areas can change the expression of factors controlling angiogenesis.

We sought to extend our earlier study to measure the changes in gene expression after laser photocoagulation long-term. The focus was on identifying genes that had a known functional relationship with angiogenesis. Ultimately, this study aimed to identify novel targets for diabetic retinopathy therapy whereby a directed change in expression would lead to a reduction in neovascularisation without the need to use lasers. Changes in gene expression at 90 days post-laser photocoagulation in the normal mouse were measured by micro-array analysis and further examined by real-time PCR (Binz *et al.*, 2005). At 90 days post laser photocoagulation, 107 genes were identified as differentially expressed compared to unlasered controls. Of these 107 genes, 34 had previously been identified as differentially expressed at three days post-treatment. Therefore, these genes demonstrated a true long-term change in expression due to laser photocoagulation.

Inducers of angiogenesis such as VEGF, PDGF, or bFGF were not differentially expressed, indicating that beneficial effects of laser photocoagulation do not stem from long-term changes to this pathway. However, this study identified genes previously associated with cytoskeletal and structural remodelling as differentially expressed at 90 days post-treatment and the level of protein measured corroborated this large increase. Interestingly, some of these genes were not differentially expressed at three days, suggesting there was a late stage of remodelling in the retina following initial wound healing and scar formation. Alternatively, these gene products could be functioning via a novel mechanism contributing to the beneficial effects of laser photocoagulation.

4. CONCLUSIONS

The goal of prevention and treatment of diabetic retinopathy requires the knowledge of factors and events that reduce or prevent neovascularisation. One approach to achieve this goal is to identify genes differentially expressed following successful laser photocoagulation. With the identification of genes initially affected by laser treatment and those whose expression remains changed long-term, we can now apply this knowledge to the diabetic retina. Ultimately, this will enable the development of therapeutic targets for long-term protection and prevention of vision impairment caused by chronic conditions such as diabetes.

5. ACKNOWLEDGEMENTS

We thank the Foundation for Fighting Blindness for C.E. Graham's Young Investigator Award to attend the RD2004 Conference. The authors gratefully acknowledge financial support from the Juvenile Diabetes Research Foundation International, the Australian National Health and Medical Research Council and Westpac Foundation. This work is part of the research effort of the Diabetic Retinopathy Consortium, Perth, Western Australia.

6. REFERENCES

Binz, N., Graham, C. E., Simpson, K., Lai, Y. K. Y., Shen, W.-Y., Lai, C.-M., Speed, T. P. and Rakoczy, P. E., 2005, Long-term effect of therapeutic laser photocoagulation on gene expression in the eye, *EMBO*:Submitted.

Humphrey, M. F., Chu, Y., Mann, K. and Rakoczy, P., 1997, Retinal GFAP and bFGF expression after multiple argon laser photocoagulation injuries assessed by both immunoreactivity and mRNA levels, *Exp Eye Res.* **64**:361-9.

Lewis, G. P., Erickson, P. A., Guerin, C. J., Anderson, D. H. and Fisher, S. K., 1992, Basic fibroblast growth factor: a potential regulator of proliferation and intermediate filament expression in the retina, *J Neurosci.* **12**:3968-78.

Mainster, M. A., 1999, Decreasing retinal photocoagulation damage: principles and techniques, *Semin Ophthalmol.* **14**:200-9.

Oh, H., Takagi, H., Suzuma, K., Otani, A., Matsumura, M. and Honda, Y., 1999, Hypoxia and vascular endothelial growth factor selectively up-regulate angiopoietin-2 in bovine microvascular endothelial cells, *J Biol Chem.* **274**:15732-9.

Park, Y. S., Kim, N. H. and Jo, I., 2003, Hypoxia and vascular endothelial growth factor acutely up-regulate angiopoietin-1 and Tie2 mRNA in bovine retinal pericytes, *Microvasc Res.* **65**:125-31.

Petrovic, V. and Bhisitkul, R. B., 1999, Lasers and diabetic retinopathy: the art of gentle destruction, *Diabetes Technol Ther.* **1**:177-87.

Porta, M. and Allione, A., 2004, Current approaches and perspectives in the medical treatment of diabetic retinopathy, *Pharmacology & Therapeutics.* **103**:167-77.

Roider, J., El Hifnawi, E. S. and Birngruber, R., 1998, Bubble formation as primary interaction mechanism in retinal laser exposure with 200-ns laser pulses, *Lasers Surg Med.* **22**:240-8.

Wilson, A. S., Hobbs, B. G., Shen, W.-Y., Speed, T. P., Schmidt, U., Begley, C. G. and Rakoczy, P. E., 2003, Argon Laser Photocoagulation-Induced Modification of Gene Expression in the Retina, *Invest Ophthalmol Vis Sci.* **44**:1426-34.

Xiao, M., McLeod, D., Cranley, J., Williams, G. and Boulton, M., 1999, Growth factor staining patterns in the pig retina following retinal laser photocoagulation, *Br J Ophthalmol.* **83**:728-36.

Xiao, M., Sastry, S. M., Li, Z. Y., Possin, D. E., Chang, J. H., Klock, I. B. and Milam, A. H., 1998, Effects of retinal laser photocoagulation on photoreceptor basic fibroblast growth factor and survival, *Invest Ophthalmol Vis Sci.* **39**:618-30.

Yancopoulos, G. D., Davis, S., Gale, N. W., Rudge, J. S., Wiegand, S. J. and Holash, J., 2000, Vascular-specific growth factors and blood vessel formation, *Nature.* **407**:242-8.

Zhang, N. L., Samadani, E. E. and Frank, R. N., 1993, Mitogenesis and retinal pigment epithelial cell antigen expression in the rat after krypton laser photocoagulation, *Invest Ophthalmol Vis Sci.* **34**:2412-24.

CHAPTER 30

APPLYING TRANSGENIC ZEBRAFISH TECHNOLOGY TO STUDY THE RETINA

Ross F. Collery, Maria L. Cederlund, Vincent A. Smyth, and Breandán N. Kennedy*

1. THE ZEBRAFISH RETINA

During the past two decades, zebrafish (*Danio rerio*) have become established as a prolific model for biological research (for review, see Udvadia and Linney, 2003). Zebrafish exhibit many features of an ideal model organism, including small size, rapid development, and high fecundity. Since zebrafish are vertebrates, there is a conservation of physiology with higher vertebrates, including humans. Juvenile zebrafish develop rapidly and oviparously, facilitating developmental studies without invasive procedures. Zebrafish are amenable to drug discovery, having similar responses to mammals in pharmacological tests using cardiovascular, anti-angiogenic and anti-cancer drugs (for review, see Langheinrich *et al.*, 2002). To date, zebrafish have been used in genetic screens to identify genes involved in development and function of organs including the eye, brain, ear and heart (Haffter *et al.*, 1996). More recently, transgenic technologies have been developed for zebrafish. In this review, we will describe how this transgenic technology can be applied to study retinal biology.

The zebrafish retina is similar morphologically and physiologically to the human retina (Fig. 30.1a). Discrete layers stratify the retinal cell types (Schmitt and Dowling, 1994). Cell types can be distinguished by location, morphology, and detection of unique cellular markers (Fig. 30.1b, c). In contrast to the commonest models used for retinal research (rat and mouse), zebrafish, being diurnal, have abundant cone photoreceptors. These include long single cones sensitive to blue light, double cones with red-sensitive and green-sensitive types, and short single cones sensitive to ultraviolet light. Though zebrafish do not have a macula, they have a comparable number and density of cones to humans. Thus, zebrafish are particularly suited to studies of cone photoreceptor development and function. An immature eye anlage is distinguishable in the externally fertilised zebrafish embryo at 24 hours

---

*Conway Institute & Department of Pharmacology, University College Dublin, Dublin 4, Ireland.

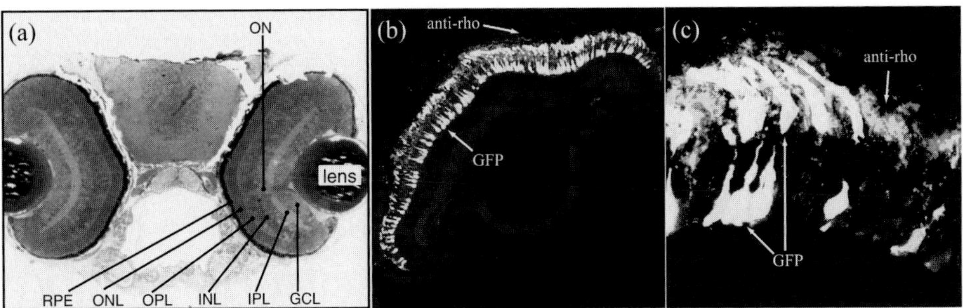

Figure 30.1. (a) Retinal histology of transverse cryosections of 5 dpf zebrafish stained with methylene blue and Azure II. ON, optic nerve; RPE, retinal pigment epithelium; ONL, outer nuclear layer; OPL, outer plexiform layer; INL, inner nuclear layer; IPL, inner plexiform layer; GCL, ganglion cell layer. (b), (c) Merged fluorescent imaging of GFP in GFP-transgenic zebrafish retinal cryosection, and immunohistochemistry with anti-rhodopsin antibody visualised with Cy3-conjugated secondary antibody.

post-fertilisation (hpf), the retina is differentiated at 72 hpf and there is a measurable visual response at 120 hpf. This rapid development makes zebrafish suitable for the study of all stages of retinal development. *Ex utero* development and larval transparency means that gene expression patterns may be observed in wholemount zebrafish larvae. Established behavioural tests exist for fast and accurate assay of visual transduction in zebrafish (Brockerhoff *et al.*, 1995; Neuhauss *et al.*, 1999). Electroretinography (ERG) though a more invasive and more time-consuming assay enables detailed characterisation of visual function (Brockerhoff *et al.*, 1995).

Zebrafish are a cost-effective model organism. Adult zebrafish, being ~3 cm in length, can be maintained at high densities in multiple tanks, which allows for the establishment of a large experimental population. This population can produce large numbers of offspring, since a given mating pair can routinely produce batches of 150 eggs every 3-4 days. Zebrafish are amenable to transgenesis, whereby foreign DNA sequences injected into developing embryos are incorporated into the genome. The first transgenic zebrafish were generated in 1988, demonstrating the inheritance of an antibiotic cassette (Stuart *et al.*, 1988). Transgenic zebrafish lines are now routinely used to study diverse aspects of zebrafish development and physiology.

2. GENERATION OF TRANSGENIC ZEBRAFISH

Transgenic zebrafish are generated by injecting transgenes into newly fertilised zebrafish eggs. These embryos are raised to adulthood (~3 months) and founder zebrafish that transmit the transgene through the germline are identified by PCR and/or reporter gene expression.

Zebrafish routinely mate at daybreak, which is artificially simulated in laboratory conditions by the onset of a light-dark cycle. The evening before embryo injection, male and female adults are placed in mating tanks. These tanks have a gridded insert to aid collection of the eggs, and an optional barrier insert to separate male and female zebrafish. This

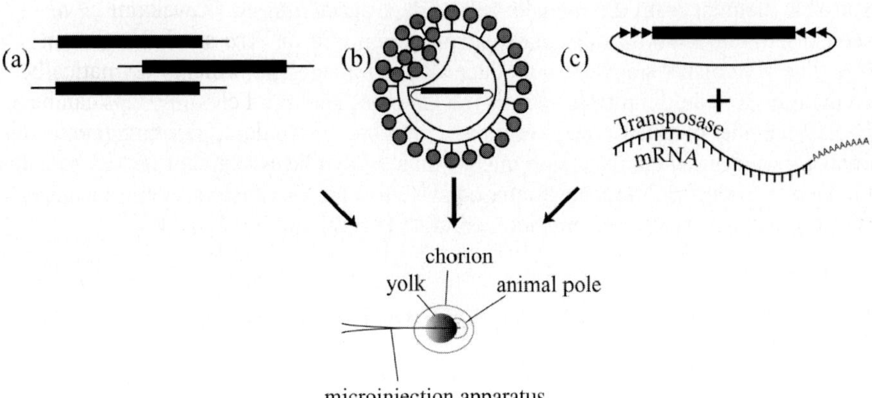

Figure 30.2. Schematic showing the three modes of delivery of a transgene for germline integration. (a) linearised DNA constructs are injected alone. (b) DNA constructs are packaged into retrovirus for infection. (c) DNA constructs are flanked by transposon inverted repeats and co-injected with in vitro transcribed transposase mRNA. The schematic shows the major components of the microinjection apparatus. Transgenes are represented by black bars, and inverted repeats by black triangles.

insert enables the investigator to delay matings so that newly fertilised eggs can be collected at several intervals. DNA is delivered for transgenesis using fine glass needles that can pass easily through an egg's chorion and yolk into the animal pole. These needles are filled with a DNA and a tracer dye, usually phenol red. To optimise transgene integration, eggs are injected at the 1- or 2-cell stage. The eggs are immobilised in an agarose trough, a micro-manipulator is used to guide the needle into the animal pole, and a microinjector used to inject the DNA solution. Between 50 and 70% of injected eggs routinely survive the procedure.

DNA injected for transgenesis can integrate into the genome by random insertion, by transposase-mediated insertion, or by retroviral insertion (Fig. 30.2). During random integration, injected DNA assembles into concatemers consisting of many head-to-tail copies of the construct. The zebrafish's DNA repair machinery targets the free double-stranded ends and incorporates the concatemers into the genome, usually, at a single, random point. The transgene can integrate into native enhancer or repressor regions, which can distort the levels and patterns of transcription driven by the promoter. The major disadvantage to this method is the low efficiency of transgene integration. It is advantageous, however, since it can be performed without labour-intensive cloning, as even PCR amplicons can be injected. Retroviral systems can also generate transgenic zebrafish (Gaiano *et al.*, 1996). The transgenic DNA is packaged by a viral packaging line, and delivered using a separate viral vehicle. Integration is brought about by infection of the zebrafish cells with the transgenic DNA. The disadvantages to this method are that the generation of high-titre viral stocks can be technically difficult and requires specialised safety procedures. The advantages are that the retrovirus inserts the transgene into the genome with high efficiency.

Most recently, technologies have been developed to generate transgenic zebrafish by transposition. In this procedure the transgene is flanked with inverted repeats from the *Tol2* transposable element from the medaka zebrafish, *Oryzias latipes* (Kawakami *et al.*, 1998). The construct is injected as a circular plasmid along with *in vitro*-synthesised transposase mRNA. The zebrafish's ribosomes produce active transposase, which enzymatically integrates the transgene into multiple sites of the genome. The initial cloning steps can be problematic, since the inverted-repeat regions are prone to random rearrangements during bacterial passage, and the transposase mRNA must be synthesised and protected from degradation. However, the *Tol2* transposase based system is highly efficient, with transgenes integrating in greater than 50% of injected zebrafish (Kawakami *et al.*, 2000).

3. TRANSGENIC ZEBRAFISH TECHNOLOGY APPLIED TO THE RETINA

There are several transgenic zebrafish lines and tissue/cell-specific promoters available to study the retina. These lines/promoters can be used to study temporal and spatial gene regulation, to characterise gene function by overexpression and to report on retinal cell/tissue integrity in genetic and pharmacological screens.

Transgenic zebrafish have been successfully applied to characterise the temporal and spatial regulation of genes *in vivo*. The patterning and timing of the expression of a gene can be characterised by fusing reporters (*e.g.* EGFP) to putative promoter regions and generating transgenic zebrafish transmitting this construct. Expression of fluorescent transgenes can be analysed in anaesthetised zebrafish by fluorescence microscopy. Phenyl-thio-urea (0.003%) added to embryo media prevents pigment formation, facilitating the analysis of fluorophores in transparent larvae up to 8 days post-fertilisation. Apart from generating stable transgenic lines, reporter constructs can also be used to test promoter activity using transient expression assays. By analysing approximately one hundred zebrafish injected with constructs containing promoter deletions/mutations, it is possible to quickly locate *cis*-acting enhancer or repressor elements that confer tissue- or temporal-specific expression (Luo *et al.*, 2004).

The line *Tg(1.2ZOP-EGFP)* expresses enhanced GFP (EGFP) under the control of 1.2 kb of promoter sequence from the zebrafish rod opsin gene (Kennedy *et al.*, 2001). EGFP is expressed exclusively in rod photoreceptors, and exhibits a similar developmental expression profile to the rod opsin protein. Similarly, the line *Tg(ZUV-GFP)* expresses GFP under the control of a UV opsin promoter, which directs expression solely to UV- cones (Luo *et al.*, 2004). This line was used to analyse proximal and distal elements of the UV opsin promoter, and showed that a UV opsin promoter element could direct ectopic expression of a chimeric rhodopsin promoter to UV cones. Promoters from other species can also be used to study the zebrafish retina, such as a *Xenopus laevis* rod opsin promoter (Perkins *et al.*, 2002).

Another application of transgenic zebrafish technology is enhancer trapping. This process takes advantage of the random nature of transgene integration. Constructs consisting of GFP coupled to a minimal promoter are injected into zebrafish embryos. Integration of the transgene near strong *cis*-acting elements can result in zebrafish expressing EGFP in spatial/temporal expression dictated by the *cis*-element and easily visualised in the transparent embryos by fluorescent microscopy.

Expression of GFP in nascent cells can be used to plot the migration of cells during development of primordial structures. Constructs highlighting the migration of presumptive cells within the eye can give insights into signalling pathways. A wave of differentiation generating ganglion and amacrine cells in the retina during neurogenesis has been imaged using GFP under the regulation of a *sonic hedgehog* promoter (Shkumatava *et al.*, 2004).

Promoters that direct expression of transgenes in the retina can also be used to characterise gene function. Expression of foreign genes in zebrafish allows novel genes to be characterised by analyzing phenotypes associated with overexpression of that gene. Transgenic zebrafish technologies can also "rescue" mutant/disease phenotypes. The *lakritz* mutant, which has a mutation in *ath5,* an eye-specific transcription factor, causes elimination of the ganglion cell layer (Kay *et al.*, 2001). Injection of a plasmid carrying the unmutated gene under control of its native promoter rescues the wild type phenotype, confirming that the mutant phenotype is caused by the mutation. Similarly, a nonsense mutation in the gene for the *rx3* transcription factor causes an eyeless phenotype, which can be rescued by the injection of constructs carrying the wild type gene (Kennedy *et al.*, 2004).

The low efficiency of transgenesis by random integration previously impeded the generation of transgenic lines. To overcome this, existing lines often took advantage of dual expression of both an effector transgene and a reporter gene for easy screening. Transgenic lines exist that express both EGFP and protein kinase A (PKA), or EGFP and glycogen synthase kinase-3β (GSK-3β), under the control of a retinal ganglion cell (RGC) promoter (Yoshida and Mishina, 2003). These lines allow characterisation of the promoter, clarification of the roles of PKA and GSK-3β, and efficient screening owing to the concomitant expression of the reporter gene. Likewise, transgenic expression of a Gap43-GFP fusion protein allows visualisation of dynamic behaviour of GFP-labeled amacrine cell neurites *in vivo* from the earliest stages of neurite outgrowth (Kay *et al.*, 2004). The recent emergence of an efficient transposase-mediated system for transgenesis eliminates the need for a reporter to be co-injected with the effector transgene.

4. LIMITATIONS AND FUTURE DEVELOPMENTS

Transgenic zebrafish technology has advanced such that transgenic zebrafish can now be routinely generated *in-house*. Transgenic zebrafish will provide powerful tools for functional genomics, for modelling human diseases and for drug discovery. However, a number of limitations exist, particularly the current inability to apply homologous recombination to generate targeted knockouts/knockins in zebrafish.

Standard approaches to understand gene function include analysing "*gain-of function*" and "*loss-of function*" phenotypes. In zebrafish the most common method to overexpress a gene (*gain-of function*) is by microinjection of embryos with *in vitro*-synthesised mRNA. However, as the mRNA is translated throughout the developing larvae, this assay may be misinformative. Also, it is difficult to limit the temporal onset of expression of injected mRNAs, which are typically degraded by 5 dpf. Thus, it is desirable to express a transgene under regulation of a promoter that drives specific temporal and spatial expression patterns. Though a few promoters have been identified that direct specific patterns of expression in the zebrafish retina, there is a need to identify specific promoters for other cell types so that these can be comprehensively studied. In addition, there is a need for inducible expression of transgenes in the retina. A heat-shock promoter can induce selective expression when coupled with highly localised laser treatment (Halloran *et al.*, 2000). In the future transgenic lines generated from retinal-specific promoters coupled with the Tet ON/OFF system will provide greater flexibility for inducible expression of transgenes in the zebrafish retina.

Analysis of phenotypes arising from the specific reduction of gene expression (*loss-of-function*) is a key method to understanding gene function. In zebrafish, loss-of-function phenotypes have been generated by morpholino "knockdown" or by mutagenesis screening (Warren and Fishman, 1998; Nasevicius and Ekker, 2000). However, neither of these approaches are designed to target selective and complete eliminate of the expression of a specific gene. The gold standard of "*loss-of-function*" analyses, *i.e.* targeted knockout by homologous recombination, is currently not feasible in zebrafish. Though cells displaying embryonic stem cell properties and of producing zebrafish germ-line chimeras from embryo cell cultures have been reported the long-term culture of zebrafish embryonic stem cells required for selection of homologous recombinants has been elusive though concerted efforts to overcome this hurdle continue ((Ma *et al.*, 2001; Fan *et al.*, 2004). Once this is achieved other variations including the Cre/*lox* binary system for conditional knockouts will become feasible. As a result, targeted knockout lines of zebrafish cannot be generated, currently. The ability to make directed knockout lines will facilitate reverse genetic studies accelerating the characterisation of genes whose function is either partially or wholly unknown.

Currently available transgenic technologies in zebrafish now enable the generation of transgenic models of human retinopathies, especially to study the approximately 50 genes associated with dominant forms of inherited human blindness (http://www.sph.uth.tmc.edu/Retnet/). Currently, promoters exist for expressing transgenes in rod photoreceptors, cone photoreceptors, or in ganglion cells, and many more promoters for other retinal cells will become available in the future. Such zebrafish models will aid in understanding the molecular pathogenesis of disease. In addition, it is envisaged that zebrafish models of human retinopathies can be used in pharmacological screens to identify drugs that modulate the disease. The features of small size, rapid development, larval transparency and large offspring numbers that are inherent to zebrafish are also beneficial to pharmacological screens.

5. ACKNOWLEDGEMENTS

The authors wish to thank Susan Brockerhoff, Tom Vihtelic, and David Hyde for providing photographic figures.

6. REFERENCES

Brockerhoff, S. E., J. B. Hurley, U. Janssen-Bienhold, S. C. Neuhauss, W. Driever and J. E. Dowling, 1995. A behavioral screen for isolating zebrafish mutants with visual system defects. *Proc Natl Acad Sci U S A* **92**(23):10545-9.

Fan, L., A. Alestrom, P. Alestrom and P. Collodi, 2004. Development of cell cultures with competency for contributing to the zebrafish germ line. *Crit Rev Eukaryot Gene Expr* **14**(1-2):43-51.

Gaiano, N., M. Allende, A. Amsterdam, K. Kawakami and N. Hopkins, 1996. Highly efficient germ-line transmission of proviral insertions in zebrafish. *Proc Natl Acad Sci U S A* **93**(15):7777-82.

Haffter, P., M. Granato, M. Brand, M. C. Mullins, M. Hammerschmidt, D. A. Kane, J. Odenthal, F. J. van Eeden, Y. J. Jiang, C. P. Heisenberg, R. N. Kelsh, M. Furutani-Seiki, E. Vogelsang, D. Beuchle, U. Schach, C. Fabian and C. Nusslein-Volhard, 1996. The identification of genes with unique and essential functions in the development of the zebrafish, *Danio rerio*. *Development* **123**:1-36.

Halloran, M. C., M. Sato-Maeda, J. T. Warren, F. Su, Z. Lele, P. H. Krone, J. Y. Kuwada and W. Shoji, 2000. Laser-induced gene expression in specific cells of transgenic zebrafish. *Development* **127**(9):1953-60.

Kawakami, K., A. Koga, H. Hori and A. Shima, 1998. Excision of the tol2 transposable element of the medaka fish, *Oryzias latipes*, in zebrafish, *Danio rerio. Gene* **225**(1-2):17-22.

Kawakami, K., A. Shima and N. Kawakami, 2000. Identification of a functional transposase of the Tol2 element, an Ac-like element from the Japanese medaka fish, and its transposition in the zebrafish germ lineage. *Proc Natl Acad Sci U S A* **97**(21):11403-8.

Kay, J. N., K. C. Finger-Baier, T. Roeser, W. Staub and H. Baier, 2001. Retinal ganglion cell genesis requires lakritz, a Zebrafish atonal Homolog. *Neuron* **30**(3):725-36.

Kay, J. N., T. Roeser, J. S. Mumm, L. Godinho, A. Mrejeru, R. O. Wong and H. Baier, 2004. Transient requirement for ganglion cells during assembly of retinal synaptic layers. *Development* **131**(6):1331-42 Epub 2004 Feb 18.

Kennedy, B. N., G. W. Stearns, V. A. Smyth, V. Ramamurthy, F. van Eeden, I. Ankoudinova, D. Raible, J. B. Hurley and S. E. Brockerhoff, 2004. Zebrafish *rx3* and *mab21l2* are required during eye morphogenesis. *Dev Biol* **270**(2):336-49.

Kennedy, B. N., T. S. Vihtelic, L. Checkley, K. T. Vaughan and D. R. Hyde, 2001. Isolation of a zebrafish rod opsin promoter to generate a transgenic zebrafish line expressing enhanced green fluorescent protein in rod photoreceptors. *J Biol Chem* **276**(17):14037-43 Epub 2001 Jan 18.

Langheinrich, U., E. Hennen, G. Stott and G. Vacun, 2002. Zebrafish as a model organism for the identification and characterization of drugs and genes affecting p53 signaling. *Curr Biol* **12**(23):2023-8.

Luo, W., J. Williams, P. M. Smallwood, J. W. Touchman, L. M. Roman and J. Nathans, 2004. Proximal and distal sequences control UV cone pigment gene expression in transgenic zebrafish. *J Biol Chem* **279**(18):9286-93 Epub 2004 Feb 13.

Ma, C., L. Fan, R. Ganassin, N. Bols and P. Collodi, 2001. Production of zebrafish germ-line chimeras from embryo cell cultures. *Proc Natl Acad Sci U S A* **98**(5):2461-6.

Nasevicius, A. and S. C. Ekker, 2000. Effective targeted gene 'knockdown' in zebrafish. *Nat Genet* **26**(2):216-20.

Neuhauss, S. C., O. Biehlmaier, M. W. Seeliger, T. Das, K. Kohler, W. A. Harris and H. Baier, 1999. Genetic disorders of vision revealed by a behavioral screen of 400 essential loci in zebrafish. *J Neurosci* **19**(19):603-15.

Perkins, B. D., P. M. Kainz, D. M. O'Malley and J. E. Dowling, 2002. Transgenic expression of a GFP-rhodopsin COOH-terminal fusion protein in zebrafish rod photoreceptors. *Vis Neurosci* **19**(4):257R-64R.

Schmitt, E. A. and J. E. Dowling, 1994. Early eye morphogenesis in the zebrafish, *Brachydanio rerio. J Comp Neurol* **344**(4):532-42.

Shkumatava, A., S. Fischer, F. Muller, U. Strahle and C. J. Neumann, 2004. *Sonic hedgehog*, secreted by amacrine cells, acts as a short-range signal to direct differentiation and lamination in the zebrafish retina. *Development* **131**(16):3849-58.

Stuart, G. W., J. V. McMurray and M. Westerfield, 1988. Replication, integration and stable germ-line transmission of foreign sequences injected into early zebrafish embryos. *Development* **103**(2):403-12.

Udvadia, A. J. and E. Linney, 2003. Windows into development: historic, current, and future perspectives on transgenic zebrafish. *Dev Biol* **256**(1):1-17.

Warren, K. S. and M. C. Fishman, 1998. "Physiological genomics": mutant screens in zebrafish. *Am J Physiol* **275**(1 Pt 2):H1-7.

Yoshida, T. and M. Mishina, 2003. Neuron-specific gene manipulations to transparent zebrafish embryos. *Methods Cell Sci* **25**(1-2):15-23.

CHAPTER 31

BMI1 LOSS DELAYS PHOTORECEPTOR DEGENERATION IN *RD1* MICE

Bmi1 loss and neuroprotection in *Rd1* mice

Dusan Zencak[1], Sylvain V. Crippa[1], Meriem Tekaya[1], Ellen Tanger[2], Daniel F. Schorderet[1,3], Francis L. Munier[1], Maarten van Lohuizen[2], and Yvan Arsenijevic[1]

1. SUMMARY

Retinitis pigmentosa (RP) is a heterogeneous group of genetic disorders leading to blindness, which remain untreatable at present. *Rd1* mice represent a recognized model of RP, and so far only GDNF treatment provided a slight delay in the retinal degeneration in these mice. Bmi1, a transcriptional repressor, has recently been shown to be essential for neural stem cell (NSC) renewal in the brain, with an increased appearance of glial cells *in vivo* in *Bmi1* knockout (*Bmi1$^{-/-}$*) mice. One of the roles of glial cells is to sustain neuronal function and survival. In the view of a role of the retinal Müller glia as a source of neural protection in the retina, the increased astrocytic population in the *Bmi1$^{-/-}$* brain led us to investigate the effect of *Bmi1* loss in *Rd1* mice. We observed an increase of Müller glial cells in *Rd1-Bmi1$^{-/-}$* retinas compared to *Rd1*. Moreover, *Rd1-Bmi1$^{-/-}$* mice showed 7-8 rows of photoreceptors at 30 days of age (P30), while in *Rd1* littermates there was a complete disruption of the outer nuclear layer (ONL). Preliminary ERG results showed a responsiveness of *Rd1-Bmi1$^{-/-}$* mice in scotopic vision at P35. In conclusion, *Bmi1* loss prevented, or rescued, photoreceptors from degeneration to an unanticipated extent in *Rd1* mice.

In this chapter, we will first provide a brief review of our work on the cortical NSCs and introduce the Bmi1 oncogene, thus offering a rational to our observations on the retina.

---

[1] Unit of Oculogenetics, Jules Gonin Eye Hospital, Lausanne, Switzerland. [2] Division of Molecular Genetics, The Netherlands Cancer Institute, The Netherlands. [3] Institute for Research in Ophthalmology, Sion, Switzerland. Corresponding author: Y. Arsenijevic, E-mail: yvan.arsenijevic@ophtal.vd.ch, Fax: +41 21 6268888.

2. BMI1 REQUIREMENT FOR NSCs

Neural stem cells (NSCs) have a primary role in brain organogenesis during development, and several studies suggest that they are responsible for the generation of certain neuronal sub-populations in adulthood (reviewed in Kuhn and Svendsen 1999). Similarly, retinal stem cells (RSCs) give rise to the different cell types composing the retina in a temporally coordinate manner (reviewed in Marquardt and Gruss 2002). Throughout development and adulthood, the stem cell pool is maintained by asymmetric divisions in order to generate one stem cell and one committed cell, while symmetric divisions giving rise to two daughter stem cells are required for the initial expansion of the stem cell population. Several pathways interact and cross-talk to control NSC renewal and commitment to specific cell fates (reviewed in Arsenijevic 2003). Likewise, RSCs require a specific regulation for a proper generation of the different retinal cell types (Hatakeyama and Kageyama 2004). For instance, transcription factors like Pax6 and Hes1 play key roles in the control of NSC renewal and commitment to neuronal versus glial differentiation, both being necessary for proper NSC renewal and to direct neural stem cells to a neuronal (Pax6, Heins et al. 2002) or glial fate (Hes1, Nakamura et al. 2000; Wu et al. 2003). The same two factors are strongly implicated in eye development and in the control of cell differentiation in the retina. Pax6 is necessary for the formation of multipotent retinal progenitors (Marquardt and Gruss 2002) and for the specification of the different layers and neuronal subtypes of the retina (Hatakeyama and Kageyama 2004). On the other hand, Hes1 is important in the development of Müller glia (Hatakeyama and Kageyama 2004). Several factors play therefore parallel roles in the brain and in the retina.

Since *Bmi1*-deficient mice present overall growth retardation and a smaller brain after two weeks of age, a profound defect in cerebellum growth and progressive neurological defects (van der Lugt et al. 1994), we focalized our attention on Bmi1 as a possible key factor in NSC renewal (Molofsky et al. 2003, Zencak et al., submitted). Bmi1 belongs to the Polycomb group of transcription factors, and controls the cell cycle by promoting entry into the S-phase through *p16ink4a* inhibition, (Jacobs et al. 1999, Figure 31.1). On the other hand, Bmi1 leads to a decrease of p53 through *p19arf* repression, leading to a decrease in

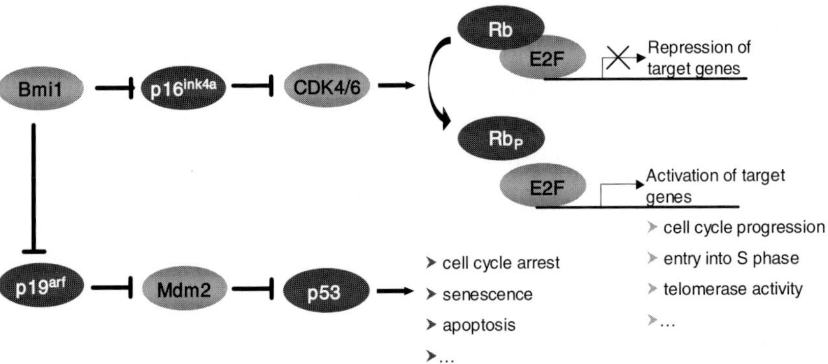

Figure 31.1. Schematic representation of the pathways controlled by Bmi1, as described above. Light gray represents promotion of the cell cycle while dark grey represents its inhibition. Rb_p = phosphorylated Rb.

senescence and apoptosis (Jacobs et al. 1999). It was demonstrated that *Bmi1* loss leads to an impairment in neural stem cell (NSC) proliferation and renewal (Molofsky et al. 2003). Another recent study showed that Bmi1 is crucial in the expansion of cerebellar granular cells both *in vivo* and in *vitro* (Leung et al. 2004). However, the effect of *Bmi1* loss in other brain regions remains to be addressed, as well as the question whether the action of Bmi1 is intrinsic to the NSCs or due to the stem cell niche.

Our recent study (Zencak et al., submitted) provided evidence that Bmi1 is expressed in neural progenitor cells (NPCs), and that the distribution of the Bmi1-positive cells *in vivo* is similar to what would be expected for NPCs and NSCs. More precisely, Bmi1+ cells were observed mostly in the sub-ventricular zone (SVZ), and more dispersed in the corpus callosum and in the cortex at birth. Because of the action of Bmi1 on the *ink4a/arf* locus, *Bmi1* loss was expected to result either in a decrease in proliferation or in an increase in apoptosis. BrdU incorporation analysis was performed on newborn brains after a short pulse injection of BrdU 30 minutes prior to sacrifice, and on adult brains (at P30) after a two-day treatment with BrdU. Interestingly, Bmi1 loss led to a decrease in proliferation in the newborn cortex, as well as in the newborn and P30 dorso-lateral corner of the SVZ, while no significant change was observed in the number of apoptotic cells (evaluated by TUNEL analysis). This observation was strongly accentuated *in vitro* with a 10-fold reduction of NSC colony formation from primary cortical cultures in conditions allowing NSC proliferation and renewal. When we tested self-renewal by replating and dissociating individual primary colonies, we observed an almost complete failure of *Bmi1*$^{-/-}$ NSCs to self-renew. In addition, by using an RNAi approach, we showed that the effect of Bmi1 loss was intrinsic to NSCs and not due to surrounding cells, such as the stem cell niche. In summary, Bmi1 is intrinsically required for NSC proliferation and self-renewal.

3. INCREASED PRESENCE OF GLIAL CELLS IN THE *BMI1*$^{-/-}$ BRAIN

The reduced proliferation *in vitro* and *in vivo* could result in an altered cell pattern *in vivo*. Surprisingly, no significant difference was observed in the distribution of neurons and oligodendrocytes. However, an increased number of astrocytes was observed at birth in the marginal zone of the cortex during the early stages of astrocyte appearance, while a massive gliosis was detected in the young adult *Bmi1*$^{-/-}$ brain. Interestingly, the increased astrocytic population appeared to proliferate normally *in vivo*, as assessed by BrdU incorporation. Taken together, our recent study showed that Bmi1 is intrinsically required for neural stem cell renewal, leading to a reduced proliferation *in vivo* and to an increased astroglial population that retains the ability to proliferate.

4. BMI1 LOSS AND NEUROPROTECTION IN *RD1* MICE

4.1. Introduction

The increased glial population in the brain led us to investigate the effect of Bmi1 loss on Müller glia in the retina. Several studies suggest a role of the Müller cells in retinal neuroprotection (Ooto et al. 2004; Garcia and Vecino 2003). In this view, we hypothesized an increased presence or activity of Müller cells in the *Bmi1*$^{-/-}$ retina and analyzed the effect of Bmi1 deletion in *rd1* mice, on an inbred FVB genetic background. *Rd1* mice are currently used as a model of human RP, a group of inherited retinal dystrophies untreatable at present. *Rd1* mice are characterized by an early severe degeneration of the outer nuclear layer (ONL) due to a mutation of the Phosphodiesterase-6-beta (Pde6b) gene. In these mice, a complete loss of rod photoreceptors is observed at 3 weeks of age, followed by a death of cone photoreceptors. To date, ciliary neurotrophic factor (CNTF) treatment slowed moderately the photoreceptor cell death in *Rd1* mice as revealed by histological analyses (LaVail et al. 1998), while glial cell line-derived neurotrophic factor (GDNF) injections could in some cases lead to recordable ERGs with a slight delay of the degeneration (Frasson et al. 1999). In the present study we tested whether an increased glial activity was present in the $Bmi1^{-/-}$ mice and if such increase could protect a degenerating retina.

4.2. Materials and Methods

4.2.1. Animals

Bmi1-knockout mice on a FVB (*Rd1*) inbred genetic background were generated and handled as previously described (van der Lugt et al. 1994; Jacobs et al. 1999). All mice carried the Pde-6b mutation characteristic of *Rd1* mice (FVB background), and were wild-type ($Bmi1^{+/+}$), heterozygous ($Bmi1^{+/-}$) or homozygous ($Bmi1^{-/-}$) for the *Bmi1*-knockout allele. In the text, $Bmi1^{+/+}$ mice are mentioned as *Rd1* mice.

4.2.2. Immunohistochemistry and Antibodies

Immunostainings were performed on eye cryostat sections from newborn (P0) or perfusion-fixed adult (P30) mice. Primary antibodies included rabbit antiserum, rabbit polyclonal anti-CRALBP (1/1000, gift of J.Saari), mouse monoclonal anti-Rho4D2 (1/40, gift of D.Hicks), rabbit polyclonal anti-Recoverin (1/500, Chemicon). They were revealed by fluorescence using the appropriate FITC- or Cy3-conjugated secondary antibody. Primary antibodies were incubated overnight at 4°C.

4.2.3. Electroretinogram Recording (ERGs)

$Rd1$-$Bmi1^{-/-}$ and control ($Rd1$-$Bmi1^{+/+ \, or \, +/-}$) mice were tested in scotopic and photopic conditions at different ages ranging from P16 to P35. They were dark-adapted overnight before recording scotopic responses and then light-adapted for 5 min. before recording the photopic responses.

4.3. Results

To analyze the presence of Müller cells, we analyzed CRALBP expression by immunohistochemistry in the *Rd1* and *Rd1-Bmi1*$^{-/-}$ retina in adult mice (P30). Confirming our hypothesis, *Rd1- Bmi1*$^{-/-}$ retinas displayed a more intense CRALBP immunoreactivity compared to *Rd1* (Figure 31.2A-B) this should be translated in a quantitative way. The DAPI dye used as counterstaining evidenced another important difference: the *Rd1-Bmi1*$^{-/-}$ retinas presented typically 7 to 8 rows of photoreceptors with the characteristic condensed chromatin (Figure 31.2C-D), while *Rd1* mice had a completely dystrophic outer nuclear layer

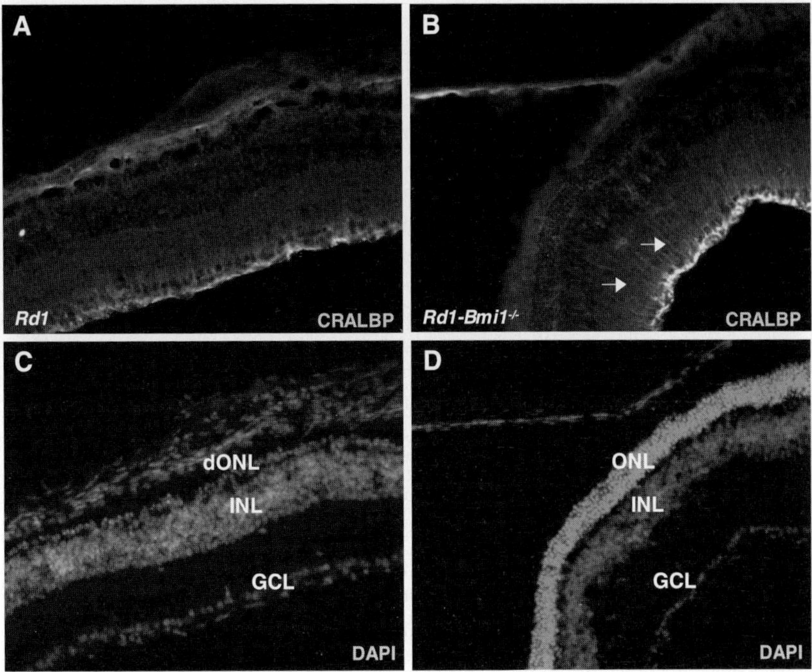

Figure 31.2. Histology of the *Rd1* and *Rd1-Bmi1*[-/-] adult retina. **A,B,** CRALBP immunohistochemistry on P30 *Rd1* and *Rd1-Bmi1*[-/-] respectively. Note the stronger CRALBP immunoreactivity in *Rd1-Bmi1*[-/-] retina compared to *Rd1*, with the characteristic pattern of Müller glia (arrows). **C,D,** DAPI counterstaining corresponding respectively to **A** and **B**. Note the presence of photoreceptors with the typically condensed chromatin in the *Rd1-Bmi1*[-/-] ONL, while in (**D**). ONL = outer nuclear layer, dONL = dystrophic outer nuclear layer, INL = inner nuclear layer, GCL = ganglion cell layer. Magnification: 200×.

(ONL). Immunohistochemical characterization of the cells present in the *Rd1-Bmi1*[-/-] ONL revealed rod photoreceptor features (data not shown). Moreover, the length and the shape of the outer segments in the *Rd1-Bmi1*[-/-] retina were similar to functional photoreceptors.

To test the visual function of *Rd1* and *Rd1-Bmi1*[-/-] mice, we recorded ERG responses in scotopic and photopic conditions at several stages from P16 to P35. As expected, *Rd1* mice displayed a slight response at P16, which disappeared with the progression of the retinal degeneration. On the contrary, ERG recordings in *Rd1-Bmi1*[-/-] mice showed a clear response at all tested ages to both single flash and flicker stimuli, improving from P16 to P35 (data not shown). Nevertheless, the shape of the ERG response was unusual, probably due to Pde6b loss of function. No significant response was observed in photopic vision in neither group.

Taken together, these results show that Bmi1 loss provided a consistent delay in the rod photoreceptor degeneration in the *Rd1* mice to an extent unattained to date, with the presence of functional rod photoreceptors at 35 days of age. The reduced viability of the mice prevented us from investigating the effect of Bmi1 in *Rd1* mice in the long term.

4.4. Discussion and Perspectives

Our work provides evidence that Bmi1 loss strongly delays rod photoreceptor degeneration in *Rd1* mice, a model of human RP. The reported observation of an ONL with functional photoreceptors at 30 and 35 days of age has never been described in *Rd1* mice so far, and may open new perspectives for the treatment of retinal degeneration once the mechanism underlying the delayed degeneration is identified.

The stronger intensity of the CRALBP staining in the absence of Bmi1 may reflect an increased activity of Müller glia, which remains to be quantified. Considering the role of Müller cells in neuroprotection, this could explain the rescue of photoreceptors. On the other hand, we can hypothesize an involvement of the Retinoblastoma (Rb) protein, a critical indirect downstream target of Bmi1 and recently described to play a critical role in rod photoreceptor development (Zhang et al. 2004). Both hypotheses require further investigation currently in progress.

5. ACKNOWLEDGEMENTS

We would like to thank Dana Hornfeld and Muriel Jaquet for editing help. This work was supported by the Swiss National Science Foundation, the ProVisu Foundation, the Velux Foundation, and the French Association Against Myopathies.

6. REFERENCES

Arsenijevic, Y., 2003, Mammalian neural stem-cell renewal: Nature versus nurture, *Mol Neurobiol* **27**:73.

Frasson, M., Picaud, S., Leveillard, T., Simonutti, M., Mohand-Said, S., Dreyfus, H., Hicks, D., and Sabel, J., 1999, Glial cell line-derived neurotrophic factor induces histologic and functional protection of rod photoreceptors in the rd/rd mouse, *Invest Ophthalmol Vis Sci* **40**:2724.

Garcia, M., and Vecino, E., 2003, Role of muller glia in neuroprotection and regeneration in the retina, *Histol Histopathol* **18**:1205.

Hatakeyama, J., and Kageyama, R., 2004, Retinal cell fate determination and bhlh factors, *Semin Cell Dev Biol* **15**:83.

Heins, N., Malatesta, P., Cecconi, F., Nakafuku, M., Tucker, K. L., Hack, M. A., Chapouton, P., Barde, Y. A., and Gotz, M., 2002, Glial cells generate neurons: The role of the transcription factor pax6., *Nat Neurosci* **5**:308.

Jacobs, J. J., Kieboom, K., Marino, S., DePinho, R. A., and van Lohuizen, M., 1999, The oncogene and polycomb-group gene bmi-1 regulates cell proliferation and senescence through the ink4a locus, *Nature* **397**:164.

Kuhn, H. G., and Svendsen, C. N., 1999, Origins, functions, and potential of adult neural stem cells, *Bioessays* **21**:625.

LaVail, M. M., Yasumura, D., Matthes, M. T., Lau-Villacorta, C., Unoki, K., Sung, C. H., and Steinberg, R. H., 1998, Protection of mouse photoreceptors by survival factors in retinal degenerations, *Invest Ophthalmol Vis Sci* **39**:592.

Leung, C., Lingbeek, M., Shakhova, O., Liu, J., Tanger, E., Saremaslani, P., Van Lohuizen, M., and Marino, S., 2004, Bmi1 is essential for cerebellar development and is overexpressed in human medulloblastomas, *Nature* **428**:337.

Marquardt, T., and Gruss, P., 2002, Generating neuronal diversity in the retina: One for nearly all., *Trends Neurosci* **25**:32.

Molofsky, A. V., Pardal, R., Iwashita, T., Park, I. K., Clarke, M. F., and Morrison, S. J., 2003, Bmi-1 dependence distinguishes neural stem cell self-renewal from progenitor proliferation, *Nature* **425**:962.

Nakamura, Y., Sakakibara, S., Miyata, T., Ogawa, M., Shimazaki, T., Weiss, S., Kageyama, R., and Okano, H., 2000, The bhlh gene hes1 as a repressor of the neuronal commitment of cns stem cells, *J Neurosci* **20**:283.

Ooto, S., Akagi, T., Kageyama, R., Akita, J., Mandai, M., Honda, Y., and Takahashi, M., 2004, Potential for neural regeneration after neurotoxic injury in the adult mammalian retina, *Proc Natl Acad Sci U S A* **101**:13654.

van der Lugt, N. M., J., D., Linders, K., van Roon, M., Robanus-Maandag, E., te Riele, H., van der Valk, M., Deschamps, J., Sofroniew, M., and van Lohuizen, M., 1994, Posterior transformation, neurological abnormalities, and severe hematopoietic defects in mice with a targeted deletion of the bmi-1 proto-oncogene, *Genes Dev* **8**:757.

Wu, Y., Liu, Y., Levine, E. M., and Rao, M. S., 2003, Hes1 but not hes5 regulates an astrocyte versus oligodendrocyte fate choice in glial restricted precursors, *Dev Dyn* **226**:675.

Zhang, J., Gray, J., Wu, L., Leone, G., Rowan, S., Cepko, C. L., Zhu, X., Craft, C. M., and Dyer, M. A., 2004, Rb regulates proliferation and rod photoreceptor development in the mouse retina, *Nat Genet* **36**:351.

CHAPTER 32

TRANSCRIPTIONAL AND POST-TRANSCRIPTIONAL REGULATION OF THE ROD cGMP-PHOSPHODIESTERASE β-SUBUNIT GENE

Recent advances and current concepts

Leonid E. Lerner[1], Natik Piri[2], and Debora B. Farber[2]

1. INTRODUCTION

In eukaryotic cells, gene expression is controlled at multiple levels that could be grouped in two large categories, transcriptional and post-transcriptional regulatory events. Regulation of gene expression at the level of transcription is the major determinant of the initiation of protein synthesis and of the level of gene expression. However, there is increasing evidence of the important contribution of the post-transcriptional control mechanisms in determining and fine-tuning the final amount of the synthesized protein product. Post-transcriptional regulatory mechanisms could be grouped into events that modulate mRNA stability, localization, translation, as well as protein stability and modifications.

Expression of the key effector enzyme in the rod phototransduction cascade, the rod-specific cGMP-phosphodiesterase (cGMP-PDE) is restricted to rod photoreceptors in the mammalian retina. cGMP-PDE is a membrane-associated, heterotetrameric enzyme composed of two catalytic α- and β-subunits and two inhibitory γ-subunits (Fung et al., 1990). Each of these subunits is essential for normal cGMP-PDE activity required for phototransduction and for the maintenance of retinal health. Mutations in the protein-coding region of the gene encoding its *β*-subunit (β-*PDE*) cosegregate with retinal degenerations leading to blindness in human (Farber and Danciger, 1997), mice (Bowes et al., 1990; Pittler and Baehr, 1991) and dogs (Farber et al., 1992; Suber et al., 1993). However, even if the exonic sequences are intact, impaired regulatory mechanisms would result in suboptimal expression of the *β-PDE* gene causing alterations in the phototransduction cascade and

[1] Leonid E. Lerner, F.M. Kirby Center for Molecular Ophthalmology, Scheie Eye Institute, University of Pennsylvania School of Medicine, Philadelphia, Pennsylvania, 19104. Natik Piri and Debora B. Farber, Jules Stein Eye Institute, UCLA School of Medicine, Los Angeles, California, 90095.

abnormally high levels of cGMP. Therefore, such events will likely result in retinal functional and structural abnormalities. Recently, genetic defects in transcriptional mechanisms that control the expression of several retina-specific genes have been linked to different types of retinal disorders (Bessant et al., 1999; Freund et al., 1997; Freund et al., 1998; Haider et al., 2000; Swain et al., 1997). Given the detrimental effect mutations in the protein-coding region of the *β-PDE* gene have on photoreceptor integrity, it is crucial to understand the molecular events that mediate the precise regulation of expression of this gene in rod photoreceptors in the human retina.

Recent advances in the field of regulation of gene expression indicate that genes are differentially expressed according to their interplay with particular sets of transcription factors. Therefore, rod photoreceptor-specific expression of a gene is likely to be regulated by a unique set of transcription factors specific for rods, rather than a single rod-specific transcription factor. Our long-term interest in understanding the mechanisms of rod-specific regulation of the *β-PDE* gene expression has led us to continue the investigation of transcriptional and, more recently, post-transcriptional control of this gene utilizing a combination of *in vitro, ex vivo* and *in vivo* approaches. The results of these studies reviewed in this chapter will contribute to our pursuit of knowledge of how to maintain the expression of the *β-PDE* gene in rod photoreceptors at physiological levels, and to decelerate or prevent the development of certain forms of retinal degenerations.

2. TRANSCRIPTIONAL STUDIES

Formation of the preinitiation transcription complex involves coordinated interactions of RNA polymerase II with an array of basal transcription factors at the basal promoter sequences in the upstream regions of genes. The basal level of transcription is supported by general transcription factors and is enhanced by the action of activator proteins that interact with specific DNA elements. In addition, activated transcription may be repressed by the action of repressor proteins that also interact with specific DNA regulatory sequences. In previous studies, we reported our initial results on the transcriptional control mechanisms that take place in the human *β-PDE* 5′-flanking region. Mutational analysis of the *β-PDE* promoter tested both *in vitro* and *ex vivo*, and confirmed by the generation of transgenic *Xenopus* expressing mutant *β-PDE* promoter/GFP fusion constructs *in vivo*, revealed a minimal promoter region, from −93 to +53, that supports high levels of rod-specific transcription (Lerner et al., 2001). Two enhancer elements were localized within this minimal promoter, βAp1/NRE and β/GC that interact with nuclear factors and activate transcription from the *β-PDE* promoter.

The functionally important β/GC element is homologous to the consensus GC box that binds members of the Sp family of transcription factors including Sp1, Sp3 and Sp4. These nuclear factors share similar structural features and have highly conserved DNA binding domains that allow them to bind with identical affinity to the consensus GC box (Hagen et al., 1992). However, while Sp1 and Sp3 are ubiquitously expressed, Sp4 is expressed predominantly in the CNS. This prompted us to further test its abundance in the adult retina, and to evaluate its activation properties on the rod-specific *β-PDE* promoter under defined conditions in direct comparison to Sp1 and Sp3.

Figure 32.1. Sp4 expression in mouse retina, brain and in Y79 human retinoblastoma cells. Western blot analysis was carried out using anti-Sp4 antibodies and nuclear extracts (60 µg/lane) prepared from (lane 1) Y79 human retinoblastoma cells, (lane 2) retina, (lane 3) brain (cortex and cerebellum), and (lane 4) liver. The predominant reactive band of 95-105 kDa was observed in retina, brain and the Y79 cells, but not in liver. (Reprinted with permission from Lerner et al., J Biol Chem, 2001).

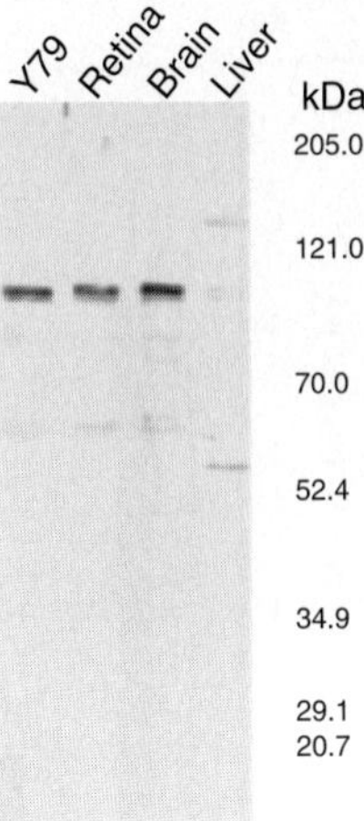

2.1. Retina-Enriched Expression of Sp4 Is Consistent with its Proposed Role as a Regulator of the Rod-Specific *β-PDE* Gene

In order to investigate the potential role for Sp4 as a transcriptional regulator of the rod-specific *β-PDE* gene, we tested whether this protein was expressed in the adult retina. Sp4-specific antibodies were used for immunoblot analysis of the nuclear extracts from adult mouse retina, brain, and liver, as well as from Y79 human retinoblastoma cells. We observed a single, high intensity band of predicted molecular weight (95-105 kDa) in lanes containing nuclear extracts from brain, retina and Y79 retinoblastoma cells (Figure 32.1). No major bands were seen in the lane containing the liver nuclear extract. The results of these experiments demonstrate that Sp4 is also relatively abundant in the adult vertebrate retina at a concentration comparable to that in the brain. This distribution is consistent with its proposed role as a regulator of the rod-specific *β-PDE* gene. In addition, co-localization of Sp4 in the retina with certain other transcriptional regulators such as Nrl and Crx suggests a combinatorial mechanism of photoreceptor-specific gene regulation.

2.2. Proximal Promoter Region of the *β-PDE* Gene Contains a Regulatory DNA Element

Previously, we observed significant reduction of transcriptional activity with the 5′-end deletion of *β-PDE* promoter from −72 to −45 (Lerner et al., 2001). However, the truncated −45 to +53 construct also showed residual promoter activity well above the control (Figure 32.2), which suggests the presence of at least one additional regulatory sequence. In order to identify the potential regulatory element located in this region, further deletions of the proximal promoter were performed. Multiple *β-PDE* promoter/*luciferase* fusion constructs (i.e. −45 to +53, −23 to +53 and +4 to +53) were transiently transfected in cultured Y79 human retinoblastoma cells, and also *ex vivo* in dissected *Xenopus* embryo heads. These human retinoblastoma cells and amphibian *in situ* transfection systems have been employed

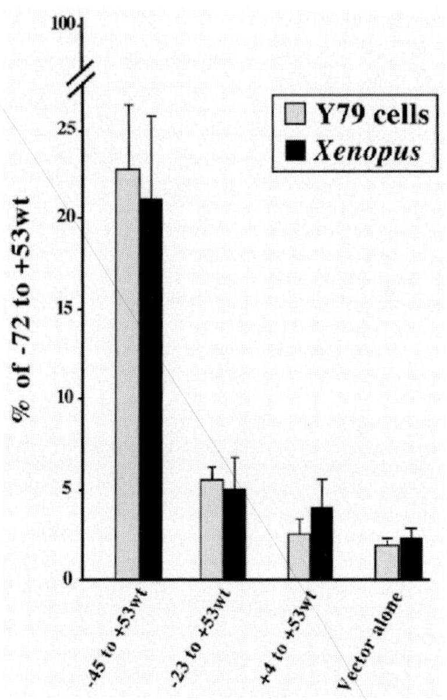

Figure 32.2. Deletion analysis of the β-*PDE* promoter suggests a regulatory DNA element in the −45 to −23 region. Serial 5′-end deletion constructs of the −45 to +53 β-*PDE* promoter were transfected in Y79 retinoblastoma cells (light bars) and in *ex vivo* dissected *Xenopus* embryonic heads (dark bars). Luciferase activity was measured in cell lysates and normalized to the corresponding β-galactosidase activity for each sample. The results are expressed as percent of the mean activity produced by the −72 to +53 β-*PDE* construct ± standard deviation. Each transfection was done in triplicate and repeated several times. (Reprinted with permission from Lerner et al., J Biol Chem, 2002).

previously for studying the regulation of photoreceptor-specific gene expression, including that of β-*PDE* (Lerner et al., 2001). Luciferase activities were measured and normalized to the β-galactosidase activities obtained with a control plasmid in Y79 cells, or expressed per embryo and averaged statistically as described previously for *Xenopus* transfections (Batni et al., 2000). When the −45 to −23 region was deleted, further reduction in promoter activity was observed in both transfection systems (Figure 32.2). The activity level of the −23 to +53 promoter was not significantly different from that observed with the promoter-less control vector when tested in Y79 cells or *Xenopus* embryos. Luciferase activity remained low when the +4 to +53 promoter construct carrying further 5′-end deletion past the major transcription start site was tested. High evolutionary conservation of the −45 to −23 region (Di Polo et al., 1996) that comprises the consensus Crx response element (*CRE*, −41/−36) and the T/A-rich sequence (β/TA) located at a consensus position for the TATA box is evident between mouse and human also suggesting its functional importance.

2.3. Functional Testing of Nucleotide Substitution Mutants in the Proximal Promoter and the 5′-Untranslated Region of the β-*PDE* Gene in Neuroretina-Related Transfection Systems

The initial proximal promoter deletion analysis described above prompted us to investigate whether the putative *CRE* (−41/−36) and its flanking sequences in the β-*PDE* proximal promoter were functionally relevant to the transcriptional regulation of the β-*PDE* gene *in vivo* in the context of a retina-relevant environment. A series of β-*PDE* promoter mutants carrying small oligonucleotide substitutions within the −45 to −23 region was transiently transfected in Y79 retinoblastoma cells and also in *Xenopus* embryos maintained *ex vivo* (summarized in Figure 32.3, top). Interestingly, the −41/−38 m mutation that completely disrupted the consensus *CRE* motif, had little effect on the β-*PDE* promoter activity in both transfection systems (Figure 32.3A). In addition, no significant changes were seen with mutations in the CRE flanking sequences (i.e. −37/−36 m, −35/−34 m and −33/32 m). The −30/−27 m mutant containing nucleotide substitutions in positions 2 through 5 of the T/A-rich β/*TA* sequence TAAGAAA to TCCTCAA) also showed no significant effect on promoter activity in transient transfections.

Although neither the mutations in consensus *CRE* or β/*TA* affected promoter activity, a cooperative interaction of transcription factors at both sites located in close proximity of each other could not be ruled out. Therefore, a double-mutant was constructed that contained both −30/−27 m and −41/−38 m. However, transient transfections of Y79 cells using the double-mutant showed no significant alterations in promoter activity compared to the wild type β-*PDE* promoter (Figure 32.3B). To search for other regulatory sequences in this *TATA*- and *Inr*-less gene, additional β-*PDE* promoter mutants (n = 14) containing nucleotide substitutions spanning the proximal 5′-flanking and the 5′-untranslated regions (−23 to +53; Figure 32.3, top) were tested in transient transfections of Y79 retinoblastoma cells. Promoter activity determined in these mutants ranged between approximately 0.5- and 1.5-fold that of the wild type control (Figure 32.3B). A 3′-end deletion mutant (−72 to +4) lacking most of the 5′-UTR showed approximately 3-fold reduction of promoter activity, which may be attributed to the deletion of certain translational control elements such as the Kozak sequence.

In summary, our results suggest that the β-*PDE* promoter may not have well-defined core elements responsible for basal transcription in Y-79 cells or *Xenopus* embryo heads. Rather, it appears that the transcription factors that interact with this region and mediate low-level β-*PDE* gene expression may not require a rigid sequence, but can accommodate a range of nucleotides.

2.4. TBP and TFIIB Bind the β-*PDE* Promoter

The possibility of an additional regulatory sequence(s) in the β-*PDE* basal promoter region or the 5′-UTR is suggested by the tight regulation of the transcriptional initiation site selection in this gene. There are only one major and one minor transcription start sites in both human and murine β-*PDE* genes (Di Polo et al., 1996). This indicates the assembly of the basal transcription machinery at a specific core promoter element rather than random binding to a variety of sequences. Sequence analysis of the β-*PDE* promoter showed that there are no known consensus core promoter elements present in this gene. However, there is the β/*TA* sequence that has a high T/A content and is located in the close proximity of the transcription start site ($^{-31}$TAAGAAA^{-25}, which is the consensus location for the TATA-box element) of the β-*PDE* promoter. However, this β/*TA* sequence is quite different from the known functional TATA-box elements (TATAAA, consensus).

Thus, we tested whether the β/TA sequence was able to bind purified TBP separately or in complex with TFIIB in GMSAs. As a control, we compared the binding of TBP, TFIIB and the TFIIB-TBP combination to the *AdML* promoter. Shifted bands were observed with the addition of either TBP alone or TFIIB alone to the β/TA probe (Figure 32.4A). Addition of the combination of TBP and TFIIB resulted in a slower migrating band with about three-fold increase in band intensity compared to TBP alone, producing a characteristic supershifted pattern described previously for the *AdML* promoter (Wolner and Gralla, 2000).

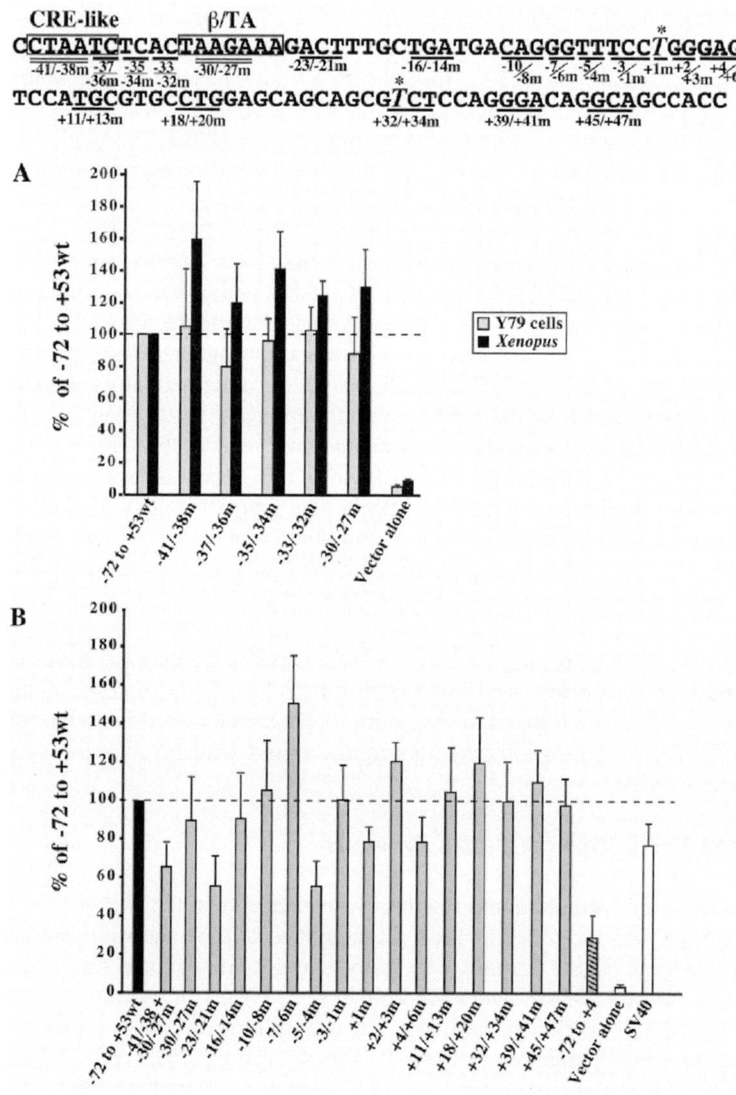

Figure 32.3. Analyses of transcriptional activity of the *β-PDE* promoters carrying various 1-4 bp nucleotide substitutions in the −41 to +53 proximal promoter and 5′-untranslated regions. A series of *β-PDE* 5′-flanking and 5′-untranslated region mutants were generated in the context of −72 to +53 *β-PDE/luciferase* fusion construct and tested in transient transfections. (Top) All mutations (1-4 bp transversions) are summarized and shown as underlined nucleotides. The two nucleotide sequences mutated in the double-mutant construct are underlined with a double line. An asterisk marks the transcription start site designated as +1. (Bottom) Transient transfections of the mutant constructs in Y79 retinoblastoma cells (light bars) and in *Xenopus* embryos *ex vivo* (dark bars). A: Constructs contained nucleotide substitutions spanning the −41 to −27 region. B: Constructs contained mutations spanning the −23 to +53 sequence. Luciferase activity produced by each mutant was normalized for each transfection system and expressed as percent of the mean activity of the −72 +53 wild-type *β-PDE* promoter ± standard deviation. Transfections were performed in triplicate and repeated at least two times. (Reprinted with permission from Lerner et al., J Biol Chem, 2002).

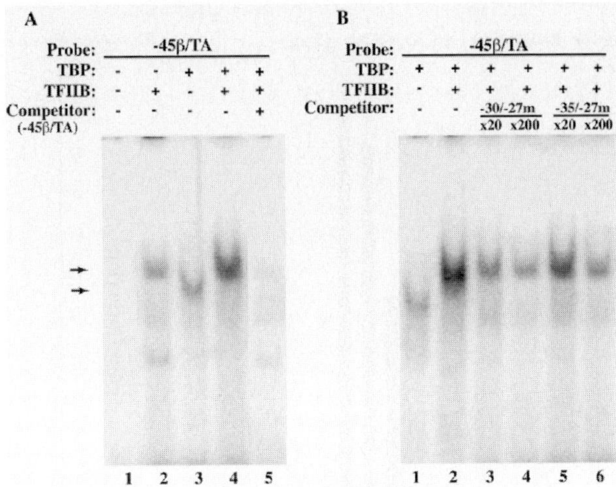

-45β/TA: 5'-GTTCCTAATCTCACTAAGAAAGACTTTGCT-3'

Figure 32.4. TBP and TFIIB bind the *β-PDE* promoter as single proteins with cooperative enhancement of binding by TFIIB-TBP complex. The −45β/TA probe sequence comprising the −45/−16 region of the *β-PDE* promoter is shown at the bottom. In mutant competitors, nucleotide transversions were introduced into the −45/16 sequence. Electrophoretic mobility shift assays: A: In the control experiment (*lane 1*), protein was not included in the binding reaction. Purified TFIIB (*lane 2*), TBP (*lane 3*), or a combination of TFIIB-TBP (*lane 4*) was added to the binding reactions. In *lane 5*, the 200-fold molar excess of unlabeled −45β/TA oligonucleotide was added to the reaction mixture identical to that resolved in *lane 4*. B: Combination of TFIIB-TBP (*lane 2*) produces a typical supershifted complex with a 3-fold increase in intensity compared to TBP alone (*lane 1*). In *lanes 3* and *4*, a 20-fold and 200-fold molar excess of the −31/−27 m mutant unlabeled competitor, and in *lanes 5* and *6*, a 20-fold and 200-fold molar excess of the −35/−27 m competitor were included in the same binding reaction as in *lane 2*. Retarded protein-DNA complexes are labeled with *arrows*. (Reprinted with permission from Lerner et al., J Biol Chem, 2002.)

These results suggest an enhanced cooperative binding by the TFIIB-TBP complex to the *β-PDE* promoter compared to TBP alone.

In contrast, when comparable protein concentrations were used, the *AdML* promoter interacted with TBP and TFIIB-TBP, but did not form a stable TFIIB-DNA complex in GMSA (data not shown), as previously demonstrated (Wolner and Gralla, 2000). Although, the addition of a 200-fold molar excess of the wild type −45/−16 competitor to the binding reaction prevented the shifted complex formation (Figure 32.4A, lane 5), the mutant −30/ −27 m and −35/−27 m competitors also showed substantial competition with the wild type sequence for TFIIB-TBP binding (Figure 32.4B). These results further corroborate our functional transfection data that a well-defined core promoter sequence could not be found in the *β-PDE* 5'-flanking region.

2.5. The *β-PDE* Promoter is a Specific Target for Activation by the Sp4 Transcriptional Regulator

The most significant finding of the present investigation was the demonstration of the functional involvement of members of the Sp family in transcriptional regulation of the *β-*

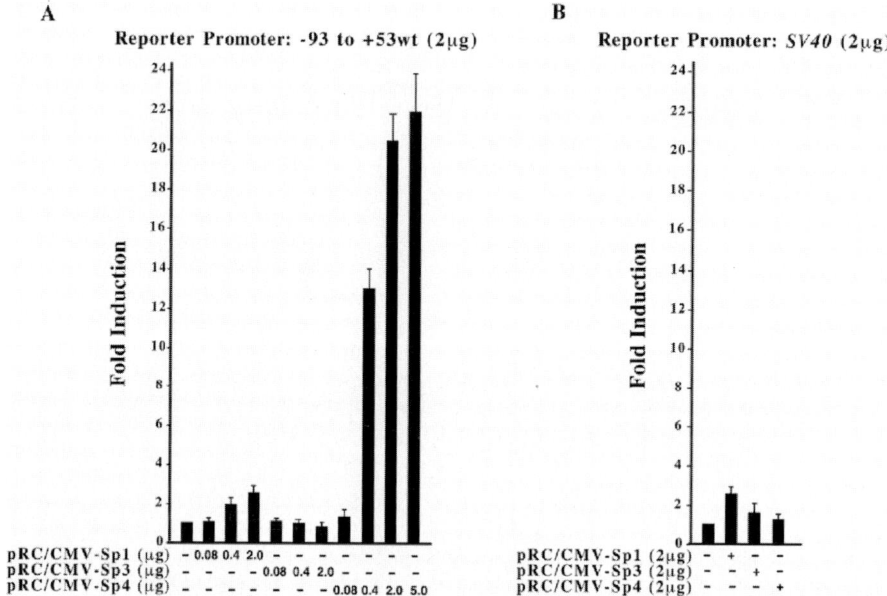

Figure 32.5. Sp4 is a potent activator of the *β-PDE* gene promoter. The fold induction of the *β-PDE* or *SV40* reporter constructs was determined relative to the uninduced reporter activity. **A:** The −93 to +53 minimal rod-specific *β-PDE* promoter (2 μg) was cotransfected with increasing amounts of Sp1, Sp3 and Sp4 plasmids and compared to the uninduced promoter contransfected with an empty plasmid. **B:** The *SV40* promoter/*luciferase* vector (2 μg, pGL2-Control, Promega®) was cotransfected with 2 μg of the plasmid containing either Sp1, Sp3 or Sp4, and compared to an empty plasmid. Luciferase activity was measured in cell lysates and normalized to the corresponding β-galactosidase activity for each sample. The results are expressed as the fold induction of the mean activity of the uninduced −93 to +53 *β-PDE* reporter construct ± SD. (Reprinted with permission from Lerner et al., J Biol Chem, 2002.)

PDE promoter. Our previous investigations of promoter activity using transient transfections of multiple *β-PDE* promoter mutants, as well as protein-DNA binding studies using the *β-PDE* promoter sequences, have revealed the *β/GC* element (−55/−46) as an important enhancer of the *β-PDE* promoter that binds different transcription factors of the Sp family (Lerner et al., 2001). Members of the Sp family bind to GC-rich DNA sequences through three zinc finger motifs. The residues involved in the determination of the target site specificity and binding affinity are highly conserved between different family members. Recently, we compared different Sp proteins (Sp1, Sp3 and Sp4) that share similar DNA-binding characteristics (Hagen et al., 1992) for their effects on transcription from the *β-PDE* promoter. Interestingly, Sp1, Sp3 and Sp4 transcription factors showed differential effects on the *β-PDE* promoter activity. Wild-type minimal rod-specific *β-PDE* promoter (−93/+53, 2 μg) was transiently co-transfected with increasing amounts of expression plasmids (0.08 μg, 0.4 μg and 2 μg) each carrying a full-length cDNA for either Sp1, Sp3 or Sp4. Compared to other members of the Sp family, Sp4 was the only transcription factor that showed significant dose-dependent effect on the *β-PDE* promoter (Figure 32.5A). Promoter-specificity of the Sp4-mediated transactivation was confirmed by comparing its effect on transcription from the β promoter (approximately 21-fold enhancement) to that on the *SV40* promoter (no significant change) relative to the uninduced transcription, respectively (Figure 32.5B). The maximum activation was observed using 2 μg of pRC/CMV-Sp4 without any further increase in promoter activity using 5 μg of pRC/CMV-Sp4, indicating a saturation effect. In contrast, neither Sp1 nor Sp3 showed any significant effect on transcription from the β-*PDE* promoter.

3. POST-TRANSCRIPTIONAL STUDIES

Regulation of expression of PDE subunits is also controlled at the post-transcriptional level. We examined the retinal steady-state mRNA and protein levels, protein biosynthesis rate, as well as the translational efficiency of rod-specific cGMP-phosphodiesterase. Our findings indicated that in mouse retina the number of mRNA molecules for β-*PDE* is approximately 5 times higher than that for α-*PDE* and that the levels of α-*PDE* and β-*PDE* transcripts in 10-day-old mice are approximately 85% of those in 30-day-old animals. The relative concentrations of two endogenously expressed β-PDE transcripts differing by the length of their 5′-UTR are similar in the developing and adult retina. At the protein level, α-*PDE* and β-*PDE* show equimolar expression in retinas of 10- and 30-day-old mice. Furthermore, we observed similar turnover rates for both subunits through pulse-chase experiments. The discordance between the mRNA and protein levels suggested that PDE expression is regulated post-transcriptionally and most likely at the translational level that is generally accepted to modulate the global synthetic activity of the cell.

To investigate whether the production of equal amounts of α-*PDE* and β-*PDE* from different amounts of the corresponding mRNAs was controlled at the level of translation, we determined the translational efficiency of α-*PDE* and β-*PDE* mRNAs and examined the role of their 5′ and 3′ UTRs as well as coding regions in the regulation of protein synthesis. Using constructs containing the full-length cDNA for α-*PDE* or β-*PDE* we were able to conclusively demonstrate that α-*PDE* mRNA is translated approximately 5 times more efficiently than its β-*PDE* counterpart (Figure 32.6).

Thus, our results indicated that the low level of α-*PDE* mRNA found in retina is counterbalanced by its efficient translation. These data also point at possible regulation of PDE expression in photoreceptor cells by the feedback mechanism: protein synthesis efficiency dictates the level of mRNA transcription.

After determining the translational efficiency of α-*PDE* and β-*PDE* mRNAs, we investigated which factors contribute to their differential translation. Since protein synthesis is controlled primarily at the initiation step and is generally dependent on the structural properties of individual mRNAs, we evaluated the role of the 5′ and 3′ UTRs, known to be involved in the regulation of protein synthesis (Day and Tuite, 1998; Kozak, 1987; Kozak, 1997; Geballe and Morris, 1994), as well as the coding sequences of α-*PDE* and β-*PDE* mRNAs on this process. Sequence analysis revealed the presence of an upstream AUG and the absence of a "strong" initiation sequence in the 5′ UTR of β-*PDE* mRNA (Piri et al., 2003). In contrast, the α-*PDE* mRNA has a consensus translation initiation sequence and no upstream AUG. Furthermore, neither of the 5′ UTRs of α-*PDE* and β-*PDE* mRNAs contains stable secondary structures that can reduce the rate of protein synthesis. On the basis of these analyses we hypothesized that these differences could account for the lower protein synthesis efficiency of β-*PDE*. Indeed, when we mutated the upstream AUG and restored

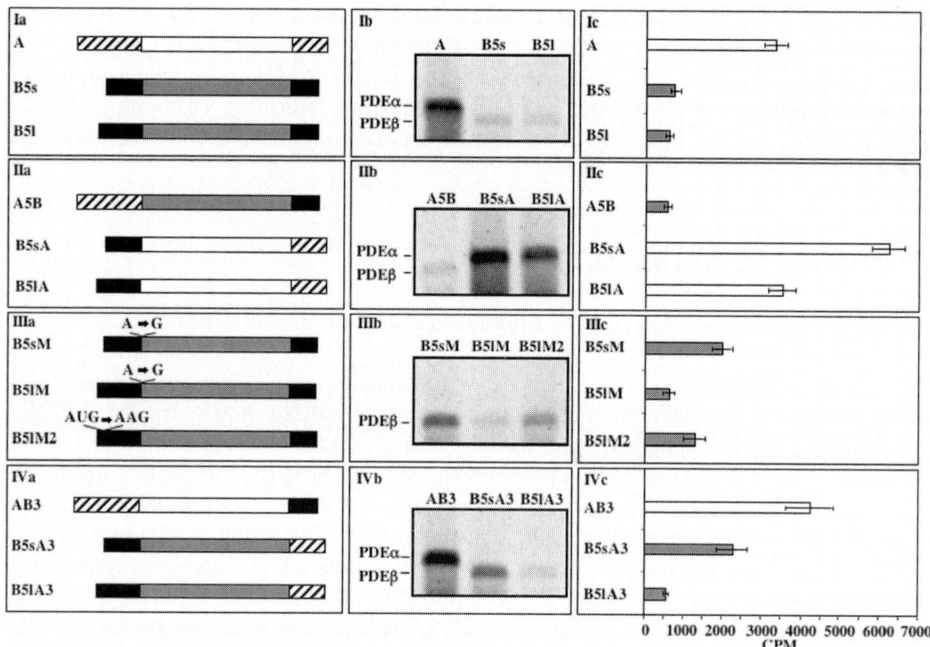

Figure 32.6. In vitro translation of full-length α-PDE and β-PDE cDNAs. Ia-IVa: The different regions of α-PDE and β-PDE mRNAs are shown in the constructs as: hatched boxes, α-PDE 5′ and 3′ UTRs; open box, α-PDE coding region; filled boxes, β-PDE 5′ and 3′ UTRs; grey box, β-PDE coding region. Mutations in the translation initiation sequence and upstream AUG are depicted. Ib-IVb: Autoradiographs and Ic-IVc: quantitative analysis of the *in vitro* synthesized α-PDE and β-PDE proteins.

the consensus translation initiation sequence in the *β-PDE* 5′ UTR, the protein production was increased. However, using several chimeric constructs we demonstrated that, in fact, the 5′ UTR of *α-PDE* leads to lower protein synthesis than the 5′ UTR of *β-PDE* mRNA (Figure 32.6). Therefore, the differential translation of the PDE subunits cannot be solely explained based upon the primary or secondary structures of their 5′ UTRs. Multiple examples describe the involvement of *cis*-elements within the 3′ UTR in translational regulation and the mechanisms by which these elements influence protein synthesis (Jackson and Standart, 1990; Conne et al., 2000; Stuart et al., 2000). We demonstrated that both *α-PDE* and *β-PDE* 3′ UTRs have a stimulatory effect on translation. The results of our studies undoubtedly implicate the involvement of the coding regions in the differential translation of *α-PDE* and *β-PDE* mRNAs. All eight constructs containing the *β-PDE* coding region resulted in lower protein expression than the four constructs containing the *α-PDE* coding region, regardless of the flanking 5′ or 3′ UTR.

In summary, our results indicate that the low level of *α-PDE* mRNA found in retina can be compensated by its more efficient translation to achieve equimolar expression with *β-PDE*. Moreover, the *α-PDE* and *β-PDE* coding regions are involved in the differential expression of these subunits, with the former producing more protein than the latter.

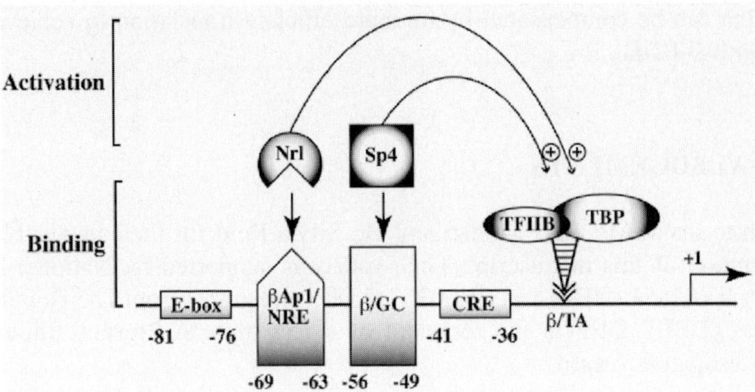

Figure 32.7. Schematic model of the molecular events required for rod-specific transcription from the minimal −93 to +53 β-*PDE* promoter. Functionally relevant DNA response elements in the β-*PDE* promoter are represented by tall graphic figures, whereas consensus binding sequences for known transcriptional regulators that do not affect transcription from this promoter are shown as narrow rectangles. Nucleotides are numbered relatively to the major transcription start site at +1. Potential protein-DNA interactions are shown as vertical arrows, whereas their functional effects on promoter activity are represented by semicircular arrows. Basal transcription factors TBP and TFIIB may interact with the β-*PDE* promoter and their higher affinity cooperative binding is indicated as a hatched double-arrow. (Reprinted with permission from Lerner et al., J Biol Chem, 2002.)

4. CONCLUSIONS

Sp4 is the least characterized member of the Sp family of transcriptional regulators, possibly because of its restricted pattern of expression that is predominant in the central nervous system. Here, we demonstrate that the rod-specific β-*PDE* gene represents the first natural target gene for Sp4. In addition, the β-*PDE* promoter seems to lack transcriptional regulation by the other related members of the Sp family, Sp1 and Sp3. The fact that Sp4 is able to specifically transactivate the β-*PDE* promoter supports our finding of this relatively restricted protein, compared to the ubiquitous Sp1 or Sp3, being abundantly expressed in the mammalian retina. The lack of other known specific Sp4 targets, combined with its regulation of a rod-restricted β-*PDE* gene, implies that this transcription factor functions in a very narrow cell type-specific manner.

In addition, Sp4 could have a more universal role in cell-specific expression of certain genes in rods and possibly other retinal cell populations by interacting with different arrays of transcription factors. We have previously shown that another nuclear factor, Nrl, regulates transcription from the β-*PDE* promoter (Lerner et al., 2001). Considering the additional β-*PDE* transcriptional mechanism described in this chapter and those from our previous studies, we can suggest that a unique combination of molecular interactions may be required for rod-specific transcription from this TATA- and Inr-less promoter (Figure 32.7). This model is consistent with the combinatorial principles of transcriptional regulation of cell-specific gene expression.

Finally, the expression of PDE subunits is also regulated at the post-transcriptional level, with α-*PDE* mRNA translated approximately 4.1- and 5.5-fold more efficiently than β-*PDE* short and long transcripts, respectively. This indicates that the low level of α-*PDE* mRNA found in retina can be compensated by its more efficient translation to achieve equimolar expression with *β-PDE*.

5. ACKNOWLEDGEMENTS

We wish to thank Ms. Lisa Mohan and Dr. Silvia Reid for their invaluable assistance in the preparation of this manuscript. This work was supported by National Institutes of Health grants EY02651 (DBF) and EY00367 (LEL), and grants from The Foundation Fighting Blindness (DBF). DBF is the recipient of a Research to Prevent Blindness Senior Scientific Investigators Award.

6. REFERENCES

Batni, S., Mani, S. S., Schlueter, C., Ji, M., and Knox, B. E., 2000, Xenopus rod photoreceptor: model for expression of retinal genes. *Methods Enzymol* **316**:50-64.

Bessant, D. A., Payne, A. M., Mitton, K. P., Wang, Q. L., Swain, P. K., Plant, C., Bird, A. C., Zack, D. J., Swaroop, A., and Bhattacharya, S. S., 1999, A mutation in NRL is associated with autosomal dominant retinitis pigmentosa. *Nat Genet* **21**:355-356.

Bowes, C., Li, T., Danciger, M., Baxter, L. C., Applebury, M. L., and Farber, D. B., 1990, Retinal degeneration in the rd mouse is caused by a defect in the β–subunit of rod cGMP-phosphodiesterase. *Nature* **347**:677-680.

Conne, B., Stutz, A., and Vassalli, J.D., 2000, The 3′ untranslated region of messenger RNA: A molecular 'hotspot' for pathology?. *Nat Med* **6**:637-641.

Day, D.A., and Tuite, M.F., 1998, Post-transcriptional gene regulatory mechanisms in eukaryotes: an overview. *J Endocrinol* **157**:361-371.

Di Polo, A., Rickman, C. B., and Farber, D. B., 1996, Isolation and initial characterization of the 5′ flanking region of the human and murine cyclic guanosine monophosphate-phosphodiesterase beta-subunit genes. *Invest Ophthalmol Vis Sci* **37**:551-560.

Farber, D. B., Danciger, J. S., and Aguirre, G., 1992, The β-subunit of cyclic GMP-phosphodiesterase mRNA is deficient in canine rod-cone dysplasia 1. *Neuron* **9**:349-356.

Farber, D. B. and Danciger, M., 1997, Identification of genes causing photoreceptor degenerations leading to blindness. *Curr Opin Neurobiol* **7**:666-673.

Freund, C. L., Gregory-Evans, C. Y., Furukawa, T., Papaioannou, M., Looser, J., Ploder, L., Bellingham, J., Ng, D., Herbrick, J. A., Duncan, A., et al., 1997, Cone-rod dystrophy due to mutations in a novel photoreceptor-specific homeobox gene (CRX) essential for maintenance of the photoreceptor. *Cell* **91**:543-553.

Freund, C. L., Wang, Q. L., Chen, S., Muskat, B. L., Wiles, C. D., Sheffield, V. C., Jacobson, S. G., McInnes, R. R., Zack, D. J., and Stone, E. M., 1998, De novo mutations in the CRX homeobox gene associated with Leber congenital amaurosis. *Nat Genet* **18**:311-312.

Fung, B. K. K., Young, J. H., Yamane, H. K., and Griswold-Prenner, I., 1990, Subunit stoichiometry of retinal rod cGMP phosphodiesterase. *Biochemistry* **29**:2657-2664.

Geballe, A.P. and Morris, D.R., 1994, Initiation codons within 5′-leaders of mRNAs as regulators of translation. *Trends Biochem Sci* **19**:159-164.

Hagen, G., Muller, S., Beato, M., and Suske, G., 1992, Cloning by recognition site screening of two novel GT box binding proteins: a family of Sp1 related genes. *Nucleic Acids Res* **20**:5519-5525.

Haider, N. B., Jacobson, S. G., Cideciyan, A. V., Swiderski, R., Streb, L. M., Searby, C., Beck, G., Hockey, R., Hanna, D. B., Gorman, S., et al., 2000, Mutation of a nuclear receptor gene, NR2E3, causes enhanced S cone syndrome, a disorder of retinal cell fate. *Nat Genet* **24**:127-131.

Kozak, M., 1987, At least six nucleotides preceding the AUG initiator codon enhance translation in mammalian cells. *J Mol Biol* **196**:947-950.

Kozak, M., 1997, Recognition of AUG and alternative codons is augmented by G in position +4 but is not generally affected by the nucleotides in positions +5 and +6. *EMBO J* **16**:2482-2492.

Jackson, R.J. and Standart, N., 1990, Do the poly(A) tail and 3′ untranslated region control mRNA translation? *Cell* **62**:15-24.

Lerner, L. E., Gribanova, Y. E., Ji, M., Knox, B. E., and Farber, D. B., 2001, Nrl and Sp nuclear proteins mediate transcription of rod-specific cGMP-phosphodiesterase β-subunit: Involvement of multiple response elements. *J Biol Chem* **276**:34999-35007.

Lerner, L. E., Gribanova, Y. E., Whitaker, L., Knox, B. E., and Farber, D. B., 2002, The rod cGMP-phosphodiesterase beta-subunit promoter is a specific target for Sp4 and is not activated by other Sp proteins or CRX. *J Biol Chem* **277**:25877-25883.

Pittler, S. J., and Baehr, W., 1991, Identification of a nonsense mutation in the rod photoreceptor cGMP phosphodiesterase β-subunit gene of the rd mouse. *Proc Natl Acad Sci USA* **88**:8322-8326.

Piri, N., Yamashita, C.K., Shih, J., Akhmedov, N.B., and Farber, D.B., 2003, Differential expression of rod photoreceptor cGMP-phosphodiesterase alpha and beta subunits: mRNA and protein levels. *J Biol Chem* **278**: 36999-37005.

Stuart, J.J., Egry L.A., Wong, G.H., and Kaspar, R.L., 2000, The 3′ UTR of human MnSOD mRNA hybridizes to a small cytoplasmic RNA and inhibits gene expression. *Biochem Biophys Res Commun* **274**:641-648.

Suber, M. L., Pittler, S., Qin, N., Wright, G., Holcombe, V., Lee, R., Craft, C., Lolley, R., Baehr, W., and Hurwitz, R., 1993, Irish setter dogs affected with rod-cone dysplasia contain a nonsense mutation in the rod cyclic GMP phosphodiesterase β subunit gene. *Proc Natl Acad Sci, USA* **90**:3968-3972.

Swain, P. K., Chen, S., Wang, Q. L., Affatigato, L. M., Coats, C. L., Brady, K. D., Fishman, G. A., Jacobson, S. G., Swaroop, A., Stone, E., et al., 1997, Mutations in the cone-rod homeobox gene are associated with the cone-rod dystrophy photoreceptor degeneration. *Neuron* **19**:1329-1336.

Wolner, B. S., and Gralla, J. D., 2000, Roles for non-TATA core promoter sequences in transcription and factor binding. *Mol Cell Biol* **20**:3608-3615.

PART IV

GENE THERAPY AND NEUROPROTECTION

CHAPTER 33

DOWN-REGULATION OF RHODOPSIN GENE EXPRESSION BY AAV-VECTORED SHORT INTERFERING RNA

Jacqueline T. Teusner, Alfred S. Lewin, and William W. Hauswirth*

1. INTRODUCTION

RNA interference (RNAi) is an evolutionarily ancient method of genome defense in many organisms. RNAi is triggered when a cell encounters a long double-stranded RNA (dsRNA) and results in the silencing of gene expression (Fire *et al.*, 1998). The dsRNA is processed by the RNAse III-like nuclease, Dicer, into 20-25 nt duplex RNA, termed short interfering RNA (siRNA, Zamore *et al.*, 2000; Hammond *et al.*, 2000). The siRNAs assemble into RNA-induced silencing complexes (RISCs) containing endoribonucleases and are unwound. It is believed that the sense strands are degraded, while the anti-sense strands activate the RISC to participate in repeated cycles of cleavage and degradation of complementary cognate mRNA, effectively silencing the gene such that no protein is made. For this reason, RNAi is seen as a useful tool for analyzing gene function. Given that introducing long dsRNA into mammalian cells initiates a potent antiviral response (Stark *et al.*, 1998), most investigators bypass this by introducing or expressing siRNAs, which are the key mediators of RNAi (Elbashir *et al.*, 2001a and b).

Short interfering RNA are now employed across many disciplines due to their great specificity and potency compared to alternative silencing methods, such as anti-sense and ribozyme-based strategies. Our group has incorporated siRNA technology into the research of autosomal dominant retinitis pigmentosa (ADRP), a progressive rod-cone dystrophy (Berson, 1996; Adler, 1996). More than 100 point mutations in the rhodopsin gene can cause ADRP, and these mutations account for about one third of cases (Naash *et al.*, 2004). The P23H mutation was among the first genetic defects to be identified as a cause of ADRP (Dryja *et al.*, 1990), and it is the most prevalent ADRP mutation in North America. The purpose of this work was to demonstrate down-regulation of P23H rhodopsin mRNA using siRNA *in vitro* prior to their use in an *in vivo* model of ADRP.

---

*Departments of Ophthalmology and Molecular Genetics, and the Powell Gene Therapy Center, University of Florida, Gainesville, FL, U.S.A. 32610.

2. METHODS

2.1. Cell Lines and Growth Conditions

HEK293 cells expressing bovine P23H rhodopsin under a tetracycline-inducible promoter were a kind donation from Dr. S. M. Noorwez. Cells were routinely grown under blasticidin and zeocin selection in Dulbecco's modified Eagle's medium (DMEM, Invitrogen Corp., Carlsbad, CA) and 10% fetal bovine serum (FBS) at 37°C in 5.0% CO_2. Expression of P23H rhodopsin was induced by the addition of $1\,\mu g/ml$ tetracycline.

2.2. siRNA Design and Screening *In Vitro*

Four pairs of complementary 21 nt RNAs that target different regions of both the bovine (accession number M12689) and human rhodopsin gene (accession NM_000539) were chemically synthesized by Dharmacon Research (Lafayette, CO). The siRNAs were all specifically designed to have low identity to mouse rhodopsin (accession number BC031766) and other known genes in the NCBI database. The nomenclature of each siRNA was chosen based on the location of its target sequence on the bovine rhodopsin transcript (121, 403, 605 and 769). To form the siRNA duplexes, the RNAs were annealed and purified according to the manufacturer's instructions.

The siRNAs were screened for silencing efficiency using the P23H-expressing cells in a six-well plate format. When the cells had reached 30-50% confluency, duplicate wells were transfected with 300 pmol siRNA using Oligofectamine™ reagent (Invitrogen Corp.) under serum-free, antibiotic-free conditions. Transfections included a mock control (without siRNA), a non-specific siRNA control (Dharmacon) and an RNA sense and anti-sense strand only control. Three hours post-transfection, the medium was changed to DMEM containing 10% FBS with antibiotics and the cells were induced with tetracycline. Cells were maintained for up to 72 hours until harvesting for RNA isolation.

Mammalian plasmids that express functional siRNAs were also constructed. Briefly, two complementary oligonucleotides for the 121, 403 and 769 siRNA target sites were chemically synthesized, annealed and ligated into pSilencer™ (Ambion Inc., Austin, TX) downstream of the H1 RNA promoter. When transcribed, each encodes a short hairpin RNA (shRNA) with a 19mer stem derived from the mRNA target and a loop that is susceptible to cleavage by Dicer to form functional siRNA.

P23H-expressing cells plated in a 12-well plate format were transfected with 400 ng of each shRNA plasmid using siPORT XP-1 reagent (Ambion Inc.) in DMEM. Transfection controls included a mock and a GFP-expressing plasmid, pTR-UF11. After 3 hours, the medium was replaced with DMEM containing 10% FBS, then at 72 hours, antibiotics, including tetracycline, were added. Cells were maintained for up to 7 days.

Finally, the shRNAs were cloned into an AAV vector and packaged into functional serotype 2 virus. Briefly, PCR primers to pSilencer were designed so as to amplify products containing the H1 RNA promoter and the oligonucleotides encoding the active hairpin. Each amplification product was cloned into the UF11GFPflip vector (GFPflip refers to the GFP coding sequence in the reverse orientation hence it is not expressed) and packaged into AAV2 as described previously (Zolotukhin *et al.*, 2002). Using a 12-well plate format, P23H cells at 80% confluency were infected with 1×10^5 particles per cell of each AAV2-shRNA in DMEM. Infection controls included a mock, the empty UF11GFPflip-AAV2 and UF11-

AAV2. Three hours post-infection, the medium was adjusted to contain 10% FBS plus antibiotics including tetracycline. After 24 hours, the medium was replaced and the cells were maintained until harvesting for RNA isolation.

2.3. RNA Isolation and Real Time PCR

Expression of mRNA was investigated by quantitative real time PCR. Total RNA was isolated from transfected cells using TRIzol® Reagent (Invitrogen Corp.), according to the manufacturer's protocol. The cDNA was synthesized using Superscript™ II reverse transcriptase (Invitrogen Corp.) and PCR amplified using Amplitaq Gold® (Applied Biosystems, Foster City, CA). Rhodopsin- and β-actin-specific primers were selected from the corresponding bovine and human sequences, respectively. Negative controls consisting of mock reverse transcribed cDNA were included with each PCR set to detect possible genomic DNA contamination. PCR products were quantified using SYBR green dye (Molecular Probes, Eugene, OR) and the Gene Amp 5700 system (Version 1.3, Perkin-Elmer Biosystems©, Boston, MA). Gel electrophoresis and dissociation curve analysis confirmed that amplification of non-specific reaction products did not occur.

3. RESULTS

Given that RNAi acts post-transcriptionally, real time PCR was used to assess the effect of the chemically-synthesized siRNAs on P23H rhodopsin mRNA. Figure 33.1 illustrates that 24 hours after cell transfection, tetracycline-induced rhodopsin mRNA levels were

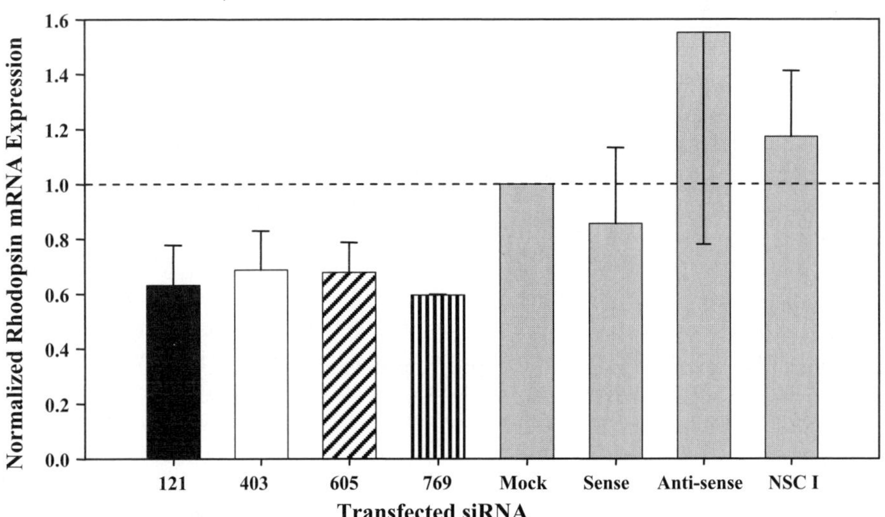

Figure 33.1. Real Time PCR analysis indicates chemically-synthesized siRNAs inhibit P23H rhodopsin mRNA expression in vitro. Endogenous β-actin was used as the active reference for RNA normalization. The normalized fold difference in bovine rhodopsin mRNA expression of each transfected sample relative to the untransfected control is shown. Real time PCR was performed in triplicate for each transfection and data represents the mean ± SD of 2 transfections.

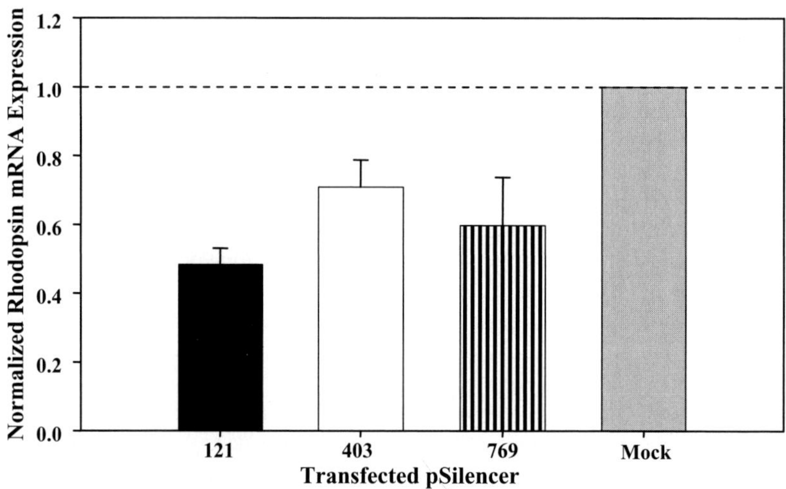

Figure 33.2. Real Time PCR analysis indicates shRNA-expressing plasmids inhibit P23H rhodopsin mRNA expression *in vitro*. Real time PCR was performed as described in Figure 33.1.

significantly down-regulated by all of the siRNAs compared to controls. By 48 hours, rhodopsin mRNA levels were reduced by 3 of the siRNA duplexes, namely, 121, 403 and 769. Duplex 605 did not show reproducible silencing (data not shown).

In the hope of obtaining longer-term suppression of the P23H rhodopsin gene, as opposed to the few days observed with chemically synthesized siRNAs, mammalian plasmids that express shRNAs for the 121, 403 and 769 target sites were constructed. Figure 33.2 shows the real time results of P23H rhodopsin mRNA expression in cells harvested 7 days post-transfection. Transfection efficiency, determined using a control GFP reporter plasmid, was found to be ≤50%. Despite this sub-optimal transfection efficiency, reduced target mRNA was observed with all three of the plasmids tested.

As a preliminary step towards using siRNA technology *in vivo*, the siRNA hairpins were also cloned and packaged into functional AAV2 and used tested once more *in vitro*. Serotype 2 virus was chosen due to its apparent infectivity towards a variety of different cell types. Using the control UF11-AAV2 that expresses the GFP reporter, we found the infection efficiency to be about 30%. Initial experiments showed down-regulation of the P23H rhodopsin transcript is not seen until the cells had been passaged about three times (data not shown). This is anticipated as viral expression of shRNAs is delayed compared with direct transfection into cells. Therefore, RNA was isolated from cells after their first passage, at which point the shRNAs are likely to be expressed, then again at P3, which corresponded to 2 weeks post-infection. Figure 33.3 shows the resulting real time analysis, with the level of P23H rhodopsin mRNA expression at P3 normalized to that seen at P1. The data suggest the target is down-regulated by 2 weeks post-infection. However, the cells are still in passage to determine how long suppression is maintained. Nevertheless, the results are promising and the vectors are being tested in an *in vivo* model of ADRP.

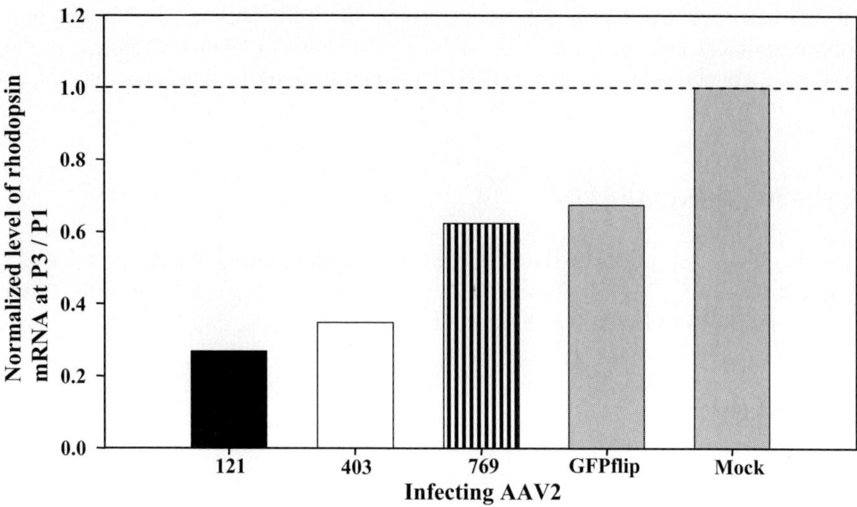

Figure 33.3. Real Time PCR analysis of shRNA-AAV2 infections *in vitro*. Endogenous β-actin was used for RNA normalization. The normalized level of bovine rhodopsin mRNA expression in each infected sample at P3 relative to that level seen at P1 is shown. The data was calculated from the mean of 3 real time PCRs.

4. DISCUSSION

The P23H mutation in rhodopsin causes a late onset form of ADRP in humans (Dryja *et al.*, 1990). We have performed careful *in vitro* screening of chemically synthesized siRNAs and DNA vectors that express shRNAs to identify three candidate sequences that down-regulate P23H rhodopsin mRNA expression. In order to effectively deplete protein levels and obtain longer-term suppression of the rhodopsin gene, the shRNAs were packaged into serotype 2 and 5 AAV. Preliminary *in vitro* experiments have indicated the AAV2-delivered shRNAs are also effective at reducing P23H target.

Transgenic mice expressing a single copy of the human rhodopsin gene containing the P23H mutation on a C57BL/6 background have been generated in our lab (Fritz *et al.*, submitted). Mice created with one copy of the human transgene on a heterozygous knockout background for endogenous mouse rhodopsin display a loss of scotopic electroretinography response by 3 months of age. Additionally, this correlates with a loss of ONL thickness, as assessed by histological analysis of the retinas. Due to their progressive retinal degeneration, these transgenic mice are considered to be a useful model for ADRP. We know from other studies in our lab that AAV-delivered siRNAs designed to target mouse rhodopsin are functional in this *in vivo* model (Dr. Marina Gorbatyuk, personal communication). Therefore, the AAV5-shRNA reagents described here, specifically designed to target bovine and human rhodopsin but not the mouse transcript, are currently being tested in this mouse line to determine if rescue of rod function is attainable. Since the siRNAs were designed to target the human rhodopsin transcript avoiding the region around the P23H mutation, these siRNA vectors have potential to treat other transgenic mouse models of human rhodopsin mutants.

RNAi, coupled to AAV vectors demonstrating efficient and long-term retinal gene expression, offers a new tool in gene therapy to potentially treat many autosomal dominant retinal diseases requiring that a defective gene be silenced. This research is the first step towards the clinical goal of using AAV–delivered siRNAs to silence defective rhodopsin, while simultaneously re-introducing a siRNA-resistant form of the gene encoding the correct protein.

5. ACKNOWLEDGEMENTS

This work was supported by the NEI, the Foundation Fighting Blindness, the Steinbach Fund and Research to Prevent Blindness Inc. The authors also thank Dr. S.M. Noorwez, Dr. J. Fritz, Dr. S. Min, Dr. M. Gorbatyuk, V. Chiodo, T. Doyle, and M. Ding.

6. REFERENCES

Adler R., 1996. Mechanisms of photoreceptor death in retinal degenerations. From the cell biology of the 1990s to the ophthalmology of the 21st century? *Arch Ophthalmol.* **114**(1):79-83.

Berson E. L., 1996. Retinitis pigmentosa: Unfolding its mystery. *Proc. Natl. Acad. Sci. USA* **93**:4526-4528.

Dryja, T. P., McGee, T. L., Reichel, E., Hahn, L. B., Cowley, G. S., Yandell, D. W., Sandberg, M. A. and Berson., E. L., 1990. A point mutation of the rhodopsin gene in one form of retinitis pigmentosa. *Nature* **343**:364-366.

Elbashir, S. M., Lendeckel, W., and Tuschl, T., 2001a. RNA interference is mediated by 21- and 22-nucleotide RNAs. *Genes & Dev.* **15**:188-200.

Elbashir, S. M, Harborth, J. Lendecknel, W., Yalcin, A., Weber, K., and Tuschl, T., 2001b. Duplexes of 21-nucleotide RNAs mediate RNA interference in mammalian cell culture. *Nature* **411**:494-498.

Fire, A., Xu, S., Montgomery, M. K., Kostas, S. A., Driver, S. E., and Mello, C. C., 1998. Potent and specific genetic interference by double-stranded RNA in *Caenorhabditis elegans. Nature* **391**:806-811.

Hammond, S. M., Bernstein, E., Beach, D., and Hannon, G. J., 2000. An RNA-directed nuclease mediates post-transcriptional gene silencing in *Drosophila* cells. *Nature* **404**:293-296.

Naash, M. I., Wu, T. H., Chakraborty, D., Fliesler, S. J., Ding, X. Q., Nour, M., Peachey, N. S., Lem, J., Qtaishat, N., Al-Ubaidi, M. R. and Ripps, H., 2004. Retinal abnormalities associated with the G90D mutation in opsin. *J Comp Neurol.* **478**(2):149-63.

Stark, G. R., Kerr, I. M., Williams, B. R., Silverman, R. H., and Schreiber, R. D., 1998. How cells respond to interferons. *Annu. Rev. Biochem.* **67**:227-264.

Zamore, P. D., Tuschl, T., Sharp, P. A. and Bartel, D. P., 2000. RNAi: double-stranded RNA directs the ATP-dependent cleavage of mRNA at 21 to 23 nucleotide intervals. *Cell* **101**:25-33.

Zolotukhin, S., Potter, M., Zolotukhin, I., Sakai, Y., Loiler, S., Fraites, T.J. Jr., Chiodo, V.A., Phillipsberg, T., Muzyczka, N., Hauswirth, W.W., Flotte, T.R., Byrne, B.J., and Snyder, R.O., 2002. Production and purification of serotype 1, 2, and 5 recombinant adeno-associated viral vectors. *Methods* **28**:158-167.

CHAPTER 34

ASSESSING THE EFFICACY OF GENE THERAPY IN *Rpe65*[-/-] MICE USING PHOTOENTRAINMENT OF CIRCADIAN RHYTHM

Chris W. Stoddart*, Meaghan J.T. Yu*, Matthew T. Martin-Iverson, Dru M. Daniels, C.-May Lai, Nigel L. Barnett, T. Michael Redmond, Kristina Narfström, and P. Elizabeth Rakoczy[†]

1. INTRODUCTION

Gene therapy (GT) can be described as the *in vivo* transfer of DNA for therapeutic purposes. In the case of congenital retinal dystrophies (RD), GT can only be considered a successful treatment option if the gene transfer results in the restoration of vision at some level. Thus, a critical component of developing GT treatments for RDs is assessing the amount of functioning vision that is produced. In mouse models of RD, visual function after gene therapy is tested using techniques such as electroretinograms (ERGs) and retinoid analysis.[1-3] A potentially useful addition to these tests would be a murine behavior-based technique,[4] like those used with the RPE65 dog model,[5-7] to demonstrate effective GT-induced visual recovery.

In the current study we tested a newly developed mouse behavioral assay[4] to determine its potential for assessing the efficacy of ocular gene therapy *in vivo*. This behavioral assay involved quantifying the circadian locomotor activity of mice under dynamic light condi-

*CWS & MJTY contributed equally to this work.

[†]CWS, MJTY, DMD, CML, PER, Centre for Ophthalmology and Visual Science, The University of Western Australia (UWA), Perth, Western Australia. CWS, MTMI, School of Medicine & Pharmacology, UWA, Perth, Western Australia. MJTY, Lions Eye Institute, Perth, Western Australia. NLB, Vision Touch & Hearing Research Centre, School of Biomedical Sciences, University of Queensland, Queensland, Australia. TMR, Laboratory of Retinal Cell & Molecular Biology, National Eye Institute, National Institutes of Health, Maryland, USA. KN, Vision Science Group, Department of Veterinary Medicine and Surgery, College of Veterinary Medicine, University of Missouri-Columbia, Columbia, Missouri, USA.

Corresponding author: Prof P. Elizabeth Rakoczy, Centre for Ophthalmology and Visual Science, AA block, QEII Medical Centre, Verdun St, Nedlands, Western Australia. Ph: 61 8 9381 0726. Fx: 61 8 9381 0700. email rakoczy@cyllene.uwa.edu.au.

tions.[4] The circadian rhythm is the biological clock about which mammalian physiology revolves, and entrainment of this rhythm to a 24 hour cycle forms the basis of the present research. Circadian rhythms are primarily entrained by the perception of light (photo-entrainment),[8-10] where higher levels of photic detection lead to faster re-entrainment.[4] In this study we assessed the ability of mice to re-entrain their locomotor circadian rhythms to a 12 hour (h) light reversal. An improved entrainment ability in these animals can be viewed as an increased level of photic detection and thus an improved ability to perceive light.[4]

The behavioral technique was assessed using the *Rpe65*[-/-] knockout mouse model, a strain of mice in which a targeted disruption of the retinal *Rpe65* gene leads to a disease phenotype reflective of LCA in humans.[11] RPE65 is a protein expressed predominantly within the retinal pigment epithelial (RPE) cells and is an essential component of the visual cycle and thus also phototransduction and visual function.[11] In the current study, *Rpe65*[-/-] mice were subjected to gene therapy by using a recombinant adeno-associated virus (rAAV) to deliver and express RPE65 in the retinas of these mice.[3,12] The effectiveness of this gene therapy approach was then assessed using the new behavioral technique.

2. METHODS

2.1. Production of rAAV-gene Constructs for Gene Therapy, Subretinal Delivery, and Expression of rAAV-gene Constructs

All procedures were approved by the University of Western Australia Animal Experimentation Ethics Committee, and were in compliance with the Association for Research in Vision and Ophthalmology Statement for the Use of Animals in Ophthalmic and Vision Research.

The construction, delivery and expression of the rAAV-RPE65 gene therapy construct has been described in detail previously.[3] Briefly, an rAAV construct carrying normal, non-mutated mouse RPE65 cDNA (rAAV.RPE65) was generated, then delivered to the target RPE cells of *Rpe65*[-/-] mice by subretinal injection.[3,12] The success of the delivery was assessed by performing ERG analysis and RPE65 immunohisto-chemistry.[3,13]

2.2. Behavioral Procedure

Three experimental groups of adult mice were assayed: a test group of rAAV.RPE65-injected *Rpe65*[-/-] knockout mice ("injected *Rpe65*[-/-]", n = 8), a control group of age-matched, sham-injected (normal saline) *Rpe65*[-/-] mice ("control *Rpe65*[-/-]", n = 8), and second control group of age-matched, normal C57Bl/6J controls ("C57" n = 8). The mice were approximately 3 months post injection at the time of testing (injections were performed soon after weaning). The mice were individually housed in Plexiglas cages accommodating infrared photocell beams for continuous monitoring of motor activity, as previously described.[4] Interruptions of the photocell beams were measured by PC computer and expressed as counts/h (bins). The mice were given free access to water and food (Glen Forrest Rodent Chow) at all times.

Procedures were performed as described previously.[4] Briefly, test and control groups were adapted to the testing environment via a 12:12h light-dark cycle for 6 days (light intensity during light cycle = 700 ± 50 lux, Sekonic L-28C photometer, NY, USA). After the 6 days, the light cycle was reversed, (12:12h dark-light). The 180° change in photoperiod (light reversal) was made such that the mice were initially exposed to 24h of dark, as a result of 2 consecutive 12h dark periods. Data was recorded until the circadian rhythms of the mice entrained to the new lighting conditions.

2.3. Data Analysis

Data bins were collected from each animal and graphed as double plot actograms for visual inspection. Hourly activity data was analyzed in blocks of 72 hourly bins (3 days). The spectral amplitudes (in natural log units) of the possible periods (and their Z score transformations) of each block (72h) were determined for each mouse independently using the periodogram method of Spectral analysis (SPSS-PC). Z scores for the 24h period were analyzed further to determine if significant 24 hour rhythms were present. Individual activity amplitudes and acrophases were generated using Halberg's cosinor regressions (with period = 24h).[14]

Parametric statistical analysis of the amplitudes, acrophases and 24h z-scores across days were performed with the general linear model (GLM) Analysis of Variance (ANOVA) procedure for repeated measures, which including simple pair-wise contrast tests for further group comparisons, using SPSS 10.0 for Windows, with $\alpha = 0.05$. For a more detailed discussion of the statistical analysis see Daniels et al. 2003[4] and associated papers.[15,16]

3. RESULTS

Periodogram analysis showed all groups of mice maintained a significant 24h period in their locomotor circadian activity (data no shown). Double plot actograms of the data collected showed that, while all 3 groups of mice were capable of re-entraining after the 12:12h light reversal,[17,18] the amount of time taken to re-entrain varied between the groups. The re-entrainment of C57 controls to the photoperiod reversal was rapid, being complete by 4 days in all mice (Fig. 34.1A), while re-entrainment of the control *Rpe65*[-/-] group took significantly longer, needing at least 13 days (Fig. 34.1B). While there was little within-subjects variance in the two control groups, the re-entrainment times of the injected *Rpe65*[-/-] group varied considerably; mice in this group taking between 8 & 14 days to re-entrain (Fig. 34.1C, 1D).

When the amplitude data for each group was compared using simple contrast tests, the control *Rpe65*[-/-] mice expressed a hypolocomotive phenotype with respect to the C57 controls (p = 0.014). Assessment of the injected *Rpe65*[-/-] group showed some of the animals (n = 3) exhibited a small amount of recovery from the *Rpe65*[-/-] hypolocomotive phenotype, however a majority of the group displayed the same phenotype as the control *Rpe65*[-/-]s (data not shown, C57 p = 0.213; control *Rpe65*[-/-] p = 0.182).

The acrophase of a circadian rhythm shows the peak of the 24 hour cycling component of activity, occurring in the dark phase of the photoperiod for motor activity in nocturnal animals. By tracking the change in acrophase following light reversal, the re-entrainment rate of the circadian rhythm can be assessed. Upon analysis, the C57 control group exhibited rapid re-entrainment, which differed significantly from both the control and injected

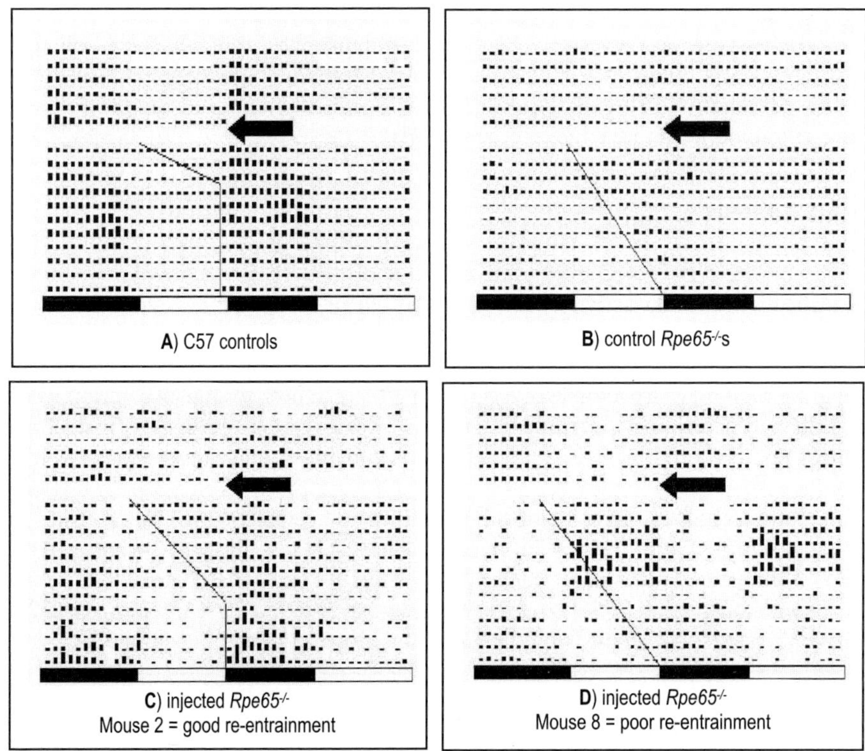

Figure 34.1. Double plotted actograms for **(A)** control C57 counts, **(B)** control *Rpe65*[-/-] counts and **(C)**-**(D)** individual injected *Rpe65*[-/-] mice. The two control groups, C57 and control *Rpe65*[-/-], are plotted as group means, while the two gene treated examples represent individual mice to demonstrate the variable nature of the results. Mouse 2 **(C)** shows a positive outcome from the gene therapy, with some re-entrainment occurring, while mouse 8 **(D)** shows no recovery after therapy.

Rpe65[-/-] groups ($F_{(2,21)}$ = 8.138, p = 0.002; Fig. 34.2A). The control *Rpe65*[-/-] and injected *Rpe65*[-/-] mice did not differ significantly from each other (p = 0.331). A small proportion (n = 3) of injected *Rpe65*[-/-] mice showed an enhanced ability to re-entrain their acrophase after light reversal (Fig. 34.2B), showing faster re-entrainment compared to control *Rpe65*[-/-] mice (p = 0.002).

4. DISCUSSION

In the current work, we test a new mouse behavioral method for its ability to asses GT efficacy in rAAV.RPE65-injected *Rpe65*[-/-] mice. Using this method, the injected mice displayed an overall phenotype indicative of dystrophic retinas, as shown by the hypolocomotive activity levels and long re-entrainment of acrophase. Both of these features were also seen in the control *Rpe65*[-/-] mice. An advantage of using a behavioral, circadian rhythm-based approach, with the collection of hourly data over a number of days, is that the

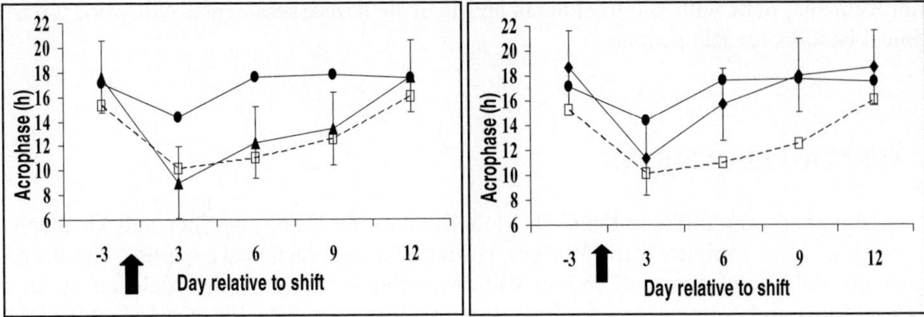

Figure 34.2. The shift in acrophase of injected *Rpe65*[-/-] mice and controls, following a 12:12 h change in photoperiod. **A)** When taken as a group, the injected *Rpe65*[-/-] mice (▲, n = 8) show very small improvements in their ability to re-entrain to the new photoperiod, relative to the control *Rpe65*[-/-] mice (□, n = 8). Their rate of entrainment remains significantly reduced relative to the C57 controls (●, n = 8). **B)** When the injected *Rpe65*[-/-] mice were analyzed individually, a small group (◆, n = 3) showed a significantly enhanced re-entrainment rate. Arrows show the time of light reversal. Error bars represent critical difference of pair wise comparisons using the multiple F-test (α = 0.05). Since critical difference calculations account for both between-subjects and within-subjects variance measures the error bars are only included on the test group and they refer to a confidence interval outside of which values are significantly different.

statistical analyses can be applied to individual mice. When the injected *Rpe65*[-/-] mice were analyzed individually, 3 of the 8 mice showed promising results. These 3 gene-treated animals re-entrained to the light reversal in a shorter time than the control *Rpe65*[-/-]s, suggesting a greater degree of light perception in these animals. However the proportion of animals showing improvements, the small effect size of the changes and relatively low power meant that these individual improvements were insufficient to show significant improvement in the injected group as a whole.

The behavioral tests were performed using the *Rpe65*[-/-] mouse model of LCA, which has previously undergone GT treatment using an rAAV-based vector (rAAV.RPE65).[3,12] Previous work with this system has shown that rAAV.RPE65 injection has some capacity to induce recovery in the mice, with distinct RPE65 expression, improved ERG signals and the presence of cone opsin immunoreactivity.[3] However, like the behavioral analysis undertaken here, these changes were small and mainly short term, and unable to make a significant impact on the long-term outlook of retina degeneration in these mice.[3] The small responses seen from the behavioral assay therefore agree with the magnitude of those seen previously,[3,12] and thus indicates that, if used as part of an overall GT assessment, the technique may be a useful addition to the already established techniques. However, the assessment in this case was limited by the small responses obtained from the *Rpe65*[-/-] mouse, and confirmation of the behavioral assay's usefulness will need to wait advancements in GT technology and/or approaches.[3,12]

In conclusion, we tested a new behavioral assay to measure the outcome of gene therapy in the *Rpe65*[-/-] knockout mouse model. This new approach produced results of similar magnitude to those seen previously with this model, and we observed some promising outcomes within a subgroup of the injected mice. However the limited success of these trials was noted, and further work will be required to confirm these results. It is hopeful that future improvements, both with GT treatments and their *in vivo* assessments, will work towards clinical benefits for RD patients.

5. ACKNOWLEDGEMENTS

The authors would like to thank Dr Mela Brankov, Dr Wei-Yong Shen and Mr Stephen Moore, Lions Eye Institute, Perth, Western Australia for their technical expertise. The *Rpe65*[-/-] knockout mice[11] and goat anti mouse RPE65 antibody[13] were kindly provided by Dr T. Michael Redmond, National Eye Institute, National Institute of Health, Bethesda, MD, USA. This project was funded in part by the National Health and Medical Research Council of Australia and Retina Australia.

6. REFERENCES

1. P. Gouras, J. Kong and S. H. Tsang, Retinal degeneration and RPE transplantation in *Rpe65*[-/-] mice, *Invest. Ophthalmol. Vis. Sci.* **43**(10), 3307-3311 (2002).
2. J. P. Van Hooser, Y. Liang, T. Maeda, V. Kuksa, G. F. Jang, Y. G. He, F. Rieke, H. K. Fong, P. B. Detwiler and K. Palczewski, Recovery of visual functions in a mouse model of Leber Congenital Amaurosis, *J. Biol. Chem.* **277**(21), 19173-19182 (2002).
3. C. M. Lai, M. J. Yu, M. Brankov, N. L. Barnett, X. Zhou, T. M. Redmond, K. Narfstrom and P. E. Rakoczy, Recombinant adeno-associated virus type 2-mediated gene delivery into the *Rpe65*[-/-] knockout mouse eye results in limited rescue, *Genet. Vacc. Ther.* **2**(1), 3 (2004).
4. D. M. Daniels, C. W. Stoddart, M. T. Martin-Iverson, C. M. Lai, T. M. Redmond and P. E. Rakoczy, Entrainment of circadian rhythm to a photoperiod reversal shows retinal dystrophy in *Rpe65*[-/-] mice, *Physiol. Behav.* **79**(4-5), 701-711 (2003).
5. G. M. Acland, G. D. Aguirre, J. Ray, Q. Zhang, T. S. Aleman, A. V. Cideciyan, S. E. Pearce-Kelling, V. Anand, Y. Zeng, A. M. Maguire, S. G. Jacobson, W. W. Hauswirth and J. Bennett, Gene therapy restores vision in a canine model of childhood blindness, *Nat. Genet.* **28**(1), 92-95 (2001).
6. K. Narfstrom, M. L. Katz, R. Bragadottir, M. Seeliger, A. Boulanger, T. M. Redmond, L. Caro, C. M. Lai and P. E. Rakoczy, Functional and structural recovery of the retina after gene therapy in the RPE65 null mutation dog, *Invest. Ophthalmol. Vis. Sci.* **44**(4), 1663-1672 (2003).
7. K. Narfstrom, M. L. Katz, M. Ford, T. M. Redmond, E. Rakoczy and R. Bragadottir, In vivo gene therapy in young and adult RPE65-/- dogs produces long-term visual improvement, *J. Hered.* **94**(1), 31-37 (2003).
8. R. J. Lucas, M. S. Freedman, M. Munoz, J. M. Garcia-Fernandez and R. G. Foster, Regulation of the mammalian pineal by non-rod, non-cone, ocular photoreceptors, *Science* **284**(5413), 505-507 (1999).
9. I. Provencio and R. G. Foster, Circadian rhythms in mice can be regulated by photoreceptors with cone-like characteristics, *Brain Res.* **694**(1-2), 183-190 (1995).
10. C. P. Selby, C. Thompson, T. M. Schmitz, R. N. Van Gelder and A. Sancar, Functional redundancy of cryptochromes and classical photoreceptors for nonvisual ocular photoreception in mice, *Proc. Natl. Acad. Sci. U.S.A* **97**(26), 14697-14702 (2000).
11. T. M. Redmond, S. Yu, E. Lee, D. Bok, D. Hamasaki, N. Chen, P. Goletz, J. X. Ma, R. K. Crouch and K. Pfeifer, RPE65 is necessary for production of 11-cis-vitamin A in the retinal visual cycle, *Nat. Genet.* **20**(4), 344-351 (1998).
12. P. E. Rakoczy, C. M. Lai, M. J. T. Yu, D. M. Daniels, M. Brankov, B. C. Rae, C. W. Stoddart, N. L. Barnett, M. T. Martin-Iverson, T. M. Redmond, K. Narfstrom, X. Zhou and I. J. Constable, in: *Retinal degeneration mechanisms and experimental therapy*, edited by J. G. Hollyfield, M. M. La Vail (Luwer Academic / Plenum Publishers, New York, 2003), pp. 431-438.
13. T. M. Redmond and C. P. Hamel, Genetic analysis of RPE65: From human disease to mouse model, *Methods Enzymol.* **316**, 705-724 (2000).

14. F. Halberg, M. Engeli, C. Hamburger and D. Hillman, Spectral resolution of low-frequency, small-amplitude rhythms in excreted 17-ketosteroids; probable androgen-induced circaseptan desynchronization, *Acta Endocrinol. (Copenh.)* **50**, Suppl-54 (1965).
15. M. W. Vasey and J. F. Thayer, The continuing problem of false positives in repeated measures ANOVA in psychophysiology: A multivariate solution, *Psychophysiology* **24**(4), 479-486 (1987).
16. P. P. Vitaliano, Parametric statistical analysis of repeated measures experiments, *Psychoneuro-endocrinology* **7**(1), 3-13 (1982).
17. S. Ebihara and K. Tsuji, Entrainment of the circadian activity rhythm to the light cycle: Effective light intensity for a zeitgeber in the retinal degenerate C3h mouse and the normal C57Bl mouse, *Physiol. Behav.* **24**(3), 523-527 (1980).
18. C. Kopp, E. Vogel, M. C. Rettori, P. Delagrange and R. Misslin, Re-entrainment of the spontaneous locomotor activity rhythm to a daylight reversal in C57Bl/6 and C3h/he mice: Implication of melatonin, *Physiol. Behav.* **70**(1-2), 171-176 (2000).

LENTIVIRAL VECTORS CONTAINING A RETINAL PIGMENT EPITHELIUM SPECIFIC PROMOTER FOR LEBER CONGENITAL AMAUROSIS GENE THERAPY

Lentiviral gene therapy for LCA

Alexis-Pierre Bemelmans[1], Corinne Kostic[1], Dana Hornfeld[1],
Muriel Jaquet[1], Sylvain V. Crippa[1], William W. Hauswirth[2], Janis Lem[3],
Zhongyan Wang[3], Daniel F. Schorderet[1,4], Francis L. Munier[1],
Andreas Wenzel[5], and Yvan Arsenijevic[1,6]

1. INTRODUCTION

Leber congenital amaurosis (LCA) is a retinitis pigmentosa with early onset, leading to blindness in infants. There is currently no efficient therapy to treat LCA. At the present time, mutations in seven different genes have been associated with the disease (Hanein et al. 2004). In 10 to 15% of the cases LCA originates from a mutation in RPE65 (Gu et al. 1997), a gene specifically expressed in the cells of the retinal pigment epithelium layer (RPE cells). This gene encodes a 65 kD protein the function of which has been dissected in a recently published study demonstrating its crucial role as a regulator of the visual cycle and a chaperone for the chromophore of the visual pigment (Xue et al. 2004). The patients affected by a mutation in this gene could benefit from a substitutive gene therapy consisting in the transfer of a fully functional allele of the RPE65 gene in RPE cells. Furthermore, animal models of RPE65 mutations have been identified (Aguirre et al. 1998; Veske et al. 1999) or genetically produced (Redmond et al. 1998) and thus provide the necessary tools to set up the conditions of such a strategy before a clinical trial can be started. The proof

[1] Oculogenetics Unit, Hôpital Ophtalmique Jules Gonin, Lausanne, Switzerland; [2] Dept. of Ophthalmology and Powell Gene Therapy Center, University of Florida, Gainesville, USA; [3] Dept. of Ophthalmology and Program in Genetics, Tufts University School of Medicine, Boston, USA; [4] IRO, Institut de Recherche en Ophtalmologie, Sion, Switzerland; [5] Laboratory for Retinal Cell Biology, Dept. of Ophthalmology, University of Zürich, Zürich, Switzerland; [6] Corresponding author: 15 avenue de France, Case Postale 133, CH-1000 Lausanne 7, Switzerland; Tel: +41 21 626 8260; Fax: +41 21 626 8888; yvan.arsenijevic@ophtal.vd.ch.

of feasibility of this approach has indeed already been established in dogs bearing a spontaneous mutation in the RPE65 gene (Acland et al. 2001; Narfström et al. 2003), as well as in knock-out mice (Dejneka et al. 2004; Lai et al. 2004). These studies have shown that an adeno-associated virus (AAV)-derived vector is able to deliver the RPE65 gene to RPE cells and thus to restore vision at least partially. Nevertheless, before a clinical trial can take place, a great effort must be provided to assess the bio-safety of the procedure. In particular, transgene expression has to be tightly controlled to achieve the following criteria: (i) expression should occur only in RPE cells; (ii) expression should reach the therapeutic level without disturbing the homeostasis of the target cells.

Adenovirus, AAV and lentivirus-derived vectors have been successfully used to transfer genes into retinal cells *in vivo*. Although adenoviruses are able to transduce RPE cells with a high efficiency, and photoreceptors to a lesser extent (Bennett et al. 1994; Li et al. 1994), they trigger an immune response which leads to the rejection of the transduced cells (Hoffman et al. 1997; Kumar-Singh and Farber 1998; Reichel et al. 1998). AAV-derived vectors are able to transduce RPE cells and photoreceptors, or photoreceptors alone, depending on the serotype, and allow for long-lasting transgene expression (Ali et al. 1996; Bennett et al. 1997; Flannery et al. 1997). Nevertheless, the delay that has been occasionally observed between AAV administration and transgene expression (Bennett et al. 1999) could become a serious hurdle in the case of LCA where treatment efficiency is desirable as early as in neonate. Furthermore, in the case of RPE65 mutations, the treatment by gene transfer will certainly have to remain active for the entire lifespan, a prerequisite that will be more likely fulfilled with an integrative vector. The lentivirus-derived vector is such a candidate, because its integrating properties make it particularly interesting for a "life-long treatment". This type of vector is well known to target RPE cells in a highly predominant fashion after administration by injection into the subretinal space (Miyoshi et al. 1997; Kostic et al. 2003).

To develop a vector specifically designed for the gene therapy of RPE65 mutations in human, we have constructed a lentivirus in which the expression of the transgene is driven by a 0.8 kb proximal fragment of the human RPE65 promoter (LV-R0.8). We report here the transgene expression pattern of this vector in the mouse eye as well as the ability of this vector to trigger therapeutic levels of RPE65 protein in the RPE65 knock-out mouse model.

2. METHODS

2.1. Construction of Lentiviral Vectors

The lentiviral backbones used in this study were derived from the Hlox-EFS-GFP vector described in (Salmon et al. 2000). In the control vector (LV-R0.8-GFP), the EFS promoter was replaced by a fragment of 0.8 kb of the human RPE65 promoter. In the therapeutic vector (LV-R0.8-RPE65), after addition of the R0.8 promoter, the GFP coding sequence was replaced by the mouse RPE65 cDNA. Viral particles were produced by transient transfection of 293T cells as previously described (Naldini et al. 1996). Briefly, 293T cells were co-transfected by the vector plasmid, an HIV-1 packaging plasmid and an envelope plasmid encoding VSV-G. Two days after transfection, recombinant viral particles were harvested in the supernatant and concentrated by ultracentrifugation. Viral stocks were then stored in small aliquots at −80°C until use. The concentration of total viral particles was determined using ELISA quantification of the p24 capsid protein. Infectious activity was assessed on infected 293T cells by flow cytometry detection of GFP-positive cells for the LV-R0.8-GFP vector and by western blot detection of RPE65 for the LV-R0.8-RPE65 vector.

2.2. Animal Treatment

All the mice used in this study were from the RPE65 knock-out line (Redmond et al. 1998). For vector injection, animals were sedated by volatile anesthesia; the temporal part of the sclera was carefully freed from the conjunctive tissue and perforated with a thin needle. A 31G Hamilton syringe was then inserted in the subretinal space and 1 µl of virus was injected.

For measurement of the corneal electroretinogram (ERG), mice were dark-adapted overnight and, under dim red light, anesthetized using ketamine and xylazine and placed in a Ganzfeld bowl. Pupils were dilated and thin silver wires were used as electrodes. Mice were then subjected to light stimuli as described in Grüter et al. (Gene Ther, in press).

2.3. Histology

To evidence transgene expression, animals were sacrificed by injection of pentobarbital, the eyes were then enucleated and fixed by immersion in PBS containing 4% paraformaldehyde. Eyes were then cut at 14 µm thickness on a cryostat and sections were collected on slides. Control transgene expression was evidenced by visualization of direct GFP fluorescence or GFP immunolabeling. RPE65 transgene expression was detected using a polyclonal rabbit antibody. Immunolabeling was performed as previously described (Kostic et al. 2003).

3. RESULTS AND DISCUSSION

3.1. *In Vitro* Testing of Lentiviral Vectors

We first assessed that our lentiviral vectors were able to transduce cells *in vitro*. To that aim, we used 293T cells, in which the R0.8 promoter in the lentiviral context is active. Three days after infection by LV-R0.8-GFP, cells were harvested and analyzed by flow cytometry. This allowed us to evidence a population of GFP expressing cells (data not shown), reflecting a titer of 5.10exp8 transducing units per ml of viral solution (TU/ml).

The activity of the LV-R0.8-RPE65 vector was assessed by Western blot analysis. We detected a band corresponding to RPE65 in cytosolic extracts of infected 293T cells, as well as in those of 293T cells transfected by the vector plasmid, the higher intensity of the band in the latter probably reflecting the higher copy number of transgenes per cell with the transfection method (Fig. 35.1).

3.2. Activity of the R0.8 Promoter in the Mouse Retina

We then investigated the pattern of transgene expression of the LV-R0.8-GFP vector in the eye of adult RPE65 knock-out mice. After subretinal injection into the adult, GFP was exclusively detected in the RPE cells across a wide span of the retina (Fig. 35.2a),

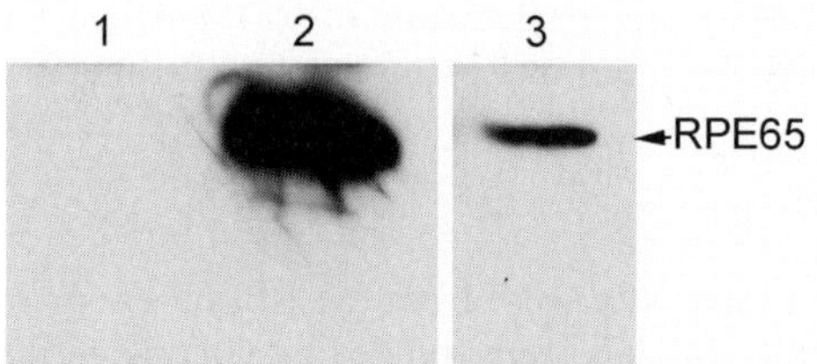

Figure 35.1. Western blot analysis of RPE65 expression. 293T cells were infected with LV-R0.8-RPE65 and cytosolic extracts were prepared three days later. For each condition, 25 µg proteins were loaded on SDS-PAGE and subsequently transferred to a PVDF membrane. RPE65 was detected using a rabbit polyclonal antibody, a HRP-linked secondary antibody (Amersham), and the ECL[+] detection kit (Amersham). 1: uninfected cells; 2: cells transfected with the vector plasmid; 3: cells infected with the LV-R0.8-RPE65 vector.

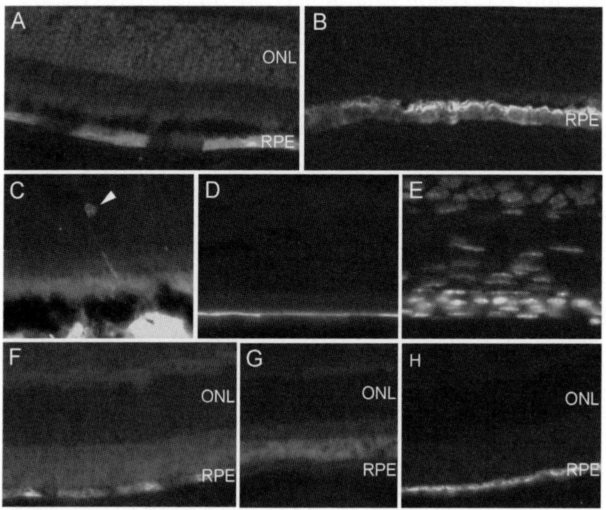

Figure 35.2. Immunolabeling of GFP and RPE65 one week after injection of LV-R0.8-GFP (A-E) or LV-R0.8-RPE65 (F) into the subretinal space of adult (A and F) or P5 (B-E) mouse retina. A and B show examples of GFP-expressing RPE cells. Dapi counterstaining in A allows visualization of the outer nuclear layer. Arrowhead in C indicates the nucleus of a photoreceptor expressing GFP; note the presence of GFP in the corresponding segment. D shows an example of GFP expression located in cells of the corneal epithelium (E: dapi counterstaining of D). F: immunolabeling of RPE65 in an adult RPE65 knock-out mouse one week after injection of LV-R0.8-RPE65. Expression of the transgenic protein was restricted to RPE cells. G: non-injected knock-out mouse. H: non-injected wild type control. ONL: outer nuclear layer; RPE: retinal pigment epithelium.

confirming previous studies that have demonstrated the high capacity of lentiviral vectors to transduce RPE cells (Miyoshi et al. 1997; Bainbridge et al. 2001; Kostic et al. 2003).

We also performed intravitreal injections of LV-R0.8-GFP into the eye of mouse pups of 5 days of age (P5). This led to high GFP expression in the RPE cells (Fig. 35.2b). Surprisingly, the external surface of the corneal epithelium also expressed a high level of GFP (Fig. 35.2d-e). Although the injection procedure surely resulted in the diffusion of the vector on the eye surface, it was unlikely to observe GFP here with the R0.8 promoter. The reason for this expression remains unclear, but a transient activation of the RPE65 promoter in these regions cannot be excluded, knowing that RPE65 is expressed in human keratinocytes (Hinterhuber et al. 2004). Occasionally, we also observed some cells expressing GFP in the photoreceptor layer (Fig. 35.2c). The differences in expression pattern between the adult and neonates illustrate the fact that to target a specific cell type, one has to consider the combination of the vector tropism, the injection site, and the specificity of the promoter used. In the present study, the cause of the absence of GFP in photoreceptors after gene transfer in the adult is probably due to an absence of infection of these cells – the vector being known to poorly transduce adult photoreceptors (Kostic et al. 2003). In addition, endogenous RPE65 is not expressed in rods, which represent more than 95% of the mouse photoreceptors, and its expression in cones is unclear (Seeliger et al. 2001; Znoiko et al. 2002).

3.3. Therapeutic Level of RPE65 in the Retina

We next injected the LV-R0.8-RPE65 therapeutic vector into the subretinal space of 2.5 months old RPE65–/– mice. At this age the ERG of these mice is nearly flat (Seeliger et al. 2001). One week after injection, a few mice were sacrificed to assess expression of the transgene. This allowed us to detect significant levels of RPE65 protein in the RPE cells of injected animals (Fig. 35.2f), whereas no RPE65 could be detected in non-injected RPE65–/– mice (Fig. 35.2g). Nevertheless, the level of expression of RPE65 appeared to be weaker in LV-R0.8-RPE65 treated RPE65–/– mice than in wild type controls (compare Fig. 35.2f and 2h).

To assess that the expression of RPE65 was sufficient to reach a therapeutic level, we measured the ERG of the treated animals at three months, i.e. two weeks after injection of the lentiviral vector. In some animals we detected an ERG response characterized by a b-wave amplitude that could reach as much as 50% of wild type controls (Fig. 35.3).

4. CONCLUSION

In the present study we report the first successful attempt to restore visual function in RPE65 knockout mice by injection of a lentivirus encoding RPE65, thus demonstrating that lentiviral vectors could be valuable tools to treat LCA. Moreover, we report that the R0.8 promoter fragment was efficient to drive the expression of therapeutic levels of RPE65 protein. The use of the GFP reporter gene revealed that in the context of a lentiviral-mediated gene transfer, R0.8 activity was specific for RPE cells in adult mice. In newborn mice, however, we detected additional expression in the corneal epithelium and in few photoreceptors. Nevertheless, this phenomenon occurred only after injection in neonates, for which (i) the injection procedure is difficult to control and (ii) the differences with the adult extracellular matrix could explain the transduction of photoreceptors.

There is now accumulating body of evidence that RPE65 gene transfer using AAV and the chicken beta-actin promoter in animal models of LCA is beneficial over the long term for visual function (Bennett 2004). For the treatment of human patients, it is nevertheless mandatory to test alternative promoters and vectors in animal models of RPE65 mutations, to achieve safe, sufficient and long-lasting expression of RPE65. Lentiviral vector and RPE65 promoter might provide such an alternative.

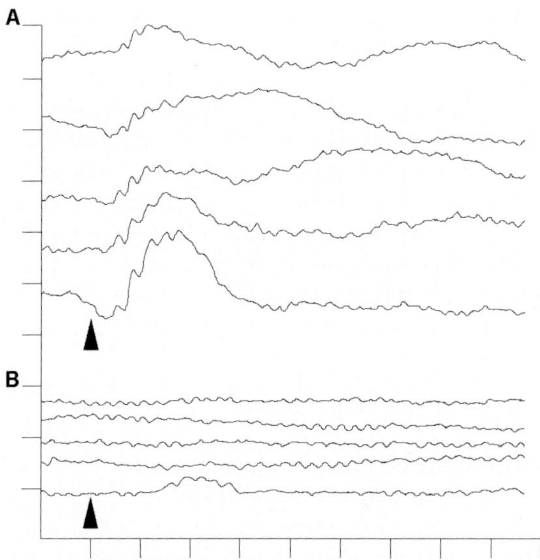

Figure 35.3. Representative single-flash ERG of RPE65-/- mice recorded in scotopic condition. The arrowheads indicate the time of stimulation. From top to bottom: single flash recordings with the stimulus intensity increasing from 0.3 to 25 cd.s.m^{-2}. Vertical scale: 200 μV/div; horizontal scale: 40 msec/div. A: recording of a three months old RPE65-/- mouse two weeks after subretinal injection of LV-R0.8-RPE65 vector. The vehicle-injected contralateral control eye showed no recordable ERG (data not shown). B: recording of a 2.5 months old RPE65-/- mouse just before surgery. The electroretinogram response is nearly absent at this stage.

5. ACKNOWLEDGEMENTS

This work was supported by the ProVisu Foundation.

6. REFERENCES

Acland, G. M., Aguirre, G. D., Ray, J., Zhang, Q., Aleman, T. S., et al., 2001, Gene therapy restores vision in a canine model of childhood blindness, *Nat Genet* **28**:92.

Aguirre, G. D., Baldwin, V., Pearce-Kelling, S., Narfström, K., Ray, K., and Acland, G. M., 1998, Congenital stationary night blindness in the dog: Common mutation in the RPE65 gene indicates founder effect, *Mol Vis* **4**:23.

Ali, R. R., Reichel, M. B., Thrasher, A. J., Levinsky, R. J., Kinnon, C., Kanuga, N., Hunt, D. M., and Bhattacharya, S. S., 1996, Gene transfer into the mouse retina mediated by an adeno-associated viral vector, *Hum Mol Genet* **5**:591.

Bainbridge, J. W., Stephens, C., Parsley, K., Demaison, C., Halfyard, A., Thrasher, A. J., and Ali, R. R., 2001, In vivo gene transfer to the mouse eye using an hiv-based lentiviral vector; efficient long-term transduction of corneal endothelium and retinal pigment epithelium, *Gene Ther* **8**:1665.

Bennett, J., 2004, Gene therapy for leber congenital amaurosis, *Novartis Found Symp* **255**:195.

Bennett, J., Duan, D., Engelhardt, J. F., and Maguire, A. M., 1997, Real-time, noninvasive in vivo assessment of adeno-associated virus-mediated retinal transduction, *Invest Ophthalmol Vis Sci* **38**:2857.

Bennett, J., Maguire, A. M., Cideciyan, A. V., Schnell, M., Glover, E., et al., 1999, Stable transgene expression in rod photoreceptors after recombinant adeno-associated virus-mediated gene transfer to monkey retina, *Proc Natl Acad Sci U S A* **96**:9920.

Bennett, J., Wilson, J., Sun, D., Forbes, B., and Maguire, A., 1994, Adenovirus vector-mediated in vivo gene transfer into adult murine retina, *Invest Ophthalmol Vis Sci* **35**:2535.

Dejneka, N. S., Surace, E. M., Aleman, T. S., Cideciyan, A. V., Lyubarsky, A., et al., 2004, In utero gene therapy rescues vision in a murine model of congenital blindness, *Mol Ther* **9**:182.

Flannery, J. G., Zolotukhin, S., Vaquero, M. I., LaVail, M. M., Muzyczka, N., and Hauswirth, W. W., 1997, Efficient photoreceptor-targeted gene expression in vivo by recombinant adeno-associated virus, *Proc Natl Acad Sci U S A* **94**:6916.

Gu, S. M., Thompson, D. A., Srikumari, C. R., Lorenz, B., Finckh, U., et al., 1997, Mutations in RPE65 cause autosomal recessive childhood-onset severe retinal dystrophy, *Nat Genet* **17**:194.

Hanein, S., Perrault, I., Gerber, S., Tanguy, G., Barbet, F., et al., 2004, Leber congenital amaurosis: Comprehensive survey of the genetic heterogeneity, refinement of the clinical definition, and genotype-phenotype correlations as a strategy for molecular diagnosis, *Hum Mutat* **23**:306.

Hinterhuber, G., Cauza, K., Brugger, K., Dingelmaier-Hovorka, R., Horvat, R., Wolff, K., and Foedinger, D., 2004, RPE65 of retinal pigment epithelium, a putative receptor molecule for plasma retinol-binding protein, is expressed in human keratinocytes, *J Invest Dermatol* **122**:406.

Hoffman, L. M., Maguire, A. M., and Bennett, J., 1997, Cell-mediated immune response and stability of intraocular transgene expression after adenovirus-mediated delivery, *Invest Ophthalmol Vis Sci* **38**:2224.

Kostic, C., Chiodini, F., Salmon, P., Wiznerowicz, M., Deglon, N., et al., 2003, Activity analysis of housekeeping promoters using self-inactivating lentiviral vector delivery into the mouse retina, *Gene Ther* **10**:818.

Kumar-Singh, R., and Farber, D. B., 1998, Encapsidated adenovirus mini-chromosome-mediated delivery of genes to the retina: Application to the rescue of photoreceptor degeneration, *Hum Mol Genet* **7**:1893.

Lai, C. M., Yu, M. J., Brankov, M., Barnett, N. L., Zhou, X., Redmond, T. M., Narfström, K., and Rakoczy, P. E., 2004, Recombinant adeno-associated virus type 2-mediated gene delivery into the RPE65−/− knockout mouse eye results in limited rescue, *Genet Vaccines Ther* **2**:3.

Li, T., Adamian, M., Roof, D. J., Berson, E. L., Dryja, T. P., Roessler, B. J., and Davidson, B. L., 1994, In vivo transfer of a reporter gene to the retina mediated by an adenoviral vector, *Invest Ophthalmol Vis Sci* **35**:2543.

Miyoshi, H., Takahashi, M., Gage, F. H., and Verma, I. M., 1997, Stable and efficient gene transfer into the retina using an hiv-based lentiviral vector, *Proc Natl Acad Sci U S A* **94**:10319.

Naldini, L., Blömer, U., Gallay, P., Ory, D., Mulligan, R., Gage, F. H., Verma, I. M., and Trono, D., 1996, In vivo gene delivery and stable transduction of nondividing cells by a lentiviral vector, *Science* **272**:263.

Narfström, K., Katz, M. L., Bragadottir, R., Seeliger, M., Boulanger, A., Redmond, T. M., Caro, L., Lai, C. M., and Rakoczy, P. E., 2003, Functional and structural recovery of the retina after gene therapy in the RPE65 null mutation dog, *Invest Ophthalmol Vis Sci* **44**:1663.

Redmond, T. M., Yu, S., Lee, E., Bok, D., Hamasaki, D., Chen, N., Goletz, P., Ma, J. X., Crouch, R. K., and Pfeifer, K., 1998, RPE65 is necessary for production of 11-cis-vitamin a in the retinal visual cycle, *Nat Genet* **20**:344.

Reichel, M. B., Ali, R. R., Thrasher, A. J., Hunt, D. M., Bhattacharya, S. S., and Baker, D., 1998, Immune responses limit adenovirally mediated gene expression in the adult mouse eye, *Gene Ther* **5**:1038.

Salmon, P., Oberholzer, J., Occhiodoro, T., Morel, P., Lou, J., and Trono, D., 2000, Reversible immortalization of human primary cells by lentivector-mediated transfer of specific genes, *Mol Ther* **2**:404.

Seeliger, M. W., Grimm, C., Stahlberg, F., Friedburg, C., Jaissle, G., et al., 2001, New views on RPE65 deficiency: The rod system is the source of vision in a mouse model of leber congenital amaurosis, *Nat Genet* **29**:70.

Veske, A., Nilsson, S. E., Narfström, K., and Gal, A., 1999, Retinal dystrophy of swedish briard/briard-beagle dogs is due to a 4-bp deletion in RPE65, *Genomics* **57**:57.

Xue, L., Gollapalli, D. R., Maiti, P., Jahng, W. J., and Rando, R. R., 2004, A palmitoylation switch mechanism in the regulation of the visual cycle, *Cell* **117**:761.

Znoiko, S. L., Crouch, R. K., Moiseyev, G., and Ma, J. X., 2002, Identification of the RPE65 protein in mammalian cone photoreceptors, *Invest Ophthalmol Vis Sci* **43**:1604.

CHAPTER 36

GENE DELIVERY TO THE RETINA
USING LENTIVIRAL VECTORS

Kenneth P. Greenberg[1,2], Edwin S. Lee[2,3], David V. Schaffer[2,4], and
John G. Flannery[1,2]

1. INTRODUCTION

The delivery of foreign DNA to the retina has proven to be a valuable tool for investigations of retinal disease, development, and complex cellular interactions. To achieve efficient and stable retinal gene expression with minimal unwanted side effects, viral vectors derived from AAV (adeno-associated virus) and LV (lentivirus) remain the vehicles of choice. LV vectors have gained recent attention in CNS gene delivery due in part to their large transgene capacity, however contradictory results regarding retinal transduction ability exist in the literature. We sought specifically to characterize the temporal and spatial expression pattern of LV vectors when delivered to the rodent retina.

The primary goals in the development of gene transfer vectors have been to harness a virus's innate ability to deliver a gene payload to a cell, while eliminating any pathogenic potential of that virus. Evolution has cleverly accomplished the former, while the latter has been achieved through the selective removal of wild type replication and virulence genes.[1,2] Current vectors are both safe and efficient vehicles for gene transfer to the primate retina[3,4] and phase I human clinical trials are underway.[5]

2. LENTIVIRAL VECTOR DESIGN AND BIOSAFETY

Lentiviruses are a type of retrovirus capable of infecting both dividing and nondividing cells due to their unique ability to pass through a target cell's intact nuclear membrane, an essential property for CNS gene transfer. LV vectors have many desirable features including a large transgene carrying capacity (~8 kb vs. AAV's 4.8 kb) and stable integration

---

[1] Vision Science, [2] Helen Wills Neuroscience Institute, [3] Molecular and Cell Biology, and [4] Chemical Engineering, University of California, Berkeley, CA 94720, USA; Corresponding author: K. P. Greenberg, E-mail: kgreenb@berkeley.edu

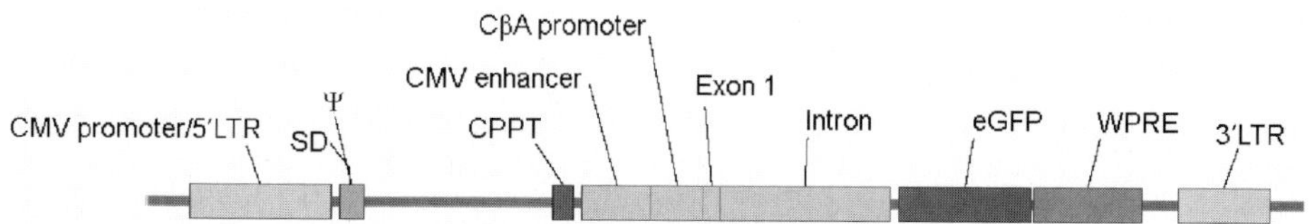

Figure 36.1. HIV-1 based lentiviral transfer vector (pFCβAGW) containing SIN LTRs, CPPT, hybrid CMV enhancer/Chicken β-Actin promoter (including exon 1 and the intron) driving eGFP, and WPRE.

leading to long-term expression. Adenovirus (Ad) vectors have an exceedingly large capacity (36 kb), however their lack of stable integration results in transient expression. LV vectors also have the apparent ability to avoid immune system inactivation, a current handicap of both Ad and AAV vectors. The most frequently used LV vectors are based on HIV-1, however vectors derived from non-human lentiviruses such as FIV (feline immunodeficiency virus), SIV (simian immunodeficiency virus), BIV (bovine immunodeficiency virus) and EIAV (equine infectious anemia virus) could theoretically provide biosafety advantages.[6-9]

An obvious concern for the delivery of HIV based vectors to humans is the potential for AIDS or an AIDS-like disease caused by a replication competent vector. This concern has been addressed by completely deleting six (env, tat, vif, vpr, vup, and nef) of the nine viral genes, keeping only those essential for gene delivery (gag, pol, and rev). Furthermore, any potential replication competent recombinant (RCR) vectors can be screened for by sensitive methods.[10,11] A second safety issue concerns the inadvertent activation of celluar oncogenes from transcriptional read-through of the HIV long terminal repeat promoter (LTR). Deletion of the 3′ LTR core promoter and replacement of the wild type U3 region of the 5′ HIV LTR with the cytomegalovirus (CMV) promoter in the current third generation of self-inactivating (SIN) vectors results is an integrated provirus with transcriptionally silent LTRs and no viral genes.[12]

Recent improvements in LV vector delivered transgene expression levels have been achieved by two methods. First, incorporation of the HIV central polypurine tract (CPPT) sequence upstream of the desired payload appears to enhance nuclear entry of the pre-integration complex and significantly increases expression levels.[13] Although the precise role of the CPPT is debated,[14,15] our results indicate that this element increases transgene expression levels in the retina (data not shown). Secondly, incorporation of the woodchuck hepatitis virus post-transcriptional regulatory element (WPRE) downstream of the transgene enhances stability of the RNA transcript and therefore increases expression levels.[16] An HIV-1 based transfer vector containing these elements is shown (Fig. 36.1).

3. TARGETING LENTIVIRAL VECTORS

Much effort is currently underway to specifically target all classes of neural, glial, and epithelial cells in the retina with viral vectors. However, photoreceptors, RPE, and ganglion cells remain the only retinal cell types successfully targeted with LV vectors.[17-19] Presently, the efficiency of delivery to photoreceptors and ganglion cells remains significantly higher with AAV vectors than LV vectors.

Methods for targeting specific cell types (altering the tropism of a virus) can be divided into two basic categories: restricting viral entry at the point of transduction or restricting transgene expression at the stage of transcription. Ideally, one could limit vector entry to a desired class of cell, therefore eliminating potential off-target effects. Although it has been attempted with LV vectors, transductional targeting has been met with limited success.[20] Ongoing research aimed at genetically inserting cell specific peptides into the envelope or discovering mutants with novel tropisms will likely resolve these difficulties.

Enveloped viruses such as lentiviruses may utilize envelope glycoproteins derived from other enveloped viruses. Pseudotyping, or the replacement of one virus's envelope glycoproteins with those from another virus, has been effective for increasing vector host cell range, increasing vector particle stability, and limiting vector entry to certain types of cells.[21-23] The majority of LV vectors used in retinal gene transfer are pseudotyped with envelope glycoproteins derived from the vesicular stomatitis virus (VSV) glycoprotein, although other pseudotypes have been tested.[24,25] Novel vector pseudotypes are frequently discovered which may have the innate ability to target certain cell types while maintaining their highly evolved and efficient delivery characteristics.[26]

In addition to pseudotyping, transcriptionally targeting specific cell types has been used with great success in the CNS. Cell specific regulatory elements inserted upstream of the transgene can direct expression to photoreceptors, hippocampal neurons, and astrocytes.[27-29] Transcriptional targeting is increasingly feasible due to the ongoing identification of cell specific promoters for virtually all classes of cells in the retina.

4. METHODS

4.1. Vector Production

All procedures involving vector production, concentration, and titration were performed in a Type IIA biosafety cabinet under strict BL2 practice. LV vectors were produced by either calcium phosphate or Lipofectamine 2000 (Invitrogen) transient transfection.

Calcium phosphate transfections were adapted from a previous protocol[1] and performed as follows. Five T-175 (Nunc) flasks were coated with poly-L-lysine (Sigma #P4832 diluted 1:10 in PBS and sterile filtered) and allowed to stand for 10 minutes before aspirating. Low passage 293T cells (ATCC #CRL-11268) were seeded at $1.2\text{-}1.5 \times 10^7$ cells per flask in 20 mL complete IMDM (IMDM + 10% FBS, 1XPen/Strep, 2 mM L-glutamine). The following day the calcium phosphate/DNA precipitate was prepared after all reagents equilibrated to room temperature. For five T-175 flasks, 158 μg transfer vector (pCS-CG or pFCβAGW), 79 μg pMDLg/pRRE, 24 μg pRSV-REV, and 55 μg pMD.G (VSVG) were mixed in a final volume of 13.9 mL sterile ddH_2O (buffered with Hepes to 2.5 mM) and 1.9 mL 2.5 M $CaCl_2$. After mixing, 15.8 mL 2X HeBS (Hepes Buffered Saline pH 7.05) was added to the $DNA/H_2O/CaCl_2$ solution and mixed by pipetting briefly. The $CaPO_4$ precipitate formed during a 1.5 minute incubation, and the reaction was quenched by adding 18.4 mL complete IMDM media. After mixing briefly, 10 mL of this solution was added to each flask which was placed in an incubator (37°C, 5% CO_2) overnight. Media was aspirated and replaced with 20 mL fresh IMDM 12 hours later. Two harvests of the cell supernatant were performed 24 hours and 48 hours after the first media change. The cell supernatant

(200 mL) was then filtered through a 0.45 μm pore PVDF Durapore filter (Millipore, Bedford, MA) and stored at 4°C until concentrated.

For Lipofectamine 2000 transfections, 293T cells were plated as described in complete IMDM lacking antibiotics. We found that for optimal transfections, the total amount of plasmid DNA can reduced by 2.25 fold, while maintaining the above ratio of four plasmids. Transfection complexes were prepared by mixing the plasmids in a final volume of 21.9 mL Opti-MEM reduced serum media (Invitrogen). In a separate reaction tube, 21.4 mL Opti-MEM media was gently mixed with 525 μL Lipofectamine 2000 reagent. Both tubes were incubated at room temperature for 5 minutes, gently mixed together, and incubated another 20 minutes. This solution was added to each of the five flasks which were placed in an incubator overnight. Transfection media was aspirated 12 hours later, cells were washed with PBS, and given 20 mL complete IMDM. The additional PBS wash was found necessary to remove transfection amine complexes which frequently caused cataracts when carried over into the injected vector preparation. Vector supernatant was harvested and filtered as described above.

4.2. Vector Concentration for *In Vivo* Use

High titer LV vector stocks were generated after two rounds of ultracentrifugation. The filtered vector supernatant (32 mL) was carefully overlaid on a 20% sucrose solution (4 mL) in six ultracentrifuge tubes (Beckman #344058) which were centrifuged at 24,000 rpm in a SW-28 rotor for 2 hours at 4°C. The supernatant was aspirated (avoiding the pellet) and 800 μL cold PBS was added to each tube and mixed by pipetting. After a 30 minute incubation on ice, the six vector/PBS tubes were pooled and overlayed on 1 mL of 20% sucrose in one ultracentrifuge tube (Beckman #344059). The vector was centrifuged in a SW-41Ti rotor at 25,000 rpm for 1.5 hours at 4°C. The supernatant was aspirated and pelleted vector was resuspended in 200 μL cold PBS. Vector was incubated on ice overnight and again mixed by pipetting. If not used immediately, vector was stored for up to one week at 4°C or flash frozen and stored at −80°C for long term.

4.3. Vector Titer Determination by Q-PCR

Both physical particle and functional biological titers may be determined by several methods including p24 ELISA, FACS, and quantitative PCR.[30,31] A particle titer estimates the amount of vector present in a preparation, however it provides no information regarding the biological function of a vector. Conversely, functional titer determination can accurately estimate the infectious ability of a vector through the quantitative detection of integrated proviral genomes by real time PCR. This method has the advantage of isolating the viral transduction event from later gene trancription and translation, which is the basis for protein expression titers (FACS). Although time consuming, one clear benefit to this approach is the ability to determine vector titer on a cell line (ie 293s) irrespective of the vector delivered promoter element. Vectors may contain cell specific promoters whose gene product is not expressed in an available cell line, and therefore titer determination based on protein expression is not feasable. Additionally, we find this method invaluable for testing vector transduction efficiency of pseudotyped or engineered vectors on primary retinal cell isolates regardless of promoter.

Functional titer was determined based on a protocol[31] by quantitative PCR as follows. Cultured 293T cells were infected with serial dilutions of vector (10^{-3}-10^{-7}) in 1.0 mL media with 8 μg/mL polybrene in a six well plate (2-5 × 10^5 cells/well). Cells were incubated for at least 4-5 days and washed multiple times to remove residual plasmid carried over from vector production. The transduced cells were then trypsinized, counted, and DNA from $1 × 10^6$ cells from each well was isolated (Gentra Puregene #D-5000A). The total amount of DNA from each sample was normalized and 5 μl was added to each Q-PCR reaction (ABI #N808-0228) containing 3.5 mM $MgCl_2$, 200 μM each DNTP, 320 nM each primer, 320 nM probe, 0.025 U/μL amplitaq, 2.5 μL reaction buffer, and ddH_2O to 25 μL. Primers (Fwd-ACCTGAAAGCGAAAGGGAAAC, Rev-CACCCATC TCTCTCCTTCTAGCC) and probe (5′FAM-AGCTCTCTCGACGCAGGACTCGGC-BHQ-3′ Biosearch Technologies) sequences are specific to the HIV-1 packaging signal (Ψ) and may be used with any HIV-1 based vectors containing this element. A standard curve was generated by amplification of a spectrophotometrically predetermined quantity (10^{10}-10^2 molecules/reaction) of transfer vector plasmid containing the HIV-1 packaging sequence.

Each reaction was performed in triplicate under the following conditions in a Stratagene Mx-3000P thermocycler: 1 cycle of 95°C for 10 minutes, 40 cycles of 95°C for 15 seconds and 60°C for 2 minutes. The thermocycler was set to detect and report fluorescence during the annealing/extension step of each cycle. A standard curve was generated by plotting theshold cycles vs. copy number and vector DNA titer in TU/mL (transducing units/mL) was determined at multiple dilutions (Fig. 36.2).

An RNA based particle titer may also be determined using Quantitative Reverse Transcriptase PCR (QRT-PCR). Serial dilutions of vector stock were prepared in PBS, RNA was extracted (QIAamp MinElute Virus Kit Qiagen #57714), and residual DNA removed while RNA was bound to the purification column (Qiagen Rnase-Free Dnase set #79254). QRT-PCR reactions (Stratagene Brilliant QRT-PCR Master Mix Kit #600551) were prepared as follows: 1X QRT-PCR Master Mix, 320 nM each primer (see above), 320 nM probe (see above), 0.375 μL of 1:500 diluted reference dye, 0.1 μL StrataScript RT/Rnase, and ddH_2O to 25 μL. Reactions were performed in triplicate and reactions lacking RT were used to deter-

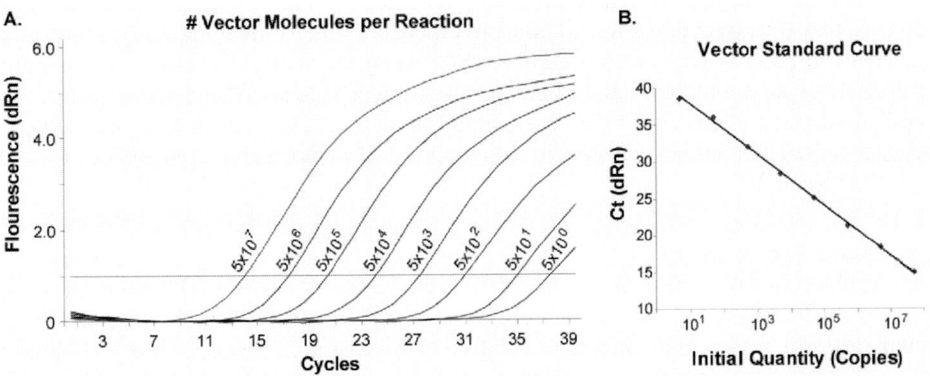

Figure 36.2. Q-PCR Amplification plot of pCS-CG plasmid (A), and standard curve (B) generated by plotting threshold cycle (Ct) against number of vector DNA molecules.

mine backgound DNA amplification. Cycling conditions were as described above with the addition of an initial 48°C RT cycle for 30 minutes. RNA titer was determined by using transfer vector plasmid as the standard after subtracting out background signal from the reactions lacking RT.

4.4. Intraocular Injection Procedure

All procedures used were in accordance with the ARVO Statement for the Use of Animals in Ophthalmic and Vision Research and were approved by the University of California, Berkeley Committee on Animal Research. C57BL/6 mice and Sprague-Dawley rats were used in all the studies. For LV time course studies, C57BL/6 mice aged between P4-P17 were used. Animals were anesthetized by intraperitoneal injection of ketamine/xylazine and eyes were dilated using 2.5% phenylephrine hydrochloride and 1% atropine sulfate. A shelving puncture was made through the sclera with a sharp 30-gauge needle, followed by a Hamilton syringe equipped with a blunt 33-gauge needle. For sub-retinal injections, the tip of the needle was advanced through the sclera, choroid, retina, and vitreous, and the needle penetrated the superior central retina to deliver the vector (0.5-3μL) into the subretinal space. We found this approach to be most successful in avoiding damage to the lens. Intravitreal injections were performed by delivering the vector (2-10μL) directly into the vitreous body. Immediately after injection, the quality (ie. lack of hemorrhage) and size of the subretinal bleb were evaluated under a stereo microscope by visualizing through a cover slip with Celluvisc (Allergan, Irvine, CA) placed on the cornea.

4.5. *In Vivo* GFP Imaging

GFP expression was evaluated *in vivo* 3-60 days after injection of LV vectors. A Retcam II (Massie Research, Pleasanton, CA) was used for fluorescent and visible light fundus imaging in live anesthetized rodents. The Retcam II is a contact fiber-optic, digital, color fundus camera originally developed for wide-field pediatric retinal imaging. The RetCam II is based around a 3 CCD medical grade digital camera providing high resolution 24 bit color images and a 20 second real time video capture mode. The Retcam II's handheld camera unit and foot controlled focus and incident light intensity made imaging the immobilized rodent retina an extremely rapid and efficient procedure. *In vivo* imaging was performed on anesthetized rodents with dilated pupils, while using Genteal gel (Novartis Opthalmics, Duluth, GA) as a contact medium between the camera lens and the rodent's cornea. The onset, duration, and extent of GFP expression was tracked in this manner. Additionally, the Retcam II was extremely useful for imaging subretinal detachments immediately after LV vector delivery.

4.6. Tissue Preparation

Eyes were enucleated from animals injected with LV-CMV-GFP or LV-CβA-GFP at 10-60 days post-injection. Eye cups were fixed in 4% paraformaldehyde in PBS for 1 hour at 4°C and washed in PBS. Eyes were cryoprotected in 15% sucrose for 2 hours followed by 30% sucrose overnight at 4°C, embedded in OCT, and flash frozen in a dry ice/ethanol slurry. Sections were cut (10μM thick) using a CM1850 cryostat (Leica, Nussloch,

Germany) and were thaw mounted on Superfrost Plus slides (Fisher Scientific). Images were acquired using a Zeiss Axiophot epifluorescence microscope (Thornwood, NY).

5. RESULTS

5.1. Q-PCR Vector Titers

DNA based functional vector titers in the cell supernatant ranged from 5×10^6-2×10^7 TU/mL before and 7×10^8-1×10^{10} TU/mL after concentration. RNA based particle titers were 3×10^8-8×10^9 particles/mL in the supernatant, and 6×10^{10}-2×10^{12} particles/mL after concentration. Taking the difference between RNA and DNA titers, we found the functional vector : inactive particle ratio to be from 1 : 100 to 1 : 1000. GFP titers were also determined by direct visualization for some vector batches and were found to be slightly lower than Q-PCR determined functional titers. Titers of vectors produced by Lipofectamine 2000 transfection were routinely higher than those produced by calcium phosphate transfection.

5.2. *In Vivo* GFP Expression

GFP expression was detectable within 3 days after injection and persisted for at least 6 months. When delivered in a subretinal injection to adult animals, we found VSV-CMV-GFP and VSV-CβA-GFP vectors to have a strong cellular tropism for the RPE. Intravitreal delivery of LV was inefficient and resulted in occasional expression in cells of the GCL (not shown). Retcam II imaging revealed GFP expression covering a large surface area after subretinal LV injection with little indication of retinal trauma (Fig. 36.3A). Of the five available interchangeable lenses, we found the Retcam II equipped with a wide angle 130 degree ROP lens to provide extremely sharp fundus images of the rat retina (Fig. 36.3B). A high contrast 80 degree lens provided satisfactory images in large mouse eyes, although shadowing was apparent around the retinal periphery in smaller mouse eyes.

Most interestingly, we found that when delivered subretinally to young mouse pups aged P4 and P7, the VSV-CMV-GFP LV vector transduced photoreceptors in addition to the RPE layer (Fig. 36.4A and 4B). Photoreceptor transduction in these young animals was present primarily at the injection site. GFP expression was highly restricted to the RPE layer in all mice aged P10, P14, and P17 (Fig. 36.4C). Mild evidence of an immune response at the injection site was observed in 2 of the 24 injected animals (Fig. 36.4D). PBS injected control animals exhibited no obvious immune response. GFP expression extended over the majority of the RPE (Fig. 36.4E).

6. DISCUSSION

We aimed to determine the cellular tropism of VSV-LV vectors when injected subretinally in rodents. An inconsistency exists in the literature where some reports show that VSV-LV efficiently transduces both photoreceptors and RPE, while others show RPE restricted expression.[17,18,32] To better understand the capability of this vector, we sought to clarify the temporal tropism for photoreceptor transduction when delivered to young mice pups.

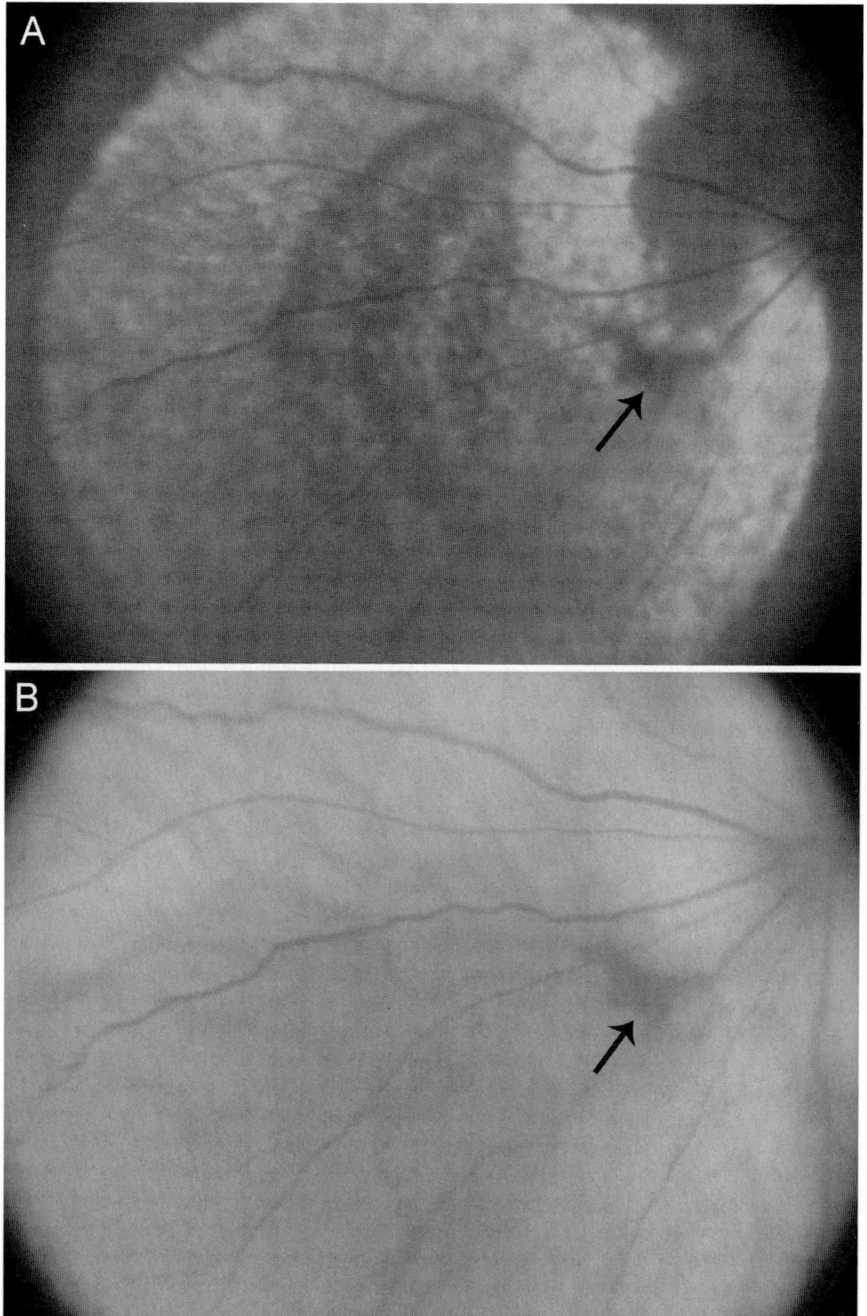

Figure 36.3. Flourescent fundus image showing widespread GFP expression in rat retina 1 week after subretinal injection of 3 μL VSV-CβA-GFP lentiviral vector (A). Fundus image of same rat under white light illumination (B). Arrows indicate small hemorrhage resulting from subretinal injection. Both images acquired with a Retcam II imaging system (Massie Research, Pleasanton, CA). See also color insert.

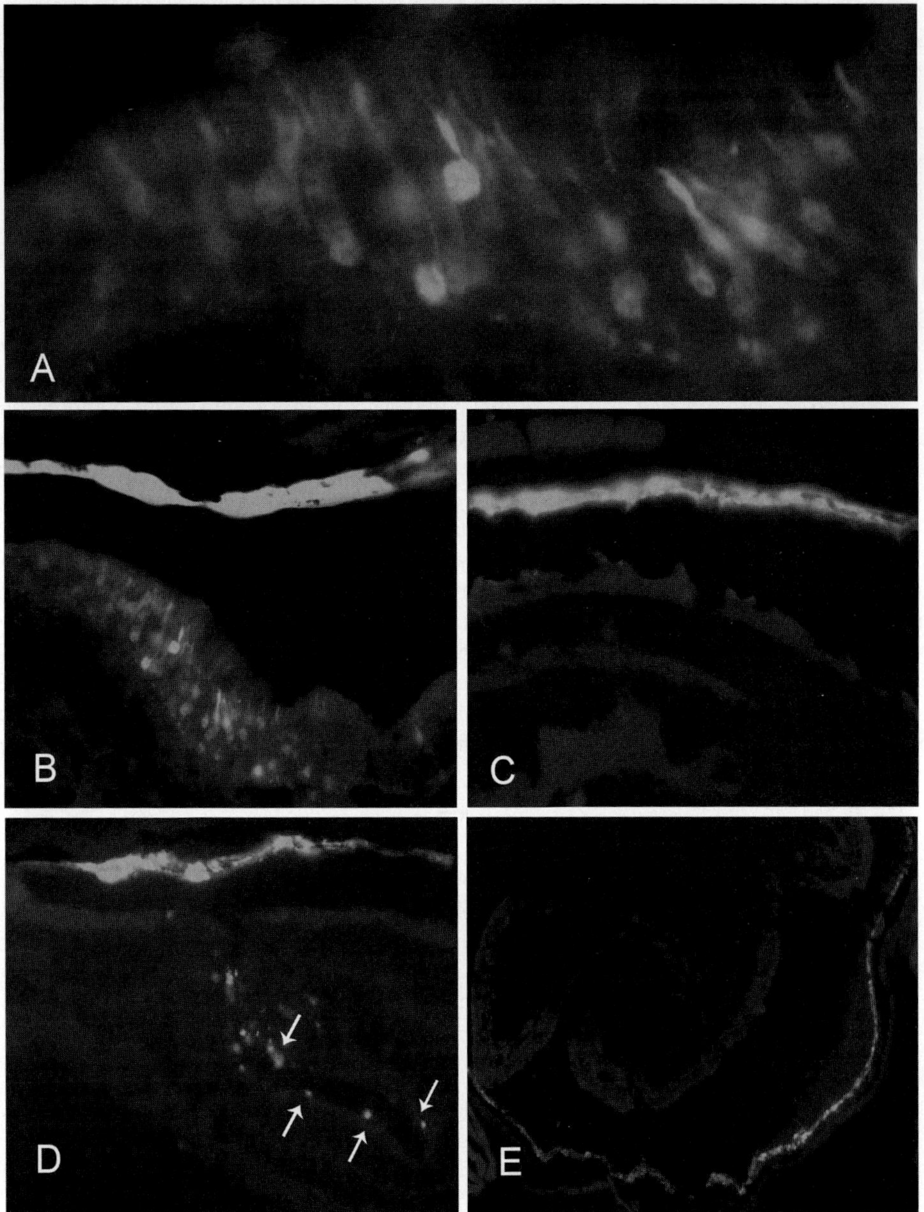

Figure 36.4. High magnification view of GFP positive photoreceptors of mouse retina injected subretinally with VSV-CMV-GFP LV vector at age P7 (A). Lower magnification view (B) of the same retina shown in (A) where RPE and photoreceptors are seen expressing GFP. Expression of GFP restricted to the RPE layer in P14 mouse retina injected with VSV-CMV-GFP LV vector (C). Injection track mark shown (arrows) and evidence of immune response from autoflourescent macrophages bordering track mark (D). Low magnification view showing extent of GFP expression along entire length of the RPE (E). See also color insert.

A developmental window was found to exist for the VSV-LV vector's ability to transduce photoreceptors when subretinally injected into C57BL/6 mice. This vector readily transduced RPE cells in mice of all ages, however transduction of photoreceptors occurred only in mice aged P7 and younger. The temporal window for photoreceptor transduction coincides with a period of rod photoreceptor neurogenesis during retinal development.[33] The onset of RPE specific transduction coincides with the completion of photoreceptor development and the beginning of normally occurring photoreceptor death.[34]

Binding and fusion of the VSV-LV vector does not appear to depend on the presence of specific cell surface receptors, therefore this tropism is unlikely the result of transient surface receptor expression on photoreceptors during development.[35] The relatively unorganized architecture of the immature mouse retina may facilitate viral access to photoreceptor cells. The mechanism for this restricted tropism is not entirely understood, however it appears to be related to the restricted access of the viral particles to photoreceptors probably due to the protein rich (Chondroitins, collagen, and fibronectin) inter-photoreceptor matrix (IPM). Enzymatic digestion of the IPM appears to improve LV vector access to photoreceptors.[36] Additionally, direct RPE phagocytosis of LV vector could play an important role in high RPE transduction.

In some cases, such as the secretion of therapeutic growth factors, ubiquitous expression may be desired in as many retinal cell types as possible. LV vectors containing "ubiquitous" promoters such as CMV, CMV-β-actin, EF1-α, PGK, and ubiquitin have been tested in the retina. Although excellent tools for strong expression in the RPE, these "ubiquitous" promoters should be regarded with care as they demonstrate specific, rather than universal spatial expression patterns when delivered subretinally by LV vectors.

Current and future efforts to target specific classes of retinal cells with LV vectors will be particularly useful for the treatment of retinal degenerative diseases through gene therapy.

7. ACKNOWLEDGEMENTS

The authors wish to thank Scott Geller, Natalie Walsh, and Josh Leonard for procedural advice, Debbie Kuo and Aaron Pham for technical assistance, and the Foundation Fighting Blindness for their travel support of KPG to attend the RD2004 meeting.

8. REFERENCES

1. L. Naldini, U. Blomer, P. Gallay, D. Ory, R. Mulligan, F. H. Gage, I. M. Verma and D. Trono, In vivo gene delivery and stable transduction of nondividing cells by a lentiviral vector, *Science.* **272**(5259):263-267 (1996).
2. R. J. Samulski, L. S. Chang and T. Shenk, Helper-free stocks of recombinant adeno-associated viruses: Normal integration does not require viral gene expression, *J Virol.* **63**(9):3822-3828 (1989).
3. A. J. Lotery, T. A. Derksen, S. R. Russell, R. F. Mullins, S. Sauter, L. M. Affatigato, E. M. Stone and B. L. Davidson, Gene transfer to the nonhuman primate retina with recombinant feline immunodeficiency virus vectors, *Hum Gene Ther.* **13**(6):689-696 (2002).
4. J. Bennett, A. M. Maguire, A. V. Cideciyan, M. Schnell, E. Glover, V. Anand, T. S. Aleman, N. Chirmule, A. R. Gupta, Y. Huang, G. P. Gao, W. C. Nyberg, J. Tazelaar, J. Hughes, J. M. Wilson and S. G. Jacobson, Stable transgene expression in rod photoreceptors after recombinant adeno-associated virus-mediated gene transfer to monkey retina, *Proc Natl Acad Sci* U S A. **96**(17):9920-9925 (1999).
5. Foundation Fighting Blindness (2003) http://www.blindness.org/content.asp?id=208

6. N. Loewen, D. A. Leske, J. D. Cameron, Y. Chen, T. Whitwam, R. D. Simari, W. L. Teo, M. P. Fautsch, E. M. Poeschla and J. M. Holmes, Long-term retinal transgene expression with fiv versus adenoviral vectors, *Mol Vis.* **10**:272-280 (2004).

7. Y. Ikeda, Y. Goto, Y. Yonemitsu, M. Miyazaki, T. Sakamoto, T. Ishibashi, T. Tabata, Y. Ueda, M. Hasegawa, S. Tobimatsu and K. Sueishi, Simian immunodeficiency virus-based lentivirus vector for retinal gene transfer: A preclinical safety study in adult rats, *Gene Ther.* **10**(14):1161-1169 (2003).

8. K. Takahashi, T. Luo, Y. Saishin, J. Sung, S. Hackett, R. K. Brazzell, M. Kaleko and P. A. Campochiaro, Sustained transduction of ocular cells with a bovine immunodeficiency viral vector, *Hum Gene Ther.* **13**(11):1305-1316 (2002).

9. L. F. Wong, M. Azzouz, L. E. Walmsley, Z. Askham, F. J. Wilkes, K. A. Mitrophanous, S. M. Kingsman and N. D. Mazarakis, Transduction patterns of pseudotyped lentiviral vectors in the nervous system, *Mol Ther.* **9**(1):101-111 (2004).

10. T. Kafri, H. van Praag, L. Ouyang, F. H. Gage and I. M. Verma, A packaging cell line for lentivirus vectors, *J Virol.* **73**(1):576-584 (1999).

11. P. Escarpe, N. Zayek, P. Chin, F. Borellini, R. Zufferey, G. Veres and V. Kiermer, Development of a sensitive assay for detection of replication-competent recombinant lentivirus in large-scale hiv-based vector preparations, *Mol Ther.* **8**(2):332-341 (2003).

12. H. Miyoshi, U. Blomer, M. Takahashi, F. H. Gage and I. M. Verma, Development of a self-inactivating lentivirus vector, *J Virol.* **72**(10):8150-8157 (1998).

13. S. C. Barry, B. Harder, M. Brzezinski, L. Y. Flint, J. Seppen and W. R. Osborne, Lentivirus vectors encoding both central polypurine tract and posttranscriptional regulatory element provide enhanced transduction and transgene expression, *Hum Gene Ther.* **12**(9):1103-1108 (2001).

14. V. Zennou, C. Petit, D. Guetard, U. Nerhbass, L. Montagnier and P. Charneau, Hiv-1 genome nuclear import is mediated by a central DNA flap, *Cell.* **101**(2):173-185 (2000).

15. A. Limon, N. Nakajima, R. Lu, H. Z. Ghory and A. Engelman, Wild-type levels of nuclear localization and human immunodeficiency virus type 1 replication in the absence of the central DNA flap, *J Virol.* **76**(23):12078-12086 (2002).

16. R. Zufferey, J. E. Donello, D. Trono and T. J. Hope, Woodchuck hepatitis virus posttranscriptional regulatory element enhances expression of transgenes delivered by retroviral vectors, *J Virol.* **73**(4):2886-2892 (1999).

17. M. Takahashi, H. Miyoshi, I. M. Verma and F. H. Gage, Rescue from photoreceptor degeneration in the rd mouse by human immunodeficiency virus vector-mediated gene transfer, *J Virol.* **73**(9):7812-7816 (1999).

18. J. W. Bainbridge, C. Stephens, K. Parsley, C. Demaison, A. Halfyard, A. J. Thrasher and R. R. Ali, In vivo gene transfer to the mouse eye using an hiv-based lentiviral vector; efficient long-term transduction of corneal endothelium and retinal pigment epithelium, *Gene Ther.* **8**(21):1665-1668 (2001).

19. B. A. van Adel, C. Kostic, N. Deglon, A. K. Ball and Y. Arsenijevic, Delivery of ciliary neurotrophic factor via lentiviral-mediated transfer protects axotomized retinal ganglion cells for an extended period of time, *Hum Gene Ther.* **14**(2):103-115 (2003).

20. V. Sandrin, S. J. Russell and F. L. Cosset, Targeting retroviral and lentiviral vectors, *Curr Top Microbiol Immunol.* **281**:137-178 (2003).

21. J. C. Burns, T. Friedmann, W. Driever, M. Burrascano and J. K. Yee, Vesicular stomatitis virus g glycoprotein pseudotyped retroviral vectors: Concentration to very high titer and efficient gene transfer into mammalian and nonmammalian cells, *Proc Natl Acad Sci* U S A. **90**(17):8033-8037 (1993).

22. Y. Kang, C. S. Stein, J. A. Heth, P. L. Sinn, A. K. Penisten, P. D. Staber, K. L. Ratliff, H. Shen, C. K. Barker, I. Martins, C. M. Sharkey, D. A. Sanders, P. B. McCray, Jr. and B. L. Davidson, In vivo gene transfer using a nonprimate lentiviral vector pseudotyped with ross river virus glycoproteins, *J Virol.* **76**(18):9378-9388 (2002).

23. N. D. Mazarakis, M. Azzouz, J. B. Rohll, F. M. Ellard, F. J. Wilkes, A. L. Olsen, E. E. Carter, R. D. Barber, D. F. Baban, S. M. Kingsman, A. J. Kingsman, K. O'Malley and K. A. Mitrophanous, Rabies virus glycoprotein pseudotyping of lentiviral vectors enables retrograde axonal transport and access to the nervous system after peripheral delivery, *Hum Mol Genet.* **10**(19):2109-2121 (2001).

24. A. Auricchio, G. Kobinger, V. Anand, M. Hildinger, E. O'Connor, A. M. Maguire, J. M. Wilson and J. Bennett, Exchange of surface proteins impacts on viral vector cellular specificity and transduction characteristics: The retina as a model, *Hum Mol Genet.* **10**(26):3075-3081 (2001).

25. G. Duisit, H. Conrath, S. Saleun, S. Folliot, N. Provost, F. L. Cosset, V. Sandrin, P. Moullier and F. Rolling, Five recombinant simian immunodeficiency virus pseudotypes lead to exclusive transduction of retinal pigmented epithelium in rat, *Mol Ther.* **6**(4):446-454 (2002).

26. D. A. Sanders, No false start for novel pseudotyped vectors, *Curr Opin Biotechnol.* **13**(5), 437-442 (2002).

27. C. Kostic, F. Chiodini, P. Salmon, M. Wiznerowicz, N. Deglon, D. Hornfeld, D. Trono, P. Aebischer, D. F. Schorderet, F. L. Munier and Y. Arsenijevic, Activity analysis of housekeeping promoters using self-inactivating lentiviral vector delivery into the mouse retina, *Gene Ther.* **10**(9):818-821 (2003).

28. H. Miyoshi, M. Takahashi, F. H. Gage and I. M. Verma, Stable and efficient gene transfer into the retina using an hiv-based lentiviral vector, *Proc Natl Acad Sci* U S A. **94**(19):10319-10323 (1997).

29. J. Jakobsson, C. Ericson, M. Jansson, E. Bjork and C. Lundberg, Targeted transgene expression in rat brain using lentiviral vectors, *J Neurosci Res.* **73**(6):876-885 (2003).

30. L. P. de Almeida, D. Zala, P. Aebischer and N. Deglon, Neuroprotective effect of a cntf-expressing lentiviral vector in the quinolinic acid rat model of huntington's disease, *Neurobiol Dis.* **8**(3):433-446 (2001).

31. L. Sastry, T. Johnson, M. J. Hobson, B. Smucker and K. Cornetta, Titering lentiviral vectors: Comparison of DNA, rna and marker expression methods, *Gene Ther.* **9**(17):1155-1162 (2002).

32. A. R. Harvey, W. Kamphuis, R. Eggers, N. A. Symons, B. Blits, S. Niclou, G. J. Boer and J. Verhaagen, Intra-vitreal injection of adeno-associated viral vectors results in the transduction of different types of retinal neurons in neonatal and adult rats: A comparison with lentiviral vectors, *Mol Cell Neurosci.* **21**(1):141-157 (2002).

33. L. D. Carter-Dawson and M. M. LaVail, Rods and cones in the mouse retina. Ii. Autoradiographic analysis of cell generation using tritiated thymidine, *J Comp Neurol.* **188**(2):263-272 (1979).

34. K. Mervin and J. Stone, Developmental death of photoreceptors in the c57bl/6j mouse: Association with retinal function and self-protection, *Exp Eye Res.* **75**(6):703-713 (2002).

35. D. A. Coil and A. D. Miller, Phosphatidylserine is not the cell surface receptor for vesicular stomatitis virus, *J Virol.* **78**(20):10920-10926 (2004).

36. O. Grüter, C. Kostic, M. Tekaya, D. F. Schorderet, L. Zografos, F. L. Munier, Y. Arsenijevic, Potential improvement of lentiviral gene transfer by weakening the extracellular matrix, ARVO poster #B97 (2004).

CHAPTER 37

POTENTIAL USE OF CELLULAR PROMOTER(S) TO TARGET RPE IN AAV-MEDIATED DELIVERY

Cellular promoters and RPE-targeting

Erika N. Sutanto[1], Dan Zhang[1,2], Yvonne K.Y. Lai[1,2], Wei-Yong Shen[1,2], and P. Elizabeth Rakoczy[1,2]

1. INTRODUCTION

Gene therapy has been reported to show potential as an alternative treatment to conventional therapy. In ocular research, recombinant adeno-associated virus (rAAV) has been chosen as a vector of interest due to its low immunogenicity, its broad host range and its ability to result in long-term transduction (Ali et al., 1997; Bennett and Maguire, 2000).

Many reports on AAV-mediated ocular gene transfer have utilized strong, ubiquitous promoters such as cytomegalovirus (CMV) and chicken β-actin (CBA), which could result in a high level of transgene expression. Nevertheless, the possibility of non-specific expression outside the target cells (Guy et al., 1999; Sanftner et al., 2001) and the silencing of viral promoter activity (Stone et al., 2000; Prosch et al., 1996; Loser et al., 1998) may limit the use of such promoters. Cellular specificity of rAAV-mediated gene delivery can be modulated either by changing viral capsid serotype (Auricchio et al., 2001; Weber et al., 2003) or by the use of cell-specific promoters. The latter approach has been tested using a photoreceptor-specific promoter, namely an opsin gene promoter, which was shown to target photoreceptor efficiently (Flannery et al., 1997; Jomary et al., 1999). With regards to the importance of retinal pigment epithelim (RPE) in maintaining health and integrity of the retina, the focus of this study was to evaluate the use of cellular promoter(s) to target RPE following subretinal injection of rAAV. The cellular promoters proposed for this study were cathepsin D (CatD) and human RPE65 proximal promoters.

In a previous immunohistochemical study, it was shown that in human eye CatD is expressed at high levels in the RPE and at a lower level in ganglion cells (Rakoczy et al.,

[1] Centre for Ophthalmology and Visual Science, The University of Western Australia; [2] Department of Molecular Ophthalmology, Lions Eye Institute, Nedlands, 6009, Western Australia, Australia. Corresponding author: P.E. Rakoczy, E-mail: rakoczy@cyllene.uwa.edu.au.

1999). Considering that following subretinal injection, the rAAV would be localized in the subretinal space in close proximity to the RPE and photoreceptors, the use of CatD proximal promoter may be ideal to target the RPE. The RPE65 promoter was chosen because it has been shown *in vitro* to have specific activity in RPE cells (Nicoletti et al., 1998) despite it being a weak promoter compared to the CMV promoter. CatD proximal promoter contains five transcription start sites, and is specifically controlled by estrogen-responsive elements (EREs), a retinoic acid-responsive element (RARE) and a major late promoter element (MLPE) (Cavailles et al., 1993; Sheikh et al., 1996; Wang et al., 1997). Human RPE65 (hRPE65) contains general transcriptional machinery elements and positive elements such as Oct-1 and E-box sites for RPE-specific expression (Boulanger et al., 2000). GAL4 is a yeast transcriptional factor that activates transcription by binding to four related dyad symmetrical sequences. Fusion of the herpes simplex virus transcriptional activator VP16 partial activation domain to GAL4 DNA binding domain resulted in enhancement of transcriptional activity of a reporter gene (Sadowski et al., 1988).

The first part of this study was to evaluate which region of the CatD proximal promoter was necessary to target high transgene expression in cultured RPE and to test the relative specificity of this region *in vivo* by subretinal injection of rAAV. The second part was to evaluate the potential use of chimeric transcriptional activator, GAL4-VP16, to enhance the weak promoter activity of hRPE65.

2. MATERIALS AND METHOD

2.1. Construction of Plasmid DNA

Plasmids pCD(L)-gfp, pCD(M)-gfp and pCD(Sm)-gfp carrying different sizes of CatD proximal promoter fragments were constructed following restriction digestion of pCatD (Figure 37.1).

Human RPE65 promoter fragment (−655 to +31) was amplified by PCR from human genomic DNA (Promega, Madison, WI) and subcloned into pGEM-T Easy vector (Promega)

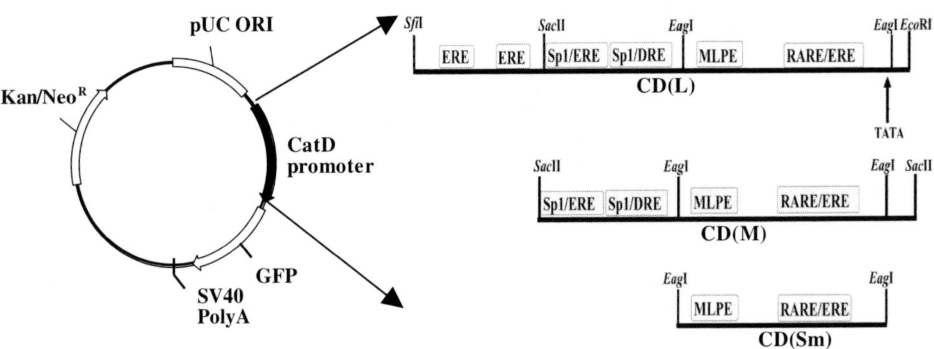

Figure 37.1. Schematic diagram of CatD proximal promoter regions used in plasmid constructions. ERE = estrogen responsive element, RARE = retinoic acid response element, MLPE = major late promoter element, DRE = dioxin responsive element, GFP = green fluorescent protein.

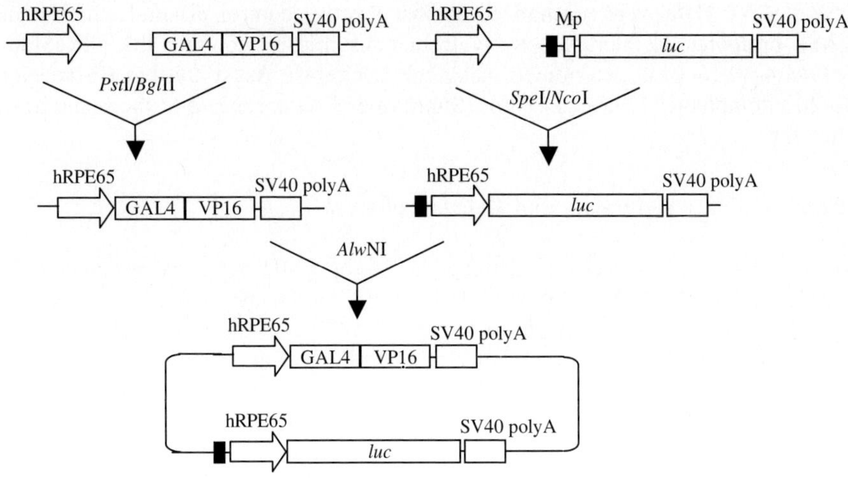

Figure 37.2. A flow-chart summarising construction of plasmids containing hRPE65 promoter with luciferase reporter gene and/or GAL4-VP16 transactivator. Mp = minimal promoter containing only of a TATA box and a transcription initiation site. ■ = GAL4 DNA binding sites.

to create the pGEM-hRPE65 plasmid. Plasmids pBIND, pACT and pG5luc which contains relevant regions of the GAL4, VP16 and a firefly luciferase reporter gene linked with GAL4 DNA binding domain, respectively, were obtained from the Checkmate™ Mammalian Two-Hybrid System (Promega). These plasmids were then used to generate phR65luc and phR65GAL4-VP16 for co-transfection as well as single construct phR65luc-GAL4-VP16 which carries both the transactivator and the luciferase gene under the control of hRPE65 promoter. A simplified flowchart summarising steps for constructing such plasmids is shown in Figure 37.2.

2.2. Cell Culture and Transient Transfection

Cell culture reagents were obtained from Invitrogen Life Technologies (Carlsbad, CA). The low-passage human RPE cells HRPE51 (established from 51-year old donor), an RPE cell line D407 (kindly given by Dr Richard Hunt, University of South Carolina, SC), and human fibroblast cell line, F2000 (Flow Laboratories, Herts, UK) were grown in Dulbecco's modified Eagle's medium, supplemented with 1% (v/v) penicillin/ streptomycin and 10% (v/v) fetal bovine serum (FBS). Cells were seeded onto 24-well plates one day prior to transfection and transiently transfected at 70% confluency using FuGene6 Transfection Reagent (Roche, Indianapolis, IN). A mixture of 2 μg DNA plasmid DNA and 3 μL FuGene 6 was prepared as described in the manufacturer's protocol. For co-transfection, a 1 : 1 molar ratio of the reporter vector and the activator vector were used.

2.3. Fluorescence Activated Cell Sorter (FACS) Analysis and Luciferase Assay

At the completion of experiment, cells transfected with plasmids carrying the GFP reporter gene were harvested and analyzed using a FACSCalibur Flow Cytometer (Becton

Dickinson, CA). Data were normalized against positive control plasmid which contains the CMV promoter. Luciferase activity from cells transfected with the phR65luc and phR65GAL4-VP16 were determined using the Luciferase Assay System (Promega) and TD-20/20 luminometer (Turner Designs, Sunnyvale, CA) according to the manufacturer's specification.

2.4. Construction, Production, and Delivery of rAAV

A gene expression cassette containing the 365 bp CatD proximal promoter (CD(L)) was subcloned into the AAV serotype-2 plasmid SSV9 to create pSSV.CD(L)-gfp. A large-scale production of rAAV.CD(L)-gfp and control rAAV.CMV-gfp were performed according to routine methodologies in our laboratory (Rolling et al., 1999). Two microliters of rAAV.CD(L)-gfp or rAAV.CMV-gfp (7.2×10^5 tu/eye) were subretinally injected into non-pigmented RCS/rdy$^+$ rats. All procedures adhered to the University of Western Australia Animal Experimentation Committee, and to the Association for Research in Vision and Ophthalmology guidelines for the Use of Animals in Ophthalmic and Vision Research.

2.5. Detection of GFP Expression

At 12 weeks post-injection, animals were euthanized, the eyes enucleated, and the retinas isolated and separated into sclera/choroid/RPE and neuroretina layers. Each layer was then flatmounted and the GFP expression was detected using fluorescence microscopy. The numbers of GFP-positive cells were then counted on each layer using ×10 objective magnification to compare transduction efficiency between constructs containing CatD and CMV promoters.

3. RESULTS AND DISCUSSION

3.1. Activity of CatD Promoter Fragments *In Vitro*

Quantification of GFP signal using FACS demonstrated that in HRPE51 cells, signal intensity varied from 45.7 ± 6.23 (pCD(L)-gfp), 17.7 ± 5.64 (pCD(M)-gfp) and 17.1 ± 1.62 (pCD(Sm)-gfp) (Figure 37.3). In D407, the intensity ranged from 48.15 ± 4.20 (pCD(L)-gfp), 23.8 ± 0.31 (pCD(M)-gfp) and 24.2 ± 1.92 (pCD(Sm)-gfp). In both the HRPE51 and D407 cultures, pCD(L)-gfp-transfected cells consistently had higher GFP signal intensity than those transfected with either pCD(M)-gfp or pCD(Sm)-gfp. The transfection of F2000 resulted in weaker GFP signal intensity with no significant difference ($p > 0.05$) between any of the constructs. There is however a marked difference ($p < 0.05$) in GFP signal intensities between the RPE and fibroblast cultures for pCD(L)-gfp-transfected cells. These results suggest that the presence of two EREs (Figure 37.1) are necessary for high CatD expression as the removal of these elements in both CD(M) and CD(Sm) fragments resulted in lower activity. In addition, the low activity of CatD proximal promoter in F2000 cultures could be due to the absence of additional factors required for activation of intrinsic elements present in the CatD promoter region. Nevertheless, the identification of such factors and the exact mechanism of how they interact are not yet elucidated.

Figure 37.3. The longest CatD proximal promoter fragment used in this study, CD(L), drives high transgene expression in cultured RPE cells.

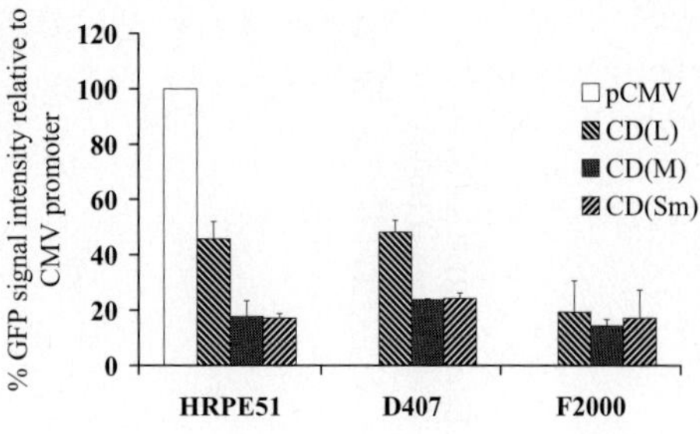

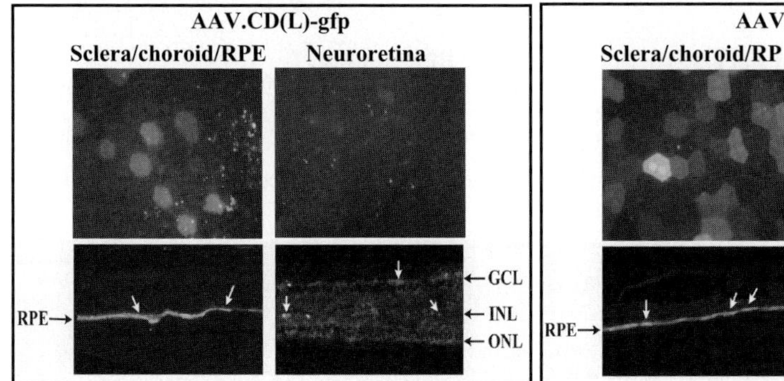

Figure 37.4. Fluorescence micrographs of separated RPE and neuroretina flatmounts, and cryosections of non-pigmented RCS/rdy[+] following AAV subretinal delivery. Original magnification: ×20.

3.2. CatD Promoter Activity *In Vivo*

At 12-week post subretinal injection, fluorescence signal was detected with equal intensity in both the RPE and neuroretina layers of AAV.CMV-gfp-injected eyes (Figure 37.4). On the other hand, in AAV.CD(L)-gfp-injected eyes, the majority of signal was observed in the RPE layer with very few GFP-positive cells present in the neuroretina.

The location of the signal was confirmed in each layer by cryosectioning, with some weak signal detected in the photoreceptors and ganglion cell layer following injection with AAV.CD(L)-gfp (Figure 37.4). Based from the cell counting results, there was approximately three times the number of photoreceptor cells being transduced than RPE cells in AAV.CMV.gfp-injected eyes. In contrast, there was similar numbers of photoreceptor and RPE cells expressing GFP in AAV.CD(L).gfp-injected eyes. Since the same viral titer was used in this study, the higher number of GFP-positive cells associated with AAV.CMV-gfp could be due to the promoter strength. This is reflected in the *in vitro* analysis where there was two-fold lower signal intensity in pCD(L)-gfp-transfected cells than in pCMV-transfected cells (Figure 37.3). Furthermore, as both virus preparations were constructed and packaged in the same serotype (serotype 2), and assuming that both of them equally transduce the same type of cells, the lower number of GFP-positive cells in the neuroretina

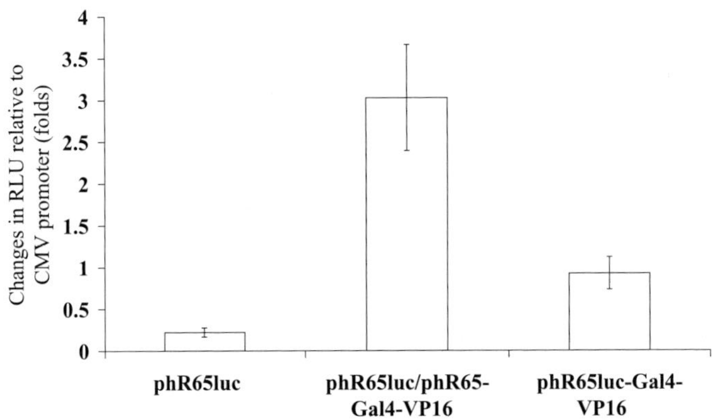

Figure 37.5. Luciferase activity of different DNA constructs in HRPE51 cells.

following AAV.CD(L)-gfp delivery verified the preferential CatD promoter activity in the RPE.

3.3. Effect of GAL4-VP16 Transactivator

An alternative gene expression system mediated by GAL4-VP16 was established and evaluated for its potential to increase hRPE65 promoter activity. The transcriptional activity was then assessed by evaluation of the luciferase level in the transfected-HRPE51 cells. In the absence of GAL4-VP16 transactivator, the relative light unit (RLU) associated to the hRPE65 promoter was very low compared to the CMV promoter (Figure 37.5). The hRPE65 promoter activity was markedly enhanced ($p < 0.01$) however in the presence of GAL4-VP16 transactivator such that there was a three-fold increase in activity compared to the CMV promoter. Interestingly, transfections of a single construct phR65luc-Gal4-VP16 resulted in lower but comparable reporter gene activity to that of the CMV promoter.

4. CONCLUSION

The study presented here demonstrated that the longest proximal CatD promoter used in this study (CD(L)) retains high promoter activity in cultured RPE cells. The presence of estrogen elements might play a role whereby upon ligand binding, their pathways might interact with those of other elements, thus contribute to higher promoter activity. Although the *in vivo* results showed that it has lower activity than the CMV promoter, CatD proximal promoter has the ability to predominantly target transgene expression to the RPE. The second part of the study showed the ability of the transactivator GAL4-VP16 to increase weak cell-specific promoter activity. It also demonstrated that significant enhancement was achieved when the hRPE65 promoter controlled both the reporter and GAL4-VP16 genes.

In conclusion, this current work suggests that there is potential to use the CatD proximal promoter to target RPE following subretinal delivery. However, a further study is required to try to increase the promoter activity, either by incorporating cell-specific enhancer or the GAL4-VP16 transactivator as used in the current work. Furthermore, this study showed that in combination with a strong transcriptional activator, the hRPE65 promoter has the capacity to result in at least similar to or even higher transgene expression than the CMV promoter.

5. ACKNOWLEDGEMENTS

We thank the Foundation Fighting Blindness for the travel award provided to ENS to attend this meeting.

6. REFERENCES

Ali RR, Reichel MB, Hunt DM and Bhattacharya SS (1997). Gene therapy for inherited retinal degeneration. *Br J Ophthalmol* **81**:795-801.

Auricchio A, Kobinger G, Anand V, *et al.* (2001). Exchange of surface proteins impacts on viral vector cellular specificity and transduction characteristics: the retina as a model. *Hum Mol Genet* **10**:3075-81.

Bennett J and Maguire AM (2000). Gene therapy for ocular disease. *Mol Ther* **1**:501-5.

Boulanger A, Liu S, Henningsgaard AA, *et al.* (2000). The upstream region of the Rpe65 gene confers retinal pigment epithelium-specific expression in vivo and in vitro and contains critical octamer and E-box binding sites. *J Biol Chem* **275**:31274-82.

Cavailles V, Augereau P and Rochefort H (1993). Cathepsin D gene is controlled by a mixed promoter, and estrogens stimulate only TATA-dependent transcription in breast cancer cells. *Proc Natl Acad Sci U S A* **90**:203-7.

Flannery JG, Zolotukhin S, Vaquero MI, *et al.* (1997). Efficient photoreceptor-targeted gene expression in vivo by recombinant adeno-associated virus. *Proc Natl Acad Sci U S A* **94**:6916-21.

Guy J, Qi X, Muzyczka N and Hauswirth WW (1999). Reporter expression persists 1 year after adeno-associated virus-mediated gene transfer to the optic nerve. *Arch Ophthalmol* **117**:929-37.

Jomary C, Chatelain G, Michel D, *et al.* (1999). Effect of targeted expression of clusterin in photoreceptor cells on retinal development and differentiation. *J Cell Sci* **112**:1455-64.

Loser P, Jennings GS, Strauss M and Sandig V (1998). Reactivation of the previously silenced cytomegalovirus major immediate-early promoter in the mouse liver: involvement of NFkappaB. *J Virol* **72**:180-90.

Nicoletti A, Kawase K and Thompson DA (1998). Promoter analysis of RPE65, the gene encoding a 61-kDa retinal pigment epithelium-specific protein. *Invest Ophthalmol Vis Sci* **39**:637-44.

Prosch S, Stein J, Staak K, *et al.* (1996). Inactivation of the very strong HCMV immediate early promoter by DNA CpG methylation in vitro. *Biol Chem Hoppe Seyler* **377**:195-201.

Rakoczy PE, Sarks SH, Daw N and Constable IJ (1999). Distribution of cathepsin D in human eyes with or without age-related maculopathy. *Exp Eye Res* **69**:367-74.

Rolling F, Shen WY, Tabarias H, *et al.* (1999). Evaluation of adeno-associated virus-mediated gene transfer into the rat retina by clinical fluorescence photography. *Hum Gene Ther* **10**:641-8.

Sadowski I, Ma J, Triezenberg S and Ptashne M (1988). GAL4-VP16 is an unusually potent transcriptional activator. *Nature* **335**:563-4.

Sanftner MLH, Abel H, Hauswirth WW and Flannery JG (2001). Glial cell line derived neurotrophic factor delays photoreceptor degeneration in a transgenic rat model of retinitis pigmentosa. *Mol Ther* **4**:622-9.

Sheikh MS, Augereau P, Chalbos D, *et al.* (1996). Retinoid regulation of human cathepsin D gene expression. *J Steroid Biochem Mol Biol* **57**:283-91.

Stone D, David A, Bolognani F, *et al.* (2000). Viral vectors for gene delivery and gene therapy within the endocrine system. *J Endocrinol* **164**:103-18.

Wang F, Porter W, Xing W, *et al.* (1997). Identification of a functional imperfect estrogen-responsive element in the 5′-promoter region of the human cathepsin D gene. *Biochemistry* **36**:7793-801.

Weber M, Rabinowitz J, Provost N, *et al.* (2003). Recombinant adeno-associated virus serotype 4 mediates unique and exclusive long-term transduction of retinal pigmented epithelium in rat, dog, and nonhuman primate after subretinal delivery. *Mol Ther* **7**:774-81.

CHAPTER 38

CYTOKINE-INDUCED RETINAL DEGENERATION: ROLE OF SUPPRESSORS OF CYTOKINE SIGNALING (SOCS) PROTEINS IN PROTECTION OF THE NEURORETINA

Charles E. Egwuagu, Cheng-Hong Yu, Rashid M. Mahdi, Marie Mameza, Chikezie Eseonu, Hiroshi Takase, and Samuel Ebong*

1. INTRODUCTION

The vertebrate retina is comprised of a collection of highly specialized cell types, with each subtype playing unique roles and functions in the reception, transduction and conversion of incident light rays into visual images. The photo-transduction mechanism is extremely sensitive and minute alterations in the relative abundance of any retinal cell type can severely compromise the quality of the visual image. Because ganglion cells and other retinal neurons are terminally differentiated cells, it has been argued that evolutionarily conserved mechanisms must exist to protect retinal neurons from injury or death caused by exposure to environmental toxins or toxic bi-products of intermediary metabolism. For example, neuroretinal cells require protection from infectious agents that occasionally colonize and kill them, leading to permanent loss of such cells. Although intraocular infections is rapidly cleared by inflammatory cells, prolonged secretion of inflammatory cytokines in the retina may induce cytopathic effects that can produce retinal degenerative changes and possibly retinal degeneration. Although much effort has been made in characterizing chromosomal mutations and other biochemical lesions that may underlie the development of retinal degeneration, few studies have addressed the role of inflammation or inflammatory mediators in pathogenic mechanisms of retinal degenerative diseases. In fact, inflammation and dysregulation of activities of proinflammatory cytokines have been implicated in pathogenesis of other human degenerative diseases including Alzheimer's disease and multiple sclerosis.

In this study, expression of the proinflammatory cytokine, interferon gamma (IFNγ), was targeted to the lens of transgenic (TR) rats and the lens was used to serve as a depot

---

*National Eye Institute, National Institutes of Health, MD, U.S.A.

for releasing IFNγ into retina. This TR rat model allows us to directly test the hypothesis that prolonged exposure of retinal cells to pro-inflammatory cytokines, as may occur during persistent chronic infection of the retina, can induce retinal disease or retinal degenerative changes. We have also investigated mechanisms that may underlie protection of retinal cells from hypoxia. Because suppressors of cytokine signaling (SOCS) proteins play important role in regulating the activation, intensity and duration of cytokine- and stress-induced signals,[1-3] we examined whether retinal cells respond to cytokines and oxidative stress by inducing SOCS expression.

2. RESULTS

2.1. IFNγ Induces Retinal Degenerative Changes and Apoptosis of Ganglion Cells

The chimeric construct used for generating the TR rats with constitutive expression of IFNγ in the eye consists of the αA-Crystallin promoter fused to the mouse IFNγ coding sequence.[4] Because of lens-specificity of the αA-Crystallin promoter,[5,6] transgene expression occurs preferentially in the lens and its effects are initially confined to this tissue. However, after the first month of postnatal life, the lens capsule begins to disintegrate, releasing lenticular material into the posterior chamber and vitreous cavity and this coincides temporally with appearance of the effects of IFNγ on the retina. Appearance of retinal infoldings is observed in adult TR rats after three months of age (Fig. 38.1A) and number and size of these folds increase with time (data not shown). In addition, growth of the ganglion cell layer is significantly inhibited and its thickness approximates one-half of that seen in WT eye (data not shown).

The inhibition of ganglion cell growth is of particular importance in view of the critical functions of these cells in the visual process. It is therefore interesting to note that *In situ* detection of apoptotic cells by the TUNEL assay revealed presence of apoptotic cells in TR but not in WT rat retina (Fig. 38.1B). It is even more remarkable that the apoptotic response is restricted to the ganglion cell layer, suggesting that ganglion cells are more sensitive to the effects of IFNγ. In addition, the morphological changes seen in the TR rat retina correlates with upregulated expression of interferon regulatory factors 1 (IRF-1), interferon

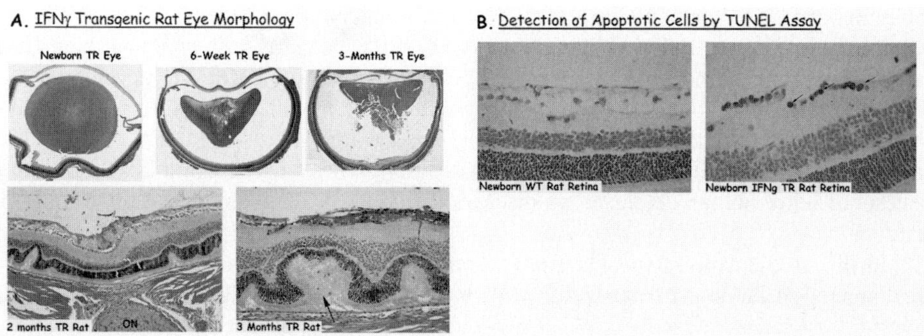

Figure 38.1. Chronic exposure of retinal cells can induce retinal degenerative changes.

consensus binding protein (ICSBP), RT1-Bα (equivalent to mouse MHC class II), ICAM-1 and TNFα genes (data not shown), suggesting that pathogenic effects of IFNγ are mediated, in part, by altering normal patterns of gene expression in the eye.

2.2. Retinal Response to Inflammatory Cytokines is Under Feedback-Regulation by SOCS

We next examined whether cytokine activities in the retina are under feedback regulation by SOCS proteins. To establish that SOCS proteins are expressed in the retina, we isolated total RNA from human or murine retina, prepared cDNAs and subjected them to 30 cycles of PCR amplification as reported previously.[7] We found that CIS, SOCS4, SOCS5, SOCS6, SOCS7 are constitutively expressed at very high levels in human retina while SOCS3 is not detectable even after 35 cycles of RT-PCR amplification. Although SOCS1 is also detected, it is at very low levels as detection required 35 cycles.

To examine whether SOCS expression is induced in retinal cells by proinflammatory cytokines it was necessary to establish that these cytokines do indeed activate gene transcription in the retina. Human retinal pigment epithelial (hRPE) or Müller cell line was stimulated with either interleukin 4 (IL-4) (10 ng/ml) or IFNγ (100 u/ml) for 15 min and transcriptional activation was assessed by gel-shift assay. Activation by IL-4 or IFNγ is mediate through STAT6 or STAT1, respectively.[8] As shown in Fig. 38.3, a retarded band is induced by the STAT6 probe in nuclear extracts from cells stimulated with IL-4 while IFNγ-

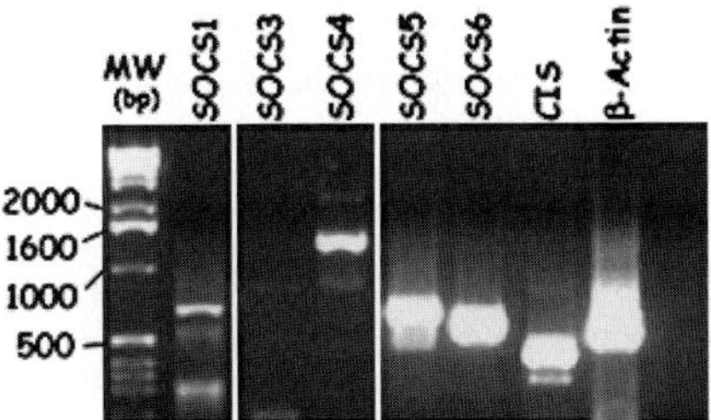

Figure 38.2. SOCS mRNA transcripts are constitutively expressed in the retina.

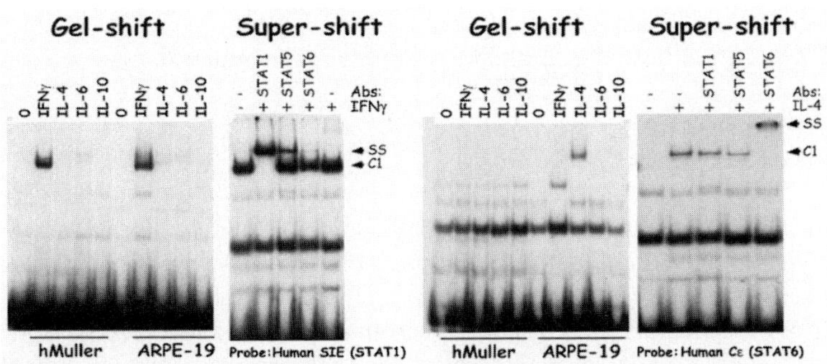

Figure 38.3. Inflammatory cytokines activate JAK/STAT signaling pathways in retinal cells.

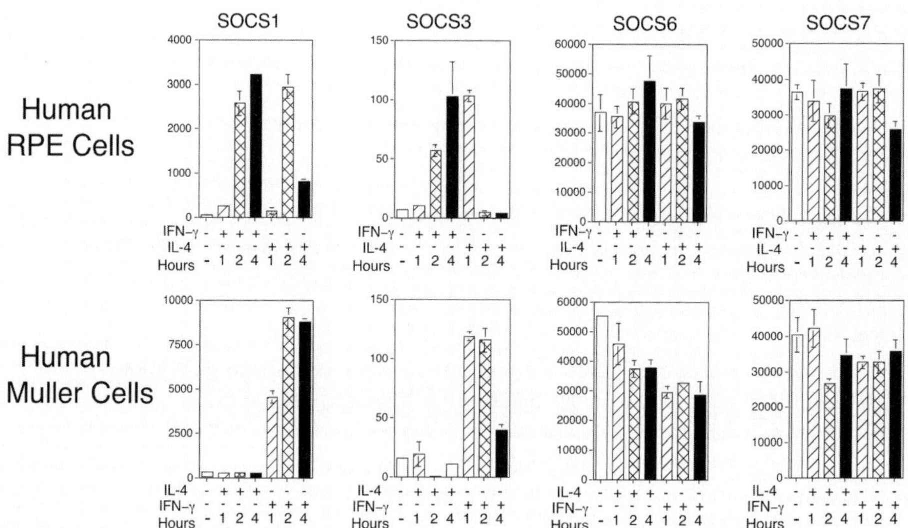

Figure 38.4. Retinal cell response to proinflammatory cytokines is under feedback regulation by SOCS proteins.

stimulated cells induce a STAT1 band-shift. Presence of STAT1 or STAT6 in the retarded band is confirmed by super-shift analysis indicated by SS.

We then examined whether SOCS proteins are induced in human retinal cells by pro-inflammatory cytokines. Müller or hRPE cells was washed, starved for 2 h before stimulation with IFNγ or IL-4 and then analyzed for induction of SOCS expression by real-time RT-PCR. As indicated in Fig. 38.4, expression of SOCS1 or SOCS3 is induced by both cytokines, although intensity or kinetics of induction is different. Although SOCS6 or SOCS7 are constitutively expressed in these cells, they are not induced in response to these inflammatory cytokines.

We next examined whether retina cells respond to hypoxia by inducing SOCS expression. Mouse retina explants were propagated for varying amounts of time under hypoxia condition. Induction of vascular endothelial growth factor (VEGF) or hypoxia-inducing factor 1 (HIF-1α) expression, two markers of hypoxia, was used to verify that the cells were indeed exposed to significant hypoxia. Induction of SOCS expression was analyzed by real-time RT-PCR and as indicated in Fig. 38.5, only SOCS3 is induced.

3. DISCUSSION AND CONCLUSION

Inflammatory cells that mediate host immunity to intraocular pathogens produce copious amounts of pro-inflammatory cytokines, IFNγ and IL-4. In this study we show that prolonged secretion of IFNγ in the neuroretina promotes formation of retinal in-foldings in the photoreceptor layer and induces apoptotic death of retinal ganglion cells. However, these results appear to be at variance with the fact that humans are constantly infected with a variety of pathogens that induce expression of this proinflammatory with no evidence of

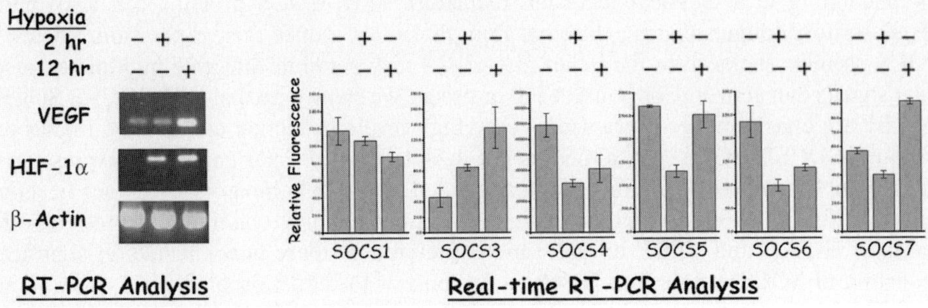

Figure 38.5. Retinal cell response to hypoxia is under feedback regulation by SOCS proteins.

similar clinical or histological manifestation. It is however of note that these symptoms occur only in older rats and is therefore consistent with age-dependent occurrence of retinal degenerative diseases in older humans. The data suggests the possibility that retinal degeneration is a slow and progressive pathogenic condition that is initiated or amplified, in part, by proinflammatory cytokines produced by inflammatory cells elicited by low-grade persistent infections. Thus, retinal degenerative diseases may occur only in individuals that are not able to adequately control activities of these inflammatory mediators.

Neurotrophic factors and neuregulatory cytokines, such as, CNTF, OSM, CT1, LIF, IGFs and FGFs are also produced during inflammation and have been shown to counteract deleterious effects of proinflammatory cytokines. Thus, retinal degenerative diseases may occur in individuals that are not able to adequately control activities of inflammatory mediators or activate protective mechanism that confer protection of retinal neuronal cells from cytokine- or stress-induced retinal damage. In the developing CNS and during spinal cord or brain injury, the steady-state levels of neuregulatory cytokines determines whether the neural progenitors would differentiate to neurons or astrocyte pathway and is therefore an important determinant of whether healing, repair or regeneration would occur.[9-11] It is of note that most proinflammatory cytokines and neurotrophic/neuregulatory factors mediate their effects through activation Janus kinase (JAK)/signal transducers and activators of transcription (STAT) pathway.[12,13] Homeostatic regulation of activities of these competing factors is essential to normal physiology of the retina and under stringent control by endogenous feedback regulators of the JAK/STAT signal transduction pathway.[14]

Recent studies on cellular mechanisms that switch off signals induced by growth factors and cytokines have uncovered existence of a family of endogenous negative feedback regulators, generically called suppressors of cytokine signaling (SOCS).[1-3] The best characterized members of the 8-member SOCS family are SOCS1, SOCS3, SOCS5 and CIS (cytokine induced SH2-domain protein) and expression one or more of these proteins is transiently induced by a wide variety of inflammatory and anti-inflammatory cytokines, including interferon IFNγ, IL-3, IL-4, IL-6, IL-12, IL-13, leukemia-inhibitory factor, stem cell factor, CNTF, GM-CSF and leptin.[1-3] Growth factors such as IGF-1, PDGF, FGFs, EGF, prolactin, growth hormone and erythropoietin also induce their expression Inhibitory effects of SOCS proteins derive from direct interactions with cytokine receptors and/or JAK kinases, thereby preventing recruitment of STATs to the signaling complex.[1-3] In addition to functioning in a classical feedback regulatory loop, SOCS proteins can also inhibit responses to cytokines that are different from those that induce their expression. Interest in SOCS proteins stems from the belief that SOCS may serve to integrate multiple extracellular signals that converge on a target cell or tissue. We show here that CIS, SOCS5, SOCS6, SOCS7 are constitutively expressed at very high levels in human and murine retinas and although SOCS1, SOCS2 or SOCS3 is not detectable in the normal retina, expression of these SOCS members is significantly upregulated by proinflammatory cytokines in retinal cells. We further show that retinas maintained under hypoxic conditions express elevated levels of HIF-1α and VEGF mRNAs and expression of these genes results in significant induction of SOCS expression. However, in contrast to induction of SOCS1, SOCS2 and SOCS3 by retinal cells in response of to proinflammatory cytokines, response to hypoxia is under feedback regulation by only SOCS3, suggesting a remarkable specificity of SOCS-mediated regulation in the retina.

In summary, we have shown in this study that similar to other neurodegenerative diseases, apoptotic death of retinal ganglion cells and retinal degeneration may result, in part, from chronic exposure of ocular cells to pro-inflammatory cytokines, as may occur in chronic inflammatory diseases of the eye. We also show that SOCS proteins are constitutively expressed in the retina and that retinal cells respond to cytotoxic cytokines or to hypoxic conditions by upregulating SOCS expression. These results are remarkable because SOCS proteins generally have a short half-life and are not detectable in many tissues. Demonstration that retinal cells respond to exposure to cytotoxic cytokines and hypoxia by upregulating SOCS expression, suggests that SOCS proteins may mitigate injurious effects of environmental, chemical or oxidative stress and should be exploited as neuroprotective agents of the mammalian retina.

4. REFERENCES

1. W. S. Alexander, D. J. Hilton, The role of suppressors of cytokine signaling (SOCS) proteins in regulation of the immune response. *Annu Rev Immunol.* **22**:503-29 (2004).
2. M. Kubo, T. Hanada, A. Yoshimura, Suppressors of cytokine signaling and immunity. *Nat Immunol.* **4**(12):1169-76 (2003).
3. A. Yoshimura, H. Mori, M. Ohishi, D. Aki, T. Hanada, Negative regulation of cytokine signaling influences inflammation. *Curr Opin Immunol.* **15**(6):704-8 (2003).
4. A. B. Chepelinsky, J. S. Khillan, K. A. Mahon, P. A. Overbeek, H. Westphal, J. Piatigorsky, Crystallin genes: lens specificity of the murine alpha A-crystallin gene. *Environ Health Perspect.* **75**:17-24 (1987).
5. C. E. Egwuagu, R. M. Mahdi, C. C. Chan, J. Sztein, W. Li, J. A. Smith, A. B. Chepelinsky, Expression of interferon-gamma in the lens exacerbates anterior uveitis and induces retinal degenerative changes in transgenic Lewis rats. *Clin Immunol.* **91**(2):196-205 (1999).
6. C. E. Egwuagu, J. Sztein, R. M. Mahdi, W. Li, C. Chao-Chan, J. A. Smith, P. Charukamnoetkanok P, A. B. Chepelinsky, IFN-gamma increases the severity and accelerates the onset of experimental autoimmune uveitis in transgenic rats. *J Immunol.* **162**(1):510-7 (1999).
7. C. E. Egwuagu, C. R. Yu, M. Zhang, R. M. Mahdi, S. J. Kim, I. Gery, Suppressors of cytokine signaling proteins are differentially expressed in Th1 and Th2 cells: implications for Th cell lineage commitment and maintenance. *J Immunol,* **168**:3181-7 (2002).
8. C. R. Yu, R. M. Mahdi, S. Ebong, B. P. Vistica, J. Chen, Y. Guo, I. Gery, C. E. Egwuagu, Cell proliferation and STAT6 pathways are negatively regulated in T cells by STAT1 and suppressors of cytokine signaling. *J Immunol.* **173**(2):737-46 (2004).
9. A. M. Turnle, C. H. Fau, R. L. Rietze, J. R. Coonan, P. F. Bartlett, Suppressor of cytokine signaling 2 regulates neuronal differentiation by inhibiting growth hormone signaling. *Nat Neurosci.* **5**(11):1155-62 (2002).

10. A. M. Turnley, R. Starr, P. F. Bartlett, SOCS1 regulates interferon-gamma mediated sensory neuron survival. *Neuroreport*. **12**(16):3443-5 (2001).

11. G. Wong, Y. Goldshmit, A. M. Turnley, Interferon-gamma but not TNF alpha promotes neuronal differentiation and neurite outgrowth of murine adult neural stem cells. *Exp Neurol*. **187**(1):171-7 (2004).

12. J. E. Darnell, Jr., I. M. Kerr, G. R. Stark, Jak-STAT pathways and transcriptional activation in response to IFNs and other extracellular signaling proteins. *Science* **264**:1415-21 (1994).

13. A. M. Turnley, P. F. Bartlett, Cytokines that signal through the leukemia inhibitory factor receptor-beta complex in the nervous system. *J Neurochem*. **74**(3):889-99 (2000).

14. H. Paradis, R. L. Gendron, LIF transduces contradictory signals on capillary outgrowth through induction of stat3 and (P41/43) MAP kinase. *J Cell Sci*. **113**(23):4331-9 (2000).

DISEASE MECHANISMS AND GENE THERAPY IN A MOUSE MODEL FOR X-LINKED RETINOSCHISIS

Laurie L. Molday, Seok-Hong Min, Mathias W. Seeliger, Winco W.H. Wu, Astra Dinculescu, Adrian M. Timmers, Andreas Janssen, Felix Tonagel, Kristiane Hudl, Bernhard H.F. Weber, William W. Hauswirth, and Robert S. Molday*

1. INTRODUCTION

X-linked retinoschisis (RS) is an inherited recessive macular degeneration that affects between 1 in 5000 and 1 in 25,000 males early in life (George et al., 1995; Sieving, 1998; Tantri et al., 2004). It is characterized by a loss in central vision, splitting of the retina with the appearance of spoke-like cystic cavities radiating from the parafoveal region of the retina, a loss in the b-wave of the electroretinogram (ERG), and progressive atrophy of the macula. In about 50% of the cases, bilateral schisis is observed in the peripheral retina with some loss in peripheral vision. During the course of the disease, complications can arise which include retinal detachment, vitreal hemorrhage and choroidal sclerosis.

The gene responsible for RS was identified in 1997 by positional cloning and shown to encode a retinal specific, discoidin domain containing protein known as RS1 or retino-schisin (Sauer et al., 1997). RS1 is primarily expressed in photoreceptors and secreted as a disulfide-linked homo-octameric complex (Molday et al., 2001; Reid et al., 1999). It is localized along the surface of photoreceptors at the level of the inner segment, outer nuclear and outer plexiform layers of the outer retina and the surface of bipolar cells in the inner retina.

*Laurie L. Molday, University of British Columbia, Vancouver, BC V6T 1Z3. Seok-Hong Min, University of Florida, Gainsville, FL 32610. Mathias W. Seeliger, University of Tubingen, D-72076 Tubingen, Germany, Winco W.H. Wu, University of British Columbia, Vancouver, BC V6T 1Z3. Astra Dinculescu, University of Florida, Gainesville, FL 32610. Adrian M. Timmers, University of Florida, Gainesville, FL 32610. Andreas Janssen, University of Regensburg, Regensburg, D-92043 Germany, Felix Tonagel, University of Tubingen, D-72076 Tubingen, Germany, Kristiane Hudl, University of Tubingen, D-72076 Tubingen, Germany. Bernhard H.F. Weber, University of Regensburg, D-92043 Regensburg, Germany, William W. Hauswirth, University of Florida, Gainsville, FL 32610. Robert S. Molday, University of British Columbia, Vancouver, BC V6T 1Z3.

A mouse deficient in *Rs1h*, the mouse ortholog of the human *RS1* gene, was generated by homologous recombination to determine the role of RS1 in retina structure and function (Weber et al., 2002). The hemizygous *Rs1h$^{-/Y}$* mouse displays a number of features found in individuals with RS including a loss in the ERG b-wave, cystic structures within the inner retina, and progressive rod and cone degeneration. Morphological studies further reveal a highly disorganized retina with displacement of photoreceptor nuclei into the outer segment layer, merging of the outer and inner nuclear layers, gaps between bipolar cells, and disruption of the photoreceptor-bipolar synapse.

In this chapter we discuss disease mechanisms responsible for RS in relation to the structure of the RS1 protein and structural and functional recovery of the *Rs1h*-deficient mouse following adeno-associated viral (AAV) vector mediated gene therapy.

2. STRUCTURAL FEATURES AND DISEASE MECHANISMS

The 224 amino acid RS1 polypeptide is organized in four modules: an N-terminal signal peptide, an Rs1 domain, a discoidin (DS) domain, and a short C-terminal segment (Figure 39.1). Each module plays a key role in the biosynthesis, structure, subunit assembly and function of RS1 as a putative cell adhesion protein. Disease-causing missense mutations have been identified in each module (Consortium, 1998). The effect of selected mutations on RS1 has been reported (Wang et al., 2002; Wu and Molday, 2003).

2.1. Signal Peptide

At the N-terminus, a 23 amino acids hydrophobic signal peptide directs the nascent RS1 polypeptide to the endoplasmic reticulum (ER) membrane as an initial step in the processing of RS1 for secretion from cells (Sauer et al., 1997; Wu and Molday, 2003). Upon translocation of the polypeptide through the ER membrane, a signal peptidase in the ER lumen removes the signal peptide to produce the mature 201 amino acid RS1 polypeptide.

The effect of two disease-linked mutations in the signal peptide on the processing of RS1 has been examined (Wang et al., 2002; Wu and Molday, 2003). The L12H and L13P mutations cause a mislocalization of RS1 to the cytoplasm and rapid degradation via proteosomes. These amino acid substitutions most likely disrupt the helical conformation of the peptide thereby preventing its incorporation into the ER membrane.

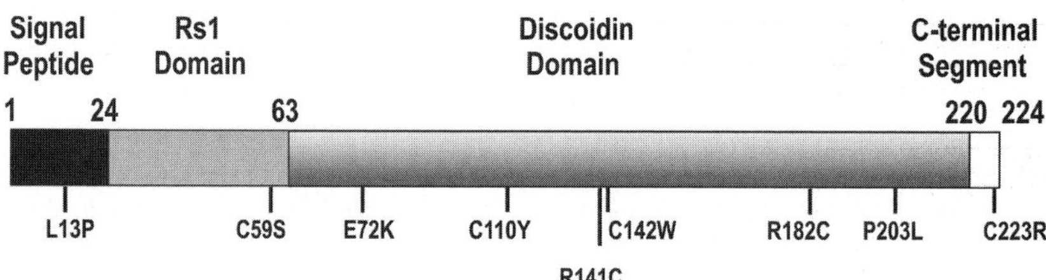

Figure 39.1. Schematic linear representation of RS1 showing the four modules and selected disease-linked missense mutations in each module. Modules include: the signal peptide important for insertion of the nascent polypeptide into the ER membrane; the discoidin domain implicated in cell adhesion, and the Rs1 domain and C-terminal segment both of which contribute to the octameric assembly of RS1.

2.2. Discoidin Domain

The dominant feature of the RS1 polypeptide is the 157 amino acid discoidin (DS) domain or F5/8 type C domain which comprises over 75% of the mature polypeptide chain. DS domains are present in a wide range of membrane and extracellular proteins where they mediate a variety of cell adhesion, cell signaling and developmental processes (Baumgartner et al., 1998; Vogel, 1999). Examples of DS domain containing proteins are Factors V and VIII involved in blood coagulation, neuropilins 1 and 2, which mediate nervous system regeneration and degeneration, discoidin domain receptors DDR1 and DDR2 implicated in cancer metastasis, and discoidin I involved in cellular adhesion during slime mold development.

High-resolution structural studies indicate that DS domains consist of eight antiparallel β-strands arranged in a barrel-like structure with several loops or spikes, projecting from one end of the core barrel (Pratt et al., 1999). The DS domain of RS1 has been modeled after the C2 DS domain of Factors V and VIII and shown to consist of the core beta barrel conformation with 3 spike regions (Wu and Molday, 2003). Conserved cysteine residues at the beginning and end of the DS domains (C63 and C219 in RS1) form an intramolecular disulfide bond important in protein folding. A second intramolecular disulfide bond, absent in DS domains of other proteins, joins C110 in spike 2 to C142 in spike 3. The function of the DS domain of RS1 is not known although it has been implicated in cell adhesion.

Most disease-linked missense mutations occur within the DS domain of RS1 with over a quarter resulting in a loss or gain of a cysteine (Consortium, 1998). These disease-linked mutant proteins are expressed in culture cells at relatively normal levels, but unlike wild-type (WT) RS1, they are not secreted from cells. Instead, they are retained in the ER as mis-folded, aggregated proteins (Wang et al., 2002; Wu and Molday, 2003).

2.3. Rs1 Domain and C-Terminal Segment

Two additional modules flank the DS domain. The 38 amino acid Rs1 domain resides just upstream of the DS domain, while a 5 amino acid C-terminal segment lies just downstream of the DS domain. Cysteine mutagenesis indicate that C59 of the Rs1 domain forms an intermolecular disulfide bond with C223 of the C-terminal segment of another RS1 subunit resulting in a disulfide-linked homo-octameric complex (Wu and Molday, 2003). Disease-causing mutations in these cysteines (C59S and C223R) are expressed and secreted from culture cells indicating that these mutant proteins fold into a native-like conformation. However, unlike WT RS1, the mutants fail to assemble into a disulfide-linked octameric complex indicating that multimeric assembly is critical for the function of RS1 as a cell adhesion protein (Wu and Molday, 2003).

3. GENE THERAPY IN THE *RS1H*-DEFICIENT MOUSE

Since female carriers of RS are asymptomatic, the lack of a functional RS1 protein and not the presence of a mutated protein is responsible for RS. Therefore, delivery of the normal *RS1* gene to retinal cells, and in particular photoreceptor cells, using established gene therapy approaches (Acland et al., 2001; Flannery et al., 1997) could lead to an improved outcome for RS patients.

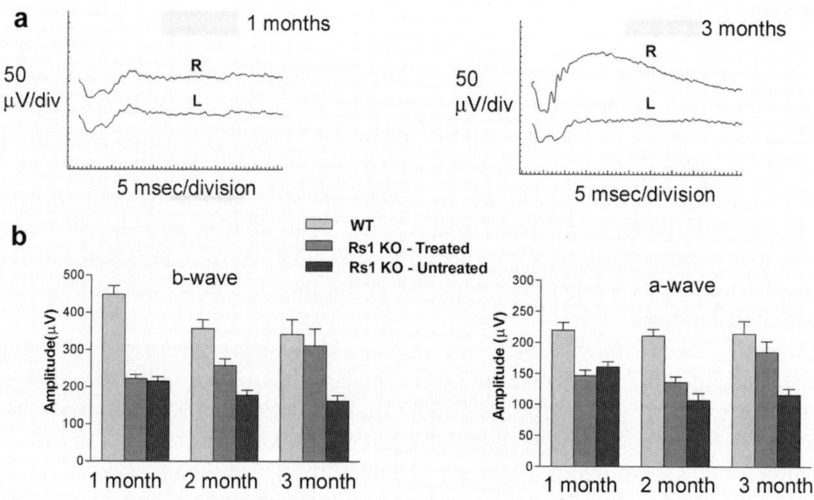

Figure 39.2. ERG recordings of AAV5-mOPs-*RS1* treated and untreated eyes. (a). Bilateral full-field scotpic ERG recordings of a treated (R) and untreated (L) *Rs1h*-deficient mouse at 1 and 3 months posttreatment. Traces are the average of 5 responses to a stimulus intensity of $0.173 \log \mathrm{cd\,m^{-2}}$. (b) Maximum ERG amplitudes of WT and treated and untreated mice. Bar is the mean amplitude $\pm$ SEM for 7 eyes.

To explore the feasibility of using gene therapy as a treatment for RS, we have delivered an AAV serotype 5 vector containing the human *RS1* cDNA under the control of the mouse opsin promoter (AAV5-mOPs-*RS1*) into the subretinal space of the right eyes of *Rs1h*-deficient 15-day old mice. The left eyes were not injected and served as controls. ERG recordings and immunocytochemical studies were carried out to determine the effect of gene delivery on the recovery of retinal structure and function.

3.1. ERG Recordings

Full-field scotopic ERG recordings of untreated and treated eyes of *Rs1h*-deficient mice were made at various times after a single subretinal injection with AAV5-mOPs-*RS1*. Figure 39.2a shows typical ERG recordings at 1 and 3 months posttreatment and Figure 39.2b displays the mean amplitudes of the a and b waves for the treated and untreated eyes at 1, 2 and 3 months. At 1 month, the amplitudes of the b-wave for the treated and untreated mice were similar and significantly reduced relative to WT mice. After 2 and 3 months there was a significant increase in b-wave amplitude for the treated eye, while the untreated eye showed a modest decline. The a-wave of the treated eye also showed an increase in amplitude at 3 months. Photopic responses were also significantly improved in the treated eye. There was a recovery of the oscillatory potentials of light-adapted single flash ERGs and a marked improvement in the responses to stimulus frequencies.

3.2. Immunocytochemical and Morphological Studies

The effect of AAV5-mOPs-*RS1* treatment on the expression and tissue distribution of RS1 was studied by immuncytochemical labeling of retinal cryosections of mice 5 month

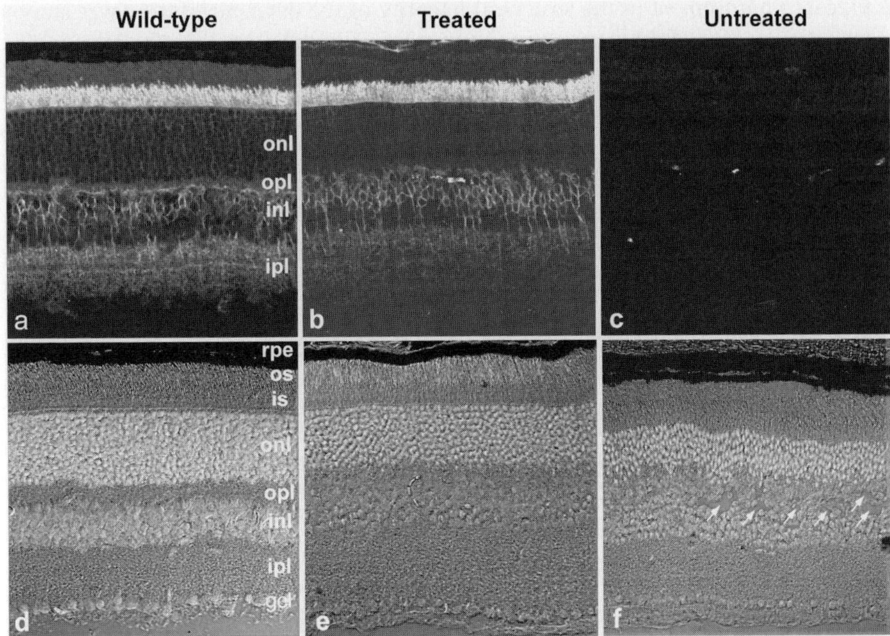

Figure 39.3. Immunoflurorescence microscopy of retinas from AAV-mOPs-*RS1* treated and untreated *Rs1h*-deficient mice 5 month posttreatment. (a-c) Cryosections labeled with the 3R10 anti-RS1 monoclonal antibody. (d-f) Sections imaged with DIC and stained with DAPI to highlight onl, inl and gcl layers. Arrows show gaps in the untreated mice. rpe, retinal pigment epithelium; os, outer segment; is, inner segment; onl, outer nuclear layer; opl, outer plexiform layer; inl, inner nuclear layer, ipl, inner plexiform layer; gcl, ganglion cell layer.

posttreatment. As shown in confocal micrographs of Figure 39.3a-c, retina tissue from a treated eye showed a RS1 immunostaining pattern similar to that of a wild-type retina. Intense RS1 labeling was present in the inner segment layer and more moderate labeling was seen in the outer nuclear and outer plexiform layers. The inner nuclear layer and to a lesser degree the inner plexiform layer was also labeled indicating that RS1 secreted from photoreceptors was able to move into the inner retina and bind to the surface of bipolar cells. Analysis of the whole retina of the treated eye showed that over 85% of the retina was labeled indicating that the RS1 protein spread laterally from the site of injection and expression. As previously reported (Weber et al., 2002), no RS1 labeling was observed for the untreated *Rs1h*-deficient retina.

The expression of RS1 coincided with a marked improvement in the structural organization of the retinal layers as visualized in DIC images merged with DAPI nuclear stain (Figure 39.3d-f). The retina was organized into characteristic layers with a distinct separation of the inner and outer nuclear layers and an absence of gaps between bipolar cells. An increased thickness of the outer nuclear layer indicative of enhanced photoreceptor survival was also seen in the treated retina. In contrast, the untreated eye showed a pronounced disorganization of the retinal cell layers, reduced outer nuclear layer thickness, shortened outer segments and gaps between bipolar cells. Immunofluorescence labeling studies using antibodies specific for the synaptic layers, bipolar cells and cone photoreceptors further showed a significant improvement in the structural integrity of the outer plexiform layer and inner nuclear layers and a 2-fold increase in the number of middle wavelength cones (data not shown).

Finally, glial fibrillary acidic protein (GFAP) expression and localization was studied to evaluate the pathological state of the retina. GFAP expression and distribution in the AAV5-mOPs-*RS1* treated retina was comparable to that of a wild-type retina with labeling restricted to the basal region of the Müller cells. In contrast, the untreated retina showed increased GFAP expression and localization throughout the Müller cells.

In one series of studies, we also evaluated the long-term effect of AAV5-mOPs-*RS1* treatment. RS1 expression and recovery of retinal structure and function persisted for at least 1 year following treatment.

Recently, Zeng et al. (2004) reported on AAV-mediated delivery of the mouse *Rs1h* gene under the control of a CMV promoter into the eye of a 13 week old *Rs1h*-deficient mouse. In this preliminary study, an improvement in the scotopic ERG b wave for the treated eye was observed, but no improvement in retina tissue organization or photoreceptor cell survival was demonstrated using this gene delivery protocol.

4. CONCLUSIONS

RS1 is a multisubunit, discoidin domain containing protein implicated in retinal cell adhesion. Disease-causing mutations cause defective protein biosynthetic processing, folding or subunit assembly resulting in an inactive protein. Delivery of the human *RS1* gene via the AAV5 vector to photoreceptor cells of *Rs1h*-deficient mice results in near normal RS1 expression and tissue localization and a substantial recovery of retinal structure and visual function over the long term. These studies suggest that AAV-mediated *RS1* gene delivery to photoreceptors may be an effective treatment for RS.

5. ACKNOWLEDGEMENTS

This work was supported by grants from the Foundation Fighting Blindness, Macular Vision Research Foundation, and the National Eye Institute.

6. REFERENCES

Acland, G.M., Aguirre, G.D., Ray, J., Zhang, Q., Aleman, T.S., Cideciyan, A.V., Pearce-Kelling, S.E., Anand, V., Zeng, Y., Maguire, A.M., Jacobson, S.G., Hauswirth, W.W. and Bennett, J. (2001) Gene therapy restores vision in a canine model of childhood blindness. *Nat. Genet.*, **28**:92-95.

Baumgartner, S., Hofmann, K., Chiquet-Ehrismann, R. and Bucher, P. (1998) The discoidin domain family revisited: new members from prokaryotes and a homology-based fold prediction. *Protein Sci.*, **7**:1626-1631.

Consortium, R. (1998) Functional implications of the spectrum of mutations found in 234 cases with X-linked juvenile retinoschisis. *Hum. Mol. Genet.*, **7**:1185-1192. (http://www.dmd.nl/rs)

Flannery, J.G., Zolotukhin, S., Vaquero, M.I., LaVail, M.M., Muzyczka, N. and Hauswirth, W.W. (1997) Efficient photoreceptor-targeted gene expression in vivo by recombinant adeno-associated virus. *Proc. Natl. Acad. Sci. U.S.A.*, **94**:6916-6921.

Fraternali, F., Cavallo, L. and Musco, G. (2003) Effects of pathological mutations on the stability of a conserved amino acid triad in retinoschisin. *FEBS Lett.*, **544**:21-26.

George, N.D., Yates, J.R. and Moore, A.T. (1995) X linked retinoschisis. *Br. J. Ophthalmol.*, **79**:697-702.

Molday, L.L., Hicks, D., Sauer, C.G., Weber, B.H. and Molday, R.S. (2001) Expression of X-linked retinoschisis protein RS1 in photoreceptor and bipolar cells. *Invest. Ophthalmol. Vis. Sci.*, **42**:816-825.

Pratt, K.P., Shen, B.W., Takeshima, K., Davie, E.W., Fujikawa, K. and Stoddard, B.L. (1999) Structure of the C2 domain of human factor VIII at 1.5 A resolution. *Nature*, **402**:439-442.

Reid, S.N., Akhmedov, N.B., Piriev, N.I., Kozak, C.A., Danciger, M. and Farber, D.B. (1999) The mouse X-linked juvenile retinoschisis cDNA: expression in photoreceptors. *Gene*, **227**:257-266.

Sauer, C.G., Gehrig, A., Warneke-Wittstock, R., Marquardt, A., Ewing, C.C., Gibson, A., Lorenz, B., Jurklies, B. and Weber, B.H. (1997) Positional cloning of the gene associated with X-linked juvenile retinoschisis. *Nat. Genet.*, **17**:164-170.

Sieving, P.A. (1998) Juvenile Retinoschisis. In Traboulsi, E.I. (ed.), *Genetic Diseases of the Eye.* Oxford University Press, New York.

Tantri, A., Vrabec, T.R., Cu-Unjieng, A., Frost, A., Annesley, W.H., Jr. and Donoso, L.A. (2004) X-linked retinoschisis: a clinical and molecular genetic review. *Surv. Ophthalmol.*, **49**:214-230.

Vogel, W. (1999) Discoidin domain receptors: structural relations and functional implications. *FASEB J.*, **13 Suppl**:S77-S82.

Wang, T., Waters, C.T., Rothman, A.M., Jakins, T.J., Romisch, K. and Trump, D. (2002) Intracellular retention of mutant retinoschisin is the pathological mechanism underlying X-linked retinoschisis. *Hum. Mol. Genet.*, **11**:3097-3105.

Weber, B.H., Schrewe, H., Molday, L.L., Gehrig, A., White, K.L., Seeliger, M.W., Jaissle, G.B., Friedburg, C., Tamm, E. and Molday, R.S. (2002) Inactivation of the murine X-linked juvenile retinoschisis gene, Rs1h, suggests a role of retinoschisin in retinal cell layer organization and synaptic structure. *Proc. Natl. Acad. Sci. U.S.A.*, **99**:6222-6227.

Wu, W.W.H. and Molday, R.S. (2003) Defective discoidin domain structure, subunit assembly, and endoplasmic reticulum processing of retinoschisin are primary mechanisms responsible for X-linked retinoschisis. *J. Biol. Chem.*, **278**:28139-28146.

Zeng, Y., Takada, Y., Kjellstrom, S., Hiriyanna, K., Tanikawa, A., Wawrousek, E., Smaoui, N., Caruso, R., Bush, R.A. and Sieving, P.A. (2004) RS-1 Gene Delivery to an Adult Rs1h Knockout Mouse Model Restores ERG b-Wave with Reversal of the Electronegative Waveform of X-Linked Retinoschisis. *Invest. Ophthalmol. Vis. Sci.*, **45**:3279-3285.

CHAPTER 40

MOLECULAR MECHANISMS OF NEUROPROTECTION IN THE EYE

Colin J. Barnstable and Joyce Tombran-Tink*

1. INTRODUCTION

Neurons are continuously subjected to fluctuating levels of oxidative stress, neuro-transmitters and other compounds that have the potential of damaging the cells. Under physiological conditions the levels of these compounds do not reach pathological levels and the cells survive. Under pathological conditions, however, the compounds reach toxic levels and apoptotic cell death can be triggered. Endogeneous neuroprotective factors are molecules which can prevent the switch from survival to cell death. These factors work in several ways, only a few of which will be considered in this chapter.

2. MITOCHONDRIAL UNCOUPLING PROTEINS

Reactive oxygen species are toxic and can reach levels that kill neurons. Cellular levels of reactive oxygen species are the sum of endogenous production and exogenous sources. Endogenous reactive oxygen species are produced within the cell as an obligatory by-product of oxidative phosphorylation. Exogenous reactive oxygen species arise from a variety of biochemical reactions and other pathological stimuli. Mitochondrial uncoupling proteins are intrinsic membrane proteins that can be activated to provide a controlled decrease in mitochondrial membrane potential and thus lower endogenous levels of reactive oxygen species. The neuroprotective effect of uncoupling proteins has been documented in a variety of degenerative models (Horvath et al., 2003a,b; Diano et al., 2003). Overexpression of this protein in transgenic mice provides protection against a variety of excitotoxins and other pathological stimuli. Retinal ganglion cells express the uncoupling protein UCP2, to date, however, we have not found clear evidence for expression of any of the five members

---

*Colin J. Barnstable, Department of Ophthalmology and Visual Science, Yale University School of Medicine, New Haven, CT 06520. Joyce Tombran-Tink, Division of Pharmaceutical Sciences, UMKC, Kansas City, MO 64110.

of the UCP gene family in photoreceptors. It is likely that the high energy requirements of photoreceptors require very tight coupling of energy sources to ATP production and that any uncoupling of oxidative phosphorylation would be deleterious to photoreceptor function. Nevertheless, molecules that activate uncoupling proteins, such as the Coenzyme Q co-factor, may be useful in preventing damage to other retinal cells such as RPE and Müller cells.

3. CNTF HAS BOTH POSITIVE AND NEGATIVE EFFECTS ON ROD PHOTORECEPTORS

Ciliary Neurotrophic Factor (CNTF), an endogeneous retinal protein (Walsh et al., 2001), has long been known to protect a variety of neurons, including rod photoreceptors, from pathological stimuli (LaVail et al., 1992; Tao et al., 2002; Bok et al., 2002). CNTF binds to a receptor complex found on a number of retinal cells types, particularly Müller glial cells. Activation of the CNTF receptor, in turn, activates two signal transduction pathways, the JAK/STAT and the Erk1/2 MAPK pathways. Within the retina, there is some evidence that activation of one of these pathways is restricted to specific cell types, although both can be activated in adult Müller glial cells. Although Müller glial cells are one of the strongest candidates to respond to and transmit CNTF signals to other cell types of the retina, there is still ongoing debate about which cells actually mediate CNTF neuroprotective actions in the retina.

We have shown that CNTF can block the formation of rod photoreceptors in retinal explants, using opsin as a specific marker to monitor rod cell number (Zhang et al., 2004). This effect depends on the activation of the STAT3 and not the MAPK pathway. Recent experiments have shown that in retinas from animals with a retina specific STAT3 knockout, there is an apparent increase in thickness of the Outer Nuclear Layer, suggesting that CNTF/STAT3 inhibition of rod formation is part of normal retinal development.

To test whether CNTF had an effect on specific rod genes we carried out a microarray analysis of retinal explants treated with CNTF for various periods of time. CNTF cause a reduction in expression of a range of genes involved in visual transduction, including cGMP-phosphodiesterase, recoverin, transducin and Abca4. On the other hand, we found that other genes such as Crx, Chx10 and Nr2E3 were not changed. The results suggest that CNTF has a very specific negative effect on rod development and may not always be an ideal therapeutic agent to combat retinal degenerations.

4. PEDF IS A POTENT NEUROPROTECTIVE FACTOR

PEDF is a novel neuroprotective factor that has proven therapeutic potential for a number of retinal diseases. PEDF is a 50 kD protein of the serpin family that was first isolated from medium conditioned by human RPE cells (Tombran-Tink and Johnson, 1989; Tombran-Tink et al., 1991). PEDF is an effective neuroprotective factor in many parts of the nervous system. In the eye, PEDF reduces apoptosis induced by H_2O_2 or light damage in rat photoreceptors (Cao et al., 1989, 2001), preserves the spatial organization, morphology, and function of photoreceptors after RPE detachment in a *Xenopus* model of retinal degeneration (Jablonski et al.) and protects retinal neurons from injuries caused by increased

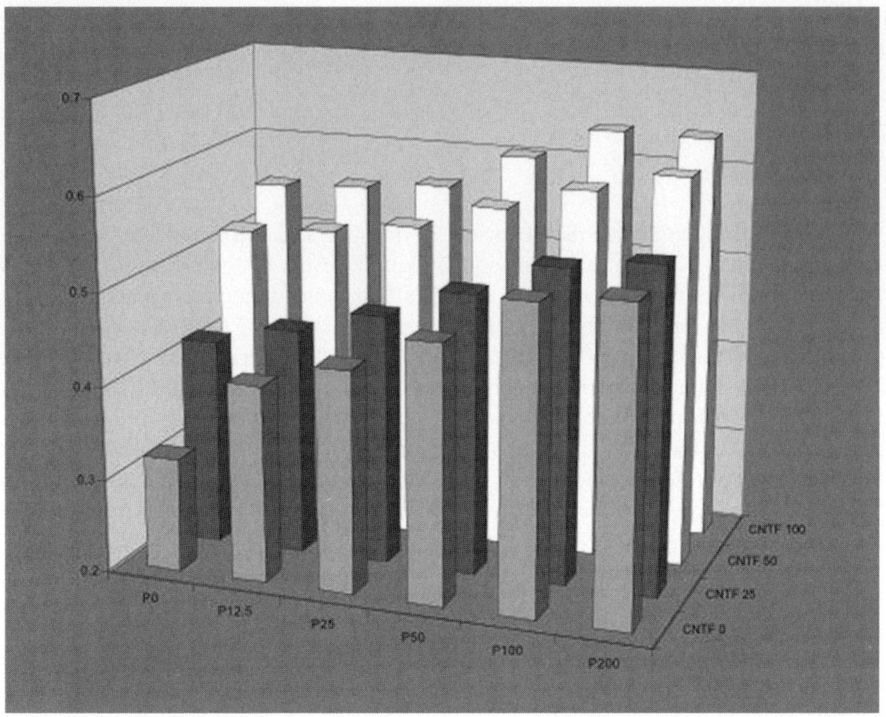

Figure 40.1. Increased survival of human RPE cells treated with H2O2. Cells were pretreated with neuroprotective factor for 24 hrs before being treated with 60 μM H2O2 for I hour. 3 hr later, cell survival was measured by labeling cells with Calcein AM and measuring fluorescence. Concentrations of PEDF (P) and CNTF (CNTF) in ng/ml are marked on the axes.

intraocular pressure from transient ischemic reperfusion.[11] In cells of other parts of the nervous system, such as cerebellar granule cells, hippocampal neurons and spinal cord motor neurons, nanogram amounts of PEDF provide protection from the damaging effects of glutamate toxicity.[12-14]

We have compared the efficacy of CNTF and PEDF to protect cells subjected to toxic levels of H2O2. As shown in the 3-D histogram of Figure 40.1, increasing concentrations of PEDF (left to right) or CNTF (front to back) increased the proportion of surviving cells. Saturation of the effect of each neuroprotective factor occurred at concentrations of approximately 100 ng/ml. Interestingly though even at saturating concentrations of one factor, addition of the other factor increases cell survival.

To test whether CNTF and PEDF might be acting in the same way at a molecular level we used a microarray analysis to compare changes in gene expression induced by treatment with each factor. Retinal explants were cultured in the presence or absence of each factor for 12, 24 or 48 hr. RNA from each group of retinas was used to prepare Cy3 and Cy5 labeled cDNA probes. These probes were hybridized to a microarray containing 12,000 spots of non-redundant retinal cDNA. Of the almost 10,000 retinal genes on these arrays CNTF induced changes in 62 genes and PEDF in only 30 genes. Very few of these genes were the same, suggesting that the two neuroprotective factors act, at least in part, via different mechanisms. The results also emphasize that neuroprotective factors have very specific actions on target cells.

5. CONCLUSIONS

There is growing evidence that a variety of intrinsic and extrinsic factors can increase a neuron's ability to withstand the episodic spikes in levels of toxic insults that occur in many neurodegenerative diseases. This suggests that neuroprotective agents can be used therapeutically for a range of complex retinal disorders such as Macular Degeneration and glaucoma, though they may be less efficacious over the long term against monogenic disorders with high penetrance such as many forms of retinitis pigmentosa. PEDF is an attractive candidate for neuroprotective therapies because it has no known harmful effects and is the only neuroprotective factor that also has antiangiogenic activity. Our findings suggest that different neuroprotective factors act via different pathways and thus that combinations of factors may be the most effective way of combating these retinal diseases.

6. ACKNOWLEDGEMENTS

We thank Drs. Samuel Shao-Min Zhang and Bing Chen for ongoing collaborations and discussions. Work in our laboratories has been supported by grants from the NIH and the David Woods Kemper Memorial Foundation.

7. REFERENCES

Bilak, M.M., Corse, A.M., Bilak, S.R., Lehar, M., Tombran-Tink, J., and Kuncl, R.W., 1999, Pigment epithelium-derived factor (PEDF) protects motor neurons from chronic glutamate-mediated neurodegeneration. *J. Neuropathol. Exp. Neurol.* **58**:719.

Bok, D., Yasumura, D., Matthes, M.T., Ruiz, A., Duncan, J.L., Chappelow, A.V., Zolutukhin, S., Hauswirth, W., and LaVail, M.M., 2002, Effects of adeno-associated virus-vectored ciliary neurotrophic factor on retinal structure and function in mice with a P216L rds/peripherin mutation. *Exp Eye Res.* **74**:719.

Cao, W., Tombran-Tink, J., Chen, W., Mrazek, D., Elias, R., and McGinnis, J.F., 1999, Pigment epithelium-derived factor protects cultured retinal neurons against hydrogen peroxide-induced cell death. *J. Neurosci. Res.* **57**:789.

Cao, W., Tombran-Tink, J., Elias, R., Sezate, S., Mrazek, D., and McGinnis, J.F., 2001, In vivo protection of photoreceptors from light damage by pigment epithelium-derived factor. *Invest Ophthalmol Vis Sci.* **42**:1646.

DeCoster, M.A., Schabelman, E., Tombran-Tink, J., and Bazan, N.G., 1999, Neuroprotection by pigment epithelial-derived factor against glutamate toxicity in developing primary hippocampal neurons. *J Neurosci Res.* **56**:604.

Diano, S., Matthews, R.T., Patrylo, P., Yang, L., Beal, M.F., Barnstable, C.J., and Horvath, T.L., 2003, Uncoupling protein 2 prevents neuronal death including that occuring during seizures: a mechanism for pre-conditioning. *Endocrinology*, **144**:5014.

Horvath, T.L., Diano, S., and Barnstable, C.J., 2003a, Mitochondrial uncoupling protein 2 in the central nervous system: neuromodulator and neuroprotector. *Biochem. Pharmacol.* **65**:1917.

Horvath, T.L., Diano, S., Leranth, C., Garcia-Segura, L.M., Cowley, M.A., Shanabrough, M., Elsworth, J.D., Sotonyi, P., Roth, R.H., Dietrich, E.H., Matthews, R.T., Barnstable, C.J., and Redmond, Jr, D.E., 2003b, Coenzyme Q induces mitochondrial uncoupling and prevents dopamine cell loss in a primate model of Parkinson's disease. *Endocrinol* **144**:2757.

Jablonski, M.M., Tombran-Tink, J., Mrazek, D.A., and Iannaccone, A., 2000, Pigment epithelium-derived factor supports normal development of photoreceptor neurons and opsin expression after retinal pigment epithelium removal. *J Neurosci.* **20**:7149.

LaVail, M.M., Unoki, K., Yasumura, D, Matthes, M.T., Yancopoulos, G.D., and Steinberg, R.H., 1992, Multiple growth factors, cytokines, and neurotrophins rescue photoreceptors from the damaging effects of constant light. Proc Natl Acad Sci USA. **89**:11249.

Ogata, N., Wang, L., Jo, N., Tombran-Tink, J., Takahashi, K., Mrazek, D., and Matsumura, M., 2001, Pigment epithelium derived factor as a neuroprotective agent against ischemic retinal injury. *Curr Eye Res.* **22**:245.

Taniwaki, T., Hirashima, N., Becerra, S.P., Chader, G.J., Etcheberrigaray, R., and Schwartz, J.P., 1997, Pigment epithelium-derived factor protects cultured cerebellar granule cells against glutamate-induced neurotoxicity. *J Neurochem.* **68**:26.

Tao, W., Wen, R., Goddard, M.B., Sherman, S.D., O'Rourke, P.J., Stabila, P.F., Bell, W.J., Dean, B.J., Kauper, K.A., Budz, V.A., Tsiaras, W.G., Acland, G.M., Pearce-Kelling, S., Laties, A.M., and Aguirre, G.D., 2002, Encapsulated cell-based delivery of CNTF reduces photoreceptor degeneration in animal models of retinitis pigmentosa. *Invest Ophthalmol Vis Sci.* **43**:3292.

Tombran-Tink, J., and Johnson, L.V., 1989, Neuronal differentiation of retinoblastoma cells induced by medium conditioned by human RPE cells. *Invest. Opthalmol. Vis. Sci.* **30**:1700.

Tombran-Tink, J., and Barnstable, C.J., 2003, PEDF: A multifaceted neurotrophic factor. *Nature Reviews Neuroscience* **4**:628.

Walsh, N., Valter, K., Stone, J. Cellular and subcellular patterns of expression of bFGF and CNTF in the normal and light stressed adult rat retina. *Exp Eye Res.* 2001 May; **72**(5):495-501.

Zhang, S.S., Wei, J., Qin ,H., Zhang, L., Xie, B., Hui, P., Deisseroth, A., Barnstable, C.J., and Fu, X.Y., 2004, STAT3-mediated signaling in the determination of rod photoreceptor cell fate in mouse retina Invest Ophthalmol Vis Sci. **45**:2407.

CHAPTER 41

RETINAL DAMAGE CAUSED BY PHOTODYNAMIC THERAPY CAN BE REDUCED USING BDNF

Jacque L. Duncan, Daniel M. Paskowitz, George C. Nune,
Douglas Yasumura, Haidong Yang, Michael T. Matthes,
Marco A. Zarbin, and Matthew M. LaVail*

1. INTRODUCTION

Age related macular degeneration (AMD) is the leading cause of blindness among the elderly in the United States (Klein et al., 1992; Klein et al., 2002), and choroidal neovascularization (CNV) accounts for the majority of severe vision loss (Ferris et al., 1984). The current standard treatment for CNV is verteporfin photodynamic therapy (PDT) (Landy and Brown, 2003), which uses a laser to activate a photosensitizing dye accumulated within the CNV. Although PDT causes less damage to the retina overlying CNV than thermal laser, in normal primate (Husain et al., 1996; Kramer et al., 1996; Reinke et al., 1999; Peyman et al., 2000), rabbit (Peyman et al., 2000) and rat (Zacks et al., 2002) models, PDT damages photoreceptors and retinal pigment epithelial (RPE) cells. Although there has been no histologic evidence of damage to normal human retinal cells after PDT (Schlotzer-Schrehardt et al., 2002), patients treated with PDT experience visual disturbances and acute severe vision loss significantly more often than patients receiving placebo (Arnold et al., 2004; Azab et al., 2004). Because neurotrophic agents, such as brain-derived neurotrophic factor (BDNF) have been proven effective in reducing retinal damage in rodents after exposure to constant light (LaVail et al., 1992; Okoye et al., 2003), we hypothesized that BDNF treatment prior to PDT might reduce collateral damage to retinal and RPE cells in normal rats.

* Jacque L. Duncan, Daniel M. Paskowitz, George C. Nune, Douglas Yasumura, Haidong Yang, Michael T. Matthes, Matthew M. LaVail, University of California, San Francisco, California, 94143. Marco A. Zarbin, Institute of Ophthalmology and Visual Science, University of Medicine and Dentistry of New Jersey, Newark, New Jersey, 07101.

2. METHODS

2.1. PDT

PDT was performed as described previously (Zacks et al., 2002) on adult Brown-Norway rats. Laser parameters used were modified from the standard protocol used to treat humans as follows. First, the 689 nm laser fluence was reduced from $50\,J/cm^2$ to $10\,J/cm^2$ because preliminary experiments demonstrated severe retinal damage to the retina after PDT laser using a fluence greater than $10\,J/cm^2$. Second, to permit interocular comparison between control and BDNF-treated eyes, the first eye of each rat received laser 3 minutes, and the second eye 4 minutes after intravenous verteporfin injection. All rats received $6\,mg/m^2$ verteporfin. The PDT laser spot size (3.0 mm) was determined by the size of the dilated rat pupil. Because there is no macula in the rat fundus, all eyes were treated superior to the optic disk for reproducible localization using fundus imaging and histology.

2.2. Intravitreal Injections

Rats were anesthetized with xylazine (13 mg/kg) and ketamine (87 mg/kg), and pupils were dilated using phenylephrine 2.5% and atropine 1%. One eye of each rat was injected with 2 µl BDNF (2 mg/ml). The contralateral eye received 2 µl phosphate buffered saline (PBS) or remained uninjected to serve as a control. Intravitreal injections were performed transsclerally using a 0.5-inch, 32-gauge beveled needle 2 days prior to PDT.

2.3. Retinal Analysis

Retinal function was evaluated using both full-field and multifocal electroretinography (ERG). Fundus photographs and fluorescein angiograms were performed 7 days after PDT. Rats were perfused, and eyes were embedded in plastic and sectioned for study using light microscopy as described previously (LaVail and Battelle, 1975).

3. RETINAL TOXICITY OF PDT IS REDUCED BY BDNF

3.1. Fundus Appearance

As described elsewhere, (Paskowitz et al., 2004) one week after PDT, a discrete circle of choroidal and RPE hypopigmentation with RPE clumping was visible superior to the optic disk in all eyes. This region demonstrated early hypofluorescence with delayed filling of choroidal vessels. A small ring of hyperfluorescence around the hypofluorescent region was visible in the late frames of the fluorescein angiogram, consistent with RPE damage surrounding the region of choriocapillaris closure. There was no significant difference in the appearance of the PDT-treated region between eyes that received BDNF and control eyes.

3.2. Histological Characteristics after Verteporfin PDT

Compared to eyes treated with 689-nm laser in the absence of verteporfin dye, eyes that received verteporfin PDT demonstrated severe damage to the retina and RPE when studied

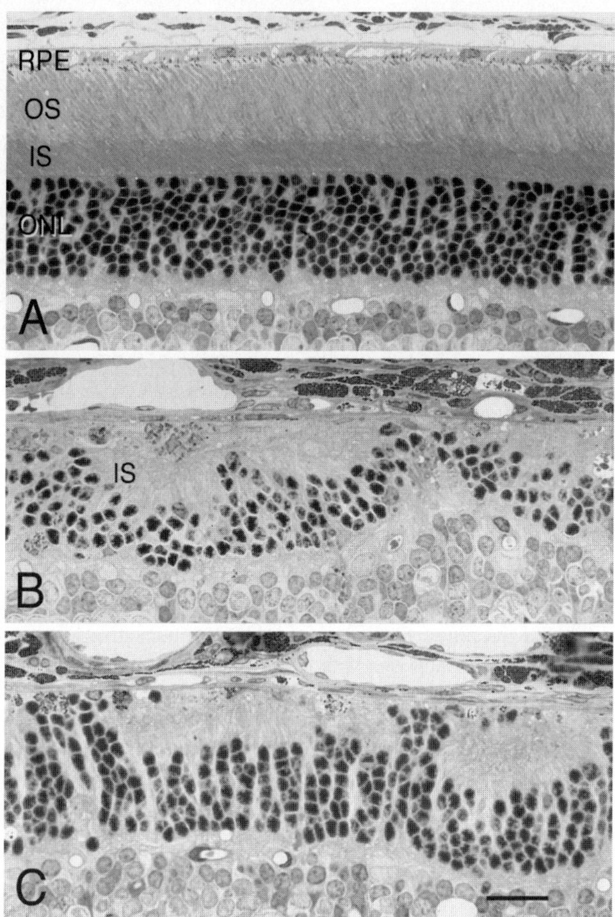

Figure 41.1. Light micrographs 1 week after PDT show the region of laser application in 3 eyes. (**A**) Retinal and RPE structure is normal after PDT laser without verteporfin dye. (**B**) PDT laser delivered 3-4 minutes after verteporfin injection produces severe retinal and RPE damage in this control eye treated with PBS 2 days prior to PDT. (**C**) The contralateral eye of the rat shown in (**B**) demonstrates greater ONL, IS and presumptive OS layer thickness after receiving BDNF 2 days prior to PDT. Epon-Araldite, 1 μm-thick sections; bar, 20 μm. Reproduced with permission of *Investigative Ophthalmology and Visual Science* in the format Other Book via Copyright Clearance Center from Paskowitz DM, Nune G, Yasumura D, Yang H, Bhisitkul RB, Sharma S, Matthes MT, Zarbin MA, LaVail MM, Duncan JL. BDNF Reduces the Retinal Toxicity of Verteporfin Photodynamic Therapy, *Invest Ophthalmol Vis Sci* 2004;45:4190-4196.

histologically 1 week after PDT (Fig. 41.1A and 41.1B). The outer nuclear layer (ONL) thickness in PDT-treated eyes was reduced from 9-10 rows in rats that had not received verteporfin PDT (Fig. 41.1A) to 3-4 irregular rows. In addition, variable photoreceptor outer segment and inner segment loss was apparent, with partial rosette formation in some areas (Fig. 41.1B). Damage to the RPE was present with pigment clumping, RPE cell attenuation and duplication, and migration of pigmented cells into the subretinal space. The retinal damage was limited to the PDT-treated region.

3.3. BDNF Improves Retinal Structure after PDT

Eyes that received BDNF 2 days prior to PDT showed less damage than contralateral control eyes. Although not normal, BDNF-treated eyes showed greater ONL thickness than controls (Fig. 41.1C). The number of surviving photoreceptors was quantified both by average ONL thickness in microns and by counting rows of nuclei in the ONL at the 5 most severely damaged regions in each eye. Significant photoreceptor rescue was observed in BDNF-treated eyes compared to control when measured both of these ways. The mean ONL thickness increased from $26.8 \pm 1.1\,\mu m$ (control) to $34 \pm 1.1\,\mu m$ (BDNF) ($P < 10^{-7}$), and the mean number of rows of ONL nuclei increased from 3.6 ± 0.6 rows (control) to 5.0 ± 0.3 rows (BDNF) ($P < 10^{-6}$, n = 17).

3.4. BDNF Improves Retinal Function after PDT

Bilateral full-field ERG recording demonstrated no significant difference between eyes treated with BDNF and control eyes, suggesting BDNF injection had no effect on global retinal function. To assess retinal function in the region that received PDT, multifocal ERG testing was performed. An infrared fundus camera imaged the retina being tested, and these images were aligned with fundus photos demonstrating RPE hypopigmentation in the PDT-treated area to identify the multifocal ERG responses produced by the PDT-treated retina.

Figure 41.2A shows averaged multifocal ERG responses from 6 normal rat eyes in a topographic pseudocolor plot. Each eye was tested with the optic disk at the same location (white circle). Figure 41.2B shows a representative multifocal ERG performed 1 week after PDT in a PBS-injected eye, demonstrating a large focal region of reduced retinal function superior to the optic disk in the PDT-treated area. The contralateral, BDNF-treated eye shows improvement in retinal function in the PDT-treated region (Fig. 41.2C). When superimposed, the traces used to generate the topographic plots demonstrate similar improved retinal function centrally in the BDNF-treated eye (red traces) compared to control (black traces) (Fig. 41.2D). Among 7 rats tested, BDNF-treated eyes showed significantly less abnormal multifocal ERG responses centrally than contralateral control eyes ($P = 0.03$).

4. CONCLUSIONS

Intravitreal BDNF injection 2 days prior to verteporfin PDT reduced retinal damage in normal rats. Future studies will investigate the effect of BDNF prior to PDT in rats with experimental CNV. Adjunctive therapy with BDNF or other neuroprotective agents may reduce undesired side effects of visual disturbances and severe vision loss among patients treated with PDT.

5. ACKNOWLEDGMENTS

The authors thank Jose Velarde, Kate Donohue-Rolfe, Shivani Sharma, Kamran Hosseini, Robert J. Lowe, Nancy Lawson and Dean Cruz for assistance, and Dr. Erich Sutter for advice in carrying out the multifocal ERG analyses. This work was supported by a Career Development Award from Research to Prevent Blindness (JLD), National Institutes of

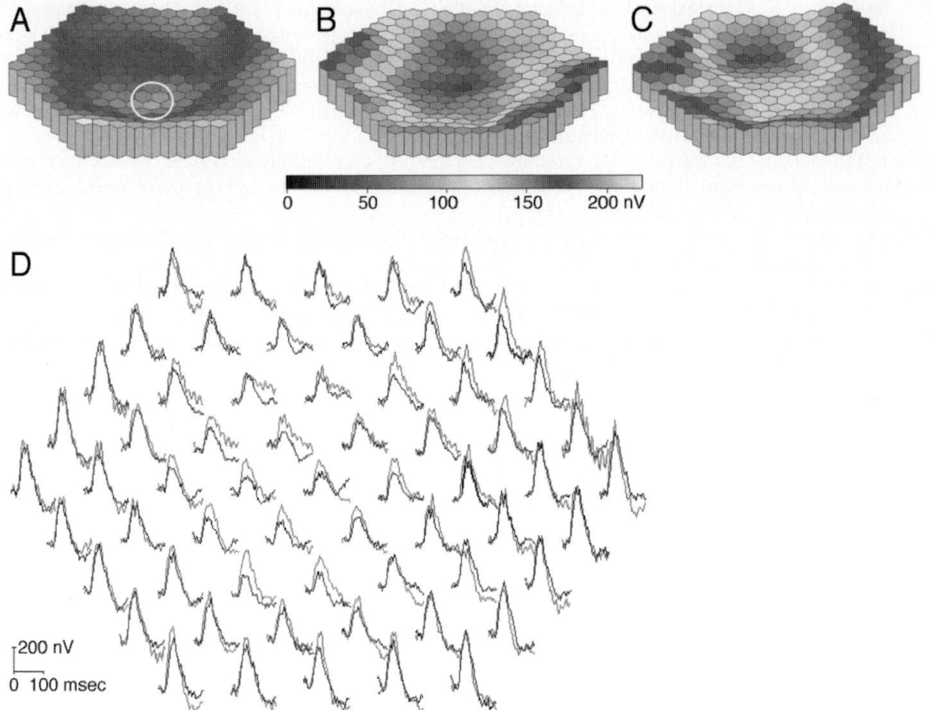

Figure 41.2. Multifocal ERG performed 1 week after PDT demonstrates reduced local retinal function in areas of PDT-treated retina. (**A**) Averaged responses from 6 normal rat eyes shown topographically demonstrate slightly reduced retinal function at the location of the optic disk (white circle). (**B**) Responses are reduced centrally in the PDT-treated area of an eye treated with PBS 2 days prior to PDT. (**C**) The contralateral, BDNF-treated eye shows improved retinal function in the PDT-treated area. (**D**) Superimposed multifocal ERG responses from the eyes shown in (**B**) (black traces) and (**C**) (red traces) show improved function centrally in the BDNF-treated eye. Reproduced with permission of *Investigative Ophthalmology and Visual Science* in the format Other Book via Copyright Clearance Center from Paskowitz et al., *Invest Ophthalmol Vis Sci* 2004;45:4190-4196. See also color insert.

Health Grants EY01919, EY02162 (MML) and EY00415 (JLD); the Bernard A. Newcomb Macular Degeneration Foundation (JLD, MML); the Foundation Fighting Blindness, Inc.; the Macular Vision Research Foundation (MML); and That Man May See, Inc. MML is a Research to Prevent Blindness Senior Scientist Investigator.

6. REFERENCES

Arnold, J. J., Blinder, K. J., Bressler, N. M., Bressler, S. B., Burdan, A., Haynes, L., Lim, J. I., Miller, J. W., Potter, M. J., Reaves, A., Rosenfeld, P. J., Sickenberg, M., Slakter, J. S., Soubrane, G., Strong, H. A. and Stur, M., 2004, "Acute severe visual acuity decrease after photodynamic therapy with verteporfin: case reports from randomized clinical trials-TAP and VIP report no. 3," *Am J Ophthalmol* **137**:683-96.

Azab, M., Benchaboune, M., Blinder, K. J., Bressler, N. M., Bressler, S. B., Gragoudas, E. S., Fish, G. E., Hao, Y., Haynes, L., Lim, J. I., Menchini, U., Miller, J. W., Mones, J., Potter, M. J., Reaves, A., Rosenfeld, P. J.,

Strong, A., Su, X. Y., Slakter, J. S., Schmidt-Erfurth, U. and Sorenson, J. A., 2004, "Verteporfin therapy of subfoveal choroidal neovascularization in age-related macular degeneration: meta-analysis of 2-year safety results in three randomized clinical trials: Treatment Of Age-Related Macular Degeneration With Photodynamic Therapy and Verteporfin In Photodynamic Therapy Study Report no. 4," *Retina* **24**: 1-12.

Ferris, F. L., Fine, S. L. and Hyman, L., 1984, "Age-related macular degeneration and blindness due to neovascular maculopathy," *Arch Ophthalmol* **102**:1640-1642.

Husain, D., Miller, J. W., Michaud, N., Connolly, E., Flotte, T. J. and Gragoudas, E. S., 1996, "Intravenous infusion of liposomal benzoporphyrin derivative for photodynamic therapy of experimental choroidal neovascularization," *Arch Ophthalmol* **114**:978-85.

Klein, R., Klein, B. E. and Linton, K. L., 1992, "Prevalence of age-related maculopathy. The Beaver Dam Eye Study," *Ophthalmology* **99**:933-43.

Klein, R., Klein, B. E., Tomany, S. C., Meuer, S. M. and Huang, G. H., 2002, "Ten-year incidence and progression of age-related maculopathy: The Beaver Dam eye study," *Ophthalmology* **109**:1767-79.

Kramer, M., Miller, J. W., Michaud, N., Moulton, R. S., Hasan, T., Flotte, T. J. and Gragoudas, E. S., 1996, "Liposomal benzoporphyrin derivative verteporfin photodynamic therapy. Selective treatment of choroidal neovascularization in monkeys," *Ophthalmology* **103**:427-38.

Landy, J. and Brown, G. C., 2003, "Update on photodynamic therapy," *Curr Opin Ophthalmol* **14**:163-8.

LaVail, M. M. and Battelle, B. A., 1975, "Influence of eye pigmentation and light deprivation on inherited retinal dystrophy in the rat," *Exp. Eye Res.* **21**:167-92.

LaVail, M. M., Unoki, K., Yasumura, D., Matthes, M. T., Yancopoulos, G. D. and Steinberg, R. H., 1992, "Multiple growth factors, cytokines and neurotrophins rescue photoreceptors from the damaging effects of constant light," *Proc. Natl. Acad. Sci. USA* **89**:11249-53.

Okoye, G., Zimmer, J., Sung, J., Gehlbach, P., Deering, T., Nambu, H., Hackett, S., Melia, M., Esumi, N., Zack, D. J. and Campochiaro, P. A., 2003, "Increased expression of brain-derived neurotrophic factor preserves retinal function and slows cell death from rhodopsin mutation or oxidative damage," *J Neurosci* **23**:4164-72.

Paskowitz, D. M., Nune, G., Yasumura, D., Yang, H., Bhisitkul, R. B., Sharma, S., Matthes, M. T., Zarbin, M. A., Lavail, M. M. and Duncan, J. L., 2004, "BDNF reduces the retinal toxicity of verteporfin photodynamic therapy," *Invest Ophthalmol Vis Sci* **45**:4190-6.

Peyman, G. A., Kazi, A. A., Unal, M., Khoobehi, B., Yoneya, S., Mori, K. and Moshfeghi, D. M., 2000, "Problems with and pitfalls of photodynamic therapy," *Ophthalmology* **107**:29-35.

Reinke, M. H., Canakis, C., Husain, D., Michaud, N., Flotte, T. J., Gragoudas, E. S. and Miller, J. W., 1999, "Verteporfin photodynamic therapy retreatment of normal retina and choroid in the cynomolgus monkey," *Ophthalmology* **106**:1915-23.

Schlotzer-Schrehardt, U., Viestenz, A., Naumann, G. O., Laqua, H., Michels, S. and Schmidt-Erfurth, U., 2002, "Dose-related structural effects of photodynamic therapy on choroidal and retinal structures of human eyes," *Graefes Arch Clin Exp Ophthalmol* **240**:748-57.

Zacks, D. N., Ezra, E., Terada, Y., Michaud, N., Connolly, E., Gragoudas, E. S. and Miller, J. W., 2002, "Verteporfin photodynamic therapy in the rat model of choroidal neovascularization: angiographic and histologic characterization," *Invest Ophthalmol Vis Sci* **43**:2384-91.

CHAPTER 42

CONTROLLING VASCULAR ENDOTHELIAL GROWTH FACTOR: THERAPIES FOR OCULAR DISEASES ASSOCIATED WITH NEOVASCULARIZATION

Robert J. Marano[1,2] and P. Elizabeth Rakoczy[2]

1. INTRODUCTION

Vascular endothelial growth factor (VEGF) is a potent stimulator of angiogenesis and is essential for normal embryonic development and many physiological events that require the growth of new blood vessels. Abnormal expression of endogenous VEGF can lead to ocular diseases including age related macular degeneration (Ohno-Matsui et al., 2001) and diabetic retinopathy (Aiello et al., 1994; Boulton et al., 1998), which are the two leading causes of blindness in the developed world. Regulation of VEGF expression occurs primarily through trans-factor interactions with cis-elements located on the 5′ and 3′ untranslated regions (UTR's) and include stabilizing and destabilizing elements in addition to enhancer regions (Coles et al., 2004; Dibbens et al., 1999; Iida et al., 2002; Levy et al., 1997; Marano et al., 2004). The prime stimuli of VEGF upregulation are hypoxic or ischemic conditions, which indirectly activates VEGF through interactions between hypoxia inducible factor 1 (HIF-1) and the hypoxia response element (HRE) located within the promoter region of the VEGF gene (Forsythe et al., 1996).

Due to the relative importance of VEGF in angiogenesis and neovascularisation, it has come under close scrutiny as a possible target for the control of both angiogenic and ischemic diseases. In the case of the former, inhibition of VEGF expression has been explored through the use of sense oligonucleotides and siRNA's (Garrett et al., 2001; Marano et al., 2003; Tolentino et al., 2004). In addition, the posttranslational activity of VEGF has been inhibited by interfering with protein:receptor interactions using VEGF antibodies and a truncated version of the VEGF receptor Flt-1 (Adamis et al., 1996). In the case of the ischemic diseases, in an effort to maintain blood supply to oxygen starved tissue, angiogenesis has been induced by elevating VEGF protein levels using gene transfer to introduce and expressible VEGF gene construct into cells, or injecting tissues with VEGF protein

---

[1] Department of Molecular Ophthalmology, Lions Eye institute, 2 Verdun St. Nedlands, WA, Australia.
[2] Centre for Ophthalmology and Visual Sciences, University of Western Australia, Australia.

directly (de Boer et al., 2001). However, in most cases, the temporal control in addition to tissue localization remains a limiting factor for this method of therapy.

In this chapter we will discuss several methods used in controlling VEGF expression at both the transcriptional and translational levels in addition to post translational inhibition of VEGF activity and how these may be exploited as possible therapies for neovascular diseases of the eye.

2. OLIGONUCLEOTIDE THERAPY

Oligonucleotides consist of short sequences of nucleotides generally between 15 and 30 base pairs in lengths and possess a phosphodiester (P) backbone in the native form. Many chemical modifications can be made to improve the biological half-life of oligonucleotides in vivo and include substituting the P backbone for phophorothioate (PS), which results in the oligonucleotide being less susceptible to the effects of nucleases. Oligonucleotides used in expressional control of genes consist of three main forms; sense oligonucleotides, which as the name suggests resembles a protion of the sense stand of duplex DNA; antisense oligonucleotides, which resemble a protion of the complimentary stand of duplex DNA; and aptamers, whose activity relies on the secondary structure formed by folded oligonucleotides. Depending on which form is used, oligonucleotides are able to control gene expression at either the transcriptional or translational level in addition to affecting downstream, post translation activities.

All three oligonucleotide types have been explored as a possible control for increased VEGF expression associated with neovascularization in aged related macular degeneration (AMD). In our laboratory, we have examined the 5′-UTR sequence of the VEGF gene for possible oligonucleotide target sites. From this study, a sense oligonucleotide (DS-085) was found to significantly down-regulate VEGF expression in cultured cells (Garrett et al., 2001). Subsequent experiments later demonstrated that *in vivo*, intravitreal administration of DS-085 resulted in retinal cells being transfected, in addition, DS-085 mediated a reduced angiogenic response in a laser photocoagulation induced choroidal neovascularisation (CNV) rodent model (Marano et al., 2003). The most likely mechanism of downregulation in this case was thought to occur through hybridization of the oligonucleotide in the major groove of duplex DNA forming of a triple helix. This would result in polymerase arrest and lead to a reduction in the quantity of transcript available for translation.

In addition to sense oligonucleotides, an antisense oligonucleotide has been developed to target the VEGF-R2 receptor (KDR/Flk) to control neovascularization of the cornea (Berdugo et al., 2003). Antisense oligonucleotides bind to their compliment sequence on the single stranded mRNA. From this point, regulation of the gene may occur in one of three ways; through polymerase arrest; through occupational inhibition, whereby the bound oligonucleotide occupies a functional site such as the 5′ end capping sequence; and finally through RNAase recruitment caused by the presence of the bound oligonucleotide, which results in the rapid degradation of the mRNA strand.

While these results are promising, it was an aptamer (EYE001) that first progressed to the clinical trial stage as a means of controlling neovascularization of retina (The Eyetech Study Group, 2002; The Eyetech Study Group, 2003). Aptamers are produced by exploiting an inherent natural phenomena of nucleic acids i.e. depending on their sequence, they are able to fold and form complex secondary structures. The folded oligonucleotide then acts similar to an antibody in recognizing and binding to specific elements on a protein. In the case of EYE001, the aptamer binds to VEGF and inhibits its interaction with its cell surface receptor molecules, preventing any further downstream activity.

3. GENE THERAPY

A primary drawback in the use of oligonucleotides as a therapeutic has been the relative biological instability of the compound. Oligonucleotides are generally short lived *in vivo*, and while the use of modified chemistries has led to improved half lives, longevity remains a limiting issue and re-administration is required on a regular basis. In this respect, gene therapy, whereby a modified, recombinant gene is inserted into the genome, is a superior method of controlling gene function. Once inserted, the new gene remains functional for an indefinite period, producing the therapeutic agent for the life of the cell. In addition to longevity, specific cell types may be targeted for gene expression through the use of cell and tissue specific promoters, which control gene expression. This approach may be useful for systemic delivery of gene constructs as it ensures that the gene is not expressed in tissues where its presence may prove detrimental. For the control of VEGF associated with ocular angiogenesis, this approach of secretion gene therapy has been explored on several occasions. Constructs containing the von Hippel-Lindau gene, (Akiyama et al., 2004) pigment epithelium-derived factor gene, (Auricchio et al., 2002) and genes expressing the soluble VEGF receptor sFlt-1, (Bainbridge et al., 2002; Lai et al., 2001; Lai et al., 2002) have been used to transform retinal cells *in vivo* to produced recombinant proteins and control the expression and/or the downstream effects of VEGF. In these cases, the gene construct was delivered to cells using a modified virus, which has been engineered to be unable to replicate. The virus genomes containing the recombinant gene constructs are packaged into virus particles *in vitro*, which are then injected into the eye. The virus particles attaches to the cell surface and injects the nucleic acid material into the host cell, where it is incorporated into the genome and expressed using the hosts metabolic processes.

Several limitations exist in this system of gene therapy. Using viral vectors to insert genes into the host's genome runs the risk of causing insertional inactivation of other genes within the cell, which can be highly damaging depending on the gene that is inactivated (Baum et al., 2004). In addition, while it is possible to exert some crude control over the level of expression through the use of weak or strong promoters, a more refined, temporal control system is more difficult to achieve. In the case of angiogenic diseases of the eye, VEGF concentrations may fluctuate over time; therefore, maintaining high levels of an anti-VEGF compound may prove detrimental in the long term. Attempts have been made to address this issue, and involve incorporating the same elements that control the expression of endogenous VEGF (such as the hypoxia control element), into the anti-VEGF transgene construct (Bainbridge et al., 2003). In addition to these logistical issues involved in the science and 'workability' of gene therapy, the use of viruses as a delivery vehicle for modified genetic material has not received wide public acceptance at this point in time.

4. RNA INTERFERENCE (RNAi)

RNAi through the use of short interfering RNA (siRNA) molecules has also proven efficient at controlling laser induced CNV in mouse and primate models (Reich et al., 2003; Tolentino et al., 2004). The mechanism of siRNA occurs through the activity of a multicomponent ribonuclease-protein complex termed RISC (RNA-induced silencing complex), which recruits siRNA and guides them to the homologous portion on the mRNA transcript. This triggers the degradation of the mRNA strand, once again reducing the quantity of transcript available for translation. Several advantages of RNAi include high specificity for the target gene and as it utilizes a naturally occurring pathway for mRNA degradation, there are relatively few side effects. In addition, siRNA may be administered directly as duplex RNA strands for immediate recruitment by the RISC, alternatively, siRNA can be introduced as a gene therapy style construct whereby the sense sequence of the siRNA is cloned into a vector upstream of the reverse compliment sequence separated by a short spacer sequence. When the construct is transcribed, the resulting mRNA strand will fold on itself forming a duplex strand with a hairpin loop at one end. This structure can then be processed by the dicer protein, which removes the hairpin and facilitates recruitment by the RISC (Arenz and Schepers, 2003).

5. CONCLUSION

The use of anti-VEGF therapies are becoming a popular method for controlling diseases where angiogenesis is one of the major pathologies. For angiogenic diseases affecting the eye, several such methods are currently being explored, each with their own merits and limitations. At present, efficient delivery and longevity of the biological effect continue to be the major issues facing novel therapeutics. However, with the dearth of techniques and methods aimed at controlling VEGF expression currently being developed, a decision will need to be made on which is the most suitable as a clinical treatment. Future research may eventually indicate that the most effective treatment is a combination of two or more methods i.e. one treatment to achieve a rapid early response followed by a second, alternative treatment which provides a longer effect.

6. REFERENCES

Adamis, A. P., Shima, D. T., Tolentino, M. J., Gragoudas, E. S., Ferrara, N., Folkman, J., D'Amore, P. A., Miller, J. W., 1996, Inhibition of vascular endothelial growth factor prevents retinal ischemia-associated iris neovascularization in a nonhuman primate, *Arch Ophthalmol,* **114:**66-71.

Aiello, L. P., Avery, R. L., Arrigg, P. G., Keyt, B. A., Jampel, H. D., Shah, S. T., Pasquale, L. R., Thieme, H., Iwamoto, M. A., Park, J. E., et al., 1994, Vascular endothelial growth factor in ocular fluid of patients with diabetic retinopathy and other retinal disorders, *N Engl J Med,* **331:**1480-1487.

Akiyama, H., Tanaka, T., Itakura, H., Kanai, H., Maeno, T., Doi, H., Yamazaki, M., Takahashi, K., Kimura, Y., Kishi, S., Kurabayashi, M., 2004, Inhibition of ocular angiogenesis by an adenovirus carrying the human von Hippel-Lindau tumor-suppressor gene in vivo. *Invest Ophthalmol Vis Sci,* **45:**1289-1296.

Arenz, C., Schepers, U., 2003, RNA interference: from an ancient mechanism to a state of the art therapeutic application? *Naturwissenschaften,* **90:**345-359.

Auricchio, A., Behling, K. C., Maguire, A. M., O'Connor, E. M., Bennett, J., Wilson, J. M., Tolentino, M. J., 2002, Inhibition of retinal neovascularization by intraocular viral-mediated delivery of anti-angiogenic agents, *Mol Ther,* **6:**490-494.

Bainbridge, J. W., Mistry, A., Binley, K., De Alwis, M., Thrasher, A. J., Naylor, S., Ali, R. R., 2003, Hypoxia-regulated transgene expression in experimental retinal and choroidal neovascularization. *Gene Ther,* **10:**1049-1054.

Bainbridge, J. W., Mistry, A., De Alwis, M., Paleolog, E., Baker, A., Thrasher, A. J., Ali, R. R., 2002, Inhibition of retinal neovascularisation by gene transfer of soluble VEGF receptor sFlt-1. *Gene Ther,* **9:**320-326.

Baum, C., von Kalle, C., Staal, F. J., Li, Z., Fehse, B., Schmidt, M., Weerkamp, F., Karlsson, S., Wagemaker, G., Williams, D. A., 2004, Chance or necessity? Insertional mutagenesis in gene therapy and its consequences. *Mol Ther,* **9:**5-13.

Berdugo, M., Valamanesh, F., Andrieu, C., Klein, C., Benezra, D., Courtois, Y., Behar-Cohen, F., 2003, Delivery of antisense oligonucleotide to the cornea by iontophoresis. *Antisense Nucleic Acid Drug Dev,* **13:**107-114.

Boulton, M., Foreman, D., Williams, G., McLeod, D., 1998, VEGF localisation in diabetic retinopathy. *Br J Ophthalmol,* **82:**561-568.

Coles, L. S., Bartley, M. A., Bert, A., Hunter, J., Polyak, S., Diamond, P., Vadas, M. A., Goodall, G. J., 2004, A multi-protein complex containing cold shock domain (Y-box) and polypyrimidine tract binding proteins forms on the vascular endothelial growth factor mRNA. Potential role in mRNA stabilization. *Eur J Biochem,* **271:**648-660.

de Boer, R. A., Siebelink, H. J., Tio, R. A., Boomsma, F., van Veldhuisen, D. J., 2001, Carvedilol increases plasma vascular endothelial growth factor (VEGF) in patients with chronic heart failure. *Eur J Heart Fail,* **3:**331-333.

Dibbens, J. A., Miller, D. L., Damert, A., Risau, W., Vadas, M. A., Goodall, G. J., 1999, Hypoxic regulation of vascular endothelial growth factor mRNA stability requires the cooperation of multiple RNA elements. *Mol Biol Cell,* **10:**907-919.

Forsythe, J. A., Jiang, B. H., Iyer, N. V., Agani, F., Leung, S. W., Koos, R. D., Semenza, G. L., 1996, Activation of vascular endothelial growth factor gene transcription by hypoxia-inducible factor 1. *Mol Cell Biol,* **16:**4604-4613.

Garrett, K. L., Shen, W. Y., Rakoczy, P. E., 2001, In vivo use of oligonucleotides to inhibit choroidal neovascularisation in the eye. *J Gene Med,* **3:**373-383.

Iida, K., Kawakami, Y., Sone, H., Suzuki, H., Yatoh, S., Isobe, K., Takekoshi, K., Yamada, N., 2002, Vascular endothelial growth factor gene expression in a retinal pigmented cell is up-regulated by glucose deprivation through 3′ UTR. *Life Sci,* **71:**1607-1614.

Lai, C. M., Brankov, M., Zaknich, T., Lai, Y. K., Shen, W. Y., Constable, I. J., Kovesdi, I., Rakoczy, P. E., 2001, Inhibition of angiogenesis by adenovirus-mediated sFlt-1 expression in a rat model of corneal neovascularization. *Hum Gene Ther,* **12:**1299-1310.

Lai, Y. K., Shen, W. Y., Brankov, M., Lai, C. M., Constable, I. J., Rakoczy, P. E., 2002, Potential long-term inhibition of ocular neovascularisation by recombinant adeno-associated virus-mediated secretion gene therapy. *Gene Ther,* **9:**804-813.

Levy, N. S., Goldberg, M. A., Levy, A. P., 1997, Sequencing of the human vascular endothelial growth factor (VEGF) 3′ untranslated region (UTR): conservation of five hypoxia-inducible RNA-protein binding sites. *Biochim Biophys Acta,* **1352:**167-173.

Marano, R. J., Wimmer, N., Kearns, P. S., Thomas, B. G., Toth, I., Brankov, M., Rakoczy, P. E., 2003, Inhibition of in vitro VEGF expression and choroidal neovascularization by synthetic dendrimer peptide mediated delivery of a sense oligonucleotide. *Experimental Eye Research,* **Accepted**.

Marano, R. J., Brankov, M., Rakoczy, P. E., 2004, Discovery of a novel control element within the 5′UTR of VEGF: Regulation of expression using sense oligonucleotides. *J Biol Chem,* **279**(36):37808-37814.

Ohno-Matsui, K., Morita, I., Tombran-Tink, J., Mrazek, D., Onodera, M., Uetama, T., Hayano, M., Murota, S. I., Mochizuki, M., 2001, Novel mechanism for age-related macular degeneration: an equilibrium shift between the angiogenesis factors VEGF and PEDF. *J Cell Physiol,* **189:**323-333.

Reich, S. J., Fosnot, J., Kuroki, A., Tang, W., Yang, X., Maguire, A. M., Bennett, J., Tolentino, M. J., 2003, Small interfering RNA (siRNA) targeting VEGF effectively inhibits ocular neovascularization in a mouse model. *Mol Vis,* **9:**210-216.

The Eyetech Study Group, 2002, Preclinical and phase 1A clinical evaluation of an anti-VEGF pegylated aptamer (EYE001) for the treatment of exudative age-related macular degeneration, *Retina,* **22:**143-152.

The Eyetech Study Group, 2003, Anti-vascular endothelial growth factor therapy for subfoveal choroidal neovascularization secondary to age-related macular degeneration: phase II study results. *Ophthalmology,* **110:**979-986.

Tolentino, M. J., Brucker, A. J., Fosnot, J., Ying, G. S., Wu, I. H., Malik, G., Wan, S., Reich, S. J., 2004, Intravitreal injection of vascular endothelial growth factor small interfering RNA inhibits growth and leakage in a nonhuman primate, laser-induced model of choroidal neovascularization. *Retina,* **24:**132-138.

CHAPTER 43

INTRAVITREAL INJECTION OF TRIAMCINOLONE ACETONIDE FOR MACULAR EDEMA DUE TO RETINITIS PIGMENTOSA AND OTHER RETINAL DISEASES

Changguang Wang[1,2], Jianbin Hu[1,2], Paul S. Bernstein[1], Michael P. Teske[1], Marielle Payne[1,2], Zhenglin Yang[1,2], Chumei Li[1,2], David Adams[1,2], Jennifer H. Baird[1,2], and Kang Zhang[1,2]

1. INTRODUCTION

Macular edema is a swelling of the macula that can result in decreased visual acuity. Because the macula is extensively surrounded by blood vessels, any resulting leakage can lead to macular edema and subsequent visual loss. Such leakage can be secondary to retinitis pigmentosa, and other retinal diseases including diabetic retinopathy, retinal vein occlusion, inflammatory processes such as uveitis, or can be a result of ocular surgery, referred to as Irvine-Gass Syndrome.

Triamcinolone acetonide (TA) is a slow-dissolving depot corticosteroid suspension previously used for arthritis as an intra-articular injection. TA has also used as a periocular injection to treat ocular inflammatory diseases. (Yoshikawa et al., 1995; Zamir et al., 2002) More recently, TA has been given as an intravitreal injection to treat macular edema with encouraging results. Macular edema affects a significant number of RP and approximately 29% of diabetic patients (Klein et al., 1984) with disease duration of 20 years or more and consequently is a significant cause of reduced visual acuity. Intravitreal TA has been shown to be effective in reducing macular thickening caused by diabetic clinically significant macular edema (CSME) refractory to other treatments including laser photocoagulation (Martidis et al., 2002; Massin et al., 2004). Focal macular edema, characterized by focal leakage from microaneurysms, is more responsive to laser photocoagulation than diffuse leakage from the posterior retinal capillary bed. (Bresnick, 1983; Bresnick, 1986) TA is becoming an increasingly important therapeutic method for improving visual acuity in those

[1] Moran Eye Center, Department of Ophthalmology and Visual Science, and [2] Program in Human Molecular Biology & Genetics, Eccles Institute of Human Genetics, University of Utah, Salt Lake City, UT.

with diffuse macular edema refractory to laser treatment. Intravitreal injection of TA has been shown to reduce retinal thickening, improve blood retinal barrier function and improve visual acuity in patients with diffuse macular edema. (Jonas et al., 2003; Martidis et al., 2002) TA has also been shown to reduce vitreal levels of Vascular Endothelial Growth Factor (VEGF) and Stromal-Derived Factor 1 (SDF1), angiogenic growth factors that potentially play an important role in the pathogenesis of diffuse macular edema. (Brooks et al., 2004; Funatsu et al., 2003)

Central retinal vein occlusion (CRVO) is another common retinal vascular disorder that has shown promising results following treatment with TA. Experimental studies in monkeys have shown that a hypoxic environment is produced in the retina following venous occlusion. (Hockley et al., 1979) This hypoxia causes functional structural changes in the retinal capillaries resulting in increased permeability and retinal edema, potentially through factors including VEGF and SDF1. Although both ischemic and nonischemic types of CRVO show anatomical improvement following treatment with TA, visual acuity does not improve as well in ischemic CRVO. (Ip et al., 2004)

Cystoid macular edema (CME) can result from cases of long standing retinitis pigmentosa (RP), uveitis, or Irvine-Gass Syndrome. The exact mechanism of CME occurrence in patients with RP is not known. The high prevalence of antiretinal autoantibodies in those with CME associated with RP suggests an inflammatory, autoimmune process. (Heckenlively et al., 1999) There has been one previous case report of a patient with CME and RP treated with intravitreal injections of TA. (Sallum et al., 2003; Saraiva et al., 2003) Intravitreal TA not only allows improvement in visual acuity with minimal systemic side-effects, but has also been shown to be effective in reducing visual loss in those with long-term refractory inflammatory CME. (Antcliff et al., 2001)

Complications of corticosteroids administered ocularly include intraocular pressure elevation, retinal detachment, vitreous hemorrhage, cataractogenesis, endophthalmitis, and potential cytotoxicity to photoreceptors and the retinal pigment epithelium. TA is a minimally water-soluble steroid that is injected in a suspension form. The decreased water solubility contributes to its prolonged duration of action. After intravitreal injection, the duration of effect has been reported to last between 4 weeks and 9 months. (Beer et al., 2003; Jonas et al., 2004)

2. METHODS

A retrospective chart review was performed for the historical period of June through November 2003. During this period, 51 eyes of 50 patients were treated for macular edema with intravitreal injections of TA at the Moran Eye Center. This includes 18 patients with chronic diffuse macular edema due to the following conditions: retinitis pigmentosa (two patients), CSME (seven patients), Irvine-Gass Syndrome (four patients), and central retinal vein occlusion (five patients). Of the eyes treated for macular edema, 20 eyes were measured before and after TA injection with optical coherence tomography (OCT) on a Zeiss-Humphrey Stratus OCT version 3.0. The data for these eyes were analyzed for this study. Pre-injection OCT measurements were taken between 47 days and zero days before TA injection, with a mean of seven days and a median of zero days. The time of the first post-injection measurement ranged from 21 days to 62 days with a mean of 35 days and a median of 33 days.

One-time injections of triamcinolone acetonide were performed as follows: eyes were anesthetized with proparacaine or tetracaine and sterilized with topical 5% povidone-iodine solution. Four mg of TA was injected through a 27-gauge needle 3.5 mm temporal to the limbus in phakic patients and 3.0 mm temporal from the limbus in pseudophakic patients.

Three parameters were analyzed between the pre-injection and first post-injection visit: LogMAR best-corrected visual acuity, OCT foveal thickness in microns, and Goldmann or Tono-Pen XL applanation pressure in mm Hg. The differences in these values were analyzed with a one-tailed Wilcoxon Signed-Rank analysis for statistical significance.

3. RESULTS

Patients treated for macular edema were classified into four categories: RP, diabetic, Irvine-Gass, and vascular. Foveal thickness, measured by OCT, improved proportionately in all categories, but was greatest in RP patients (Figure 43.1). Visual acuity also showed improvement in all categories, but was greatest in Irvine-Gass and vascular patients and least in diabetics (Figure 43.2). Intraocular pressures increased in all categories, as expected, but the overall magnitude of that change was only 2.5 mm Hg. Long-term OCT data are available on too few patients to achieve statistical relevance, but the overall trend showed fluid reaccumulation starting 60 days post-injection without return to pre-injection levels.

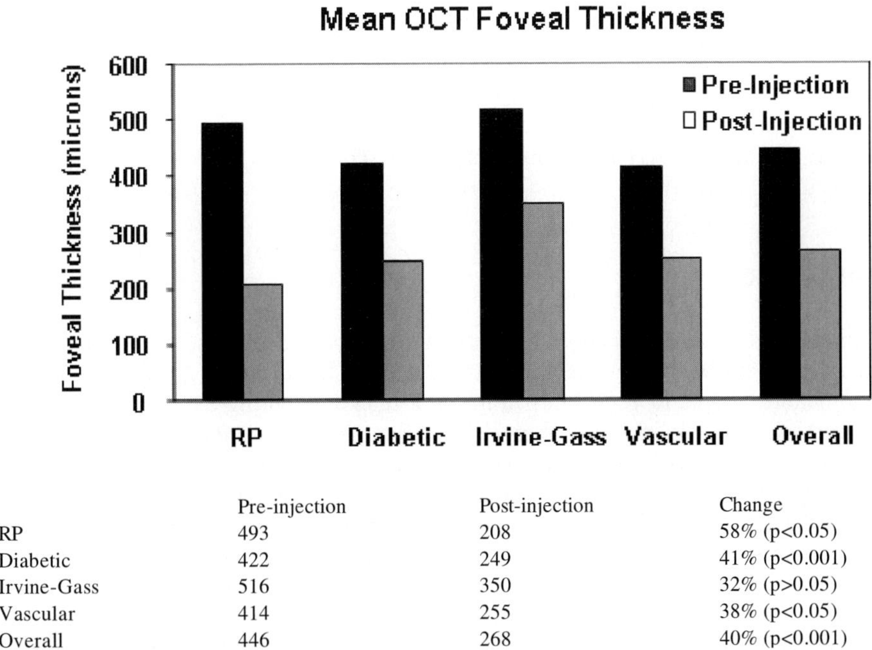

	Pre-injection	Post-injection	Change
RP	493	208	58% (p<0.05)
Diabetic	422	249	41% (p<0.001)
Irvine-Gass	516	350	32% (p>0.05)
Vascular	414	255	38% (p<0.05)
Overall	446	268	40% (p<0.001)

Figure 43.1. Mean Foveal Thickness, measured by OCT, is broken down into RP, diabetic, Irvine-Gass, and vascular categories. Thickness improved in all categories, although most improvement was seen in RP patients.

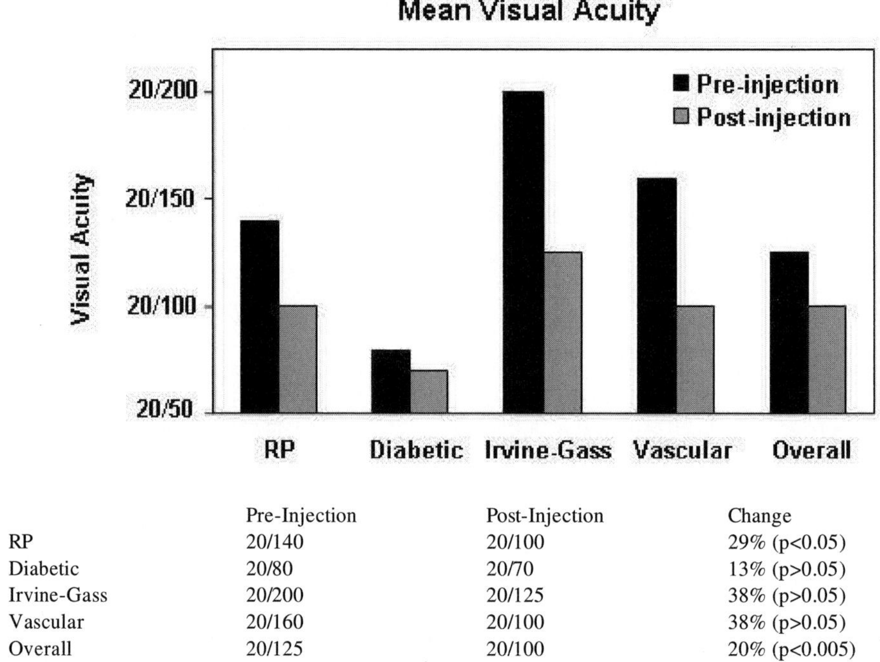

	Pre-Injection	Post-Injection	Change
RP	20/140	20/100	29% (p<0.05)
Diabetic	20/80	20/70	13% (p>0.05)
Irvine-Gass	20/200	20/125	38% (p>0.05)
Vascular	20/160	20/100	38% (p>0.05)
Overall	20/125	20/100	20% (p<0.005)

Figure 43.2. Visual Acuity improved in all categories. Improvement was greatest in Irvine-Gass and vascular patients and least in diabetics.

Four eyes of two RP patients were treated for CME. In all cases, there was a marked reduction of CME on OCT measurements (Figure 43.1 & 43.3) and some improvement of visual acuity (Figure 43.2).

4. DISCUSSION

Intravitreal TA injection appears to be a promising treatment option for macular edema from various retinal diseases including RP. Our OCT foveal thickness data show definite, and sometimes dramatic, reductions in macular thickness during the expected duration of action of TA. Pressure increases were relatively modest, 2.5 mm Hg on average and were reversed by topical medications (data not shown). Certainly, there is a greater pressure rise in the subset of steroid-responding patients. Even so, the highest IOP response at 3-9 week follow-up was 8 mm Hg. Unfortunately, visual acuity did not consistently increase with decreased foveal thickness. The improvement in visual acuity was most impressive in the Irving-Gass class of patients, perhaps in keeping with their shorter duration of edema. The diabetic patients in our series showed only a modest improvement in visual acuity. Perhaps this reflects the fact that the diabetic patients selected for intravitreal TA injections at our center predominantly had long-standing chronic diffuse macular edema. One limitation to intravitreal TA treatment is apparent even in this limited study: the long-term follow-up OCT

Figure 43.3a. OCT scans of right eye of patient with CME resulting from RP. Top image is before TA injection, notice large intraretinal cysts; bottom image is 10 weeks after injection. Foveal thickness improved from 672 µM to 192 µM. Foveal contour was restored.

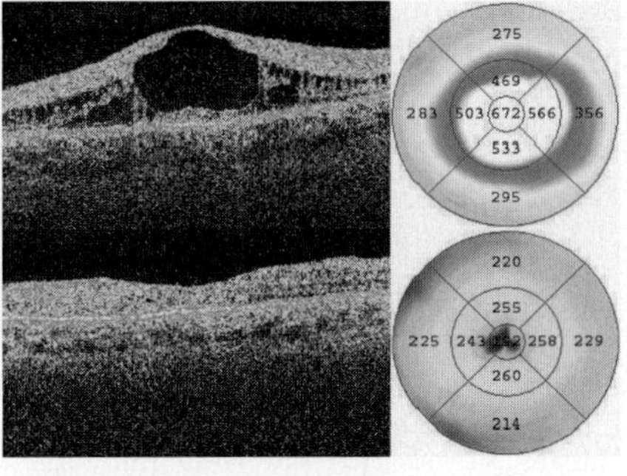

Figure 43.3b. OCT scans of left eye of patient with CME resulting from RP. Top image is before TA injection, notice large intraretinal cysts; bottom image is 14 weeks after injection. Foveal thickness improved from 661 µM to 246 µM. Notice foveal contour was restored, yet there were still intraretinal cysts.

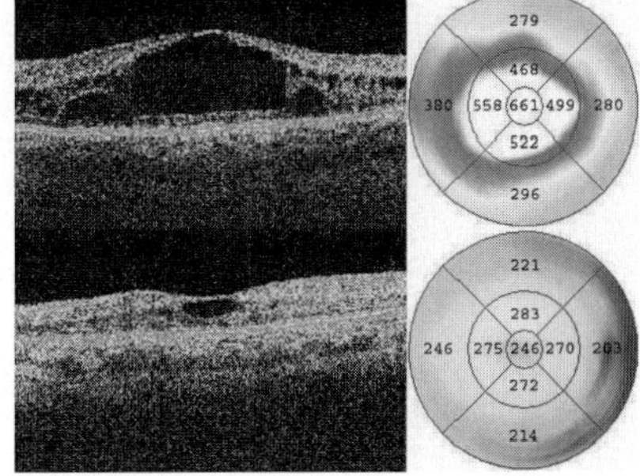

numbers seem to reach their minimum around 60 days, and the effect was gone by 150 days in our longest follow-up. This study indicates further investigation with more patients and longer duration are warranted.

5. REFERENCES

Antcliff, R.J., Spalton, D.J., Stanford, M.R., Graham, E.M., Ffytche, T.J. & Marshall, J. (2001). Intravitreal triamcinolone for uveitic cystoid macular edema: an optical coherence tomography study. *Ophthalmology*, **108**:765-72.

Beer, P.M., Bakri, S.J., Singh, R.J., Liu, W., Peters, G.B., 3rd & Miller, M. (2003). Intraocular concentration and pharmacokinetics of triamcinolone acetonide after a single intravitreal injection. *Ophthalmology*, **110**:681-6.

Bresnick, G.H. (1983). Diabetic maculopathy. A critical review highlighting diffuse macular edema. *Ophthalmology*, **90**:1301-17.

Bresnick, G.H. (1986). Diabetic macular edema. A review. *Ophthalmology*, **93**:989-97.

Brooks, H.L., Jr., Caballero, S., Jr., Newell, C.K., Steinmetz, R.L., Watson, D., Segal, M.S., Harrison, J.K., Scott, E.W. & Grant, M.B. (2004). Vitreous levels of vascular endothelial growth factor and stromal-derived factor 1 in patients with diabetic retinopathy and cystoid macular edema before and after intraocular injection of triamcinolone. *Arch Ophthalmol*, **122**:1801-7.

Funatsu, H., Yamashita, H., Ikeda, T., Mimura, T., Eguchi, S. & Hori, S. (2003). Vitreous levels of interleukin-6 and vascular endothelial growth factor are related to diabetic macular edema. *Ophthalmology*, **110**:1690-6.

Heckenlively, J.R., Jordan, B.L. & Aptsiauri, N. (1999). Association of antiretinal antibodies and cystoid macular edema in patients with retinitis pigmentosa. *Am J Ophthalmol*, **127**:565-73.

Hockley, D.J., Tripathi, R.C. & Ashton, N. (1979). Experimental retinal branch vein occlusion in rhesus monkeys. III. Histopathological and electron microscopical studies. *Br J Ophthalmol*, **63**:393-411.

Ip, M.S., Gottlieb, J.L., Kahana, A., Scott, I.U., Altaweel, M.M., Blodi, B.A., Gangnon, R.E. & Puliafito, C.A. (2004). Intravitreal triamcinolone for the treatment of macular edema associated with central retinal vein occlusion. *Arch Ophthalmol*, **122**:1131-6.

Jonas, J.B., Degenring, R.F., Kamppeter, B.A., Kreissig, I. & Akkoyun, I. (2004). Duration of the effect of intravitreal triamcinolone acetonide as treatment for diffuse diabetic macular edema. *Am J Ophthalmol*, **138**:158-60.

Jonas, J.B., Kreissig, I., Sofker, A. & Degenring, R.F. (2003). Intravitreal injection of triamcinolone for diffuse diabetic macular edema. *Arch Ophthalmol*, **121**:57-61.

Klein, R., Klein, B.E., Moss, S.E., Davis, M.D. & DeMets, D.L. (1984). The Wisconsin epidemiologic study of diabetic retinopathy. IV. Diabetic macular edema. *Ophthalmology*, **91**:1464-74.

Martidis, A., Duker, J.S., Greenberg, P.B., Rogers, A.H., Puliafito, C.A., Reichel, E. & Baumal, C. (2002). Intravitreal triamcinolone for refractory diabetic macular edema. *Ophthalmology*, **109**:920-7.

Massin, P., Audren, F., Haouchine, B., Erginay, A., Bergmann, J.F., Benosman, R., Caulin, C. & Gaudric, A. (2004). Intravitreal triamcinolone acetonide for diabetic diffuse macular edema: preliminary results of a prospective controlled trial. *Ophthalmology*, **111**:218-24; discussion 224-5.

Sallum, J.M., Farah, M.E. & Saraiva, V.S. (2003). Treatment of cystoid macular edema related to retinitis pigmentosa with intravitreal triamcinolone acetonide: case report. *Adv Exp Med Biol*, **533**:79-81.

Saraiva, V.S., Sallum, J.M. & Farah, M.E. (2003). Treatment of cystoid macular edema related to retinitis pigmentosa with intravitreal triamcinolone acetonide. *Ophthalmic Surg Lasers Imaging*, **34**:398-400.

Yoshikawa, K., Kotake, S., Ichiishi, A., Sasamoto, Y., Kosaka, S. & Matsuda, H. (1995). Posterior sub-Tenon injections of repository corticosteroids in uveitis patients with cystoid macular edema. *Jpn J Ophthalmol*, **39**:71-6.

Zamir, E., Read, R.W., Smith, R.E., Wang, R.C. & Rao, N.A. (2002). A prospective evaluation of subconjunctival injection of triamcinolone acetonide for resistant anterior scleritis. *Ophthalmology*, **109**:798-805; discussion 805-7.

CHAPTER 44

CONE SURVIVAL: IDENTIFICATION OF RdCVF

Olivier Lorentz, José Sahel, Saddek Mohand-Saïd, and Thierry Leveillard*

1. INTRODUCTION

The foremost cause of irreversible blindness in major retinal diseases is photoreceptor degeneration. In animal models as well as in human retinal hereditary dystrophies, the mutations described since 1990 affect mainly coding sequences for structural proteins (peripherine, Rom 1) or components of the phototransduction cascade (rhodopsin, cGMP-dependent phosphodiesterase) found in the rod outer segments.[1,2,3] The mechanisms leading to programmed cell death of these cells are still hypothetical.[4] In addition to this direct rapid rod loss, delayed cone loss is seen in clinical situations and was described in 1978 in the "retinal degeneration" (rd) mouse model.[5] Their loss is responsible for the major visual handicap because cones are essential for diurnal, colour and central vision.[6] This secondary loss of cone photoreceptors does not have any obvious explanation since cones are generally not directly affected by the genetic anomaly found in these diseases.

In several models leading to selective rod loss, such as transgenic mice[7] or mice carrying a spontaneous mutation,[8] secondary cone loss is observed whereas the causal abnormality is not directly incriminated in their degeneration. In certain studies the link between rod loss and cone drop-out is still hypothetical. The cellular interactions involved in cone survival have never been the subject of a systematic experimental approach, and can be amply justified through the major perspectives in fundamental neurobiology and therapeutic outcomes. Taking into account the multiple cone functions, preservation of this population would open an original avenue of therapeutic investigation which would enable a considerable limitation of functional consequences for the patients.

2. ROD SECRETE FACTOR(S) ABLE TO PROMOTE CONE SURVIVAL

Our approach is based on the observation that in retinal pathologies (mutant animal models and patients) rod degeneration precedes that of cones. Our efforts initially focused

* Olivier Lorenta, Ph.D., INSERM U592, Bâtiment Kourilsky, HôpitalSt-Antoine, 184 rue du Faubourg St-Antoine, 75571 Paris CEDEX 12.

315

on selective rod replacement, obtained by use of a vibratome sectioning method. Our first results demonstrated that when pure photoreceptor layers were transplanted to the subretinal space in 5 week old *rd1* mice [at this stage very few rods remain (<0.02%) whereas about 30% of the cones are still present] the transplant induces a significant increase in the number of surviving cones (an average increase of 30%, p < 0.001) compared to that of non-treated congenic retina.[9] This trophic effect can be detected distant from the transplant, suggesting the existence of a diffusible factor liberated by the transplanted cells. Tests carried out *in vitro* on *rd1* retina and wild type retina co-cultures confirmed this hypothesis.[10] This study showed that about 40% of cones normally lost during the sixth postnatal week are saved when cultured in the presence of rod-rich samples. This effect is photoreceptor specific since transplants of inner retina exert no beneficial effects on cone survival.[11,12] The neuroprotective activity (40-50% increase in viability) on cones is heat labile and has an apparent molecular weight larger than 15 kDa.[13] These experiments indicate that the activity is carried out by protein(s): the Rod-dependent Cone Viability Factors (RdCVFs).

Taken together, these assays show the existence of at least one trophic factor secreted by rods that is essential for cone survival. Rod degeneration in *rd1* mice and RP patients might hence lead to survival factor deprivation and consequently progressive cone loss. A key implication is that prevention of apoptotic rod death could lengthen cone survival. Such a mechanism, which has never been proposed prior to our work, provides a justification for the numerous strategies aimed at preserving non-functional rods, since these will protect cones.[14-19] In humans, diagnosis is often made at late stages, and in the absence of residual rods only PR transplantation (mainly rods) could restore the expression of these factors allowing to block or reduce secondary cone degeneration.[19] The potential clinical importance of such factors is obvious, cone loss representing the main cause of visual handicap in RP and AMD. Of importance, since cone loss is a late-onset downstream event and occurs independently of the specific mutation expressed by rods, a potentially broad number of patients could benefit from such therapy.

The observed trophic effect of rod transplants on cone survival suggests that non-functional rod protection could be sufficient to preserve cone viability.[10,12,19,20]

3. IDENTIFICATION OF RdCVF

Our systematic approach to characterize factors involved in cone viability has led to identification of a new gene (RdCVF) encoding a secreted protein for which we will explore its functions.

We postulate that the degeneration of rods of the *rd1* mouse retina is leading to the loss of expression of secreted protein factor(s) essential for cone viability. This mechanism of cone degeneration is also likely in human retinas affected with RP.[25] The identification of the genes encoding these Rod-dependent Cone Viability Factors is a prerequisite to a therapy aimed at preventing the secondary loss of cones and of vision.

To identify the RdCVF genes, we used a systematic strategy based on a functional assay using cone–enriched cultures. We developed a high throughput cone-enriched culture system using chicken retina.[26,27] Contrarily to the mammals, birds have retinas dominated by cones. In these cultures, the primary postmitoting cells (60-80% cones) are degenerating over a period of few days. An increase in cell survival was observed when cultured in the presence of conditioned media isolated from wild-type mouse[13]. The viability activity on chicken cone, as for RdCVF, is heat labile has an apparent molecular weight larger than 15 kDa. The chicken embryo retinal culture system is an easy, reproducible and high throughput cone viability assay.[27]

We constructed an expression library from wild type mouse retina and tested all the genes for their potential to promote chicken cone survival by expression cloning methods. Briefly, pooled by 100 clones from the expression library were transfected into a cell line (COS-1). The conditioned media from the COS-1 tranfected cell were added to primary chicken cone cells seeded into 96 well-plates. After 7 days, viable cell counts from the cone-enriched cultures were measured using in house high content screening methods and compared to that of empty library vector. Twenty-one hundred pools, corresponding to 210,000 individual clones, were screened. A Pool (number 939) contained twice as many living cells as the negative controls. By limiting dilution clone 939.09.08 was isolated and shown to contain a 502-bp insert with an open reading frame encoding a putative 109-amino acid polypeptide. We named this gene Rod-derived Cone Viability Factor. The novel gene carries many characteristics of the postulated therapeutic gene:

1) Purified recombinant RdCVF protected chicken cones in a dose dependent manner.
2) RdCVF, when transfected into COS-1 cells exerts its survival activity on cones from *rd1* retinal explants, the model of the degenerative disease.
3) Purified recombinant RdCVF protected mouse cones when injected into the subretinal space on the *rd1* mouse.
4) RdCVF protein is detected in conditioned media from wild-type retina and COS-transfected cells.
5) RdCVF messenger RNA and protein expression is largely decreased in a rod-less mouse retina (the degenerated *rd1* retina).
6) RdCVF is expressed by pure cultures of photoreceptors from mouse (97% rods).
7) Immunohistochemistry demonstrated that RdCVF localized in the photoreceptor layer of the retina with a more intense staining in the extracellular matrix surrounding cone cells.
8) RdCVF antibodies are able to block the neuroprotective effect generated by wild-type conditioned media.

In addition, RdCVF expression was found to be restricted to the retina. RdCVF encodes for two polypetides of 17 and 34 kDa by alternative splice, the longer form being extended in its C-terminal region. Both forms have a limited homology with the thioredoxin family and the gene was named accordingly Txnl6 (Thioredoxin-like-6) in the databases. We could not demonstrate any thiol-oxidoreductase activity for the isolated polypeptide (the 17 kDa form). The founder member of the thioredoxin family, Trx-1 has been isolated originally as the adult T-cell leukemia-derived factor[28] a factor secreted by cells by a mechanism that does not involve a signal peptide sequence,[29] a signal also absent in RdCVF.

4. CONCLUSION

Mutation-independent therapies offer a means of slowing down or even stopping photoreceptor degeneration process in the medium term. They would be applicable to most patients regardless of the genetic defect and limit their handicap. Such therapies are based on the use of either transfer of antiapoptotic genes to the photoreceptors or, more readily, delivery of neuroprotective factors. The endogenous rod-derived cone viability factors (RdCVFs) are some of the most appropriate neuroprotective candidates. In addition, the development of eye delivery methods using genetically engineered and encapsulated cells implanted directly into the vitreous offers a new means of long-term controllable and reversible neuroprotective treatment. Finally, progress in the understanding of endogenous neurogenesisn will perhaps provide new possibilities of treatment in the late stages of RP.

5. AKNOWLEDGEMENTS

This work was supported by the Institut National pour la Santé et la Recherche Médicale (INSERM), Louis Pasteur University (Strasbourg), Pierre et Marie Curie University (Paris VI), the French Ministère des Sciences et des Technologies, The Association Française contre les Myopathies (AFM), the Fondation Bétancourt, the Fédération des Aveugles de France and the European Commission: PROAGERET (# QLK6-2001-00385) and PRORET (# QLK6-2001-00569).

6. REFERENCES

1. K. Kajiwara, E. L. Berson, and T. P. Dryja. Digenic retinitis pigmentosa due to mutations at the unlinked peripherin/RDS and ROM1 loci, *Science.* **264**(5165):1604-8 (1994).
2. M. E. McLaughlin, M. A. Sandberg, E. L. Berson, and T. P. Dryja. Recessive mutations in the gene encoding the beta-subunit of rod phosphodiesterase in patients with retinitis pigmentosa, *Nat Genet.* **4**(2):130-4 (1993).
3. P. J. Rosenfeld, L. B. Hahn, M. A. Sandberg, T. P. Dryja, and E. L. Berson. Low incidence of retinitis pigmentosa among heterozygous carriers of a specific rhodopsin splice site mutation, *Invest Ophthalmol Vis Sci.* **36**(11):2186-92 (1995).
4. G. Q. Chang, Y. Hao, and F. Wong. Apoptosis: final common patxhway of photoreceptor death in rd, rds, and rhodopsin mutant mice, *Neuron.* **11**(4):595-605 (1993).
5. L. D. Carter-Dawson, M. M. LaVail, and R. L. Sidman. Differential effect of the rd mutation on rods and cones in the mouse retina, *Invest Ophthalmol Vis Sci.* **17**(6):489-98 (1978).
6. J. E. Dowling, *The Retina. Ed.Belknap press of Harvard University Press,Cambridge,Massachussets and London* (1987).
7. M. A. McCall, R. G. Gregg, K. Merriman, Y. Goto, N. S. Peachey, and L. R. Stanford. Morphological and physiological consequences of the selective elimination of rod photoreceptors in transgenic mice, *Exp Eye Res.* **63**(1):35-50 (1996).
8. C. Bowes, T. Li, M. Danciger, L. C. Baxter, M. L. Applebury, and D. B. Farber. Retinal degeneration in the rd mouse is caused by a defect in the beta subunit of rod cGMP-phosphodiesterase, *Nature.* **347**(6294):677-80 (1990).
9. S. Mohand-Said, D. Hicks, M. Simonutti, D. Tran-Minh, A. Deudon-Combe, H. Dreyfus, M. S. Silverman, J. M. Ogilvie, T. Tenkova, and J. Sahel. Photoreceptor transplants increase host cone survival in the retinal degeneration (rd) mouse, *Ophthalmic Res.* **29**(5):290-7 (1997).
10. S. Mohand-Said, A. Deudon-Combe, D. Hicks, M. Simonutti, V. Forster, A. C. Fintz, T. Leveillard, H. Dreyfus, and J. A. Sahel. Normal retina releases a diffusible factor stimulating cone survival in the retinal degeneration mouse. *Proc Natl Acad Sci U S A.* **95**, 8357-62 (1998).
11. S. Mohand-Said, D. Hicks, H. Dreyfus, and J. A. Sahel. Selective transplantation of rods delays cone loss in a retinitis pigmentosa model. *Arch. Ophthalmol.* **118**:807-811 (2000).
12. J. A. Sahel, S. Mohand-Said, T. Leveillard, D. Hicks, S. Picaud, and H. Dreyfus Rod-cone interdependence: implications for therapy of photoreceptor cell diseases. *Prog Brain Res* **131**:649-61 (2001).
13. A. C. Fintz, I. Audo, D. Hicks, S. Mohand-Said, T. Leveillard, and J. A. Sahel. Partial characterization of retina-derived cone neuroprotection in two culture models of photoreceptor degeneration. *Invest Ophthalmol Vis Sci.* **44**:818-25 (2003).

14. E. G. Faktorovich, R. H. Steinberg, D. Yasumura, M. T. Matthes, and M. M. LaVail. Photoreceptor degeneration in inherited retinal dystrophy delayed by basic fibroblast growth factor. *Nature.* **347**(6288):83-6 (1990).

15. M. M. LaVail, K. Unoki, D. Yasumura, M. T. Matthes, G. D. Yancopoulos, and R. H. Steinberg. Multiple growth factors, cytokines, and neurotrophins rescue photoreceptors from the damaging effects of constant light. *Proc Natl Acad Sci U S A.* **89**(23):11249-53 (1992).

16. C. K. Chen, M. E. Burns, M. Spencer, G. A. Niemi, J. Chen, J. B. Hurley, D. A. Baylor, and M. I. Simon. Abnormal photoresponses and light-induced apoptosis in rods lacking rhodopsin kinase. *Proc Natl Acad Sci U S A.* **96**(7):3718-22 (1999).

17. M. Frasson, J. A. Sahel, M. Simonutti, H. Dreyfus, and S. Picaud. Retinitis pigmentosa: rod photoreceptor rescue by a Ca^{2+} channel blocker in the rd mouse. *Nature Medicine,* **5**: 1183-7 (1999).

18. M. Frasson, S. Picaud, T. Léveillard, M. Simonutti, S. Mohand-Said, H. Dreyfus, D. Hicks, and J. A. Sahel. Glial cell line-derived neurotrophic factor induces histologic and functional protection of rod photoreceptors in the rd/rd mouse. *Invest Ophthalmol Vis Sci.,* **40**:2724-34 (1999).

19. D. Hicks, and J. A. Sahel. The implications of Rod-Dependent Cone Survival for Basic and Clinical Research. *Invest Ophthalmol Vis Sci.* **40**:3071-4 (2001).

20. S. Mohand-Said, D. Hicks, T. Leveillard, S. Picaud, F. Porto, and J. A. Sahel. Rod-cone interactions: development and clinical significance. *Progress in retinal and eye research.* **20**:451-67 (2001).

21. S. Jing, D. Wen, Y. Yu, P. L. Holst, Y. Luo, M. Fang, R. Tamir, L. Antonio, Z. Hu, R. Cupples, J. C. Louis, S. Hu, B. W. Altrock, and G. M. Fox. GDNF-induced activation of the ret protein tyrosine kinase is mediated by GDNFR-alpha, a novel receptor for GDNF. *Cell.* **85**(7):1113-24 (1996).

22. M. N. Delyfer, T. Leveillard, S. Mohand-Saïd, D. Hicks, S. Picaud, and J.A. Sahel. Inherited retinal degenerations : therapeutic prospects. *Biol. Cell.* **96**: 261-9 (2004).

23. J. C. Blanks, A. M. Adinolfi, and R. N. Lolley. Photoreceptor degeneration and synaptogenesis in retinal-degenerative (rd) mice. *J Comp Neurol.* **156**(1):95-106 (1974).

24. E. Banin, A. V. Cideciyan, T. S. Aleman, R. M. Petters, F. Wong, A. H. Milam, and S. G. Jacobson. Retinal rod photoreceptor-specific gene mutation perturbs cone pathway development. *Neuron.* **23**(3):549-57 (1999).

25. A. V. Cideciyan, D. C. Hood, Y. Huang, E. Banin, Z. Y. Li, E. M. Stone, A. H. Milam, and S. G. Jacobson. Disease sequence from mutant rhodopsin allele to rod and cone photoreceptor degeneration in man. *Proc Natl Acad Sci U S A.* **95**(12):7103-8 (1998).

26. R. Adler, and M. Hatlee. Plasticity and differentiation of embryonic retinal cells after terminal mitosis. *Science* **243**:391-3 (1989).

27. T. Leveillard, S. Mohand-Said, O. Lorentz, D. Hicks, A. C. Fintz, E. Clerin, M. Simonutti, V. Forster, N. Cavusoglu, F. Chalmel, P. Dolle, O. Poch, G. Lambrou, and J. A. Sahel. Identification and characterization of rod-derived cone viability factor. *Nat Genet.* **36**(7):755-9 (2004).

28. N. Wakasugi, Y. Tagaya, H. Wakasugi, A. Mitsui, M. Maeda, J. Yodoi, and T. Tursz. Adult T-cell leukemia-derived factor/thioredoxin, produced by both human T-lymphotropic virus type I- and Epstein-Barr virus-transformed lymphocytes, acts as an autocrine growth factor and synergizes with interleukin 1 and interleukin 2. *Proc Natl Acad Sci U S A.* **87**(21):8282-6 (1990).

29. W. Nickel. The mystery of nonclassical protein secretion. A current view on cargo proteins and potential export routes. *Eur. J. Biochem.* **270**:2109-19 (2003).

NEUROPROTECTION OF PHOTORECEPTORS IN THE RCS RAT AFTER IMPLANTATION OF A SUBRETINAL IMPLANT IN THE SUPERIOR OR INFERIOR RETINA

Machelle T. Pardue[1], Michael J. Phillips[1], Brett Hanzlicek[2], Hang Yin[1], Alan Y. Chow[3], and Sherry L. Ball[2,4]

1. INTRODUCTION

The artificial silicon retina (ASR™) consists of an array of photodiodes on a silicon disk that responds to incident light in a gradient fashion (Peyman et al., 1998; Chow et al., 2001, 2002). This device is designed to be placed in the subretinal space and serve as a replacement for degenerating photoreceptors. Two possible mechanisms for the ASR device to improve visual function include 1) direct activation of the remaining inner retinal neurons and subsequent activation of visual centers in the brain or 2) a delay in photoreceptor loss due to a neurotrophic effect from subretinal electrical stimulation. Initial results of ongoing FDA trials with the ASR device suggest that subretinal electrical stimulation could elicit a neurotrophic effect (Chow et al., 2004). Ten advanced retinitis pigmentosa (RP) patients implanted with the ASR device have increased central visual fields and improved visual acuity and color vision (Chow et al., 2004). These improvements cannot be easily explained by direct activation since the implant was placed 20° from the macula. To determine whether neuroprotection results from subretinal electrical stimulation, the RCS rat model of RP was implanted with an ASR device. Subretinal implantation of an ASR device into the superior retina of the Royal College of Surgeons (RCS) rat resulted in preservation of photoreceptors (Pardue et al., 2004). However, the RCS rat is known to have delayed photoreceptor degeneration in the superior region of the retina (LaVail and Battelle, 1975). To determine whether the superior retina is a "privileged" site in the RCS rat, ASR devices were subretinally implanted in the superior and inferior retina.

---

[1]Rehab R&D, Atlanta VA Medical Center, Decatur, GA 30033, USA and Department of Ophthalmology, Emory University, Atlanta GA 30322, USA; [2]Research Service, Cleveland VA Medical Center, Cleveland, OH 44106, USA; [3]Optobionics, Corp, Naperville, IL 60563; [4]Department of Psychology, Case Western Reserve University, Cleveland, OH 44106.

2. METHODS

2.1. Experimental and Implant Design

RCS rats were obtained from an in-house breeding colony at the Cleveland VA Medical Center which originated from Dr. Matthew M. LaVail at University of California, San Francisco. Three RCS rats were implanted at 21 days of age such that one eye was implanted superiorly and the other inferiorly. Three other RCS rats served as unoperated controls. All animals were followed until 8 weeks post-implantation or 11 weeks of age. All procedures were carried out in accordance with the Association for Research in Vision and Ophthalmology statement concerning the use of animals in ophthalmic and vision research.

Each implant consisted of a series of $20\,\mu m$ by $20\,\mu m$ microphotodiodes on a silicon disk that was 1 mm in diameter and $\sim30\,\mu m$ thick. Each implant was backed with a uniform layer of iridium oxide to serve as the electrode. Implants were fabricated by Optobionics Corp.

2.2. Surgical Procedures

As previously described (Ball et al., 2000), each RCS rat was sedated and anaesthetizing drops were applied to the corneal surface. One eye was rotated inferiorly with a suture and an incision made through the eye cup into the vitreous approximately 3 mm from the superior limbus. The implant was then gently placed in the superior subretinal space. The eye was rotated back to the normal position and antibiotic ointment applied. The same surgical procedures were performed in the opposite eye, except the eye was rotated superiorly and the implant was placed 3 mm from the inferior limbus.

2.3. Electroretinography (ERG)

ERGs were recorded every 1 or 2 weeks until 8 weeks post-op using the LKC ERG system (Gaithersburg, MD). Rats were dark-adapted overnight, anesthetized, and their pupils dilated. ERGs were recorded from both eyes simultaneously with silver wire electrode loops contacting the corneal surface through a layer of 1% methylcellulose. Platinum needle electrodes placed in the cheek and tail served as reference and ground, respectively.

To record a dark-adapted response series, stimuli were presented in order of increasing luminance from 0.001 to 137 cd sec/m^2; interstimulus intervals increased from 15 sec to 1 min with increasing flash intensity. Cone-mediated responses were isolated by superimposing stimuli upon a rod-desensitizing adapting field (30 cd/m^2), and presenting the stimulus at 2.1 Hz after a 10 min period of light adaptation.

2.4. Histology

After eight weeks of implantation or at 11 weeks of age, rats were sacrificed by anesthetic overdose, and their eyes enucleated. Each eye was marked for orientation and immersed in 2% paraformaldehyde/ 2.5% glutaraldehyde overnight. Eyes were then dehydrated through a graded alcohol series and embedded in Embed 812/Der736. Blocks were cut at $0.5\,\mu m$ on a Reichert Ultramicrotome and stained with toluidine blue.

Retinal morphometry was analyzed from sections taken from both operated and un-operated groups. From each retinal section, six 0.5 mm regions were analyzed as indicated in Figure 45.3A. Digital images were taken of each retinal region and photoreceptor cells counted with samplings from five sections averaged for each region.

3. RESULTS

3.1. Electroretinography

Figure 45.1A shows the dark-adapted ERG responses recorded from RCS rat eyes in the three treatment groups: superior-implantation, inferior-implantation, and unoperated. Each waveform is the maximal dark-adapted response recorded at each post-implantation time indicated for a single eye. At 2 weeks after implantation, retinal responses from all eyes are similar. However, by 4 weeks, eyes implanted with a subretinal device in either the superior or inferior retina have larger amplitude b-waves than unoperated eyes. At 5 weeks after implantation, the unoperated eyes have begun to develop the negative scotopic thresh-old response (STR; arrow) while the implanted eyes do not develop the STR until 7 or 8 weeks post-implantation. At 8 weeks post-op, the unoperated eyes have a larger amplitude STR response than the implanted eyes.

Figure 45.1B summaries the average maximal dark-adapted b-wave or STR amplitude response at each post-operative timepoint. Eyes with a superiorly or inferiorly placed sub-retinal implant have very similar responses and have significant functional preservation com-pared to unoperated control eyes. Responses of implanted eyes were significantly greater

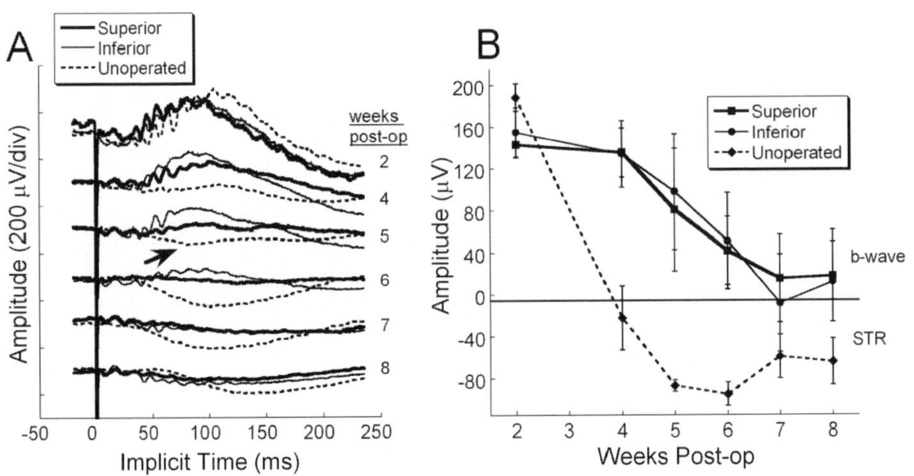

Figure 45.1. Dark-adapted ERG responses from RCS rats implanted with a subretinal implant in the superior or inferior retina versus unoperated rats. A) Maximal dark-adapted response from a single eye in each treatment group elicited by a 137 cd sec/m^2 flash at each post-operative week indicated. The arrow indicates the appearance of the STR in the unoperated eye. B) Average maximal dark-adapted b-wave across time (± standard error). The horizontal line marks the division between the positive b-wave and negative STR that are plotted on a continuum.

than unoperated controls at 4 to 6 weeks post-op (Student's t-test, p < 0.005). While response amplitudes decrease over the 8 weeks of follow-up, only three of six implanted eyes develop the STR response at 7 and 8 weeks post-op. In contrast, the unoperated eyes have a faster rate of retinal function loss with only the STR visible in the majority of animals by 4 weeks post-op. Light-adapted b-wave responses from implanted eyes also had significantly higher amplitudes than unoperated eyes from 2 to 6 weeks post-op (Student's t-test, p < 0.05; data not shown).

3.2. Histology

Figure 45.2 presents photomicrographs of the RCS retina implanted superiorly (A) or inferiorly (B) or age-matched unoperated RCS rats (C). In eyes implanted with an ASR device, 5-7 layers of photoreceptor nuclei are still present at 8 weeks after implantation. In the unoperated retina, a sparse single row of photoreceptor nuclei is visible along with a large layer of photoreceptor segment debris.

These results are summarized in Figure 45.3 which plots the number of photoreceptor cells in each retinal region across the retina for each treatment group. While unoperated eyes have very few photoreceptors with a small increase in the superior retina, both the superior- and inferior-implanted eyes have significantly more photoreceptors across all the regions of the retina examined (Repeated measures ANOVA for main effect of treatment; superior vs unoperated $F_{(1,4)} = 91.5$, p = 0.001; inferior vs unoperated $F_{(1,4)} = 51.0$, p = 0.002). Photoreceptor preservation peaks in regions directly overlying the implant (regions 2 and 7). A direct comparison of the number of photoreceptors in the inferior and superior regions of the eye (i.e. overlaying the curves) reveals no significant differences (Repeated measures ANOVA, $F_{(5,20)} = 0.34$, p = 0.88).

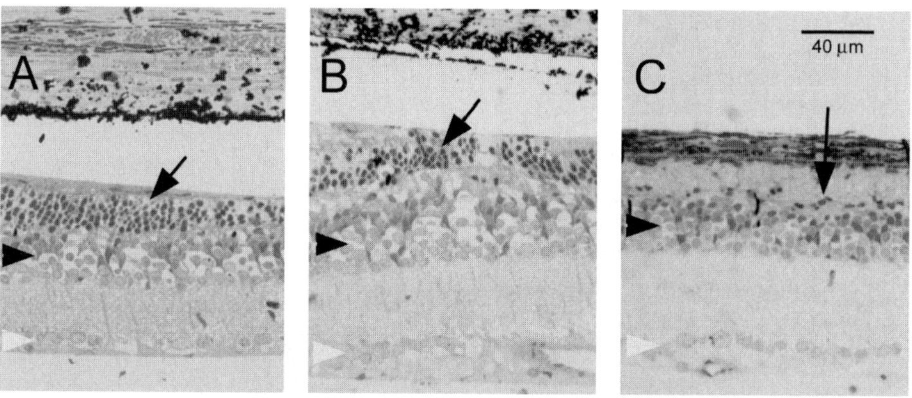

Figure 45.2. Photomicrograph of RCS retina implanted with a subretinal implant in the superior (A), inferior (B) retina or an unoperated control (C). The larger number of photoreceptors in the implanted eyes is indicated by the arrows. The black material on the RPE side of the retina is the remnants of the implant after sectioning. The separation of the retina from the implant is an artifact of tissue processing. The black arrowhead indicates the inner nuclear layer and the white arrowhead indicates the ganglion cell layer.

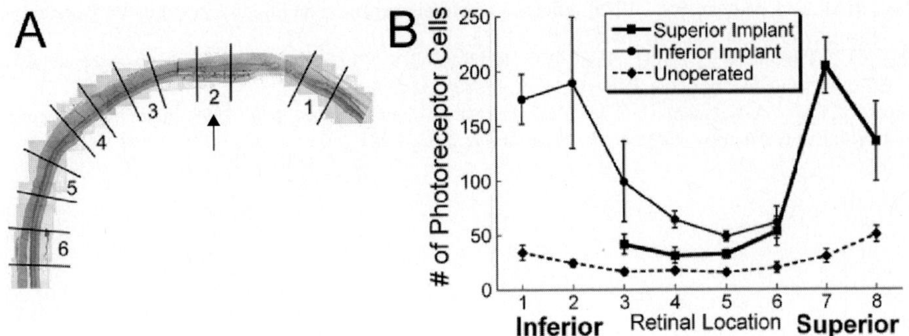

Figure 45.3. A) Location of the six regions examined across the retina in each implanted retina. In eyes implanted inferiorly, the region nearest to the inferior limbus was indicated as region 1. In eyes implanted superiorly, the region nearest to the superior limbus was indicated as region 8. The arrow indicates the implant. B) Average photoreceptor counts (± standard error) for each retinal location. A greater number of photoreceptors was measured directly over the implant in both the inferior and superior retina. Note that the regions only overlap in areas 3-6 in the two implanted eyes.

4. DISCUSSION

Implantation of an ASR device significantly preserved retinal function and photoreceptor cells in the RCS rat implanted at 21 days of age, as previously reported (Pardue et al., 2004). The amount of functional or anatomical preservation was not regionally dependent. Eyes implanted in either the superior or inferior region both had significant preservation of retinal function. In addition, both the superior and inferior regions of the retina had similar photoreceptor counts. As reported previously (LaVail & Battelle, 1975), more photoreceptor nuclei were counted in the superior retina of unoperated rats. However, this delay in photoreceptor degeneration in the superior region of the RCS rat did not influence the amount of neuropreservation caused by the subretinal implant. These data provide additional evidence that subretinal implantation of an ASR device in the RCS rat results in significant preservation of retinal function and morphology.

5. REFERENCES

Ball, S.L., Pardue, M.T., Chow, A.Y., Chow, V.Y., and Peachey, N.S., 2001, Subretinal implantation of photodiodes in rodent models of photoreceptor degeneration, in: *New Insights into Retinal Degenerative Diseases,* R.E. Anderson RE, M.M. LaVail MM, J.G. Hollyfield, eds., Kluwer/Plenum, New York, pp.175-182.

Chow, A.Y., Pardue, M.T., Chow, V.Y., Peyman, G.A., Liang, C., Perlman, J.I., and Peachey, N.S., 2001, Implantation of silicon chip microphotodiode arrays into the cat subretinal space. *IEEE Trans Neural Syst Rehabil Eng.* **9**:86-95.

Chow, A.Y., Pardue, M.T., Perlman, J.I., Ball, S.L., Chow, V.Y., Hetling, J.R., Peyman G.A., Liang, C., Stubbs, E.B, Jr., and Peachey, N.S., 2002, Subretinal implantation of semiconductor-based photodiodes: durability of novel implant designs. *J Rehabil Res Dev.* **39**:313-321.

Chow, A.Y., Chow, V.Y., Peyman, G.A., Packo, K.H., Pollack, J.S., and Schuchard, R., 2004, The artificial silicon retina™ (ASR™) chip for the treatment of vision loss from retinitis pigmentosa. *Arch Ophthalmol.* **122**:460-469.

LaVail, M.M., and Battelle, B.A., 1975, Influence of eye pigmentation and light deprivation on inherited retinal dystrophy in the rat. *Exp. Eye Res.* **21**:167-192.

Pardue, M.T., Phillips, M.J., Yin, H., Sippy, B.D., Webb-wood, S., Chow, A.Y., Ball, S.L, in press, Neuroprotective effect of subretinal implants in the RCS rat. *Invest Ophthalmol Vis Sci.*

Peyman, G., Chow, A.Y., Liang, C., Chow, V.Y., Perlman, J.I., and Peachey, N.S., 1998, Subretinal semiconductor microphotodiode array. *Ophthalmic. Surg. Lasers* **29**:234-241.

GLUTAMATE TRANSPORT MODULATION: A POSSIBLE ROLE IN RETINAL NEUROPROTECTION

Nigel L. Barnett, Kei Takamoto, and Natalie D. Bull*

1. INTRODUCTION

The regulation of extracellular glutamate levels in the retina, under physiological and pathophysiological conditions, is essential for the prevention of excitotoxic neurodegeneration. Glial and neuronal high-affinity glutamate transporters (excitatory amino acid transporters, EAATs) facilitate the rapid removal of glutamate from the extracellular space, thereby terminating the excitatory signal and reducing the possibility of excitotoxic neuronal damage. Failure or reversal of these transport systems leads to raised levels of extracellular glutamate and contributes to the development of excitotoxic retinal degeneration.

Five distinct human EAATs and their rodent homologues (i.e. GLAST, GLT-1, EAAC-1, EAAT4 and EAAT5) have been cloned and localised by immunohistochemistry in the rodent retina.[1-3] GLAST and EAAT4 are associated with the Müller and astroglial cells respectively.[1,4] Uptake studies indicate that Müller cells dominate normal retinal glutamate transport, utilising GLAST.[5,6] GLT-1 is associated with cones and cone bipolar cells,[7] EAAC-1 is localised to amacrine cells plus horizontal cells and ganglion cells[8] while EAAT5 is associated with photoreceptors.[3] We have shown that GLAST activity is more susceptible to an ischaemic attack than the other classes of retinal glutamate transporters,[6] with its failure contributing to the dangerous ischaemia-induced elevation of extracellular glutamate and consequent neurodegeneration. Thus, posing the question, can the activity of retinal glutamate transporters be enhanced during an ischaemic episode to provide neuroprotection? In this study we modulated the activity of rat retinal glutamate transporters pharmacologically during an acute ischaemic attack in an attempt to protect the retina from degeneration and dysfunction.

---

* Vision, Touch & Hearing Research Centre, School of Biomedical Sciences, The University of Queensland, Brisbane, 4072, Australia. Email: n.barnett@uq.edu.au.

2. METHODS

2.1. Pharmacological Modulation of Ischaemic Glutamate Transport

Glutamate transport activity was assessed immunohistochemically, by tracking the accumulation of the non-endogenous, non-metabolisable glutamate transporter substrate, D-aspartate, as previously described.[9,10] D-Aspartate antiserum was generously provided by Prof. D. Pow (University of Newcastle, Australia). We have previously shown that inhibitors of protein kinase C (PKC)[9] and activators of protein kinase A (PKA)[11] can inhibit retinal glutamate transport. In an attempt to stimulate retinal glutamate transport we investigated the effect of the PKCδ/ε activator, ingenol (300 μM) and the protein kinase A inhibitor, KT5720 (1 μM).

2.2. Acute Ischaemic Insult

All procedures were approved by the University of Queensland Animal Experimentation Committee and were conducted in accordance with the ARVO Statement for the Use of Animals in Ophthalmic and Vision Reseach. Adult Dark Agouti rats (200-250 g) were anaesthetised with ketamine (100 mg/kg) and xylazine (12 mg/kg). Twenty minutes after a 2 μl intravitreal injection of a pharmacological agent (or saline control), retinal ischaemia was induced by elevating intraocular pressure to 110 mmHg for 60 minutes as previously described.[12] Following the ischaemic insult, reperfusion was permitted for 7 days.

2.3. Histological Analysis

Semi-thin (500 μm) sections obtained from either the nasal, temporal, superior or inferior regions of the retina approximately 2 mm from the optic disk were stained with toluidine blue, viewed on a Zeiss Axioskop microscope and digitally imaged. Morphometric analysis of the inner plexiform layer (IPL), the inner nuclear layer (INL), the outer nuclear layer (ONL) and the total retinal thickness was performed using Adobe Photoshop software to quantify retinal damage following the ischaemic insult. Cell count analysis was also performed on the same sections. To do this, 100 μm-length sections of the INL and ONL were cropped out of each image and the number of cell bodies in the cropped area was manually counted. The retinas of four rats for each experimental condition were analysed and a one-way ANOVA with Tukey *post hoc* test was used to compare values obtained from protein kinase modulator-treated eyes with those from control eyes.

2.4. Electroretinography

Following overnight dark adaptation, full field flash electroretinograms were recorded over a 4.8 log unit intensity range (-3 to 1.8 log cd.s.m^{-2}) to assess retinal function as previously described.[13] The a- and b-wave amplitudes were measured and expressed as the mean wave amplitude $\pm$ SEM. Two-way repeated measures ANOVA was performed on log transformed data to compare the responses from the control and protein kinase treated retinas. A *post-hoc* Bonferroni test was used to isolate significant differences ($p < 0.05$) between control and experimental responses at each stimulus intensity. All rats were subjected to the same conditions for ERG measurements.

3. RESULTS

3.1. Pharmacological Modulation of Ischaemic Glutamate Transport

Under non-ischaemic conditions, the Müller cells dominate the uptake of D-aspartate, via their glutamate transporter GLAST.[9] Figure 46.1 shows that an *in vitro* simulated ischaemic insult of 15 minutes altered the pattern of retinal D-aspartate uptake. Uptake by Müller cells was greatly reduced. Following treatment with the PKCδ/ε activator, ingenol, there was a slight increase in the uptake of D-aspartate by the Müller cells. Ischaemic Müller cell uptake of D-aspartate was even more pronounced following the inhibition of PKA by KT5720.

3.2. Histological Analysis

Morphometric analysis of post ischaemic retinas did not reveal any differences between saline-treated ischaemic tissue and protein kinase modulated ischaemic tissue ($p > 0.05$ for the inner plexiform layer, inner nuclear layer, outer nuclear layer and total retinal thickness). However, both ingenol and KT5720 appeared to reduce the severity of histological damage caused by the ischaemic insult. Cellular disruption was less pronounced as was swelling of neurones and Müller cell processes. The neuroprotective property of ingenol was confirmed by cell counts which revealed that significantly more cells in the ONL remained 7 days after ischaemia than in non-treated ischaemic retinas (214 ± 7 cells/100 μm compared with 189 ± 5 cells/100 μm, n = 4, $p < 0.05$).

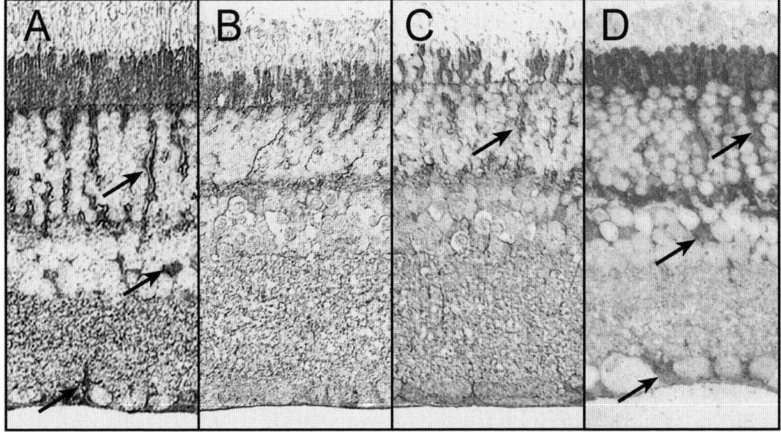

Figure 46.1. Glutamate transporter activity in the retina reflected by the accumulation of the glutamate analogue, D-aspartate. (A) GLAST-mediated uptake by Müller cells (arrows) in the non-ischaemic retina. (B) Greatly reduced glutamate transport under ischaemic conditions. (C) PKCδ/ε activation during ischaemia slightly increases glutamate transport. (D) Inhibition of PKA with KT5720 during ischaemia significantly increases glutamate uptake by the Müller cells.

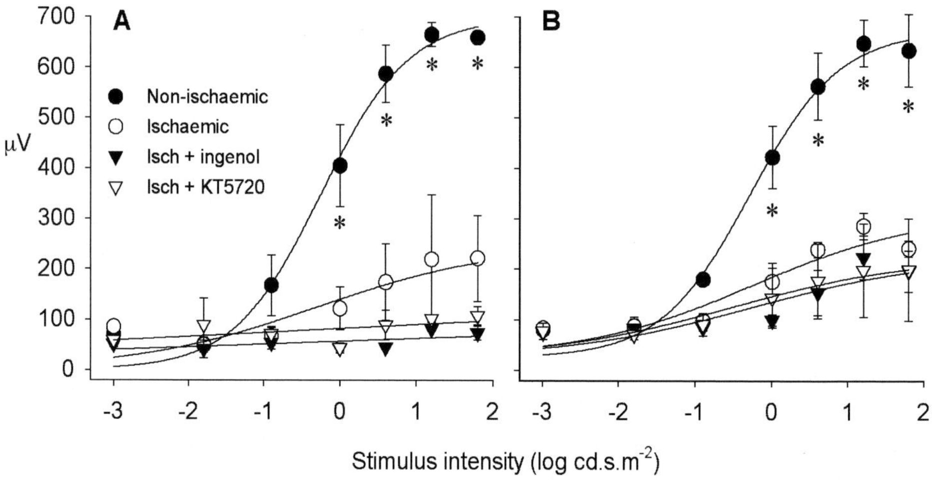

Figure 46.2. Stimulus-response characteristics of the ERG b-wave amplitude recorded (A) 1 day or (B) 7 days after an acute ischaemic insult. Data are expressed as mean amplitude ± SEM and fitted with a Naka-Rushton plot. The non-ischaemic b-wave was significantly greater than the ischaemic response at both time points ($* p < 0.05$). Stimulation of glial glutamate uptake with ingenol (300 µM) or KT5720 (1 µM) did not significantly ameliorate the ischaemia-induced suppression of the ERG.

3.3. Electroretinography

ERGs were recorded 1 and 7 days after the ischaemic insult. Intensity-response curves showed a significant suppression of the a-wave (data not shown) and b-wave amplitudes both 1 day (Figure 46.2A) and 7 days (Figure 46.2B) after acute retinal ischaemia ($p < 0.05$ compared with non-ischaemic). Neither ingenol nor KT5720 ameliorated the ischaemia-induced attenuation of the ERG at either time point. The respective intensity-response curves for ingenol and KT5720 treated retinas were not significantly different ($p > 0.05$) from the response characteristics of the saline-treated ischaemic retinas (Figure 46.2).

4. DISCUSSION

The activity of excitatory amino acid transporters is regulated either directly or indirectly by protein kinase activity.[14,15] The modulation of PKA and PKC has been shown to alter retinal glutamate transport.[9,11] However, it is not yet clear whether the pharmacological manipulation of glutamate transport could offer protection to the retina during an excitotoxic event in which extracellular levels of glutamate are expected to be elevated. In this study we showed that ingenol, a PKCδ/ε-selective activator, slightly enhanced the uptake of a glutamate analogue into Müller cells during an ischaemic insult. Given that PKCδ is localised to Müller cells and that antagonists of PKCδ inhibit Müller cell glutamate uptake,[9] this is not surprising. More surprising was the pronounced stimulation of ischaemic Müller cell uptake following the inhibition of PKA with KT5720. In cortical cultures, cyclic AMP analogues, not inhibitors, indirectly stimulate glial glutamate transport.[16] However, this effect is believed to be due to increased trafficking of the transporter to the membrane rather than a direct effect upon transporter activity.[17] Given that ischaemic failure of GLAST activity may contribute to raised extracellular glutamate concentrations and subsequent retinal degeneration, it is possible that stimulation of glial glutamate transport may be neuroprotective. Observation of ischaemic retinal sections suggests that this is the case. Following treatment with either ingenol or KT5720, ischaemia-induced disruption of retinal morphology was reduced. In particular, there was reduced swelling of neurones and Müller cell processes, suggesting decreased activation of AMPA-kainate receptors.[18] Moreover, photoreceptor cell counts 7 days after ischaemia revealed a significant reduction in neurodegeneration following PKCδ/ε activation. Thus, such intervention could be classed as neuroprotective.

Electroretinographic analyses of post-ischaemic retinas did not demonstrate any alleviation of retinal dysfunction by the pharmacological stimulation of Müller cell glutamate uptake. Scotopic a-wave and b-wave amplitudes were greatly suppressed despite the apparent rescue of retinal neurones by protein kinase modulation. A mismatch between retinal morphology and function is not uncommon following retinal ischaemia.[19] A likely explanation of the disparity between anatomical and functional results is the action of trophic factors on the retina. In particular, that of CNTF which is released following retinal stress such as ischaemia.[20] CNTF has been shown to protect photoreceptors from degeneration and suppresses the ERG.[21] Therefore, it would be of interest to assess long-term retinal function following glutamate transport stimulation during ischaemia to determine whether the observed cell rescue is maintained and is translated into a significant restoration of retinal function.

5. ACKNOWLEDGEMENTS

We thank the National Health & Medical Research Council (Australia) for funding this study and Retina Australia for the travel award provided to NLB to attend this meeting.

6. REFERENCES

1. M. M. Ward, A. I. Jobling, T. Puthussery, L. E. Foster, and E. L. Fletcher, Localization and expression of the glutamate transporter, excitatory amino acid transporter 4, within astrocytes of the rat retina, *Cell Tissue Res.* 315(3):305-310 (2004).
2. T. Rauen, J. D. Rothstein, and H. Wässle, Differential expression of three glutamate transporter subtypes in the rat retina, *Cell Tissue Res.* 286(3):325-336 (1996).
3. D. V. Pow and N. L. Barnett, Developmental expression of excitatory amino acid transporter 5: a photoreceptor and bipolar cell glutamate transporter in rat retina, *Neurosci. Lett.* 280(1):21-24 (2000).
4. D. V. Pow and N. L. Barnett, Changing patterns of spatial buffering of glutamate in developing rat retinae are mediated by the Müller cell glutamate transporter GLAST, *Cell Tissue Res.* 297(1):57-66 (1999).
5. N. L. Barnett and D. V. Pow, Antisense knockdown of GLAST, a glial glutamate transporter, compromises retinal function, *Invest. Ophthalmol. Vis. Sci.* 41(2):585-591 (2000).
6. N. L. Barnett, D. V. Pow, and N. D. Bull, Differential perturbation of neuronal and glial glutamate transport systems in retinal ischaemia, *Neurochem. Int.* 39(4):291-299 (2001).
7. T. Rauen and B. I. Kanner, Localization of the glutamate transporter GLT-1 in rat and macaque monkey retinae, *Neurosci. Lett.* 169(1-2):137-140 (1994).
8. K. Schultz and W. K. Stell, Immunocytochemical localization of the high-affinity glutamate transporter, EAAC1, in the retina of representative vertebrate species, *Neurosci. Lett.* 211(3):191-194 (1996).

9. N. D. Bull and N. L. Barnett, Antagonists of protein kinase C inhibit rat retinal glutamate transport activity in situ, *J. Neurochem.* 81(3):472-480 (2002).

10. N. D. Bull and N. L. Barnett, Retinal glutamate transporter activity persists under simulated ischemic conditions, *J. Neurosci. Res.* 78(4):590-599 (2004).

11. K. Takamoto and N. L. Barnett. Modulation of rat retinal glutamate transport by protein kinase A, *Proc. Austr. Neurosci. Soc.* 15:131 (2004).

12. N. L. Barnett and S. D. Grozdanic, Glutamate transporter localization does not correspond to the temporary functional recovery and late degeneration after acute ocular ischemia in rats, *Exp. Eye Res.* 79(4):513-524 (2004).

13. C. M. Lai, M. J. Yu, M. Brankov, N. L. Barnett, X. Zhou, T. M. Redmond, K. Narfstrom, and P. E. Rakoczy, Recombinant adeno-associated virus type 2-mediated gene delivery into the Rpe65-/- knockout mouse eye results in limited rescue, *Genet Vaccines Ther.* 2(1):3 (2004).

14. G. E. Gochenauer and M. B. Robinson, Dibutyryl-cAMP (dbcAMP) up-regulates astrocytic chloride-dependent L-[^{3}H]glutamate transport and expression of both system xc(-) subunits, *J. Neurochem.* 78(2):276-286 (2001).

15. L. A. Dowd and M. B. Robinson, Rapid stimulation of EAAC1-mediated Na+-dependent L-glutamate transport activity in C6 glioma cells by phorbol ester, *J. Neurochem.* 67(2):508-516 (1996).

16. B. D. Schlag, J. R. Vondrasek, M. Munir, A. Kalandadze, O. A. Zelenaia, J. D. Rothstein, and M. B. Robinson, Regulation of the glial Na+-dependent glutamate transporters by cyclic AMP analogs and neurons, *Mol Pharmacol.* 53(3):355-369 (1998).

17. E. G. Hughes, J. L. Maguire, M. T. McMinn, R. E. Scholz, and M. L. Sutherland, Loss of glial fibrillary acidic protein results in decreased glutamate transport and inhibition of PKA-induced EAAT2 cell surface trafficking, *Brain Res Mol Brain Res.* 124(2):114-123 (2004).

18. O. Uckermann, L. Vargova, E. Ulbricht, C. Klaus, M. Weick, K. Rillich, P. Wiedemann, A. Reichenbach, E. Sykova, and A. Bringmann, Glutamate-evoked alterations of glial and neuronal cell morphology in the guinea pig retina, *J. Neurosci.* 24(45):10149-10158 (2004).

19. N. L. Barnett and N. N. Osborne, Prolonged bilateral carotid artery occlusion induces electrophysiological and immunohistochemical changes to the rat retina without causing histological damage, *Exp. Eye Res.* 61:83-90 (1995).

20. G. Chidlow, K. G. Schmidt, J. P. Wood, J. Melena, and N. N. Osborne, Alpha-lipoic acid protects the retina against ischemia-reperfusion, *Neuropharmacol.* 43(6):1015-1025 (2002).

21. F. C. Schlichtenbrede, A. MacNeil, J. W. Bainbridge, M. Tschernutter, A. J. Thrasher, A. J. Smith, and R. R. Ali, Intraocular gene delivery of ciliary neurotrophic factor results in significant loss of retinal function in normal mice and in the Prph2Rd2/Rd2 model of retinal degeneration, *Gene Ther.* 10(6):523-527 (2003).

CHAPTER 47

ACTIVATION OF CELL SURVIVAL SIGNALS IN THE GOLDFISH RETINAL GANGLION CELLS AFTER OPTIC NERVE INJURY

Yoshiki Koriyama, Keiko Homma, and Satoru Kato*

1. SUMMARY

Generally, nerve injury of adult mammalian CNS neurons leads to a retrograde neuronal degeneration and cell death. The retinal ganglion cells (RGCs) of rat fail to regenerate and become apoptotic after optic nerve injury. In contrast, goldfish RGCs can survive and regrow their axons after injury. Focusing on this different response of RGCs in both species to optic nerve injury, we compared cell death and cell survival signals in the rat and goldfish RGCs after optic nerve injury. In goldfish retina, levels of phospho-Akt (p-Akt) and phospho-Bad (p-Bad) first rapidly increased at 3-5 days after optic nerve injury. Subsequently, levels of Bcl-2 increased and caspase-3 activity decreased at 10 days after nerve injury. In rat retina, levels of p-Akt and p-Bad first rapidly decreased at 1-2 days after optic nerve injury. Subsequently, levels of Bax and caspase-3 activity increased 6 days after optic nerve crush. These changes after optic nerve injury were all morphologically localized only in the RGCs. The data suggest that goldfish RGCs are warranted the cell survival by rapid p-Akt and subsequent Bcl-2 activations during the optic nerve regeneration, whereas rat RGCs are made a progress of the cell death by rapid inactivation of p-Akt and subsequent activation of Bax after optic nerve crush.

2. INTRODUCTION

Adult mammalian optic nerve cannot regenerate after nerve injury. Retinal ganglion cells (RGCs) in rat initiate a sprouting reaction at the damaged nerve endings, but this growth is abortive and the cells soon begin to die (Bähr and Bonhoeffer, 1994). In contrast, fish and amphibian RGCs can regenerate their optic nerves. Outgrowth of the leading axons

*Saturo Kato, Department of Molecular Neurobiology, Graduate School of Medicine, University of Kanazawa, 13-1 Takara-machi, Kanazawa 920-8640, Japan, Tel: +81 76 265 2450; E-mail: satoru2med.kanazawa-u.ac.jp.

proceeds at 0.3 mm/day after axtomy (McQuarrie et al., 1981). The regenerating fibers first arrive to the optic lobe at 2 weeks after nerve lesion and the plexiform layer formed by the regenerating fivers is visible in all areas of the optic tectum at 3-4 weeks after nerve lesion (Kato et al., 1999). At view of this point, one question is raised why can goldfish RGCs but not rat RGCs regrow their axons after optic nerve injury. Therefore, we compared cell death and cell survival signals in both species RGCs after optic nerve injury. From these comparative study with rat, we clearly demonstrate for the first time that rapid activation of phospho-Akt (active form of serine-threonine protein kinase) and subsequent activation of Bcl-2 enable goldfish RGCs to survive and regenerate their optic nerve.

3. METHODS

3.1. Caspase-3 Activity in the Retina

The optic nerves of common goldfish (*Carrassius auratus*, 6-7 cm body length) and Sprague-Dawley rats (250-300 g in body weight) were crushed by tweezers 1 mm from the eyeball for 10 sec. Retinal samples at indicated time points after optic nerve crush were incubated with a fluorogenic substrate Ac-DEVD-7-amino-4-methylcoumarin (AMC, 20 µM), specific for caspase-3, at 37°C for 1 h. Cleavage of the substrate by caspase-3 was measured by the fluorescent product, AMC, using a spectrofluorometer (excitation at 355 nm and emission at 460 nm).

3.2. Western Blotting and Immunohistochemistry

Retinal extracts from rat and goldfish were prepared at the indicated times after optic nerve crush. The retina was carried out Western blot analysis as described previously (Koriyama et al., 2003). And the expression pattern of cell death and survival signal proteins in the rat and goldfish retinas after optic nerve crush was investigated by immunohistochemical staining as described previously (Liu et al., 2002). The primary antibodies used were anti-mouse Bcl-2 (1:200; recognized as a 26 kDa protein band), Bax (1:500; recognized as a 23 kDa protein band), which were purchased from Santa Cruz (CA). Anti-Akt antibody react the absolute Akt including p-Akt. Anti-mouse p-Akt antibody (1:500; recognized as a 56 kDa protein band) and phospho-Bad (1:500; recognized as a 23 kDa protein band) were purchased from Sigma-Aldrich (USA). An appropriate anti-IgG antibody was used as the secondary antibody (1:200-1000).

4. RESULTS

4.1. The Time Course of Cell Death Signals in the Rat RGCs after Optic Nerve Crush

We measured caspase-3 activity, which is the most downstream molecule of the apoptotic signal cascade, in the rat retinas following optic nerve injury. The caspase-3 activity was significantly increased by 25% of the control value 6-8 days after nerve injury (Figure 47.1). Next, we measured levels of Akt and p-Akt proteins which are activated by trophic

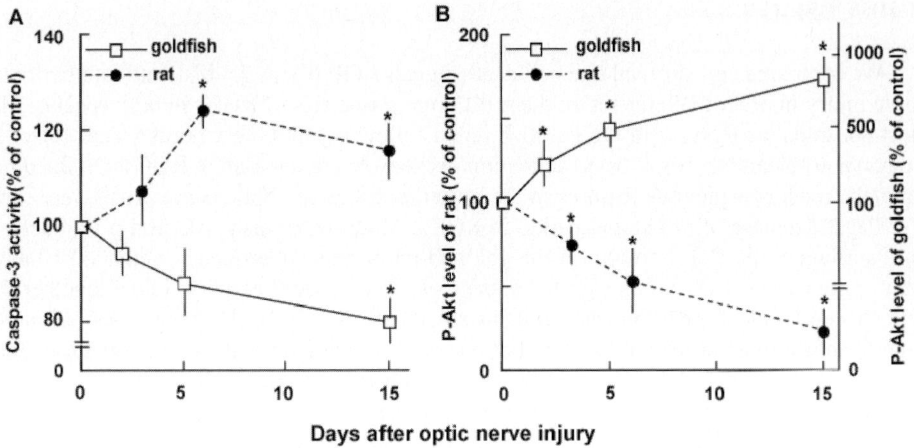

Figure 47.1. The time course of cell survival and cell death signals after optic nerve injury. (A) Caspase-3 activity. (B) Phospho-Akt protein level. The value indicates the mean ± S.E.M. (n = 3). *p < 0.01 as compared with control.

factors via phosphatidylinositol-3-kinase (PI3K). The p-Akt protein levels rapidly decreased in the rat retina 1-2 days after nerve injury (Figure 47.1), whereas total Akt protein levels did not change in this period. The decrease of p-Akt levels in rat was localized in RGCs by immunohistochemistry. We further measured levels of Bcl-2, Bax, and p-Bad protein in the rat retina following optic nerve injury. The Bcl-2 protein levels slightly increased 1.3 fold in the rat retina 6 days after injury and then decreased rather than control value by 12 days after nerve injury. The Bax protein levels drastically increased 3-4 fold 6-12 days after optic nerve crush. The p -Bad protein levels rapidly decreased less than 50% of control value at 1-2 days after nerve injury.

4.2. The Time Course of Cell Survival Signals in the Goldfish RGCs after Optic Nerve Crush

In goldfish retina, no significant reduction of cell number in the ganglion cell layer could be seen over 40 days after optic nerve crush. We measured caspase-3 activity in the goldfish retina after nerve injury. The caspase-3 activity was significantly decreased by 20% of the control value at 10-20 days after nerve crush (Figure 47.1) and then returned to the control value 30 days later. Next, we measured levels of Akt and p-Akt proteins in the goldfish retina following optic nerve injury. The p-Akt protein levels were rapidly increased 3-5 days and peaked 8 fold at 10-20 days and then decreased by 40 days after nerve crush (Figure 47.1). The Akt protein levels did not change in this period. The increase of p-Akt in goldfish was localized in the RGCs by immunohistochemistry. We further measured Bcl-2, Bax and p-Bad protein levels in the goldfish retina following optic nerve injury. The Bcl-2 protein levels significantly increased 1.7 fold 10-20 days and then decreased 30 days after injury. The Bax protein levels did not change during this 30 days of period. The p-Bad protein levels in the goldfish retina were rapidly increased 3-5 days and peaked 8-fold at 10-20 days and gradually decreased by 40 days after optic nerve injury.

5. DISCUSSION

We compared cell survival and cell death signals in RGCs in goldfish and rat after optic nerve injury in order to characterize these different properties of RGCs in both species. The most conspicuous features in the goldfish retina following optic nerve crush were early (3-5 days) and sustained (10-30 days) activation of the p-Akt and p-Bad in RGCs. On the other hand, the most conspicuous features in the rat retina following optic nerve crush were early (1-2 days) decrease of p-Akt and p-Bad in RGCs. Along with this p-Akt and p-Bad activation, anti-apoptotic Bcl-2 protein in the goldfish retina was subsequently activated 10 days after optic nerve crush, with a significant decrease in caspase-3 activity. Taken together the data strongly indicate that the early and sustained activation of p-Akt is an essential molecular event for maintaining cell survival of fish RGCs during optic nerve regeneration. Subsequent activation of Bcl-2 is a reinforcement phenomenon to prevent RGCs from cell death. Phospho-Akt plays a central role in the cell survival system via PI3K/Akt system and inactivates apoptotic Bad and caspase-9 (Datta et al., 1997; Cardone et al., 1998). Furthermore, p-Akt directly activates cyclic AMP responsive element binding protein (CREB) and NF-kappaB leading to activation of Bcl-2 expression (Romashkova and Makarov, 1999). In the goldfish retina, initial activation of p-Akt 3-5 days after optic nerve crush follows inactivation of Bad at 3-5 days, and finally activation of Bcl-2 at 10-20 days after nerve injury.

Figure 47.2 illustrates a schematic drawing of the major role of p-Akt in goldfish and rat RGCs after optic nerve crush. The long time (3-30 days) activation of p-Akt suggests that p-Akt works upon not only cell survival, but also axonal elongation via direct cell survival and indirect anti-apoptotic and neurite promoting mechanisms mentioned above. Lavie et al. (1997) reported that explant culture experiment in goldfish, inhibition of the PI3K system by wortmannin produced a dose-dependent suppression of neurite outgrowth although the exact mechanism is not yet clarified. On the other hand, overexpression of Bcl-

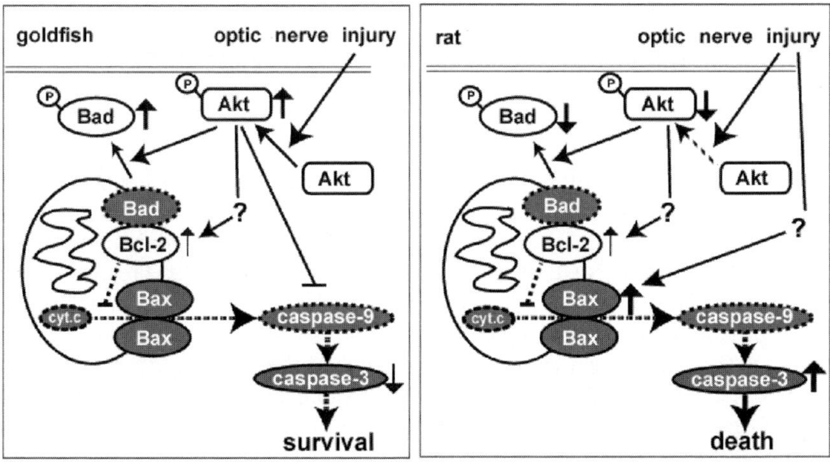

Figure 47.2. Signal cascade for cell survival and cell death through activation of the p-Akt system. In goldfish (left) and rat (right) RGCs after optic nerve crush. Molecules in the solid line were examined in this study but molecules surrounded in the dotted line were not examined.

2 into RGCs in the transgenic mouse prevented cell death but did not regenerate their axons (Lodovichi et al., 2001). These molecular studies also support our hypothesis that p-Akt activation in RGCs might be a determining factor for which goldfish optic nerve can regenerate their axons after nerve injury. Our goldfish optic nerve regeneration system offers some therapeutic clues to overcome the difficulties of mammalian CNS regeneration.

6. REFERENCES

Bähr, M., and Bonhoeffer, F., 1994, Perspectives on axonal regeneration in the mammalian CNS. Trends Neurosci. 17:473-479.

Cardone, M. H., Roy, N., Stennicke, H. R., Salvesen G. S., Franke T. F., Stanbridge E., Frisch S., and Reed, J. C., 1998, Regulation of cell death protease caspase-9 by phosphorylation. Science 282:1318-1321.

Datta, S. R., Dudek, H., Tao, X., Masters, S., Fu, H., Gotoh, Y., and Greenberg, M.E., 1997, Akt phosphorylation of BAD couples survival signals to the cell-intrinsic death machinery. Cell 91:231-241.

Kato, S., Devadas, M., Okada, K., Shimada, Y., Ohkawa, M., Muramoto, K., Takizawa, N., and Matsukawa, T., 1999, Fast and slow recovery phases of goldfish behavior after transection of the optic nerve revealed by a computer image processing system. Neuroscience 93:907-914.

Koriyama, Y., Chiba, K., and Mohri, T., 2003, Propentofylline protects beta-amyloid protein-induced apoptosis in cultured rat hippocampal neurons. Eur. J. Pharm. 458:235-241.

Lavie, Y., Dybowski, J., and Agranoff, B. W., 1997, Wortmannin blocks goldfish retinal phosphatidylinositol 3-kinase and neurite outgrowth. Neurochem. Res. 22:373-378.

Liu, Z. W., Matsukawa, T., Arai, K., Devadas, M., Nakashima, H., Tanaka, M., Mawatari, and K., Kato, S., 2002, Na,K-ATPase alpha3 subunit in the goldfish retina during optic nerve regeneration. J. Neurochem. 80:763-770.

Lodovichi, C., Di Cristo, G., Cenni, M.C., and Maffei, L., 2001, Bcl-2 overexpression per se does not promote regeneration of neonatal crushed optic fibers. Eur. J. Neurosci. 13:833-838.

McQuarrie, I. G., and Grafstein, B., 1981, Effect of a conditioning lesion on optic nerve regeneration in goldfish. Brain Res. 216:253-264.

Romashkova, J. A., and Makarov, S. S., 1999, NF-kappaB is a target of AKT in anti-apoptotic PDGF signalling. Nature 401:86-90.

PART V

USHER SYNDROME

CHAPTER 48

ROLES AND INTERACTIONS OF USHER 1 PROTEINS IN THE OUTER RETINA

Concepción Lillo, Junko Kitamoto, and David S. Williams*

1. INTRODUCTION

Usher syndrome (USH) describes a group of inherited blindness-deafness disorders, resulting from retinal degeneration and cochlear dysfunction (Usher, 1914). There are three subtypes of Usher syndrome, Usher syndrome type 1 is the most severe of the three. Seven Usher 1 genes have been mapped (Usher 1A-G), and all show the same clinical phenotype in humans. The most common form is Usher 1B, which accounts for at least 50% of Usher 1 cases (Astuto et al., 2000). Usher 1B is caused by mutations in the gene, *MYO7A*, which encodes an unconventional myosin, myosin VIIa (Weil et al., 1995). Usher 1C has been shown to be caused by defects in the gene, *USH1C*, which encodes harmonin, a scaffold protein with PDZ domains (Verpy et al., 2000; Bitner-Glindzicz et al., 2000). PDZ proteins, such as harmonin, form multiprotein complexes that are localized in specific subcellular domains, such as the microvilli of epithelial cells, synaptic terminals and the tight junctions (Sheng and Sala, 2001). Alternative splicing of the *USH1C* gene results in multiple harmonin isoforms, named a, b and c (Verpy et al., 2000). The short isoform a is the most abundant of the three and is present in most tissues. Both harmonin and myosin VIIa have been found in the stereocilia of the hair cells in the inner ear and the microvilli of other epithelial cells (Wolfrum et al., 98; Verpy et al., 2000; Boëda et al., 2002). Recent studies have shown that these two proteins interact to shape the stereocilia bundle in the inner ear (Boëda et al., 2002).

In the retina, myosin VIIa is present in the connecting cilium of the photoreceptor cells and the apical processes of the pigmented epithelium (Hasson et al., 1995; El-Amraoui et al., 1996; Liu et al., 1997, 1999). One recent work (Reiners et al., 2003) has identified harmonin in the mouse retina.

For the present study, we have generated antibodies against specific domains for harmonin. Our goal is to determine whether harmonin colocalizes with myosin VIIa and thus whether it is feasible for them to interact in the retina.

*C. Lillo, J. Kitamoto and D.S. Williams, Departments of Pharmacology and Neurosciences, UCSD School of Medicine, La Jolla, CA, 92092-0983, U.S.A.

2. MATERIALS AND METHODS

2.1. Generation of Antibodies

We raised domain-specific harmonin antibodies in rabbits. H1 antibody was made against the N-terminus and PDZ1 domain, H2 antibody against the PDZ2 domain and they both recognize all harmonin isoforms. H3 is specific for the PDZ3 domain and it will only recognize a and b isoforms (Fig. 48.1). The three different antibodies were affinity purified against its corresponding antigen.

The myosin VIIa polyclonal antibody pAb2.2, was affinity purified using the bacterially expressed antigen (Liu et al., 1997) coupled to an NHS-Sepharose column (Amersham, CA). Alternatively, pAb2.2 antiserum was purified by repeated depletion against western blots of retinal tissue from homozygous $Myo7a^{4626SB}$ mice (which are null for myosin VIIa).

2.2. Immunohistochemistry and Immunoelectron Microscopy

For immunofluorescence microscopy, eyecups were fixed in 4% paraformaldehyde in phosphate-buffered saline (PBS), cryoprotected and cryosectioned. After quenching autofluorescence with 0.1% sodium borohydride in PBS, sections were incubated with primary antibody overnight at 4°C in blocking solution (2% goat serum, 0.1% Triton X-100 in PBS), and secondary antibody for 1 hour at room temperature using the Alexa 594 nm goat anti-rabbit IgG (Molecular Probes, OR, USA). Nuclei were labeled with DAPI (diluted 1 : 10,000) (Molecular Probes, OR, USA). Images were collected with a BioRad 1024 laser scanning confocal microscope.

For immunoelectron microscopy, we followed the protocol described in more detail in Gibbs et al. 2004. Briefly, eyecups were fixed in 0.1% glutaraldehyde + 2% paraformaldehyde in 0.1 M cacodylate buffer, and processed for embedment in LR White resin. After blocking, ultrathin sections were incubated with purified myosin VIIa or harmonin antibody in TBS + 1% BSA + 1% Tween 20 overnight at 4°C, washed, and incubated with goat anti-rabbit IgG conjugated to 10 nm gold (Amersham, Arlington Heights, IL, USA) for 1 hour. Appropriate negative controls, processed at the same time, were done.

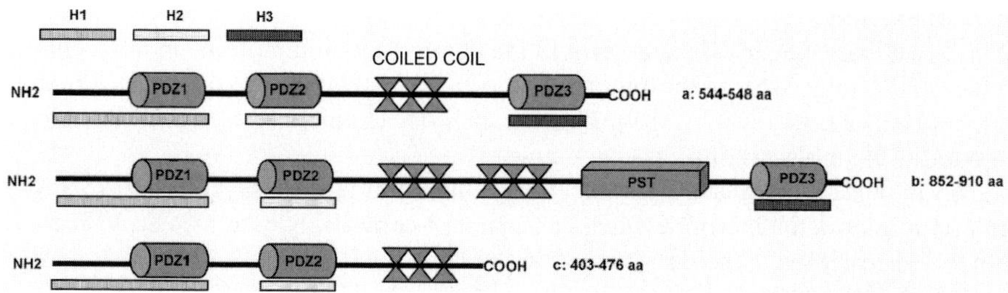

Figure 48.1. Cartoon showing the domains of the different harmonin isoforms and the regions recognized by the three different antibodies (H1, H2 & H3).

3. RESULTS

3.1. Myosin VIIa Localization

We immunodetected myosin VIIa in mouse and human retinas by immunofluorescence and immunoelectron microscopy. In the RPE cells, myosin VIIa is found in the apical processes (Fig. 48.2A, B). To elucidate more precisely the distribution of myosin VIIa in the RPE cells and the organelles with which it is associated, we performed immunoelectron microscopy labeling of mouse and human retinas (Fig. 48.3A, B). By quantification of immunolabel, we determined that most of the myosin VIIa (75%) in the apical RPE of mouse and human retinas is associated with the membrane of melanosomes (Gibbs et al., 2004).

In the photoreceptor cells, myosin VIIa is present only in the connecting cilium in mouse (Fig. 48.2B, C) and human retinas (Fig. 48.2D). We could not detect myosin VIIa in any

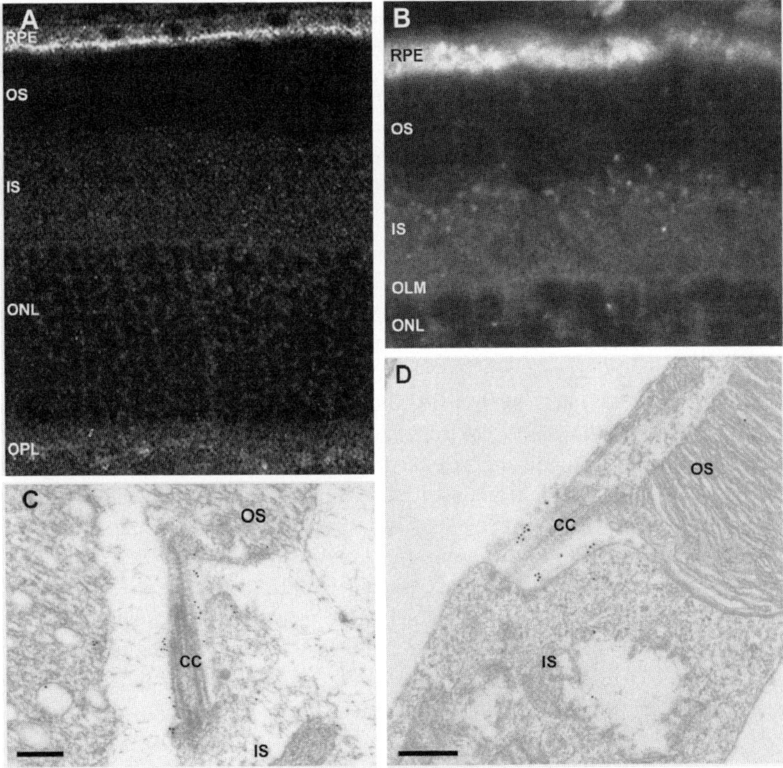

Figure 48.2. Immunofluorescence and electron microscopy images of myosin VIIa labeling in the mouse and human outer retina. A: Myosin VIIa is localized in the apical region of the RPE cells in the mouse retina. B: In a detailed image we detect the presence of punctate labeling in the outer segment (OS)/ inner segment (IS) interface, where the connecting cilium is located. The immunoelectron microscopy labeling of Myosin VIIa in mouse (C) and human (D) photoreceptor cells shows that this protein is localized around the connecting cilium membrane. RPE: retinal pigmented epithelium, OS: outer segment, IS: inner segment, CC: connecting cilium, ONL: outer nuclear layer, OLM: outer limiting membrane, OPL: outer plexiform layer. Scale bars: 300 nm.

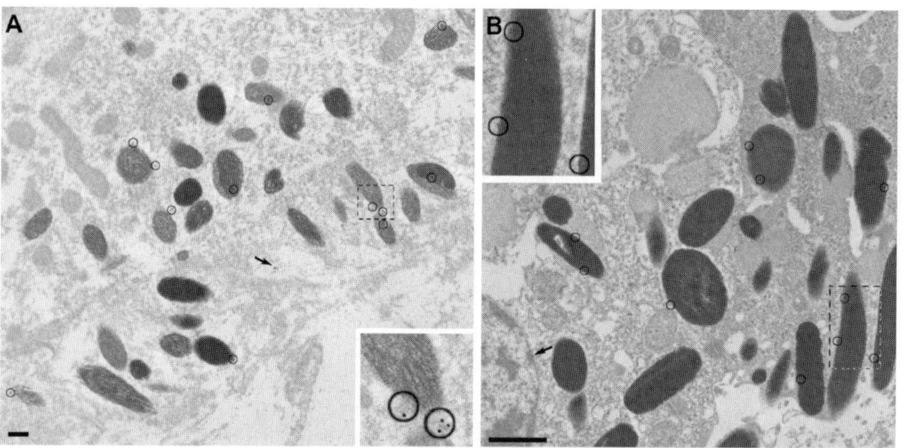

Figure 48.3. Electron micrographs of mouse (A) and human (B) RPE cells labeled with Myosin VIIa antibody. The apical region is lower right and the basal region is upper left. Myosin VIIa is associated with the melanosome membrane (circles) and elsewhere in the cytoplasm (arrows). Insets: magnification from the square in the picture. Scale bar: 300 nm.

other location or organelle in the photoreceptor cell. This finding is consistent with previous reports from our lab (Liu et al., 1997, 1999).

3.2. Harmonin Localization

A battery of antibodies raised against the harmonin protein shows that the distribution of this protein in the mouse outer retina is completely different from that of the myosin VIIa. The antibody H1, which recognizes the N terminus of harmonin and its PDZ1 domain, detects this protein in the inner segments of the photoreceptor cells and in the photoreceptor synaptic terminals in the outer plexiform layer (Fig. 48.4A). H2, which recognizes the PDZ2 domain of harmonin, labels the outer segments of the photoreceptor cells (Fig. 48.4B). H3 was made against the PDZ3 domain of harmonin, and labels the inner segments, outer limiting membrane (OLM) and outer plexiform layer in the mouse retina (Fig. 48.4C, D). None of the antibodies showed any harmonin labeling in the RPE cells (Fig. 48.4A, B). These data are supported by biochemical assay of cellular subfractions (not shown).

By immunoelectron microscopy of mouse and human retinas, using the H1 antibody, harmonin was detected mostly in the mitochondria in the inner segments of the photoreceptor cells (Fig. 48.5A, 48.6A). In the outer plexiform layer, harmonin seems to be associated with the cell membrane and mitochondria of the photoreceptors synaptic terminals (Fig. 48.5B, 48.6B). In contrast to myosin VIIa, we did not detect any labeling in the photoreceptor connecting cilium (Fig. 48.5C).

4. DISCUSSION

Our results demonstrate that in mouse and human RPE cells, most of myosin VIIa is associated with the melanosomes membrane, suggesting a role in their transport in this

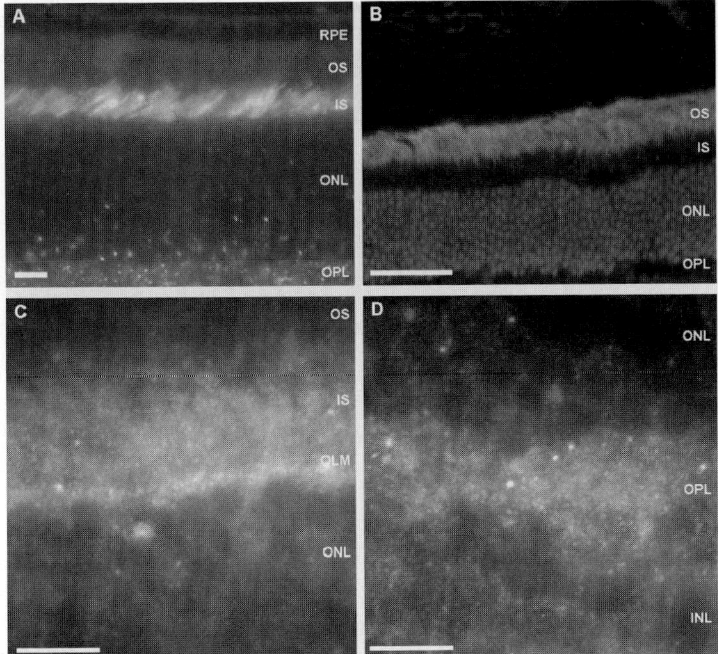

Figure 48.4. Immunofluorescence images of mouse outer retina labeled with harmonin antibodies. H1 (A) and H3 (C, D) antibodies detect harmonin in the photoreceptors inner segments and in the outer plexiform layer. B: H2 antibody detects harmonin in the outer segments of the photoreceptor cells; nuclei in ONL are stained with DAPI. Scale bars, A, C and D: 10 μm., B: 50 μm. RPE: Retinal pigmented epithelium, OS: outer segments, IS: inner segments, OLM: outer limiting membrane, ONL: outer nuclear layer, OPL: outer plexiform layer, INL: inner nuclear layer.

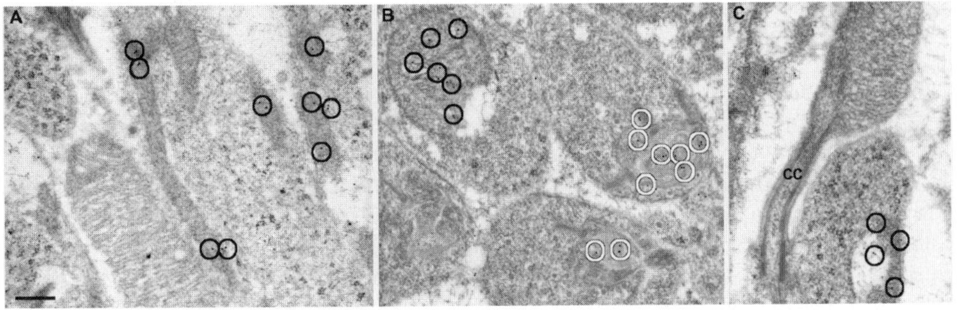

Figure 48.5. Electron micrographs of mouse photoreceptor cells labeled with harmonin H1 antibody. A: Harmonin is present in the mitochondria, both, in the inner segments (black circles, A, C) and in the synaptic terminal (black circles, B). It is also associated with the membrane in the synaptic terminal (white arrows, B). CC: connecting cilium. Scale bar = 300 nm.

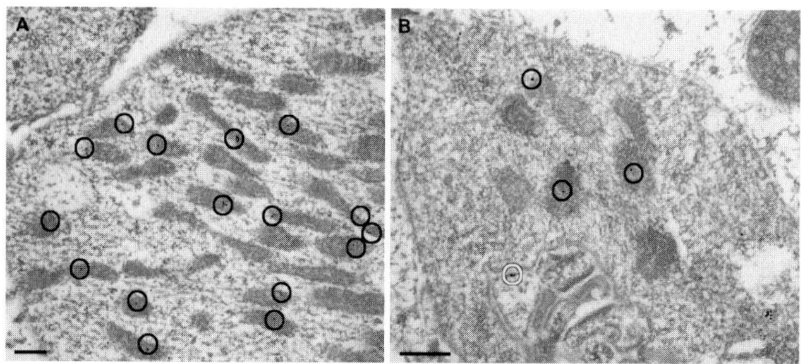

Figure 48.6. Electron micrographs of human photoreceptor cells labeled with harmonin H1 antibody. Also in human photoreceptor cells, harmonin is localized in the mitochondria in both, the inner segment (black circles, A) and in the synaptic terminal (black circles, B). In the synaptic terminal, harmonin is also associated with the cell membrane (white circles, B). Scale bars = 300 nm.

specific cell type (Gibbs et al., 2004). In fact, mice lacking myosin VIIa, show mislocalization and abnormal motility of melanosomes in the RPE cells (Liu et al., 1998, Gibbs et al., 2004). In the photoreceptor cells, myosin VIIa is only present in the connecting cilium, where it has been shown to play a role in the opsin transport from the inner segment to the outer segment (Liu et al., 1999).

Here we show that harmonin, the product of the *USH1C* gene, is not present in the RPE cells, and, in the photoreceptor cells, its distribution does not overlap with myosin VIIa.

In humans, defects in any one of the seven Usher 1 genes cause the same clinical phenotype in the inner ear and the retina. This observation suggests that these proteins either could be interacting with each other forming a multiprotein complex, or could be involved in the same functional pathway. Some studies have shown that this hypothesis is valid for some of these proteins, such as harmonin, myosin VIIa and cadherin 23 (the product of *USH1D*) in the inner ear, where they may interact during the stereociliary morphogenesis, and, after development, to maintain stereociliary integrity (Boëda et al., 2002, Siemens et al., 2002). Nevertheless, our results show that, in the mouse and human retina, the distributions of the two proteins do not overlap, thus preventing any interaction.

In the photoreceptor cells, harmonin was localized mainly with membrane of mitochondria and in the synaptic terminal. Due to the nature of PDZ domains, known to serve as scaffolds for the formation of large molecular complexes (Sheng and Sala, 2001), we suggest that harmonin may form protein complexes together with other structural proteins either to shape the synaptic terminal, or to help in the delivery or trafficking of synaptic vesicles. Several studies have demonstrated that in general, pre- and post-synaptic terminals are very complex structures composed of several anchoring proteins, many of them containing PDZ domains (Kim and Sheng, 2004). Harmonin might be one of them.

In mitochondria, there is at least one PDZ domain protein, called outer membrane protein 25 (OMP25) that functions by recruiting cytoplasmic proteins to the mitochondria membrane (Nemoto and De Camilli, 1999). A role for harmonin might be in the localization of mitochondria within the cell, especially with respect to the actin cytoskeleton, due to the ability of harmonin b to bind directly to actin filaments (Boëda et al., 2002).

5. SUMMARY

Our studies demonstrate that harmonin and myosin VIIa are not localized in the same compartments in the mouse and human retinas, indicating that they do not interact in this organ, contrary to what has been shown in the inner ear.

The enrichment of harmonin in the photoreceptor synapses indicates that this protein may form multiple complexes with others to maintain the synaptic structure or to mediate in the release of synaptic vesicles.

6. ACKNOWLEDGMENTS

We thank Piotr Kazmierczak and Ulrich Mueller for their generous gift of the H2 and H3 harmonin antibodies. This research was supported by NIH R01 grant EY07042, NIH core grant EY EY12598, and a grant from the Foundation Fighting Blindness (to DSW).

7. REFERENCES

Astuto, L. M., Weston, M. D., Carney, C. A., Hoover, D. M., Cremers, C., Wagenaar, M., Moller, C., Smith, R. J. H., Pieke-Dahl, S., Greenberg, J., Ramesar, R., Jacobson, S. G., Ayuso, C., Heckenlively, J. R., Tamayo, M., Gorin, M. B., Reardon, W., and Kimerling, W. J., 2000, Genetic heterogeneity of Usher syndrome: Analysis of 151 families with Usher type I. *Amer. J. Hum. Genet.* **67**:1569-1574.

Bitner-Glindzicz, M., Lindley, K. J., Rutland, P., Blaydon, D., Smith, V. V., Milla, P. J., Hussain, K., Furth, Lavi, J., Cosgrove, K. E., Shepherd, R. M., Barnes, P. D., O'Brien, R. E., Farndon, P. A., Sowden, J., Liu, X. Z., Scanlan, M. J., Malcolm, S., Dunne, M. J., Aynsley-Green, A. and Glaser, B., 2000, A recessive contiguous gene deletion causing infantile hyperinsulinism, enteropathy and deafness identifies the Usher type 1C gene. *Nat. Genet.* **26**(1):56-60.

Boëda, B., El-Amraoui, A., Bahloul, A., Goodyear, R., Daviet, L., Blanchard, S., Perfettini, I., Fath, K. R., Shorte, S., Reiners, J., Houdusse, A., Legrain, P., Wolfrum, U., Richardson, G. and Petit, C., 2002, Myosin VIIa, harmonin and cadherin 23, three Usher I gene products that cooperate to shape the sensory hair cell bundle. *EMBO J.* **21**(24):6689-6699.

Dodd, P. R., Hardy, J. A., Oakley, A. E., Edwardson, J. A., Perry, E. K. and Delaunoy, J. P., 1981, A rapid method for preparing synaptosomes: comparison, with alternative procedures. *Brain Res.* **226**(1-2):107-118.

El-Amraoui, A., Sahly, I., Picaud, S., Sahel, J., Abitbol, M. and Petit, C., 1996, Human Usher 1B/mouse shaker-1: the retinal phenotype discrepancy explained by the presence/absence of myosin VIIA in the photoreceptor cells. *Hum. Mol. Genet.* **5**(8):1171-1178.

Gibbs, D., Azarian, S. M., Lillo, C., Kitamoto, J., Klomp, A. E., Steel, K. P., Libby, R. T. and Williams, D. S., 2004, Role of myosin VIIa and Rab27a in the motility and localization of RPE melanosomes. *J. Cell. Sci.* **117**:6473-6483.

Gibbs, D., Kitamoto, J., and Williams, D. S., 2003, Myosin VIIa in RPE phagocytosis and a potential cause of blindness in Usher syndrome 1B. *Proc. Natl. Acad. Sci. USA.* **100**(11):6481-6486.

Hasson, T., Heintzelman, M. B., Santos-Sacchi, J., Corey, D. P., and Mooseker, M. S., 1995, Expression in cochlea and retina of myosin VIIa, the gene product defective in Usher syndrome type 1B. *Proc. Natl. Acad. Sci. USA* **92**:9815-9819.

Kim, E. and Sheng, M., 2004, PDZ domains proteins of synapses. *Nat. Rev. Neurosci.* **5**(10):771-781.

Liu, X., Ondek, B., and Williams, D. S., 1998, Mutant myosin VIIa causes defective melanosome distribution in the RPE of shaker-1 mice. *Nat. Genet.* **19**:117-118.

Liu, X., Udovichenko, I. P., Brown, S. D. M., Steel, K. P., and Williams, D. S., 1999, Myosin VIIa participates in opsin transport through the photoreceptor cilium. *J. Neurosci.* **19**:6267-6274.

Liu, X., Vansant, G., Udovichenko, I. P., Wolfrum, U., and Williams, D. S., 1997, Myosin VIIa, the product of the Usher 1B syndrome gene, is concentrated in the connecting cilia of photoreceptor cells. *Cell Motil. Cytoskel.* **37**:240-252.

Nemoto, Y and De Camilli, P., 1999, Recruitment of an alternatively spliced form of synaptojanin 2 to mitochondria by the interaction with the PDZ domain of a mitochondrial outer membrane protein. *EMBO J.* **18**(11):2991-3006.

Reiners, J., Reidel, B., El-Amraoui, A., Boeda, B., Huber, I., Petit, C. and Wolfrum, U., 2003. Differential distribution of harmonin isoforms and their possible role in Usher-1 protein complexes in mammalian photoreceptor cells. *IOVS* **44**(11):5006-5015.

Sheng, M. and Sala, C., 2001, PDZ domains and the organization of supramolecular complexes. *Annu. Rev. Neurosci.* **24**:1-29.

Siemens, J., Kazmierczak, P., Reynolds, A., Sticker, M., Littlewood-Evans, A. and Muller, U., 2002, The Usher syndrome proteins cadherin 23 and harmonin form a complex by means of PDZ-domain interactions. *Proc. Natl. Acad. Sci. USA* **99**(23):14946-14951.

Usher, C., 1914, On the inheritance of retinitis pigmentosa with notes of cases. *R. Lond. Ophthalmol. Hosp. Rep.* **19**:130-236.

Verpy, E., Leibovici, M., Zwaenepoel, I., Liu, X. Z., Gal, A., Salem, N., Mansour, A., Blanchard, S., Kobayashi, I., Keats, B. J., Slim, R.and Petit, C., 2000, A defect in harmonin, a PDZ domain-containing protein expressed in the inner ear sensory hair cells, underlies Usher syndrome type 1C. *Nat. Genet.* **26**(1):51-55.

Weil, D., Blanchard, S., Kaplan, J., Guilford, P., Gibson, F., Walsh, J., Mburu, P., Varela, A., Levilliers, J., Weston, M. D., Kelley, P. M., Kimberling, W. J., Wagenaar, M., Levi-Acobas, F., Larget-Piet, D., Munnich, A., Steel, K. P., Brown, S. D. M., and Petit, C., 1995, Defective myosin VIIA gene responsible for Usher syndrome type 1B. *Nature* **374**:60-61.

Wolfrum, U., Liu, X., Schmitt, A., Udovichenko, I. P. and Williams, D. S., 1998, Myosin VIIa as a common component of cilia and microvilli. *Cell Motil. Cytoskeleton* **40**(3):261-271.

CHAPTER 49

MOLECULAR ANALYSIS OF THE SUPRAMOLECULAR USHER PROTEIN COMPLEX IN THE RETINA

Harmonin as the key protein of the Usher Syndrome

Jan Reiners and Uwe Wolfrum*

1. Introduction

Human Usher syndrome (USH) is the most common form of deaf-blindness and also the most frequent case of recessive *retinitis pigmentosa*. According to the degree of the clinical symptoms, three different types of the Usher syndrome are distinguished: USH1, USH2 and USH3 (Davenport and Omenn, 1977). USH is genetically heterogeneous with eleven chromosomal loci, which can be assigned to the three USH types (USH1A-G, USH2A-C, USH3A) (Petit, 2001). Out of these, USH1 is the most severe form, characterized by profound congenital deafness, constant vestibular dysfunction and prepubertal-onset *retinitis pigmentosa*. USH2 patients show a milder congenital deafness, a slightly later onset of *retinitis pigmentosa* and no vestibular dysfunction. The rarest Usher type 3 shows a late onset of *retinitis pigmentosa* and a progressing hearing loss. So far the different USH subtypes have been grouped into one disease basically on the same phenotype of the patients, although the clinical symptoms of the individual differ noticeably. The protein harmonin, responsible for USH1C, is of special interest, since it contains three PDZ domains, known for protein-protein interactions. We have gathered evidence that the different USH proteins are molecularly linked essentially via the scaffold protein harmonin. Harmonin interacts hereby not only with USH1 proteins, but also with USH2 proteins. Thus, this is the first evidence for a molecular linkage between USH1 and USH2, beyond the shared phenotype.

*Jan Reiners and Uwe wolfrumm, Instiutut für Zoologie, Universität Mainz, 55099 Mainz, Germany; E-mail: wolfrum@uni-mainz.de.

2. HARMONIN INTERACTS WITH ALL USH1 PROTEINS

All USH proteins belong into different protein classes. Among the USH1 proteins there are myosin VIIa (USH1B), two cadherins (cadherin 23/USH1D and protocadherin 15/USH1F) and with harmonin (USH1C) and SANS (USH1G) two scaffold proteins. Among these proteins harmonin is of special interest since it contains three PDZ domains. Recent results showed that all known USH1 proteins interact with harmonin via its PDZ domains (Boëda et al., 2002; Siemens et al., 2002; Weil et al., 2003; Adato et al., 2004; Reiners et al., submitted). Thereby, the USH-cadherins cadherin 23 and protocadherin 15 interact with the PDZ2 domain of harmonin. Myosin VIIa and SANS on the other hand bind to harmonin's PDZ1 and PDZ1/PDZ3 respectively.

In addition, USH1 proteins also exhibit homomeric interactions. Harmonin has been shown to initiate homomeric interactions via PDZ1 and the C-terminal of the major harmonin isoform a1 which provides the basis for polymeric protein chains (Siemens et al., 2002; Adato et al., 2004; Reiners et al., unpublished). Homodimers were also demonstrated for the USH1 proteins, myosin VIIa and SANS (Inoue and Ikebe, 2003; Adato et al., 2004). Moreover, dimerization is commonly found in cadherins and seems likely for the USH1 cadherins, cadherin 23 and protocadherin 15 (Bolz et al., 2002). Chains of harmonin may connect these dimers and integrate them into a protein network. In summary, the harmonin scaffold may integrate USH1 proteins and their dimers into USH1 networks and complexes.

3. HARMONIN INTERACTS WITH ALL USH2 PROTEINS

To date three USH2 genes have been identified. The first isolated USH2 gene was the most common form of the Usher syndrome, USH2A. It encodes for Usherin previously depicted as an extracellular matrix protein (Eudy et al., 1998). Recently, a splice variant of Usherin has been described which contains a transmembrane domain and a cytoplasmatic part including a PDZ binding motif (Van Wijk et al., 2004). At nearly the same time Weston et al. (2004) identified the gene defective in USH2C patients. The isoform causing USH2C is called "very large G-protein coupled receptor 1b" (VLGR1b), a member of the GPCR-superfamily. Its cytoplasmatic C-terminal tail contains a PDZ binding motif, as well. The affected gene in patients of USH2B was suggested to be the sodium bicarbonate transporter NBC3 (Bok et al., 2003). This prediction was based on following observations: the gene encoding for NBC3 is located in the human USH2B locus and mice lacking the murine ortholog of NBC3 show the USH phenotype. In previous studies, this co-transporter was localized in the kidney, where it interacts with the PDZ-protein NHERF-1 via its C-terminal (Pushkin et al., 1999; Pushkin et al., 2003). To summarize, although the USH2 proteins are members of very distinct families of transmembrane proteins, they have promising PDZ binding motifs of the class I (Nourry et al., 2003) at their C-terminus in common.

The interaction of these three proteins with the PDZ1 domain of harmonin was demonstrated in GST-pull down assays and using the yeast two-hybrid system (Reiners et al. in prep.). While the USH1 proteins rather function as cell adhesion and scaffold proteins, the USH2 proteins seem to be large "functional proteins", in a physiological point of view for the cell. These functional proteins may be positioned and anchored via harmonin in the USH1 protein network to form a supramolecular Usher protein complex.

4. FURTHER INTERACTING PARTNERS OF HARMONIN

Harmonin is expressed in many tissues and has been shown to interact with further proteins, which are not directly related to the Usher syndrome. The protein MCC2 was reported to interact with the PDZ1 domain of harmonin (Ishikawa et al., 2001). In pancreatic cells, the protein HARP, a protein with a large homology to SANS, binds to harmonin (Góñez et al., 2001; Johnston et al., 2004).

Recently, we identified further interacting partners of harmonin by yeast two-hybrid screening of a retinal cDNA-library: one of these proteins is the actin-binding protein filamin A. In the cellar environment, filamin A forms homodimers, stabilizes three-dimensional branching of actin filaments and links membrane proteins to the actin cytoskeleton (Gorlin et al., 1990; van der Flier and Sonnenberg, 2001). The interaction of filamin A with harmonin was confirmed by GST-pull down-assays and immuno-precipitations (Reiners et al., in prep.). Like actin-associated motor myosin VIIa (USH1B), filamin A provides a connection between the Usher protein complex and the actin cytoskeleton.

5. SUPRAMOLECULAR USHER PROTEIN COMPLEXES IN THE RETINA

To understand the cellular function of the USH proteins and their complexes respectively, it was essential to determine their subcellular localization. For this purpose, specific antibodies against the different interaction partners were generated and used for subcellular localization in the mammalian retina (Reiners et al., 2003; Wolfrum and Reiners, 2004; Reiners et al., submitted; Reiners et al., in prep.). Immunocytochemical analyses revealed that the partner molecules are localized in several distinct compartments of retinal photoreceptor cells. However, the co-localization of all USH proteins and the other complex partners – a necessary prerequisite for the assembly of a supramolecular complex – was determined in the outer plexiform layer. In this retinal layer the synaptic terminals of photoreceptor cells are sited. Together with our binding studies, this indicates that the identified Usher protein complex partners assemble in the photoreceptor cell synapses.

In the photoreceptor synaptic terminals, USH complexes may play fundamental roles in the structural and functional integrity of this synaptic junction. The scaffold protein harmonin bridges the activity of integral membrane USH2 proteins with the actin cytoskeleton (including filamin A and myosin VIIa) and the cell-cell adhesion sites generated by cadherin 23 and protocadherin 15. It is very likely that there are even more "functional" proteins, which are integrated into this USH complex. This hypothesis is outlined in figure 49.1. Defects of any of the USH-complex partners should result in synaptic dysfunction which in turn may cause *retinitis pigmentosa*, the clinical phenotype in the retina of USH patients.

6. ACKNOWLEDGEMENTS

This work was supported by Deutsche Forschungsgemeinschaft, Forschung contra Blindheit Initative Usher Syndrom e.V., ProRetina Deutschland e.V., and the FAUN-Stiftung, Nürnberg, Germany. The authors thank Boris Reidel, Tina Märker, Karin Jürgens, Gabi

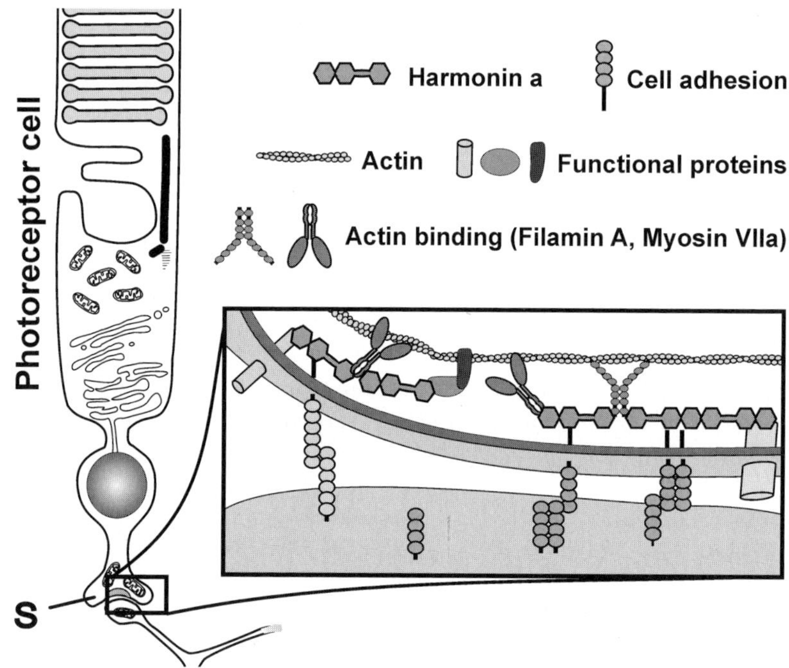

Figure 49.1. Scheme of the supramolecular Usher protein complex at photoreceptor synapse (S). Harmonin (USH1C) may act as the scaffold between the USH-cadherins (cadherin 23 / USH1D, protocadherin 15 / USH1F) and the actin binding proteins (myosin VIIa / USH1B, filamin A). In this fundament further "functional" proteins (e.g. NBC3 / USH2B) may be anchored and positioned.

Stern-Schneider, Elisabeth Sehn, Karla Kubicki and Jürgen Harf for their contribution to this project.

7. REFERENCES

Adato, A., Michel, V., Kikkawa, Y., Reiners, J., Alagramam, K. N., Weil, D., Yonekawa, H., Woifrum, U., El Amraoui, A., and Petit, C., 2004, Interactions in the network of Usher syndrome type 1 proteins, *Hum Mol Genet.*

Boëda, B., El Amraoui, A., Bahloul, A., Goodyear, R., Daviet, L., Blanchard, S., Perfettini, I., Fath, K. R., Shorte, S., Reiners, J., Houdusse, A., Legrain, P., Wolfrum, U., Richardson, G., and Petit, C., 2002, Myosin VIIa, harmonin and cadherin 23, three Usher I gene products that cooperate to shape the sensory hair cell bundle, *EMBO J.* **21**:6689-6699.

Bok, D., Galbraith, G., Lopez, I., Woodruff, M., Nusinowitz, S., BeltrandelRio, H., Huang, W., Zhao, S., Geske, R., Montgomery, C., Van, S., I, Friddle, C., Platt, K., Sparks, M. J., Pushkin, A., Abuladze, N., Ishiyama, A., Dukkipati, R., Liu, W., and Kurtz, I., 2003, Blindness and auditory impairment caused by loss of the sodium bicarbonate cotransporter NBC3, *Nat. Genet.* **34**:313-319.

Bolz, H., Reiners, J., Wolfrum, U., and Gal, A., 2002, Role of cadherins in Ca^{2+}-mediated cell adhesion and inherited photoreceptor degeneration, *Adv. Exp. Med. Biol.* **514**:399-410.

Davenport, S. L. H. and Omenn, G. S., 1977, The heterogeneity of Usher syndrome, *Vth Int. Conf. Birth Defects, Montreal.*

Eudy, J. D., Weston, M. D., Yao, S., Hoover, D. M., Rehm, H. L., Ma-Edmonds, M., Yan, D., Ahmad, I., Cheng, J. J., Ayuso, C., Cremers, C., Davenport, S., Moller, C., Talmadge, C. B., Beisel, K. W., Tamayo, M., Morton, C. C., Swaroop, A., Kimberling, W. J., and Sumegi, J., 1998, Mutation of a gene encoding a protein with extracellular matrix motifs in Usher syndrome type IIa, *Science* **280**:1753-1757.

Góñez, L. J., Johnston, A. M., Naselli, G., Braakhuis, A. J., Niwa, H., and Harrison, L. C., 2001, Islet cell precursors, growth and differentiation, *The Walter and Eliza Hall Institute of Medical Research Annual Report 2000/2001* 60.

Gorlin, J. B., Yamin, R., Egan, S., Stewart, M., Stossel, T. P., Kwiatkowski, D. J., and Hartwig, J. H., 1990, Human endothelial actin-binding protein (ABP-280, nonmuscle filamin): a molecular leaf spring, *J. Cell Biol.* **111**:1089-1105.

Inoue, A. and Ikebe, M., 2003, Characterization of the motor activity of mammalian myosin VIIA, *J.Biol.Chem.* **278**:5478-5487.

Ishikawa, S., Kobayashi, I., Hamada, J., Tada, M., Hirai, A., Furuuchi, K., Takahashi, Y., Ba, Y., and Moriuchi, T., 2001, Interaction of MCC2, a novel homologue of MCC tumor suppressor, with PDZ-domain Protein AIE-75, *Gene* **267**:101-110.

Johnston, A. M., Naselli, G., Niwa, H., Brodnicki, T., Harrison, L. C., and Gonez, L. J., 2004, Harp (harmonin-interacting, ankyrin repeat-containing protein), a novel protein that interacts with harmonin in epithelial tissues, *Genes Cells* **9**:967-982.

Nourry, C., Grant, S. G., and Borg, J. P., 2003, PDZ domain proteins: plug and play!, *Sci. STKE.* 2003: RE7.

Petit, C., 2001, Usher syndrome: from genetics to pathogenesis, *Annu. Rev. Genomics Hum. Genet.* **2**:271-297.

Pushkin, A., Abuladze, N., Lee, I., Newman, D., Hwang, J., and Kurtz, I., 1999, Cloning, tissue distribution, genomic organization, and functional characterization of NBC3, a new member of the sodium bicarbonate cotransporter family, *J. Biol. Chem.* **274**:16569-16575.

Pushkin, A., Abuladze, N., Newman, D., Muronets, V., Sassani, P., Tatishchev, S., and Kurtz, I., 2003, The COOH termini of NBC3 and the 56-kDa H+-ATPase subunit are PDZ motifs involved in their interaction, *Am. J. Physiol Cell Physiol* **284**:C667-C673.

Reiners, J., Märker, T., Reidel, B., and Wolfrum, U., 2005, Retinal expression and interaction of the Usher syndrome type 1 protein protocadherin 15 (USH1F) with harmonin (USH1C) via the PDZ2-domain., submitted.

Reiners, J., Reidel, B., El Amraoui, A., Boeda, B., Huber, I., Petit, C., and Wolfrum, U., 2003, Differential distribution of harmonin isoforms and their possible role in Usher-1 protein complexes in mammalian photoreceptor cells, *Invest Ophthalmol. Vis. Sci.* **44**:5006-5015.

Siemens, J., Kazmierczak, P., Reynolds, A., Sticker, M., Littlewood-Evans, A., and Muller, U., 2002, The Usher syndrome proteins cadherin 23 and harmonin form a complex by means of PDZ-domain interactions, *Proc. Natl. Acad. Sci. U.S.A* **99**:14946-14951.

van der Flier, A. and Sonnenberg, A., 2001, Structural and functional aspects of filamins, *Biochim. Biophys. Acta* **1538**:99-117.

Van Wijk, E., Pennings, R. J., Te, B. H., Claassen, A., Yntema, H. G., Hoefsloot, L. H., Cremers, F. P., Cremers, C. W., and Kremer, H., 2004, Identification of 51 Novel Exons of the Usher Syndrome Type 2A (USH2A) Gene That Encode Multiple Conserved Functional Domains and That Are Mutated in Patients with Usher Syndrome Type II, *Am. J. Hum. Genet.* **74**:738-744.

Weil, D., El Amraoui, A., Masmoudi, S., Mustapha, M., Kikkawa, Y., Laine, S., Delmaghani, S., Adato, A., Nadifi, S., Zina, Z. B., Hamel, C., Gal, A., Ayadi, H., Yonekawa, H., and Petit, C., 2003, Usher syndrome type I G (USH1G) is caused by mutations in the gene encoding SANS, a protein that associates with the USH1C protein, harmonin, *Hum. Mol. Genet.* **12**:463-471.

Weston, M. D., Luijendijk, M. W., Humphrey, K. D., Moller, C., and Kimberling, W. J., 2004, Mutations in the VLGR1 gene implicate G-protein signaling in the pathogenesis of Usher syndrome type II, *Am. J. Hum. Genet.* **74**:357-366.

Wolfrum, U. and Reiners, J., IOVS eLetters (26. May 2004), Myosin VIIa in Photoreceptor Cell Synapses May Contribute to an Usher 1 Protein Complex in the Retina; http://www.iovs.org/cgi/eletters/44/11/5006.

PART VI

STEM CELLS, TRANSPLANTATION AND RETINAL REPAIR

LIMITED NEURAL DIFFERENTIATION OF RETINAL PIGMENT EPITHELIUM

Ryosuke Wakusawa*, Toshiaki Abe, Yoko Saigo, and Makoto Tamai

1. INTRODUCTION

The retinal pigment epithelial (RPE) cell shares its origin with the neural retina as an anterior neural plate derivative. Recently RPE cell was reported on its capacity of transdifferentiating into a neuron-like cell in mammals.[1,2] Neurotrophic factors have reported to play essential roles. Some of these factors are basic fibroblast growth factor (bFGF) for transdifferentiation of RPE into neural cells[2,3] or epithelial growth factor (EGF) for proliferation of neural progenitor cells.[4]

In this study, we investigated whether adult human RPE cell line ARPE-19 could acquire neural property in the presence with the neurotrophic factors.

2. MATERIALS AND METHODS

2.1. Epithelial Culture (Non-Transdifferentiation Culture)

ARPE-19 cells were kindly given to us by Dr. Hjelmeland (Department of Ophthalmology, Section of Molecular and Cellular biology, University of California). The cells were cultured in Dulbecco's modified Eagle's medium-Ham's F12 (DMEM/F12; Gibco BRL) with 10% fetal bovine serum (FBS; ThermoTrace Melbourne Australia) in a humidified incubator at 37°C in 5% CO_2. The medium was changed every 2 to 3 days.

*Ryosuke Wakusawa, Department of Ophthalmology and Visual Science, Tohoku University, Graduate School of Medicine, 1-1 Seiryomachi Aobaku, Sendai Miyagi, 980-8574, Japan. Fax: 81 22 717 7298; Tel: 81 22 717 7294.

2.2. Transdifferentiation Culture

The cells were cultured in DMEM/F12 supplemented with B27 (Gibco), 1%FBS, 20 ng/ml bFGF (Genzyme, Cambridge, MA) and 20 ng/ml EGF (Genzyme) on the laminin-coated plastic plates for 14 days.

2.3. Immunocytochemistry

The monolayer cells were fixed in ethanol for 30 min and blocked with 10% goat serum in PBS for 60 min at RT. After removal of the blocking solution, the cells were incubated with primary antibodies overnight at 4°C. Primary antibodies were used at the following concentrations: mouse monoclonal anti-pancytokeratin (CK) (1 : 500; Sigma), mouse mono-clonal anti-tubulin-beta (TUB) 3 (1 : 500, Sigma), mouse monoclonal anti-microtubule asso-ciated protein (MAP) 5 (1 : 1000; Chemicon, Temecula, CA), mouse monoclonal anti-neurofilament (NF) 200 (1 : 1000, Sigma). After washing with PBS, the cells were further incubated with fluorescein isothiocyanate (FITC)-labeled goat anti-mouse immunogloblin (1 : 100, Dako) for 30 min at RT. Cell nuclei were counterstained with 4', 6'-deamino-2-phenylindole, dihydrochloride (DAPI) (1 microgm/ml).

Double staining of NF200 and CK antibody was also performed. The cells were incubated with Alexa Fluor 488 (Molecular Probes)-conjugated mouse monoclonal anti-panCK (1 : 500) and Alexa Fluor 546-conjugated mouse monoclonal anti-NF200 (1 : 1000) for 60 min at 37°C.

Photographs were taken by fluorescein microscopy (Leica DMIRE2, Leica Micro-systems Imaging Solutions Ltd., Cambridge UK).

2.4. Flowcytometry (FACS)

After cultivation, ARPE-19 cells were collected in cold PBS, and 1×10^5 cells/ml were incubated with each anti-neuron marker (TUB, MAP5, NF200, and rod-opsin) and CK in 3% FBS- phosphate buffered saline (PBS) on ice for 30 min. After washing with 3% FBS-PBS, the cells were incubated with FITC-labeled anti-mouse IgG (Jackson ImmunoTesearch Lab.) for 30 min on ice. Each sample was analyzed by flowcytometer. The cells that reacted with the each neuron marker showed stronger FITC fluorescence and shifted to the intense fluorescence area (M2) when compared to that of control cells. The percentage of positive shift was calculated and compared.

2.5. Extraction of mRNA, cDNA Generation, and Reverse-Transcriptase Polymerase Chain Reaction (RT-PCR)

mRNA was extracted from each cells using oligo dT cellulose and cDNA was gener-ated according to the manufacturer's instruction (Pharmacia Biotech Inc., Uppsala, Sweden).

RT-PCR was carried out for beta-actin, CK isotype 8, TUB3, MAP5 and NF200 in 50 μl of reaction mixture (reaction cycles were 30 or 35). Nested PCR was carried out for NF200 with 1 μl of the primary PCR product. The annealing temperatures depended on each primer sets for 2 minutes.

Primer sequences and amplification length were as follows:

Table 50.1. Sequences of each primer set.

Primer sequences	Amplification length (bp)	
Beta-actin	5′-CTACAATGAGCTGCGTGTGG-3′	313
	5′-CGGTGAGGATCTTCATGAGG-3′	
CK isotype 8	5′-AGATGCTGGAGACCAAGTGG-3′	323
	5′-GAGGAAGTTGATCTCGTCGG-3′	
TUB3	5′-GCTGCAATAAGACAGAGACAGG-3′	253
	5′-CGAGATGTACGAAGACGACG-3′	
MAP5	5′-AGCTCGAGGAAGAACAGTCC-3′	348
	5′-TGTTGGTACCAGTCCACTGC-3′	
NF200	5′-GTGAACACAGACGCTATGCG-3′	251
	5′-AGCAGGTCCTGGTATTCTCG-3′	
NF200, nested	5′-GGAGATAACTGAGTACCGGC-3′	189
primer	5′-CATCTCCCACTTGGTGTTCC-3′	

3. RESULTS

3.1. Immunocytochemistry

In the condition of non-transdifferentiation culture, ARPE-19 cells showed a flat and polygonal epithelial-like morphology. More than 98% of cells were positive for CK and TUB3. Some of them were positive for MAP5 but none for NF200 (figure 50.1. A and B).

In the condition of transdifferentiation culture, a small number of cells were spindle-shaped and extended multiple processes. These cells showed immunoreactivity for MAP5 and NF200 (figure 50.1. C and D). The cells expressing NF200 were also positive for CK by double staining.

3.2. FACS

In the condition of transdifferentiation culture, ARPE-19 cells were positive about 79% in TUB3, 5.6% in MAP5, 4.8% in NF200 and 83.4% in CK by FACS.

3.3. RT-PCR

RT-PCR also followed these results. CK, TUB3 and MAP5 were expressed in the condition of non-transdifferentiation culture. In the transdifferentiation culture, not only CK, TUB3 and MAP5, but also NF200 were expressed (figure 50.1. E).

4. DISCUSSION

RPE could dedifferentiate or transdifferentiate in response to several microenvironmental changes. Neurotrophic factors, such as bFGF played important roles in the control of the transdifferentiation processes of vertebrate pigment epithelial cells.[1,3,5]

It was reported that cultured adult human RPE cells, regardless of the presence of exogenous neurotrophic factors, expressed TUB3, one of the neural cell markers.[2,6] In the

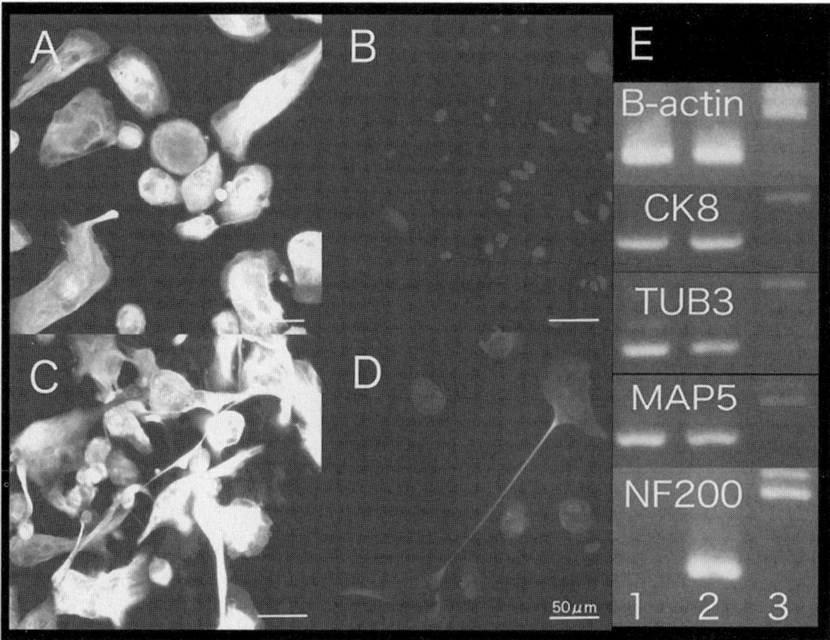

Figure 50.1. (A-D) Immunocytochemistry. Anti-panCK (A) and anti-NF200 (B) staining of the cells in the condition of non-transdifferentiation culture. Anti-panCK (C) and anti-NF200 (D) staining of the cells in the condition of transdifferentiation culture. (E) RT-PCR of each primer set. Lane 1: non-transdifferentiation culture. Lane 2: transdifferentiation culture. Lane 3: 100 bp marker.

present study, we also showed TUB3 expression on ARPE-19 cells at the condition of non-transdifferentiation. Some of them also expressed another neural cell marker, MAP5. Together with the previous studies, our results showed that RPE could have a partial neuronal property.

It is supposed that FGF signaling inhibits RPE differentiation and activates neural retina specification in eye development.[7] In embryonic rat, RPE cells transdifferentiated into neural retina in the presence of bFGF, although this ability was restricted until age E13.[1] Present study exhibited NF200 expression on ARPE-19 in the presence of bFGF and EGF, which was not observed in the standard culture condition. This result was quantitatively analyzed by FACS and confirmed its expression by RT-PCR and immunocytochemistry. However, immunocytochmistry revealed that the RPE cells expressing NF200 were also positive for CK. This result may show that some of the RPE cells were transdifferentiated into neuron partially by the stimulation of bFGF and EGF, but the partial transdifferentiated RPE itself still maintained epithelial property. For adult RPE cells, additional stimuli other than bFGF and EGF may be required for dedifferentiation from epithelium and transdifferentiation into more mature neuron.

5. REFERENCES

1. S. Zhao, S.C. Thornquist, and C.J. Barnstable, 1974, In vitro transdifferentiation of embryonic rat retinal pigment epithelium to neural retina, Brain Res. 677:300-310.
2. K. Amemiya, M. Haruta, M. Takahashi, M. Kosaka, and G. Eguchi, 2004, Adult human retinal pigment epithelial cells capable of differentiating into neurons, Biochem Biophys. Res. Commun. 316:1-5.
3. D. S. Sakaguchi, L.M. Janick, and T.A. Reh, 1997, Basic fibroblast growth factor (FGF-2) induced transdifferentiation of retinal pigment epithelium: Generation of retinal neurons and glia, Dev. Dynamics 209:387-398.
4. V. Tropepe, M. Sibilia, B.G. Ciruna, J. Rossant, E.F. Wagner, and D. van der Kooy, 1999, Distinct neural stem cells proliferate in response to EGF and FGF in the developing mouse telencephalon, Dev. Biol. 208:166-188.
5. R. Kodama, and G. Eguchi, From lens regeneration in the newt to in vitro transdifferentiation of vertebrate pigmented epithelial cells, 1995, Semin. Cell. Biol. 6:143-149.
6. S.A. Vinores, N.L. Derevjanik, J. Mahlow, S.F. Hackett, J.A. Haller, E. deJuan, A. Frankfurter, and P.A. Campochiaro, 1995, Class III beta-tubulin in human retinal pigment epithelial cells in culture and in epire tinalmembranes, Exp. Eye Res. 60:385-400.
7. J.R. Martinez-Morales, I. Rodrigo, and P. Bavolentra, 2004, Eye development: a view from the retina pigment epithelium, Bioessays 26:766-777.

CHAPTER 51

RETINAL PIGMENT EPITHELIAL CELLS FROM THERMALLY RESPONSIVE POLYMER-GRAFTED SURFACE REDUCE APOPTOSIS

Toshiaki Abe*, Masayoshi Hojo, Yoko Saigo, Masahiko Yamato, Teruo Okano, Ryosuke Wakusawa, and Makoto Tamai

1. INTRODUCTION

Brain-derived neurotrophic factor (BDNF) has reported to show photoreceptor protection for retinal degeneration either genetically programmed,[1] or light induced experimental retinal degeneration.[2] Genetically modified BDNF gene expressing cell transplantation in the subretinal space also rescued light induced photoreceptor degeneration.[3] When we consider the genetically modified BDNF gene expressing cell transplantation, the fate or the behavior of the cell at the subretinal regions has still unclear, especially at the diseased subretinal lesions. The cells placed at deeper layer of bruch membrane tend to be affected by apoptosis.[4] The fate of the transplanted cell may be one of the important factors for success of the transplantation. Cells cultured on poly-(N-isopropylacrylamide (PIPAAm)-grafted plates were easily detached from the culture plates as cell sheet by reducing the temperature from 37° C to 20° C without using enzymes.[5] We further cultured the cell and examined the degree of apoptosis by comparing with those of cells collected by trypsin treatment.

2. MATERIALS AND METHODS

2.1. Construction of Vector DNA with Rat BDNF cDNA and Transfection

Rat BDNF cDNA was kindly given from Dr. Atsushi Takeda, at Department of Neurology Tohoku University and was inserted into the plasmid Topo TA expression vector (CLONTECH Laboratories, Inc., CA. USA) and transduced into the cells by lipofections. These cells were selected by Zeocin (25 mg/ml) as we reported previously.[3]

* Toshiaka Abe, Division of Clinical Cell therapy, Tohoku University, School of Medicine, 1-1 Seiryomachi Aobaku, Sendai Miyagi, 980-8574, Japan. Fax: 81 22 717 8234; E-mail: toshi@oph.med.tohoku.ac.jp.

2.2. Preparation of the RPE Cell Sheet from PIPAAm-Grafted Plates

RPE cultured either on PIPAAm-grafted or non-grafted plates was started with same cell concentration (3×10^5 cells/ml) in 35 mm culture plates. Changing the temperature of the culture plate from 37° C to 20° C, we could collect the cultured RPE as cell sheet as reported previously.[5]

2.3. Quantification of Collagen Type IV and BDNF Expression

The amount of collagen type IV was determined by enzyme-linked imunosorbent assay (ELISA). The amount of BDNF (Promega Co.) protein were also quantified according to the manufacture's instruction.

2.4. Western Blot Analysis

Western blot analysis was performed using anti-chondroitin sulfate A (Seikagaku corporation, Tokyo, Japan) as we reported previously.[6]

2.5. Apoptosis of RPE and Flowcytometry

When RPE cultured on PIPAAm plate was detached from the plates, RPE cultured on normal culture plates was collected by 0.25% trypsin. These collected cells were cultured in the same medium coated with 1.5% agarose culture plate, which unable the cells to attach the culture plates, as reported previously.[7] Rat RPE collected as cell sheet and cultured with unattached condition for 3 hr was recovered. These cells were used for FACS analysis to detect early apoptosis using annexin V. After 24 hr culture in the same culture medium and plates, these cells were also collected and analyzed by FACSCalibur Hg flowcytometer. The % of cells showing the fluorescein intensity at M1 region was also calculated and was considered as positive % of apoptotic cells (Fig. 51.2).[8]

2.6. Caspase-3 Activity

At the indicated time, caspase-3 activity was assayed by caspase-3 activity assay kit (Oncogene (Red-DEVD-FMK) with FACS.

2.7. Extraction of mRNA, cDNA Generation, Reverse-Transcriptase Polymerase Chain Reaction (RT-PCR), Sequencing and Real-Time PCR

Extraction of mRNA and cDNA generatgion was performed. PCR, and Real-time PCR were carried out as reported previously.[6] We examined the gene expression of p-53 and β-actin.

2.8. Microscopic and Electron Microscopic Examination

After 24 hour culture, the sample was dissected, postfixed in phosphate-buffered 1% osmium tetroxide (pH 7.4), and dehydrated in an ascending series of ethanol solutions, and observed with a TEM-100CX.

2.9. Statistical Analysis

Statistical significance was determined by Fisher's protected least significant difference (PLSD) test.

3. RESULTS

When we examined the amount of chondroitin sulfate by western blotting analysis, cells collected as cell sheet from PIPAAm-grafted plates showed more chondroitin sulfate than that of cells collected by trypsin (Fig. 51.1). The amount of collagen type IV, the amount was also higher in the cells collected as cell sheet.

When we examined the early apoptosis from the cells collected as cell sheet, we found about 10% less apoptosis in the RPE with or without BDNF than those collected as cell suspension. When we collected the cell by trypsin 24 hour after culture, the ratio of the cell death was also about 10% less in the cells collected as cell sheet than those of cells collected as cell suspension in both BDNF transduced or non-transduced RPE.

P-53 gene expression was increased within 3-hour culture, in every cell examined. Further the gene was more expressed in the cells collected as cell suspension.

When we examined the caspase-3 activity, we found prominent increase of the activity after 10-hour culture in every cell type examined. The activity was less in the BDNF-transduced RPE cell collected as cell sheet at 3 and 10 hour after floating culture than those of RPE cells collected as cell suspension, although statistical analysis was not significant.

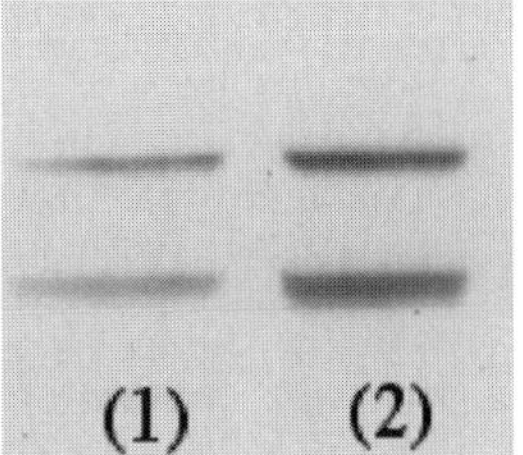

Figure 51.1. Western blotting for chondroitin sulfate.

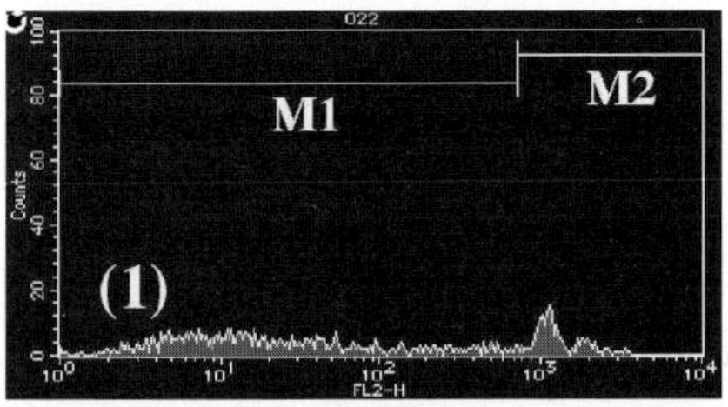

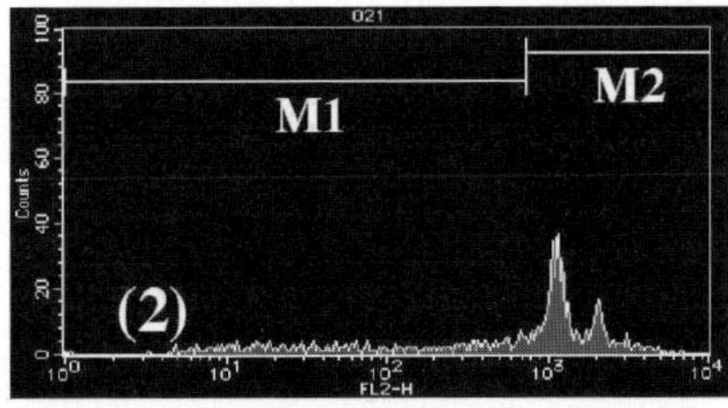

Figure 51.2. Late cell death analysis after fixation of the RPE cells. (1) Cell suspension, (2) Cell sheet.

4. DISCUSSION

The signal transduction from extracellular matrix and some soluble factors play an important role for epithelilal cell survival.[9] Further the extracellular matrix derived from the RPE is reported to be superior to other material such as tissue culture plastic plate or ligands including gelatin or fibronectin.[4] When we collected the RPE from PIPAAm-grafted culture plates, we found more chondroitin sulfate and collagen type IV.

When we examined the RPE as cell sheet from PIPAAm-grafted culture plate, early apoptotic cell death and late cell death were less than those of cells from standard culture plates. Our results may show that rat RPE from PIPAAm-grafted plates as cell sheet may offer better RPE for transplantation than that of standard culture.

5. ACKNOWLEDGMENT

This work was supported in part by grants from Grant-in-Aid for Scientific Research 14370553 (Dr. T. Abe); from the Ministry of Education and Culture of the Japanease Government, Tokyo (Dr. Tamai).

6. REFERENCES

1. LaVail MW, Yasumura D, Matthes MT, et al. Protection of mouse photoreceptors by survival factors in retinal degenerations. *Invest Ophthalmol Vis Sci.* 1998; **39**:592-602.
2. LaVail MM, Unoki K, Yasumura D, et al. Multiple growth factors, cytokines, and neurotrophins rescue photoreceptors from the damaging effects of constant light. *Proc Natl Acad Sci U S A.* 1992; **89**:11249-11253.
3. Kano T, Abe T, Tomita H, et al. Protective Effect against Ischemia and Light Damage of Iris Pigment Epithelial Cells Transfected with the BDNF Gene. *Invest Ophthalmol Vis Sci.* 2002; **43**:3744-3753.
4. Tezel TH, Del Priore LV. Repopulation of different layers of host human Bruch's membrane by retinal pigment epithelial cell grafts. *Invest Ophthalmol Vis Sci.* 1999; **40**:767-774.
5. von Recum H, Kikuchi A, Yamato M, et al. Growth factor and matrix molecules preserve cell function on thermally responsive culture surfaces. *Tissue Eng.* 1999; **5**:251-265.
6. Abe T, Sugano E, Saigo Y, Tamai M. Interleukin-1beta and Barrier Function of Retinal Pigment Epithelial Cells (ARPE-19): Aberrant Expression of Junctional Complex Molecules. *Invest Ophthalmol Vis Sci.* 2003; **44**:4097-4104.
7. Tezel TH, Del Priore LV. Reattachment to a substrate prevents apoptosis of human retinal pigment epithelium. *Graefes Arch Clin Exp Ophthalmol.* 1997; **235**:41-47.
8. Telford WG, King LE, Fraker PJ. Evaluation of glucocorticoid-induced DNA fragmentation in mouse thymocytes by flow cytometry. *Cell Prolif.* 1991; **24**:447-59.
9. Prince JM, Klinowska TC, Marshman E, et al. Cell-matrix interactions during development and apoptosis of the mouse mammary gland in vivo. *Dev Dyn.* 2002; **223**:497-516.

RETINAL TRANSPLANTATION

A treatment strategy for retinal degenerative diseases

Biju B. Thomas[1], Robert B. Aramant[2], SriniVas R. Sadda[1], and Magdalene J. Seiler[1,3,*]

1. ABSTRACT

Retinal transplantation is one among the various treatment strategies aimed to prevent and restore visual loss. Sheets of fetal retina with or without retinal pigment epithelium (RPE) are transplanted into the subretinal space. Retinal transplants have been shown to substantially improve visual responses in rat retinal degeneration models following retinal transplantation, based on behavior and electrophysiology. The transplantation effects may be influenced by several factors such as the age of the recipient at transplantation and the type of species used. Modified functional evaluation techniques permit better understanding of the physiological mechanisms underlying visual improvement in animal models.

2. INTRODUCTION

Retinal degenerations, such as retinitis pigmentosa (RP) (Santos et al., 1997; Humayun et al., 1999) and age related macular degeneration (AMD) (Yates and Moore, 2000), are devastating causes of progressive vision loss and blindness. These diseases primarily affect the photoreceptors or the retinal pigment epithelium (RPE). A variety of approaches to preserve or restore vision are under investigation. Treatment strategies for retinal degeneration are aimed at either preventing photoreceptor loss or restoring vision by replacing the lost photoreceptors and/or the RPE. One of these approaches, retinal transplantation, is discussed in detail in this chapter.

[1]Dept. Ophthalmology, Doheny Retina Institute, Keck School of Medicine, at the University of Southern California, Los Angeles, CA; [2]Dept. Anatomical Sciences & Neurobiology, Univ. of Louisville, Louisville, KY; [3]Dept. Cell and Neurobiology, Keck School of Medicine, at the University of Southern California, Los Angeles, CA. *Corresponding author Magdalene J. Seiler, e-mail: mseiler@doheny.org.

Retinal transplantation is based on the hypothesis that the degenerated retina can be repaired by newly introduced normal RPE and photoreceptor cells that may develop appropriate connections with the still functional part of the host retina (Aramant and Seiler, 2002; 2004).

3. RODENT MODELS OF RETINAL DEGENERATION

Retinal degeneration models have been used to test the effects of retinal transplants. The Royal College of Surgeon (RCS) rat expresses a recessive mutation in the receptor tyrosine kinase, Mertk (D'Cruz et al., 2000) which causes RPE dysfunction with subsequent photoreceptor death. In the rd mouse, a mutation of the rod-specific cGMP-phosphodiesterase (Bowes et al., 1990; Pittler and Baehr, 1991) results in extensive and rapid loss of incompletely formed rods.

Many transgenic rodent models for various retinal degenerative diseases have been developed. La Vail et al., have created different lines of transgenic rats, carrying either the P23H or the S334ter rhodopsin mutation (Steinberg et al., 1997; Liu et al., 1999; Lee et al., 2003). Many more transgenic mouse models are available for various retinal degenerative diseases (Fauser et al., 2002; Wilson and Wensel, 2003).

Excessive absorption of photons by the visual pigment rhodopsin triggers damage to the photoreceptors (Boulton et al., 2001). Moderate continuous blue light exposure for 2-4 days effectively destroys photoreceptors in albino rats while keeping the RPE intact (Seiler et al., 2000).

4. FUNCTIONAL EVALUATION OF VISUAL RESPONSES

Functional evaluation of visual responses is important for assessment of the efficacy of various treatment strategies in animal models of retinal degeneration.

4.1. Fullfield ERG

One of the first methods used to assess visual function has been the ERG (Jiang and Hamasaki, 1994). However, because the ERG reflects the response of the entire retina, it cannot be recommended for evaluating local retinal responses. As a mass response, the ERG averages the rescued area with much larger areas of nonfunctional retina making it difficult to distinguish the response from the rescued area. For various therapeutic interventions, however, functional recovery may be restricted to specific retinal areas.

4.2. Multifocal ERG

The mfERG has been developed to obtain local retinal responses in patients, and has also been used to detect local functional differences in rats (Ball and Petry, 2000) and mice (Nusinowitz et al., 1999). A more advanced mfERG system is now available powered with a fundus camera that may help to locate responses of specific areas in the retina.

4.3. Recording from Visual Brain Centers

Electrophysiological recording from the visual centers such as superior colliculus (SC) and visual cortex can provide an in-depth analysis of response properties at various levels of the visual centers in the central nervous system (CNS). Recording from the SC may be considered as most effective for the evaluation of the visual responses. The surface of the superior colliculus can be easily exposed and maintained for several hours of recording. Different areas of the retina are represented by corresponding areas on the surface of the SC in a retino-collicular map (Siminoff et al., 1966). In rodents, the SC responses represent mostly direct input from the retina and reflect overall ganglion cell output (Lund et al., 2001). Evaluation of response characteristics such as the response onset latency (Thomas et al., 2004a) may reflect the physiological status of the neural retina. Recording from higher visual areas such as the visual cortex (Girman et al., 1999; Coffey et al., 2002; Girman et al., 2003b) may elucidate the adaptive mechanisms that may occur following visual loss and after therapeutic events aimed at improving visual function.

4.4. Visual Behavioral Tests

Behavioral testing for visual responses which are mediated through the central neural circuitry is an equally important approach for evaluating visual function. Various tests have been employed in rodents to evaluate progression of visual loss following retinal degeneration, and to assess the functional effects of various therapeutic interventions (Lund et al., 2001). These include simple startle reflex (del Cerro et al., 1995) and orientation tests (Hetherington et al., 2000), as well as more complex light discrimination and maze tests (Little et al., 1998; Kwan et al., 1999; Prusky et al., 2000; Coffey et al., 2002). Another particularly effective test of visual performance measures an animal's ability to track moving stimuli. This head-tracking (HT) test is based on the optokinetic response, a compensatory eye movement in the direction of the movement of a stimulus. By scoring the total time spent tracking the movements, visual acuity can be measured (Coffey et al., 2002). However this technique is not reliable in albino rats due to their abnormal visual sensory apparatus (Precht and Cazin, 1979).

5. RETINAL TRANSPLANTATION

Retinal transplantation, the transplantation of new photoreceptors either as dissociated cells or retinal sheets into the subretinal space, is aimed at replacing the lost photoreceptors (reviewed in Aramant and Seiler, 2002; Lund et al., 2003; Aramant and Seiler, 2004). Intact sheets of fetal retina, with or without RPE, can be transplanted to the subretinal space in various rat models of retinal degeneration and morphologically repair an area of a damaged retina. (Seiler and Aramant, 1998; Aramant et al., 1999) These transplanted photoreceptors can also respond to light (Seiler et al., 1999).

5.1. Restoration of Visual Responses in the SC by Retinal Sheet Transplants in Various Rodent Models of Retinal Degeneration

It was shown in RCS rats that after transplantation of retina together with RPE, visually evoked multi-unit responses could be recorded in the superior colliculus (SC) from 66%

of the transplanted rats, but only from 46% of sham surgery controls (Woch et al., 2001). Transplant responses were only found in a small area of the SC corresponding to the retinal placement of the transplant. Thus if a rescue effect was the mechanism, it appeared to be highly localized. None of the age-matched RCS rats without surgery showed visual responses. It is important to note that a sham surgery effect in RCS rats has been reported previously (Wen et al., 1995; Humphrey et al., 1997). Visual responses of the transplant rats, however, were more robust, and had shorter latencies and higher amplitudes compared with the sham animals. This study indicated that the visual responses recorded from the transplanted rats originated from the transplant area; the possibility of functional connections between the transplant and the host retina was suggested (Woch et al., 2001).

Because the decline in visually driven activity in the RCS rat is slow (Sauve et al., 2001), another degeneration model was selected to better distinguish the transplant-induced recovery/restoration (Sagdullaev et al., 2003). In *S334ter*-line-3 rats, at the age of transplantation (21-28 days), no visual responses could be recorded within the area of the SC that represents transplant placement. The percentage of transgenic rats with transplants that recovered visual activity (64%) was similar to the percentage reported for RCS rats (66%). Again, visual responses were recorded only from a small area of the SC corresponding to the retinal placement of the graft. None of the age matched control rats including the sham surgery rats showed visual responses in the SC. Qualitative anatomical comparisons suggested that one important predictor of functional outcome is the morphological integrity of the transplant. The lack of a sham surgery effect in S334ter-line 3 rats may be attributed to the difference in the mechanism underlying retinal degeneration in different degeneration models (RCS and line-3 transgenic rats).

To study the long term effect of retinal transplantation, visual responses were recorded from *S334ter*-line-5 rats, a model for slow retinal degeneration (Thomas et al., 2004a). Visual responses can persist for a substantial period of time after neural retinal transplantation because these responses are preserved in the transplanted area of line-5 rats up to 254 days of age. However, none of the "poor" or disorganized transplants had any effect. Further, all the rats with "good" transplants had relatively "short" latency responses suggesting better inner retinal function.

Retinal transplantation studies conducted in *rd* mice (Arai et al., 2004) yielded a 43% success rate (positive responses were recorded from three out of seven eyes studied). Interestingly, the organization of the graft did not appear to correlate as expected with the electrophysiology results in this model. Eyes with well-organized, laminated grafts showed no response whereas the three light-responsive eyes had rosetted or disorganized grafts. All three light-responsive eyes demonstrated much higher levels of recoverin immunoreactivity in the host retina overlying the graft compared with untreated age-matched *rd/rd* mice. Thus, the mechanism of visual restoration after retinal sheet transplantation in this experiment appeared to derive at least in part from a rescue effect on host cones. The lack of an apparent functional rescue effect in the animals with laminated grafts may have been due to a glial barrier between transplant and host, and the loss of cones in the host retina.

5.2. Better Head Tracking Response in Transplanted *S334ter*-Line-3 Rats

The functional capability of retinal transplants in S334ter-line-3 rats was further evaluated by means of optokinetic behavioral tests, a modified optokinetic head tracking apparatus (Thomas et al., 2004b). This apparatus consists of a striped rotating drum, specifically

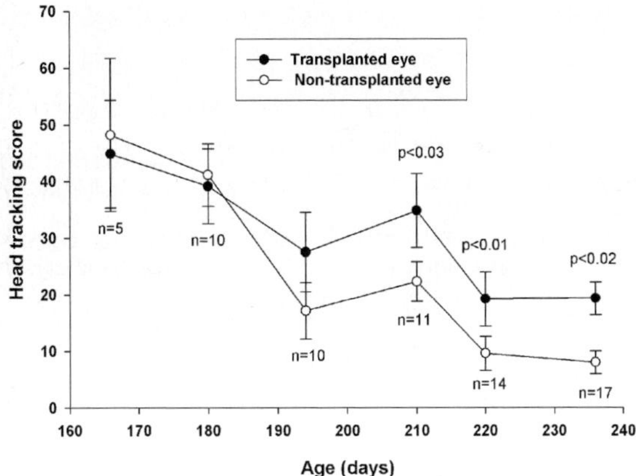

Figure 52.1. Optokinetic Testing (modified method to test each eye separately) of S334-ter-line 3 rats, 166-236 days old. The head tracking score is compared between left and right eye among transplanted S334ter-3 rats. Significant preservation of visual responses was seen in the transplanted eyes at later stages of retinal degeneration. Among non-transplanted rats the progression of visual loss is more symmetrical in both eyes. Taken from Figure 5A of Thomas et al., 2004b. Optokinetic test to evaluate visual acuity of each eye separately. (Reprinted from Journal of Neuroscience Methods, Vol 138, Thomas et al., Optokinetic test to evaluate visual activity acuity of each eye separately, 7-13; 2004, with permission from Elsevier).

modified to measure vision in each eye separately for evaluation of monocular treatments. In line-3 rats with retinal transplants, visual responses are significantly preserved in transplanted eyes at late stages of retinal degeneration (**Fig. 52.1**). The result of the behavioral evaluation corroborates our electrophysiological findings and demonstrates that the visual sensitivity of transplant is more than just light perception.

5.3. Transplanted *S334ter*-Line-3 Rats Show Improved Visual Sensitivity to Low Light

The light stimulus for the above SC recordings in S334ter-line-3 rats (Sagdullaev et al., 2003) consisted of a full-field stimulus with bright light of $1300\,cd/m^2$ ($3.11\,\log cd/m^2$). In the following experiments, the recording stimulus was modified to evaluate the sensitivity of the visual responses (visual threshold). Light stimuli of 100 ms to 1 second duration and intensity -8 to $1\,\log cd/m^2$ were presented on a full field white screen. $-6\,\log cd/m^2$ corresponds to the rod threshold (overcast night sky), $-1\,\log cd/m^2$ corresponds to the cone threshold (dawn). Visual responses from the SC were recorded from pigmented normal as well as retinal degenerate rats (with or without transplants). In normal pigmented rats, the dark adapted visual threshold was $-5.52\,\log cd/m^2$ which is almost similar to previously reported observations in pigmented Long Evans rats (Herreros de Tejada et al., 1992). In transgenic retinal degenerate rats, at 45 days of age, the threshold was elevated to $-0.09\,\log cd/m^2$. On the other hand, the visual threshold in the SC was considerably improved in transplanted line-3 rats (**Fig. 52.2**) in the area corresponding to the placement of the graft in the retina.

Girman et al. (2003a) demonstrated that the mesopic range of normal pigmented rats is at around $-4\,\log cd/m^2$. Once it is possible to determine the scotopic and photopic range of dark adapted rats, it is also possible to assess whether the visual responses originate from

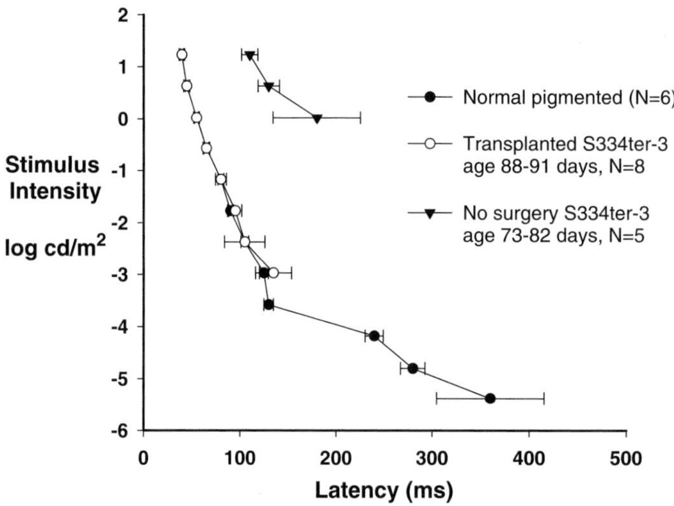

Figure 52.2. Visual responses recorded from the superior colliculus of transplanted rats after low light stimulation. Transplanted rats show similar latency responses as normals down to a stimulus intensity of $-3 \log \text{cd/m}^2$ in the SC area corresponding to the graft placement. No surgery S334ter-3 rats, which contain no rods, show only long latency responses to bright light, and no responses to low light (below $0 \log \text{cd/m}^2$).

the host photoreceptors because the rods are only found in the transplants and not in the host retina. Preliminary data suggest that visual threshold recorded in the transplanted line-3 rats is above the level of a clear rod response. However, the possibility of rod contribution cannot be ruled out for several reasons. The physiological capability of the transplanted photoreceptors may not be expected to be same as normal rods. Also, the neural circuitry formed between the transplant and host retina may influence the light threshold level of the transplant. Also, it should be noted that in a non-transplant rat, the visual threshold at a very early age (45 days) is much higher ($-0.09 \log \text{cd/m}^2$) than the threshold recorded from a transplanted rat. This suggests that restored visual response is not due to photoreceptor rescue but a more direct contribution of the transplanted photoreceptors.

6. ADVANTAGES OF FETAL RETINAL SHEET TRANSPLANTS

Although there are other important treatment strategies aimed at improving visual function, retinal transplantation remains a major hope for those who have lost the photoreceptors. Fetal retinal sheet transplants with or without RPE, can morphologically integrate into a degenerated retina. This has been demonstrated in various retinal degeneration rat models (Seiler and Aramant, 1998; Aramant et al., 1999; Woch et al., 2001; Sagdullaev et al., 2003; Thomas et al., 2004a). Significant improvement in visual sensitivity has been reported following transplantation of a fetal retinal sheet together with its RPE in a human patient (Radtke et al., 2004). This patient improved from hand motion vision to being able to read enlarged letters from a computer screen.

The various advantages of fetal retinal sheet transplantation technique are briefly described here.

6.1. Transplants Reconstruct the Damaged Retina

Using a special implantation instrument and procedure developed by Aramant and Seiler (Aramant and Seiler, 2002), it is possible to transplant fresh intact sheets consisting of 2 tissues: RPE together with the neuroblastic retina. The photoreceptors of intact-sheet transplants develop both inner and outer segments morphologically resembling a normal retina and remain healthy for many months (Seiler and Aramant, 1998; Aramant et al., 1999; Seiler et al., 1999). These transplanted sheets of retinal neuroblastic progenitor cells already have a primordial circuitry established.

6.2. Transplants are Physiologically Active

It has been shown that normal phototransduction processes takes place in the transplanted photoreceptors (Seiler et al., 1999) which are hence capable of transforming light into electrical signals. There are also indications to prove establishment of neural connections between the transplant and the host retina (Seiler et al., in press).

6.3. Trophic Effects of Fetal Retinal Sheets

It has been well established that trophic factors can be used as a treatment strategy to protect photoreceptors from degeneration (Delyfer et al., 2004). Most of the trophic factors are naturally present in a normal healthy retinal environment. Following cell loss, the natural resources for these trophic factors will be depleted and subsequently a degenerating retina is subjected to large-scale secondary changes. Supplementation with one single factor may not be of sufficient help to reverse these changes. The fetal retinal sheets, especially when transplanted together with its RPE can be a good source of many different trophic factors. Hence, fetal retinal sheet transplantation may help to regain the normal retinal homeostasis by protecting the inner retina from further damage.

6.4. Permanence of the Transplants

For any therapeutic strategy, it is very important that the functional recovery is long-lasting. A significant advantage of the fetal retinal transplants is that they can remain safely in the host environment for an extended period of time. This may be due to the relative immunological privilege of the subretinal space. The safety of the transplants in the subretinal space has been shown in animal models (Thomas et al., 2004a) as well as in human patients (Radtke et al., 2002; Radtke et al., 2004). The functional effects of transplants can be maintained through a large part of the life span of the experimental animals (Woch et al., 2001; Sagdullaev et al., 2003; Thomas et al., 2004a). A report from a clinical trial showed improvement in visual sensation in some patients even years after surgery (Radtke et al., 2004).

6.5. Transplants Provide Mechanical Support to the Degenerating Retina

There are recent reports on elaborate changes taking place in the retina following degeneration (Jones et al., 2003; Marc and Jones, 2003; Marc et al., 2003). The loss of photoreceptors and the subsequent change in the retinal thickness can be one of the factors that trigger this global remodeling. The fetal retinal sheet transplants as a natural replacement for lost photoreceptors may provide the most suitable mechanical support to prevent retinal collapse.

7. FUTURE DIRECTIONS

Fetal retinal sheet transplantation has been shown to have a rescue and restoration effect on visual function in various rodent models with retinal degeneration. Clinical studies indicate that fetal retinal transplants with its RPE can survive without apparent rejection and show improvement in visual acuity in some patients. Additional work is in progress attempting to confirm the reproducibility of these results and enhance the magnitude of the effect. Current and future studies are aiming at improving the interconnections between the transplant and the host retina. Elimination of the inner limiting membrane on the surface of the donor retina that can act as a barrier might lead to improvements in the connectivity between transplant and host. Connectivity could be also improved using appropriate factors that promote neurogenesis and reduce glial scars and trauma.

8. ACKNOWLEDGEMENTS

Supported by Foundation Fighting Blindness, Foundation for Retinal Research, Fletcher Jones Foundation, NIH EY03040 and Private Funds, and an NEI travel grant to BT. For space reasons, only some of the people involved in the studies discussed in this chapter can be mentioned. We thank Zhenhai Chen, Xiaoji Xu, Lilibeth Lanceta, and Betty Nunn for technical assistance. Guanting Qiu (Doheny Eye Institute, Los Angeles, CA); Shinichi Arai (Niigata University, Niigata, Japan); Botir T. Sagdullaev (Washington University, St. Louis); Norman D. Radtke, Heywood M. Petry, Maureen A. McCall, Peng Yang (University of Louisville); Gustaw Woch (Hershey Medical School, Hershey, PA); and Sherry L. Ball (Cleveland University, Cleveland, OH) contributed to the work presented in this paper. We thank Matthew M. LaVail, UCSF, for the founder breeding pairs of transgenic S334ter rats, and Eric Sandgren, University of Wisconsin, for founder breeders of transgenic hPAP rats.

Robert Aramant and Magdalene Seiler have a proprietary interest in the implantation instrument and procedure.

9. REFERENCES

Arai, S., Thomas, B. B., Seiler, M. J., Aramant, R. B., Qiu, G., Mui, C., De Juan, E. and Sadda, S. R., 2004. Restoration of visual responses following transplantation of intact retinal sheets in rd mice. *Exp Eye Res*, **79**:331-341

Aramant, R. B. and Seiler, M. J., 2002. Retinal transplantation–advantages of intact fetal sheets. *Prog Retin Eye Res*, **21**:57-73

Aramant, R. B. and Seiler, M. J., 2004. Progress in retinal sheet transplantation. *Prog Retin Eye Res*, **23**:475-494

Aramant, R. B., Seiler, M. J. and Ball, S. L., 1999. Successful cotransplantation of intact sheets of fetal retina with retinal pigment epithelium. *Invest Ophthalmol Vis Sci*, **40**:1557-1564

Ball, S. L. and Petry, H. M., 2000. Noninvasive assessment of retinal function in rats using multifocal electroretinography. *Invest Ophthalmol Vis Sci*, **41**:610-617

Boulton, M., Rozanowska, M. and Rozanowski, B., 2001. Retinal photodamage. *J Photochem Photobiol B*, **64:** 144-161

Bowes, C., Li, T., Danciger, M., Baxter, L. C., Applebury, M. L. and Farber, D. B., 1990. Retinal degeneration in the rd mouse is caused by a defect in the beta subunit of rod cGMP-phosphodiesterase. *Nature*, **347:**677-680

Coffey, P. J., Girman, S., Wang, S. M., Hetherington, L., Keegan, D. J., Adamson, P., Greenwood, J. and Lund, R. D., 2002. Long-term preservation of cortically dependent visual function in RCS rats by transplantation. *Nat Neurosci*, **5:**53-56

D'Cruz, P. M., Yasumura, D., Weir, J., Matthes, M. T., Abderrahim, H., LaVail, M. M. and Vollrath, D., 2000. Mutation of the receptor tyrosine kinase gene Mertk in the retinal dystrophic RCS rat. *Hum Mol Genet*, **9:** 645-651

del Cerro, M., DiLoreto, D., Jr., Cox, C., Lazar, E. S., Grover, D. A. and del Cerro, C., 1995. Neither intraocular grafts of retinal cell homogenates nor live non-retinal neurons produce behavioral recovery in rats with light-damaged retinas. *Cell Transplant*, **4:**133-139

Delyfer, M. N., Leveillard, T., Mohand-Said, S., Hicks, D., Picaud, S. and Sahel, J. A., 2004. Inherited retinal degenerations: therapeutic prospects. *Biol Cell*, **96:**261-269

Fauser, S., Luberichs, J. and Schuttauf, F., 2002. Genetic animal models for retinal degeneration. *Surv Ophthalmol*, **47:**357-367

Girman, S. V., Lu, B. and Lund, R. D., 2003a. Light Adaptation Study in RCS Rats, Untreated and with Subretinal Graft of Human RPE Cells. *ARVO abstract*:program #482

Girman, S. V., Sauve, Y. and Lund, R. D., 1999. Receptive field properties of single neurons in rat primary visual cortex. *J Neurophysiol*, **82:**301-311

Girman, S. V., Wang, S. and Lund, R. D., 2003b. Cortical visual functions can be preserved by subretinal RPE cell grafting in RCS rats. *Vision Res*, **43:**1817-1827

Herreros de Tejada, P., Green, D. G. and Munoz Tedo, C., 1992. Visual thresholds in albino and pigmented rats. *Vis Neurosci*, **9:**409-414

Hetherington, L., Benn, M., Coffey, P. J. and Lund, R. D., 2000. Sensory capacity of the royal college of surgeons rat. *Invest Ophthalmol Vis Sci*, **41:**3979-3983

Humayun, M. S., Prince, M., de Juan, E., Jr., Barron, Y., Moskowitz, M., Klock, I. B. and Milam, A. H., 1999. Morphometric analysis of the extramacular retina from postmortem eyes with retinitis pigmentosa. *Invest Ophthalmol Vis Sci*, **40:**143-148

Humphrey, M. F., Chu, Y., Mann, K. and Rakoczy, P., 1997. Retinal GFAP and bFGF expression after multiple argon laser photocoagulation injuries assessed by both immunoreactivity and mRNA levels. *Exp Eye Res*, **64:**361-369

Jiang, L. Q. and Hamasaki, D., 1994. Corneal electroretinographic function rescued by normal retinal pigment epithelial grafts in retinal degenerative Royal College of Surgeons rats. *Invest Ophthalmol Vis Sci*, **35:** 4300-4309

Jones, B. W., Watt, C. B., Frederick, J. M., Baehr, W., Chen, C. K., Levine, E. M., Milam, A. H., Lavail, M. M. and Marc, R. E., 2003. Retinal remodeling triggered by photoreceptor degenerations. *J Comp Neurol*, **464:** 1-16

Kwan, A. S., Wang, S. and Lund, R. D., 1999. Photoreceptor layer reconstruction in a rodent model of retinal degeneration. *Exp Neurol*, **159:**21-33

Lee, D., Geller, S., Walsh, N., Valter, K., Yasumura, D., Matthes, M., LaVail, M. and Stone, J., 2003. Photoreceptor degeneration in Pro23His and S334ter transgenic rats. *Adv Exp Med Biol*, **533:**297-302

Little, C. W., Cox, C., Wyatt, J., del Cerro, C. and del Cerro, M., 1998. Correlates of photoreceptor rescue by transplantation of human fetal RPE in the RCS rat. *Exp Neurol*, **149:**151-160

Liu, C., Li, Y., Peng, M., Laties, A. M. and Wen, R., 1999. Activation of caspase-3 in the retina of transgenic rats with the rhodopsin mutation s334ter during photoreceptor degeneration. *J Neurosci*, **19:**4778-4785

Lund, R. D., Kwan, A. S., Keegan, D. J., Sauve, Y., Coffey, P. J. and Lawrence, J. M., 2001. Cell transplantation as a treatment for retinal disease. *Prog Retin Eye Res*, **20:**415-449

Lund, R. D., Ono, S. J., Keegan, D. J. and Lawrence, J. M., 2003. Retinal transplantation: progress and problems in clinical application. *J Leukoc Biol*, **74:**151-160

Marc, R. E. and Jones, B. W., 2003. Retinal remodeling in inherited photoreceptor degenerations. *Mol Neurobiol*, **28:**139-147

Marc, R. E., Jones, B. W., Watt, C. B. and Strettoi, E., 2003. Neural remodeling in retinal degeneration. *Prog Retin Eye Res*, **22:**607-655

Nusinowitz, S., Ridder, W. H., 3rd and Heckenlively, J. R., 1999. Rod multifocal electroretinograms in mice. *Invest Ophthalmol Vis Sci*, **40:**2848-2858

Pittler, S. J. and Baehr, W., 1991. Identification of a nonsense mutation in the rod photoreceptor cGMP phospho-diesterase beta-subunit gene of the rd mouse. *Proc Natl Acad Sci U S A*, **88:**8322-8326

Precht, W. and Cazin, L., 1979. Functional deficits in the optokinetic system of albino rats. *Exp Brain Res*, **37:** 183-186

Prusky, G. T., West, P. W. and Douglas, R. M., 2000. Behavioral assessment of visual acuity in mice and rats. *Vision Res*, **40:**2201-2209

Radtke, N. D., Aramant, R. B., Seiler, M. J., Petry, H. M. and Pidwell, D., 2004. Vision change after sheet transplant of fetal retina with retinal pigment epithelium to a patient with retinitis pigmentosa. *Arch Ophthalmol*, **122:**1159-1165

Radtke, N. D., Seiler, M. J., Aramant, R. B., Petry, H. M. and Pidwell, D. J., 2002. Transplantation of intact sheets of fetal neural retina with its retinal pigment epithelium in retinitis pigmentosa patients. *Am J Ophthalmol*, **133:**544-550

Sagdullaev, B. T., Aramant, R. B., Seiler, M. J., Woch, G. and McCall, M. A., 2003. Retinal transplantation-induced recovery of retinotectal visual function in a rodent model of retinitis pigmentosa. *Invest Ophthalmol Vis Sci*, **44:**1686-1695

Santos, A., Humayun, M. S., de Juan, E., Jr., Greenburg, R. J., Marsh, M. J., Klock, I. B. and Milam, A. H., 1997. Preservation of the inner retina in retinitis pigmentosa. A morphometric analysis. *Arch Ophthalmol*, **115:** 511-515

Sauve, Y., Girman, S. V., Wang, S., Lawrence, J. M. and Lund, R. D., 2001. Progressive visual sensitivity loss in the Royal College of Surgeons rat: perimetric study in the superior colliculus. *Neuroscience*, **103:**51-63

Seiler, M. J. and Aramant, R. B., 1998. Intact sheets of fetal retina transplanted to restore damaged rat retinas. *Invest Ophthalmol Vis Sci*, **39:**2121-2131

Seiler, M. J., Aramant, R. B. and Ball, S. L., 1999. Photoreceptor function of retinal transplants implicated by light-dark shift of S-antigen and rod transducin. *Vision Res*, **39:**2589-2596

Seiler, M. J., Liu, O. L., Cooper, N. G., Callahan, T. L., Petry, H. M. and Aramant, R. B., 2000. Selective photoreceptor damage in albino rats using continuous blue light. A protocol useful for retinal degeneration and transplantation research. *Graefes Arch Clin Exp Ophthalmol*, **238:**599-607

Seiler, M. J., Sagdullaev, B. T., Woch, G., Thomas, B. B. and Aramant, R. B., in press. Transsynaptic virus tracing from host brain to subretinal transplants. *Eur J Neurosci*

Siminoff, R., Schwassmann, H. O. and Kruger, L., 1966. An electrophysiological study of the visual projection to the superior colliculus of the rat. *J Comp Neurol*, **127:**435-444

Steinberg, R. H., Matthes, M. T., Yasumura, D., Lau-Villacorta, C., Nishikawa, S., Cao, W., Flannery, J. G., Naash, M., Chen, J. and LaVail, M. M., 1997. Slowing by survival factors of inherited retinal degeneration in transgenic rats with mutant opsin genes. *ARVO abstract:*S226

Thomas, B. B., Seiler, M. J., Sadda, S. R. and Aramant, R. B., 2004a. Superior colliculus responses to light – preserved by transplantation in a slow degeneration rat model. *Exp Eye Res*, **79:**29-39

Thomas, B. B., Seiler, M. J., Sadda, S. R., Coffey, P. J. and Aramant, R. B., 2004b. Optokinetic test to evaluate visual acuity of each eye independently. *J Neurosci Methods*, **138:**7-13

Wen, R., Song, Y., Cheng, T., Matthes, M. T., Yasumura, D., LaVail, M. M. and Steinberg, R. H., 1995. Injury-induced upregulation of bFGF and CNTF mRNAS in the rat retina. *J Neurosci*, **15:**7377-7385

Wilson, J. H. and Wensel, T. G., 2003. The nature of dominant mutations of rhodopsin and implications for gene therapy. *Mol Neurobiol*, **28:**149-158

Woch, G., Aramant, R. B., Seiler, M. J., Sagdullaev, B. T. and McCall, M. A., 2001. Retinal transplants restore visually evoked responses in rats with photoreceptor degeneration. *Invest Ophthalmol Vis Sci*, **42:**1669-1676

Yates, J. R. and Moore, A. T., 2000. Genetic susceptibility to age related macular degeneration. *J Med Genet*, **37:** 83-87

MICROARRAY ANALYSIS REVEALS RETINAL STEM CELL CHARACTERISTICS OF THE ADULT HUMAN EYE

For contributed volumes

Brigitte Angénieux[1], Lydia Michaut[2], Daniel F. Schorderet[3], Francis L. Munier[1], Walter Gehring[2], and Yvan Arsenijevic[1,4]

1. INTRODUCTION

In western countries, retinitis pigmentosa (RP) affects 1/3,500 individuals and age related macula degeneration (AMD) affects 1% to 3% of the population aged over 60. *In vitro* generation of retinal cells is thus a promising tool to screen protective drugs and to provide an unlimited cell source for transplantation. However, one main limitation is the amount of cells available. Stem cells, that can generate unlimited quantity of cells, could overcome this hurdle. Indeed, stem cells are defined by three characteristics: the ability to produce a large population of cells (expansion) and the potency, to produce the differentiated cells composing the organ from which the stem cells are originated. They are also able to self-renew indefinitely: for instance haematopoietic stem cells, located in the bone marrow, can expand, divide and generate differentiated cells into the diverse lineages throughout the life, the stem cells conserving its status (Till *et al*, 1961). Intestinal stem cells also are able to regenerate the intestine all along life (Potten *et al*, 1975). The other stem cells properties are the ability to produce a large population of cells (expansion) and as well as the differentiated cells composing the organ from which they originated.

Some species, such as the salamanders, can regenerate their retina (Haynes *et al*, 2004) which is not the case for the human retina. Whether this lack of regeneration is an evolutionary capacity loss, due to a blocking mechanism or a missing signal, is an important question to address. Nevertheless, production of retinal cells can be observed even in species

[1] Oculogenetics Unit, Hôpital Ophtalmique Jules Gonin, Lausanne, Switzerland; [2] Dpt of Cell Biology, Biozentrum, Basel, Switzerland; [3] IRO, Institut de Recherche en Ophtalmologie, Sion, Switzerland; [4] Corresponding author: 15 avenue de France, Case Postale 133, CH-1000 Lausanne 7, Switzerland; Tel: +41 21 626 8260; Fax: +41 21 626 8888; yvan.arsenijevic@ophtal.vd.ch.

where no regeneration occurs. The eye of some fishes and amphibians continues growing during adulthood due to the persistent activity of retinal stem cells (RSCs). In fishes, the RSCs are located in the ciliary margin zone (CMZ) at the periphery of the retina, at the proximity of the iris. Although the adult mammalian eye does not grow during adult life, Tropepe *et al* (2000) have shown that the adult mouse eye contains retinal stem cells in the homologous zone (termed the ciliary margin zone), in the pigmented epithelium and not in the neuroretina. We demonstrated that the adult human eye also contains retinal stem cells in the same region (*i.e.* the *pars plicata* and the *pars plana*, Coles *et al*, 2004). These RSCs meet the criteria of stem cells *i.e.* they can differentiate *in vitro* into all retinal cells, expand and self-renew. Here, we further characterize the human adult retinal stem cells to investigate whether the adult retinal stem cells share common characteristics with other stem cell populations.

2. ISOLATION OF HUMAN RETINAL STEM CELLS

Stem cells and human retinal progenitors were obtained from organ donors in accordance to their wishes or those of their families. The obtaining and the use of the tissues agreed with the guidelines provided by the Ethical Committee of the Lausanne University School of Medicine. The tissues were processed 10 to 24 hours *post-mortem.*

The *pars plana* (figure 53.1: white square) and the *pars plicata* (figure 53.1: grey square) both gave rise to clonal spheres *in vitro* in the presence of either EGF, FGF-2 or only with insulin after one week (Figure 53.2A sphere formed after one week in the presence of EGF). Stem cells can be isolated regardless of the age of the donor (from 2 to 82 years old, n = 21, figure 53.2B) suggesting that the stem cell pool is conserved throughout life. We expanded the cells by placing one clonal single sphere into a 24-wells plate in the presence of EGF and 10% FBS. After 3 days, the cells began to spread out of the sphere leading to a monolayer culture in less than one month (Figure 53.2C). The monolayer culture contains highly proliferative progenitor cells that can generate over 300 million cells within one month (Figure 53.2D). During proliferation, 89 +/− 0,02% (n = 3) of the cells express nestin (Figure 53.2E, n = 3), a marker of undifferentiated cells (Tohyama *et al.*, 1992).

3. HUMAN RETINAL STEM CELLS CHARACTERIZATION

To further characterize this cell population, we performed an expression analysis using the Affymetrix® U133 Plus 2.0 GeneChip. We analyzed the global transcriptional expres-

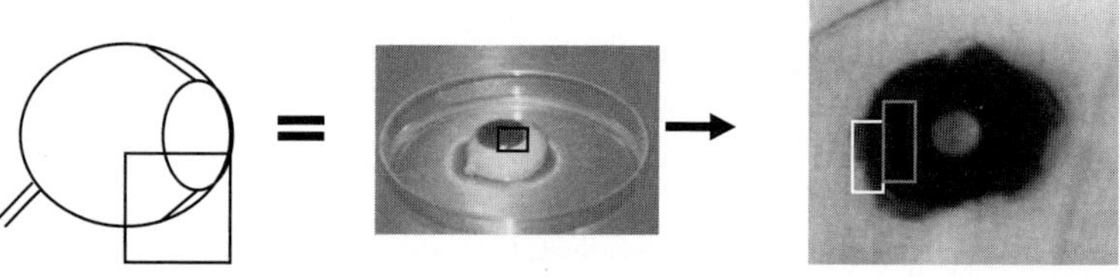

Figure 53.1. Location of the *pars plicata (white square)* and *plana (red square)* in the human eye.

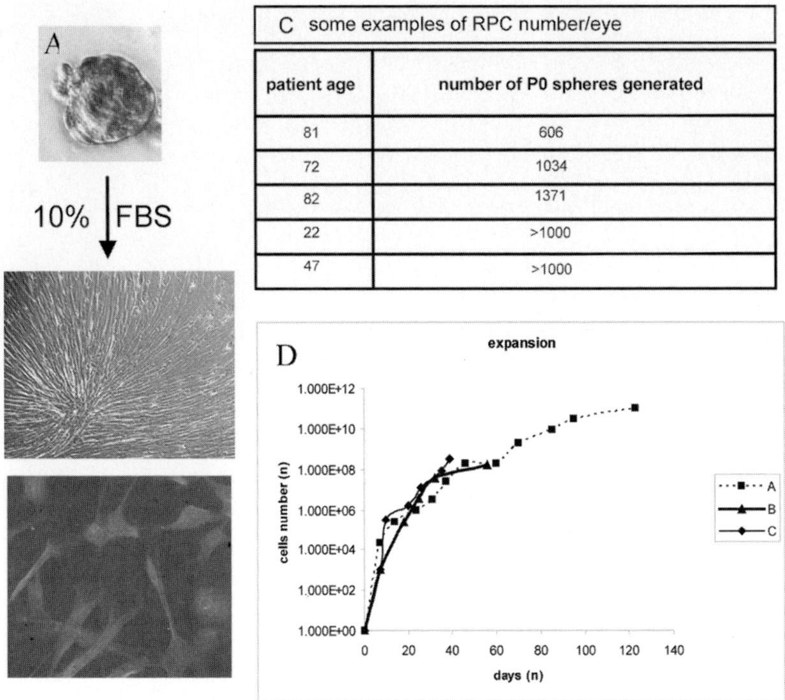

Figure 53.2. The adult human eye contains highly proliferative undifferentiated cells.

sion of expanded and differentiated hRPCs from two different donors (E and L). During expansion (*i.e.* in the presence of EGF and 10% FBS), RSCs expressed markers present in neural and other stem cells: as evidenced by immunostaining (Figure 53.2E), nestin was detected at a high expression level during the expansion, validating the microarray approach. We also observed the expression of ABCG2 which is present in hematopoietic stem cells (Zhou *et al*, 2003), or Bmi1 which is required for hematopoietic stem cell (Park *et al*, 2003, Lessard *et al*, 2003) and neural stem cell renewal (Molofsky *et al*, 2003), or nucleostemin which is necessary for NSC renewal (Tsai and McKay, 2002). The expression of Bmi1 was confirmed by RT-PCR (data not shown). Thus, the highly proliferative cells derived from the adult human eye share common characteristics with other stem cells.

The ability of these cells to differentiate into the different cell types composing the organ from which they have been isolated is one of the stem cell characteristics. Thus, to portray the progeny of the RSCs we also analysed their gene expression profile after differentiation (*i.e.* after withdrawal of FBS and stimulation by EGF). The cells express genes that have different functions in neurons such as genes coding for the skeleton (Map2, Tau), for the synaptic vesicles proteins (SNAPAP). Moreover specific markers of retinal cells are also expressed as shown by the presence of RPE65 (protein are expressed in retinal pigmented cells), opsin, and peripherin (specific to photoreceptors). In one of the cell line, we also detected by immunostaining, cells that have differentiated into retinal neurons and which expressed specific proteins such as calbindin for horizontal cells (Figure 53.3A), syn-

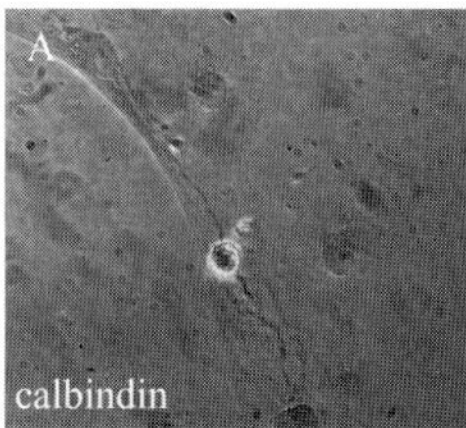

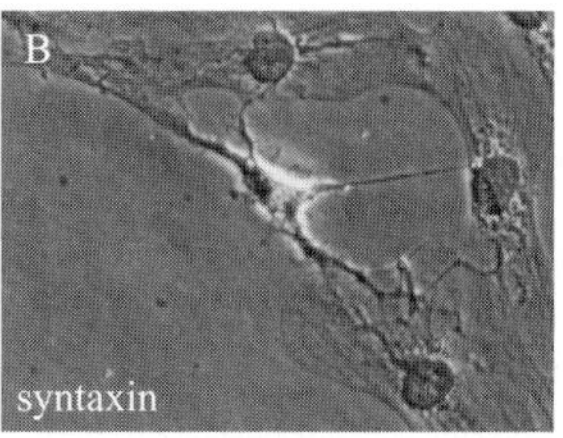

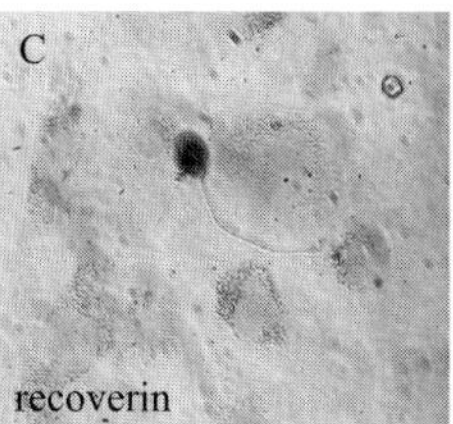

Figure 53.3. The RSCs differentiate into retinal neurons.

taxin for amacrine cells (Figure 53.3B), recoverin for photoreceptor and some bipolar cells (Figure 53.3C) and vimentin for glial cells (data not shown).

4. CONCLUSION

We showed that adult human eye contains cells that can easily be expanded (one cell giving rise to more than 100 billion cells) and share common characteristics with other stem cell populations. Furthermore, after differentiation, those cells expressed markers of several retinal cell types such as pigmented epithelium (RPE65), glia (vimentin) or retinal neurons (opsin, recoverin). The potential of generating retinal neurons *in vitro* is a promising tool. Indeed the human macula is composed of 200,000 cells which represent only 1/1,000,000 of the total number of cells that we can generate from one RSC. Thus, RSCs could circumvent the problem of the cell number limitation for transplantation studies and could serve to dissect the mechanisms leading to the generation of neurons and their survival.

5. REFERENCES

Coles BL, Angenieux B, Inoue T, Del Rio-Tsonis K, Spence JR, McInnes RR, Arsenijevic Y, van der Kooy: Facile isolation and the characterization of human retinal stem cells. *PNAS*. 2004, **101**(44).

Lessard J, Sauvageau G.: Bmi-1 determines the proliferative capacity of normal and leukaemic stem cells. *Nature* 2003, **423**(6937).

Molofsky AV, Pardal R, Iwashita T, Park IK, Clarke MF, Morrison SJ.: Bmi-1 dependence distinguishes neural stem cell self-renewal from progenitor proliferation. *Nature* 2003, **425**(6961).

Park IK, Qian D, Kiel M, Becker MW, Pihalja M, Weissman IL, Morrison SJ, Clarke MF.: Bmi-1 is required for maintenance of adult self-renewing haematopoietic stem cells. *Nature* 2003, **423**(6937).

Potten CS, Hendry JH.: Differential regeneration of intestinal proliferative cells and cryptogenic cells after irradiation, *Int J Radiat Biol Relat Stud Phys Chem Med*. 1975, **27**(5).

Till J, and McCulloch E.: A direct measurement of the radiation sensitivity of normal mouse bone marrow cells., *Radiat. Res* 1961, **14**(213).

Tohyama T, Lee VM, Rorke LB, Marvin M, McKay RD, Trojanowski JQ.: Nestin expression in embryonic human neuroepithelium and in human neuroepithelial tumor cells. *Lab Invest* 1992, **66**(3).

Tsai RY, McKay RD.: A nucleolar mechanism controlling cell proliferation in stem cells and cancer cells. *Genes Dev.*, 2002 **16**(23).

Tropepe V, Coles BL, Chiasson BJ, Horsford DJ, Elia AJ, McInnes RR, van der Kooy D.: Retinal stem cells in the adult mammalian eye. *Science* 2000, **287**(5460).

Zhou S, Schuetz JD, Bunting KD, Colapietro AM, Sampath J, Morris JJ, Lagutina I, Grosveld GC, Osawa M, Nakauchi H, Sorrentino BP.: The ABC transporter Bcrp1/ABCG2 is expressed in a wide variety of stem cells and is a molecular determinant of the side-population phenotype. *Nat Med* 2001, **7**(9).

CHAPTER 54

USING STEM CELLS TO REPAIR THE DEGENERATE RETINA

Stem cells in the context of retinal degenerations

Christine M. Hall[1,3], Anthony Kicic[2], Chooi-May Lai[1,3], and P. Elizabeth Rakoczy[1,3]

1. INTRODUCTION

It is conceivable that in the future, stem cells will be used in the treatment of retinal degenerative conditions. They could be transplanted to replace the photoreceptor victims of retinal degeneration or to function as vehicles for the provision of survival or regenerative factors. Their full potential will, most likely, only be realized with thorough investigations of both embryonic and adult stem cells. Studies into the use of stem cells for the treatment of retinal degenerations have primarily involved retinal and neural stem cells. A few laboratories, including our own, have investigated allografts of stem cells from more distant sites like the bone marrow to the retina. Still other groups have investigated the application of embryonic stem cells for treatment of retinal degenerations. In this mini-review we will compare adult versus embryonic stem cells for use in the retina and summarize the recent investigations using stem cells to repair the degenerating retina. Due to space limitations, we are unable to cite all relevant papers so have chosen to focus on key articles that have made significant contributions to this field.

2. STEM CELLS FOR THE TREATMENT OF RETINAL DEGENERATIONS

Stem cells may be the ideal candidates for cell replacement therapy. Stem cells (SC) are highly plastic cells that are capable of becoming different types of cells and retain the

[1] Centre for Ophthalmology and Visual Science, The University of Western Australia; [2] Department of Respiratory Medicine, Princess Margaret Hospital for Children; [3] Stem Cell Unit, Department of Molecular Ophthalmology, Lions Eye Institute, Nedlands, 6009, Western Australia, Australia. Corresponding author: P.E. Rakoczy, E-mail: rakoczy@cyllene.uwa.edu.au

Table 54.1. Comparison of embryonic vs. adult stem cell characteristics.

Embryonic Stem Cells	Adult Stem Cells
Pluripotent	Multipotent
Unlimited proliferative capacity	Limited proliferative capacity
Ethically complicated	Ethically valid
Derived from 2 sources – inner cell mass of blastocyst or gonads of foetus	Derived from many sources – e.g. brain, eye, bone marrow
Less likely to contain DNA abnormalities	May contain more DNA abnormalities
Can cause teratomas when transplanted	Less evidence of teratoma formation when transplanted
Problem of immune rejection	Can be derived autologously so less likelihood of immune rejection
Can be rapidly expanded *in vitro*	Some sites contain very small numbers of stem cells

capacity for self-renewal. There are advantages and disadvantages associated with the use of both embryonic (ESC) and adult stem cells (ASC) in cell-based therapeutic strateies to treat degenerative disease and these have been summarised in Table 54.1. It is beyond the scope of this article to comprehensively review these aspects in detail and we therefore refer the reader to several excellent articles and reviews specifically dealing with these topics (Daar et al., 2004; Hochedlinger et al., 2004).

The potential of stem cells in a cell replacement strategy for the treatment of degenerative disease has been recognised from the results of both *in vitro* and *in vivo* analyses. *In vitro* analyses of photoreceptor sheets, retinal explants, primary retinal pigment epithelium (RPE), cultured RPE, and of stem cells themselves have contributed to the belief that stem cells may have a role in the treatment of retinal degenerations. Photoreceptor differentiation has been analyzed by co-culture of stem cells with retinal cells from the embryonic retina (Belliveau et al., 2000), neonatal retina (Ahmad et al., 1999; Ahmad et al., 2000), RPE (Chiou et al., 2005), and stromal cells (Hirano et al., 2003). Other studies inducing the differentiation of marrow stromal cells (MSC) into mesodermal, neuroectodermal and endodermal derivatives have also substantiated the ability of stem cells to differentiate into the lineages required for cell replacement in the degenerate retina (Jiang et al., 2002; Woodbury et al., 2002). Perhaps one of the most informative studies demonstrating the potential of stem cells for the treatment of retinal degenerative disease is that of Zhao et al. (2002) where they showed that ESC-derived neural cells express photoreceptor regulatory genes.

Investigations of the viability, integration, and functionality of transplanted retinal tissue have shown promise for possible treatment of retinal degenerations. Transplanted tissue has included fetal (Sagdullaev et al., 2003), and adult sheets or dissociated photoreceptors (Gouras et al., 1991), RPE (Gouras et al., 1989; Wang et al., 2004), full-thickness neuroretina (Ghosh et al., 2004), and microaggregates of neural retina (Gouras and Tanabe, 2003). In a recent study, the activity in the superior colliculus was analysed following transplant of embryonic rat retinal sheets to the retina of a transgenic rat with photoreceptor degeneration where the transplant was shown to improve visual function (Sagdullaev et al., 2003). Using a mouse model, similar results were found to be associated with the rescue of host cones in the transplant recipients (Arai et al., 2004). Whilst such results are promising, there is evidence that rescue of visual function may only be possible if the RPE cells are transplanted prior to the major loss of photoreceptors (Li and Turner, 1991; Castillo et al., 1997). The small number of cells harvested from fetal ocular tissue, and the ethical issues surrounding the use of fetal tissue, present major obstacles to the clinical application of such cell replacement therapy.

3. RECENT INVESTIGATIONS USING STEM CELLS TO REPAIR THE DEGENERATE RETINA

3.1. Transplantation of Embryonic Stem Cells

A number of different laboratories have studied the effects of transplanting ESC into the subretinal space (Schraermeyer et al., 2001; Arnhold et al., 2004; Haruta et al., 2004) and into the vitreous of the eye (Hara et al., 2004; Meyer et al., 2004). Transplantation of undifferentiated murine and primate ESC into the dystrophic RCS rat (Schraermeyer et al., 2001; Haruta et al., 2004) both resulted in establishing a multilayered photoreceptor outer nuclear layer (ONL) compared to a non-existent ONL in the sham-injected animals. Haruta and associates (2004) also presented evidence for a concomitant improvement in visual function. Recently, Meyer et al. (2004) transplanted retinoic acid-induced, ESC-derived, neural precursors intravitreally into five week old *rd1* mice. A small proportion of the transplanted cells were able to cross the inner limiting membrane, incorporate into the neural retina and to differentiate into cells expressing neural cell markers whilst resembling neural cell morphology. In the age-matched normal C57BL/6J mouse model, the donor cells remained in the vitreous distant from the retinal surface. This suggests that chemo-attractant or migration factors are present in the degenerate retinas that are not present in the normal model. Arnhold et al. (2004) also differentiated the ESC into neural precursors prior to subretinal transplantation into rhodopsin-knockout mice. The transplanted cells did not appear to disrupt the cellular laminae until 8 weeks post-transplant when teratomas were seen to affect the choroid, retina, and vitreous of host eyes. Hara et al. (2004) also saw teratoma formation after transplantation of ESC intravitreally in mice and this appears to be a major obstacle to the application of ESC in cell replacement therapy.

3.2. Transplantation of Adult-Derived Stem Cells

3.2.1. Retinal-Derived Stem Cells

As with ESC, the subretinal transplantation of retinal stem cells (RSC) into rats with disease or injury has shown the survival, integration and differentiation into cells expressing markers of the neural retina, including photoreceptors (Akagi et al., 2003; Chacko et al., 2003; Klassen et al., 2004a). Injection into normal animal models appears to inhibit the incorporation of such cells into the retina (Chacko et al., 2000; Chacko et al., 2003). Intravitreal injection of RSC into immunocompetent mice (Coles et al., 2004) and rats (Seigel et al., 1998) have shown that RSC survive, migrate from the vitreous and integrate into the neural retina. The characteristics of retinal and neural stem cells (NSC), their developmental roles, molecular mechanisms known to control their specification, and transplantation of NSC into the retina are all described in a comprehensive review by Klassen et al. (2004b).

3.2.2. Brain-Derived Stem Cells

Most of the NSC transplant studies in the retina have focused on the survival, integration, and differentiation of the transplanted cells. Again, a lack of integration and differentiation was noted when NSC were transplanted into normal rat retinas (Kurimoto et al., 2001). Intravitreal injection of adult rat NSC into mechanically injured rat retinae differentiated into cells expressing markers of astrocytes and neurons together with some evidence of synapse-like formations between graft and host cells but did not express markers for photoreceptors (Nishida et al., 2000). Similar results were obtained by Young et al. (2000), subsequent to intravitreal injection of NSC into the dystrophic RCS rat retina. Pressmar et al. (2001) transplanted NSC into the retinae of β2/β1 knock-in mice that serve as a model of photoreceptor apoptosis and into normal mice. Transplanted cells were shown to differentiate into astrocytes and oligodendrocytes and the mutant mice were seen to contain more grafted cells than the wild type mice. Transplantation of mouse NSC into Brazilian opossum retinae (Van Hoffelen et al., 2003) also showed that the age of the host determined the fate of the differentiating cells *in vivo*.

3.2.3. Bone Marrow-Derived Stem Cells

Intravitreal injection of MSC into injured adult Brown Norway rats showed incorporation, mostly in the ONL in the vicinity of the site of injury, and differentiation into retinal neural cells (Tomita et al., 2002). Otani et al. (2002) demonstrated that MSC transplant into the eye rescued the retinal vasculature from degeneration. The rescued vasculature was seen to be responsible for the preservation of retinal nuclear layer thickness and improved ERG recordings following intravitreal injection of MSC into mouse models of retinal degeneration, *rd1* and *rd10* (Otani et al., 2004). The rescued retina was found to consist of mostly cones as opposed to the normal retina which is composed of mostly rods, and microarray analysis of the rescued retinae found up-regulation of many anti-apoptotic genes. Additional substantiation of MSC suitability for treatment of neural degenerative disease is provided by the fact that Priller et al. (2001) found bone marrow-derived fully differentiated Purkinje neurons in the brains of lethal irradiated C57BL/6J mice that had received tail vein injections of MSC.

4. ADVANTAGES AND DISADVANTAGES OF MSC FOR THE TREATMENT OF RETINAL DEGENERATIONS

The pool of MSC is significantly larger and more accessible than the pools of SC in the eye or brain hence our laboratory chose to analyze MSC in more depth. As outlined in section two, there exists both benefits and disadvantages for the use of ESC and ASC. The less controversial nature of using adult-derived SC, the potential for autologous grafts, and the lack of evidence for tumor formation after transplant into the eye made the study of adult-derived MSC more attractive for initial investigations. It is possible that ESC may be more malleable than ASC in their ability to respond to extrinsic cues due to the evidence of various levels of competence noted during development (Cepko et al., 1996; Belliveau and Cepko, 1999) and therefore it is also important to investigate ESC in this setting.

To assess the feasibility of MSC for cell replacement in retinal degenerative disease, our laboratory performed both *in vitro* and *in vivo* analyses. The *in vitro* analyses of rat CD90+ MSC have involved the use of activin A and taurine to induce differentiation and resulted in the expression of rhodopsin, opsin and recoverin (Kicic et al., 2003a). The rat MSC were transduced with rAAV.GFP prior to transplantation into the subretinal space of 4-5 week old normal RCS rats in order to demonstrate the viability of combining gene therapy with cell replacement (Kicic et al., 2003a). This was performed because autologous transplantation may necessitate the correction of the genes responsible for the retinal degenerative condition in order to prevent the transplanted cells from degenerating. The transplanted cells showed no morphological differentiation into photoreceptors and thus we suspect that the rhodopsin positive transplanted cells are immature photoreceptors. Despite little morphological resemblance, transplanted MSC were able to attract synapses, which provides evidence for some level of functional restoration.

Having established the safety of grafting MSC into the subretinal space and demonstrating that some signals inducing photoreceptor differentiation did exist in the RCS retina we next questioned whether the characteristics of MSC differed between species. The majority of studies have been performed in both mouse and rat models of normal and degenerate retinae and therefore it seemed natural to question whether results obtained with one species could be extrapolated to the other. Investigations of stem cell differences are also important when assessing the clinical application of cell replacement therapy in humans. The population profiles of CD45, CD11b and CD90 expression for MSC were compared between the rat, (RCS rdy+p+) and the mouse, (C57BL/6J) (Kicic et al., 2003b). The rat MSC population appeared to sustain a high level of sole CD90 expression, a marker of undifferentiated cells (Woodbury et al., 2000) through passages 1-8, however, the mouse MSC population did not show significant levels of sole CD90+ cells until passage 8. These results were comparable to those seen by Woodbury et al. (2000) in the rat and by Phinney et al. (1999) in the mouse. Differences between these two species were also observed with respect to their ability to differentiate into photoreceptors following activation with 100 ng/mL activin A with a larger proportion of rat MSC expressing rhodopsin than mouse MSC. However MSC from mice and rats do show some similarities in that they can be transduced with comparable efficiencies by both adenovirus and adeno-associated virus vehicles encoding GFP (Kicic et al., 2003b).

Having revealed differences in the characteristics of MSC between species it was then necessary to investigate if differences exist between the MSC of the healthy animal model compared with the MSC from a model with retinal degenerative disease (Kicic et al., In Press). Again, the population profiles of MSC from C57BL/6J normal mice and C3H/HeJ retinal degenerative mouse model were compared and both were shown to have similar profiles at early passage as reported for the mouse model above. In contrast to the normal model, the MSC derived from the retinal degenerate model showed a more rapid increase in the proportion of sole CD90+ cells with increasing time in culture. Marrow stromal cells from the healthy and disease origins were both found to respond to inductive cues for adipocytes and photoreceptor differentiation as determined by expression of cell-specific genes with RT-PCR and their associated protein products with Western blot analysis. There were no significant differences observed in the efficiency of adenovirus and adeno-associated virus to transduce MSC from disease and healthy origins. Collectively these results demonstrate that there do not appear to be limitations to the retinal transplant of autologous MSC in retinal degenerative animal models.

These studies have proven that MSC may be ideal candidates for cell replacement therapy due to their stem cell characteristics, their ease of isolation, their accessibility and the large pool of cells available requiring less time in culture. The safety of MSC transplant in the retina and the ability of MSC to differentiate into cells of retinal lineage have both been demonstrated. Variations exist between MSC from different species and further investigations are required to examine whether differences exist between SC from human and animal models as currently thought (Ginis et al., 2004; Rao et al., 2004). There do appear to be intrinsic differences in the MSC from disease and healthy models which affect their ability to respond to extrinsic cues however, we have shown that the MSC from retinal degenerative models do still have the potential for photoreceptor differentiation. Therefore it is feasible that autologous MSC may be used in the future for cell replacement therapy in the treatment of retinal degenerative disease.

5. CONCLUSION

The prevalence of retinal degenerative disease and the lack of available treatments have prompted investigation into the feasibility of cell replacement therapy. Studies using allogeneic and xenogeneic transplants have supported the notion that the photoreceptor layer may be repaired or regenerated if suitable signals and support cells are present in the eye. The lack of suitably matched tissue available for transplant severely limits the clinical application of such methods. Stem cells provide a source of cells that could be used to enhance sight in retinal degenerative patients. The potential of stem cells for the treatment of retinal disease is substantiated by *in vitro* studies indicating the capacity of both adult and embryonic stem cells to differentiate into photoreceptors and *in vivo* studies where transplantation of stem cells into injured or diseased retinae results in the survival, integration, and a level of differentiation. Many questions still exist regarding the behavior of stem cells after transplantation in the retina and these factors are important for translating current knowledge into a safe and effective treatment for retinal degenerative diseases.

6. ACKNOWLEDGEMENTS

We thank the National Eye Institute for supporting the participation of Christine Hall at the XIth International Symposium on Retinal Degenerations.

7. REFERENCES

Ahmad, I., Dooley, C. M., Thoreson, W. B., Rogers, J. A., and Afiat, S., 1999, In vitro analysis of a mammalian retinal progenitor that gives rise to neurons and glia, *Brain Res* **831**(1-2):1-10.

Ahmad, I., Tang, L., and Pham, H., 2000, Identification of neural progenitors in the adult mammalian eye, *Biochem Biophys Res Commun* **270**:517-521.

Akagi, T., Haruta, M., Akita, J., Nishida, A., Honda, Y., and Takahashi, M., 2003, Different characteristics of rat retinal progenitor cells from different culture periods, *Neurosci Lett* **341**(3):213-6.

Arai, S., Thomas, B. B., Seiler, M. J., Aramant, R. B., Qiu, G., Mui, C., de Juan, E., and Sadda, S. R., 2004, Restoration of visual responses following transplantation of intact retinal sheets in rd mice, *Exp Eye Res* **79**(3):331-41.

Arnhold, S., Klein, H., Semkova, I., Addicks, K., and Schraermeyer, U., 2004, Neurally selected embryonic stem cells induce tumor formation after long-term survival following engraftment into the subretinal space, *Invest Ophthalmol Vis Sci* **45**(12):4251-5.

Belliveau, M. J., and Cepko, C. L., 1999, Extrinsic and intrinsic factors control the genesis of amacrine and cone cells in the rat retina, *Development* **126**(3):555-66.

Belliveau, M. J., Young, T. L., and Cepko, C. L., 2000, Late retinal progenitor cells show intrinsic limitations in the production of cell types and the kinetics of opsin synthesis, *J Neurosci* **20**(6):2247-54.

Castillo, B. V., Jr., del Cerro, M., White, R. M., Cox, C., Wyatt, J., Nadiga, G., and del Cerro, C., 1997, Efficacy of nonfetal human RPE for photoreceptor rescue: a study in dystrophic RCS rats, *Exp Neurol* **146**(1):1-9.

Cepko, C. L., Austin, C. P., Yang, X., Alexiades, M., and Ezzeddine, D., 1996, Cell fate determination in the vertebrate retina, *Proc Natl Acad Sci U S A* **93**(2):589-95.

Chacko, D. M., Rogers, J. A., Turner, J. E., and Ahmad, I., 2000, Survival and differentiation of cultured retinal progenitors transplanted in the subretinal space of the rat, *Biochem Biophys Res Commun* **268**(3):842-6.

Chacko, D. M., Das, A. V., Zhao, X., James, J., Bhattacharya, S., and Ahmad, I., 2003, Transplantation of ocular stem cells: the role of injury in incorporation and differentiation of grafted cells in the retina, *Vision Res* **43**(8):937-46.

Chiou, S. H., Kao, C. L., Peng, C. H., Chen, S. J., Tarng, Y. W., Ku, H. H., Chen, Y. C., Shyr, Y. M., Liu, R. S., Hsu, C. J., Yang, D. M., Hsu, W. M., Kuo, C. D., and Lee, C. H., 2005, A novel in vitro retinal differentiation model by co-culturing adult human bone marrow stem cells with retinal pigmented epithelium cells, *Biochem Biophys Res Commun* **326**(3):578-585.

Coles, B. L., Angenieux, B., Inoue, T., Del Rio-Tsonis, K., Spence, J. R., McInnes, R. R., Arsenijevic, Y., and van der Kooy, D., 2004, Facile isolation and the characterization of human retinal stem cells, *Proc Natl Acad Sci U S A* **101**(44):15772-7.

Daar, A. S., Bhatt, A., Court, E., and Singer, P. A., 2004, Stem cell research and transplantation: science leading ethics, *Transplant Proc* **36**(8):2504-6.

Ghosh, F., Wong, F., Johansson, K., Bruun, A., and Petters, R. M., 2004, Transplantation of full-thickness retina in the rhodopsin transgenic pig, *Retina* **24**(1):98-109.

Ginis, I., Luo, Y., Miura, T., Thies, S., Brandenberger, R., Gerecht-Nir, S., Amit, M., Hoke, A., Carpenter, M. K., Itskovitz-Eldor, J., and Rao, M. S., 2004, Differences between human and mouse embryonic stem cells, *Dev Biol* **269**(2):360-80.

Gouras, P., Lopez, R., Kjeldbye, H., Sullivan, B., and Brittis, M., 1989, Transplantation of retinal epithelium prevents photoreceptor degeneration in the RCS rat, *Prog Clin Biol Res* **314**:659-71.

Gouras, P., Du, J., Kjeldbye, H., Kwun, R., Lopez, R., and Zack, D. J., 1991, Transplanted photoreceptors identified in dystrophic mouse retina by a transgenic reporter gene, *Invest Ophthalmol Vis Sci* **32**(13):3167-74.

Gouras, P., and Tanabe, T., 2003, Survival and integration of neural retinal transplants in rd mice, *Graefes Arch Clin Exp Ophthalmol* **241**(5):403-9.

Hara, A., Niwa, M., Kunisada, T., Yoshimura, N., Katayama, M., Kozawa, O., and Mori, H., 2004, Embryonic stem cells are capable of generating a neuronal network in the adult mouse retina, *Brain Res* **999**(2):216-21.

Haruta, M., Sasai, Y., Kawasaki, H., Amemiya, K., Ooto, S., Kitada, M., Suemori, H., Nakatsuji, N., Ide, C., Honda, Y., and Takahashi, M., 2004, In Vitro and In Vivo Characterization of Pigment Epithelial Cells Differentiated from Primate Embryonic Stem Cells, *Invest Ophthalmol Vis Sci* **45**(3):1020-1025.

Hirano, M., Yamamoto, A., Yoshimura, N., Tokunaga, T., Motohashi, T., Ishizaki, K., Yoshida, H., Okazaki, K., Yamazaki, H., Hayashi, S., and Kunisada, T., 2003, Generation of structures formed by lens and retinal cells differentiating from embryonic stem cells, *Dev Dyn* **228**(4):664-71.

Hochedlinger, K., Rideout, W. M., Kyba, M., Daley, G. Q., Blelloch, R., and Jaenisch, R., 2004, Nuclear transplantation, embryonic stem cells and the potential for cell therapy, *Hematol J* **5 Suppl 3**:S114-7.

Jiang, Y., Jahagirdar, B. N., Reinhardt, R. L., Schwartz, R. E., Keene, C. D., Ortiz-Gonzalez, X. R., Reyes, M., Lenvik, T., Lund, T., Blackstad, M., Du, J., Aldrich, S., Lisberg, A., Low, W. C., Largaespada, D. A., and Verfaillie, C. M., 2002, Pluripotency of mesenchymal stem cells derived from adult marrow, *Nature* **418**(6893):41-9.

Kicic, A., Shen, W. Y., Wilson, A. S., Constable, I. J., Robertson, T., and Rakoczy, P. E., 2003a, Differentiation of marrow stromal cells into photoreceptors in the rat eye, *J Neurosci* **23**(21):7742-9.

Kicic, A., Shanley, A. C., Hall, C. M., and Rakoczy, P. E., 2003b, Marrow stromal cells (MSC): a species comparison, *Adv Exp Med Biol* **533**:407-14.

Kicic, A., Hall, C. M., Shen, W. Y., Rakoczy, P. E., Are stem cell characteristics altered by disease state?, *Stem Cells Dev.* In Press.

Klassen, H. J., Ng, T. F., Kurimoto, Y., Kirov, I., Shatos, M., Coffey, P., and Young, M. J., 2004a, Multipotent retinal progenitors express developmental markers, differentiate into retinal neurons, and preserve light-mediated behavior, *Invest Ophthalmol Vis Sci* **45**(11):4167-73.

Klassen, H., Sakaguchi, D. S., and Young, M. J., 2004b, Stem cells and retinal repair, *Prog Retin Eye Res* **23**(2): 149-81.

Kurimoto, Y., Shibuki, H., Kaneko, Y., Ichikawa, M., Kurokawa, T., Takahashi, M., and Yoshimura, N., 2001, Transplantation of adult rat hippocampus-derived neural stem cells into retina injured by transient ischemia, *Neurosci Lett* **306**(1-2):57-60.

Li, L., and Turner, J. E., 1991, Optimal conditions for long-term photoreceptor cell rescue in RCS rats: the necessity for healthy RPE transplants, *Exp Eye Res* **52**(6):669-79.

Meyer, J. S., Katz, M. L., Maruniak, J. A., and Kirk, M. D., 2004, Neural differentiation of mouse embryonic stem cells in vitro and after transplantation into eyes of mutant mice with rapid retinal degeneration, *Brain Res* **1014**(1-2):131-44.

Nishida, A., Takahashi, M., Tanihara, H., Nakano, I., Takahashi, J. B., Mizoguchi, A., Ide, C., and Honda, Y., 2000, Incorporation and differentiation of hippocampus-derived neural stem cells transplanted in injured adult rat retina, *Invest Ophthalmol Vis Sci* **41**(13):4268-74.

Otani, A., Kinder, K., Ewalt, K., Otero, F. J., Schimmel, P., and Friedlander, M., 2002, Bone marrow-derived stem cells target retinal astrocytes and can promote or inhibit retinal angiogenesis, *Nat Med* **8**(9):1004-10.

Otani, A., Dorrell, M. I., Kinder, K., Moreno, S. K., Nusinowitz, S., Banin, E., Heckenlively, J., and Friedlander, M., 2004, Rescue of retinal degeneration by intravitreally injected adult bone marrow-derived lineage-negative hematopoietic stem cells, *J Clin Invest* **114**(6):765-74.

Phinney, D. G., Kopen, G., Isaacson, R. L., and Prockop, D. J., 1999, Plastic adherent stromal cells from the bone marrow of commonly used strains of inbred mice: variations in yield, growth, and differentiation, *J Cell Biochem* **72**(4):570-85.

Pressmar, S., Ader, M., Richard, G., Schachner, M., and Bartsch, U., 2001, The fate of heterotopically grafted neural precursor cells in the normal and dystrophic adult mouse retina, *Invest Ophthalmol Vis Sci* **42**(13):3311-9.

Priller, J., Persons, D. A., Klett, F. F., Kempermann, G., Kreutzberg, G. W., and Dirnagl, U., 2001, Neogenesis of cerebellar Purkinje neurons from gene-marked bone marrow cells in vivo, *J Cell Biol* **155**(5):733-8.

Rao, M., 2004, Conserved and divergent paths that regulate self-renewal in mouse and human embryonic stem cells, *Dev Biol* **275**(2):269-86.

Sagdullaev, B. T., Aramant, R. B., Seiler, M. J., Woch, G., and McCall, M. A., 2003, Retinal transplantation-induced recovery of retinotectal visual function in a rodent model of retinitis pigmentosa, *Invest Ophthalmol Vis Sci* **44**(4):1686-95.

Schraermeyer, U., Thumann, G., Luther, T., Kociok, N., Armhold, S., Kruttwig, K., Andressen, C., Addicks, K., and Bartz-Schmidt, K. U., 2001, Subretinally transplanted embryonic stem cells rescue photoreceptor cells from degeneration in the RCS rats, *Cell Transplant* **10**(8):673-80.

Scigel, G. M., Takahashi, M., Adamus, G., and McDaniel, T., 1998, Intraocular transplantation of E1A-immortalized retinal precursor cells, *Cell Transplant* **7**(6):559-66.

Tomita, M., Adachi, Y., Yamada, H., Takahashi, K., Kiuchi, K., Oyaizu, H., Ikebukuro, K., Kaneda, H., Matsumura, M., and Ikehara, S., 2002, Bone marrow-derived stem cells can differentiate into retinal cells in injured rat retina, *Stem Cells* **20**(4):279-83.

Van Hoffelen, S. J., Young, M. J., Shatos, M. A., and Sakaguchi, D. S., 2003, Incorporation of murine brain progenitor cells into the developing mammalian retina, *Invest Ophthalmol Vis Sci* **44**(1):426-34.

Wang, H., Yagi, F., Cheewatrakoolpong, N., Sugino, I. K., and Zarbin, M. A., 2004, Short-term study of retinal pigment epithelium sheet transplants onto Bruch's membrane, *Exp Eye Res* **78**(1):53-65.

Woodbury, D., Schwarz, E. J., Prockop, D. J., and Black, I. B., 2000, Adult rat and human bone marrow stromal cells differentiate into neurons, *J Neurosci Res* **61**(4):364-70.

Woodbury, D., Reynolds, K., and Black, I. B., 2002, Adult bone marrow stromal stem cells express germline, ectodermal, endodermal, and mesodermal genes prior to neurogenesis, *J Neurosci Res* **69**(6):908-17.

Young, M. J., Ray, J., Whiteley, S. J., Klassen, H., and Gage, F. H., 2000, Neuronal differentiation and morphological integration of hippocampal progenitor cells transplanted to the retina of immature and mature dystrophic rats, *Mol Cell Neurosci* **16**(3):197-205.

Zhao, X., Liu, J., and Ahmad, I., 2002, Differentiation of embryonic stem cells into retinal neurons, *Biochem Biophys Res Commun* **297**(2):177-84.

CHAPTER 55

OPTIC NERVE REGENERATION: MOLECULAR PRE-REQUISITES AND THE ROLE OF TRAINING

Restoring vision after optic nerve injury

Lyn D. Beazley[1,2], Jennifer Rodger[1,2], Carolyn E. King[1,2], Carole A. Bartlett[1], Andrew L. Taylor[1], and Sarah A. Dunlop[1,2,*]

1. INTRODUCTION

The vertebrate visual system is a valuable model for examining recovery after injury to the central nervous system (CNS). It is a relatively "simple" part of the CNS having one major class of projection neuron, the retinal ganglion cells (RGCs), which make topographic connections within well defined visual nuclei, thus recreating visual space within the brain. Topographic maps can be readily assessed electrophysiologically and anatomically and are a critical template for useful visually guided behaviour which can be examined behaviourally. Furthermore, the optic nerve is accessible, an extra-foramenal crush injury severing all RGC axons but leaving the meningeal sheath intact as a conduit for regeneration and preventing gross axonal mis-routing. The procedure also leaves the blood supply to the eye patent, avoiding ischaemic-induced RGC death.

The optic nerve injury model allows examination of the response of one class of central neuron to axotomy that may have wider implications for CNS neurotrauma. Restoration of useful function in any CNS region will require the fulfilment of several pre-requisites. First, cell death must be prevented and axons encouraged to regrow to their target tissue. Topographic maps must then be re-created in tandem with an appropriate balance of excitation and inhibition to allow the restoration of useful function. Whereas neuroprotection and neuroregeneration have been widely studied, restoration of topography and appropriate functional connections has received less attention.

[1] Sarah A. Dunlop, et al., School of Animal Biology, [2] The Western Australian Institute for Medical Research, The University of Western Australia, Nedlands, Western Australia, 6009. *Corresponding author.

2. SUCCESSFUL OPTIC NERVE REGENERATION IN FISH AND AMPHIBIA

Comparative studies of optic nerve regeneration across the vertebrate classes have provided powerful insights whereby the pre-requisites for neural repair are fulfilled (reviewed: Beazley, 2000; Beazley and Dunlop, 2000). In fish and amphibia, RGC survival is robust; almost 100% of RGCs survive in goldfish and up to 50-70% in amphibia with RGC axons regaining the major visual centre, the optic tectum, by approximately 2 weeks (Beazley and Humphrey, 1985; Murray and Edwards, 1982). The restoration of topography appears to be biphasic requiring an initial activity-independent phase followed by a phase that is critically dependent on normal patterns of neural activity (Fig. 55.1).

During the initial phase, a coarse topographic map is restored in which receptive fields map approximately correctly within the optic tectum but are somewhat larger than normal (Schmidt and Edwards, 1983). The anatomical correlate is that regenerating RGC axonal arbors are more widespread than normal and are located within both topographically appropriate and inappropriate tectal regions (Schmidt et al., 1988). The initial phase appears to be activity-independent since it occurs if animals are raised in the dark, a condition permitting only spontaneous activity, or in a stroboscopic environment which synchronises all activity and prevents normal patterned form vision (Schmidt and Edwards, 1983; Schmidt et al., 1983). Similarly, a coarse map forms if sodium channel-mediated activity is blocked with tetrodotoxin or if glutamatergic NMDA receptor-mediated activity is prevented with

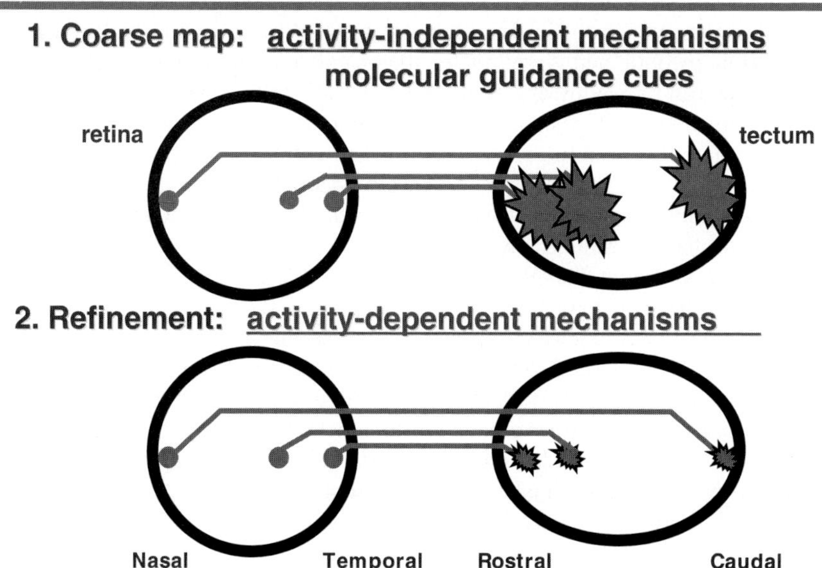

Figure 55.1. Diagram of two overlapping phases involved in restoration of retinotectal topography. During the activity-independent phase (left panel), RGC axons (dots in retinal outline) project axons with wide-spread terminal arbors in the optic tectum (jagged shapes in tectal outline). In the second phase, activity-dependent mechanisms focus terminal arbors and produce precise topography. See also color insert.

the antagonist APV (Schmidt, 1990). In the second phase, the map is refined via activity-dependent mechanisms, whereby incorrectly located terminal arbors, or parts thereof, are pruned or silenced by inhibition or are inactive (Lin et al., 1998); absence of appropriate activity results in failure to refine.

Here we describe some of our experiments in different models of optic nerve regeneration which throw new light on mechanisms underpinning both phases. We summarise evidence which implicates a role for molecular guidance cues in the first, activity-independent phase. We also highlight the importance, during the second phase, of normal patterns of neural activity, which can be induced by training on a specific visual task, to restore topography and therefore useful vision.

3. MOLECULAR PRE-REQUISITES FOR RESTORATION OF TOPOGRAPHY

In his "chemoaffinity hypothesis", the Nobel Laureate Roger Sperry predicted that matching gradients of guidance molecules would be involved in conferring positional identity to both RGCs in the eye and their postsynaptic target cells within the optic tectum, allowing the establishment of topographic connections between retinal and tectal cells with matched identity (Sperry, 1963). The hypothesis received widespread experimental support although the molecular basis remained elusive until the identification of an orphan tyrosine kinase Eph receptor with a graded expression pattern in the developing visual system (Cheng et al., 1995). Subsequent studies identified classes of Eph receptors and their ligands, the ephrins, which were involved in the development of retino-tectal topography (Flanagan and Vanderhaeghen, 1998).

The "A" family of Eph/ephrins is involved in the establishment of the retinal temporo-nasal to the tectal rostro-caudal axis. EphA receptors are expressed as temporal^{high} to nasal^{low} gradients on RGCs and their axons. Within the tectum, ephrin-As are expressed in a rostral^{low} to caudal^{high} gradients. EphA/ephrin-A binding is primarily repulsive. Developing temporal axons, with high EphA receptor expression, are repulsed by high ephrin-A expression in caudal tectum and map rostrally. Conversely, nasal axons with low EphA receptor expression are less sensitive to high ephrin-A expression and map caudally. At maturity, such gradients are down-regulated. Functional involvement of EphA/ephrin-As is evident from ephrin-A2/-A5 double knock-out mice and EphA3 knock-in mice in which topography is disrupted (Feldheim et al., 2000; Brown et al., 2000). *In vitro* experiments demonstrate that temporal and nasal growth cones display opposite behaviours when presented with high ephrin-expressing tectal cells; temporal axons are repelled whilst nasal ones are not.

Given the well-established role for EphA/ephrin-As in the development of retinotectal topography, we investigated whether such developmental guidance molecules were involved in the restoration of topography after injury to the adult visual system. In normal goldfish, EphA3 and EphA5 have a uniform profile across the temporo-nasal retinal axis (King et al., 2003). However, during the time that a coarse map is restored, EphA3 and EphA5 are upregulated as a temporal^{high} to nasal^{low} gradient. Upregulation is transient, returning to normal levels once topography is refined. Similarly, within the optic tectum, ephrin-A2 is transiently up-regulated as an ascending rostral^{low} to caudal^{high} gradient as the coarse map is restored (Rodger et al., 2000). We have also demonstrated a functional role for EphA/ephrin-As during optic nerve regeneration (Rodger et al., 2003). Blocking EphA/ephrin-A interactions *in vivo* by intracranial injection of recombinant EphA3 (EphA3-AP) resulted in a degradation of retinotectal topography as assessed electrophysiologically. Furthermore, *in vitro*, goldfish temporal RGC axons were repulsed by ephrin-A5 whereas nasal ones were not.

4. RESTORING TOPOGRAPHY – THE NEED FOR A NOVEL MODEL

Although spontaneous optic nerve regeneration in mammals is largely abortive (Zeng et al., 1995), RGC axon regeneration can be assisted by, for example, grafting a piece of peripheral nerve (PN) between the back of the eye and the superior colliculus (SC; mammalian homologue of the optic tecum in lower vertebrates). After PN grafting, approximately 10% of RGCs survive but only a small proportion regenerate their axons along the graft, less than 1% enter the SC and precise topography is absent (Sauvé et al., 2001). However, the number of RGC axons entering the SC falls well below the 20% minimum required to restore topography, suggesting that once sufficient neuroprotection and neuroregeneration can be stimulated in mammals, further steps may be required to restore topography and therefore useful vision.

In searching for models lacking topography in visual projections, we turned our attention to reptiles, the class of vertebrate phylogenetically intermediate between the fish and amphibia and the birds and mammals. Although the response to axotomy is variable (reviewed: Dunlop et al., 2004), we identified a lizard, *Ctenophorus ornatus*, which was similar to fish and amphibia with large numbers of RGC axons regaining the tectum. However, the outcome of optic nerve regeneration mimicked peripheral nerve-assisted mammalian optic nerve regeneration in that topography was lacking (Beazley et al., 1997; Stirling et al., 1999; Dunlop et al., 2000).

5. NEURAL ACTIVITY AND VISUAL TRAINING TO RESTORE TOPOGRAPHY AND USEFUL VISION

In lizard, anatomical tracing throughout optic nerve regeneration revealed that RGC axons failed to restore topography (Beazley et al., 1997; Dunlop et al., 2000). Nevertheless, electrophysiological recording indicated the presence of a coarse topographic map at an intermediate stage. However, the map was transient, breaking down in the long term with blindness persisting throughout. Furthermore, whereas normal animals displayed purely AMPA-mediated glutamatergic fast secure excitatory synaptic neurotransmission accompanied by low levels of GABA-ergic inhibition, during optic nerve regeneration, responses were weak and habituated readily, were both NMDA- and AMPA-mediated and displayed high levels of inhibition (Dunlop et al., 2003). Taken together, the data indicated that low levels of visually elicited activity from widely spread RGC arbors resulted in non-correlated firing of postsynaptic partner cells and therefore weak NMDA receptor activation. As a consequence, AMPA-mediated activity presumably decreased while GABA-ergic inhibition increased, thus preventing the maturation of fast, secure synaptic transmission (Shi et al., 1997) and the restoration of topography.

The corollary that low levels of activity will delay synaptic maturation is that high levels will accelerate it. Extensive evidence suggests that neural activity, either spontaneous or elicited, influences synaptic circuitry not only during development but also in both the

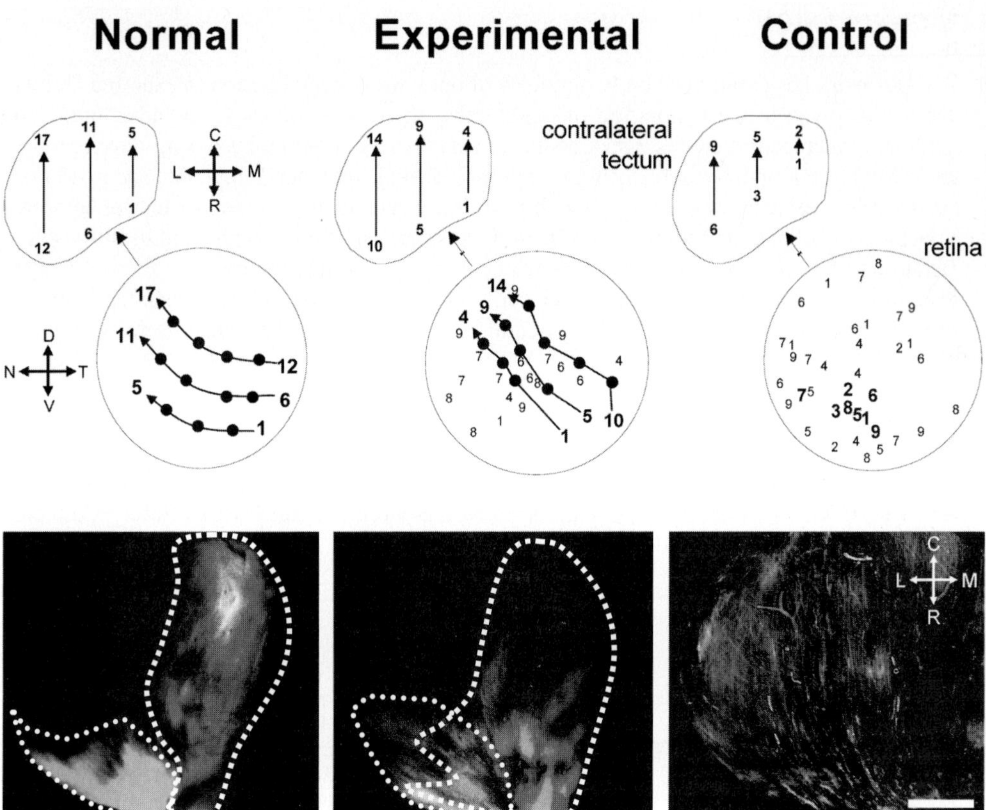

Figure 55.2. Visual training restores topography. Top panel: retinotectal maps assessed electrophysiologically. Numbers in tectal outlines (top) indicate electrode positions and those in retinal outlines (bottom) indicate the location of receptive fields. In normal and trained animals (left and centre) the map is topographic (rows). In untrained animals (right), topography is lacking. Bold numbers represent robust responses, non-bold weak responses. Bottom panel: retinotectal topography as assessed anatomically. Dorsal view of the optic tectum after placements of the carbocyanine dyes DiI (square dotted lines) and DiAsp (round dotted lines) respectively in dorsal or ventral retina. A retinotopic map is observed in normal and trained, but not untrained (complete overlap of dye labelling), animals. (Beazley et al., Journal of Neurotrauma, 2003, Vol 20, No. 11, 1263–1270, by the permission of Mary Ann Liebert, Inc. Publishers). See also color insert.

normal and damaged adult brain (reviewed, Dunlop and Steeves, 2003). Enhanced neural activity in the form of specific training improves functional outcome in a wide range of situations including peripheral nerve injury, stroke, spinal cord and head injury. We therefore assessed the influence of specific visual training on the outcome of optic nerve regeneration in lizard (Beazley et al., 2003). Food items were presented to the monocular field of the experimental eye in trained animals and to the unoperated eye in untrained animals twice weekly throughout the course of optic nerve regeneration. A topographic map was restored (Fig. 55.2) as well as glutamatergic excitation that was predominantly AMPA-mediated together with low levels of GABA-ergic inhibition (Beazley et al., 2003; Dunlop et al., 2003). Crucially, animals responded to and fed on prey items presented to the experimental eye indicating a return of useful vision.

7. CONCLUSION

Our work has capitalised on two models of optic nerve regeneration to examine factors that are involved in the restoration of topographic maps after injury to the adult brain. In goldfish, guidance molecules instrumental in establishing topography during development are involved functionally in restoring topography during optic nerve regeneration in adults. Furthermore, in lizard, a model in which RGC axon regeneration is robust but topography is lacking, specific visual training results in the restoration of topography and useful vision. Taken together, the work suggests that restoration of vision in mammals will require the recapitulation of developmental guidance cues during the restoration of a coarse topographic map and appropriate levels of relevant neural activity to ensure the return useful vision.

8. ACKNOWLEDGEMENTS

We acknowledge the financial support provided by the National Health and Medical Research Council (Australia), the Neurotrauma Research Program (Road Safety Council, Western Australia) and Woodside Energy Limited.

9. REFERENCES

Beazley, L. D., 2000, Optic nerve regeneration in the CNS of amphibians and reptiles. In: *Axonal regeneration in the central nervous system*, N. Ingoglia and M. Murray Eds., Marcell Dekker Press, pp67-105.

Beazley, L. D., and Dunlop, S. A., 2000, Evolutionary hierarchy of optic nerve regeneration: implications for cell survival, axon outgrowth and map making. In: *Degeneration and Regeneration of the Nervous System*, Saunders, N. R. & Dziegielewska, K. M. Eds., Harwood Academic Publishers, pp119-152.

Beazley, L. D., Sheard, P., Tennant, M., Starac, D., and Dunlop, S. A., 1997, The optic nerve regenerates but does not restore retinotopic projections in the lizard *Ctenophorus ornatus*, *J. Comp. Neurol.* **377**:105.

Beazley, L. D., Rodger, J., Chen, P., Stirling, R. V., Taylor, A. L., Tee, L. B. G., and Dunlop, S. A., 2003, Training on a visual task improves the outcome of optic nerve regeneration, *J. Neurotrauma* **20**:1263.

Brown, A., Yates, P. A., Burrola, P., Ortuno, D., Vaidya, A., Jessell, T. M., Pfaff, S. L., O'Leary, D. D., and Lemke, G., 2000, Topographic mapping from the retina to the midbrain is controlled by relative but not absolute levels of EphA receptors for signaling, *Cell* **102**:77.

Cheng, H-J., Nakamoto, M., Beregeman, A., and Flanagan, J. D. 1995, Complementary gradients in expression and binding of ELF-1 and Mek4 in development of the retinotectal projection map, *Cell*, **82**:371.

Dunlop, S. A., Tran, N., Tee, L. B. G., Papadimitriou, J., and Beazley, L. D., 2000, Retinal projections throughout optic nerve regeneration in the lizard, *Ctenophorus ornatus*, *J. Comp. Neurol.* **416**:188.

Dunlop, S. A., and Steeves, J. D., 2003, Neural activity and facilitated recovery following CNS injury: Implications for rehabilitation. In: *Topics in Spinal Cord Rehabilitation*, Ed J. C. Bresnahan. A Thomas Land, St Louis, pp92-103.

Dunlop, S. A., Stirling, R. V., Rodger, J., Symonds, A. C. E., Bancroft, W. J., Tee, L. B. G., and Beazley, L. D., 2003, Failure to form a stable topographic map during optic nerve regeneration: Abnormal activity-dependent mechanisms, *Exp. Neurol.* **184**:805.

Dunlop, S. A., Tee, L. B. G., Stirling, R. V., Taylor, A. L., Runham, P. B., Barber, A. B., Bartlett, C. A., Kuchling, G., Rodger, J., Roberts, J. D., Harvey, A. R., and Beazley, L. D., 2004, Failure to restore vision after optic nerve regeneration in reptiles: varying responses to axotomy, *J. Comp. Neurol.* **478**:292.

Feldheim, D. A., Kim, A. Y.-I., Bergeman, D., Frisen, J., Barbacid, M., and Flanagan, J. G., 2000, Genetic analysis of ephrn-A2 and ephrin-A5 shows their requirement in multiple aspects of retinocollicular mapping, *Neuron* **21**:1303.

Flanagan, J. D., and Vanderhaeghen, P., 1998, The Eph receptors and ephrins in neural development, *Ann. Rev. Neurosci.* **21**:309.

Humphrey, M. F., and Beazley, L. D., 1985, Retinal ganglion cell death during optic nerve regeneration in the frog *Hyla moorei*, *J. Comp. Neurol.* **236**:382.

King, C. E., Wallace, A., Bartlett, C. A., Beazley, L. D., and Dunlop, S. A., 2003, Transient up-regulation of retinal EphA3 and EphA5 but not ephrin-A2 coincides with topographic map restoration during optic nerve regeneration, *Exp. Neurol.* **183**:593.

Lin, S.-Y., and Constantine-Paton, M., 1988, Suppression of sprouting: an early function of NMDA receptors in the absence of AMPA/kainate receptor activity, *J. Neurosci.* **15**:3725.

Murray, M., and Edwards, M. A., 1982, A quantitative study of the reinnervation of the goldfish optic tectum following optic nerve crush, *J. Comp. Neurol.* **209**:363.

Rodger, J., Bartlett, C. A., Beazley, L. D., and Dunlop, S. A., 2000, Ephrin-A2 is transiently upregulated during optic nerve regeneration in the goldfish, *Exp. Neurol.* **166**:196.

Rodger, J., Vitale, P. N., King, C. E., Bartlett, C. A., Brennan, C. O'Shea, J. E., Dunlop, S. A., and Beazley, L. D., 2004, EphA/ephrin interactions are required for restoration of topography during optic nerve regeneration, *Mol. Cell Neurosci.* **25**:56.

Sauvé, Y., Sawai, H., and Rasminsky, M., 2001, Topological specificity in reinnervation of the superior colliculus by regenerated retinal ganglion cell axons in adult hamsters, *J. Neurosci.* **21**:951.

Schmidt, J. T., 1990, Long-term potentiation and activity-dependent retinotopic sharpening in the regenerating retino tectal projection of goldfish: Common sensitive period and sensitivity to NMDA blockers, *J. Neurosci.* **10**:233.

Schmidt, J. T., and Edwards, M. A., 1983, Activity sharpens the map during the regeneration of the retinotectal projection in goldfish, *Brain Res.* **269**:29.

Schmidt, J. T., Edwards, D. L., and Stuermer, C. A. O., 1983, The re-establishment of synaptic transmission by regenerating optic axons in goldfish: time course and effects of blocking activity by intraocular injection of tetrodotoxin, *Brain Res.* **269**:15.

Schmidt, J. T., Turcote, J. C., Buzzard, M., and Tieman, D. G., 1988, Staining of regenerated optic arbors in goldfish tectum: progressive changes in immature arbors and a comparison of mature regenerated arbors with normal arbors, *J. Comp. Neurol.* **269**:565.

Shi, J., Aamodt, S. M., and Constantine-Paton, M., 1997, Temporal correlations between functional and molecular changes in NMDA receptors and GABA neurotransmission in the superior colliculus, *J. Neurosci.* **17**:6264.

Stirling, R. V., Dunlop, S. A., and Beazley, L. D. 1999, Electrophysiological evidence for a transient physiological topographic organisation of retinotectal projections during optic nerve regeneration in the lizard, *Ctenophorus ornatus*, *Vis. Neurosci.* **16**:682.

Sperry, R. W., 1963, Chemoaffinity in the orderly growth of nerve fibre patterns and connections, *Proc. Nat. Acad. Sci.* **50**:703.

Zeng, B-Y., Anderson, P. N., Campbell, M. G., and Lieberman, A. R., 1995, Regenerative and other responses to injury in the retinal stump of the optic nerve in albino rats: transection of the intracranial optic nerve, *J. Anat.* **186**:495.

CHAPTER 56

RETINAL GANGLION CELL REMODELLING IN EXPERIMENTAL GLAUCOMA

James E. Morgan, Amit V. Datta, Jonathan T. Erichsen, Julie Albon, and Michael E. Boulton*

1. INTRODUCTION

Retinal ganglion cell (RGC) death is the key pathological event in glaucoma and the biological basis for the loss of vision. Although significant advances have been made in the medical and surgical treatment of glaucoma, the disease remains the most common cause, worldwide, of irreversible vision loss (Quigley, 1996). Our understanding of the role played by elevated intraocular pressure (IOP) in the initiation of glaucoma has recently been advanced by evidence from clinical trials that IOP levels within the normal range can influence the degree of retinal ganglion cell death (AGIS, 2000). For those patients with advanced glaucoma damage, a reduction in IOP, even within the normal range, can reduce the risk of progressive vision loss.

In spite of this, many patients will suffer glaucoma damage even when IOPs have been reduced to appropriate therapeutic levels. In order to address this limitation, there has been considerable research into the processes that initiate pathological RGC death. There is strong evidence that RGCs are lost through programmed cell death (Quigley et al., 1995), but relatively little about the changes that occur in the RGC population prior to this event. Early descriptions, based on changes in human and experimental glaucoma (Quigley et al., 1988, 1989), suggested that cells proceeded to apoptosis with little in the way of morphological changes. This interpretation of the data, based on cell and axon size changes in glaucoma, was taken as evidence for the selective loss of larger RGCs in the early stages of glaucoma. Since larger RGCs are more common in the magnocellular pathway, this supported the hypothesis that diagnostic tests based on the properties of these cells could be used to detect glaucoma at an early stage (Johnson, 1994).

Critical appraisal of these data as well as more recent evidence has challenged this view (Morgan, 2002) and suggests that retinal ganglion cells display morphological changes that

---

* School of Optometry and Vision Sciences, Redwood Building, King Edward VII Ave, Cathays Park, Cardiff CF10 3NB, Wales, UK.

are characteristic of neurons in chronic neurodegenerative diseases. Work in the primate model of experimental glaucoma has revealed remodelling at the level of the cell soma and the dendritic tree (Weber et al., 1997). Furthermore, other studies have failed to find evidence supporting the selective loss of magnocellular retinal ganglion cells in experimental glaucoma (Morgan et al., 2000). It is important to note that these observations have only been made in the primate glaucoma model. As such, they remain contentious and further work is needed to establish whether changes in RGC morphology are a general feature of other glaucoma models and of human disease. Although the primate has been a popular model for studying these changes in the past, because of its great cost represent it is not a feasible model for the study of the early pathophysiology of RGC death.

Therefore, we have adopted the rodent glaucoma model in which moderate increases in intraocular pressure can be produced by the injection of hypertonic saline into the episcleral venous system (Morrison et al., 1997). The principal aims of our work are to establish whether morphological changes are occurring in the retinal ganglion cell population prior to the onset of cell death and to develop techniques that will allow us to evaluate strategies for reversing any changes that occur. In this paper, we outline techniques for the analysis of RGC morphology in the rat model and report our preliminary findings.

2. METHODS

Unilateral ocular hypertension was induced in adult Norwegian Brown rats (retired male breeders, weight range: 335 to 465 gm). All experiments were conducted in accordance with Home Office (UK) regulations. Animals were maintained in a constant low light environment (40-60 lux) to minimise diurnal fluctuations in IOP. IOPs were measured at least every other day using a factory-calibrated Tonopen XL (Mentor) with the IOP taken as the mean of 10 readings. All measurements were made in awake animals in which the cornea was anaesthetised using topical benoxinate drops (0.5%).

Ocular hypertension was induced in left eyes with the right eye acting as an unoperated control. In each case, a single episcleral vein was exposed by conjunctival dissection and injected using a glass microcannula (outside diameter 15-35 μm) with sufficient hypertonic (1.75 M) saline (approximately 50-75 μl) to blanch the vessel and clear blood from the episcleral venous system. We obtained unilateral elevation of IOP within 24-48 hours in over 90% of animals following a single injection. A repeat injection was given to those animals that failed to show an increase after 7 days.

Following defined periods of sustained elevation of IOP, animals were sacrificed and the eyes dissected out rapidly for culture in oxygenated Ames medium. Retinal ganglion cells were labelled using carbocyanine dyes delivered ballistically using a Gene Gun (Bio-Rad) in which tungsten microparticles (1.7 μm diameter) were coated in dye and injected under high pressure directly into retinal ganglion cells (Sun et al., 2002). When injected in viable tissue, the dyes are rapidly transported within the cell membrane and reveal neuronal structure. Labelled cells can be viewed by fluorescence microscopy within minutes of injection to determine the extent of labelling and further injections administered as required. Images were captured at high resolution at a series of focal planes through the dendritic tree and then compressed in the z-axis for the analysis of dendritic structure in a 2 dimensional view. Two methods were used to determine changes in dendritic structure. Firstly, the number of dendritic branches was estimated using a modified Sholl analysis (Sholl, 1953) in which a series of concentric rings centred on the cell soma are drawn at regular intervals to the outermost boundary of the dendritic field. The number of dendritic branches intersecting with each ring was then determined to provide an index of dendritic complexity. Secondly, the maximum diameter of the dendritic tree was measured based on a perimeter that connected the tips of the terminal dendrites.

The determination of retinal ganglion cell loss in this paradigm can be problematic. Conventionally, retinal ganglion cells are labelled prior to the induction of ocular hypertension by retrograde labelling from injections made in the superior colliculus (Laquis et al., 1996). While this method can provide robust estimates of the population of surviving RGCs, it has the potential limitation that injections into the superior colliculus can compromise retinal ganglion cell function as a result of direct damage to retinal ganglion cell axons (Leahy et al., 2004). We therefore adopted an alternative strategy in which retinal ganglion cells were identified immunohistochemically using an antibody (TUJ1) against a neuron-specific β-3 tubulin (Covariance, UK) which can be used to distinguish RGC's and amacrine cells within the retinal ganglion cell layer (Cui et al., 2003). Use of immunohistochemical stains in the retina in which cells have been labelled intracellularly with a suitable fluorescent dye can be complicated by dispersion of dye following permeabilisation of cells during immunohistochemistry. We overcame this technical difficulty by using DiI which has been modified by the inclusion of a thiol reactive chloromethyl group (CM-DiI, Molecular Probes, OR) to increase binding to the plasma membrane and to diminish the dispersion of fluorophore within the retina following permeabilisation. Once labelled RGCs had been analysed in detail, the surrounding retinal ganglion cell population was labelled immunohistochemically to determine the degree of local RGC loss.

3. RESULTS

Consistent and moderate increases in intraocular pressure were obtained using the episcleral vein injection model. Typical intraocular pressure elevation profiles are shown in Figure 56.1.

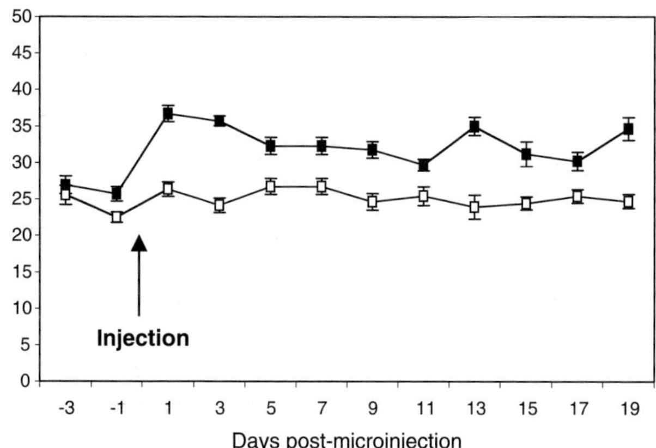

Figure 56.1. Plot showing the change in intraocular pressure following injection of hypertonic saline into an episcleral vein.

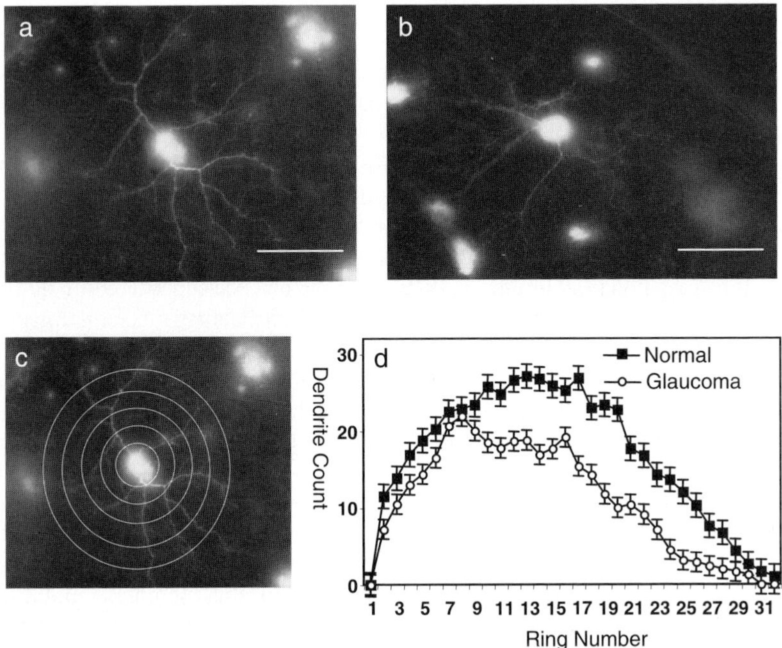

Figure 56.2. DiI labelled RGCs. (a) a normal eye. (b) glaucomatous eye. Scale bar 100 microns. (c) Sholl rings overlying a labelled retinal ganglion cell. (d) Plot showing reduction in the Sholl count for 8 cells from 4 glaucomatous eyes. The number at each ring indicates the number of dendrites crossing that ring. SEMs are indicated.

The Gene Gun method provided extensive labelling of retinal ganglion cells. However, excessive labelling can reduce the numbers of cells in which the dendritic tree can be isolated and accurately reconstructed. We therefore focused on cells in areas of retina with sparser labelling in which the dendritic tree could clearly be distinguished from those of adjacent cells. In Figure 56.2, sample cells are shown from normal and glaucomatous flat-mounted retinae that indicate pruning of the dendritic tree with reduction in overall dendritic area and in the complexity of the dendritic tree. This shrinkage and remodelling has been a consistent feature of RGCs labelled in glaucomatous retinae. The mean reduction in dendritic tree diameter for RGCs from the glaucomatous retinae was 25.7% compared with control eyes (P < 0.001, ANOVA).

Immunohistochemical labelling with TUJ1 allowed us to determine the degree of RGC loss around cells labelled with DiI. RGC identification could be confirmed by the presence of axonal labelling. An example of RGC labelling with this technique is shown in Figure 56.3.

4. DISCUSSION

Analysis of intracellularly labelled cells in this rodent supports the hypothesis that changes in dendritic morphology precede the onset of retinal ganglion cell death in exper-

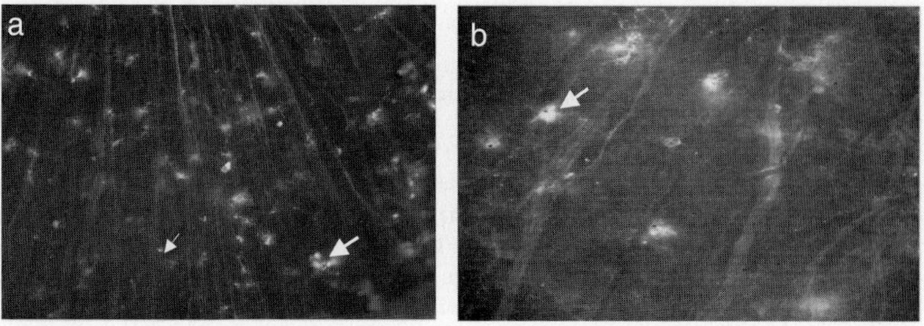

Figure 56.3. RGC's labelled immunohistochemically with TUJ1. (a) Small arrow: RGC labelled with TUJ1 alone. Large arrow: RGC labelled with DiI and TUJ1. (b). TUJ1 labelled RGCs in a glaucomatous retina (large arrow). Note the reduction in number of labelled cells in (b).

imental glaucoma. Our observations are important because they suggest that widespread remodelling occurs in the retinal ganglion cell population prior cell loss. In the primate, these changes occur in step with reduction in cell soma size (Weber et al., 1998) and similar changes appear to occur in the rodent model. An important implication of these findings is that the apparent selective loss of larger retinal ganglion cells in glaucoma could also be explained by a reduction in retinal ganglion cell soma area and dendritic tree area rather than the selective loss of one particular class of retinal ganglion cells (Morgan, 2002). These observations are consistent with clinical physical studies which have failed to demonstrate selective damage to magnocellular or parvocellular retinogeniculate pathways (Graham et al., 1996, Ansari et al., 2002) in early glaucoma.

Perhaps of greater significance for our understanding of the early changes in retinal ganglion cells in glaucoma is that our data indicate that subtle changes occur in the retinal ganglion cell population prior to the onset of retinal ganglion cell death. We hypothesise a model in which large populations of RGCs are diffusely affected by elevated IOP which predisposes to programmed cell death. It is interesting to note that in the rodent glaucoma model changes in neurofilament mRNA levels are seen early in the disease process (Johnson et al., 2000). Neurofilaments are important for neuronal structural integrity and reduced expression of NFL (the light neurofilament subtype) can correlate with reductions in dendritic complexity (Zhang et al., 2002). Changes in neurofilament expression would provide a plausible link between ocular hypertension and changes in RGC structure.

In conclusion, our data support the hypothesis that retinal ganglion cells undergo structural changes prior to the onset of cell death in experimental glaucoma. These observations have far reaching implications in terms of our understanding of the processes that precede cell death in this disease.

5. ACKNOWLEDGEMENTS

Support: National Eye Research Centre (UK), Allergan Inc.

6. REFERENCES

F. A. Ahmed, P. Chaudhary, S. C. Sharma, Effects of increased intraocular pressure on rat retinal ganglion cells, *Int. J. Dev. Neurosci.* **19**(2):209-18 (2001).

AGIS Investigators, The advanced glaucoma intervention study (AGIS): 7. The relationship between control of intraocular pressure and visual field deterioration, *Am. J. Ophthalmol.* **130**(4):429-40 (2000).

E. A. Ansari E, J. E. Morgan, R. J. Snowden. Psychophysical characterisation of early functional loss in glaucoma and ocular hypertension, *B. J. Ophthalmol.* **86**(10):1131-5 (2002).

Q. Cui, H. K. Yip, R. C. Zhao et al., Intraocular elevation of cyclic AMP potentiates ciliary neurotrophic factor-induced regeneration of adult rat retinal ganglion cell axons, *Mol. Cell. Neurosci.* **22**(1):49-61 (2003).

S. L. Graham, S. M. Drance, B. C. Chauhan et al., Comparison of psychophysical and electrophysiological testing in early glaucoma, *Invest. Ophthalmol. Vis. Sci.* **37**(13):2651-62 (1997).

C. A. Johnson, Selective versus non-selective losses in glaucoma, *J. Glaucoma* **3** (Suppl: 1):S32-4 (1994).

E. C. Johnson, L. Jia, W. Cepurna et al., Elevated intraocular pressure affects the levels of neurofilament mRNA in the retina, *Invest. Ophthalmol. Vis. Sci.* (2000).

S. Laquis, E. Garcia-Valenzuela, S. C. Sharma, The patterns of retinal ganglion cell death in hypertensive eyes, *Invest. Ophthalmol. Vis. Sci.* **37**(3):S826 (1996).

K. M. Leahy, R. L. Ornberg, Y. Wang et al., Quantitative ex vivo detection of rodent retinal ganglion cells by immunolabeling Brn-3b, *Exp. Eye Res.* **79**(1):131-40 (2004).

J. E. Morgan, H. Uchida, J. Caprioli, Retinal ganglion cell death in experimental glaucoma, *B. J. Ophthalmol.* **84**(3):303-10 (2000).

J. E. Morgan, Retinal ganglion cell shrinkage in glaucoma, *J. Glaucoma* **11**(4):365-70 (2002).

J. C. Morrison, C. G. Moore, L. M. Deppmeier et al., A rat model of chronic pressure-induced optic nerve damage, *Exp. Eye Res.* **64**(1):85-96 (1997).

H. A. Quigley, Number of people with glaucoma worldwide, *B. J. Ophthalmol.* **80**(5):389-93 (1996).

H. A. Quigley, G. R. Dunkelberger, W. R. Green, Retinal ganglion cell atrophy correlated with automated perimetry in human eyes with glaucoma, *Am. J. Ophthalmol.* **107**(5):453-64 (1989).

H. A. Quigley, R. W. Nickells, L. A. Kerrigan et al., Retinal ganglion cell death in experimental glaucoma and after axotomy occurs by apoptosis, *Invest. Ophthalmol. Vis. Sci.* **36**(5):774-86 (1995).

D. A. Sholl, Dendritic organization in the neurons of the visual and motor cortices of the cat, *J. Anat.* **87**(4):387-406 (1953).

W. Sun, N. Li, S. He, Large-scale morphological survey of rat retinal ganglion cells, *Vis. Neurosci.* **19**(4):483-93 (2002).

A. J. Weber, P. L. Kaufman, W. C. Zhang Hubbard, Morphology of single retinal ganglion cells in the glaucomatous primate retina, *Invest. Ophthalmol. Vis. Sci.* **39**(12):2304-20 (1998).

Z. Zhang, D. M. Casey, J. P. Julien et al., Normal dendritic arborization in spinal motoneurons requires neurofilament subunit L, *J. Comp. Neurol.* **450**(2):144-52 (2002).

PART VII

INDUCED RETINAL DEGENERATIONS

NEURAL PLASTICITY REVEALED BY LIGHT-INDUCED PHOTORECEPTOR LESIONS

Bryan W. Jones, Robert E. Marc, Carl B. Watt, Dana K. Vaughan, and Daniel T. Organisciak*

1. INTRODUCTION

The retina has long been assumed to remain in stasis after photoreceptor degeneration effectively deafferents the neural retina (Zrenner, 2002). However, a growing literature reveals the more insidious details of retinal degeneration and evidence of early plasticity. Retinal degenerations typically undergo three phases. Early changes observed in phase one are triggered by photoreceptor stress and include misrouting of rhodopsin to the inner segments of photoreceptors (Milam et al., 1998) followed by rhodopsin delocalization to processes extending down in fascicles projecting into the inner nuclear and ganglion cell layers (Li et al., 1995; Milam et al., 1996). Phase two is characterized by active photoreceptor cell death eventually deafferenting bipolar cell populations and eliminating light mediated signaling to the neural retina. Also observed in phase two is the formation of the Müller cell (MC) seal, entombing or walling off the remnant neural retina from what is left of the retinal pigment epithelium and vascular choroid (Jones et al., 2001; Jones et al., 2003; Marc et al., 2003). Formation of the Müller cell seal is likely due to collapse of distal elements of Müller cells, but is also possibly due to hypertrophic processes. Before completion of phase two, all dendritic elements of bipolar cells have retracted and horizontal cells typically have sent axonal processes into the inner plexiform layer (IPL). (Strettoi and Pignatelli, 2000; Park et al., 2001; Strettoi et al., 2002; Strettoi et al., 2003). The final stage of remodeling, phase three, was originally described in the GHL mouse (Jones et al., 2001), however at the time the extent of remodeling across models and the implications for vision rescue was not appreciated. Subsequent work in naturally occurring and genetic models (Jones et al., 2003) revealed extensive remodeling in response to photoreceptor degeneration. This remodeling involves the evolution of processes from all classes of neurons into

---

*B.W. Jones, R.E. Marc, C.B. Watt, Ophthalmology, Univ Utah/Moran Eye Center, Salt Lake City, UT; D.T. Organisciak, Biology, Univ of Wisconsin Oshkosh, Oshkosh, WI; D.K. Vaughan, Biochemistry and Molecular Biology, Wright State Univ, Dayton, OH.

fascicles that may run for >100 microns in addition to elaboration of new "tufts" of IPL (microneuromas) that form outside the boundaries of the normal stratification of the IPL. These microneuromas are populated with synaptic contacts corruptive of normal visual processing (Marc et al., 2003). Finally, migration of adult neuronal phenotypes throughout the vertical axis of the retina is observed with all cell classes participating. It is believed that in order to maintain normal gene expression, neurons will sprout processes to seek lost glutamatergic signaling. Failing to achieve synaptic contact may result either in cell death or cellular soma migration to other regions of the retina. Amacrine cells are commonly observed translocating to the ganglion cell layer with ganglion cells also migrating up into the inner nuclear layer.

The work with naturally occurring human, and natural and genetically engineered animal models has revealed plasticity and retinal neural remodeling as the typical response to apparent sensory deafferentation. Therefore, while light damage has long been recognized as a way to "kill" photoreceptors, this study was designed to use the light damage methodology (LD) to assess the nature and scope of plasticity in the neural retina in response to deafferentation in an environmental rather than a genetic model of deafferentation.

2. METHODS

Over 90 albino Sprague-Dawley rat retinas were exposed to light (Organisciak et al., 1998) of varying durations, pre-adaption states, circadian phases and survival times. Post-euthanasia, enucleated eyes were rapidly fixed in glutaraldehyde, resin-embedded and thin sections (250 nm) were serially probed with IgGs generated against aspartate, glutathione, glutamate, glutamine, glycine, GABA, and taurine, key retinal metabolites and cell specific markers. Primary immunohistochemical labeling was followed by silver intensification with a secondary goat anti-rabbit IgG adsorbed to 1 nm gold particles and visualized with silver intensification (Kalloniatis and Fletcher, 1993). All images of immunoreactivity were captured as 8-bit greyscale images and registered to <250 nm root-mean-square error. Computational molecular phenotyping (Marc and Jones, 2002) was then employed to identify neurons. EM overlay was employed to identify signatures at the ultrastructural level (Marc and Liu, 2000).

3. RESULTS

In the LD model, Müller cells undergo dramatic transformations in areas with complete rod and cone loss. In these regions, plasticity ensues as evidenced by neuronal migration via formation of hypertrophic MC columns throughout the axis of the retina, with some neurons migrating from the retina into the remnant choroid. Synaptic remodeling is demonstrated with neurites engaging in novel, corrupt circuitry via GABAergic, glycinergic and glutamatergic synapses. Within 14 days of even a brief, 3 hr LD treatment, focal photoreceptor loss was accompanied by irregular 2-4 fold increases in RPE glutamine and rod aspartate levels, perhaps presaging cell death. The onset of MC remodeling (formation of a fibrotic glial seal in regions of extensive rod/cone cell death) is accompanied by a dramatic >10-fold increase in MC glutamine. This occurs only in MCs engaged in seal formation; MCs a mere 0.1 mm away are normal (Figure 57.1). By 60 days post exposure, most

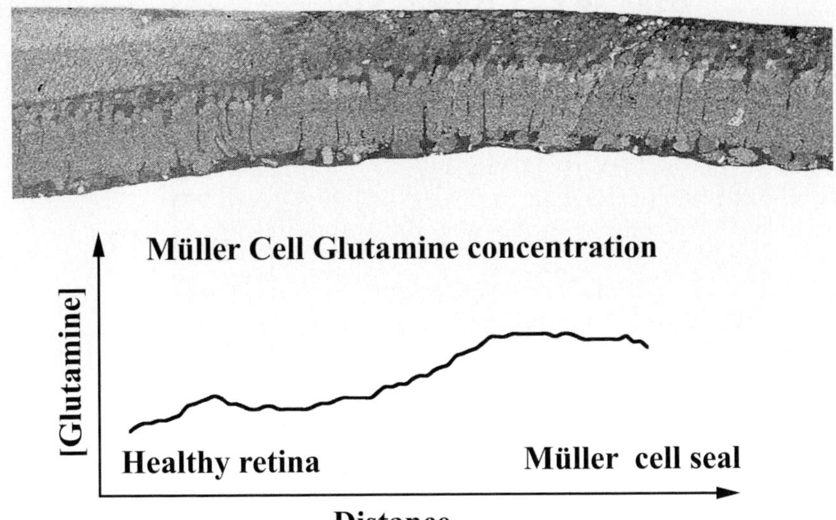

Figure 57.1. Early in the degenerative process, Müller cells exhibit large increases in glutamine concentration (often >10-fold increases) in areas where photoreceptor death is complete allowing Müller cells to begin seal formation.

photoreceptors have died and the glial seal has become complete in those areas with complete photoreceptor loss. Small microneuromas have begun to form, originating from sprouting bipolar cells and amacrine cells. Aditionally, fluid channels begin to form, likely originating from the Müller cell seal walling off of the neural retina from the vascular choroid. This seal likely impairs transretinal water flow (Bringmann et al., 2004), resulting in the formation of fluid channels or cysts. Those bipolar cells that have not sprouted and found targets to contact have begun the process of dying. For the most part however, at 60 days post exposure the normal lamination of the IPL is intact and most populations of cells other than the photoreceptors appear in approximately normal numbers. At approximately 120-240 days post-LD, when both neuronal migration on hypertrophic MC columns and synaptic remodeling are initiated, other more dramatic changes ensue. Synaptic remodeling is evinced by neuropil arising from new neurites in the remnant distal retina containing GABAergic, glycinergic, and glutamatergic synapses in novel circuits (Figure 57.2). Distal migration of MC nuclei, MC hypertrophy and disorganization of the inner nuclear layer, including cell loss, match remodeling processes in advanced genetic forms of retinal degeneration, including human retinitis pigmentosa. Neuronal migration throughout the axis of the retina is common. All classes of neurons participate including glycinergic amacrine cells migrating into the ganglion cell layer and ganglion cells can be observed migrating into the inner nuclear layer. By 240 days post-LD, the RPE has been obliterated, the vascular choroid has been compromised and there is extensive emigration of MCs and neurons from the neural retina proper into the remnant choroid, similar to that described by Sullivan et al. (Sullivan et al., 2003) for the aged ambient-LD rat. These neurons posess signatures unchanged from their signatures in the retina proper.

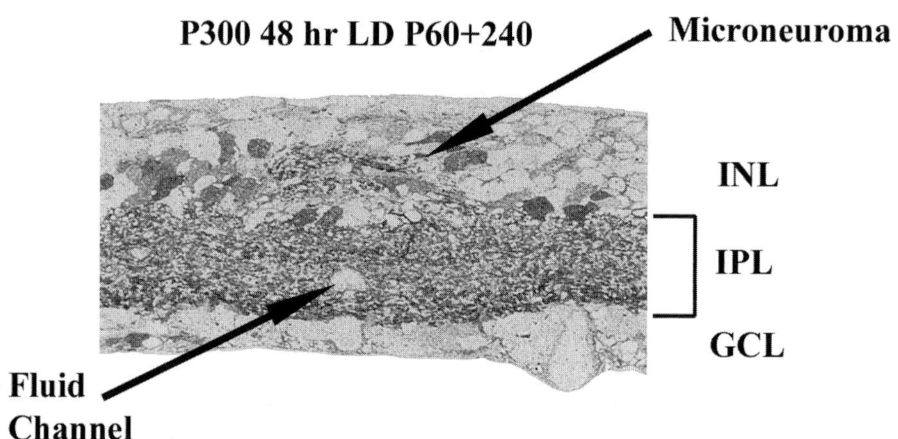

Figure 57.2. A 300 day old rat retina harvested 240 days after light damage at 60 days of age immunohisto-chemically labeled for GABA. Elaboration of microneuromas outside the normal lamination of the IPL are seen composed of tangles processes labeling for GABA, glutamate and glycine from amacrine cells, horizontal cells, bipolar cells and ganglion cells (Jones et al., 2003). Also observed in this image is the beginning formation of an aqueous fluid channel.

Confirming the exit from the neural retina proper required ultrastructural analysis. Therefore, electron microscopy (EM) with light microscopy overlay (Marc and Liu, 2000) was employed to demonstrate emigration of cells with mature neuronal phenotypes through Brüch's membrane. Figure 57.3 shows one such neuron: a GABAergic neuron that has completely passed through a small hole in Brüch's membrane demonstrating adult neuronal phenotypes remain stable when migrating through the retina and into the remnant choroid.

4. CONCLUSIONS

All insults that kill photoreceptors represent sensory deafferentations that trigger retinal remodeling akin to CNS plasticities, including neuronal loss, growth of new neurites, formation of new synapses, and reorganization of the neuronal and glial somatic positions. These data show LD that models exhibit plasticity that mirrors pathology observed in other models of retinal degeneration. LD is a fast, effective trigger of large-scale remodeling (perhaps due to the high temporal coherence of the insult) and enables study of circuitry defects emergent from remodeling.

Sparing of the ventral retina allows for a "built in" control, allowing us to compare within the same preparation both normal and remodeled portions of tissue. Furthermore, the LD model, with the possible exception of the conditional genetic knockouts, is the only model in which we know there are no developmental abnormalities with respect to circuitry and genetics throughout development making the LD model possibly the best model available for Age Related Macular Degeneration (AMD) and AMD like disorders. Cells outside the boundaries of the neural retina have escaped. Furthermore, they appear to have normal amacrine and bipolar cell signatures indicating their metabolic status appears to be stable.

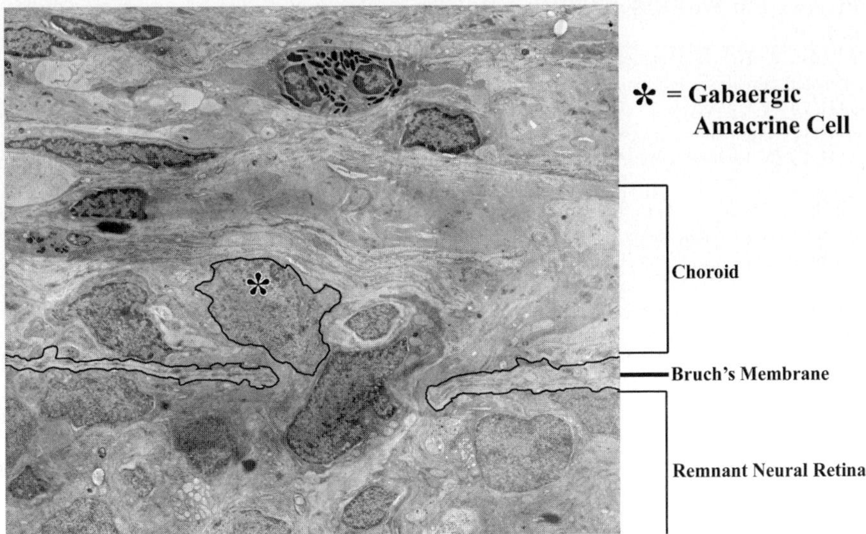

Figure 57.3. An electron micrograph of a GABAergic amacrine cell (confirmed by CMP EMoverlay (Marc and Liu, 2000)). Bruch's Membrane has been breached allowing neurons to escape the remnant neural retina into the choroid.

Many of these neurons have rewired and apparently have established some form of connectivity keeping them alive. Furthermore, the glial cells are also emigrating, leaving an abandoned retina which, for all intents and purposes may be dead. Rescue at this point is impossible.

These data show light-damaged models exhibit plasticity that mirrors pathology observed in other models of retinal degeneration. We suggest that all insults resulting in loss of photoreceptors represent sensory deafferentations, triggering retinal remodeling akin to CNS plasticities, including neuronal loss, growth of new neurites, formation of new synapses, and reorganization of neuronal and glial somatic positions and finally, adult neurons can migrate without first de-differentiating.

5. REFERENCES

Bringmann A, Reichenbach A, Wiedemann P. 2004. Pathomechanisms of cystoid macular edema. Ophthalmic Res **36**:241-249.

Jones BW, Baehr W, Frederick JM, Marc RE. 2001. Aberrant remodeling of the neural retina in the GHL transgenic mouse. In: Invest Ophthalmol Vis Sci.

Jones BW, Watt CB, Frederick JM, Baehr W, Chen CK, Levine EM, Milam AH, LaVail MM, Marc RE. 2003. Retinal remodeling triggered by photoreceptor degenerations. Journal of Comparative Neurology **464**:1-16.

Kalloniatis M, Fletcher EL. 1993. Immunocytochemical localization of the amino acid neurotransmitters in the chicken retina. J Comp Neurol **336**:174-193.

Li ZY, Kljavin IJ, Milam AH. 1995. Rod photoreceptor neurite sprouting in retinitis pigmentosa. Journal of Neuroscience **15**:5429-5438.

Marc RE, Jones BW. 2002. Molecular phenotyping of retinal ganglion cells. Journal of Neuroscience **22**:412-427.

Marc RE, Jones BW, Watt CB, Strettoi E. 2003. Neural Remodeling in Retinal Degeneration. Prog Ret Eye Res **22**:607-655.

Marc RE, Liu W. 2000. Fundamental GABAergic amacrine cell circuitries in the retina: nested feedback, concatenated inhibition, and axosomatic synapses. Journal of Comparative Neurology **425**:560-582.

Milam AH, Li ZY, Cideciyan AV, Jacobson SG. 1996. Clinicopathologic effects of the Q64ter rhodopsin mutation in retinitis pigmentosa. Invest Ophthalmol Vis Sci **37**:753-765.

Milam AH, Li ZY, Fariss RN. 1998. Histopathology of the human retina in retinitis pigmentosa. Prog Ret Eye Res **17**.

Organisciak DT, Darrow RM, Barsalou L, Darrow RA, Kutty RK, Kutty G, Wiggert B. 1998. Light history and age-related changes in retinal light damage. Invest Ophthalmol Vis Sci **39**:1107-1116.

Park SJ, Kim IB, Choi KR, Moon JI, Oh SJ, Chung JW, Chun MH. 2001. Reorganization of horizontal cell processes in the developing FVB/N mouse retina. Cell Tissue Res **306**:341-346.

Strettoi E, Pignatelli V. 2000. Modifications of retinal neurons in a mouse model of retinitis pigmentosa. Proc Natl Acad Sci **97**:11020-11025.

Strettoi E, Pignatelli V, Rossi C, Porciatti V, Falsini B. 2003. Remodeling of second-order neurons in the retina of rd/rd mutant mice. Vision Res **43**:867-877.

Strettoi E, Porciatti V, Falsini B, Pignatelli V, Rossi C. 2002. Morphological and functional abnormalities in the inner retina of the rd/rd mouse. Journal of Neuroscience **22**:5492-5504.

Sullivan R, Penfold P, Pow DV. 2003. Neuronal migration and glial remodeling in degenerating retinas of aged rats and in nonneovascular AMD. Invest Ophthalmol Vis Sci **44**:856-865.

Zrenner E. 2002. Will Retinal Implants Restore Vision? Science **295**:1022-1025.

CHAPTER 58

FACTORS UNDERLYING CIRCADIAN DEPENDENT SUSCEPTIBILITY TO LIGHT INDUCED RETINAL DAMAGE

Ruby Grewal[1], Daniel Organisciak[2], and Paul Wong[1,3]

1. INTRODUCTION

Retinal cell loss in diseases such as Retinitis Pigmentosa occurs through an apoptotic process.[1] The mechanism of this cell loss is not completely understood. Models that allow for the study of conditions in which the retina is susceptible or resistant to retinal damage help to elucidate the mechanism underlying the cell death. One model that is used to study retinal cell loss is the light induced retinal degeneration (LIRD) model.[2] Intense light exposure leads to rhodopsin bleaching[2] and is the trigger for the subsequent photoreceptor cell degeneration, as blocking the regeneration of rhodopsin prevents photoreceptor cell death.[3] Oxidative stress is also involved in retinal degeneration,[2,4-7] and the administration of natural or synthetic antioxidants prior to light exposure prevents the subsequent cell loss.[7-13] Many factors can influence the extent of light induced damage including prior light history of the animals, age, genetics, and diet.[4,8,14] The extent of LIRD is also dependent on the time of light exposure initiation. Animals exposed to light during the dark phase of the dark-light cycle suffer greater retinal damage than rats exposed to light during the day.[15] More recently, it was reported by Organisciak *et al* that relatively brief intense light exposure commencing at 0100h lead to a 2-4 fold greater loss of photoreceptor cells in rat retina than light exposure beginning at 1700h.[16] A fundamental question then is to ask what differences exist in the retina at various times of the day. It was suggested that endogenous factors regulated in a circadian manner might be involved in the observed difference in susceptibility to LIRD.[16] Circadian rhythms cue an organism about night/day changes, have a period of approximately 24 hours, and persist in constant darkness or constant light. These rhythms

[1] Department of Biological Sciences, University of Alberta, Edmonton, Alberta, Canada T6G 2E7; [2] Petticrew Research Laboratory, Department of Biochemistry and Molecular Biology, Wright State University, Dayton, Ohio, 45435, USA; [3] Departement of Ophthalmology, Emory University Eye Center, Atlanta, Georgia, 30322, USA. Corresponding author: R. Grewal, E-mail: rubyg@ualberta.ca.

411

are entrained by light, but can also be entrained by temperature[17] and feeding.[18] Circadian rhythms underlie many physiological processes that can lead to a stress tolerant or susceptible environment. There are several prominent circadian regulated physiological processes in the retina. Melatonin and dopamine are synthesized in a circadian manner or in response to light-dark cues, and may be involved in entraining the circadian cycle. They are the most obvious candidates for involvement in increased retinal susceptibility to light. Phototransduction protein levels are also important candidates as they are directly involved in modulating the retinas response to light. A fairly new candidate is metabolic activity of the retina. Increased metabolic activity results in decreased pH levels which may also play a key role as it has been shown to be involved in inducing apoptosis.[19] Any or all of these factors may be involved in the circadian dependent susceptibility to light damage. This review covers four important candidates that may play a role in circadian dependent susceptibility to light damage. A complete review of all the factors that may be involved in this phenomenon is beyond the scope of this paper.

2. MELATONIN

Melatonin produced by the pineal is thought to act as an endocrine hormone, entraining overall circadian rhythms.[20] In the retina, photoreceptor cells produce melatonin where it acts as a paracrine hormone.[21,22] The highest levels of melatonin are present during the night with levels decreasing after light onset.[23] Studies have reported that the Mel_{1b} and Mel_{1c} melatonin receptor RNA or protein are expressed in Xenopus retina[24,25] and that the melatonin MT1 receptor protein is expressed in human retina.[26] It has been suggested that the function of melatonin expression at night is to increase the sensitivity of the visual system, and in this way facilitate dark adaptation.[27] Exogenous melatonin applied to the retina has been shown to increase light induced retinal damage.[28] In contrast, the application of the melatonin receptor antagonist luzindole prevents light induced damage.[29] These results suggest that melatonin binding to receptors in the retina results in the increased retinal susceptibility to light damage. However, melatonin has also been shown to be a direct free radical scavenger, and to act as an indirect antioxidant by inducing endogenous antioxidative enzymes.[30] This creates a paradox as to the role that melatonin plays in retinal susceptibility to light damage. Although melatonin has been shown to increase sensitivity through receptor binding, oxidative stress also plays a key role in cell loss. The observation that melatonin has antioxidative properties would suggest that it would decrease susceptibility to light damage. To date, the role of melatonin in LIRD is unclear. A study by Wiechmann[31] examined melatonin induced gene expression changes by microarray analysis. 14 genes were found to change in expression in the retina, and 17 genes changed in expression in the RPE. Among the genes that changed in expression were gamma crystallin, CREB protein, CED 6, and NGF induced transcription factor.[31] Further examination of these genes may help clarify melatonin function in the retina.

3. DOPAMINE

Dopamine is synthesised in the amacrine or interplexiform cells of the retina depending on the species.[32] Studies have shown that in a number of species there is a higher concentration of dopamine during the light phase of the light cycle than in the dark phase.[32,33] Studies with the chick retina have shown rhythmic changes in dopamine levels for several days in chicks kept in constant darkness, and an increase in dopamine levels in chicks that are exposed to light during the dark phase of the light cycle.[33] Photoreceptor inner segments have D2 receptors[34] and receive paracrine input from the dopamine producing cells. Dopamine has been shown to induce apoptosis in a number of cultured neuronal and non-neuronal cell types including chick embryo sympathetic neurons and rat PC12 cells.[35,36] However it's role in retinal survival or cell loss is unclear.

4. PHOTOTRANSDUCTION PROTEINS

A number of studies have shown that many proteins involved in phototransduction have a light-dependent expression patterns. However many of the observed changes tend to be species specific. The following section will only address the mouse and rat retina.

Opsin mRNA in the mouse retina is highest just before light onset.[37] Whole eye rhodopsin protein levels in the rat retina are higher at night than in the day in both the dark reared and cyclic light reared rat.[16] However the levels are significantly higher in only cyclic-light reared rats.[16] Therefore differences in rhodopsin levels are not likely to be responsible for the difference in light susceptibility in dark-reared rats. It has been suggested that there may be a difference in rhodopsin regeneration between night and daytime.[38] However, it was found that RPE65 protein levels, and rhodopsin regeneration kinetics, did not change between night and day.[38] It has also been suggested that rhodopsin phosphorylation patterns, which are a measure of the activity of the protein, may vary at different times of the day.[39] However, it was again found that the phosphorylation patterns do not change between night and day.[39]

Transducin is a G protein, activated by rhodopsin, which propagates the visual signal.[40] Alpha-transducin mRNA levels in rat retina are highest immediately after lights on, and alpha-transducin protein is highest in the inner segments during the day, until it is transported to the outer segments at night.[41] The presence of transducin in the outer segments may increase the sensitivity of the photoreceptor rod cells to light exposure at night.

Recoverin is involved in the recycling of activated rhodopsin. An increase in the length of time that rhodopsin is in the activated state has been suggested to increase LIRD in the 'equivalent to light' hypothesis.[42] Recoverin delays the termination of phototransduction by inhibiting rhodopsin kinase activity in photoreceptor cells.[40] Recoverin mRNA and protein levels in the rat retina are high during the light period, and then decrease sharply after the onset of darkness. Levels again increase throughout the dark period, with a sharp increase in protein levels late at night, approximately 4 hours before light onset.[43] These results suggest that low recoverin levels at night may increase sensitivity to light exposure. However the sharp increase in levels before light onset makes this uncertain.

5. METABOLISM

Significant differences in the metabolic activity exist in the retina at different times of the day. Increased metabolism appears to take place at night in several species including the rabbit[44] and goldfish retina.[45] Several studies have shown an *in vitro* decrease in the extracellular pH (pH_o) of the retina at night. The change in pH_o appears to occur at dusk, when a rapid drop is observed. After dusk pH_o remains stable, until lights on, when it again increases.[46] The extracellular pH of the *in vitro* rabbit retina varies within the retinal layers, and is the lowest at the outer limiting membrane, which is located near the inner segments of photoreceptor cells. Because the pH_o is lowest at the outer limiting membrane, photo-receptor cells are likely the primary source of acid production in the retina.[46]

It was also observed that removing glucose from the solution surrounding the retina, and thereby inhibiting energy metabolism, lead to a decrease in acid production and a reduction in the difference of pH between the retina and the surrounding solution.[43] Inhibition of ATP utilization, through the inhibition of the Na^+/K^+ ATPase also reduced acid production. Therefore the observed change in pH levels was likely due to an increase in metabolism, and an increase in proton production at night. The observed change in pH_o between night and day is not due to an increase in glycolysis over oxidative phosporylation, as both processes occur at the same proportion during the day and night.[43] There is evidence that cellular acidification can initiate apoptosis. It has been shown that cellular acidification increases susceptibility of cells to heat damage.[47] It has also been shown that neutrophils in tissue culture that undergo spontaneous apoptosis have an increase in acid production before the onset of cell death.[19] These studies suggest that cellular pH levels can affect cell response to subsequent stress stimuli, and cellular acidification can lead to an induction of apoptosis. Therefore the increased acidification of the retina at night may lead to an increased susceptibility to a stress signal, such as intense light exposure.

6. CONCLUSION

The retinal environment changes in response to differing physiological requirements during the night and day. These changes may lead to subsequent differences in response to light exposure. Previous research has established that the retina is more susceptible to light damage during the dark phase of the light cycle. Although many circadian changes in hormone, protein, and gene expression levels have been documented, the mechanism underlying the circadian difference in susceptibility is not understood. Current studies examining the molecular environment of the retina at different times of the day using array screening may help elucidate some of the processes that result in the difference in retinal susceptibility. Preliminary results from our lab suggest that there are a number of genes that are differentially expressed at different times of the day that have not been characterized as such previously (unpublished results). An understanding of the conditions under which the retina is susceptible or resistant to retinal damage may help us understand the mechanism underlying cell death in retinal eye diseases.

7. ACKNOWDELGMENTS

This work was supported by NSERC (PW, RG); RP Foundation Fighting Blindness (PW); Alberta Ingenuity (RG); NIH grant EY-01959 (DTO); and M Petticrew Springfield, OH (DTO). We thank the Foundation Fighting Blindness for the travel award provided to RG to attend this meeting.

8. REFERENCES

1. A. H. Milam, Z. Y. Li, R. N. Fariss, Histopathology of the human retina in retinitis pigmentosa, *Prog Retin Eye Res*, **17**(2):175-205 (1998).
2. W. K. Noell, V. S. Walker, B. S. Kang, Retinal damage by light in rats, *Invest Ophthalmol Vis Sci*, **5**(5):450-473 (1966).
3. C. Grimm, A. Wenzel, F. Hafezi, S. Yu, M. Redmond, C. E. Reme, Protection of Rpe65-deficient mice identifies rhodopsin as a mediator of light-induced retinal degeneration, *Nat Genet*, **25**(1):63-66 (2000).
4. W. K. Noell, D. T. Organisciak, H. Ando, M. A. Braniecki, C. Durlin, Ascorbate and dietary protective mechanisms in retinal light damage of rats: electrophysiological, histological, and DNA measurements, *Prog Clin Biol Res*, **247**:469-483 (1987).
5. D. T. Organisciak, R. K. Kutty, M. Leffak, P. Wong, S. Messing, B. Wiggert, R. M. Darrow, G. J. Chader, Oxidative damage and responses in retinal nuclei arising from intense light exposure, *Degenerative Disease of the Retina*, edited by R. E. Anderson, M. M. LaVail, J. G. Hollyfield (Plenum Press, New York, 1995), pp. 9-17.
6. R. K. Kutty, G. Kutty, B. Wiggert, G. J. Chader, R. M. Darrow, D. T. Organisciak, Induction of heme oxygenase 1 in the retina by intense visible light: suppression by the antioxidant dimethylthiourea, *PNAS USA*, **92**(4):1177-1181 (1995).
7. D. T. Organisciak, R. A. Darrow, L. Barsalou, R. M. Darrow, L. A. Lininger, Light-induced damage in the retina: differential effects of dimethylthiourea on photoreceptor survival, apoptosis and DNA oxidation, *Photochem Photobiol*, **70**(2):261-268 (1999).
8. D. T. Organisciak, B. S. Winkler, Retinal light damage: practical and theoretical considerations, *Prog Retin Eye Res*, **13**:1-30 (1994).
9. D. T. Organisciak, H. M. Wang, Z. Y. Li, M. O. Tso, The protective effect of ascorbate in retinal light damage of rats, *Invest Ophthalmol Vis Sci*, **26**(11):1580-1588 (1985).
10. D. T. Organisciak, R. M. Darrow, Y. I. Jiang, G. E. Marak, J. C. Blanks, Protection by dimethylthiourea against retinal light damage in rat, *Invest Ophthalmol Vis Sci*, **33**(5):1599-1609 (1992).
11. S. Lam, M. O. Tso, D. H. Gurne, Amelioration of retinal photic injury in albino rats by dimethylthiourea, *Arch Ophthalmol*, **108**(12):1751-1757 (1990).
12. J. Li, D. P. Edward, T. T. Lam, M. O. Tso, Amelioration of retinal photic injury by a combination of flunarizine and dimethylthiourea, *Exp Eye Research*, **56**(1):71-78 (1993).
13. I. Ranchon, J. M. Gorrand, J. Cluzel, M. T. Droy-Lefaix, M. Doly, Functional protection of photoreceptors from light-induced damage by dimethylthiourea and Ginkgo biloba extract, *Invest Ophthalmol Vis Sci*, **40**(6):1191-1199 (1999).
14. D. T. Organisciak, R. M. Darrow, L. Barsalou, R. A. Darrow, R. K. Kutty, G. Kutty, B. Wiggert, Light history and age-related changes in retinal light damage, *Invest Ophthalmol Vis Sci*, **39**(7):1107-1116 (1998).
15. T. E. Duncan, W. K. O'Steen, The diurnal susceptibility of rat retinal photoreceptors to light-induced damage, *Exp Eye Res*, **41**(4):497-507 (1985).
16. D. T. Organisciak, R. M. Darrow, L. Barsalou, R. K. Kutty, B. Wiggert, Circadian-dependent retinal light damage in rats, *Invest Ophthamol Vis Sci*, **41**(12):3694-3701 (2000).
17. R. Ben-Shlomo, C. P. Kyriacou, Circadian rhythm entrainment in flies and mammals, *Cell Biochem Biophys*, **37**(2):141-56 (2002).
18. Z. Boulos, M. Terman, Food availability and daily biological rhythms, *Neurosci Biobehav Rev*, **4**(2):119-131 (1980).
19. R. A. Gottlieb, H. A. Giesing, J. Y. Zhu, R. L. Engler, B. M. Babior, Cell acidification in apoptosis: granulocyte colony-stimulating factor delays programmed cell death in neutrophils by up-regulating the vacuolar H+-ATPase, *Proc Natl Acad Sci. USA.* **92**(13):5965-5968, (1995).
20. R. J. Reiter, Melatonin: the chemical expression of darkness, *Mol Cell Endocrinol*, **79**(1):C153-C158 (1991).
21. J. C. Besharse, D. A. Dunis, Methoxyindoles and photoreceptor metabolism: activation of rod shedding, *Science*, **219**(4590):1341-1342 (1983).
22. M. L. Dubocovich, Melatonin is a potent modulator of dopamine release in the retina, *Nat*, **306**(5945):782-784 (1983).
23. S. F. Pang, H. S. Yu, H. C. Suen, G. M. Brown, Melatonin in the retina of rats: a diurnal rhythm, *J Endocrinol*, **87**(1):89-93 (1980).
24. A. F. Wiechmann, A. R. Smith, Melatonin receptor RNA is expressed in photoreceptors and displays a cyclic rhythm in *Xenopus* retina, *Mol Brain Res*, **91**:104-111 (2001).

25. A. F. Wiechmann, C. R. Wirsig-Wiechmann, Multiple cell targets for melatonin in *Xenopus laevis* retina: distribution of melatonin receptor immunoreactivity, *Vis Neurosci*, **18**(5):694-702 (2001).
26. J. Scher, E. Wankiewicz, G. M. Brown, H. Fujieda, MT (1) melatonin receptor in the human retina: expression and localization, *Invest Ophthamol Vis Sci*, **43**(3):889-897 (2002).
27. A. F. Wiechmann, X. L. Yang, S. M. Wu, J. G. Hollyfield, Melatonin enhances horizontal cell sensitivity in salamander retina, *Brain Res*, **453**(1):377-380 (1988).
28. A. F. Wiechmann, W. K. O'Steen, Melatonin increases photoreceptor susceptibility to light-induced damage, *Invest Ophthamol Vis Sci*, **33**(6):1894-1902 (1992).
29. T. Suguwara, P. A. Sieving, P. M. Iuvine, R. A. Bush, The melatonin receptor antagonist luzindole protects photoreceptors from light damage in rats, *Invest Ophthalmol Vis Sci*, **39**:2458-2465 (1998).
30. R. J. Reiter, D. Tan, J. C. Mayo, R. M. Sainz, J. Leon, Z. Czarnocki, Melatonin as an antioxidant: biochemical mechanism and pathophysiological implication in humans, *Acta Bio Pol*, **50**(4):1129-1146 (2003).
31. A. F. Wiechmann, Regulation of gene expression by melatonin: a microarray survey of the retina, *J Pineal Res.* **33**(3):178-185 (2002).
32. M. A. Djamgoz, H. J. Wagner, Localization and function of dopamine in the adult vertebrate retina, *Neurochem. Int*, **20**(2):139-191 (1992)
33. J. B. Zawilska, A. Bednarek, M. Berezinska, J. Z. Nowak, Rhythmic changes in metabolism of dopamine in the chick retina: the importance of light versus biological clock, *J Neurochem*, **84**(4):717-724 (2003).
34. Z. Muresan, J. C. Besharse, D2-like dopamine receptors in amphibian retina: localization with fluorescent ligands, *J Comp Neurol*, **331**(2):149-160 (1993).
35. I. Ziv, E. Melamed, N. Nardi, D. Luria, A. Achiron, D. Offen, A. Barzilia, Dopamine induces apoptosis like cell death in cultured chick sympathetic neurons a possible novel pathogenetic mechanism in Parkinson's disease, *Neurosci Lett*, **170**:136-140 (1994).
36. D. Offen, I. Ziv, H. Panet, L. Wasserman, R. Stein, E. Melamed, A. Barzilai, Dopamine-induced apoptosis is inhibited in PC12 cells expressing Bcl-2, *Cell Mol Neurobiol*, **17**(3):289-304 (1997).
37. C. Bowes, T. van Veen, D. B. Farber, Opsin, G-protein and 48-kDa protein in normal and rd mouse retinas: developmental expression of mRNAs and proteins and light/dark cycling of mRNAs, *Exp Eye Res*, **47**(3):369-90 (1988).
38. J. Beatrice, A. Wenzel, C. E. Reme, C. Grimm, Increased light damage susceptibility at night does not correlate with RPE65 levels and rhodopsin regeneration in rats, *Exp Eye Res*, **76**(6):695-700 (2003).
39. Z. Ablonczy, R .M. Darrow, D. R. Knapp, D. T. Organisciak, R. K. Crouch, Rhodopsin Phosphorylation in Rats Exposed to Intense Light. *Photochem Photobiol.* 2004 Aug 1; [Epub ahead of print].
40. C. W. Oyster. *The human eye structure and function* (Sinauer Associates Inc, Massachusetts, 1999).
41. M. R. Brann, L. V. Cohen, Diurnal expression of transducin mRNA and translocation of transducin in rods of rat retina, *Science*, **235**(4788):585-587 (1987).
42. G. L. Fain, J. E. Lisman, Photoreceptor degeneration in vitamin A deprivation and retinitis pigmentosa: the equivalent light hypothesis, *Exp Eye Res*, **57**(3):335-340 (1993).
43. A. F. Wiechmann, M. K. Sinacola, Diurnal expression of recoverin in the rat retina, *Mol Brain Res*, **45**(2):321-324 (1997).
44. A. V. Dmitriev, S. C. Mangel, Circadian clock regulation of pH in the rabbit retina, *J Neurosci*, **21**(8):2897-2902 (2001).
45. A. V. Dmitriev, S. C. Mangel, A circadian clock regulates the pH of the fish retina, *J Pysiol*, **522**(1):77-82 (2000).
46. A. V. Dmitriev, S. C. Mangel, Retinal pH reflects retinal energy metabolism in the day and night, *J Neurophysiol*, **91**(6):2404-2412 (2004).
47. C. W. Song, J. C. Lyons, R. J. Griffin, C. M. Makepeace, E. J. Cragoe, Increase in thermosensitivity of tumor cells by lowering intracellular pH, *Cancer Res*, **53**(7):1599-1601.

CHAPTER 59

SPACE FLIGHT ENVIRONMENT INDUCES DEGENERATION IN THE RETINA OF RAT NEONATES

Joyce Tombran-Tink and Colin J. Barnstable*

1. INTRODUCTION

Retinal degenerations can be promoted by many factors including ageing, ischemia, fluctuation in oxygen tension, oxidative stress, and increased intraocular pressure. We present new evidence that the environment encountered in space shuttle flight can also disrupt normal retinal development and mimic stimuli that induce retinal degenerations on earth. There is experimental evidence linking anomalies in visual perception with space flights since the Apollo missions (Phillpot et al., 1978; Newberg and Alavi, 1998). There is also strong evidence that pathological stimuli that disrupt retinal structure and function on earth are encountered in the space shuttle environment as well. Orbital space flights cause physiological disturbances in humans including cephalad fluid shift (Hoffler et al., 1977; Drummer, 2000), increased intraocular pressure (Mader et al., 1990; Draeger, 2000) disruption of cardiovascular function (Wang et al., 1996) and stress on the musculoskeletal system (Lane and Feedback, 2002; LeBlanc, 2000).

Animal models provide a unique opportunity to study the effects of space hazards and mechanisms of adaptation of the central nervous system to the space environment. In this paper, we report the physiological effects of space travel on the retina of rodents (NIH.R3 experiment) at various stages of postnatal development during orbital flight on Mission STS-72, which was launched in 1996.

*Joyce Tombran-Tink, Division of Pharmaceutical Sciences, UMKC, Kansas City, MO 64110. Colin J. Barnstable, Department of Ophthalmology and Visual Science, Yale University School of Medicine, New Haven, CT 06520.

2. MATERIALS AND METHODS

2.1. STS-72 In-Cabin Payloads

STS-72 was launched at the Kennedy Space Center on a 9 day Mission on January 11, 1996. The space shuttle carried the NIH.R3 Life Science Payload, a proof-of-concept study designed to test whether the Animal Enclosure Module (AEM-NF) nursing facility was capable to support nursing rats and neonates, to evaluate maternal behavior of rat dams in the cages, and to verify the retrieval of clinically healthy animals post flight. Six litters of Sprague-Dawley rats were used in the experiment. Each litter consisted of one nursing dam and 10 neonates. Two identically age and weight-matched litters were launched for each neonate ages at post-natal days 5 (PN5), 8 (PN8), and 15 (PN15). Similar groups were house on earth as ground controls.

2.2. In-Flight Rodent Conditions

All animals were fed normal chow or nursed and exposed to normal on/off light cycles. In flight activities consisted of daily health checks, water refill, and videotaping the animals. The rats were in flight for 9 days. STS-72 orbital data were: Altitude: 250 nautical miles; (288 statute miles) Inclination: 28.45 degrees. Number of Orbits: 142; Duration: 8 days, 22 hours, 01 minutes, 47 seconds; Distance travelled: 3.7 million miles.

2.3. Post Flight Evaluation of the Rodents

Post-flight, animals were removed from the AEM-NF and assessed for survival rate and health. It was reported that 6 out of 20 PN5 neonates, 19 out of 20 PN8 neonates, and all of the PN15 neonates survived the 9 day Space Mission. All flight groups weighed less than the corresponding ground controls, possible due to cage effect and independent of gravity or microgravity.

All surviving flight animals were in good condition as described by the inspecting veterinarian. The AEM-NF supported rat neonates aged PN8 or older at launch, but was unsuitable for the support of PN5 or younger rodents at launch. Because of the sample size of only 2 dams, it was difficult to determine whether the poor survival of the PN5 group was due to poor dam behavior.

2.4. Dissection and Preservation of the Rodent Eyes Post Flight

Healthy rodents were sacrificed immediately after the shuttle returned to earth and the eyes from both experimental and ground controls (caged and uncaged) were dissected and placed in 4% paraformaldehyde. The samples were archived at the Ames Research Laboratories until histological and morphometric studies were performed.

2.5. Preparation of Retinal Tissue for Histological Examination

The NIH.R3 neonatal rodent eyes were obtained from the Ames Research Laboratories, after inspection by Drs. Paul Callaghan and Alison French. Age and weight matched controls were also obtained for comparison in our study to determine the effects of space flight conditions on retinal cytoarchitecture.

The anterior segment of each eye was removed with a sharp razor blade and the posterior segment cryoprotected overnight in 30% sucrose. The eyecups were place in freezing molds containing 100% OCT and frozen overnight at −80°. Samples were cryosectioned at 15 µm and thawed onto subbed slides. Sections were stained with hematoxylin/eosin for five minutes at room temperature. After rinsing, sections were dehydrated, cleared, and mounted. Sections were viewed with brightfield and DIC optics using a Zeiss microscope.

2.6. Immunocytochemistry, Lectin, and Morphometric Studies

Sections were rehydrated in PBS, preincubated in 5% normal goat serum, 0.1% Triton X-100 in PBS (G-PBS) to block non-specific binding and then incubated in primary antibody overnight at 4°C. After washing in PBS, sections were incubated for 45 min at room temperature in Cy3-labeled secondary antibody diluted in G-PBS. After washing with PBS, sections were mounted in VectaShield with DAPI, coverslipped and viewed with a Zeiss widefield microscope equipped with DIC and epifluorescence optics. Control sections were incubated in normal mouse or rabbit serum diluted to an equivalent immunoglobulin concentration as test antibodies.

Calibrated images of sections were collected from several regions of the central retina of each section. The thickness of various retinal layers was measured in microns. At least three non-adjacent sections of each eye were measured.

3. RESULTS

The focus of this study was to examine the histological appearance of the retinas of rodents exposed to space environment. The results presented below represent the first findings of photoreceptor outer segment loss and other disruption of normal retinal development in the orbiting eyes. The results were obtained after examination of a large number of retinal sections obtained from at least 3 neonate rats at 3 different stages of development flown on orbital flight (STS-72) for 9 days.

3.1. Loss of Photoreceptor Outer Segments and Disruption of the RPE Monolayer in the Retina of Neonatal Rodents During Orbital Space Flights

The most striking difference among the space eyes and controls was the decreased length of rod photoreceptor outer segments. In normal rat retinal development, the outer segments of photoreceptor cells start to develop at about PN5 and reach their full length by about PN28 (Obata and Usukura, 1992; Bumsted et al., 2001). In figure 1, we show the length of the photoreceptor outer segments in rodents at PN5 launch (Fig 59.1A), PN8 launch (Fig 59.1B) and PN15 launch (Fig 59.1C). At all three postnatal days, the photoreceptor outer segments (OS) were either absent or shortened in animals exposed to 9 days of space travel environment. In addition, we observe that immature rod cell bodies, labeled with the RET P1 opsin antibody, present in the inner nuclear layer (INL) of both ground controls, are absent in the experimental animals. The histological preparations (Fig 59.2) show disruption of the retinal pigment epithelium (RPE). In most cases the RPE layer is

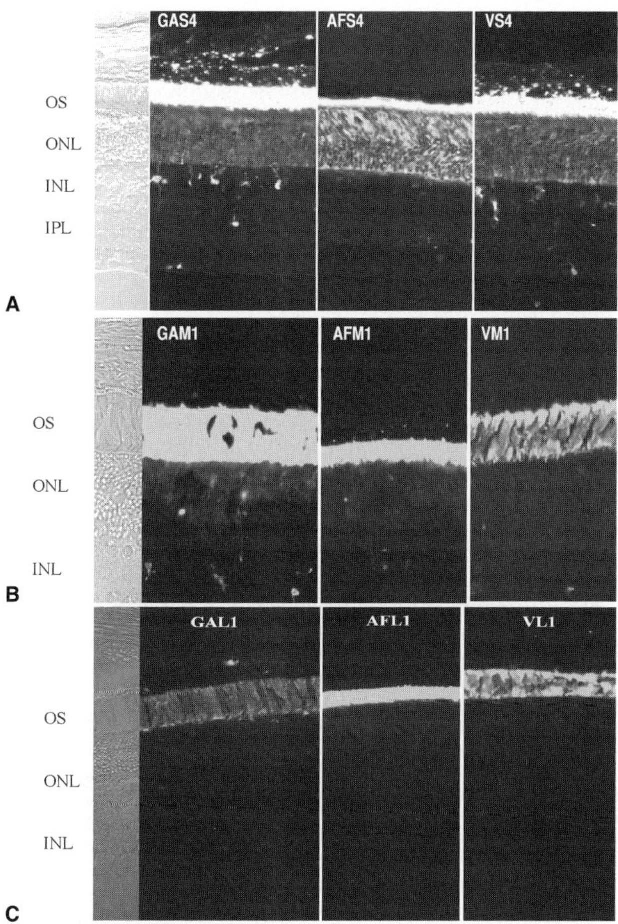

Figure 59.1. RET-P1 labeling showing differences in length of photoreceptor outer segments in female Sprague Dawley rats. **A.** PN5 launch, dissected PN14 GAS1:ground control (no cage), wt-10.9 g. AFS1: flight animals, wt-10.1 g; VS4: control (ground-cage), wt-10.7 g. **B.** PN8 launch, dissected PN17. GAM1:ground control (no cage), wt-15.6 g. AFM1: flight animals, wt-16.5 g; VM1: control (ground-cage), wt-16.2 g. **C.** PN15 launch, dissected PN24. GAL1:ground control (no cage), wt-29.7 g. AFL1: flight animals, wt-29.2 g; VL1: control (ground-cage), wt-30 g.

discontinuous or is not seen attached to the retina in the flight animals. There is little difference in the length of the inner segments of rod photoreceptors in the ground control and in flight rodents. Measurements of the average length of photoreceptor outer segments in control and experimental groups for each age are presented in Figure 59.3.

3.2. Disruption of Normal IPL Development

The IPL contains the synapses that link the bipolar cells to the amacrine and ganglion cells of the inner retina. In normal development conventional synapses in the IPL occurs

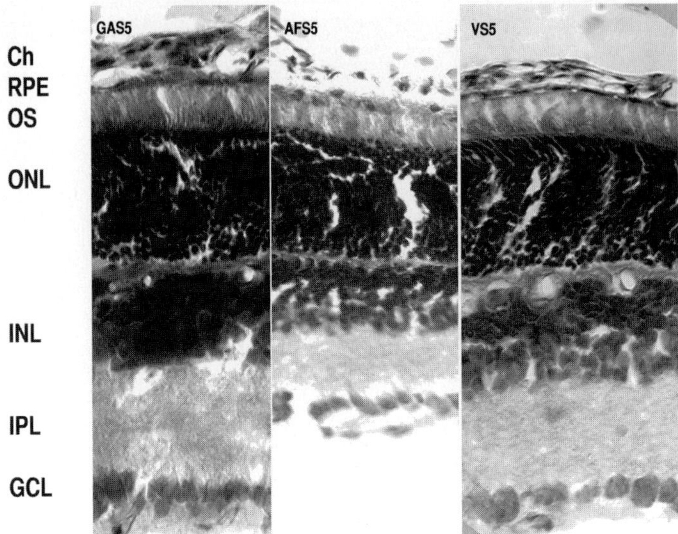

Figure 59.2. Hematoxylin and Eosin staining of the PN5-PN14 animal series. Note disrupted choroid (Ch) and choroidal vessels, lack of distinct RPE layer (RPE), and thinner IPL between ground control and flight tissue.

between PN3-PN10 and ribbon synapses between PN11-15 (Weidman and Kuwabara, 1968; Robinson, 1991). PN15 marks the time of eye opening in the animals and at this time there is a sharp reduction in both types of synapses. The widths of the IPL in ground controls and in flight rat neonates as well as those for the IS and OS are given in Figure 59.3. At PN5, PN8, and PN 15, the average width of the IPL in ground controls are approximately 48 μm, 58 μm, and 61 μm, respectively. In neonates exposed to the space flight environment, however, there is a significant decrease in the IPL at all three postnatal time points with the most dramatic reduction seen at PN15. The thinner IPL is obvious in the H&E stained micrographs of Fig 59.2. The micrographs also show large spaces in the IPL in the retina of the inflight animals suggesting that there is shrinkage and degeneration of the neuropil comprising the IPL.

3.3. Loss of Ganglion Cells

As many as 25 different ganglion cell types are found in the mammalian retina. These are the first cells to differentiate in the retina and vary in cell body size, dendritic arborization, and laminar branching patterns. The appearance of the cells in the ganglion cell layer of the retina of ground-control rodents are morphologically distinct from those seen in the retina of the matched in flight neonates (Figure 59.2). The ganglion cell layer of the control animals had more cells that were round with abundant cytoplasm. In the retinas of the space flight animals, the ganglion cells are elongated and form a discontinuous layer. The ganglion cells have scanty cytoplasm, appear detached from the neuropil of the IPL, and generally appear unhealthy. This suggests that the space flight animals have extensive damage to the ganglion cell layer.

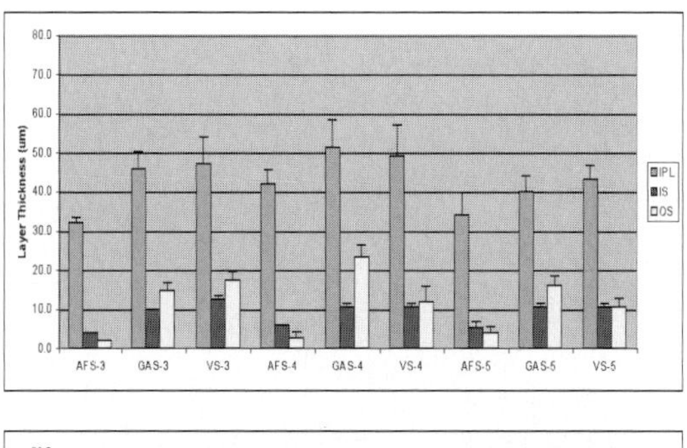

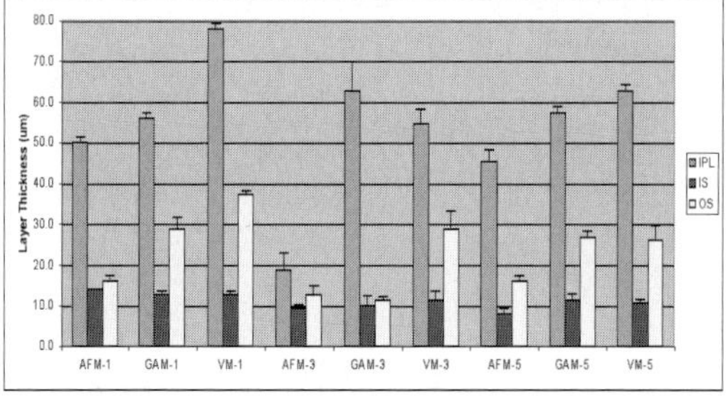

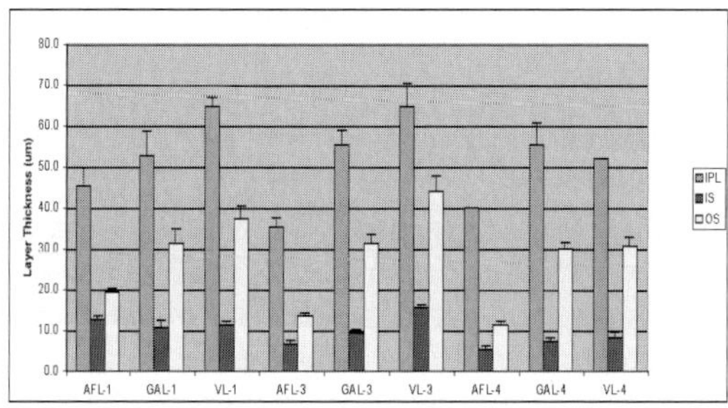

Figure 59.3. Thickness of retinal layers in the three sets of animals examined. There is a general shortening of outer segments and thinning of the inner plexiform layer in space flight animals. IPL, inner plexiform layer; IS, rod inner segments; OS, rod outer segments. Top panel, PN5 launch; AFS-3, AFS-4 and AFS-5 are the flight animals. Middle panel, PN8 launch; AFM-1, AFM-3 and AFM-5 are the flight animals. Bottom panel, PN15 launch; AFL-1, AFL-3 and AFL-4 are the flight animals. All other animals are ground controls.

4. DISCUSSION

Experimental studies have linked the perception of "light flashes" and hallucinations by astronauts in flight to the penetration of ionizing nuclei through the nervous tissue. There is some evidence that cosmic rays induced cell death in the outer nuclear layer of rats flown in space (Philpott et al., 1978), that microgravity induces intraocular pressure and vascular changes in the eye (Mader et al., 1993), and promotes apoptosis in astrocytes (Uva et al., 2002).

This study shows abnormal development of the retina and loss of photoreceptor outer segments but has not allowed us to determine whether the effects we observed in the neonatal rodent retina are transient, reversible, or dependent on flight duration. Nor does it answer the question as to whether these changes are triggered by a warp in gravitational force, solar radiation, impact of launching or reentry into the earth's atmosphere, or fluctuations in oxygen level.

It is possible that neonatal retinas do not adapt as well to the hazards encountered in the space travel environment as adult retinas. The impact on the neural retina in neonates, however, highlights the importance of developing more rigorous research efforts to study how the eyes adapt or respond to various space related assaults since the duration of manned space journeys will be significantly increased in the near future and could have irreversible adverse effects on human health and performance.

Knowledge of how this unusual environment affects molecular mechanisms and pathways of the CNS are key to accelerating development of appropriate physiological risk mitigation measures to remove biological barriers that could impede the astronauts' ability to survive and function in future long-term human space exploration. Such studies could lead to information that is important to understanding and treating similar earth-based retinal disorders as well.

5. ACKNOWLEDEMENTS

We thank Drs. Paul Callahan and Alison French (Ames Res Center, CA) and Dr. Louis Ostrach (NASA, Washington, DC) for making this study possible. Supported by NIH, the Connecticut Lions and RPB Inc.

6. REFERENCES

Bumsted KM, Rizzolo LJ, Barnstable CJ. Defects in the MITF mi/mi apical sirface are associated with a failure of outer segment elongation. Exp. Eye Res. 2001; **73**:383-392.

Draeger J, Michelson G, Rumberger E. Continuous assessment of intraocular pressure - telematic transmission, even under flight- or space mission conditions. Eur. J. Med. Res. 2000; Jan 26;**5**(1):2-4.

Drummer C, Gerzer R, Baisch F, Heer M. Body fluid regulation in micro-gravity differs from that on Earth: an overview. Pflugers Arch. 2000; 441(2-3 Suppl):R66-72.

Hoffler GW, Bergman SA, Nicogossian AE. In-flight lower limb volume measurement. In Nicogossian, A.E. (ed.): The Apollo-Soyuz Test Project Medical Report (NASA SP-411). Washington, D.C., U.S. Government Printing Office, 1977; pp. 63-68.

LeBlanc A, Lin C, Shackelford L, Sinitsyn V, Evans H, Belichenko O, Schenkman B, Kozlovskaya I, Oganov V, Bakulin A, Hedrick T, Feeback D. Muscle volume, MRI relaxation times (T2), and body composition after spaceflight. J. Appl. Physiol. 2000; Dec;**89**(6):2158-2164.

Mader TH, Gibson CR, Caputo M, Hunter N, Taylor G, Charles J, Meehan RT. Intraocular pressure and retinal vascular changes during transient exposure to microgravity. American Journal of Ophthalmology, 1993; March 15, v115 n3p347 (4).

Mader TH, Taylor GR, Hunter N, Caputo M, Meehan RT. Intraocular pressure, retinal vascular, and visual acuity changes during 48 hours of 10 degree head-down title. Aviat. Space Environ. 1990; Med. **61**:810.

Newberg AB, Alavi A. Changes in the Central Nervous System During Long-Duration Space Flight: Implications for Neuro-Imaging. Advanced Space Research, 1998; **22**:185-196.

Obata S, Kuwabara J. Morphogenesis of the photoreceptor outer segment during postnatal development in the mouse (BALB/c) retina. Cell Tiss. Res. 1992; **269**:39-48.

Philpott DE, Corbett R, Turnbill C, Harrison G, Leaffer D, Black S, Sapp W, Klein G, Savik LF. Cosmic ray effects on the eyes of rats flown on Cosmos No. 782, experimental K-007. Aviat Space Environ. Med. 1978; **49**(1 Pt 1):19-28.

Robinson, SR. Development of the mammalian retina. In Dreherand, B., Robinson S.R. (eds.): Neuroanatomy of the Visual Pathways and Their Development. London: Macmillan, 1991; pp. 69-128.

Uva BM, Masini MA, Sturla M, Tagliafierro G, Strollo F. Microgravity-induced programmed cell death in astrocytes. J. Gravit Physiol. 2002; Jul;**9**(1):P275-P276.

Wang M, Hassebrook L, Evans J, Varghese T. An Optimized Index of Human Cardiovascular Adaptation to Simulated Weightlessness. Transactions on Biomedical Engineering, 1996; **43**:502-511.

Weidman TA, Kuwabara T. Postnatal development of the rat retina. Arch. Ophthalmol. 1968; **79**:470-484.

CHAPTER 60

TOXICITY OF HYPEROXIA TO THE RETINA: EVIDENCE FROM THE MOUSE

Scott Geller, Renata Krowka, Krisztina Valter, and Jonathan Stone*

1. INTRODUCTION

Photoreceptors are vulnerable to both a lack and an excess of oxygen. Hypoxia causes a photoreceptor-specific degeneration during the critical period of normal photoreceptor development (Maslim et al., 1997), in the naturally occurring degeneration of the RCS rat (Valter et al., 1998), in direct hypoxia (induced by low inhaled pO_2) of the adult retina (Mervin and Stone, 2002b), and in the detached retina (Mervin et al., 1999). Hyperoxia causes photoreceptor degeneration at the edge of the normally developing retina (Mervin and Stone, 2002b, Stone et al., 2004) and brief reports available for the rabbit (Noell, 1955) and mouse (Yamada et al., 2001, Walsh et al., 2004a) indicate that photoreceptors degenerate when oxygen-enriched air is inhaled.

Clinically, hyperoxia-induced pathology is rare in any tissue. An oxygen-induced dysplasia of the bronchopulmonary epithelium is a complication of prolonged oxygen therapy for chronic lung disease. The lung epithelium's vulnerability to hyperoxia presumably results from its direct exposure to inspired gas. Most body tissues are protected from clinically used hyperoxia by the autoregulatory mechanisms of the capillary bed, which limit the rise of tissue pO_2. The exception is the retina, which is vulnerable because the flow of blood through the choroid, from which oxygen diffuses to the outer retina, is not controlled by autoregulatory mechanisms (reviewed in (Chan-Ling and Stone, 1993, Stone and Maslim, 1997, Stone et al., 1999)). As a consequence, hyperoxia is a factor in the induction of retinopathy of prematurity. Even this effect (reviewed in (Chan-Ling and Stone, 1993, Stone and Maslim, 1997)) is considered, however, to be due to oxygen regulation of angiogenesis, rather than a direct cellular toxicity. Prolonged hyperoxia downregulates the expression of angiogenic factors (such as vascular endothelial growth factor), which are essential for both normal

---

*Correspondence to: Jonathan Stone, Research School of Biological Sciences, The Australian National University, ACT 2601, Australia. +61 2 6125 3841 (phone), +61 2 6125 0758 (fax), e-mail: Jonathan.Stone@anu.edu.au.

vessel formation (Stone et al., 1995) and for the maintenance of adult vessels (Alon et al., 1995).

This paper explores the toxic impact of hyperoxia on photoreceptors in a mature mammalian retina. The question is important because partial depletion of the photoreceptor layer (the situation in most human retinal degenerations at diagnosis) leads to a chronic rise in tissue oxygen levels in outer retina. This rise has been demonstrated in three models of photoreceptor degeneration, the RCS rat (Linsenmeier et al., 2000, Yu et al., 2000, Yu et al., 2004), the Abyssinian cat (Linsenmeier et al., 2000), and the rhodopsin-mutant transgenic P23H rat (Yu et al., 2004). Further, there is an established basis for the understanding of this rise in analyses of oxygen gradients and consumption in the retina of rats (Yu and Cringle, 2001), cats (Haugh et al., 1990) and monkeys (Ahmed et al., 1993); the depletion-induced hyperoxia in outer retina is not likely to be specific either to species or to particular causes of depletion. For this reason, we have proposed ('the oxygen toxicity hypothesis') that depletion-induced hyperoxia is toxic to surviving photoreceptors and may be a key factor in the progress of retinal degenerations (Stone et al., 1999).

The present study explores the oxygen toxicity hypothesis by applying several regimes of hyperoxia to the non-depleted, non-degenerative retina of the C57BL/6J mouse.

2. METHODS

2.1. Animals, Oxygen Exposure

Experiments were conducted according to protocols approved by the Animal Ethics Committee of the The University of Sydney and The Australian National University. C57BL/6J mice were raised from birth in dim cyclic light (12 h 5 lux, 12 h darkness). Animals were exposed to hyperoxia by placing them, in their cages, into a plexiglass chamber in which the concentration of oxygen was increased from normal (21%) to 75%, by a feedback controlled device (OXYCYCLER, Biospherix, Redfield, NY). The periods of exposure used were 0 w (controls), 1 w, 2 w, 3 w, 4 w, 5 w and 6 w. In addition, several animals were exposed to 75% oxygen for 3 w or 4 w and then returned to room air for 1 w or 2 w. Animals were euthanised with an overdose of anaesthetic (sodium pentobarbitone, 60 mg/kg) or of halothane, followed by cervical dislocation. Eyes were immersion-fixed in 4% paraformaldehyde for 2 h, then cryo-protected by immersion in 15% sucrose overnight and were cryo-sectioned at 20 μm. Sections were placed onto glass slides, allowed to dry and stored frozen until use.

2.2. TUNEL-Labeling

Several sections from each experimental animal were labeled with the TUNEL (terminal deoxynucleotidyl transferase UTP nick end labeling) technique, as described previously (Maslim et al., 1997). The TUNEL technique detects the fragmentation of nuclear DNA characteristic of apoptosis. To allow fragmenting DNA of apoptotic cells to be seen against a background of non-degenerative cells (Fig. 60.1), the sections were counterstained with a DNA-specific dye, bisbenzamide, as described previously (Mervin and Stone, 2002a, Bravo-Nuevo et al., 2004).

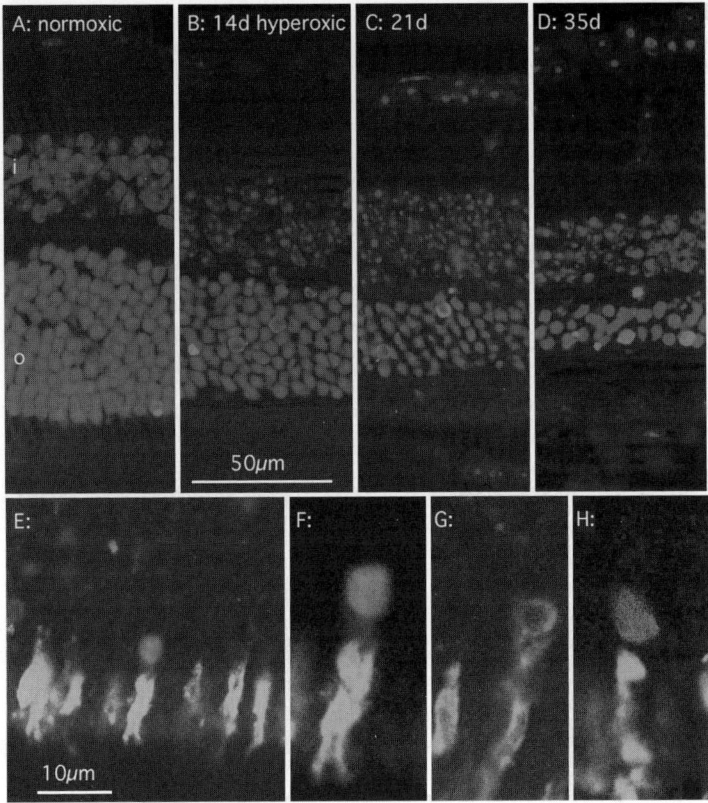

Figure 60.1. Rod and cone death induced by hyperoxia. Blue is bisbenzamide labeling of DNA, showing (A-D) the major layers of the retina (i – inner nuclear layer, o - outer nuclear layer). Red is TUNEL-labeling. Green in E - H is PNA labeling of cone sheaths. A: Young adult C57BL/6J retina, kept in room air (normoxic control). B, C, D: Young adult C57BL/6J retina after 14d, 21d and 35d in hyperoxia. These images show the most affected region of retina, at each exposure time. TUNEL labeling is restricted to the ONL up to 14d of exposure, but labeling of the INL is evident at 21d and 35d. E-H: Regions at the outer surface of the ONL. Occasional TUNEL+ nuclei were closely associated with cone sheaths (labeled green with PNA lectin). F is part of E, at higher power. The scale bar in E represents 10 μm in E, and 5 μm in F, G, H. See also color insert.

2.3. Immunolabeling

Several sections from each eye were labeled with antibodies to FGF-2, GFAP or cytochrome oxidase (CO), following previously published protocols (Maslim et al., 1997, Mervin et al., 1999). Several sections were also labeled with antibodies to c-Jun (Santa Cruz Biotechnology Inc., diluted 1:50, incubation 1h at 37°C), and further labeled with a lectin (peanut agglutinin), which labels cone sheaths (Blanks and Johnson, 1984).

2.4. Quantification of TUNEL-Labeling

The frequency of TUNEL+ profiles in a tissue section provides an estimate of the rate of cell death in the population of cells at the time of tissue fixation. To quantify cell death rates, TUNEL-labeled sections of retina were examined by fluorescence microscopy. Sections were analyzed which included or were close to the optic disc. Each section was scanned from one edge to the other in consecutive segments of 400 μm. The number of TUNEL+ profiles per segment, was recorded separately for the ONL and INL. For each animal we examined 1-4 sections, averaging the results over the full length of each section.

3. RESULTS

3.1. Hyperoxia-Induced Cell Death (HICD) in C57BL/6J Retina

3.1.1. Specificity to Photoreceptors

In retinas not exposed to hyperoxia, TUNEL+ cells were rare in any layer of the retina (Fig. 60.1A). HICD was detected as above-control frequencies of TUNEL+ profiles (red in Fig. 60.1A-D). Up to 2 w exposure, virtually all of these profiles were located in ONL (Figs. 60.2A). By 5 weeks (Figure 60.1D), with the ONL reduced to a fraction of its control thickness, some TUNEL+ profiles were observed in the INL and the layer appeared thinner, indicating cell death in the INL.

Considering ONL cell death as a function of exposure to hyperoxia (Figs. 60.2B, C), a small increase in the frequency of TUNEL+ profiles was apparent in all 4 animals examined at 1 w; this increase approached statistical significance (P = 0.0548 on a 1-tailed t-test). At 2 w exposure, 3 of the 4 animals examined showed marked increases in the frequency of TUNEL+ profiles; and, at 3 w, all 4 animals examined showed clear increases. Averaged over the 4 animals in each group (Fig. 60.2C), the increases in TUNEL+ frequencies at 2 w and 3 w were significantly (P < 0.05 on a 1-tailed t-test) higher than in the controls.

3.1.2. Cones and Rods Affected

In the hyperoxia treated animals, TUNEL+ profiles were distributed throughout the thickness of the ONL (Fig. 60.1B, C) suggesting, since cone cell nuclei are restricted to the outermost aspect of the ONL, that most dying cells were rods. When the PNA lectin was used to identify cone sheaths (Blanks and Johnson, 1984) (green in Figs. 60.1E-H) it was possible to demonstrate a close association between TUNEL+ nuclei and cones sheaths (examples in Figs. 60.1E-H). This close association suggests that hyperoxia induced death in cones as well as rods.

3.1.3. HICD after Exposure to Hyperoxia

We show below that hyperoxia up-regulates the expression of FGF-2 in the retina, a factor considered protective to photoreceptors. To test whether the photoreceptor population becomes resistant to hyperoxia after the surge of photoreceptor death (Figs. 60.2B,C) and increased FGF-2 expression, the period of exposure was extended to 4 w (2 animals), 5 w (3 animals) and 6 w (2 animals). The frequency of TUNEL+ profiles in the ONL of these 7 animals was well above control levels (13–22/mm vs. 0.21/mm (Fig. 60.3A)). On a t-test, this difference was statistically significant (P < 0.001, n = 7 for both exposed and control groups).

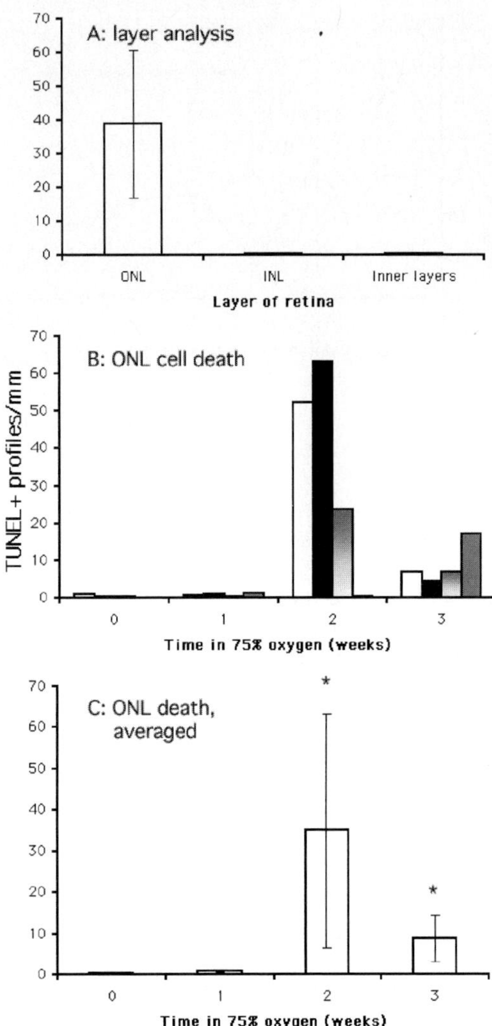

Figure 60.2. Patterns of cell death in retinas of C57BL/6J mice exposed to hyperoxia. A: TUNEL+ nuclei were mostly in the ONL. These are data from 4 animals exposed to hyperoxia for 14d. Error bars are ±1 standard deviation among the 4 animals. B: Frequency of TUNEL+ nuclei in the ONL at a series of exposures to hyperoxia. Four animals were examined at each of 1 w, 2 w, 3 w and 4 w. Values for individual animals are shown for each exposure time. C: Means and standard deviations for the data in B.

3.1.4. Photoreceptor Death after Return to Room Air

We have predicted previously (Stone et al., 1999) that depletion of the photoreceptor population will destabilize the photoreceptor population, causing photoreceptor death to continue after the stress which induces depletion is removed. To test this hypothesis, 11 animals were kept in hyperoxia for 3 w and then returned to room air (5 for 1 w, 3 for 2 w, 3 for 3 w). At all 3 survival times, TUNEL+ cells were detected in the retinas,

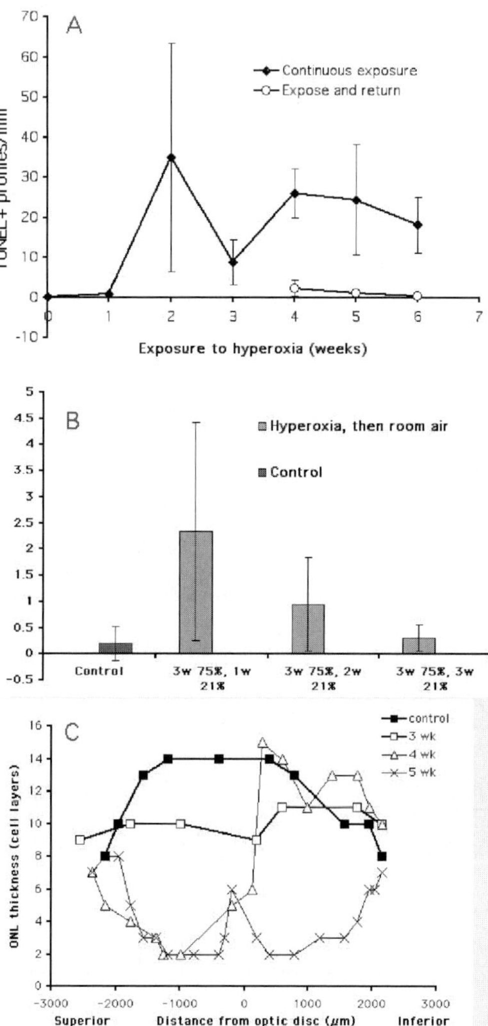

Figure 60.3. Time course of photoreceptor death. The error bars show ±1 standard deviation. A: Frequency of TUNEL+ profiles in the ONL, as a function of duration of exposure to 75% oxygen. Numbers of animals examined at each exposure time were: 0 w (control), n = 7; 1 w, n = 4; 2 w, n = 4; 3 w n = 4; 4 w, n = 2; 5 w, n = 3; 6 w n = 2. Numbers at the longer exposures were limited by morbidity. Also shown are data for animals kept in 75% oxygen for 3 w, then returned to room air for 1-3 w (n = 5 at 1 w, n = 3 at 2 w and 3 w). B: Comparison of TUNEL+ (dying) cell frequency in the ONL, between controls and animals kept in 75% oxygen for 3 w, then returned to room air for 1-3 w. Numbers of animals as for A. C: Thickness of the ONL, expressed as layers of cells, as a function of position along the retina, from superior to inferior edge, for C57BL/6J mice exposed to hyperoxia for 0 w (control), 3 w, 4 w, 5 w.

overwhelmingly (>90%) located in the ONL (data not shown). The frequency of TUNEL+ profiles, averaged over the 3 animals, was higher at all 3 survival times than in the control group (never exposed to hyperoxia) (Figs. 60.3A,B). Taking the data from 3 post-exposure times together, the frequency of TUNEL+ profiles was significantly lower than in animals remaining in hyperoxia ($P < 0.001$ on a t-test), suggesting that photoreceptor death slows on removal from hyperoxia; and significantly higher ($P < 0.02$ on a 1-tailed t-test) than in normoxic controls, suggesting that degeneration continues for some time after removal. Considering the 3 post-exposure times separately, however, the frequency of TUNEL+ profiles was significantly ($P < 0.01$ on a 1-tailed t-test) above control levels only at 1w survival. There was evidence of a trend for TUNEL+ frequency to fall with survival time in room air (Fig. 60.3B).

3.1.5. Thinning of ONL

The toxicity of hyperoxia to the mouse retina was first observed (Yamada et al., 2001) as a thinning of the ONL, in a central region of retina, away from the retinal edge. We noted a similar thinning of the ONL, for example at 2w exposure (Figs. 60.1A-D; also Figs. 60.4A-C). This thinning confirms the assumption made above that the fragmentation of DNA detected as TUNEL-labeling after exposure to hyperoxia leads to cell death. When the thinning of the ONL was mapped as a function of position in the retina, and over several weeks' exposure to hyperoxia (Figure 60.3C), several trends were noted. First, the thinning was first prominent in a central region of retina, leaving the edges of retina unaffected (see the 4w data in Figure 60.3C).

Second, this vulnerable region was located superior to the optic disc (to the left in Fig. 60.3C) and then expanded towards the edges of the retina. Finally, even at the longest exposures in the present study (6w), the edges of the retina remained less affected. It is suggested in Discussion that the lesser vulnerability of inferior retina may arise from the lighting conditions used during rearing.

3.2. Up-Regulation of FGF2, GFAP, c-Jun Expression

In control retinas, as previously reported (Mervin and Stone, 2002b), FGF-2 was prominent only in the somas of Müller cells in the INL (Fig. 60.4A), and in the nuclei of astrocytes (not shown but see (Walsh et al., 2001)). Hyperoxia caused an increase in the level of FGF2 protein in the ONL and of GFAP in Müller cells, which co-varied with TUNEL+ profile frequency both temporally and topographically. A limited up-regulation of FGF-2 in ONL somas was detected in retinas exposed to hyperoxia for 7d, in the same area of retina as a weak up-regulation of GFAP in Müller cell processes (Fig. 60.4B). After 14d exposure to hyperoxia, the up-regulation of FGF-2 in ONL somas and of GFAP in Müller cell processes was prominent (Fig. 60.4C).

Immunolabeling for c-Jun (red in Figs. 60.4D-I) was localized to foci in the inner/outer segment region of photoreceptors. In material labeled for cytochrome oxidase (green in Figs. 60.4D,E,G,H), large and small c-Jun+ foci could be distinguished. The larger seemed to be located at the outer end of cone inner segments and, when cone sheaths were labeled with the PNA lectin (Figs. 60.4F,I), the larger foci were seen to be located within cone sheaths. The smaller foci were more numerous and were located 5-10μm external to the larger (cone-

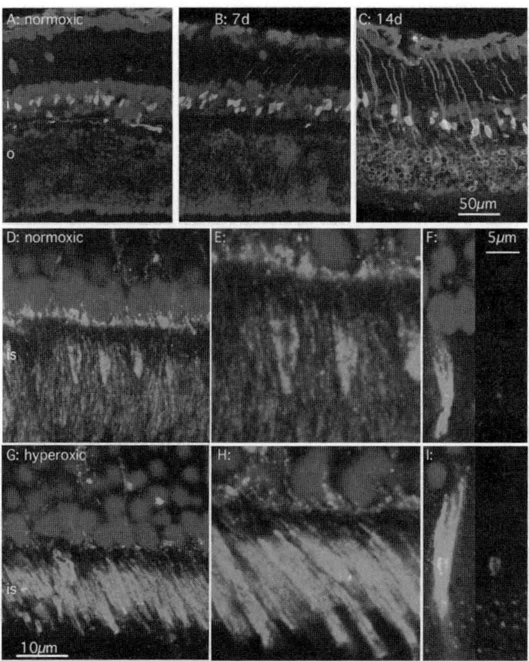

Figure 60.4. Protein up-regulation induced by hyperoxia. Blue is bisbenzamide labeling of normal DNA, showing nuclei of the major cell layers (*i* – inner nuclear layer, *o* – outer nuclear layer). Red in A – C is immunolabeling for GFAP; green is immunolabeling for FGF-2. In D – I, blue is bisbenzamide labeling, red is labeling for c-Jun. while green is labeling for either cytochrome oxidase (D, E, G, H) or PNA (F, I). See also color insert.

related) foci, suggesting that they are associated with rods, and are located at the outer end of the rod inner segment. When normoxic (Figs. 60.4D-F) and hyperoxic (Figs. 60.4G-I) retinas were compared, the c-Jun foci were consistently brighter and larger in the hyperoxic material.

3.3. Sequence of Events Following Onset of Hyperoxia

As already noted, the cell death induced by hyperoxia spread over a period of weeks, from superior-central retina towards the edges. Where photoreceptor death was localized, the up-regulation of FGF-2 and GFAP co-localized with photoreceptor death (Fig. 60.5A, B). At the edge of this region the up-regulation of FGF-2 and GFAP was graded, fading towards the edge of the retina (arrows in Fig. 60.5). The frequency of TUNEL+ profiles also fell towards the edge of the retina. It was consistently observed that TUNEL+ profiles were found more peripherally than the up-regulation of either FGF-2 or GFAP. It is argued below that these gradients occur at the edge of a spreading process, and that the fragmentation of nuclear DNA labeled by the TUNEL technique (and characteristic of apoptotic cells) occurs earlier in this process than the up-regulation of either FGF-2 or GFAP.

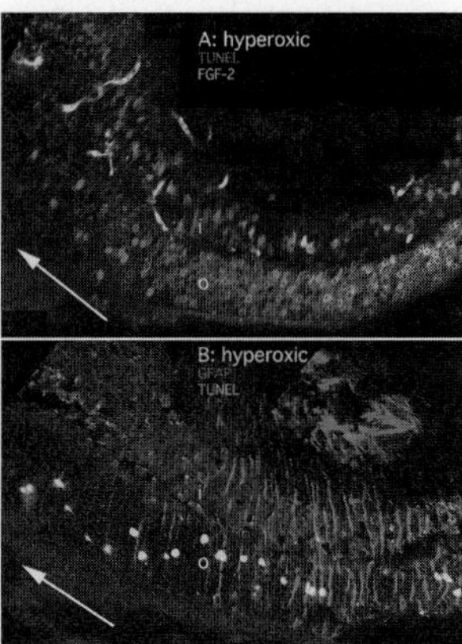

Figure 60.5. Evidence of the sequence of events induced by hyperoxia. A: Region of retina from a C57BL/6J mouse exposed to hyperoxia for 2 w, labeled for the TUNEL reaction and for FGF-2. Both labels were concentrated in a central region of retina. At the edge of the labeled region, the frequency of TUNEL-labeled profiles and the density of FGF-2 labeling in the ONL (o) both decreased towards the edge of the retina (direction of the arrow). TUNEL+ profiles extended further peripherally than FGF-2 labeling in the ONL. FGF-2 labeling in the INL (i) extended throughout the retina, as in control retinas. B: A neighboring section of retina, labeled for the TUNEL reaction and GFAP. Again, TUNEL+ profiles in the ONL (o) extended further peripherally than the up-regulation of GFAP in the radial processes of Müller cells.

4. DISCUSSION

Present results suggest that, in the mouse retina, high inhaled pO_2 is directly toxic to photoreceptors. This observation confirms previous brief reports of oxygen-induced photoreceptor death in the rabbit (Noell, 1955) and mouse (Yamada et al., 2001, Walsh et al., 2004b) and adds several new observations. First, the toxicity is initially specific to photoreceptors, affecting inner layer neurons only at later stages of the degeneration. Second, both rods and cones are affected. Third, photoreceptor death was followed by an up-regulation of stress- or protection-related proteins (FGF-2, GFAP, c-Jun). Perhaps because of the up-regulation of protective factors, the rate of hyperoxia-induced photoreceptor death peaked at 2 w, but continued significantly higher than in control retinas through the longest period of exposure examined (6 w). Fourth, reducing oxygen levels reduces the rate of photoreceptor death.

4.1. Specificity to Photoreceptors

For the first 2 w of exposure to hyperoxia, the cell death induced in the retina was largely restricted to photoreceptors. The reason for this initial specificity can be suggested from analyses of the effect of hyperoxia on oxygen levels in the rat retina (Yu and Cringle, 2001). These workers showed that high levels of inhaled oxygen increase pO_2 at the choroid, and in the outer layers of retina (outer segment, inner segment and outer nuclear layers) but not internal to the outer plexiform layer. That is, the hyperoxia initially affects only the photoreceptor layers. Once there has been significant depletion of photoreceptors, however, it is possible that the rise in pO_2 caused in the retina by high inhaled oxygen will extend to inner layers of retina, and cause death of neurons of the INL, as suggested by Figures 60.1C,D.

4.2. Significance: Understanding the Progression of Retinal Degenerations, and Potential for Therapy

As reviewed previously (Stone et al., 1999), chronic hyperoxia of the outer (photo-receptor) layers of the retina is a feature common to probably all retinal degenerations. Models of retinal oxygen consumption (Alder et al., 1990, Haugh et al., 1990, Yu and Cringle, 2001) predict a rise in pO_2 as the population of photoreceptors is depleted, and a corresponding rise has been demonstrated empirically in the RCS rat (Yu et al., 2000), the Abyssinian cat (Linsenmeier et al., 2000) and the P23H rat (Yu et al., 2004). Measurement of pO_2 has not been attempted in the human, but evidence that the human retina undergo-ing photoreceptor degeneration is hyperoxic comes from two observations. First, vessels thin in retinitis pigmentosa (Heckenlively, 1988); in animal models this thinning is reversed by hypoxia (Penn et al., 2000), suggesting that it is caused by higher than normal levels of oxygen. Second, retinitis pigmentosa is protective against hypoxic retinal disease, such as diabetic retinopathy (Sternberg et al., 1984, Arden, 2001, Lahdenranta et al., 2001).

Much attention has been given to the issue of why most photoreceptor degenerations are relentlessly progressive, involving cones although the initial degeneration may be rod-specific, and vice-versa. Evidence has been gained (reviewed in (Hicks and Sahel, 1999)) that rods produce a factor on which cone survival is dependent. The present observations give support to our earlier (Stone et al., 1999) suggestion that, when photoreceptors die, they 'leave behind' a toxin, excess oxygen, which destabilizes surviving photoreceptors.

One feature (at least) of the present data is hopeful for the management of photoreceptor degenerations. The rate of photoreceptor death slowed markedly after hyperoxia was reduced (Figure 60.3B), in the present experiments by returning the animals to room air. If, as presently suggested, depletion-induced hyperoxia of outer retina is a toxic factor in the late stages of degeneration, then reduction of oxygen levels may slow photoreceptor death. Oxygen levels in outer retina can be reduced in two ways, by reducing exposure to light (Linsenmeier, 1986), and by direct hypoxia.

4.3. Topography of HICD

HICD begins in superior retina and spreads towards the edges of the retina (Fig. 60.3C, Fig. 60.5). This temporal pattern resembles the sequence of photoreceptor death induced by bright light (Rapp and Williams, 1980, Duncan and O'Steen, 1985, Bowers et al., 2001). We have suggested (Stone et al., 1999) that the pattern of light induced cell death is deter-mined by prior light experience of the retina. Where, as in our study, the animals are raised in rooms with the light source in the ceiling, inferior retina is more exposed to light, and photoreceptors there are more resistant to light stress. Conversely, superior retina is more naïve and its photoreceptors are more vulnerable. The present results suggest that factors which make light-experienced photoreceptors more resistant to light stress also make them resistant to the stress of hyperoxia.

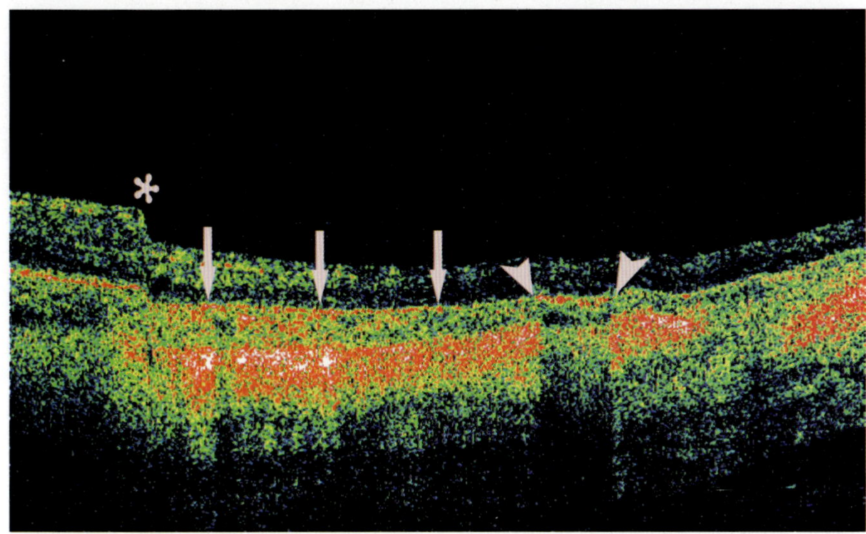

Figure 9.3. Ocular coherence tomography (OCT) through the optic nerve (arrowheads) and macula of the right eye shows an abrupt demarcation (asterisk) between the island of intact retina and the area of atrophy. The outer retina and RPE are absent, but a thin layer of inner retina and Bruch's membrane (arrows) are intact within the area of atrophy, consistent with histopathological samples from other patients with choroideremia (Compare to Figure 9.4). Increased signal from the choroid underlying the atrophic area could be consistent with fibrosis of the choroid.

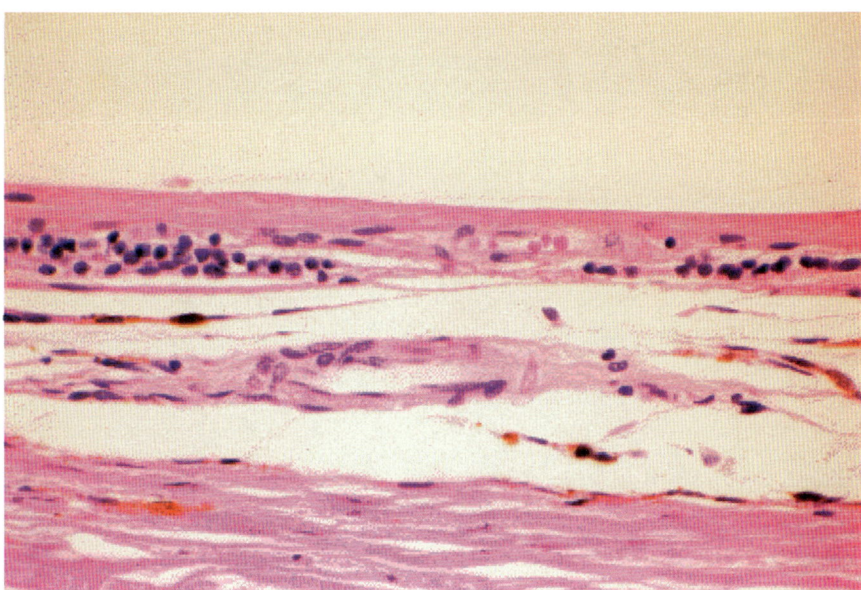

Figure 9.4. Histopathology of choroideremia in a specimen taken from another patient. There is fibrosis of the choroid and only a single choroidal artery remains. The retinal pigment epithelium and outer nuclear layers are absent. The inner nuclear layer rests against Bruch's membrane. H&E ×330. Reprinted from: Spencer WH, Ophthalmic Pathology, An Atlas and Textbook (CD-ROM), Figure 9-723, 1997, with permission from Elsevier.

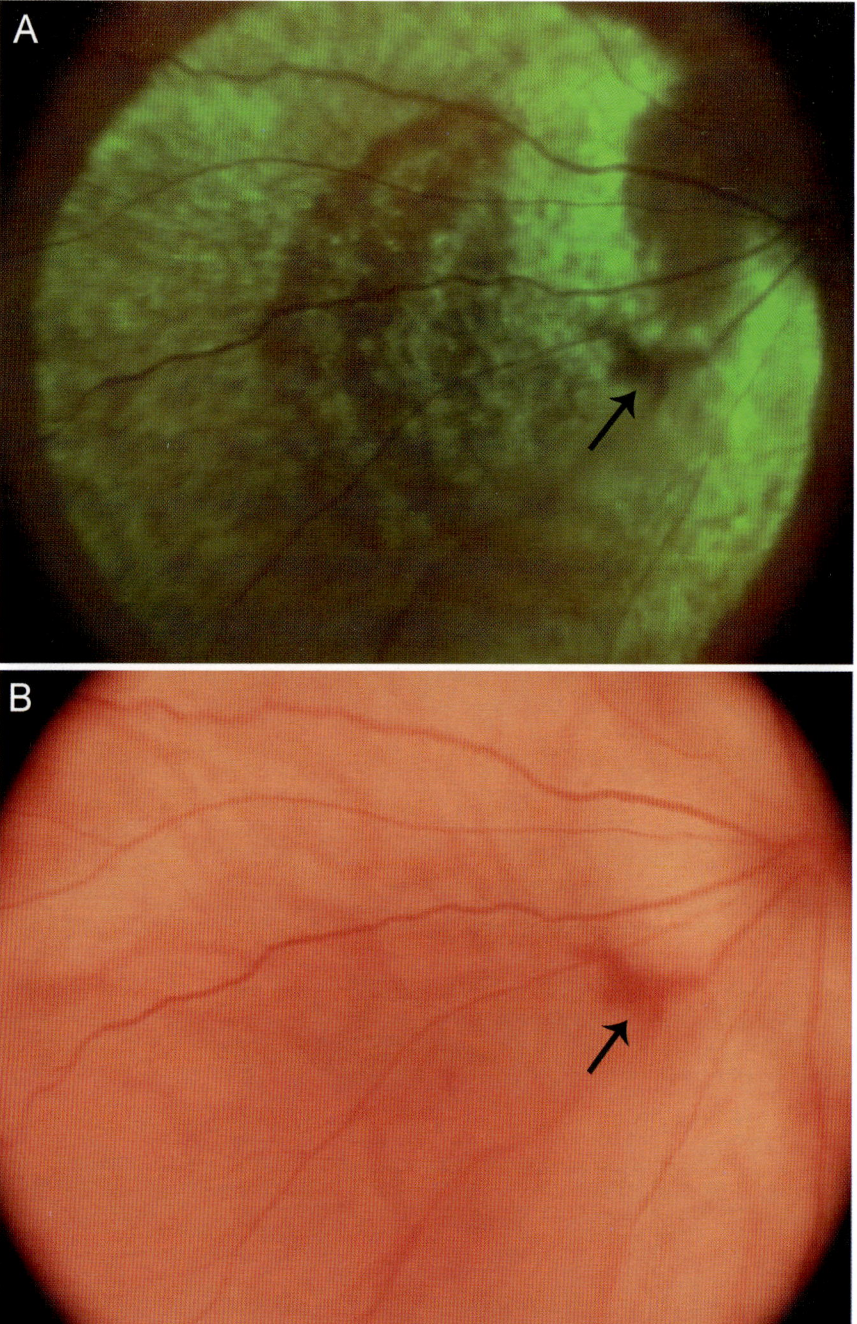

Figure 36.3. Flourescent fundus image showing widespread GFP expression in rat retina 1 week after subretinal injection of 3 μL VSV-CβA-GFP lentiviral vector (A). Fundus image of same rat under white light illumination (B). Arrows indicate small hemorrhage resulting from subretinal injection. Both images acquired with a Retcam II imaging system (Massie Research, Pleasanton, CA).

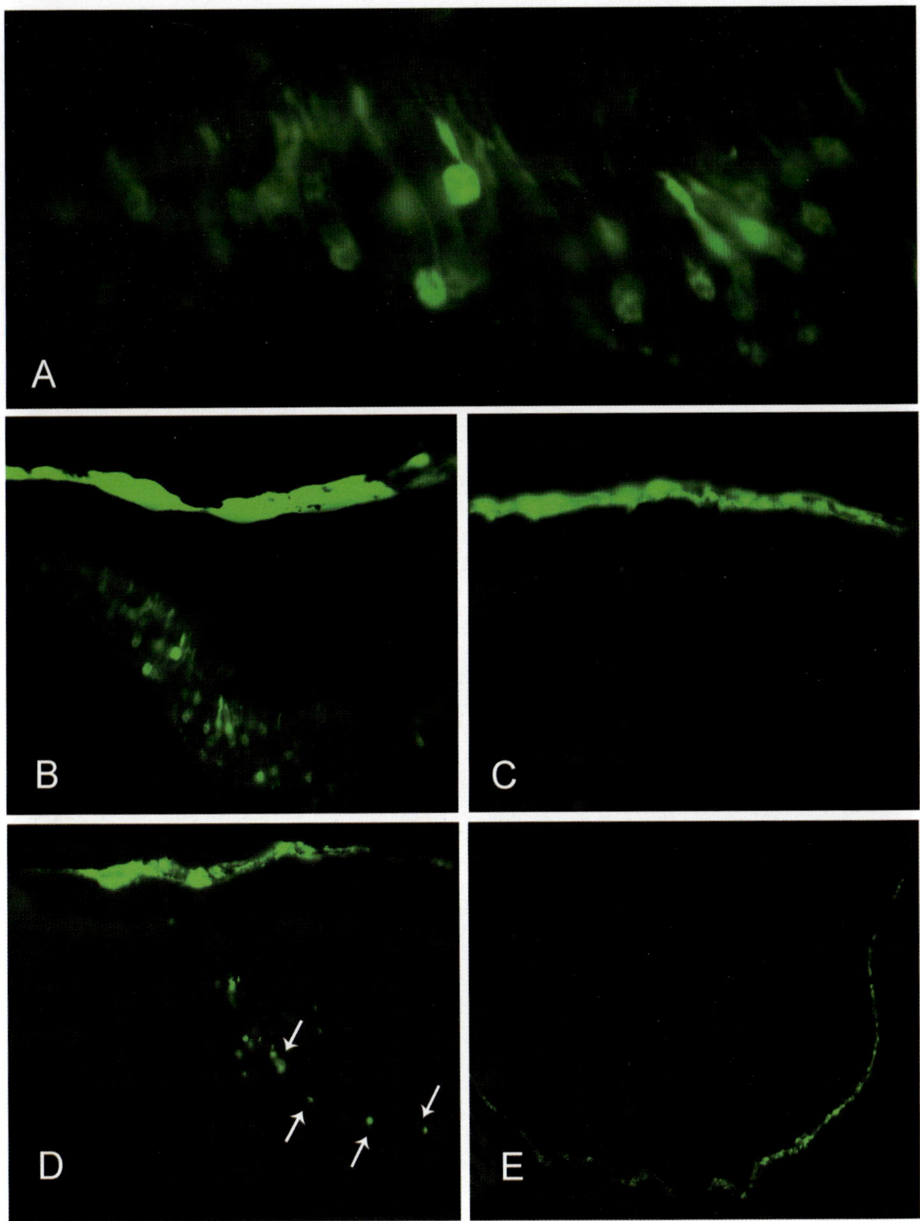

Figure 36.4. High magnification view of GFP positive photoreceptors of mouse retina injected subretinally with VSV-CMV-GFP LV vector at age P7 (A). Lower magnification view (B) of the same retina shown in (A) where RPE and photoreceptors are seen expressing GFP. Expression of GFP restricted to the RPE layer in P14 mouse retina injected with VSV-CMV-GFP LV vector (C). Injection track mark shown (arrows) and evidence of immune response from autoflourescent macrophages bordering track mark (D). Low magnification view showing extent of GFP expression along entire length of the RPE (E).

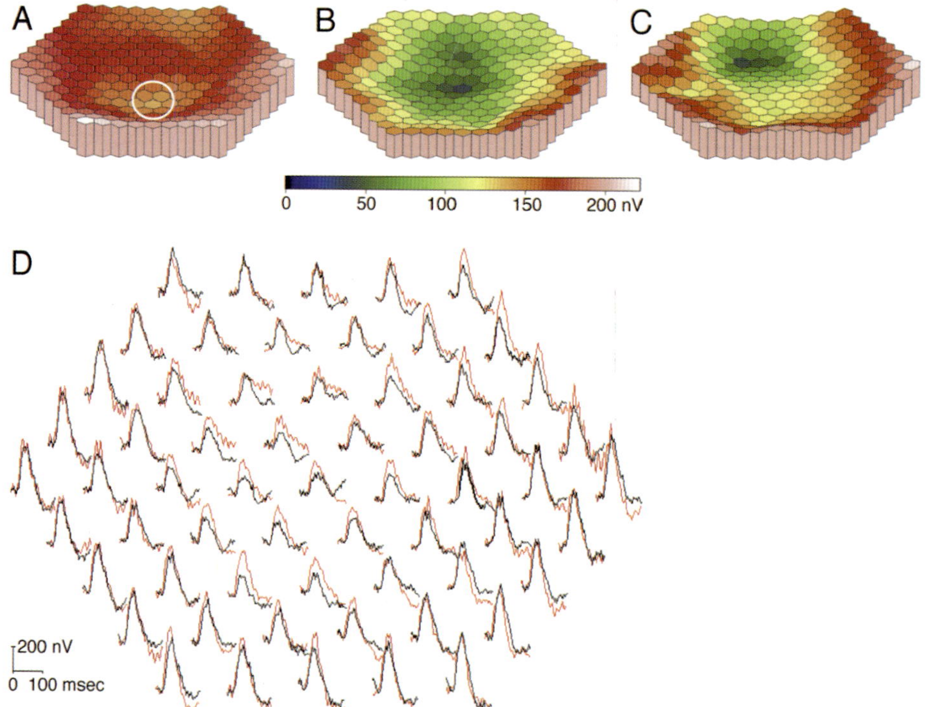

Figure 41.2. Multifocal ERG performed 1 week after PDT demonstrates reduced local retinal function in areas of PDT-treated retina. (**A**) Averaged responses from 6 normal rat eyes shown topographically demonstrate slightly reduced retinal function at the location of the optic disk (white circle). (**B**) Responses are reduced centrally in the PDT-treated area of an eye treated with PBS 2 days prior to PDT. (**C**) The contralateral, BDNF-treated eye shows improved retinal function in the PDT-treated area. (**D**) Superimposed multifocal ERG responses from the eyes shown in (**B**) (black traces) and (**C**) (red traces) show improved function centrally in the BDNF-treated eye. Reproduced with permission of *Investigative Ophthalmology and Visual Science* in the format Other Book via Copyright Clearance Center from Paskowitz et al., *Invest Ophthalmol Vis Sci* 2004;45:4190-4196.

Making a map – a two step process

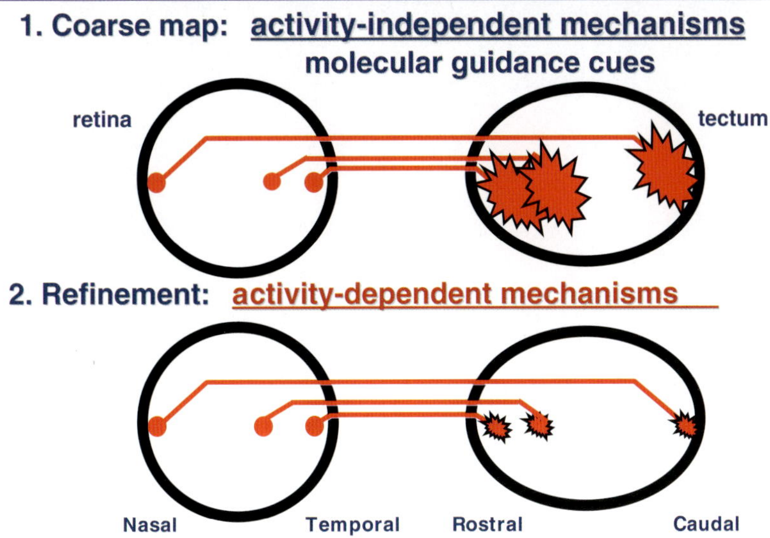

Figure 55.1. Diagram of two overlapping phases involved in restoration of retinotectal topography. During the activity-independent phase (left panel), RGC axons (dots in retinal outline) project axons with wide-spread terminal arbors in the optic tectum (jagged shapes in tectal outline). In the second phase, activity-dependent mechanisms focus terminal arbors and produce precise topography.

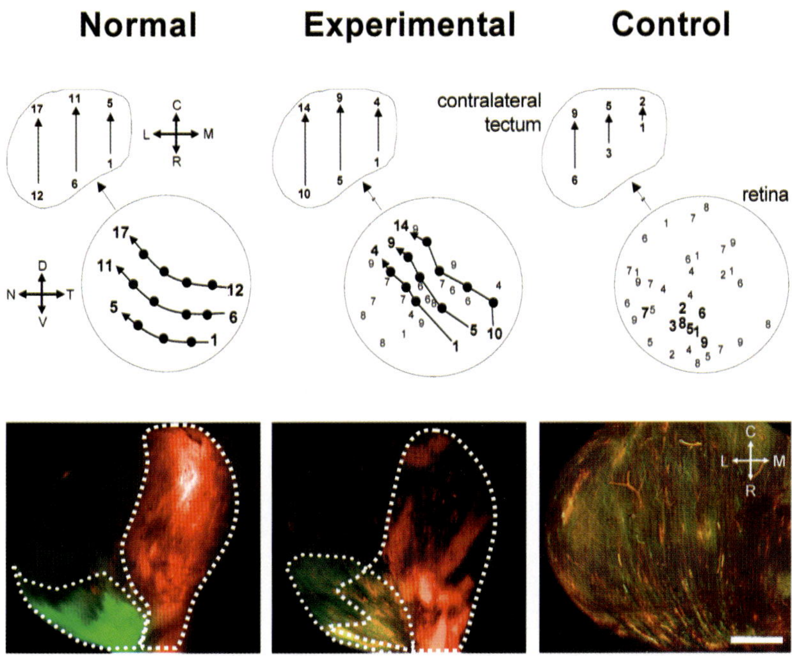

Figure 55.2. Visual training restores topography. Top panel: retinotectal maps assessed electrophysiologically. Numbers in tectal outlines (top) indicate electrode positions and those in retinal outlines (bottom) indicate the location of receptive fields. In normal and trained animals (left and centre) the map is topographic (rows). In untrained animals (right), topography is lacking. Bold numbers represent robust responses, non-bold weak responses. Bottom panel: retinotectal topography as assessed anatomically. Dorsal view of the optic tectum after placements of the carbocyanine dyes DiI (square dotted lines) and DiAsp (round dotted lines) respectively in dorsal or ventral retina. A retinotopic map is observed in normal and trained, but not untrained (complete overlap of dye labelling), animals. Reproduced with permission from J. Neurotrauma.

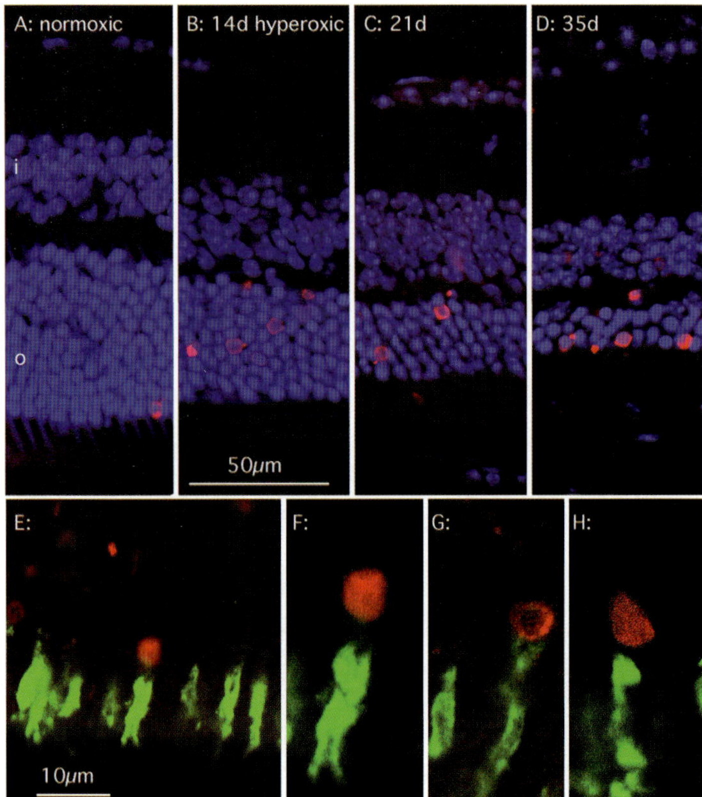

Figure 60.1. Rod and cone death induced by hyperoxia. Blue is bisbenzamide labeling of DNA, showing (A-D) the major layers of the retina (i – inner nuclear layer, o - outer nuclear layer). Red is TUNEL-labeling. Green in E - H is PNA labeling of cone sheaths. A: Young adult C57BL/6J retina, kept in room air (normoxic control). B, C, D: Young adult C57BL/6J retina after 14d, 21d and 35d in hyperoxia. These images show the most affected region of retina, at each exposure time. TUNEL labeling is restricted to the ONL up to 14d of exposure, but labeling of the INL is evident at 21d and 35d. E-H: Regions at the outer surface of the ONL. Occasional TUNEL+ nuclei were closely associated with cone sheaths (labeled green with PNA lectin). F is part of E, at higher power. The scale bar in E represents 10μm in E, and 5μm in F, G, H.

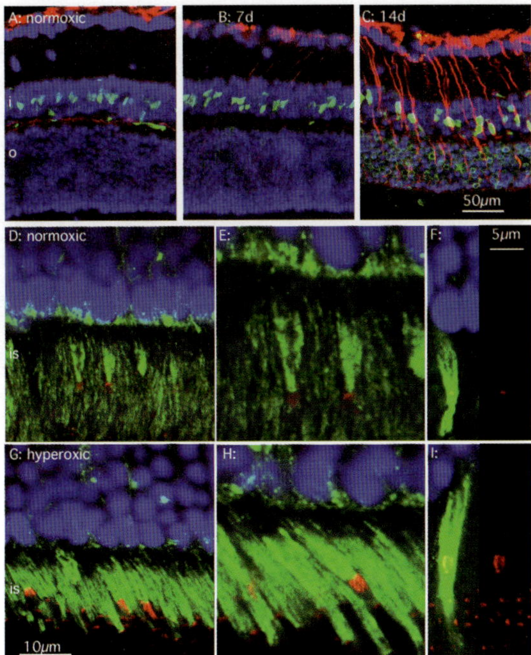

Figure 60.4. Protein up-regulation induced by hyperoxia. Blue is bisbenzamide labeling of normal DNA, showing nuclei of the major cell layers (*i* – inner nuclear layer, *o* – outer nuclear layer). Red in A – C is immunolabeling for GFAP; green is immunolabeling for FGF-2. In D – I, blue is bisbenzamide labeling, red is labeling for c-Jun. while green is labeling for either cytochrome oxidase (D, E, G, H) or PNA (F, I).

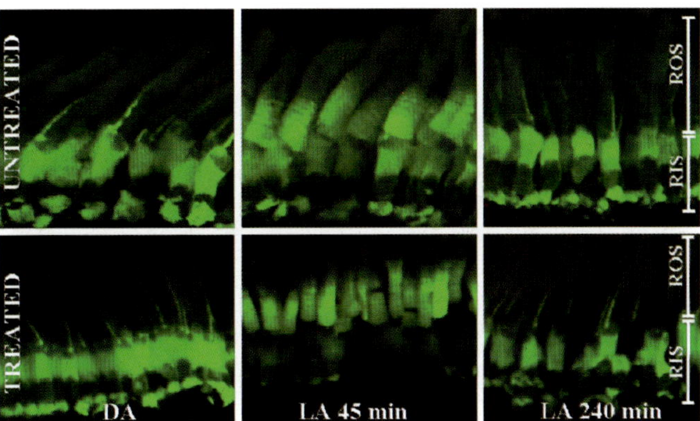

Figure 63.2. Arrestin-GFP translocation in rods from transgenic tadpoles treated or not treated with CHI. Contralateral eyes from the animals used in Figure 63.1 were cryosectioned and the endogenous fluorescence of the arrestin/GFP fusion protein visualized using confocal microscopy. Tadpoles were dark adapted overnight (DA), exposed to light for 45 min (LA 45 min), or to light for 240 min (LA 240 min).

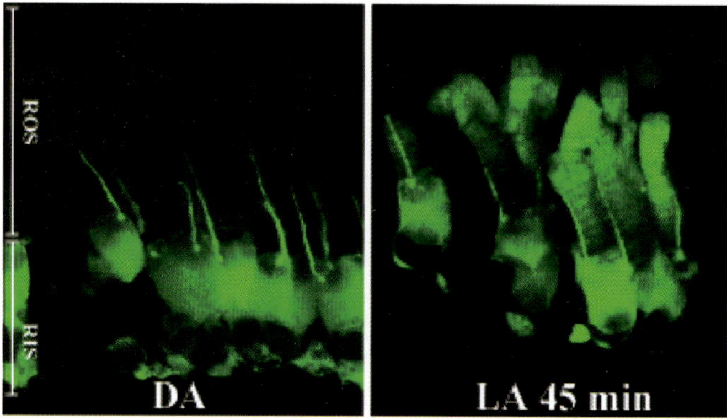

Figure 63.4. Translocation of myc-tagged arrestin-GFP fusion in transgenic tadpoles. Confocal images were obtained from cryosections of xAr-myc-GFP transgenic tadpoles that were dark-adapted overnight (DA) or exposed to light for 45 min (LA 45 min). The endogenous fluorescence of the GFP protein was imaged.

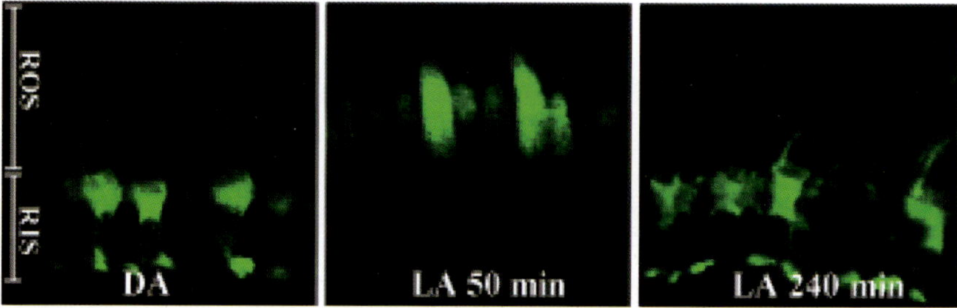

Figure 63.6. Confocal images of xAr-4K $\rightarrow$ Q-GFP translocation in rod photoreceptors. Transgenic tadpoles were dark-adapted overnight (DA), light adapted for 50 min (LA 50 min), or light adapted for 240 min (LA 240 min). Images show the endogenous fluorescence of the GFP in the fusion protein.

5. THE SEQUENCE OF PHOTORECEPTOR DEATH AND PROTECTION

If HICD spreads across the retina, then the images in Figure 60.5 suggest that photoreceptor death spreads ahead of protein (FGF-2, GFAP) up-regulation, and therefore occurs (at any one place in the retina) before protein up-regulation. This argument is indirect because it infers a temporal sequence from a spatial array, but it is consistent with previous (Walsh and Stone, 2001) evidence that, in light stress, photoreceptor death precedes the up-regulation of protective proteins. These observations suggest that photoreceptors are vulnerable to stresses, which induce apoptosis more rapidly than they up-regulate protective mechanisms. They also make clear why the up-regulation of protective factors prior to acute stress, induced by a 'pre-conditioning' exposure to bright light (Cao et al., 1997, Liu et al., 1997) or by optic nerve damage (Bush and Williams, 1991, Kostyk et al., 1994, Casson et al., 2004) is strongly protective to photoreceptors exposed to acute stress.

5.1. Questions Raised

The present data demonstrate the toxicity to retinal photoreceptors of high levels of oxygen (75%) in inspired air. In rats, hyperoxia of this level induces a marked rise in pO_2 in the outer retina, from ~20 mm Hg to ~150 mm Hg (see Fig. 60.8 in (Yu and Cringle, 2001)). Partial degeneration of the photoreceptor layer, for example in the young adult P23H-3 rat (Yu et al., 2004) also increases the pO_2 in the ONL, but to a lesser extent (~40 mmHg). The question arises whether this latter level of raised pO_2 is also toxic. Our preliminary, unpublished observations suggest a photoreceptor-specific toxic effect of 45% inspired oxygen, but long periods of exposure at these lower oxygen levels will be required. The issue is important for understanding the role of raised oxygen levels in the course of retinitis pigmentosa in humans, in which the degenerative process commonly extends over decades.

6. REFERENCES

Ahmed, J., Braun, R., Dunn, J., R. and Linsenmeier, R., 1993. Oxygen distribution in the macaque retina. Invest Ophthalmol Vis Sci. **34**:516-521.

Alder, V., Ben-Nun, J. and Cringle, S., 1990. PO2 profiles and oxygen consumption in cat retina with an occluded retinal circulation. Invest Ophthalmol Vis Sci. **31**:1029-1034.

Alon, T., Hemo, I., Itin, A., Pe'er, J., Stone, J. and Keshet, E., 1995. Vascular endothelial growth factor acts as a survival factor for newly formed retinal vessels and has implications for retinopathy of prematurity. Nature Medicine. **1**:1024-1028.

Arden, G. B., 2001. The absence of diabetic retinopathy in patients with retinitis pigmentosa: implications for pathophysiology and possible treatment. Brit J Ophthalmol. **85**:366-370.

Blanks, J. and Johnson, L., 1984. Specific binding of peanut lectin to a class of retinal photoreceptor cells. Invest Ophthalmol Vis Sci. **25**:546-557.

Bowers, F., Valter, K., Chan, S., Walsh, N., Maslim, J. and Stone, J., 2001. Effects of oxygen and bFGF on the vulnerability of photoreceptors to light damage. Invest Ophthalmol Vis Sci. **42**:804-815.

Bravo-Nuevo, A., Walsh, N. and Stone, J., 2004. Photoreceptor degeneration and loss of retinal function in the C57BL/6-C(2J) mouse. Invest Ophthalmol Vis Sci. **45**:2005-2012.

Bush, R. A. and Williams, T. P., 1991. The effect of unilateral optic nerve section on retinal light damage in rats. Exp Eye Res **52**:139-153.

Cao, W., Li, F., LaVail, M. and Steinberg, R., 1997. Development of injury-induced gene expression of bFGF, FGFR-1, CNTF and GFAP in rat retina. Invest Ophthalmol Vis Sci. **38**:S604.

Casson, R. L., Chidlow, G., Wood, J., Vidal-Sanz, M. and Osborne, N., 2004. The Effect of Retinal Ganglion Cell Injury on Light-Induced Photoreceptor Degeneration. Invest Ophthalmol Vis Sci. **45**:685-693.

Chan-Ling, T. and Stone, J., 1993. Retinopathy of prematurity: Origins in the architecture of the retina. Prog Ret Res. **12**:155-176.

Duncan, T. and O'Steen, W., 1985. The diurnal susceptibility of rat retinal photoreceptors to light-induced damage. Exp Eye Res. **41**:497-507.

Haugh, L., Linsenmeier, R. and Goldstick, T., 1990. Mathematical models of the spatial distribution of retinal oxygen tension and consumption, including changes upon illumination. Annal Biomed Engin. **18**:19-36.

Heckenlively, J., 1988. Retinitis Pigmentosa. Lippincott, Philadelphia.

Hicks, D. and Sahel, J., 1999. The implications of rod-dependent cone survival for basic and clinical research. Invest Ophthalmol Vis Sci. **40**:3071-3074.

Kostyk, S., D'Amore, P., Herman, I. and Wagner, J., 1994. Optic nerve injury alters basic fibroblast growth factor localisation in the retina and optic tract. J Neurosci. **14**:1441-1449.

Lahdenranta, J., Pasqualini, R., Schlingemann, R. O., Hagedorn, M., Stallcup, W. B., Bucana, C. D., Sidman, R. L. and Arap, W., 2001. An anti-angiogenic state in mice and humans with retinal photoreceptor cell degeneration. Proc Natl Acad Sci U S A. **98**:10368-10373.

Linsenmeier, R. A., 1986. Effects of light and dark on oxygen distribution and consumption in the cat retina. J Gen Physiol. **88**:521-542.

Linsenmeier, R. A., Padnick-Silver, L., Derwent, J. K., Ramirez, U. and Narfstrom, K., 2000. Changes in photoreceptor oxidative metabolism in Abyssinian cats with heredity rod/cone degeneration. Invest Ophthalmol Vis Sci. **41**:S886.

Liu, C., Peng, M. and Wen, R., 1997. Pre-exposure to constant light protects photoreceptor from subsequent light damage in albino rats. Invest Ophthalmol Vis Sci. **38**:S718.

Maslim, J., Valter, K., Egensperger, R., Hollander, H. and Stone, J., 1997. Tissue oxygen during a critical developmental period controls the death and survival of photoreceptors. Invest Ophthalmol Vis Sci. **38**:1667-1677.

Mervin, K. and Stone, J., 2002a. Developmental death of photoreceptors in the C57BL/6Jmouse: association with retinal function and self-protection. Exp Eye Res. **75**:703-713.

Mervin, K. and Stone, J., 2002b. Regulation by oxygen of photoreceptor death in the developing and adult C57BL/6J mouse. Exp Eye Res. **75**:715-722.

Mervin, K., Valter, K., Maslim, J., Lewis, G., Fisher, S. and Stone, J., 1999. Limiting photoreceptor death and deconstruction during experimental retinal detachment: the value of oxygen supplementation. Am J Ophthalmol. **128**:155-164.

Noell, W. K., 1955. Visual cell effects of high oxygen pressures. American Physiology Society: Federation Proceedings. **14**:107-108.

Penn, J. S., Li, S. and Naash, M. I., 2000. Ambient hypoxia reverses retinal vascular attenuation in a transgenic mouse model of autosomal dominant retinitis pigmentosa. Invest Ophthalmol Vis Sci. **41**:4007-4013.

Rapp, L. M. and Williams, T. P., 1980. A parametric study of retinal light damage in albino and pigmented rats. The effects of constant light on visual processes. Plenum Press, New York, pp. 135-159.

Sternberg, P., Landers, M. and Wolbarsht, M., 1984. The negative coincidence of retinitis pigmentosa and proliferative diabetic retinopathy. Am J Ophthalmol. **97**:788-789.

Stone, J., Itin, A., Alon, T., Pe'er, J., Gnessin, H., Chan-Ling, T. and Keshet, E., 1995. Development of retinal vasculature is mediated by hypoxia-induced vascular endothelial growth factor (VEGF) expression by neuroglia. J Neurosci. **15**:4738-4747.

Stone, J. and Maslim, J., 1997. Mechanisms of retinal angiogenesis. Prog Ret Eye Res. **16**:157-181.

Stone, J., Maslim, J., Valter-Kocsi, K., Mervin, K., Bowers, F., Chu, Y., Barnett, N., Provis, J., Lewis, G., Fisher, S., Bisti, S., Gargini, C., Cervetto, L., Merin, S. and Pe'er, J., 1999. Mechanisms of photoreceptor death and survival in mammalian retina. Prog Ret Eye Res. **18**:689-735.

Stone, J., Mervin, K., Walsh, N., Valter, K., Provis, J. and Penfold, P., 2004. Photoreceptor stability and degeneration in mammalian retina: lessons from the edge. In: Penfold, P. and Provis, J. (Eds.), Macular Degeneration: Science and Medicine in Practice. Springer Verlag, pp. (in press).

Valter, K., Maslim, J., Bowers, F. and Stone, J., 1998. Photoreceptor dystrophy in the RCS rat: Roles of oxygen, debris and bFGF. Invest Ophthalmol Vis Sci. **39**:2427-2442.

Walsh, N., Bravo-Nuevo, A., Geller, S. and Stone, J., 2004a. Resistance of photoreceptors in the C57BL/6-C2J, C57BL/6J and BALB/CJ mouse strains to oxygen stress: evidence of an oxygen phenotype. Curr Eye Res. In press.

Walsh, N. and Stone, J., 2001. Timecourse of bFGF and CNTF expression in light-induced photoreceptor degeneration in the rat retina. In: Anderson, R. E. et al. (Eds.), New Insights into Retinal Degenerative Diseases. Kluwer Academic Plenum Publishers, New York, pp. 111-118.

Walsh, N., Valter, K. and Stone, J., 2001. Cellular and subcellular patterns of expression of bFGF and CNTF in the normal and lightstressed adult rat retina. Exp Eye Res,. **72**:495-501.

Walsh, N., Van Driel, D., Lee, D. and Stone, J., 2004b. Multiple vulnerability of photoreceptors to mesopic ambient light in the P23H transgenic rat. Brain Research. **1013**:194-203.

Yamada, H., Yamada, E., Ando, A., Esumi, N., Bora, N., Saikia, J., Sung, C. H., Zack, D. J. and Campochiaro, P. A., 2001. Fibroblast growth factor-2 decreases hyperoxia-induced photoreceptor cell death in mice. Am J Pathol. **159**:1113-1120.

Yu, D. and Cringle, S. J., 2001. Oxygen distribution and consumption within the retina in vascularised and avascular retinas and in animal models of retinal disease. Prog Ret Eye Res. **20**:175-208.

Yu, D.-Y., Cringle, S., Valter, K., Walsh, N., Lee, D. and Stone, J., 2004. Photoreceptor death, trophic factor expression, retinal oxygen status, and photoreceptor function in the P23H rat. Invest Ophthalmol Vis Sci. **45**:2013-2019.

Yu, D. Y., Cringle, S. J., Su, E. N. and Yu, P. K., 2000. Intraretinal oxygen levels before and after photoreceptor loss in the RCS Rat. Invest Ophthalmol Vis Sci. **41**:3999-4006.

CHAPTER 61

TREATMENT WITH CARBONIC ANHYDRASE INHIBITORS DEPRESSES ELECTRORETINOGRAM RESPONSIVENESS IN MICE

Yves Sauvé[1], Goutam Karan[1,2], Zhenglin Yang[1,2], Chunmei Li[1,2], Jianbin Hu[1,2], and Kang Zhang[1,2,3]

1. INTRODUCTION

We showed that a functional complex of CA4 and Na^+/bicarbonate co-transporter (NBC1) is specifically expressed in the choriocapillaris and that mutations in CA4 disrupt NBC1-mediated HCO_3^- transport leading to acidification of the retina. This finding (Yang et al., 2005) point to the importance of a functional CA4 for the survival of photoreceptors and imply that CA inhibitors may have long-term adverse effects on vision.

Carbonic anhydrase inhibitors, (CAIs) have a wide range of clinical applications; indications include glaucoma, macular edema (a frequent complication of RP), acute mountain sickness, seizure, increased intracranial pressure, and fluid reduction/diuresis (Weisbecker et al., 2002). Acetazolamide is one of the most extensively used CAIs, and is given frequently as an oral dosage of 1000 mg once daily. At this dosing regimen, it produces a serum trough and peak range of 12-30 µg/ml, or 54 µM to 135 µM in concentration (Friedland et al., 1977). The IC50 of the membrane-associated carbonic anhydrase type 4 (CA4), is 4 µM (Ives, 1998). The K_i of acetazolamide for CA4, is 70 nM (Ilies et al., 2003), therefore, current carbonic anhydrase inhibitors will almost fully inhibit CA4 enzymatic activity. A study of healthy volunteers showed that a single dose of 500 mg of acetazolamide causes demonstrable changes in tests of color vision (Leys et al., 1996). In addition, light-adapted electroretinograms (ERG) a-waves in humans are attenuated about 14% by acetazolamide (Odom et al., 1994). Consequently, the use of CAIs may impair photoreceptor function.

---

[1] Department of Ophthalmology and Visual Science, University of Utah Health Science Center, Salt Lake City, UT 84132, USA.
[2] Program in Human Molecular Biology and Genetics, Eccles Institute of Human Genetics, University of Utah, Salt Lake City, UT 84112, USA.
[3] Department of Neurobiology and Anatomy, University of Utah, Salt Lake City, UT 84112, USA.

We investigated the effect of acetazolamide on retina function by measuring the ERG in mice treated with acetazolamide. We observed that ERG findings in acetazolamide treated mice were analogous to those in human patients with the autosomal dominant rod-cone dystrophy RP17 (Yang et al., 2005). Therefore we caution that long term use of carbonic anhydrase inhibitors may have potential detrimental effects on photoreceptor cells and vision.

2. METHODS

2.1. Animals and Treatment

Mice were housed and handled with the authorization and supervision of the Institutional Animal Care and Use Committee from the University of Utah. Every procedure conformed to the ARVO Statement for the Use of Animals in Ophthalmic and Vision Research. Blk6 mice of 2-3 month of age were given intraperitoneal injections of either normal saline (control group) or acetazolamide at doses of 5, 10, 20, and 30 mg/100 g body weight at a volume of 150 µl at day 0, and day 4 (n = 6 per group). ERG was recorded at day 7.

2.2. ERG Recordings

Under anesthesia with a mixture of ketamine (150 mg/kg i.p.) and xylazine (10 mg/kg i.p.), the mouse head was secured with a stereotaxic head holder and the body temperature monitored through a rectal thermometer and maintained at 38°C using a homeothermic blanket. Pupils were dilated using equal parts of topical phenylephrine (2.5%) and tropicamide (1%). Bupivacaine 0.5% was used as a topical anesthetic to avoid blinking and a drop of 0.9% saline was frequently applied on the cornea to prevent its dehydration and allow electrical contact with the recording electrode (gold wire loop). A platinum subdermal needle (Grass Telefactor, F-E2) inserted under the scalp, between the two eyes, served as the reference electrode. Amplification (at 1-1000 Hz bandpass, without notch filtering), stimulus presentation, and data acquisition were provided by the UTAS-3000 system from LKC Technologies (Gaithersburg, MD).

For dark-adapted ERG recordings, tests consisted of single flash presentations (10 µsec duration), repeated 3 to 5 times to verify the response reliability and improve the signal-to-noise ratio, if required. Stimuli were presented at sixteen increasing intensities in one log unit steps varying from −3.6 to 1.4 log cds/m^2 in luminance. To minimize the potential bleaching of rods, inter-stimulus intervals were increased as the stimulus luminance was elevated from 10 sec at lowest stimulus intensity up to 2 minutes at highest stimulus intensity.

Following dark-adapted recordings, the animals were light adapted for 15 minutes to assure maximal cone output. Photopic intensity responses (30 cds/m^2 background) ranged from −1.6 to 2.9 log cds/m^2 (−1.6, −0.6, 0.4, 1.4, 2.4 and 2.9 log cds/m^2). Criteria responses were set at 20 µV for a- and b-waves (under scotopic and photopic adaptation). For flicker ERGs, stimuli consisted of white flashes provided by a xenon bulb (luminance of 1.37 log cds/m^2), projected on a ganzfeld with a background luminance of 30 cds/m^2 presented at 20 Hz. The stimulus was presented during 3 seconds prior to data collection. This guaranteed that the first few responses (not preceded by repeated stimuli and of potentially greater amplitude) were not included in the average. A total of 40 responses were averaged.

The amplitude of the flicker response was defined as the average of the difference between consecutive negative and positive deflections.

2.3. Statistical Analysis

Error values accompanying averages were expressed as standard errors of the mean (SEM). Comparisons between two groups were made using Mann-Whitney U-test. The probability level at which the Null Hypothesis was rejected is represented as the value "p"; statistical significance was set at $p < 0.05$.

3. RESULTS

Acetazolamide treatment resulted in a depression of both cone- and rod-mediated ERG responses, in a dose dependent manner. However, dose dependence differed for cone and rod-related responses; cone related responses were affected at lower dose regimens than rod related responses.

The two lowest dose regimens, studied here, led to depressed cone mediated responses. Photopic b-wave amplitudes at high luminance stimuli under photopic adaptation (Fig. 61.1), as well as cone-mediated 20 Hz flicker amplitudes (Fig. 61.2) were significantly reduced following 5 and 10 mg/100 g acetazolamide treatment. There were tendencies for amplitude reductions at 20 and 30 mg/100 g doses, especially for maximal photopic b-wave amplitudes, but these reductions did not reach statistical significance. Optimal amplitude reductions occurred for the 10 mg/100 g dosage, both for photopic b-waves and 20 Hz flicker. At this dosage, statistically significant reductions in photopic b-wave amplitudes were seen for stimuli of 0.88 log cds/m^2 and higher. Photopic b-wave thresholds remained unaffected by acetazolamide, regardless of its dose, and b- amplitudes constantly reached a plateau for stimuli of 1.89 log cds/m^2 and of higher luminances.

The results for scotopic ERGs are presented in Fig. 61.3. Statistically significant reductions in scotopic ERG responses (for both maximal a- and b-wave amplitudes) were only obtained with the highest dose regimen, i.e. 30 mg/100 g acetazolamide (Figure 61.3A). In two animals treated at this dosage (scotopic intensity response traces from one of which are presented in Figure 61.3B), the thresholds for a- and b-waves were higher than in PBS injected animals. However, there were no statistically significant changes between acetazolamide (30 mg/100 g) and PBS treated groups. Statistically significant reductions in a- and b-wave amplitudes were seen for stimuli higher than 1.37 log cds/m2 and 0.88 log cds/m^2, respectively. There was a significant drop in b-wave amplitudes at the highest luminance tested (326 ± 34 μV at 2.86 log cds/m^2) compared with the luminance giving the maximal b-wave amplitude (434 ± 36 at 1.89 log cds/m^2). This indication of bleaching was not seen in PBS treated animals.

4. DISCUSSION

This study shows that mice treated with acetazolamide have reduced rod and cone related ERG responsiveness. Treatment with two i.p. injections of 5 or 10 mg/100g (at day 0 and day 4) led to a reduction at day 7 in cone related ERG responses, as reflected by

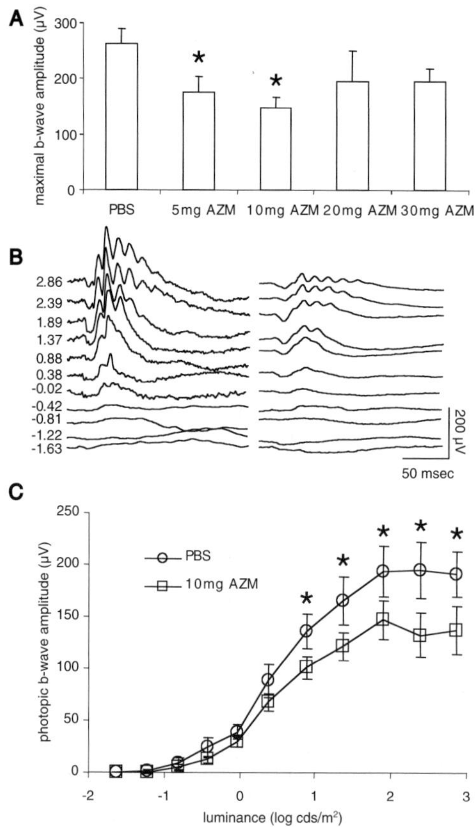

Figure 61.1. Effect of acetazolamide treatment on cone-related ERG responses using single flashes with background adaptation ($30\,\mathrm{cd/m^2}$) that saturates rods. Panel A gives the maximal photopic b-wave amplitude at 7 days following treatment with PBS and various doses of acetazolamide (AZM). Examples of intensity response traces for mice treated with PBS (left side) or $10\,\mathrm{mg/100\,g}$ AZM (right side) are illustrated in panel B; flash intensities are indicated as log $\mathrm{cds/m^2}$ on the right of the traces. Finally, photopic intensity response curves for saline and acetazolamide ($10\,\mathrm{mg/100\,g}$) treated mice are presented in panel C. Statistically significant differences are indicated by an asterisk (*).

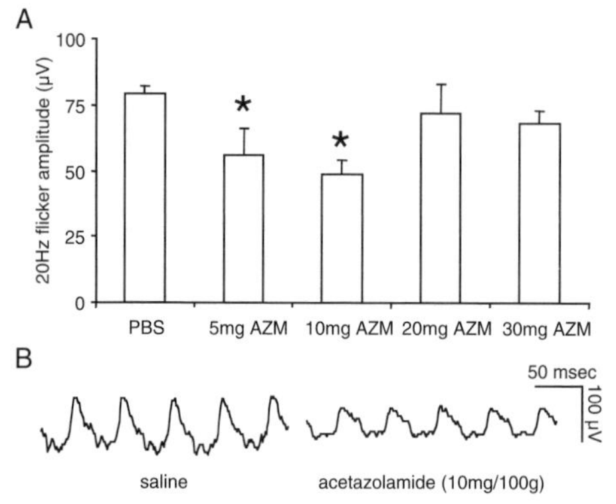

Figure 61.2. Effect of acetazolamide treatment on cone-related ERG responses using $20\,\mathrm{Hz}$ flicker stimuli with background adaptation ($30\,\mathrm{cd/m^2}$) that saturates rods. Panel A gives the maximal amplitude at 7 days following treatment with PBS and various doses of acetazolamide (AZM). Examples of $20\,\mathrm{Hz}$ flicker traces are illustrated in panel B; amplitudes for traces from saline and acetazolamide ($10\,\mathrm{mg/100\,g}$) treated mice are 94 and $41\,\mu\mathrm{V}$ respectively. Statistically significant differences are indicated by an asterisk (*).

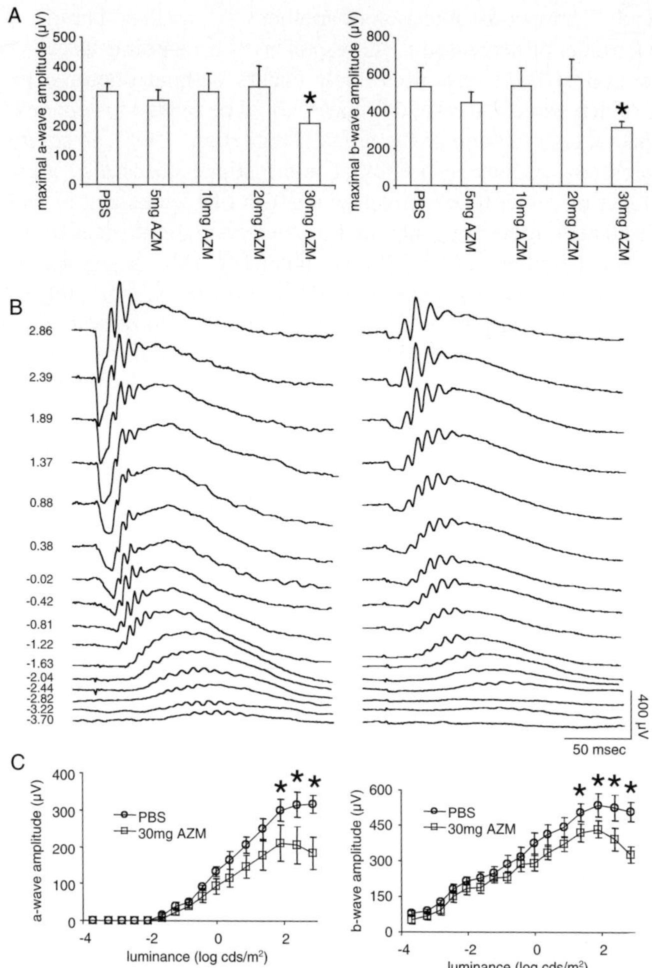

Figure 61.3. Effect of acetazolamide treatment on mixed scotopic ERG responses. Panel A gives the maximal photopic a- (left side) and b- (right side) wave amplitudes at 7 days following treatment with PBS and various doses of acetazolamide (AZM). Examples of intensity response traces for mice treated with PBS (left side) or 30 mg/100 g AZM (right side) are illustrated in panel B; flash intensities are indicated as log cds/m^2 on the right of the traces. Finally, scotopic intensity response curves for saline and acetazolamide (30 mg/100 g) treated mice are presented in panel C; a-wave on left and b-wave on right handside. Statistically significant differences are indicated by an asterisk (*).

diminutions in maximal photopic b-wave and 20 Hz flicker amplitudes. Reduction in rod-related ERG responses, as reflected by diminutions in maximal a- and b-wave amplitudes of dark-adapted scotopic ERGs, was only seen at higher doses, i.e. 30 mg/100 g. These results partially replicate previous results obtained in similar studies done in rats (Findl et al., 1995) where a treatment regimen of 5 mg/100 g i.p. at day 0 led to decrease in rod and cone related ERG responses at day 3. Our findings are also in accord with previous findings of Broeders et al. (1988) showing decreased rod and cone-mediated a- and b-wave amplitudes already 5 hours post-injection of another CAI, methazolamide, at 5 mg/100 g i.p. in rabbits. The findings of decreased ERG responses at time points as early as 5 hours indicate that diminution in ERG responses likely occurs without photoreceptor degeneration and, in the case of depressed rod responsiveness, could be related to acute metabolic changes such as involving an acidification of the retina (Findl et al., 1995). In addition, our findings of optimal depression of cone responses at intermediate doses also argue for a specific change in cell function rather than a direct toxic effect of acetazolamide for the 7 days treatment regimen used here: toxicity should not be expected to diminish as the dosage increases. Still, based on recent findings of an RP form involving a dysfunction of CA4, it is a reasonable assumption that long-term inhibition of this enzyme (and possibly other CAs) might lead to photoreceptor degeneration. Direct injections of acetazolamide in the vitreous of rabbits have indicated that doses of 1 mg or higher not only depressed the b-wave but also resulted in damaged outer segments (Borhani et al., 1994). Assuming that this effect is related to the inhibitory action of acetazolamide on CAs, then a complete inhibition should be expected to have deleterious effects. For instance, topical dorzolamide 3%, which is the optimal dose for treating glaucoma in experimental rabbit, produces a concentration of 35 μM in ocular tissues (Sugrue, 1996), which is higher than the IC50 of 4 μM for CA4 (Ives, 1998), implying a complete inhibition of CA4 enzymatic activity.

CAs are expressed in ocular tissues i.e. in choriocapillaris (CA4) and retina (CA2), their enzymatic action involves catalyzing the reaction of carbon dioxide and water to carbonic acid, which dissociates to bicarbonate and hydrogen ions. Therefore, CAs play a crucial role in regulating tissue pH via their ability to increase the effectiveness of the bicarbonate buffer system. The inhibition of CAs reduces the buffering of the acid reflux from photoreceptors resulting in a lowering of the extracellular pH around these cells. Hydrogen ions have been shown to suppress the dark current of rods in frogs (Liebmann et al., 1984), and acidification was shown to depress photoreceptor response in isolated rat retinae (Winkler, 1972). Hydrogen ions, being elevated in concentration following CAI treatment, might attenuate the amplitude of ERG components by competing with Ca^{++} for binding sites in rods (Moody, 1984) and/or decreasing the Na^+/Ca^{++} exchange in these cells (Hodgkin and Nunn, 1987), both resulting in elevated intracellular Ca^{++}, which is known to downregulate the light sensitive Na^+ conductance in rod outer segments via cGMP gated Na^+ channels (Yau, 1994). CAIs might also lead to acidification of the retina by acting on Muller cells, which have been shown to play a crucial role in regulating the extracellular pH (Oakley and Wen, 1989). In brief, the depression of rod photoreceptor response following CA inhibition in mice treated in the present study is likely to involve an acidification of the retina.

Our results indicate that cone function is also affected by acetazolamide treatment; furthermore, cone related ERG responses are depressed at lower doses, 10 mg/100 g, instead of 30 mg/100 g required to diminish scotopic responses. This CAI-induced depression of cone ERG might involve mechanisms other than acidification. Psychophysical tests in human treated with the CAI methazolamide revealed a disturbance of color discrimination which was not correlated with the degree of acidosis (Widengard, 1995). CAIs might act directly on cones themselves and only indirectly on rods (via acidification as explained above). Histochemical analysis of human retinae indicates that cones (with the exception of blue cones) contain CA activity, while rods do not (Nork et al., 1990). In addition, Muller cells, which also contain CAs, are involved in the generation of b-waves via K^+-evoked depolarization (Wen and Oakley, 1990), and CA inhibition has been shown to modify the extracellular K^+ concentration in isolated salamander Muller cells (Newman, 1996). It is therefore possible that CAIs might also affect the light induced depolarization of Muller cells, and consequently have an effect on the b-wave amplitude. Finally, acetazolamide decreases the ability of Muller cells for buffering the extracellular fluid compartment (Wen and Oakley, 1990). Changes in fluid balances, such as reported in studies involving mice deficient in the water channel aquaporin-4 (expressed by Muller cells) have been shown to lead to b-wave depression by 10 months of age (Li et al., 2002).

5. CONCLUSIONS

Acetazolamide treatment in mice leads to a depression of both rod and cone related ERG responses as seen in patients with the RP17 rod-cone dystrophy. Therefore, the chronic use of CAIs, particularly pertinent in the treatment of glaucoma, may lead to visual loss, cone function being affected at lower doses than rod function.

6. ACKNOWLEDGEMENTS

This research was supported by National Institutes of Health Grants R01EY14428, R01EY14448 and GCRC M01-RR00064, the Ruth and Milton Steinbach Fund, Ronald McDonald House Charities, the Macular Vision Research Foundation, the Research to Prevent Blindness, Inc., Knights Templar Eye Research Foundation, Grant Ritter Fund, American Health Assistance Foundation, the Karl Kirchgessner Foundation, Val and Edith Green Foundation, and the Simmons Foundation.

7. REFERENCES

Borhani, H., Rahimy, M.H., Peyman, G.A., 1994, Vitreoretinal toxicity of acetazolamide following intravitreal administration in the rabbit eye, *Ophthalmic Surg.* **25**:166-169.

Broeders, G.C., Parmer, R., Dawson, W.W., 1988, Electroretinal changes in the presence of a carbonic anhydrase inhibitor, *Ophthalmologica.* **196**:103-110.

Findl, O., Hansen, R.M., Fulton, A.B., 1995, The effects of acetazolamide on the electroretinographic responses in rats, *Invest Ophthalmol Vis Sci.* **36**:1019-1026.

Friedland, B.R., Mallonee, J., Anderson, D.R., 1977, Short-term dose response characteristics of acetazolamide in man. *Arch Ophthalmol.* **95**: 1809-1812.

Hodgkin, A.L., Nunn, B.J., 1987, The effect of ions on sodium-calcium exchange in salamander rods, *J Physiol.* **391**:371-398.

Ilies, M.A., Vullo, D., Pastorek, J., Scozzafava, A., Ilies, M., Caproiu, M.T., Pastorekova, S., Supuran, C.T., 2003, Carbonic anhydrase inhibitors. Inhibition of tumor-associated isozyme IX by halogenosulfanilamide and halogenophenylaminobenzolamide derivatives, *J Med Chem.* **46**:2187-2196.

Ives, H.E., 1998, Diuretic Agents. In Basic & Clinical Pharmacology, B.G.Katzung, ed. (Stamford, CT: Appleton & Lange), pp. 246-248.

Leys, M.J., van Slycken, S., Nork, T.M., Odom, J.V., 1996, Acetazolamide affects performance on the Nagel II anomaloscope, *Graefes Arch Clin Exp Ophthalmol.* **234** Suppl 1:S193-197.

Li, J., Patil, R.V., Verkman, A.S., 2002, Mildly abnormal retinal function in transgenic mice without Muller cell aquaporin-4 water channels, *Invest Ophthalmol Vis Sci.* **43**:573-579.

Liebman, P.A., Mueller, P., Pugh, E.N. Jr., 1984, Protons suppress the dark current of frog retinal rods, *J Physiol.* **347**:85-110.

Moody, W. Jr., 1984, Effects of intracellular H+ on the electrical properties of excitable cells, *Annu Rev Neurosci.* 7:257-278.

Newman, E.A., 1996, Acid efflux from retinal glial cells generated by sodium bicarbonate cotransport, *J Neurosci.* 16:159-168.

Nork, T.M., McCormick, S.A., Chao, G.M., Odom, J.V., 1990, Distribution of carbonic anhydrase among human photoreceptors, *Invest Ophthalmol Vis Sci.* 31:1451-1458.

Oakley, B. 2nd, Wen, R., 1989, Extracellular pH in the isolated retina of the toad in darkness and during illumination, *J Physiol.* 419:353-378.

Odom, J.V., Nork, T.M., Schroeder, B.M., Cavender, S.A., van Slycken, S., Leys, M., 1994, The effects of acetazolamide in albino rabbits, pigmented rabbits, and humans, *Vision Res.* 34:829-8237.

Sugrue, M.F., 1996, The preclinical pharmacology of dorzolamide hydrochloride, a topical carbonic anhydrase inhibitor, *J Ocul Pharmacol Ther.* 12:363-376.

Weisbecker, C.A., Fraunfelder, F.T., Tippermann, R., 2002, Physicians' desk reference for ophthalmic medicines, Montvale, N.J.: Medical Economics Company, Inc.

Wen, R., Oakley, B. 2nd, 1990, K(+)-evoked Muller cell depolarization generates b-wave of electroretinogram in toad retina, *Proc Natl Acad Sci U S A.* 87:2117-2121.

Widengard, I., Mandahl, A., Tornquist, P., Wistrand, P.J., 1995, Colour vision and side-effects during treatment with methazolamide, *Eye.* 9:130-135.

Winkler, B.S., 1972, The electroretinogram of the isolated rat retina, *Vision Res.* 12:1183-1198.

Yang, Z., Alvarez, B.V., Chakarova, C., Jiang, L., Karan, G., Frederick, J.M., Zhao, Y., Sauve, Y., Li, X., Zrenner, E., Wissinger, B., Hollander, A.I., Katz, B., Baehr, W., Cremers, F.P., Casey, J.R., Bhattacharya, S.S., Zhang, K., 2005, Mutant carbonic anhydrase 4 impairs pH regulation and causes retinal photoreceptor degeneration, *Hum Mol Genet.* 14:255-265.

Yau, KW., 1994, Phototransduction mechanism in retinal rods and cones. The Friedenwald Lecture, *Invest Ophthalmol Vis Sci.* 35:9-32.

INJURY-INDUCED RETINAL GANGLION CELL LOSS IN THE NEONATAL RAT RETINA

Kirsty L. Spalding, Qi Cui, Arunasalam M. Dharmarajan, and
Alan R. Harvey[1]

1. INTRODUCTION

In this Chapter we shall briefly review our recent studies in neonatal rats regarding the mechanisms involved in retinal ganglion cell death (RGC) after loss of central visual target areas in the brain. We also describe the influence of neurotrophins on RGC survival and consider the apparently symbiotic relationship between these retinal neurons and the cells they innervate in the developing brain.

2. NEUROTROPHINS AND RETINAL GANGLION CELL VIABILITY

During normal rat visual system development there is loss of a large number of RGCs (Perry et al., 1983). This wave of naturally occurring cell death, or programmed cell death (PCD), is a widespread phenomenon in developing nervous systems. It is generally believed that immature RGCs compete for limited amounts of trophic factors that are expressed by central target structures. Only those RGCs that receive adequate trophic support survive into adulthood (Clarke et al., 1998). Consistent with this, injury to the optic nerve (ON), or removal of central target areas such as the superior colliculus (SC), induces rapid loss of RGCs in neonatal rats (Carpenter et al., 1986; Harvey and Robertson, 1992). Six hours following SC ablation in the postnatal day 4 (P4) rat, RGC death is already twice normal PCD levels, and by 24 hrs post lesion (PL) RGC death has increased about ten-fold (Harvey and Robertson, 1992; Harvey et al., 1994; Cui and Harvey, 1995). PCD and injury-induced RGC death is decreased by exogenous application of brain-derived neurotrophic factor (BDNF)

---

[1] Alan R. Harvey, School of Anatomy & Human Biology, The University of Western Australia, Crawley, WA 6009, Australia. Kirsty L. Spalding (present address) Cell and Molecular Biology, Medical Nobel Institute, Karolinska Institute, Stockholm, Sweden.

or neurotrophin-4/5 (NT-4/5), applied intravitreally (Cui and Harvey, 1994; 1995) or distally to RGC terminals/axons (Ma et al., 1998; Spalding et al., 1998). However, the protective effect of these molecules is transient, delaying but not preventing RGC loss (Cui and Harvey, 1995).

To understand why neurotrophins only temporarily reduce neonatal RGC death after target ablation we analysed changes in neurotrophin receptor expression and possible changes in growth factor dependency (Spalding et al., 2005a). Neurotrophins mediate many of their effects via receptor tyrosine kinases (trkA, trkB and trkC), BDNF and NT-4/5 signalling primarily through the trkB receptor (Huang and Reichardt, 2003). Consistent with RGC sensitivity to BDNF and NT-4/5, these neurons express the trkB receptor (Jelsma et al., 1993; Perez and Caminos, 1995; Vecino et al., 2002).

In unlesioned rats, trkB immunohistochemical analysis revealed no change in the number of trkB positive cells in the RGC layer 24 hrs after intraocular NT-4/5 injection. However, after SC lesions there were less immunoreactive cells and even fewer cells in NT-4/5 injected eyes (Spalding et al., 2005a). Semi-quantitative confocal analysis of immunofluorescence intensity revealed an increase in trkB staining in the RGC layer in unlesioned rats 24 hrs after NT-4/5 injection, whereas in SC-lesioned animals exposed to NT-4/5 there was a significant *decrease* in staining. In summary, application of neurotrophins caused a down-regulation of the cognate trkB receptor, presumably altering the long-term responsiveness of neonatal RGCs to exogenous neurotrophins.

Injured neonatal RGCs in P4/P5 retinas do not appear to shift their trophic dependence to other survival factors (Meyer-Franke et al., 1995). Different doses of ciliary neurotrophic factor (CNTF) were given intraocularly, either alone or combined with NT-4/5. We also tested an SC-derived chondroitin sulfate proteoglycan that has been reported to promote adult RGC survival after injury (Huxlin et al., 1995). None of these interventions reduced lesion-induced RGC death 24 hrs or 36 hrs after SC ablation (Spalding et al., 2005a).

3. SOURCES OF TROPHIC SUPPORT FOR RETINAL GANGLION CELLS

Consistent with the neurotrophic hypothesis of RGC dependence on target-derived factors, BDNF is produced in the SC and is retrogradely transported by RGCs (Ma et al., 1998). However using ^{125}I-labelled peptides we recently showed that there was also substantial and rapid anterograde transport of BDNF and, to a lesser extent, neurotrophin-4/5 (NT-4/5) to central visual target areas in the neonatal rat brain (Spalding et al., 2002). Six hours after unilateral intraocular injection, all retino-recipient regions in the thalamus and midbrain were heavily labelled. Neonatal eye removal results in increased cell death in the SC (Lund et al., 1973), suggesting an anterograde trophic influence on tectal cells. We found that, 24 hrs after intraocular application of physiologically relevant doses of neurotrophin, there was significantly decreased neuronal death in the contralateral SC. Our new data thus support the proposal that BDNF and NT-4/5 can be anterograde survival factors for postsynaptic cells in the developing rat nervous system (cf. Caleo et al., 2000).

BDNF and NT-4/5 protein and mRNA have now been identified not only in central sites such as the superficial layers of the SC but also in the retina itself. Indeed, BDNF levels in neonatal rodents are greater in the SC than in the retina (Ma et al., 1998; Frost et al., 2001; Seki et al., 2003) and cells of the Xenopus and chick RGC layer have been reported to receive BDNF trophic support predominantly from intra-retinal, rather than collicular sources (Cohen-Cory et al., 1996; Herzog and von Bartheld, 1998). In mammals there is now evidence that RGCs may be supported not only by target-derived neurotrophins but also by retinally-derived neurotrophins (de Araujo and Linden, 1993; Cohen-Cory et al., 1996; Ary-Pires et al., 1997).

To determine whether neonatal rat RGC viability depends on intra-retinally derived neurotrophins, we measured RGC death 24 hrs following injections of a mixture of BDNF and NT-4/5 blocking antibodies into the eye. RGC death was also assessed 24 hrs and 48 hrs after injection of these same antibodies into the SC (Spalding et al., 2004). It was found that collicular injections of BDNF and NT-4/5 blocking antibodies significantly increased RGC death in the neonatal rat 24 hrs post injection, death rates returning to normal by 48 hrs. The increase in death was greatest following SC injections, but death was also significantly increased 24 hrs following *intravitreal* antibody injection, providing further evidence for a survival-promoting role for intraretinally derived neurotrophic factors.

The mechanisms whereby RGCs respond differentially to target-derived versus locally supplied neurotrophins are unclear, although there is evidence that intracellular signalling pathways activated by growth factors can differ depending upon where the neurotrophin binds, e.g. at the cell body, dendrite, or distally at the nerve terminal (Heerssen and Segal, 2002). There is also evidence that trophic factors derived from local and target sources have differing effects on RGC development (Lom et al., 2002). Taken together, many issues concerning the development of retinofugal connections remain to be resolved. BDNF and NT-4/5 are produced in the retina and in central targets, they are transported in both directions along RGC axons, and these neurotrophins increase the viability of both retinal and target neurons. Understanding the bi-directional relationship between developing RGCs and central neurons, and how these cells distinguish between local paracrine/autocrine neurotrophin expression and factors derived from more distant sources, are clearly important areas for future research.

4. MECHANISMS OF NEONATAL RETINAL GANGLION CELL DEATH

Compared to injury in neonates, ON injury in adult rodents results in RGC death but at a much slower rate (Misantone et al., 1984; Villegaz-Perez et al., 1993). Nonetheless, injury-induced RGC death in both the adult (Berkelaar et al., 1994; Garcia-Venezuela et al., 1994; Isenmann et al., 1997) and neonatal rat (Harvey et al., 1994; Rabacchi et al., 1994; Cui and Harvey, 1995) has primarily been ascribed to apoptosis. There is evidence in adults that RGC death after ON injury is due to caspase activation. In vivo studies have shown that caspases-3, -8, and -9 become activated in adult rat RGCs following optic nerve axotomy and intraocular injection of relevant caspase inhibitors can reduce RGC death, at least for a certain period of time (Kermer et al., 1998; Chaudhary et al., 1999; Kermer et al., 1999; Weishaupt et al., 2003).

We recently examined whether blocking caspases in vivo reduces neonatal RGC death after SC lesions (Spalding et al., 2005b). Surprisingly, intraocular injection of general and specific caspase inhibitors did not increase neonatal RGC survival 6 hrs and 24 hrs after SC ablation. These inhibitors were, however, effective in blocking caspases in another well-defined in vitro apoptosis model, the ovarian corpus luteum. Retinal caspase-3 protein and mRNA levels were assessed 3, 6 and 24 hrs after SC removal using immunohistochemistry, western and northern blots, and quantitative real-time PCR. Terminal deoxynucleotidyl transferase-mediated dUTP nick-end labelling (TUNEL) was used to independently monitor retinal cell death. The PCR data showed a small but not significant increase in caspase-3 mRNA in retinas 24hrs PL, however Western blot analysis did not reveal a significant shift to cleaved (activated) caspase-3 protein. There was a small increase in the number of cleaved caspase-3 immunolabelled cells in the ganglion cell layer 24hrs after SC removal, but this represented only a fraction of the death revealed by TUNEL.

In summary, the inability of caspase inhibitors to reduce lesion-induced RGC death 6hrs or 24hrs after neonatal SC ablation, and the lack of major activation of caspase-3 mRNA and production of cleaved protein 24hrs PL, strongly suggests that most lesion-induced RGC death in immature rats is not caspase-dependent. Within a given neuronal population, it appears that different cell death cascades can be initiated at different maturational stages during the lifetime of a cell.

5. REFERENCES

Ary-Pires, R., Nakatani, M., Rehen, S. K., and Linden, R., 1997, Developmentally regulated release of intraretinal neurotrophic factors in vitro. *Int J Dev Neurosci.* **15**:239-255.

Berkelaar, M., Clarke, D. B., Wang, Y. C., Bray, G. M., and Aguayo, A. J., 1994, Axotomy results in delayed death and apoptosis of retinal ganglion cells in adult rats. *J Neurosci.* **14**:4368-4374.

Caleo, M., Menna, E., Chierzi, S., Cenni, M. C., and Maffei, L., 2000, Brain-derived neurotrophic factor is an anterograde survival factor in the rat visual system. *Curr Biol.* **10**:1155-1161.

Carpenter, P., Sefton, A. J., Dreher, B., and Lim, W., 1986, Role of target tissue in regulating the development of retinal ganglion cells in the albino rat: effects of kainate lesions in the superior colliculus. *J Comp Neurol.* **251**:240-259.

Chaudhary, P., Ahmed, F., Quebada, P., and Sharma, S. C., 1999, Caspase inhibitors block the retinal ganglion cell death following optic nerve transection. *Mol Brain Res.* **67**:36-45.

Clarke, P. G., Posada, A., Primi, M. P., and Castagne, V., 1998, Neuronal death in the central nervous system during development. *Biomed Pharmacother.* **52**:356-362.

Cohen-Cory, S., Escandon, E., and Fraser, S. E., 1996. The cellular patterns of BDNF and trkB expression suggest multiple roles for BDNF during Xenopus visual system development. *Dev Biol* **179**:102-115.

Cui, Q., and Harvey, A. R., 1994, NT-4/5 reduces naturally occurring retinal ganglion cell death in neonatal rats. *Neuroreport* **5**:1882-1884.

Cui, Q., and Harvey, A. R., 1995, At least two mechanisms are involved in the death of retinal ganglion cells following target ablation in neonatal rats. *J Neurosci.* **15**:8143-8155.

De Araujo, E. G., and Linden, R., 1993, Trophic factors produced by retinal cells increase the survival of retinal ganglion cells in vitro. *Eur J Neurosci.* **5**:9181-9188.

Frost, D. O., Ma, Y. T., Hsieh, T., Forbes, M. E., and Johnson, J. E., 2001, Developmental changes in BDNF protein levels in the hamster retina and superior colliculus. *J Neurobiol.* **49**:173-187.

Garcia-Valenzuela, E., Gorczyca, W., Darzynkiewicz, Z., and Sharma, S. C., 1994, Apoptosis in adult retinal ganglion cells after axotomy. *J Neurobiol.* **25**:431-438.

Harvey, A. R., Cui, Q., and Robertson, D., 1994, The effect of cycloheximide and ganglioside GM1 on the viability of retinotectally projecting ganglion cells following ablation of the superior colliculus in neonatal rats. *Eur J Neurosci.* **6**:550-557.

Harvey, A. R., and Robertson, D., 1992, Time-course and extent of retinal ganglion cell death following ablation of the superior colliculus in neonatal rats. *J Comp Neurol.* **325**:83-94.

Heerssen, H. M., and Segal, A. R., 2002, Location, location, location: a spatial view of neurotrophin signal transduction. *Trends in Neurosci.* **25**:160-165.

Herzog, K-H., and von Bartheld, C. S., 1998, Contributions of the optic tectum and the retina as a source of brain-derived neurotrophic factor for retinal ganglion cells in the chick embryo. *J Neurosci.* **18**:2891-2906.

Huang, E. J., and Reichardt, L. F., 2003, Trk receptors: roles in neuronal signal transduction. *Annu Rev Biochem.* **72**:609-642.

Huxlin, K. R., Dreher, B., Schulz, M., Sefton, A. J., and Bennett, M. R., 1995, Effect of collicular proteoglycan on survival of adult rat retinal ganglion cells following axotomy. *Eur J Neurosci.* **7**:96-107.

Isenmann, S., Wahl, C., Krajewski, S., Reed, J. C., and Bähr, M., 1997, Up-regulation of Bax protein in degenerating retinal ganglion cells precedes apoptotic cell death after optic nerve lesion in the rat. *Eur J Neurosci.* **8**:1763-1772.

Jelsma, T. N., Friedman, H. H., Berkelaar, M., Bray, G. M., and Aguayo, A. J., 1993, Different forms of the neurotrophin receptor trkB mRNA predominate in rat retina and optic nerve. *J Neurobiol.* **24**:1207-1214.

Kermer, P., Klöcker, N., Labes, M., and Bähr, M., 1998, Inhibition of CPP32-like proteases rescues axotomized retinal ganglion cells from secondary cell death in vivo. *J Neurosci.* **18**:4656-4662.

Kermer, P., Klöcker, N., Labes, M., Thomsen, S., Srinivasan, A., and Bähr, M., 1999, Activation of caspase-3 in axotomized rat retinal ganglion cells in vivo. *FEBS Lett.* **453**:361-364

Lom, B., Cogen, J., Sanchez, A. L., Vu, T., and Cohen-Cory, S., 2002, Local and target-derived brain-derived neurotrophic factor exert opposing effects on the dendritic arborization of retinal ganglion cells in vivo. *J Neurosci.* **22**:7639-7649.

Lund, R. D., Cunningham, T. J., and Lund, J. S., 1973, Modified optic projections after unilateral eye removal in young rats. *Brain Behav Evol.* **8**:27-50.

Ma, Y-T., Hsieh. T., Forbes, M. E., Johnson, J. E., and Frost, D. O., 1998, BDNF injected into the superior colliculus reduces developmental retinal ganglion cell death. *J Neurosci.* **18**:2097-2107.

Meyer-Franke, A., Kaplan, M. R., Pfrieger, F. W., and Barres, B. A., 1995, Characterization of the signaling interactions that promote the survival and growth of developing retinal ganglion cells in culture. *Neuron* **15**: 805-819.

Misantone, L. J., Gershenbaum, M., and Murray, M., 1984, Viability of retinal ganglion cells after optic nerve crush in adult rats. *J Neurocytol.* **13**:449-465.

Perez, M. T. R., and Caminos, E., 1995, Expression of brain-derived neurotrophic factor and of its functional receptor in neonatal and adult rat retina. *Neurosci Lett.* **183**:96-99.

Perry, V. H., Henderson, Z., and Linden, R., 1983, Postnatal changes in retinal ganglion cell and optic axon populations in the pigmented rat. *J Comp Neurol.* **219**:356-368.

Rabacchi, S. A., Bonfanti, X-H., and Maffei, L., 1994, Apoptotic cell death induced by optic nerve lesion in the neonatal rat. *J Neurosci.* **14**:5292-5301.

Seki, M., Nawa, H., Fukuchi, T., Abe, H., and Takei, N., 2003, BDNF is upregulated by postnatal development and visual experience: quantitative and immunohistochemical analyses of BDNF in the rat retina. *Invest Ophthalmol Vis Sci.* **44**:3211-3218.

Spalding, K. L., Cui, Q., and Harvey, A. R., 1998, The effects of central administration of neurotrophins or transplants of fetal tectal tissue on retinal ganglion cell survival following superior colliculus removal in neonatal rats. *Dev Brain Res.* **107**:133-142.

Spalding, K. L., Cui, Q., and Harvey, A. R., 2005a, Retinal ganglion cell neurotrophin receptor levels and trophic requirements following target ablation in the neonatal rat. *Neuroscience* (in press)

Spalding, K. L. Dharamarajan, A. M., and Harvey, A. R., 2005b, Caspase-independent retinal ganglion cell death following target ablation in the neonatal rat. *Eur J Neurosci.* (in press).

Spalding, K. L., Rush, R. A., and Harvey, A. R., 2004, Target-derived and locally-derived neurotrophins support retinal ganglion cell survival in the neonatal rat retina. *J Neurobiol.* **60**:319-327.

Spalding, K. L, Tan, M. M. L., Hendry, I. A., and Harvey, A. R., 2002, Anterograde transport and trophic actions of BDNF and NT-4/5 in the developing rat visual system. *Mol Cell Neurosci.* **19**:485-500.

Vecino, E., Garcia-Grespo, D., Garcia, M., Martinez-Millan, L., Sharma, S. C., and & Carrascal, E., 2002, Rat retinal ganglion cells co-express brain derived neurotrophic factor (BDNF) and its receptor TrkB. *Vision Res.* **42**:151-157.

Villegas-Perez, M. P., Vidal-Sanz, M., Raminsky, M., and Aguayo, A. J., 1993, Rapid and protracted phases of retinal ganglion cell loss follow axotomy in the optic nerve of adult rats. *J Neurobiol.* **24**:23-36.

Weishaupt, J. H., Diem, R., Kermer, P., Krajewski, S., Reed, J. C., and Bähr, M., 2003, Contribution of caspase-8 to apoptosis of axotomized rat retinal ganglion cells in vivo. *Neurobiol Dis.* **13**:124-135.

PART VIII

BASIC SCIENCE UNDERLYING RETINAL DEGENERATION

ARRESTIN TRANSLOCATION IN ROD PHOTORECEPTORS

W. Clay Smith*, James J. Peterson*, Wilda Orisme*, and Astra Dinculescu*

1. INTRODUCTION

The vertebrate photoreceptor is the epitome of polarized neurons, containing two specialized compartments—the outer segment and the inner segment, connected by a narrow non-motile cilium. The outer segment of rod and cone photoreceptors is principally dedicated to capturing light and converting the energy of a photon into a change in membrane potential. The primary function of the inner segment is to provide the metabolic and synthetic demands of the photoreceptors. In order to maintain this high degree of specialization, molecules are routinely targeted to their appropriate compartment during protein synthesis. However, in addition to this relatively slow transport process, photoreceptors have a much more rapid process whereby some molecules are rapidly moved between the inner segment and outer segment through the connecting cilium in response to the light adaptational state of the eye. This translocation process has been conclusively demonstrated for two molecules involved in the phototransduction cascade—transducin and arrestin (Broekhuyse *et al.* 1985; Mangini and Pepperberg 1988; Whelan and McGinnis 1988; Sokolov *et al.* 2002; Peterson *et al.* 2003).

Recent studies have shown that the light-driven movement of arrestin is independent of activation of the phototransduction cascade since translocation is normal in transgenic mice that are deficient for transducin (Mendez *et al.* 2003; Zhang *et al.* 2003). Various ideas have been proposed regarding the potential function of protein translocation. These hypotheses include a role in the regulation of photoreceptor sensitivity (McGinnis *et al.* 1991; Sokolov *et al.* 2002), and a protective function in preventing light damage (Elias *et al.* 2004).

In this study, we address some of the mechanistic questions regarding arrestin translocation. Specifically we ask whether the light-driven redistribution of arrestin is a conse-

*Department of Ophthalmology, University of Florida, Gainesville, FL 32610-0284. Corresponding author W.C. Smith, Department of Ophthalmology, University of Florida, Box 100284 JHMHC, Gainesville, FL 32610-0284. Tel: (352) 392-0476; Fax: (352) 392-0573; E-mail: csmith@eye.ufl.edu.

quence of arrestin movement, or whether *de novo* protein synthesis coupled with arrestin degradation plays a significant role in the apparent redistribution of arrestin. In addition, we utilize a variant of arrestin that is unable to bind to rhodopsin to investigate the role of arrestin binding to light-activated phosphorhodopsin in the translocation of arrestin. Finally, we address whether vertebrate arrestins rely on an affinity for phosphoinositol lipids for translocation.

2. METHODS AND MATERIALS

2.1. Cycloheximide Treatment of Transgenic Tadpoles

Transgenic *Xenopus* expressing a fusion of GFP at the C-terminus of *Xenopus* arrestin (xAr-GFP) were obtained from breeding pairs of adult transgenic *Xenopus* as previously described (Peterson *et al.* 2003) and dark-adapted overnight. Tadpoles were placed in 0.1x tadpole Ringers with or without 100 μM cycloheximide (CHI) for 1 h. Tadpoles then either remained in the dark, or were exposed to laboratory lighting for 45 min or for 240 min. At this point, tadpoles were euthanized in 0.025% benzocaine, one eye removed, the cornea punctured with a scalpel, and placed in 100 μL 1x tadpole Ringers with 20 mM glucose and 70 μCi ^{35}S-labeled methionine and cysteine (Amersham) for 2 h at room temperature. After rinsing, the eye was disrupted by vigorous pipetting and vortexing in 50 μL 1x Laemmli sample buffer (Laemmli 1970), and viscosity reduced by the addition of 25 units of benzonase (Novagen). An aliquot of the extract was separated on 12% SDS-PAGE, the gel stained with Coomassie Brilliant blue, and the dried gel exposed to x-ray film to detect incorporated radiolabeled proteins.

The contralateral eye was fixed in methanolic formaldehyde as described (Peterson *et al.* 2003), and processed for confocal microscopy to detect the localization of the arrestin/GFP fusion using the endogenous fluorescence of GFP.

2.2. Myc-Tagged Arrestin

The ten amino acid myc tag (EQKLISEEDL) was incorporated into *Xenopus* arrestin using overlapping PCR products that introduced the cDNA for the myc-tag between the codons for Leu-76 and Thr-77. The overlapping products were combined and the complete cDNA amplified using primers against the 5′ and 3′ ends of the arrestin cDNA, incorporating *Xho*I and *Not*I sites, respectively. This product was cloned into the *Xho*I and *Not*I sites of the XOPS1.3 vector under the control of the 1.3 kb *Xenopus* rod opsin promoter (Tam *et al.* 2000). The cDNA for GFP was inserted immediately prior to the arrestin stop codon at an introduced *Nhe*I site. Transgenic animals expressing this myc-tagged xAr-GFP (xAr-myc-GFP) were prepared by nuclear transplantation (Kroll and Amaya 1996).

For *in vitro* studies, a His$_{(6)}$ tag was incorporated by PCR at the 5′ end of the cDNA immediately after the initiating ATG for xAr-GFP and xAr-myc-GFP, and then cloned into the shuttle vector pPIC-ZA for heterologous expression in *Pichia pastoris* (Dinculescu *et al.* 2002). Expressed proteins were purified over nickel-agarose (Ni-NTA, Qiagen), followed by heparin agarose. *In vitro* binding assays with rhodopsin were performed using unphosphorylated and phosphorylated rod disc membranes prepared from *Rana catesbeiana* retinas as described for bovine disc membranes (McDowell 1993).

2.3. Four Lysine Mutants of Arrestin

Multiple alignment of *Drosophila*, bovine, and *Xenopus* arrestins was performed using ClustalX (Thompson *et al.* 1997). Two of the three lysines (Lys-228 and Lys-231) identified as important for promoting an association with inositol phospholipids in *Drosophila* (Lee *et al.* 2003) are conserved in bovine arrestin and were changed to alanines by site-directed mutagenesis. Because sequence conservation is not perfect, two additional lysines were substituted with alanines to insure disruption of the potential inositol phospholipid binding site, creating the four lysine mutant K232A/K235A/K236A/K238A (4K $\rightarrow$ A) in bovine visual arrestin. The 4K $\rightarrow$ A bovine arrestin mutant was expressed in *Pichia*, purified to homogeneity, and tested for binding to inositol phospholipids immobilized on nitrocellulose strips (PIP Strips, Echelon, Inc.). PIP-Strips were blocked for 60 min with 1% gamma globulin-free horse serum in phosphate-buffered saline (PBS). Wild-type arrestin, 4K $\rightarrow$ A arrestin, and the PH domain of phospholipase C δ1 fused to glutathione-S-transferase (GST-GRIP; Echelon, Inc) (0.5 μg/mL) were then incubated with the PIP strips for 4 h. After washing with 0.05% Tween-20 in PBS, the blots were immunoprobed to detect binding using anti-arrestin monoclonal SCT-128 for bovine arrestin and anti-GST monoclonal for the GRIP positive control. Binding of the primary antibody was detected using an anti-mouse antibody conjugated to alkaline phosphatase with nitroblue tetrazolium/5-bromo-4-chloro-3-indoyl phosphate as the substrate.

Relative affinity of bovine arrestin and 4K $\rightarrow$ A arrestin for phytic acid was assessed by competitive elution from heparin agarose. Arrestin and 4K $\rightarrow$ A (1 mg) was immobilized on a 1 mL heparin sepharose column (Amersham) in 10 mM HEPES/30 mM NaCl, pH 7.0. Aliquots of phytic acid (1 μM – 10 mM) in the same mobile phase were added to the column and the amount of arrestin eluted in each aliquot measured by absorbance at 278 nm.

Homologous mutations were also created in *Xenopus* arrestin by PCR site-directed mutagenesis, substituting lysines 232, 235, 236, and 267 with glutamine (4K $\rightarrow$ Q). This substituted arrestin was fused with GFP as previously described and used to create transgenic *Xenopus* tadpoles. Translocation of the Ar(4K $\rightarrow$ Q)-GFP was assessed by confocal microscopy using tadpoles that were dark-adapted for 3 days, then exposed to laboratory lighting for 50 min, or for 240 min.

3. RESULTS AND DISCUSSION

3.1. Does Arrestin Translocate from the Outer Segments?

Our previously published work clearly demonstrates that arrestin translocates from the inner segments (RIS) of dark-adapted rod photoreceptors, moving to the rod outer segments (ROS) upon exposure to light (Peterson *et al.* 2003). In these studies, it also appears that arrestin leaves the rod outer segments and returns to the inner segments in response to either dark adaptation or extended light adaptation. However, another equally tenable explanation is that the arrestin that translocates to the outer segments is proteolyzed and that the arrestin subsequently seen in the inner segment is a result of newly synthesized arrestin (Azarian *et al.* 1995).

To discriminate between these two mechanisms, transgenic Ar-GFP tadpoles were treated with cycloheximide (CHI) to inhibit protein synthesis (Obrig *et al.* 1971), and

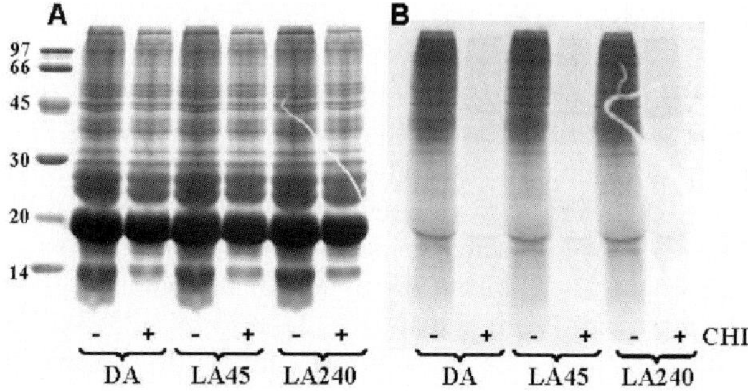

Figure 63.1. Inhibition of protein synthesis in the eyes of tadpoles treated with $100\,\mu M$ cycloheximide (CHI). Treated (+) or untreated (-) tadpoles were dark adapted overnight (DA) or exposed to light for 45 min (LA45) or for 240 min (LA240). One eye was removed and new protein synthesis assessed as described in Methods. Proteins extracted from eye homogenates were separated by 12% SDS-PAGE and stained with Coomassie blue (A). Newly synthesized proteins that incorporated radiolabeled cysteine and methionine are revealed in the autoradiograph of the same gel (B).

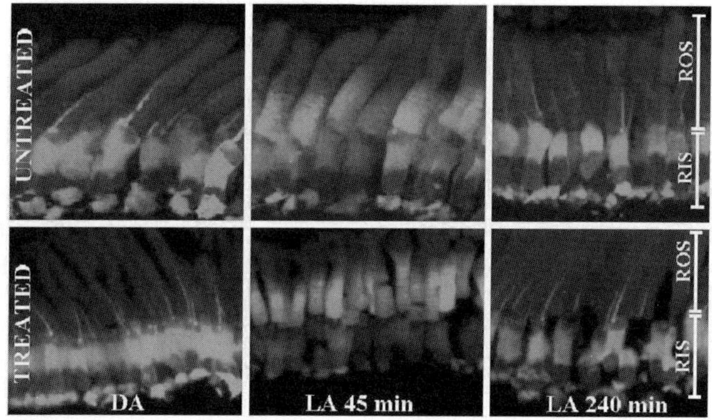

Figure 63.2. Arrestin-GFP translocation in rods from transgenic tadpoles treated or not treated with CHI. Contralateral eyes from the animals used in Figure 63.1 were cryosectioned and the endogenous fluorescence of the arrestin/GFP fusion protein visualized using confocal microscopy. Tadpoles were dark adapted overnight (DA), exposed to light for 45 min (LA 45 min), or to light for 240 min (LA 240 min). See also color insert.

translocation subsequently assessed. Figure 63.1 shows total homogenates prepared from one eye from each of these tadpoles, separated on 12% SDS-PAGE. Protein staining of total homogenates prepared from eyes treated with $100\,\mu M$ CHI revealed a similar profile as the untreated animals, although the total protein content extracted was slightly less in the treated animals (Figure 63.1A). Autoradiography of this same gel indicates that there was substantial incorporation of the radiolabeled cysteine and methionine in the untreated animals, but that new protein synthesis was largely blocked in the tadpoles treated with cycloheximide (Figure 63.1B).

The contralateral eyes from the same tadpoles used in Figure 63.1 were fixed in formaldehyde and processed for confocal microscopy to show the distribution of the Ar-GFP fusion protein, using the endogenous fluorescence of GFP (Figure 63.2). In both the untreated and CHI-treated tadpoles, the Ar-GFP localizes to the RIS and axonemes in dark-adapted animals. In response to 45 min of light exposure, the Ar-GFP almost completely translocates to the outer segments. If the cycloheximide-treated tadpoles are kept in light for 4 h or returned to the dark for 1 h (data not shown), the Ar-GFP again concentrates in the RIS and axonemes for both treated and untreated tadpoles. Because we have demonstrated that treatment of the tadpoles with CHI blocks new protein synthesis (Figure 63.1), we conclude that arrestin translocates both to and from outer segments, and that at least the vast majority of the redistribution of arrestin is a consequence of translocation and not proteolysis followed by replacement by newly synthesized protein.

These results are consistent with quantitative studies showing that the mass of arrestin-GFP is conserved during translocation from the RIS to the ROS in response to light (Peet *et al.* 2004). This study did not quantify the translocation from ROS to RIS during dark adaptation. A similar study using cycloheximide to inhibit protein synthesis in mouse retinas shows the light-driven redistribution of transducin is also largely a consequence of transducin translocation and not synthesis of new protein (Sokolov *et al.* 2002).

3.2. Is Binding to Light-Activated Rhodopsin Required for Arrestin Translocation?

When the first observations were made showing that arrestin redistributes to the outer segments during light adaptation, it was assumed that this translocation was a consequence of arrestin binding to light-activated phosphorylated rhodopsin (Wilden *et al.* 1986; Mangini *et al.* 1994). However, in mice that are deficient for rhodopsin phosphorylation, either lacking rhodopsin kinase or in which the C-terminus of rhodopsin has been mutated to remove the phosphorylation sites, light-driven translocation of arrestin between the inner segment and outer segments occurs in a manner indistinguishable from that in wild-type mice (Mendez *et al.* 2003; Zhang *et al.* 2003). To corroborate and extend these findings, we analyzed the translocation of arrestin in transgenic *Xenopus* that expressed an arrestin that is unable to bind light-activated phosphorhodopsin (R*P). In previous studies, we demonstrated that binding of arrestin to R*P could be blocked by inserting a ten amino acid myc tag (EQKLISEEDL) in a loop of bovine arrestin between Leu-77 and Ser-78 (Dinculescu *et al.* 2002). Reasoning that this variant of arrestin could be used to address the necessity of arrestin binding to R*P for translocation, we created a homologous substitution in our *Xenopus* arrestin-GFP fusion, inserting the myc tag between Leu-76 and Thr-77 (xAr-myc-GFP).

To determine if this insertion in *Xenopus* arrestin has a similar effect as that documented for bovine arrestin, both xAr-GFP and xAr-myc-GFP were heterologously expressed, purified, and used in a centrifugation binding assay with rhodopsin in ROS disk membranes obtained from bullfrog retinas. As previously documented, xAr-GFP retains its selectivity for light activated phosphorhodopsin (Figure 63.3). In contrast, xAr-myc-GFP has essentially no binding to the rhodopsin-containing disc membranes for any of the four forms of rhodopsin.

In transgenic tadpoles expressing this same transgene, xAr-myc-GFP translocation is indistinguishable from xAr-GFP (Figure 63.4). In these animals, the xAr-myc-GFP clearly translocates to the ROS in response to light in a manner that is qualitatively and quantitatively identical to xAr-GFP (compare to untreated animals in Figure 63.2). Clearly the translocation of arrestin is independent of binding to rhodopsin.

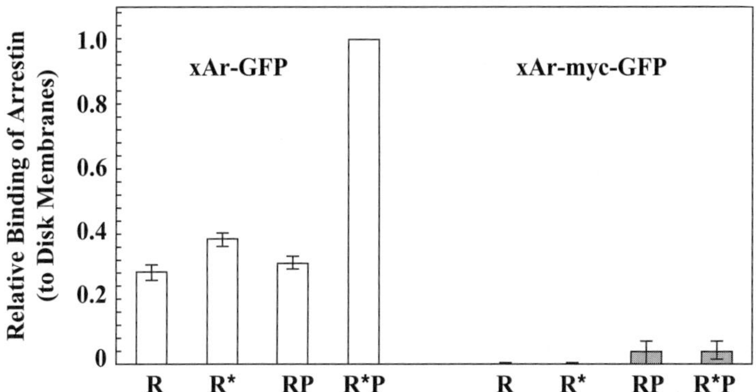

Figure 63.3. Binding of arrestin-GFP fusions to frog rhodopsin. *Xenopus* arrestin-GFP fusion protein (xAr-GFP) and arrestin-GFP fusion with at myc tag inserted between Leu-76 and Thr-77 (xAr-myc-GFP) were purified and bound to rhodopsin in disc membranes obtained from bull frog retinas in an *in vitro* centrifugation assay. The rhodopsin was either kept in the dark (R), light activated for 2 min (R*), phosphorylated but not activated (RP), or light activated and phosphorylated (R*P). Proteins bound to the disc membranes were analyzed by western blotting with an anti-GFP polyclonal antibody and quantified by scanning densitometry. Bound arrestin was normalized to the binding of xAr-GFP to R*P. Error bars represent SEM (n = 3).

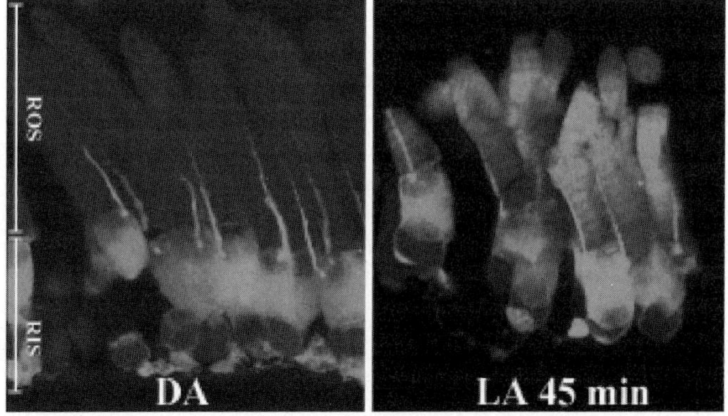

Figure 63.4. Translocation of myc-tagged arrestin-GFP fusion in transgenic tadpoles. Confocal images were obtained from cryosections of xAr-myc-GFP transgenic tadpoles that were dark-adapted overnight (DA) or exposed to light for 45 min (LA 45 min). The endogenous fluorescence of the GFP protein was imaged. See also color insert.

3.3. A Role for Inositol Phospholipids in Arrestin Translocation?

Translocation of arrestin in *Drosophila* photoreceptors has been reported to be dependent upon an association with inositol phospholipids, particularly phosphoinositol trisphosphate (PIP_3) (Lee *et al.* 2003). Mutation of three lysines in *Drosophila* arrestin (Lys-228, Lys-231, and Lys-257) disrupts this association and significantly slows the translocation of arrestin from the cell body to the rhabdomere in response to light through a NINAC myosin III-dependent mechanism (Lee and Montell 2004). Accordingly, we investigated if a similar mechanism was involved in the translocation of vertebrate arrestin, particularly since bovine

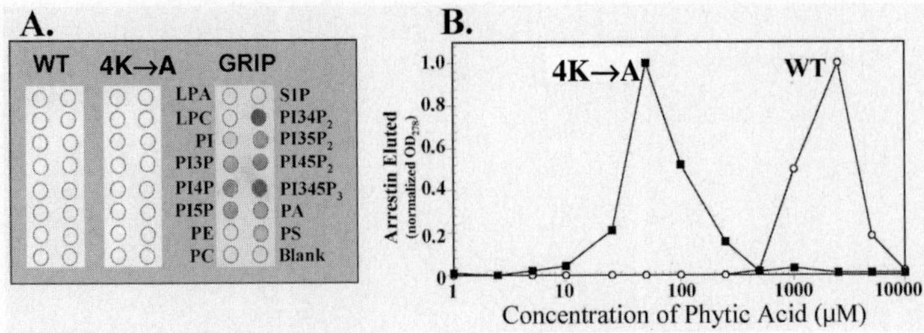

Figure 63.5. *In vitro* binding of arrestin to phospholipids. A. Native arrestin (WT), arrestin with four lysines substituted with alanine (4K → A), and GST-GRIP positive control were used to probe phospholipids immobilized on nitrocellulose. Binding was detected using anti-arrestin monoclonal antibody or anti-GST antibody for the GST-GRIP control with alkaline phosphatase-conjugated secondary antibody. Spotted lipids are lysophosphatidic acid (LPA), lysophosphocholine (LPC), phosphatidyl inositol (PI), phosphatidyl inositol 3-phosphate (PI3P), phosphatidyl inositol 4-phosphate (PI4P), phosphatidyl inositol 5-phosphate (PI5P), phosphatidylethanolamine (PE), phosphatidylcholine (PC), sphingosine-1-phosphate (S1P), phosphatidyl-3,4-bisphosphate (PI34P$_2$), phosphatidyl-3,5-bisphosphate (PI35P$_2$), phosphatidyl-4,5-bisphosphate (PI45P$_2$), phosphatidyl-3,4,5-trisphosphate (PI345P$_3$), phosphatidic acid (PA), and phosphatidylserine (PS). B. Relative affinity of WT arrestin (open circles) and 4K → A arrestin (closed squares) for inositol hexaphosphate was assessed by binding the arrestins to heparin agarose and competitively eluting with an increasing concentration of inositol hexaphosphate (phytic acid). Eluted protein was quantified by measuring absorbance at 278 nm.

arrestin has a demonstrated affinity for inositol phosphates (Palczewski *et al.* 1991). Alignment of bovine and *Drosophila* arrestins shows that two of the three mutated lysines are conserved. From an earlier study we had prepared a mutant of arrestin that coincidentally substituted these two lysines with alanines along with the two nearby lysines (K232A/K235A/K236A/K238A, used herein as 4K → A). Although the sequences of *Drosophila* and bovine arrestins are not perfectly conserved, the placement of the amino acids in three-dimensional projections based on the crystal structure of bovine arrestin (Hirsch *et al.* 1999) allows us to assign conformational homology. In addition, there are no other lysine residues that project near this grouping of lysines, giving us confidence that we have identified any lysines that might potentially bind inositol phospholipids in a similar location in vertebrate arrestins as identified in *Drosophila* arrestin.

Utilizing the procedure used in *Drosophila* (Lee *et al.* 2003), we assessed the binding of wild-type arrestin and 4K → A arrestin to different types of phospholipids spotted on nitrocellulose strips (Figure 63.5A). Unlike *Drosophila* arrestin, which showed significant binding to many forms of phosphoinositol bisphosphate and trisphosphate in this assay, we were unable to detect any binding of wild-type bovine arrestin or 4K → A arrestin to the phospholipid spots. Positive controls using the PH domain of phospholipase C-δ1 (GST-GRIP) or arrestin from house flies (*Musca domestica*, not shown), show easily detectable binding. These results suggest that the affinity of bovine arrestin for inositol phospholipids is significantly lower than that of *Drosophila* arrestin. This conclusion is supported by previous studies, showing an IC$_{50}$ of 10 μM for phyticacid (inositol hexaphosphate) for bovine arrestin (Palczewski *et al.* 1991), compared to 0.6 μM for *Drosophila* arrestin (Lee *et al.* 2003).

To assess whether the four lysine substitutions in bovine arrestin affected its affinity for phytic acid, WT and 4K → A arrestins were immobilized on heparin agarose, and competed off with increasing concentrations of phytic acid. Figure 63.5B shows that the 4K → A

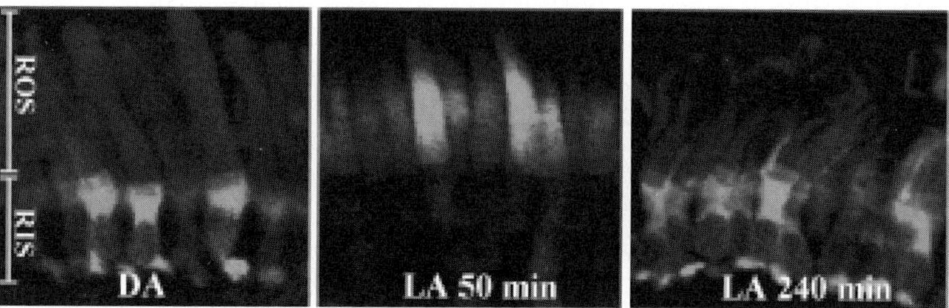

Figure 63.6. Confocal images of xAr-4K → Q-GFP translocation in rod photoreceptors. Transgenic tadpoles were dark-adapted overnight (DA), light adapted for 50 min (LA 50 min), or light adapted for 240 min (LA 240 min). Images show the endogenous fluorescence of the GFP in the fusion protein. See also color insert.

arrestin required a 50-fold lower concentration of phytic acid to elute the bound protein. This result indicates that the lysine substitutions had the predicted effect of reducing the affinity for inositol phospholipids and that we had correctly identified at least some of the lysines that participate in the association of arrestin with inositol hexaphosphate. These substitutions did not affect the specificity of 4K → A arrestin for R*P (data not shown).

Using these data, we prepared homologous substitutions of lysines in *Xenopus* arrestin to assess the impact of this reduced affinity for inositol phospholipids on arrestin translocation *in vivo*. Lysines 232, 235, 236, and 267 were substituted with glutamine (4K → Q), the same substitution used in *Drosophila* arrestin to disrupt translocation of *Drosophila* arrestin. In tadpoles expressing 4K → Q arrestin fused to the N-terminus of GFP, the translocation is indistinguishable from that of wild-type arrestin/GFP (compare to untreated tadpoles in Figure 63.2). In the dark-adapted state, 4K → Q arrestin localizes to the inner segments and axonemes of the rod photoreceptors. In response to 50 minutes of light adaptation, 4K → Q arrestin translocates to the outer segments qualitatively and quantitatively equal to wild-type arrestin. Like the native arrestin, 4K → Q arrestin also returns to the inner segments and axonemes during extended light adaptation, or if the tadpoles are subsequently dark adapted (data not shown). The similarity of these translocation results and the lower affinity of vertebrate arrestins for inositol phospholipids lead us to conclude that inositol phospholipids do not appear to play a role in the translocation of vertebrate arrestins, at least not through a direct interaction with arrestin as has been demonstrated for *Drosophila*.

4. CONCLUSIONS

These studies on selected aspects of arrestin translocation reveal three important findings. First, the light-driven redistribution of arrestin is a consequence of arrestin translocation and not a result of arrestin proteolysis with subsequent replacement by newly synthesized protein. Second, our results further substantiate that the translocation of arrestin to the ROS following light is not a result of arrestin's affinity for photoactivated, phosphorylated rhodopsin. And finally, we show that unlike *Drosophila* arrestin, vertebrate visual arrestins do not appear to rely on an affinity for phospholipids to promote translocation.

Future studies will focus on identifying the mechanism of arrestin translocation and identify how arrestin interacts with this translocation machinery.

5. ACKNOWLEDGMENTS

This research was supported by grants from the National Eye Institute (EY08571 and EY06225), the Howard Hughes Medical Institute, and Research to Prevent Blindness.

6. REFERENCES

Azarian, S. M., A. J. King, M. A. Hallett and D. S. Williams, 1995, Selective proteolysis of arrestin by calpain: molecular characteristics and its effect on rhodopsin dephosphorylation., *J. Biol. Chem.* **270**:24375-24384.

Broekhuyse, R. M., E. F. J. Tolhuizen, A. P. M. Janssen and H. J. Winkens, 1985, Light induced shift and binding of S-antigen in retinal rods., *Curr. Eye Res.* **4**:613-618.

Dinculescu, A., J. H. McDowell, S. A. Amici, D. R. Dugger, N. Richards, P. A. Hargrave and W. C. Smith, 2002, Insertional mutagenesis and immunochemical analysis of visual arrestin interaction with rhodopsin, *J. Biol. Chem.* **277**:11703-11708.

Elias, R., S. Sezate, W. Cao and J. F. McGinnis, 2004, Temporal kinetics of the light/dark translocation and compartmentation of arrestin and alpha-transducin in mouse photoreceptor cells., *Molec. Vis.* **10**:672-681.

Hirsch, J. A., C. Schubert, V. V. Gurevich and P. B. Sigler, 1999, The 2.8 angstrom crystal structure of visual arrestin: A model for arrestin's regulation, *Cell* **97**:257-269.

Kroll, K. L. and E. Amaya, 1996, Transgenic *Xenopus* embryos from sperm nuclear transplantations reveal FGF signaling requirements during gastrulation., *Development* **122**:3173-3183.

Laemmli, U. K., 1970, Cleavage of structural proteins during the assembly of the head of bacteriohage T4., *Nature* **227**:680-685.

Lee, S. J. and C. Montell, 2004, Light-dependent translocation of visual arrestin regulated by the NINAC myosin III., *Neuron* **43**:95-103.

Lee, S. J., H. Xu, L. W. Kang, L. M. Amzel and C. Montell, 2003, Light adaptation through phosphoinositide-regulated translocation of *Drosophila* visual arrestin, *Neuron* **39**:121-132.

Mangini, N. J., G. L. Garner, T. L. Okajima, L. A. Donoso and D. R. Pepperberg, 1994, Effect of hydroxylamine on the subcellular distribution of arrestin (S-antigen) in rod photoreceptors., *Vis. Neurosci.* **11**:561-568.

Mangini, N. J. and D. R. Pepperberg, 1988, Immunolocalization of 48K in rod photoreceptors., *Invest. Opthalmol. Vis. Sci.* **29**:1221-1234.

McDowell, J. H., 1993, Preparing rod outer segment membranes, regenerating rhodopsin, and determining rhodopsin concentration., *Meth. Neurosci.* **15**:123-130.

McGinnis, J. F., P. L. Stepanik, V. Lerious and J. P. Whelan, 1991, Molecular mechanisms regulating cellular and subcellular concentrations of gene products in mouse visual cells., *Retinal Degenerations (Hollyfield, J., Anderson, R.E., LaVail, M.M., eds)* CRC Press: Boca Raton, FL: 503-516.

Mendez, A., J. Lem, M. Simon and J. Chen, 2003, Light-dependent translocation of arrestin in the absence of rhodopsin phosphorylation and transducin signaling, *J. Neurosci.* **23**:3124-3129.

Obrig, T. G., W. J. Culp, W. L. McKeehan and B. Hardesty, 1971, The mechanism by which cycloheximide and related glutarimide antibiotics inhibit peptide synthesis on reticulocyte ribosomes., *J. Biol. Chem.* **246**:174-181.

Palczewski, K., A. Pulvermuller, J. Buczylko, C. Gutmann and K. P. Hofmann, 1991, Binding of inositol phosphates to arrestin., *FEBS Lett.* **295**:195-199.

Peet, J. A., A. Bragin, P. D. Calvert, S. S. Nikonov, S. Mani, X. Zhao, J. C. Besharse, E. A. Pierce, B. E. Knox and E. N. Pugh, 2004, Quantification of the cytoplasmic spaces of living cells with EGFP reveals arrestin-EGFP to be in disequlibrium in dark adapted rod photoreceptors., *J. Cell Sci.* **117**:3049-3059.

Peterson, J. J., B. M. Tam, O. L. Moritz, C. L. Shelamer, D. R. Dugger, J. H. McDowell, P. A. Hargrave, D. S. Papermaster and W. C. Smith, 2003, Arrestin migrates in photoreceptors in response to light: a study of arrestin localization using an arrestin-GFP fusion protein in transgenic frogs, *Exp. Eye Res.* **76**:553-563.

Sokolov, M., A. L. Lyubarsky, K. J. Strissel, A. B. Savchenko, V. I. Govardovskii, E. N. Pugh and V. Y. Arshavsky, 2002, Massive light-driven translocation of transducin between the two major compartments of rod cells: A novel mechanism of light adaptation, *Neuron* **34**:95-106.

Tam, B. M., O. L. Moritz, L. B. Hurd and D. S. Papermaster, 2000, Identification of an outer segment targeting signal in the COOH terminus of rhodopsin using transgenic *Xenopus* laevis, *J. Cell Biol.* **151**:1369-1380.

Thompson, J. D., T. J. Gibson, F. Plewniak, F. Jeanmougin and D. G. Higgins, 1997, The ClustalX windows interface: flexible strategies for multiple sequence alignment aided by quality analysis tools., *Nucleic Acids Res.* **24**:4876-4882.

Whelan, J. P. and J. F. McGinnis, 1988, Light-dependent subcellular movement of photoreceptor proteins., *J. Neurosci. Res.* **20**:263-270.

Wilden, U., S. W. Hall and H. Kühn, 1986, Phosphodiesterase activation by photoexcited rhodopsin is quenched when rhodopsin is phosphorylated and binds the intrinsic 48-kDa protein of rod outer segments., *Proc. Natl. Acad. Sci. USA* **83**:1174-1178.

Zhang, H. B., W. Huang, H. K. Zhang, X. M. Zhu, C. Craft, W. Baehr and C. K. Chen, 2003, Light-dependent redistribution of visual arrestins and transducin subunits in mice with defective phototransduction, *Molec. Vis.* **9**:231-237.

CHAPTER 64

BINDING OF N-RETINYLIDENE-PE TO ABCA4 AND A MODEL FOR ITS TRANSPORT ACROSS MEMBRANES

Robert S. Molday, Seelochan Beharry, Jinhi Ahn, and Ming Zhong*

1. INTRODUCTION

ABCA4, also known as ABCR or the rim protein, is a member of the ABCA subfamily of ATP binding cassette transporters (Allikmets et al., 1997b; Azarian and Travis, 1997; Illing et al., 1997). It is organized into two tandem arranged halves with each half consisting of a transmembrane segment followed by a large extracellular domain, a multi-spanning membrane domain and a cytoplasmic nucleotide binding domain (Figure 64.1) (Bungert et al., 2001). ABCA4 is localized along the rims and incisures of rod and cone photoreceptor outer segment discs where it is thought to play a role in the visual cycle (Molday et al., 2000; Sun et al., 1999; Weng et al., 1999). Mutations in the gene encoding ABCA4 are responsible for a variety of autosomal recessive retinal degenerative diseases including Stargardt macular dystrophy, cone-rod dystrophy, and retinitis pigmentosa (Allikmets, 2000; Allikmets et al., 1997b; Cremers et al., 1998). Individuals who are heterozygous for selected Stargardt-causing mutations are at a high risk for developing age-related macular degeneration (Allikmets et al., 1997a).

ABCA4 has been implicated in the movement of retinoids across the disc membrane following the photobleaching of rhodopsin or cone opsin in photoreceptor outer segments. This is based on two independent studies. First, 11-*cis* and all-*trans* retinal stimulate the ATPase activity of immunoaffinity purified and reconstituted ABCA4 by up to 4-fold (Ahn et al., 2000; Sun et al., 1999). In a number of cases, substrates that are transported by ABC transporters stimulate the ATPase activity of these proteins. Second, *abca4* knockout mice have been reported to accumulate all-*trans* retinal, N-retinylidene-phosphatidylethanolamine (retinylidene-PE) and phosphatidylethanolamine (PE) in outer segments and the diretinal pyridinium compound A2E in retinal pigment epithelial (RPE) cells in a light-dependent manner (Weng et al., 1999). Like individuals with Stargardt

---

*Robert S. Molday, University of British Columbia, Vancouver, BC V6T 1Z3. Seelochan Beharry, University of British Columbia, Vancouver, BC V6T 1Z3, Jinhi Ahn, University of British Columbia, Vancouver, BC V6T 1Z3, Ming Zhong, University of British Columbia, Vancouver, BC V6T 1Z3.

465

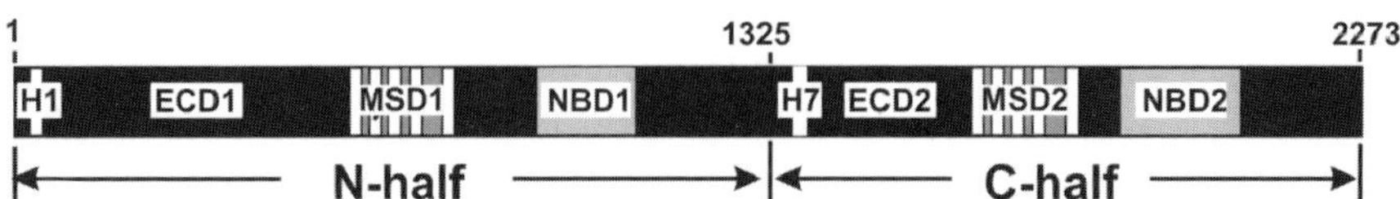

Figure 64.1. Diagram showing the organization of human ABCA4. The 2273 amino acid polypeptide chain is organized into a N-half and C-half, each of which contains a single transmembrane segment (H1 and H7) followed by an extracellular domain (ECD1 and ECD2), a multispanning domain (MSD1 and MSD2) containing 5 putative transmembrane segments, and a nucleotide binding domain (NBD1 and NBD2).

disease, *abca4* knockout mice also showed a delayed recovery of dark adaptation (Weng et al., 1999).

These results are consistent with a role of ABCA4 in the removal of all-*trans* retinal derivatives from disc membranes following the photobleaching of rhodopsin in rods and cone opsin in cones. In Stargardt disease or *abca4* knockout mice, loss in ABCA4 retinoid transport activity leads to an accumulation of all-*trans* retinal and N-retinylidene-PE in disc membranes. All-*trans* retinal can interact with opsin to activate the visual cascade, thereby causing a delay in dark adaptation. It can also react with N-retinylidene-PE to form the diretinal pyridinium phosphatidylethanolamine derivative called A2PE (Eldred and Lasky, 1993; Parish et al., 1998). Upon phagocytosis of outer segments, A2PE is hydrolyzed and the product A2E accumulates in RPE cells as fluorescent lipofuscin deposits which over the long term are toxic to these cells (Mata et al., 2000; Sun et al., 1999; Weng et al., 1999).

Although these studies strongly implicate ABCA4 in the removal of all-*trans* retinal from disc membranes and provide a molecular and cellular basis for Stargardt disease, the identity of the substrate transported by ABCA4 has not been determined experimentally and the mechanism of transport remains to be established. In particular, it is unclear whether ABCA4 acts as a flippase to translocate N-retinylidene-PE from the lumen to the cytoplasmic side of the disc membrane or if ABCA4 extrudes all-*trans* retinal from disc membranes (Sun et al., 1999; Weng et al., 1999). Some ABC transporters are known to flip phospholipids across membranes, while other ABC transporters such as P-glycoprotein actively extrude hydrophobic drugs and other compounds from cell membranes (Gottesman et al., 1995).

In this chapter, we briefly describe recent studies showing that ABCA4 preferentially binds N-retinylidene-PE in the absence of ATP. Addition of ATP releases N-retinylidene-PE from ABCA4 suggesting that ABCA4 actively transports this retinoid across disc membranes. These studies together with previous studies on the interaction of nucleotides with the nucleotide binding domains (NBD) of ABCA4 (Ahn et al., 2003) provide new insight into a possible mechanism for ABCA4-mediated retinoid transport.

2. BINDING OF RETINOIDS TO ABCA4

2.1. N-Retinylidene-PE is the Preferred Substrate for ABCA4

Recently, we have developed a solid phase assay to identify retinoids that bind to ABCA4 (Beharry et al., 2004). In this procedure, an immunoaffinity matrix consisting of

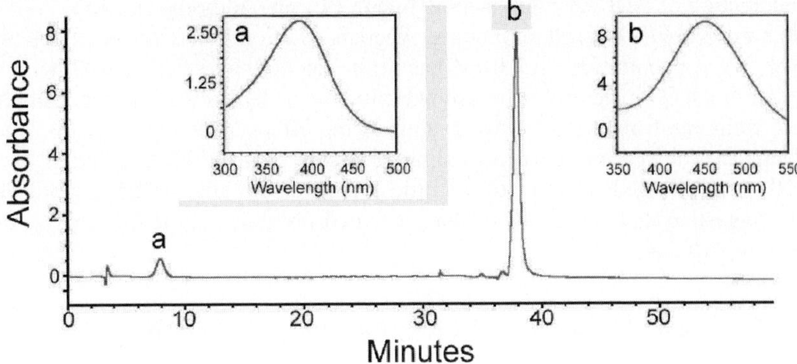

Figure 64.2. HPLC chromatography of retinoids bound to ABCA4 after the addition of all-*trans* retinal to ABCA4 in the presence of phospholipid. Peak **a** is identified as all-*trans* retinal and peak **b** as N-retinylidene-PE on the basis of retention times and spectral properties (inset).

the anti-ABCA4 Rim 3F4 monoclonal antibody covalently bound to Sepharose 2B is used to isolate ABCA4 from CHAPS detergent solubilized rod outer segments (ROS). Retinoid compounds are then added to immobilized ABCA4 in the presence of CHAPS buffer containing PE. After extensive washing to remove unbound retinoid, the bound retinoids are extracted with an organic solvent such as hexane for identification and quantification by reverse phase HPLC or radioisotope counting if radiolabeled retinoids are used.

The application of this solid phase method in the HPLC analysis of retinoids bound to ABCA4 is shown in Figure 64.2. When all-*trans* retinal is added to immobilized ABCA4 in the presence of PE, two retinoid compounds corresponding to all-*trans* retinal and N-retinylidene-PE are observed at retention times of 8 and 38 minutes, respectively. Quantitative analysis indicates that N-retinylidene-PE accounts for over 75% of the total bound retinoid suggesting that this compound is the preferred substrate for ABCA4. N-retinylidene-PE binds tightly to ABCA4 with a dissociation constant (Kd) of ~ 2µM (Beharry et al., 2004).

The binding specificity is confirmed in a series of studies (Beharry et al., 2004). First, N-retinyl-PE, the reduced form of N-retinylidene-PE, quantitatively binds to ABCA4 and inhibits the binding of N-retinylidene-PE to ABCA4. This suggests that these structurally related compounds compete for the same binding site in ABCA4. In contrast, all-*trans* retinol, the reduced form of all-*trans* retinal, does not bind to ABCA4, even at relatively high concentrations. This is in agreement with the inability of all-*trans* retinol to effectively activate the ATPase activity of ABCA4 (Ahn et al., 2000; Sun et al., 1999). Finally, N-retinylidene-PE and all-*trans* retinal do not interact with the immunoaffinity matrix in the absence of ABCA4, indicating that these retinoids bind to ABCA4 and not the immunoaffinity matrix.

2.2. ATP Releases Bound N-Retinylidene-PE from ABCA4

Since the energy needed to transport substrates across membranes by ABC transporters is generated from ATP hydrolysis, we investigated the effect of ATP and other nucleotides on the interaction of ABCA4 with N-retinylidene-PE. The addition of 50 µM ATP or GTP results in complete loss of retinoid binding, whereas addition of the dinucleotides, ADP and GDP, even at a concentration of 500 µM has little effect (Beharry et al., 2004).

To determine if nucleotide triphosphate binding or hydrolysis is responsible for the release of retinoids from ABCA4, we examined the effect of nonhydrolyzable nucleotide triphosphate derivatives on retinoid binding (Beharry et al., 2004). Addition of 500 µM AMP-PNP or GMP-PNP had relatively little effect on the amount of retinoid bound to ABCA4, suggesting that nucleotide triphosphate hydrolysis is required for release of bound retinoid from ABCA4.

2.3. ATP Binding and Hydrolysis Occurs within the NBD2 of ABCA4

The binding of adenine nucleotides to the NBD within the N and C halves of ABCA4 was evaluated using 8-azido-labeled photoaffinity derivatives (Ahn et al., 2003). When ABCA4 in ROS disc membranes was photo-affinity labeled with 8-azido-ATP or 8-azido-ADP, only NBD2 in the C-half of ABCA4 bound these nucleotides. This was also the case when membranes from culture cells co-expressing and co-assembling the C-half and N-half of ABCA4 were photo-affinity labeled.

The inability of NBD1 to bind exogenously added nucleotide could be due to the inaccessibility of the site or the presence of a tightly bound nucleotide. To delineate between these possibilities, ABCA4 was purified from ROS disc membranes and the presence of bound nucleotide was measured (Ahn et al., 2003). Results of these studies clearly indicated that ABCA4 contains one tightly bound ADP. This bound nucleotide can not be exchanged with ATP or GDP, nor can it be removed using treatments known to displace tightly bound nucleotides from other nucleotide-binding proteins. From these studies, we conclude that NBD1 contains a tightly bound, nonexchangeable ADP, whereas NBD2 binds and hydrolyzes ATP as part of the transport activity of ABCA4.

3. A WORKING MODEL FOR ABCA4 MEDIATED TRANSPORT OF N-RETINYLIDENE-PE ACROSS DISC MEMBRANES

Our results showing that N-retinylidene-PE is the preferred substrate for ABCA4 together with studies showing that only NBD2 binds and hydrolyzes ATP leads to a working mechanistic model for ABCA4 mediated transport of N-retinylidene-PE across the disc membrane. As depicted in Figure 64.3, N-retinylidene-PE trapped on the lumen leaflet of the disc bilayer following the photobleaching of rhodopsin binds to a high affinity site on ABCA4. This site is most likely located within the bundle of helical segments formed by the transmembrane segments of the C and N halves of ABCA4. The subsequent binding and hydrolysis of ATP in NBD2 provides the energy needed to induce a protein conformational change that converts the high affinity site to a low affinity site. This conformational change may involve an interaction or dimerization of NBD2 with NBD1 as suggested for other ABC transporters including P-glycoprotein, as well as changes within the transmembrane segments that promote a flipping or movement of N-retinylidene-PE. Dissociation of N-retinylidene-PE from its low affinity binding site together with the release of ADP from NBD2 returns ABCA4 to its original state.

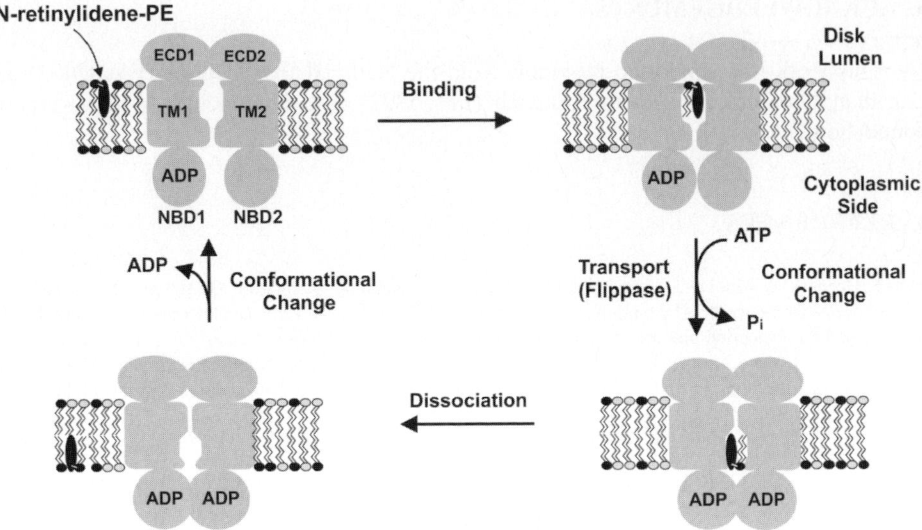

Figure 64.3. A working model for the transport of N-retinylidene-PE across disc membranes by ABCA4. N-retinylidene-PE binds to ABCA4 containing a nonexchangeable ADP in NBD1. ATP hydrolysis occurs in NBD2 resulting in a protein conformational change that initiates the flipping of N-retinylidene-PE across the disc membrane. Finally, N-retinylidene-PE and ADP dissociate returning ABCA4 its initial state.

4. SOME UNRESOLVED PROBLEMS

Although considerable progress has been made in understanding the role of ABCA4 in photoreceptor function and disease, several issues remain to be clarified. The direction of retinoid transport has not yet been determined experimentally. The current model favors the translocation of N-retinylidene-PE from the lumen to the cytoplasmic side of the disc membrane. This is based on the premise that N-retinylidene-PE trapped on the lumen side has to be transported or flipped to the cytoplasmic side of the disc membrane so that after dissociation, all-*trans* retinal can be reduced to all-*trans* retinol by retinol dehydrogenase and channeled into the visual cycle. Although this is a logical model, most eukaryotic ABC proteins transport substrates in the reverse direction i.e. from the cytoplasmic to the lumen or extracellular side of membranes. Hence, it is important to determine experimentally the direction and mechanism of retinoid transport by ABCA4. Additionally, it is important to know if ABCA4 interacts with other photoreceptor proteins and if so whether these proteins regulate the transport activity of ABCA4. The role of the various domains, including the two large extracellular domains upstream of the multispanning membrane domains and cytoplasmic segments downstream of the NBDs, on the structure, function and subcellular targeting of ABCA4 needs to be investigated. Finally, it is important to more fully understand how specific disease-causing mutations, affect the retinoid transport activity of ABCA4.

5. ACKNOWLEDGEMENTS

This work was supported by grants from the National Eye Institute (EY 02422), the Canadian Institutes for Health Research (MT 5822) and the Macular Vision Research Foundation.

6. REFERENCES

Ahn, J., Beharry, S., Molday, L.L. and Molday, R.S. (2003) Functional interaction between the two halves of the photoreceptor-specific ATP binding cassette protein ABCR (ABCA4). Evidence for a non-exchangeable ADP in the first nucleotide binding domain. *J Biol Chem,* **278**:39600-39608.

Ahn, J., Wong, J.T. and Molday, R.S. (2000) The effect of lipid environment and retinoids on the ATPase activity of ABCR, the photoreceptor ABC transporter responsible for Stargardt macular dystrophy. *J Biol Chem,* **27**:20399-20405.

Allikmets, R. (2000) Simple and complex ABCR: genetic predisposition to retinal disease. *Am J Hum Genet,* **67**:793-799.

Allikmets, R., Shroyer, N.F., Singh, N., Seddon, J.M., Lewis, R.A., Bernstein, P.S., Peiffer, A., Zabriskie, N.A., Li, Y., Hutchinson, A., Dean, M., Lupski, J.R. and Leppert, M. (1997a) Mutation of the Stargardt disease gene (ABCR) in age-related macular degeneration. *Science,* **277**:805-1807.

Allikmets, R., Singh, N., Sun, H., Shroyer, N.F., Hutchinson, A., Chidambaram, A., Gerrard, B., Baird, L., Stauffer, D., Peiffer, A., Rattner, A., Smallwood, P., Li, Y., Anderson, K.L., Lewis, R.A., Nathans, J., Leppert, M., Dean, M. and Lupski, J.R. (1997b) A photoreceptor cell-specific ATP-binding transporter gene (ABCR) is mutated in recessive Stargardt macular dystrophy. *Nat Genet,* **15**:236-246.

Azarian, S.M. and Travis, G.H. (1997) The photoreceptor rim protein is an ABC transporter encoded by the gene for recessive Stargardt's disease (ABCR). *FEBS Lett,* **409**:247-252.

Beharry, S., Zhong, M. and Molday, R.S. (2004) N-retinylidene-phosphatidylethanolamine is the preferred retinoid substrate for the photoreceptor-specific ABC transporter ABCA4 (ABCR). *J Biol Chem.*

Bungert, S., Molday, L.L. and Molday, R.S. (2001) Membrane topology of the ATP binding cassette transporter ABCR and its relationship to ABC1 and related ABCA transporters: identification of N- linked glycosylation sites. *J Biol Chem,* **276**:23539-23546.

Cremers, F.P., van de Pol, D.J., van Driel, M., den Hollander, A.I., van Haren, F.J., Knoers, N.V., Tijmes, N., Bergen, A.A., Rohrschneider, K., Blankenagel, A., Pinckers, A.J., Deutman, A.F. and Hoyng, C.B. (1998) Autosomal recessive retinitis pigmentosa and cone-rod dystrophy caused by splice site mutations in the Stargardt's disease gene ABCR. *Hum Mol Genet,* **7**:355-362.

Eldred, G.E. and Lasky, M.R. (1993) Retinal age pigments generated by self-assembling lysosomotropic detergents. *Nature,* **361**:724-726.

Gottesman, M.M., Hrycyna, C.A., Schoenlein, P.V., Germann, U.A. and Pastan, I. (1995) Genetic analysis of the multidrug transporter. *Annu Rev Genet,* **29**:607-649.

Illing, M., Molday, L.L. and Molday, R.S. (1997) The 220-kDa rim protein of retinal rod outer segments is a member of the ABC transporter superfamily. *J Biol Chem,* **272**:0303-10310.

Mata, N.L., Weng, J. and Travis, G.H. (2000) Biosynthesis of a major lipofuscin fluorophore in mice and humans with ABCR-mediated retinal and macular degeneration. *Proc Natl Acad Sci U S A,* **97**:154-7159.

Molday, L.L., Rabin, A.R. and Molday, R.S. (2000) ABCR expression in foveal cone photoreceptors and its role in Stargardt macular dystrophy. *Nat Genet,* **25**:57-258.

Parish, C.A., Hashimoto, M., Nakanishi, K., Dillon, J. and Sparrow, J. (1998) Isolation and one-step preparation of A2E and iso-A2E, fluorophores from human retinal pigment epithelium. *Proc Natl Acad Sci U S A,* **95**:14609-14613.

Sun, H., Molday, R.S. and Nathans, J. (1999) Retinal stimulates ATP hydrolysis by purified and reconstituted ABCR, the photoreceptor-specific ATP-binding cassette transporter responsible for Stargardt disease. *J Biol Chem,* **274**:269-8281.

Weng, J., Mata, N.L., Azarian, S.M., Tzekov, R.T., Birch, D.G. and Travis, G.H. (1999) Insights into the function of rim protein in photoreceptors and etiology of Stargardt's Disease from the phenotype in *abcr* knockout mice. *Cell,* **98**:3-23.

THE CHAPERONE FUNCTION OF THE LCA PROTEIN AIPL1

AIPL1 chaperone function

Jacqueline van der Spuy and Michael E. Cheetham*

1. THE AIPL1 HOMOLOGUE AIP IS A MOLECULAR CO-CHAPERONE

Mutations in the aryl hydrocarbon receptor interacting protein-like 1 (AIPL1) cause the devastating blinding disease Leber's congenital amaurosis (LCA) (Sohocki et al., 2000a). Up to 12% of recessive LCA is caused by mutations in AIPL1 (Sohocki et al., 2000b). In addition to AIPL1, LCA-causing mutations have also been identified in RetGC1, RPE65, CRX, LRAT, CRB1 and RPGRIP1, and a further two loci have been identified on 14q24 and 6q11-16 (www.retina-international.org/sci-news/mutation.htm). Although the function of AIPL1 is unknown, AIPL1 shares 49% identity with the human aryl hydrocarbon receptor (AhR)-interacting protein (AIP), also named XAP2 or ARA9 (Sohocki et al., 2000a). AIP in turn shares similarity with members of the immunophilin family of proteins including the co-chaperones FK506-binding protein (FKBP) 51 and 52 (reviewed in Chapple et al., 2001; van der Spuy and Cheetham, 2004a). Both AIP and the FKBP co-chaperones exist in a cytosolic ternary complex with the molecular chaperone Hsp90 and a specific cognate receptor, and have been shown to regulate the nuclear translocation and transactivation of the associated receptor. At the primary structural level, the tetratricopeptide repeat (TPR) motif is conserved in AIPL1, AIP and FKBP51/52. The TPR motif is an evolutionary and functionally conserved but degenerate motif found in a number of structurally unrelated proteins and mediates the binding of specific protein-interaction partners. The TPR motif in both AIP and FKBP51/52 form a TPR carboxylate clamp that mediates their interaction with the C-terminal MEEVD TPR acceptor site of Hsp90. The similarity of AIPL1 to AIP has led to suggestions that AIPL1 could function in a similar manner to AIP in facilitating protein translocation and as a component of chaperone complexes.

---

*Jacqueline van der Spuy and Michael E. Cheetham, Division of Pathology, Institute of Ophthalmology, UCL, London, EC1V 9EL, UK.

2. AIPL1 ASSOCIATES WITH THE CELL CYCLE REGULATOR NUB1

2.1. Expression of AIPL1 and NUB1 in the Human Retina

AIPL1 interacts with the NEDD8 ultimate buster protein 1 (NUB1) (Akey et al., 2002). NUB1 in turn interacts with the small ubiquitin-like protein NEDD8 and downregulates the targeted proteasomal degradation of NEDD8 and substrates that are conjugated to NEDD8 in a manner analogous to ubiquitination and sentrinization (Kamitani et al., 2001; Kito et al., 2001). All the known substrates for neddylation are members of the family of cullin (Cul) proteins, which are components of an SCF ubiquitin E3 ligase comprising Skp1, cullin, F-box protein, and ROC1 (Zheng et al., 2002). Specific substrates are targeted for SCF-mediated ubiquitination and subsequent degradation depending on the identity of the cullin component. Ubiquitination of IκBα, β-catenin, cyclin D proteins, p27 (KIP1) and p21 (CIP1/WAF1) is catalysed by the Cul-1 SCF ubiquitin E3 ligase, the activity of which is dependent on NEDD8 conjugation of the cullin component. Ubiquitination and proteasomal degradation of hypoxia-inducible factor-1α (HIF1α) and cyclin E are catalysed by the Cul-2 and Cul-3 SCF complexes respectively. Hence, the conjugation of NEDD8 to cullins has been implicated in many important biological events including cell cycle regulation and cell signalling, and it has been suggested that the interaction of NUB1 with NEDD8 implicates NUB1 and AIPL1 in the regulation of these events.

The spatiotemporal distribution of NUB1 in the developing and adult human retina has been examined in parallel with that of AIPL1 (van der Spuy et al., 2002; van der Spuy et al., 2003). In the adult human retina, AIPL1 was localized specifically in the connecting cilia, inner segments, cell bodies, axons and spherules of the rod photoreceptors but could not be detected in the cone photoreceptors. However, LCA is characterised by a flat or severely attenuated scotopic and photopic electroretinogram (ERG) within the first year of life suggesting the severe and early degeneration or impaired function of the rod and cone photoreceptors. A spatiotemporal examination of AIPL1 expression in the developing human retina detected AIPL1 in a single layer of cells in the central retina at fetal week 11.8 corresponding to early presumptive cone photoreceptors, differentiation of which precedes that of rod photoreceptors. At this age, AIPL1 was not detected in mid-peripheral or peripheral regions of the presumptive retinal outer nuclear layer (ONL). As retinal development proceeded, the expression of AIPL1 in the developing photoreceptor ONL spread gradually from the central to peripheral retina, closely following the centroperipheral gradient of rod and cone photoreceptor differentiation. During development, AIPL1 was detected in both rod and cone photoreceptors, and co-localized with both short wavelength (S)-cone and long/medium wavelength (L/M)-cone opsin expression. These data suggested that while AIPL1 is important for normal rod and cone development, it is required for the maintenance of rod photoreceptors only in adults, and that a developmental switch in AIPL1 function occurs.

Unlike AIPL1, NUB1 was not photoreceptor-specific but was expressed ubiquitously in all tissues examined and in all the retinal cell types. NUB1 was detected predominantly in the nuclei of all cells. NUB1 was expressed in all retinal cell types during human retinal development and the spatiotemporal expression of NUB1 did not follow a centroperipheral gradient. Rather, NUB1 was detected equally from central to peripheral retina at each age examined. During development a potential gradient of high to low NUB1 expression was observed from the inner to outer retina that coincided with retinal cell differentiation. The nuclear predominance of NUB1 was less evident in the developing photoreceptor cell layer. The subcellular distribution and fractionation of recombinant and endogenous AIPL1 and NUB1 in cell lines has demonstrated that while AIPL1 is predominantly cytoplasmic, NUB1 is predominantly nuclear due to a functional nuclear localisation signal near the C terminus. Hence, we wanted to test the hypothesis that AIPL1 could modulate the nuclear translocation of NUB1 in a manner analagous to the AIP-mediated regulation of the AhR.

2.2. AIPL1 Functions as a Chaperone for NUB1

AIPL1 is able to modulate the subcellular distribution of GFP-NUB1 in a concentration dependent manner (Figure 65.1) (van der Spuy and Cheetham, 2004b). Increasing amounts of AIPL1 resulted in a shift in the subcellular distribution of GFP-NUB1 from predominantly nuclear to predominantly cytoplasmic and in co-localization of GFP-NUB1 with

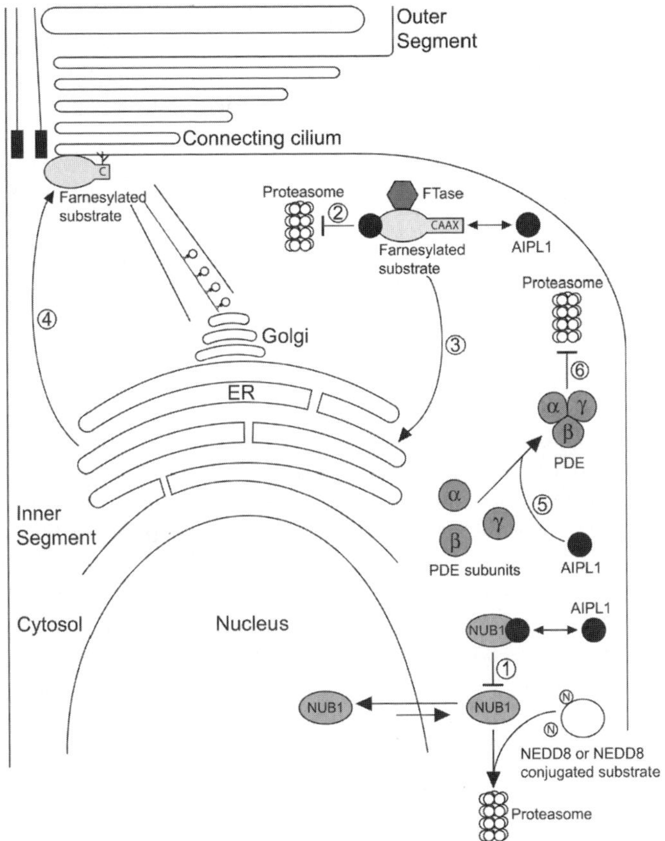

Figure 65.1. Potential AIPL1 photoreceptor functions: Modulation of NUB1 nuclear translocation (1); protection of farnesylated proteins from proteasomal degradation (2); facilitated targeting of farnesylated proteins to the ER (3); facilitated transport of farnesylated proteins to other target membranes (4); facilitated biosynthesis or assembly of PDE (5); and stabilization of PDE to proteasomal degradation (6).

AIPL1 in the cytoplasm. The closely related AIPL1 homologue AIP was unable to interact with or modulate GFP-NUB1 nucleocytoplasmic distribution, suggesting that whilst the similarity between AIPL1 and AIP correlates with a conserved function in the modulation of nuclear translocation, the specificity for the client protein differs in each case.

In addition to its role in protein translocation, AIPL1 was also able to suppress inclusion formation by GFP-NUB1 N- and C-terminal fragments. GFP-NUB1-N (residues 1 – 306) encompassed two coiled-coil (cc) domains and the ubiquitin-like (UBL) domain in the N-terminus of GFP-NUB1, and in the absence of AIPL1 formed multiple, small inclusions that decorated the cytoplasmic side of the nucleus in a perinuclear fashion. GFP-NUB1-C (residues 347-601) encompassed two ubiquitin-associated (UBA) domains, an NLS and a PEST sequence towards the C-terminus of GFP-NUB1, and formed large, intranuclear inclusions in the absence of AIPL1. Increasing amounts of AIPL1 resulted in efficient suppression of GFP-NUB1-N and GFP-NUB1-C inclusion formation, and the redistribution and co-localisation of these fragments with AIPL1 in the cytoplasm. This effect was specific for the GFP-NUB1 fragments, as AIPL1 had no effect on the formation of inclusions by unrelated aggregation-prone proteins, GFP-Huntingtin-exon 1-Q103 and P23H mutant opsin-GFP. Whilst these aggregation-prone proteins were SDS insoluble, the GFP-NUB1 fragments were SDS soluble and recruited both ubiquitin and Hsp70, suggesting that they were targeted for proteasomal degradation. The AIPL1 homologue AIP had no effect on the formation of GFP-NUB1-N and -C inclusions.

Hence, AIPL1 was able to behave in a chaperone-like manner to modulate the nuclear translocation of GFP-NUB1 and suppress the formation of inclusions by GFP-NUB1 fragments. This function of AIPL1 was compromised by certain mutations. All of the AIPL1 disease-associated and engineered mutants generated were soluble and similar in their subcellular distribution to AIPL1, with the exception of the disease-associated mutation W278X. AIPL1(W278X) formed multiple, cytoplasmic inclusions that were SDS insoluble, suggesting that this mutant protein undergoes misfolding and aggregation and is non-functional. Two additional C-terminal truncation mutants not yet associated with disease, AIPL1(E317X) and AIPL1(Q329X) were also severely compromised, suggesting that C-terminal sequences in AIPL1 are necessary for the modulation of NUB1 nuclear localization and inclusion suppression. Some of the pathological mutations, including R302L and W278X, were less efficient that wild-type protein in modulating the subcellular distribution and nuclear translocation of NUB1 suggesting a possible mechanism for disease in these patients affecting NUB1-related events involving the regulation of cell signalling and cell growth. However, other pathological AIPL1 mutants including A197P, C239R, G262S and P351Δ12 were not defective in their ability to modulate the subcellular distribution of NUB1 or suppress the formation of inclusions by NUB1 fragments, suggesting that the basis for disease in these patients may involve an alternative protein interaction and related function for AIPL1.

3. AIPL1 CHAPERONE-LIKE FUNCTIONS IN POST-TRANSLATIONAL MODIFICATION AND POST-TRANSCRIPTIONAL REGULATION

It has been demonstrated that AIPL1 is specifically able to interact with and enhance the post-translational farnesylation of proteins in the retina (Ramamurthy et al., 2003). Protein prenylation facilitates protein-protein and protein-membrane interactions, and is important in the maintenance of retinal cytoarchitecture and photoreceptor structure. It was shown that the ability of AIPL1 to interact with and enhance the processing of farnesylated proteins was severely compromised by certain pathogenic mutations including M79T and the non-functional W278X. It was suggested that AIPL1 may interact with the C-terminal prenylation motif in the cytosol dependent on the presence of a farnesyl transferase and either protect the farnesylated protein form proteasomal degradation in the cytosol, facilitate targeting of the protein to the ER for further processing, or chaperone the farnesylated protein to the target membrane (Figure 65.1). The AIPL1 mutants A197P and C239R, which were partially defective in their ability to interact with and facilitate the processing of farnesylated proteins, were functional with respect to their effect on NUB1. Another AIPL1 mutation, R302L, did not show any defect in protein farnesylation but was compromised in NUB1 function. Hence, the mechanisms of disease pathogenesis in patients with AIPL1 mutations may depend on the specific interacting partner and functional pathway affected.

Recently, mouse models of LCA with either complete or partial inactivation of AIPL1 expression have suggested that AIPL1 may also function as a potential chaperone for cGMP phosphodiesterase (PDE) (Figure 65.1) (Liu et al., 2004; Ramamurthy et al., 2004). In both models, normal retinal histology and morphological photoreceptor development were observed at birth, although no recordable photofunction could be detected in AIPL1$^{-/-}$ mice and the photoresponse onset and recovery was delayed in the rod photoreceptors of the AIPL1 hypomorphic mutant. Photoreceptor degeneration proceeded rapidly shortly after birth in the absence of AIPL1 but was significantly slowed in the presence of reduced levels of AIPL1. In both mouse models, all three subunits of the cGMP PDE holoenzyme (α, β and γ) were reduced by a post-transcriptional mechanism before the onset of photoreceptor degeneration, suggesting that AIPL1 was necessary for the biosynthesis, assembly, or stabilization of PDE to proteasomal degradation. The PDE-α subunit is farnesylated and mutations that block farnesylation cause degradation of PDE-α protein in cultured cells (Qin and Baehr, 1994). However, LCA is more severe than retinitis pigmentosa (RP) caused by mutations in the PDE subunits.

4. CONCLUSIONS

In conclusion, it has been shown that AIPL1 is able to function in a chaperone-like manner to regulate the nuclear translocation of the cell cycle regulator NUB1, enhance the processing of farnesylated proteins and post-transcriptionally regulate the levels of PDE. The severity of disease in LCA patients with mutations in AIPL1 cannot be accounted for by each of these functions on their own, suggesting that multiple and complex AIPL1-dependent mechanisms may be involved or that a single, as yet unidentified AIPL1-dependent mechanism may underlie each of these chaperone-like functions. It is tempting to speculate that these processes could converge at the level of regulating PDE, for example the interaction of AIPL1 with farnesylated components of the PDE complex could be important for the stabilization of the PDE enzyme complex, similarly the interaction of AIPL1 with NUB1 could regulate a switch in the ubiquitin proteasome machinery and protection of PDE from degradation (Figure 65.1). Alternatively, AIPL1, like AIP and FKBP51/52, may act as a part of chaperone heterocomplex in several unrelated processes. Further experimentation is needed to define the critical roles of AIPL1 and their importance for the development and function of the retina.

5. REFERENCES

Akey, D.T., Zhu, X., Dyer, M., Li, A., Sorensen, A., Blackshaw, S., Fukuda-Kamitani, T., Daiger, S.P., Craft, C.M., Kamitani, T., and Sohocki, M.M., 2002, The inherited blindness associated protein AIPL1 interacts with the cell cycle regulator protein NUB1, *Hum Mol Genet.* **15**:2723.

Chapple, J.P., Grayson, C., Hardcastle, A.J., Saliba, R.S., van der Spuy, J., and Cheetham, M.E., 2001, Unfolding retinal dystrophies: a role for molecular chaperones? *Trends Mol Med.* **9**: 414.

Kamitani, T., Kito, K., Fukuda-Kamitani, T., and Yeh, E.T., 2001, Targeting of NEDD8 and its conjugates for proteasomal degradation by NUB1, *J Biol Chem.* **276**: 46655.

Kito, K., Yeh, E.T., and Kamitani, T., 2001, NUB1, a NEDD8-interacting protein, is induced by interferon and down-regulates the NEDD8 expression, *J Biol Chem.* **276**: 20603.

Liu, X., Bulgakov, O.V., Wen, X.H., Woodruff, M.L., Pawlyk, B., Yang, J., Fain, G.L., Sandberg, M.A., Makino, C.L., and Li, T., 2004, AIPL1, the protein that is defective in Leber congenital amaurosis, is essential for the biosynthesis of retinal rod cGMP phosphodiesterase, *Proc Natl Acad Sci U S A.* **101**:13903.

Qin, N., and Baehr, W., 1994, Expression and mutagenesis of mouse rod photoreceptor cGMPphosphodiesterase, *J Biol Chem.* **269**:3265.

Ramamurthy, V., Roberts, M., van den Akker, F., Niemi, G., Reh, T.A., and Hurley, J.B., 2003, AIPL1, a protein implicated in Leber's congenital amaurosis, interacts with and aids in processing of farnesylated proteins, *Proc Natl Acad Sci U S A.* **100**:12630.

Ramamurthy, V., Niemi, G.A., Reh, T.A., and Hurley, J.B., 2004, Leber congenital amaurosis linked to AIPL1: a mouse model reveals destabilization of cGMP phosphodiesterase, *Proc Natl Acad Sci U S A.* **101**:13897.

Sohocki, M.M., Bowne, SJ, Sullivan, L.S., Blackshaw, S., Cepko, C.L., Payne, A.M., Bhattacharya, S.S., Khaliq, S., Qasim Mehdi, S., Birch, D.G., Harrison, W.R., Elder, F.F., Heckenlively, J.R., and Daiger, P., 2000a, Mutations in a new photoreceptor-pineal gene on 17p cause Leber congenital amaurosis, *Nat Genet.* **24**:79.

Sohocki, M.M., Perrault, I., Leroy, B.P., Payne, A.M., Dharmaraj, S., Bhattacharya, S.S., Kaplan, J., Maumenee, I.H., Koenekoop, R., Meire, F.M., Birch, D.G., Heckenlively, J.R., and Daiger, S.P., 2000b, Prevalence of AIPL1 mutations in inherited retinal degenerative disease, *Mol Genet Metab.* **70**:142.

van der Spuy, J., Chapple, J.P., Clark, B.J., Luthert, P.J., Sethi, C.S., and Cheetham, M.E., 2002, The Leber congenital amaurosis gene product AIPL1 is localized exclusively in rod photoreceptors of the adult human retina, *Hum Mol Genet.* **11**:823.

van der Spuy, J., Kim, J.H., Yu, Y.S., Szel, A., Luthert, P.J., Clark, B.J., and Cheetham, M.E., 2003, The expression of the Leber congenital amaurosis protein AIPL1 coincides with rod and cone photoreceptor development, *Invest Ophthalmol Vis Sci.* **44**:5396.

van der Spuy, J., and Cheetham, M.E., 2004a, Role of AIP and its homologue the blindness-associated protein AIPL1 in regulating client protein nuclear translocation, *Biochem Soc Trans.* **32**:643.

van der Spuy, J., and Cheetham, M.E., 2004b, The leber congenital amaurosis protein AIPL1 modulates the nuclear translocation of NUB1 and suppresses inclusion formation by NUB1 fragments, *J Biol Chem.* **279**:48038.

Zheng, N., Schulman, B.A., Song, L., Miller, J.J., Jeffrey, P.D., Wang, P., Chu, C., Koepp, D.M., Elledge, S.J., Pagano, M., Conaway, R.C., Conaway, J.W., Harper, J.W., and Pavletich, N.P., 2002, Structure of the Cul1-Rbx1-Skp1-F boxSkp2 SCF ubiquitin ligase complex, *Nature.* **416**:703.

CHAPTER 66

CRALBP LIGAND AND PROTEIN INTERACTIONS

Zhiping Wu[1], Sanjoy K. Bhattacharya[1], Zhaoyan Jin[1], Vera L. Bonilha[1], Tianyun Liu[2], Maria Nawrot[3], David C. Teller[2], John C. Saari[2,3], and John W. Crabb[1]

1. INTRODUCTION

The visual cycle is the complex enzymatic retinoid-processing involved in regenerating bleached rod and cone visual pigments.[1] Central to visual cycle physiology is the cellular retinaldehyde-binding protein (CRALBP), a 36 kDa cytosolic protein with high affinity for 11-*cis*-retinal and 11-*cis*-retinol. CRALBP is expressed in retinal pigment epithelium (RPE) and Müller cells, as well as in ciliary epithelium, iris, cornea, pineal gland and a subset of oligodendrocytes of the optic nerve and brain.[2] Its function outside the RPE is not known, although a recent behavioral genetic study suggests that CRALBP may contribute to ethanol preference in mice.[3] In the RPE, CRALBP serves as an 11-*cis*-retinol acceptor in the visual cycle isomerization step and as a substrate carrier for 11-*cis*-retinol dehydrogenase.[4-8] These functions require the rapid association and release of retinoid from the CRALBP ligand-binding pocket and involve critical protein interactions. To better understand the visual cycle, we are characterizing CRALBP ligand and protein interactions and retinoid trafficking within the RPE.

2. RETINAL DISEASES CAUSED BY DEFECTIVE CRALBP

Six recessive defects in the *RLBP1* gene encoding human CRALBP have been found to cause retinal pathology, including missense mutations R150Q, M225K and R233W. Retinal dystrophies associated with CRALBP gene mutations now include retinitis pigmentosa, retinitis punctata albescens, Bothnia dystrophy, fundus albipunctatas, and Newfoundland rod-cone dystrophy and have been detected in pedigrees from Europe, the middle

---

[1] Cole Eye Institute, Cleveland Clinic Foundation, Cleveland, OH 44195 USA; Departments of [2]Biochemistry and [3]Ophthalmology, University of Washington, Seattle, WA 98195 USA.

east, Newfoundland, and India.[9-13] CRALBP gene defects can tighten or abolish retinoid interactions.[6]

3. CRALBP LIGAND INTERACTIONS

3.1. Site Directed Mutagenesis

The lack of covalent interactions between the recombinant CRALBP (rCRALBP) and retinoid was demonstrated and residues Q210 and K221 associated with retinoid-binding by a combination of protein chemical modifications, site-directed mutagenesis, and UV-visible and fluorescence spectroscopy.[14] Similar methods combined with heteronuclear single quantum correlation NMR and enzymatic assays with purified recombinant 11-*cis*-retinol dehydrogenase were used to demonstrate that rCRALBP residues W165, M208, M222, M225, and W244 influence retinoid-binding and substrate carrier function.[15] Disease-causing mutations R150Q and M225K abolish ligand-binding and R233W was found to significantly tighten rCRALBP retinoid affinity.[6,9]

3.2. Photo-Labeling of the CRALBP Ligand-Binding Pocket

Photo-labeling of rCRALBP with 3-diazo-4-keto-11-*cis*-retinal yielded covalent labeling of eight residues that were identified by mass spectrometry, namely Y179, F197, C198, M208, K221, M222, V223 and M225.[16] Four of these photo-adducted residues were independently associated with ligand interactions as described above, supporting specific labeling of the ligand-binding cavity. An unexpected outcome was that each of the photoaffinity modified residues in rCRALBP exhibited one or more different adduct masses.[16] We suspect that upon UV-irradiation, the resulting carbene radical moves freely throughout the conjugated polyene structure of the retinoid analogue and fragments at the time of attachment to the protein.

3.3. Structural Modeling of the CRALBP Retinoid-Binding Domain

The homology between CRALBP and other CRAL-TRIO family members spans about 185 amino acids and includes their respective ligand binding pockets.[17] A model (Figure 66.1) of the CRALBP ligand binding domain[16] was constructed based on crystal structures of homologues α-tocopherol transfer protein (αTTP), yeast Sec14, and supernatant protein factor. Five of the CRALBP residues associated with the retinoid binding pocket by biochemical analyses (W165, Y179, F197, M222, and M225) align directly with components identified in the ligand cavities of the CRAL-TRIO crystal structures[16] and are very close to ligand in our model (average distance ~4.4 Å). Four other residues (C198, Q210, K221, and V223) align in relatively close proximity (average model distance from ligand ~7.6 Å). All nine residues line the ligand binding cavity in the model. Residues M208, R233W and W244[6,15] are more distant from ligand in the structural model (average distance from ligand ~14.3 Å). M208 and W244 may be located at the entrance/exit to the ligand cavity and involved in conformational changes necessary for ligand binding and release.

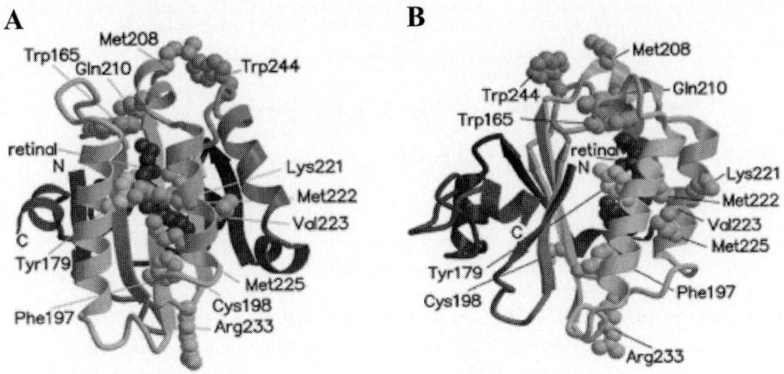

Figure 66.1. Structural Model of the CRALBP Ligand Binding Domain (residues 143-301). The twelve residues implicated in ligand interactions by biochemical analyses are labeled; 11-*cis* retinal occupies the ligand pocket. A. View parallel to the beta sheet. B. View perpendicular to that of A.

3.4. The rCRALBP Ligand-Binding Cavity from Topological Analyses

Hydrogen/deuterium (H/D) exchange detected by mass spectrometry was used to measure the effect of protein-bound 11-*cis*-retinal on the solvent accessibility of amide hydrogens.[16] N-terminal residues 5-42, C-terminal residues 282-316, and residues 41-71, 80-94, 127-137 and 262-275 all incorporated more deuterium in holo-rCRALBP and therefore are more solvent exposed in the holo-protein. With the exception of residues 127-137, the central region of CRALBP (residues 111-197) incorporated low levels of deuterium with or without ligand and appears to be buried in both the apo and holo protein structures. The only region that incorporated significantly more deuterium in the absence of bound 11-*cis*-retinal encompasses amino acids 198-255 and most of the identified ligand cavity residues. The results are also consistent with an earlier lower resolution topological analysis of CRALBP using antibodies and proteases.[18]

4. CRALBP PROTEIN INTERACTIONS

4.1. CRALBP Interacts with 11-*cis*-Retinol Dehydrogenase (RDH5)

Early studies with crude extracts from bovine RPE microsomes support a substrate carrier interaction between CRALBP and 11-*cis*-RDH.[4,19] Kinetic analyses performed with purified recombinant RDH5 and purified rCRALBP demonstrate a direct, functional interaction between rRDH5 and rCRALBP[6] and are consistent with the notion that CRALBP affects the activity of RDH5 by 'channeling' retinoids to the enzyme. Immunoprecipitation experiments with the purified recombinant proteins and different anti-peptide CRALBP antibodies[18] support a structural interaction in a C-terminal region of CRALBP (Figure 66.2). How retinoid is released from the high affinity CRALBP binding pocket for export from the RPE for visual pigment regeneration remains unresolved. Interactions with RDH5 may be involved in the release mechanism.

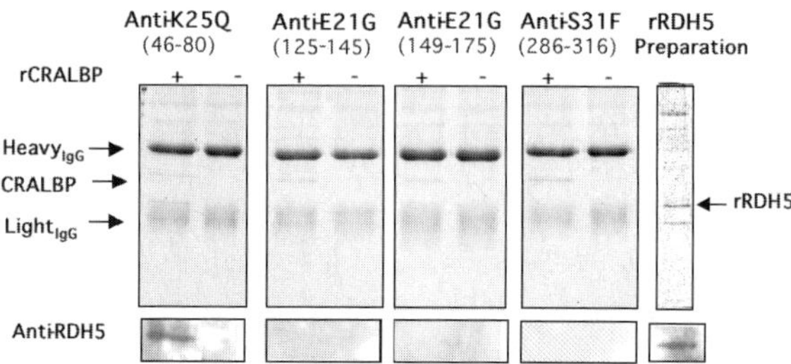

Figure 66.2. Immunoprecipitation of RDH5 with anti-CRALBP. Purified recombinant RDH5 and CRALBP were mixed together then exposed to different agarose bead-immobilized anti-peptide antibodies recognizing the CRALBP residues in parentheses. Coomassie blue stained SDS-PAGE is shown above and the Western analyses below with anti-RDH5 indicate only the N-terminally directed anti-K25Q pulled down rRDH5, supporting an interaction with a C-terminal region of CRALBP.

4.2. CRALBP Interacts with EBP50

An overlay assay was used to detect interactions of CRALBP with components of RPE microsomes.[20] Interacting proteins were separated by 2D-PAGE and identified by LC MS/MS. Protein interactions were characterized by affinity chromatography, peptide competition, and recombinant expression of protein domains. CRALBP bound to a 54 kD protein in RPE microsomes identified as ERM (ezrin, radixin, moesin)-binding phosphoprotein 50 (EBP50). EBP50 is also known as NHERF-1 (sodium/hydrogen exchanger regulatory factor type-1). EBP50 was found in multiple adjacent 2D gel spots, suggesting phosphorylaton may play a role in regulating retinoid trafficking. CRALBP bound to both recombinant PDZ domains of EBP50 but not to the C-terminal ezrin-binding domain. In outer retina, EBP50 and ezrin were localized to RPE and Müller apical processes. CRALBP was distributed throughout both RPE and Müller cells including their apical processes. ERM proteins are multivalent linkers that connect plasma membrane proteins with the cortical actin cytoskeleton. EBP50 interacts with an N-terminal domain of the ERM proteins and binds other targets through its PDZ domains, thus contributing to an apical localization of target proteins. These results support a retinoid-processing complex in the apical RPE[20] and are consistent with observations of retinoid processing proteins in reciprocal immunoprecipitations of RPE microsomes using antibodies to visual cycle proteins.[21] The C-terminus of CRALBP interacts with the PDZ domains in EBP50[20] and a separate C-terminal domain in EBP50 binds ezrin or another ERM family member. Notably, a fraction of RDH5 activity appears to be associated with the RPE plasma membrane.[22] Interactions with EBP50 and ezrin may help localize the apical plasma membrane where interaction with RDH5 could facilitate the release of 11-*cis*-retinal from CRALBP for export to the photoreceptors for visual pigment regeneration.

4.3. Retinoid-Processing Proteins in the Apical RPE

To further test the hypothesis of a retinoid-processing complex in apical RPE, we pursued proteomic analysis of RPE apical processes and plasma membranes. RPE apical membranes were isolated from mouse eyecups on lectin-coated agarose beads. Morpho-

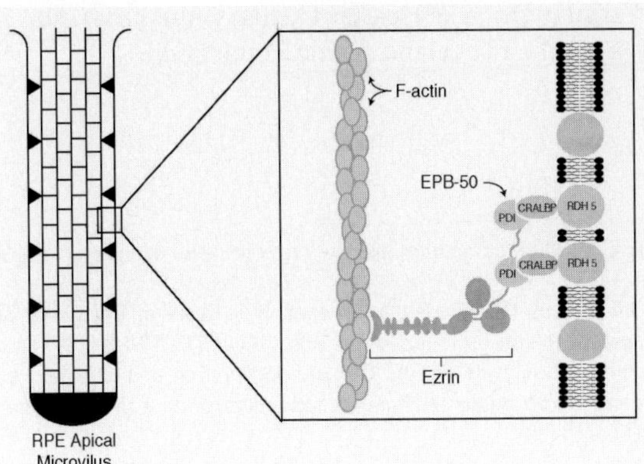

Figure 66.3. Hypothesis for a retinoid-processing complex in apical RPE. Left: A model representation of apical RPE process filled with actin fibers cross-linked to each other and to the plasma membrane. Right: Enlarged view of the model in the left with identified protein components labeled.

logical analyses of the bead-bound RPE apical membranes support a highly purified preparation. Proteins were recovered from the beads, separated by SDS-PAGE, gel bands excised, digested *in situ* with trypsin and proteins identified by LC MS/MS. CRALBP, EBP50, RDH5 and ezrin were among the proteins bound to the lectin beads.[23] These findings further support the proposed retinoid-processing complex in the RPE. Many other proteins were identified from the apical RPE membranes. The information should provide insights into processes occurring at this critical interface, which are important for the support and maintenance of vision.[24]

5. CONCLUSIONS

Considerable progress has been made in understanding the biochemical basis and enzymology of the visual cycle and associated retinal diseases. However, many aspects of the visual cycle remain unclear. The retinoid isomerization chemistry is controversial because the responsible enzymes have yet to be isolated. Less is known about cone visual pigment regeneration but the process is different than that for rhodopsin and may involve 11-*cis*-retinoid synthesis in Müller cells. In the RPE, the interaction of cytosolic proteins with membrane-associated enzymes is largely uncharacterized and little is known about mechanisms of retinoid trafficking, and spatial-compartmental organization of visual cycle components. Identification of visual cycle proteins associated with CRALBP will promote a better understanding of these processes and will advance the development of therapies for visual disorders.

6. ACKNOWLEDGEMENTS

This study was supported in part by NIH grants EY6603, EY14239, EY015638, GM63020, EY01730, EY02317, an unrestricted award from Research to Prevent Blindness,

Inc. (University of Washington), a Research Center Grant from The Foundation Fighting Blindness, and funds from the Cleveland Clinic Foundation.

7. REFERENCES

1. T.D. Lamb & E.N. Pugh, Jr., Dark adaptation and the retinoid cycle of vision, *Prog Retin Eye Res* **23**:307-80 (2004).
2. J.C. Saari, J. Huang, D.E. Possin, R.N. Fariss, J. Leonard, G.G. Garwin, J.W. Crabb & A.H. Milam, CRALBP is expressed by oligodendrocytes in optic nerve and brain, *Glia* **21**:259-68 (1997).
3. J.A. Treadwell, K.B. Pagniello & S.M. Singh, Genetic segregation of brain gene expression identifies retinaldehyde binding protein 1 and syntaxin 12 as potential contributors to ethanol preference in mice, *Behav Genet* **34**:425-39 (2004).
4. J.C. Saari, D.L. Bredberg & N. Noy, Control of substrate flow at a branch in the visual cycle, *Biochemistry* **33**:3106-12 (1994).
5. J.C. Saari, M. Nawrot, B.N. Kennedy, G.G. Garwin, J.B. Hurley, J. Huang, D.E. Possin & J.W. Crabb, Visual cycle impairment in cellular retinaldehyde binding protein (CRALBP) knockout mice results in delayed dark adaptation, *Neuron* **29**:739-48 (2001).
6. I. Golovleva, S. Bhattacharya, Z. Wu, N. Shaw, Y. Yang, K. Andrabi, K.A. West, M.S. Burstedt, K. Forsman, G. Holmgren, O. Sandgren, N. Noy, J. Qin & J.W. Crabb, Disease-causing mutations in CRALBP tighten and abolish ligand interactions, *J Biol Chem* **278**:12397-402 (2003).
7. A. Winston & R.R. Rando, Regulation of isomerohydrolase activity in the visual cycle, *Biochemistry* **37**:2044-2050 (1998).
8. H. Stecher, M.H. Gelb, J.C. Saari & K. Palczewski, Preferential release of 11-cis-retinol from retinal pigment epithelial cells in the presence of CRALBP, *J Biol Chem* **274**:8577-85 (1999).
9. M.A. Maw, B. Kennedy, A. Knight, R. Bridges, K.E. Roth, E.J. Mani, J.K. Mukkadan, D. Nancarrow, J.W. Crabb & M.J. Denton, Mutation of the gene encoding cellular retinaldehyde-binding protein in autosomal recessive retinitis pigmentosa, *Nat Genet* **17**:198-200 (1997).
10. M.S. Burstedt, O. Sandgren, G. Holmgren & K. Forsman-Semb, Bothnia dystrophy caused by mutations in the CRALBP gene (RLBP1) on chromosome 15q26, *Invest Ophthalmol Vis Sci* **40**:995-1000 (1999).
11. H. Morimura, E.L. Berson & T.P. Dryja, Recessive mutations in the RLBP1 gene encoding cellular retinaldehyde-binding protein in a form of retinitis punctata albescens, *Invest Ophthalmol Vis Sci* **40**:1000-4 (1999).
12. N. Katsanis, N.F. Shroyer, R.A. Lewis, J.C. Cavender, A.A. Al-Rajhi, M. Jabak & J.R. Lupski, Fundus albipunctatus and retinitis punctata albescens in a pedigree with an R150Q mutation in RLBP1, *Clin Genet* **59**:424-9 (2001).
13. E.R. Eichers, J.S. Green, D.W. Stockton, C.S. Jackman, J. Whelan, J.A. McNamara, G.J. Johnson, J.R. Lupski & N. Katsanis, Newfoundland rod-cone dystrophy, an early-onset retinal dystrophy, is caused by splice-junction mutations in RLBP1, *Am J Hum Genet* **70**:955-64 (2002).
14. J.W. Crabb, Z. Nie, Y. Chen, J.D. Hulmes, K.A. West, J.T. Kapron, S.E. Ruuska, N. Noy & J.C. Saari, Cellular retinaldehyde-binding protein ligand interactions. Gln-210 and Lys-221 are in the retinoid binding pocket, *J Biol Chem* **273**:20712-20 (1998).
15. Z. Wu, Y. Yang, N. Shaw, S. Bhattacharya, L. Yan, K. West, K. Roth, N. Noy, J. Qin & J.W. Crabb, Mapping the ligand binding pocket in CRALBP, *J Biol Chem* **278**:12390-6 (2003).
16. Z. Wu, A. Hasan, T. Liu, D. Teller, J.C. Saari, & J.W. Crabb, Identification of CRALBP Ligand Interactions by Photoaffinity Labeling, Hydrogen/Deuterium Exchange and Structural Modeling, *J Biol Chem* **279**:27357-27364 (2004).
17. C. Panagabko, S. Morley, M. Hernandez, P. Cassolato, H. Gordon, R. Parsons, D. Manor & J. Atkinson, Ligand specificity in the CRAL-TRIO protein family, *Biochemistry* **42**:6467-74 (2003).
18. J.W. Crabb, V.P. Gaur, G.G. Garwin, S.V. Marx, C. Chapline, C.M. Johnson & J.C. Saari, Topological and epitope mapping of CRALBP from retina, *J Biol Chem* **266**:16674-83 (1991).
19. J.C. Saari & L. Bredberg, Enzymatic reduction of 11-cis-retinal bound to cellular retinal-binding protein, *Biochim Biophys Acta* **716**:266-72 (1982).
20. M. Nawrot, K. West, J. Huang, D.E. Possin, A. Bretscher, J.W. Crabb & J.C. Saari, Cellular retinaldehyde-binding protein interacts with ERM-binding phosphoprotein 50 in retinal pigment epithelium, *Invest Ophthalmol Vis Sci* **45**:393-401 (2004).

21. S.K. Bhattacharya, Z. Wu, Z. Jin, L. Yan, M. Miyagi, K. West, M. Nawrot, J.C. Saari & J.W. Crabb. Proteomic Approach to Identification of a Mammalian Visual Cycle Protein Complex. in *Experimental Biology 2002* Vol. 16: A14 (FASEB J, New Orleans, LA, 2002).
22. N.L. Mata, E.T. Villazanna & A.T.C. Tsin, Colocalization of 11-*cis* retinyl esters and retinyl ester hydrolase activity in retinal pigment epithelium plasma membrane, *Invest Ophthal Visual Sci* **39**:1312-19 (1998).
23. V.L. Bonilha, S.K. Bhattacharya, K.A. West, J.S. Crabb, J. Sun, M.E. Rayborn, M. Nawrot, J.C. Saari & J.W. Crabb, Support For a Proposed Retinoid-Processing Protein Complex in Apical Retinal Pigment Epithelium, *Exp Eye Res* **79**:419-22 (2004).
24. V.L. Bonilha, S.K. Bhattacharya, K.A. West, J. Sun, J.W. Crabb, M.E. Rayborn & J.G. Hollyfield, Proteomic Characterization of Isolated Retinal Pigment Epithelium Microvilli, *Mol Cell. Proteomics* **3**:1119-27 (2004).

CHAPTER 67

FUNCTIONAL STUDY OF PHOTORECEPTOR PDEδ

Houbin Zhang, Jeanne M. Frederick, and Wolfgang Baehr*

1. INTRODUCTION

Cyclic GMP phosphodiesterase 6 (PDE6), a member of a large superfamily of phosphodiesterases (Soderling et al., 1998; Beavo et al., 1994), is a key enzyme in the rod and cone phototransduction cascades (Polans et al., 1996; McBee et al., 2001). PDE6 in rod photoreceptors (henceforth called PDE) is composed of two catalytic subunits–PDEα and PDEβ–and two identical inhibitory subunits, PDEγ (Baehr et al., 1979; Fung et al., 1990; Deterre et al., 1988). Photoreceptor PDE is peripherally membrane-associated via the farnesyl and geranylgeranyl chains at C-termini of PDEα and PDEβ, respectively (Anant et al., 1992; Qin et al., 1992; Qin and Baehr, 1994). PDEδ was originally copurified with photoreceptor PDE from bovine retina and considered the fourth subunit of PDE (Gillespie et al., 1989). Functional studies indicated that PDEδ could solubilize membrane-associated PDE and decouple the activation of transducin from hydrolysis of cGMP when added to the permeabilized rod outer segments (Cook and Beavo, 2000; Florio et al., 1996). However, first evidence arguing against PDEδ being an authentic PDE subunit came from the expression profile of PDEδ. Multiple tissue northern blots indicated that PDEδ mRNA is present in all tissues examined with relatively higher level in retinas, in contrast to PDE which is only expressed in photoreceptors of the retina (Florio et al., 1996; Marzesco et al., 1998). Furthermore, PDEδ homologues were also identified in organisms such as *C. elegans* which has no eyes or retina-like structures and does not express PDE6 (Li and Baehr, 1998). Like mammalian PDEδ, recombinant PDEδ from *C. elegans* can elute PDE from bovine rod outer segments, suggesting the functional conservation of PDEδ throughout evolution.

2. PDEδ INTERACTS WITH ISOPRENYLATED PROTEIN

Clues regarding the biological function of PDEδ were generated initially from yeast two-hybrid (y2h) screening (Marzesco et al., 1998; Linari et al., 1999b; Hillig et al., 2000).

*Houbin Zhang, Jeanne M. Frederick and Wolfgang Baehr. Moran Eye Center, University of·Utah, Salt Lake City, Utah, 84112.

Table 67.1. Summary of PDEδ interacting proteins.

Method	Bait	Interacting Proteins	References
Qualitative interactions of yeast expression constructs	Arl2, Arl3, Gα$_{i1}$, H-Ras, RheB, Rho6, Rap1A	PDEδ	(Hanzal-Bayer et al., 2002)
Qualitative interactions of yeast expression constructs	Rap2B, H-Ras, Rap1A, RhoA, RhoB, Rnd1	PDEδ	(Nancy et al., 2002)
Mouse embryo cDNA library	PDEδ	Arl3	(Linari et al., 1999a)
Mouse embryo cDNA library	RPGR	PDEδ	(Linari et al., 1999b)
HeLa cDNA library	Rab13	PDEδ	(Marzesco et al., 1998)
Bovine retina cDNA library	PDEδ	PDE6α GRK1	(Zhang et al., 2004)
Qualitative interactions of yeast expression constructs	PDEδ	GRK1(SAAX)	(Zhang et al., 2004)

A group of small GTPases in the Ras superfamily were shown to interact specifically with PDEδ. Most of these GTPases have a common feature, that is, a CAAX box motif in their C-terminal amino acid sequence which signals posttranslational modification resulting in cleavage of -AAX polypeptides followed by carboxymethylation and prenylation (prenyl-thioether formation) of the cysteine residue. The last residue of the CAAX box specifies whether the added prenyl chains are either farnesyl (C_{15} moieties) or geranylgeranyl (C_{20} moieties). Prenylation is a step necessary for the targeting of proteins to membrane, and interaction of PDEδ with prenylated proteins relies on the prenyl lipid chain. The binding of PDEδ to PDE *in vitro* is mediated by its prenylated C-terminal (Cook et al., 2000). The y2h screening in our lab identified rhodopsin kinase (GRK1) as another interacting member. Mammalian GRK1s are farnesylated proteins and the prenyl group is required for interaction with PDEδ. We also showed that PDEδ interacted with GRK7, a cone pigment kinase believed to be important for the recovery of photoresponse in cone photoreceptors in some species (Chen et al., 2001; Weiss et al., 2001; Weiss et al., 1998). GRK7 is a geranylger-anylated protein. Table 67.1 provides a summary of a subset of proteins which can interact with PDEδ.

3. PDEδ BINDS ISOPRENYL CHAINS IN THE ABSENCE OF C-TERMINAL CYSTEINE

Given that PDEδ interacts with multiple isoprenylated proteins and that isoprenylation is a requisite for interaction, we sought direct evidence for the interaction between PDEδ and prenyl chain alone. By conjugating a fluo-rescent probe to either farnesyl and geranyl-geranyl isoprenoid chains, we used a technique called FRET (Fluorescence Resonance Energy Transfer) to demonstrate isoprenoid binding to PDEδ (Zhang et al., 2004). FRET (reviewed by (Centonze et al., 2003)) revealed a significant increase in fluorescence emis-sion from one of the fluorescent probes suggesting interaction between PDEδ and the isoprenoid chain. By varying ligand concentrations, the binding constants for farnesyl and geranylgeranyl chains were determined to be 0.70 μM, and 19.06 μM, respectively.

4. PDEδ INTERACTS WITH Arl2 AND Arl3

Y2h screening also identified PDEδ interacting proteins other than isoprenylated proteins, such as Arl2 and Arl3 (Hanzal-Bayer et al., 2002). So far, Arl2 and Arl3 are the only non-isoprenylated proteins shown capable of PDEδ interaction. Arl2 and Arl3 are also small GTPases belonging to the Arf family in the Ras superfamily. Arf family members play important roles in intracellular vesicular trafficking processes. Arl proteins differ from Arfs in that Arl proteins do not possess co-factor activity of cholera toxin-catalyzed ADP-ribosylation of Gαs subunits. And unlike in Arfs, the glycine residue at position 2 in some Arls cannot be myristoylated (Sharer et al., 2002). The binding of PDEδ to Arl3 depends on the guanine nucleotide species bound to Arl3, and β-sheet interaction. PDEδ only interacts with GTP-bound Arl3, but not GDP-bound Arl3, suggesting PDEδ is one of the effectors of Arl3 GTPase (Linari et al., 1999a).

5. STRUCTURE OF PDEδ

BLAST searching generated the first clue about the structure PDEδ, which suggested that PDEδ has relatively weak sequence similarity to the C-terminal part of RhoGDI (Nancy et al., 2002). RhoGDI can bind Rho GTPase through interaction with its C-terminal isoprenyl group. Shortly after the identification of sequence homology, the crystal structure of PDEδ, cocrystallized with Arl2/GTP, was published (Hanzal-Bayer et al., 2002). It confirmed the striking structural similarity between PDEδ and RhoGDI, although their primary sequences have very low homology (Figure 67.1). Both of these two proteins have a hydrophobic pocket formed between two β-propellers. This structure further confirmed that PDEδ is a prenyl binding protein. Although the binding of PDEδ to prenylated proteins usually depends on the structure of the prenyl group, the protein-protein interaction may also be important for their tight binding. One example for this notion is that excess amount of prenyl compounds can only partially compete off the binding between PDEδ and GRK7. This additional protein-protein interaction could be responsible for the binding specificity of PDEδ, a possible reason why both PDEδ and RhoGDI only bind a subset of isoprenylated proteins. As for RhoGDI, it only interacts with proteins of the Rho GTPase family, whereas PDEδ interacts with a larger number of polypeptides (PDEα, PDEβ, GRK1, and GRK7 in photoreceptors).

6. CELLULAR LOCALIZATION OF PDEδ IN PHOTORECEPTORS

Several results have been published regarding localization of photoreceptor PDEδ leading to controversy. The first result of immunocytochemistry of PDEδ in retina indicated that PDEδ is localized to the outer segments of rods, but not cones (Florio et al., 1996). Later, PDEδ was shown in dark-adapted retina to be distributed throughout the whole cell, from the synaptic termini to outer segments of rod and cone photoreceptors (Zhang et al.,

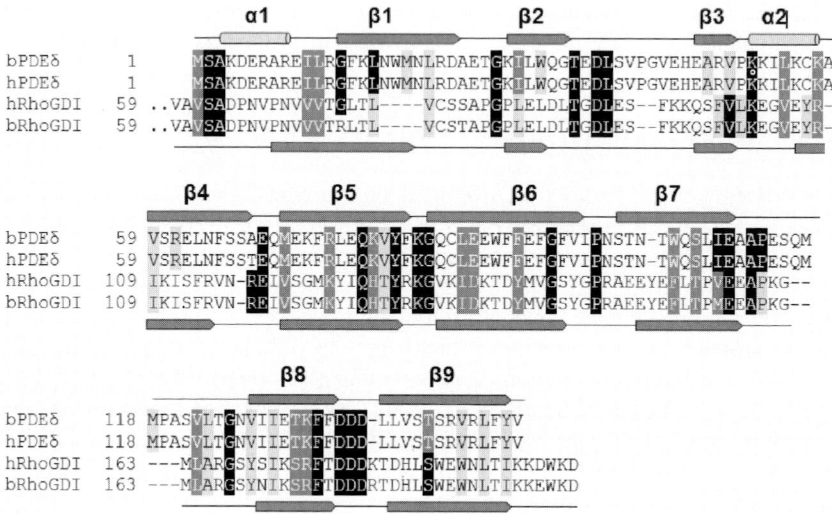

Figure 67.1. Structural homology of RhoGDI and PDEδ. The PDEδ sequence was aligned with the C-terminal portion of RhoGDI using ClustalW. Cylinders indicate two α-helices in PDEδ, and arrows depict nine β-sheets, emphasizing the structural similarity (adapted from (Hanzal-Bayer et al., 2002). Note that sequence similarity is rather limited. Residues shaded yellow represent contacts with prenyl groups inside the binding pocket.

2004). Recent data demonstrated that PDEδ in light-adapted bovine retina is localized primarily to the connecting cilium (Norton et al., 2004), a structure bridging the inner segment to the outer segment. The connecting cilium is a structure that controls the transport of phototransduction components from the inner segments, where biosynthesis takes place, to the outer segment. The localization of PDEδ to the connecting cilium is consistent with the concept that PDEδ may be involved in the intracellular trafficking of isoprenylated phototransduction proteins.

7. SUMMARY AND HYPOTHETICAL MODEL OF PDEδ FUNCTION IN PHOTORECEPTORS

Several independent assays strongly suggest that a photoreceptor polypeptide, termed "PDEδ", is not an authentic PDE subunit, but a general prenyl binding protein (PrPB). Biochemically identified binding partners include rhodopsin kinase (GRK1), cone pigment kinase (GRK7), and the PDE catalytic subunits (PDEα,β), all of which are prenylated at their C-termini. In addition to prenylated partners, PDEδ interacts with small GTPases like Arl2/3 which are not known to be prenylated. An important function of PDEδ is to solubilize prenylated proteins which are normally membrane-associated.

Based on currently available information, we propose a model in which newly synthesized prenylated protein (GRK1) interacts with PDEδ, and transports to the photoreceptor connecting cilium and the outer segment in conjunction with Arl2 charged with GTP. Hydrolysis of GTP may cause discharge of GRK1 into the OS. Arl2-GDP is recycled to Arl2-GTP by an unknown GEF most likely present in the inner segment and recombines again with PDEδ for another round of transport.

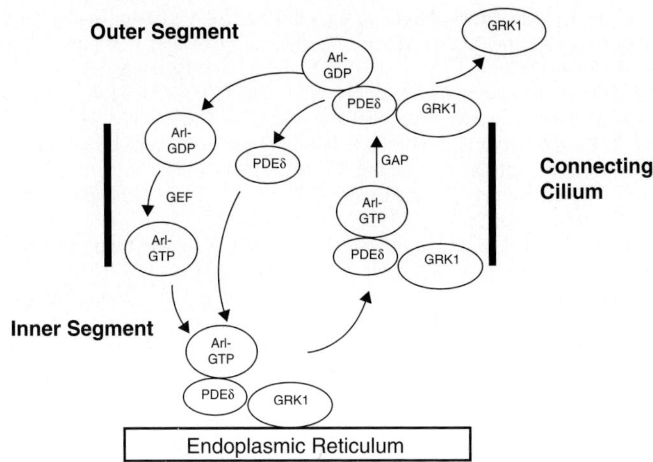

Figure 67.2. Hypothetical model of photoreceptor PDEδ in transport of rhodopsin kinase (GRK1) from the inner segment to the outer segment. PDEδ binds to GTP-bound Arl2/3, and the resulting protein complex interacts with the tail of isoprenylated GRK1 which then dissociates from the ER (site of protein isoprenylation). The Arl2/3 and PDEδ complex with GRK1 may diffuse to the connecting cilium, but more likely associate with membraneous carrier structure. At the cilium, a yet-identified GAP protein may catalyze the hydrolysis of GTP bound to ARl2/3 to GDP. Then the protein complex falls apart, and GRK1 is released after passage through the cilium. GTP-bound Arl2/3 are regenerated by an unidentified guanine nucleotide exchange factor (GEF), similarly as light activated R* promotes exchange of GDP by GTP bound to transducin.

8. FUTURE DIRECTION

Generation of a photoreceptor PDEδ knockout will be extremely useful for revealing the *in vivo* function of PDEδ in rod and cone photoreceptors. Our preliminary experiments suggested that the mouse lacking PDEδ gene had reduced levels of PDE and rhodopsin kinase, as well as shortened rod outer segments, suggesting the importance of PDEδ for the stability of its binding partners (Zhang et al., unpublished results). In the future, it will be important to study the roles Arl2 and Arl3 may play in transport of phototransduction components.

9. REFERENCES

Anant, J. S., Ong, O. C., Xie, H., Clarke, S., O'Brien, P. J., and Fung, B. K. K. (1992). In vivo differential prenylation of retinal cyclic GMP phosphodiesterase catalytic subunits. J. Biol. Chem. **267**:687-690.

Baehr, W., Devlin, M. J., and Applebury, M. L. (1979). Isolation and characterization of cGMP phosphodiesterase from bovine rod outer segments. J. Biol. Chem. **254**:1669-11677.

Beavo, J. A., Conti, M., and Heaslip, R. J. (1994). Multiple cyclic nucleotide phosphodiesterases. Mol. Pharmacol. **46**:399-405.

Centonze, V. E., Sun, M., Masuda, A., Gerritsen, H., and Herman, B. (2003). Fluorescence resonance energy transfer imaging microscopy. Methods Enzymol. **360**:542-560.

Chen, C. K., Zhang, K., Church-Kopish, J., Huang, W., Zhang, H., Chen, Y. J., Frederick, J. M., and Baehr, W. (2001). Characterization of human GRK7 as a potential cone opsin kinase. Mol. Vis. **7**:305-313.

Cook, T. A. and Beavo, J. A. (2000). Purification and assay of bovine type 6 photoreceptor phosphodiesterase and its subunits. Methods Enzymol. **315**:597-616.

Cook, T. A., Ghomashchi, F., Gelb, M. H., Florio, S. K., and Beavo, J. A. (2000). Binding of the delta subunit to rod phosphodiesterase catalytic subunits requires methylated, prenylated C-termini of the catalytic subunits. Biochemistry **39**:13516-13523.

Deterre, P., Bigay, J., Forquet, F., Robert, M., and Chabre, M. (1988). cGMP phosphodiesterase of retinal rods is regulated by two inhibitory subunits. Proc. Natl. Acad. Sci. U. S. A. **85**:2424-2428.

Florio, S. K., Prusti, R. K., and Beavo, J. A. (1996). Solubilization of membrane-bound rod phosphodiesterase by the rod phosphodiesterase delta subunit. J. Biol. Chem. **271**:24036-24047.

Fung, B. K. K., Young, J. H., Yamane, H. K., and Griswold-Prenner, I. (1990). Subunit stoichiometry of retinal rod cGMP phosphodiesterase. Biochemistry **29**:2657-2664.

Gillespie, P. G., Prusti, R. K., Apel, E. D., and Beavo, J. A. (1989). A soluble form of bovine rod photoreceptor phosphodiesterase has a novel 15-kDa subunit. J. Biol. Chem. **264**:12187-12193.

Hanzal-Bayer, M., Renault, L., Roversi, P., Wittinghofer, A., and Hillig, R. C. (2002). The complex of Arl2-GTP and PDE delta: from structure to function. EMBO J. **21**:2095-2106.

Hillig, R. C., Hanzal-Bayer, M., Linari, M., Becker, J., Wittinghofer, A., and Renault, L. (2000). Structural and biochemical properties show ARL3-GDP as a distinct GTP binding protein. Structure. Fold. Des **8**:1239-1245.

Li, N. and Baehr, W. (1998). Expression and characterization of human PDEd and its Caenorhabditis elegans ortholog CEd. FEBS Lett. **440**:454-457.

Linari, M., Hanzal-Bayer, M., and Becker, J. (1999a). The delta subunit of rod specific cyclic GMP phosphodiesterase, PDE delta, interacts with the Arf-like protein Arl3 in a GTP specific manner. FEBS Lett. **458**:55-59.

Linari, M., Ueffing, M., Manson, F., Wright, A., Meitinger, T., and Becker, J. (1999b). The retinitis pigmentosa GTPase regulator, RPGR, interacts with the delta subunit of rod cyclic GMP phosphodiesterase. Proc. Natl. Acad. Sci. U. S. A. **96**:1315-1320.

Marzesco, A. M., Galli, T., Louvard, D., and Zahraoui, A. (1998). The rod cGMP phosphodiesterase delta subunit dissociates the small GTPase Rab13 from membranes. J Biol. Chem. **273**:22340-22345.

McBee, J. K., Palczewski, K., Baehr, W., and Pepperberg, D. R. (2001). Confronting complexity: the interlink of phototransduction and retinoid metabolism in the vertebrate retina. Prog. Retin. Eye Res. **20**:469-529.

Nancy, V., Callebaut, I., El Marjou, A., and de Gunzburg, J. (2002). The delta subunit of retinal rod cGMP phosphodiesterase regulates the membrane association of Ras and Rap GTPases. J. Biol. Chem. **277**:15076-15084.

Norton, A. W., Hosier, S., Terew, J. M., Li, N., Dhingra, A., Vardi, N., Baehr, W., and Cote, R. H. (2004). Evaluation of the 17 kDa prenyl binding protein as a regulatory protein for phototransduction in retinal photoreceptors. J. Biol. Chem. *in press.*

Polans, A., Baehr, W., and Palczewski, K. (1996). Turned on by Ca^{2+}! The physiology and pathology of Ca^{2+} binding proteins in the retina. Trends Neurosci. **19**:547-554.

Qin, N. and Baehr, W. (1994). Expression and mutagenesis of mouse rod photoreceptor cGMP phosphodiesterase. J. Biol. Chem. **269**:3265-3271.

Qin, N., Pittler, S. J., and Baehr, W. (1992). *In vitro* isoprenylation and membrane association of mouse rod photoreceptor cGMP phosphodiesterase a and b subunits expressed in bacteria. J. Biol. Chem. **267**:8458-8463.

Sharer, J. D., Shern, J. F., Van, V. H., Wallace, D. C., and Kahn, R. A. (2002). ARL2 and BART enter mitochondria and bind the adenine nucleotide transporter. Mol. Biol. Cell **13**:71-83.

Soderling, S. H., Bayuga, S. J., and Beavo, J. A. (1998). Identification and characterization of a novel family of cyclic nucleotide phosphodiesterases. J Biol. Chem. **273**:15553-15558.

Weiss, E. R., Ducceschi, M. H., Horner, T. J., Li, A., Craft, C. M., and Osawa, S. (2001). Species-specific differences in expression of G-protein-coupled receptor kinase (GRK) 7 and GRK1 in mammalian cone photoreceptor cells: implications for cone cell phototransduction. J. Neurosci. **21**:9175-9184.

Weiss, E. R., Raman, D., Shirakawa, S., Ducceschi, M. H., Bertram, P. T., Wong, F., Kraft, T. W., and Osawa, S. (1998). The cloning of GRK7, a candidate cone opsin kinase, from cone- and rod-dominant mammalian retinas. Mol. Vis. **4**:27.

Zhang, H., Liu, X. H., Zhang, K., Chen, C. K., Frederick, J. M., Prestwich, G. D., and Baehr, W. (2004). Photoreceptor cGMP phosphodiesterase delta subunit (PDEdelta) functions as a prenyl-binding protein. J. Biol. Chem. **279**:407-413.

CHAPTER 68

LOCALIZATION OF THE INSULIN RECEPTOR AND PHOSPHOINOSITIDE 3-KINASE IN DETERGENT-RESISTANT MEMBRANE RAFTS OF ROD PHOTORECEPTOR OUTER SEGMENTS

Raju V.S. Rajala[1,2,3], Michael H. Elliott[1,3], Mark E. McClellan[1,3], and Robert E. Anderson[1,2,3]

1. INTRODUCTION

Lipid rafts are specialized membrane domains enriched in certain lipids, cholesterol and proteins. The existence of lipid rafts was first hypothesized in 1988 (Simons and van Meer, 1988; Simons and Ikonen, 1997), but what we know as "caveolae", flask-shaped types of lipid rafts, were observed earlier (Yamada, 1955). Three general types of rafts- caveolae, glycospingolipid enriched membranes (GEM), and polyphosphoinositide-rich rafts- have been described (Jacobson and Dietrich, 1999) and may be oriented on the "inner leaflet" (PIP_2 rich rafts and caveolae) or the "outer leaflet" (GEM). The fatty acid chains of lipids within the raft tend to be more saturated and these are more tightly packed, creating domains with higher order. It is therefore thought that rafts exist in a separate ordered phase that floats in a sea of poorly ordered lipids.

The lipid environment of rafts tends to recruit fatty acyl-modified signaling proteins (Melkonian et al., 1999) and may act as organizing centers to localize a variety of signaling molecules (Simons and Ikonen, 1997). Detergent-resistant membranes (DRMs), presumptive biochemical preparations of lipid rafts, have been isolated from bovine photoreceptor rod outer segments (ROS) and are enriched in cholesterol, caveolin-1 (Elliott et al., 2003), and saturated lipid species (Martin et al., 2005), and contain transducin, cGMP-phosphodiesterase (Seno et al., 2001), RGS9-1 (Nair et al., 2002), the p44 arrestin splice variant (Nair et al., 2004), recoverin (Senin et al., 2004), and ROM-1 (Boesze-Battaglia et al., 2002).

[1] Department of [1] Ophthalmology and [2] Cell Biology, University of Oklahoma Health Sciences Center; and [3] Dean A. McGee Eye Institute, Oklahoma City, OK 73104, USA. Corresponding author: R.V.S. Rajala, E-mail: raju-rajala@ouhsc.edu.

Cells of bovine and rat retina contain high affinity receptors for insulin (Havrankova et al., 1978; Waldbillig et al., 1987; Reiter and Gardner, 2003; Yu et al., 2004). However, little research has been done on these receptors since these early reports probably due the absence of an identified intracellular target. We have demonstrated that light stimulates tyrosine phosphorylation of the β-subunit of insulin receptor (IRβ) *in vivo*, which leads to the direct association of phosphoinositide 3-kinase (PI3K), an anti-apoptotic enzyme activity with the IRβ (Rajala et al., 2002). In this communication, we examine the localization of the insulin receptor and PI3K in DRMs prepared from bovine ROS.

2. METHODS

2.1. Preparation of Bovine ROS and Fractionation of DRMs from ROS Incubated with and without Methyl-β-Cyclodextrin (MCD)

Bovine ROS were prepared from retinas on continuous sucrose gradients (25-50%) using a modification of the method of Zimmerman and Godchaux (Godchaux, III and Zimmerman, 1979; Zimmerman and Godchaux, III, 1982) as previously described (Bell et al., 1999). DRMs were prepared according to a previously described modification (Elliott et al., 2003) of the method of Seno et al. (Seno et al., 2001). For experiments involving cholesterol depletion, ROS, in buffer B, were incubated in the presence or absence of 15 mM MCD for 1 h at 37°C prior to DRM fractionation as previously described (Elliott et al., 2003).

2.2. Tyrosine Phosphorylation of ROS and DRM

Tyrosine-phosphorylated ROS and DRMs were prepared by incubation of each in a phosphorylation buffer [50 mM Tris-HCl (pH 7.4), 100 mM NaCl, 2 mM $MgCl_2$, 1.5 mM ATP] in the presence and absence of 0.2 mM Na_3VO_3 for 15 minutes at 37°C as originally described for ROS (Rajala and Anderson, 2001). After incubation, ROS and DRM fractions were subjected to Western blot analysis.

2.3. SDS-PAGE and Western Blot Analysis

Proteins were resolved by 10% SDS-PAGE and transferred onto nitrocellulose membranes. The blots were incubated with anti-IRβ (1:250), anti-PY (1:1000), or anti-p85 (1:4000) antibodies overnight at 4°C and developed by ECL according to the manufacturer's instructions.

3. RESULTS

3.1. Localization of IRβ and p85 to DRMs Isolated from Bovine ROS

Low buoyant-density DRMs were isolated from Triton-solubilized bovine ROS by discontinuous sucrose density gradient centrifugation and fractions collected from the top of the gradients were subjected to SDS-PAGE and immunoblot analysis. Protein determinations of these fractions indicate the presence of two protein peaks, the minor DRM peak in

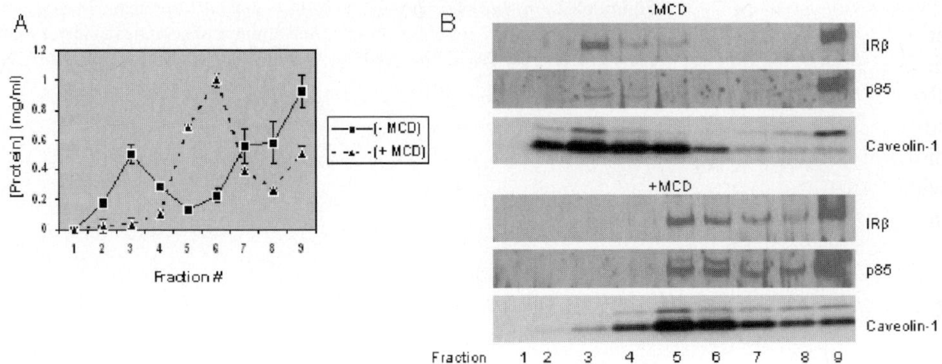

Figure 68.1. Effect of MCD on the localization of caveolin-1 and other ROS proteins in low-buoyant-density DRMs. ROS incubated in the absence (–MCD) or presence (+MCD) of 15 mM MCD were solubilized with 1% Triton X-100 and separated into low and high-density fractions by discontinuous sucrose density gradient centrifugation. Fractions collected from the top of each gradient (top to bottom shown from left to right) were subjected to either protein determination (A) or Western blot analysis (B) with antibodies against IRβ, p85 subunit of PI3K, and caveolin-1.

fraction 3 and the major peak in fraction 9 containing the majority of ROS proteins (Figure 68.1A, *solid line*). DRM fractions 2-5 were dramatically enriched in caveolin-1 with a peak in fraction 3 as previously observed (Nair et al., 2002; Elliott et al., 2003). In an attempt to determine whether DRMs isolated from ROS contained IRβ and p85, immunoblot analysis of DRM fractions was performed. The results indicate the presence of a significant pool of IRβ and p85 in the caveolin-enriched fraction 3 (Figure 68.1B). Fraction 9 was also immunoreactive for the presence of IRβ and p85 and this fraction contained the majority of ROS proteins (Figure 68.1A, *solid line*). These results indicate that portions of both IRβ and p85 are localized to photoreceptor DRMs.

To further confirm that the fractions isolated are authentic DRMs, ROS membranes were incubated with the cholesterol-sequestering agent, MCD, a treatment that disrupts cholesterol-rich DRMs. MCD treatment resulted in a dramatic loss of the low-buoyant-density band accompanied by a shift in the distribution of proteins to higher density sucrose fraction (Figure 68.1A and B). The concomitant shift in IRβ, p85 and caveolin-1 to higher density fractions following MCD treatment shows that their colocalization to DRM is disrupted by the cholesterol-depleting actions of MCD.

3.2. Absence of Phosphatase Activity in DRM Fractions

Bovine ROS prepared as described in the Methods have an endogenous tyrosine kinase activity (Bell et al., 1999). ROS and DRM fractions were subjected to *in vitro* phosphorylation resulting in the tyrosine phosphorylation of several ROS proteins (Figure 68.2). Omission of sodium vanadate resulted in the absence of tyrosine phosphorylation of ROS proteins, suggesting the inhibition of phosphatase activity in ROS by sodium vanadate. An enrichment of a tyrosine phosphorylated 80 kDa protein was observed in the DRM fraction. It is interesting to note that omission of sodium vanadate in the DRM fraction did not affect

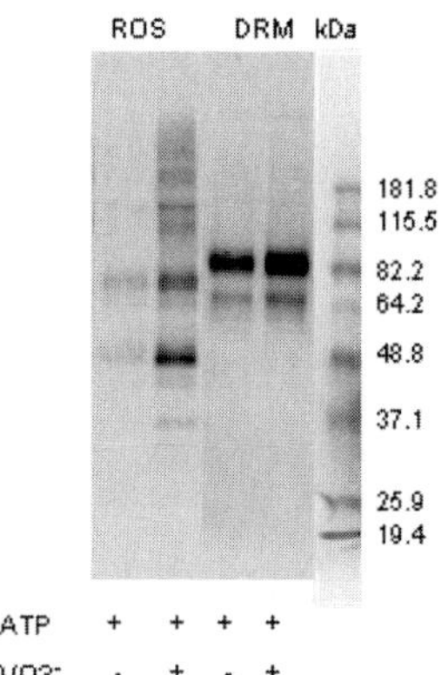

Figure 68.2. *In vitro* phosphorylation of ROS and DRM proteins. Phosphorylated ROS and DRM proteins (presence and absence of sodium vanadate) were subjected to SDS-PAGE followed by Western blot analysis with anti-PY99 antibody.

the tyrosine phosphorylation of the 80 kDa protein (Figure 68.2), suggesting the lack of phosphatase activity in DRM fractions.

4. DISCUSSION

Localization of insulin receptor in caveolae of adipocyte plasma membrane has been reported and cholesterol depletion attenuates insulin receptor signaling (Gustavsson et al., 1999). Insulin stimulation of cells prior to isolation of caveolae or insulin stimulation of the isolated caveolae fraction increased tyrosine phosphorylation of insulin receptor (Gustavsson et al., 1999). These results suggest that functional insulin receptor resides in caveolae and requires the caveolar environment for signaling. In addition, tyrosine phosphorylation of caveolin-1 was catalyzed by insulin receptor (Kimura et al., 2002). Further, caveolin-1 deficient mice show insulin resistance and defective insulin receptor protein expression in adipose tissue (Cohen et al., 2003). Baumann et al. (Baumann et al., 2000) reported that the resident lipid raft protein flotillin-1 recruits a complex of tyrosine-phosphorylated Cbl and Cbl-associated protein to rafts, and that this recruitment is required for GLUT-4 translocation in response to insulin. In addition, epidermal growth factor stimulation results in the localization of PI3K, Akt2, and PTEN to lipid rafts of intestinal cells, and this localization is important for sodium absorption and enterocyte differentiation (Li et al., 2004). Furthermore, Vero cells (Monkey, African Green kidney) stimulated with lysophosphatidic acid results in the localization of PI3K to lipid rafts (Peres et al., 2003).

We have demonstrated that light stimulates tyrosine phosphorylation of IRβ *in vivo*, which leads to the association of PI3K, an anti-apoptotic enzyme activity with the IRβ (Rajala et al., 2002). We hypothesize that an important mechanism of retinal neuroprotection is through light activation of the insulin receptor, which stimulates the anti-apoptotic PI3K/Akt pathway. The molecular mechanism behind the light-activation of IRβ is not known. In the present study, we have demonstrated the localization of IRβ and p85 subunit of PI3K to DRM fractions of bovine ROS. Perhaps light-induced IRβ activation of the anti-apoptotic PI3K/Akt pathway occurs within lipid rafts of ROS, consistent with a previous study indicating that cholesterol-rich lipid rafts mediate Akt-regulated survival in prostate cancer cells (Zhuang et al., 2002).

The insulin signaling pathway is activated by tyrosine phosphorylation of the insulin receptor and key post-receptor substrate proteins, and balanced by the action of specific protein-tyrosine phosphatases (PTPase). Inhibition of PTPase activity results in enhanced tyrosine phosphorylation of the insulin receptor (Mahadev et al., 2001). The most compelling evidence for a physiological role of phosphatase PTP1B in insulin action has been the recent demonstration of enhanced insulin sensitivity and potentiation on insulin-stimulated protein-tyrosine phosphorylation in PTP1B knockout mice (Klaman et al., 2000; Elchebly et al., 1999). Absence of phosphatase activity in DRM fractions could also stimulate insulin receptor phosphorylation. Studies are underway in our laboratory to test this hypothesis.

The phosphorylation we observed in DRMs (Figure 68.2) could be due to either receptor or non-receptor tyrosine kinases. It has been shown that the non-receptor tyrosine kinase Src phosphorylates insulin receptor on autophosphorylation sites (Yu et al., 1985; Peterson et al., 1996). Furthermore, activation of Src in rod photoreceptors cells has previously been reported (Ghalayini et al., 2002) and Src family kinases are present in detergent resistant cytoskeletal fractions (Ghalayini et al., 2002) and in DRMs isolated from bovine ROS (Martin et al., 2005). Earlier studies also demonstrated the *in vivo* tyrosine phosphorylation of caveolin by Src (Kimura et al., 2002). It is tempting to speculate that light-induced tyrosine phosphorylation of insulin receptor could be triggered in rafts by non-receptor tyrosine kinases, leading to the association with PI3K enzyme activity.

The phosphoinositide PI(4,5)P$_2$, the preferred substrate for PI3K, accumulates in membrane rafts and promotes local co-recruitment and activation of specific signaling components at the cell membrane (Caroni, 2001). Raft-localized PI(4,5)P$_2$ is regulated by lipid kinases (PI 5-Kinase) and phosphatases (eg. Synaptojanin) (Caroni, 2001; Chung et al., 1997). Localization of PI3K to the DRM fraction of ROS may have some important role in modulating PI(4,5)P$_2$ levels in lipid rafts, since the absence of protein phosphatase activity could be compensating for the presence of PI3K in the DRM fractions. Further studies, however, are required to examine whether light triggers the phosphorylation of IRβ or activation of PI3K activity in the DRM fractions.

5. ACKNOWLEDGEMENTS

This work was supported by grants from the National Institutes of Health (EY00871, EY04149, EY12190, EY15299 and RR17703); Research to Prevent Blindness, Inc. Raju V.S. Rajala is a recipient of Career Development Award from Research to Prevent Blindness, Inc.

6. REFERENCES

Baumann, C. A., Ribon, V., Kanzaki, M., Thurmond, D. C., Mora, S., Shigematsu, S., Bickel, P. E., Pessin, J. E., and Saltiel, A. R., 2000, CAP defines a second signalling pathway required for insulin-stimulated glucose transport, *Nature* **407**:202-207.

Bell, M. W., Alvarez, K., and Ghalayini, A. J., 1999, Association of the tyrosine phosphatase SHP-2 with transducin-alpha and a 97-kDa tyrosine-phosphorylated protein in photoreceptor rod outer segments, *J. Neurochem.* **73**:2331-2340.

Boesze-Battaglia, K., Dispoto, J., and Kahoe, M. A., 2002, Association of a photoreceptor-specific tetraspanin protein, ROM-1, with triton X-100-resistant membrane rafts from rod outer segment disk membranes, *J. Biol. Chem.* **277**:41843-41849.

Caroni, P., 2001, New EMBO members' review: actin cytoskeleton regulation through modulation of PI(4,5)P(2) rafts, *EMBO J.* **20**:4332-4336.

Chung, J. K., Sekiya, F., Kang, H. S., Lee, C., Han, J. S., Kim, S. R., Bae, Y. S., Morris, A. J., and Rhee, S. G., 1997, Synaptojanin inhibition of phospholipase D activity by hydrolysis of phosphatidylinositol 4,5-bisphosphate, *J. Biol. Chem.* **272**:15980-15985.

Cohen, A. W., Razani, B., Wang, X. B., Combs, T. P., Williams, T. M., Scherer, P. E., and Lisanti, M. P., 2003, Caveolin-1-deficient mice show insulin resistance and defective insulin receptor protein expression in adipose tissue, *Am. J. Physiol Cell Physiol* **285**:C222–C235.

Elchebly, M., Payette, P., Michaliszyn, E., Cromlish, W., Collins, S., Loy, A. L., Normandin, D., Cheng, A., Himms-Hagen, J., Chan, C. C., Ramachandran, C., Gresser, M. J., Tremblay, M. L., and Kennedy, B. P., 1999, Increased insulin sensitivity and obesity resistance in mice lacking the protein tyrosine phosphatase-1B gene, *Science* **283**:1544-1548.

Elliott, M. H., Fliesler, S. J., and Ghalayini, A. J., 2003, Cholesterol-dependent association of caveolin-1 with the transducin alpha subunit in bovine photoreceptor rod outer segments: disruption by cyclodextrin and guanosine 5'-O-(3-thiotriphosphate), *Biochemistry* **42**:7892-7903.

Ghalayini, A. J., Desai, N., Smith, K. R., Holbrook, R. M., Elliott, M. H., and Kawakatsu, H., 2002, Light-dependent association of Src with photoreceptor rod outer segment membrane proteins in vivo, *J. Biol. Chem.* **277**:1469-1476.

Godchaux, W., III and Zimmerman, W. F., 1979, Soluble proteins of intact bovine rod cell outer segments, *Exp. Eye Res.* **28**:483-500.

Gustavsson, J., Parpal, S., Karlsson, M., Ramsing, C., Thorn, H., Borg, M., Lindroth, M., Peterson, K. H., Magnusson, K. E., and Stralfors, P., 1999, Localization of the insulin receptor in caveolae of adipocyte plasma membrane, *FASEB J.* **13**:1961-1971.

Havrankova, J., Roth, J., and Brownstein, M., 1978, Insulin receptors are widely distributed in the central nervous system of the rat, *Nature* **272**:827-829.

Jacobson, K. and Dietrich, C., 1999, Looking at lipid rafts?, *Trends Cell Biol.* **9**:87-91.

Kimura, A., Mora, S., Shigematsu, S., Pessin, J. E., and Saltiel, A. R., 2002, The insulin receptor catalyzes the tyrosine phosphorylation of caveolin-1, *J. Biol. Chem.* **277**:30153-30158.

Klaman, L. D., Boss, O., Peroni, O. D., Kim, J. K., Martino, J. L., Zabolotny, J. M., Moghal, N., Lubkin, M., Kim, Y. B., Sharpe, A. H., Stricker-Krongrad, A., Shulman, G. I., Neel, B. G., and Kahn, B. B., 2000, Increased energy expenditure, decreased adiposity, and tissue-specific insulin sensitivity in protein-tyrosine phosphatase 1B-deficient mice, *Mol. Cell Biol.* **20**:5479-5489.

Li, X., Leu, S., Cheong, A., Zhang, H., Baibakov, B., Shih, C., Birnbaum, M. J., and Donowitz, M., 2004, Akt2, phosphatidylinositol 3-kinase, and PTEN are in lipid rafts of intestinal cells: role in absorption and differentiation, *Gastroenterology* **126**:122-135.

Mahadev, K., Zilbering, A., Zhu, L., and Goldstein, B. J., 2001, Insulin-stimulated hydrogen peroxide reversibly inhibits protein-tyrosine phosphatase 1b in vivo and enhances the early insulin action cascade, *J. Biol. Chem.* **276**:21938-21942.

Martin, R. E., Elliott, M. H., Brush, R. S., and Anderson, R. E., 2005, Detailed characterization of the lipid composition of detergent-resistant membranes from photoreceptor rod outer segment membranes, *Invest Ophthalmol Vis Sci* **In press**.

Melkonian, K. A., Ostermeyer, A. G., Chen, J. Z., Roth, M. G., and Brown, D. A., 1999, Role of lipid modifications in targeting proteins to detergent-resistant membrane rafts. Many raft proteins are acylated, while few are prenylated, *J. Biol. Chem.* **274**:3910-3917.

Nair, K. S., Balasubramanian, N., and Slepak, V. Z., 2002, Signal-dependent translocation of transducin, RGS9-1-Gbeta5L complex, and arrestin to detergent-resistant membrane rafts in photoreceptors, *Curr. Biol.* **12**:421-425.

Nair, K. S., Hanson, S. M., Kennedy, M. J., Hurley, J. B., Gurevich, V. V., and Slepak, V. Z., 2004, Direct binding of visual arrestin to microtubules determines the differential subcellular localization of its splice variants in rod photoreceptors, *J. Biol. Chem.* **279**:41240-41248.

Peres, C., Yart, A., Perret, B., Salles, J. P., and Raynal, P., 2003, Modulation of phosphoinositide 3-kinase activation by cholesterol level suggests a novel positive role for lipid rafts in lysophosphatidic acid signalling, *FEBS Lett.* **534**:164-168.

Peterson, J. E., Kulik, G., Jelinek, T., Reuter, C. W., Shannon, J. A., and Weber, M. J., 1996, Src phosphorylates the insulin-like growth factor type I receptor on the autophosphorylation sites. Requirement for transformation by src, *J. Biol. Chem.* **271**:31562-31571.

Rajala, R. V. and Anderson, R. E., 2001, Interaction of the insulin receptor beta-subunit with phosphatidylinositol 3-kinase in bovine ROS, *Invest Ophthalmol. Vis. Sci.* **42**:3110-3117.

Rajala, R. V., McClellan, M. E., Ash, J. D., and Anderson, R. E., 2002, In vivo regulation of phosphoinositide 3-kinase in retina through light-induced tyrosine phosphorylation of the insulin receptor beta-subunit, *J. Biol. Chem.* **277**:43319-43326.

Reiter, C. E. and Gardner, T. W., 2003, Functions of insulin and insulin receptor signaling in retina: possible implications for diabetic retinopathy, *Prog. Retin. Eye Res.* **22**:545-562.

Senin, I. I., Hoppner-Heitmann, D., Polkovnikova, O. O., Churumova, V. A., Tikhomirova, N. K., Philippov, P. P., and Koch, K. W., 2004, Recoverin and rhodopsin kinase activity in detergent-resistant membrane rafts from rod outer segments, *J. Biol. Chem.* **279**:48647-48653.

Seno, K., Kishimoto, M., Abe, M., Higuchi, Y., Mieda, M., Owada, Y., Yoshiyama, W., Liu, H., and Hayashi, F., 2001, Light- and guanosine 5′-3-O-(thio)triphosphate-sensitive localization of a G protein and its effector on detergent-resistant membrane rafts in rod photoreceptor outer segments, *J. Biol. Chem.* **276**:20813-20816.

Simons, K. and Ikonen, E., 1997, Functional rafts in cell membranes, *Nature* **387**:569-572.

Simons, K. and van Meer, G., 1988, Lipid sorting in epithelial cells, *Biochemistry* **27**:6197-6202.

Waldbillig, R. J., Fletcher, R. T., Chader, G. J., Rajagopalan, S., Rodrigues, M., and LeRoith, D., 1987, Retinal insulin receptors. 1. Structural heterogeneity and functional characterization, *Exp. Eye Res.* **45**:823-835.

Yamada, E., 1955, The fine structure of the gall bladder epithelium of the mouse, *J. Biophys. Biochem. Cytol.* **1**:445-458.

Yu, K. T., Werth, D. K., Pastan, I. H., and Czech, M. P., 1985, src kinase catalyzes the phosphorylation and activation of the insulin receptor kinase, *J. Biol. Chem.* **260**:5838-5846.

Yu, X., Rajala, R. V., McGinnis, J. F., Li, F., Anderson, R. E., Yan, X., Li, S., Elias, R. V., Knapp, R. R., Zhou, X., and Cao, W., 2004, Involvement of insulin/phosphoinositide 3-kinase/Akt signal pathway in 17 beta-estradiol-mediated neuroprotection, *J. Biol. Chem.* **279**:13086-13094.

Zhuang, L., Lin, J., Lu, M. L., Solomon, K. R., and Freeman, M. R., 2002, Cholesterol-rich lipid rafts mediate akt-regulated survival in prostate cancer cells, *Cancer Res.* **62**:2227-2231.

Zimmerman, W. F. and Godchaux, W., III, 1982, Preparation and characterization of sealed bovine rod cell outer segments, *Methods Enzymol.* **81**:52-57.

CHAPTER 69

MERTK ACTIVATION DURING RPE PHAGOCYTOSIS *IN VIVO* REQUIRES αVβ5 INTEGRIN

Silvia C. Finnemann[1,2] and Emeline F. Nandrot[1]*

1. INTRODUCTION

Daily phagocytosis of shed photoreceptor outer segment fragments (POS) is a key task of the retinal pigment epithelium (RPE) in the retina. Lack or inefficiency of daily POS clearance causes early onset, rapid, and complete retinal degeneration in experimental animals and likely contributes to human blinding diseases such as retinitis pigmentosa and age-related macular degeneration (Dowling and Sidman, 1962, Gal et al., 2000). The phagocytic mechanism of the RPE belongs to a group of conserved non-inflammatory clearance pathways that mediate recognition and engulfment of apoptotic cells in both non-professional and professional phagocytic cells, such as fibroblasts and macrophages, respectively (Finnemann and Rodriguez-Boulan, 1999). These pathways share the use of phagocyte cell surface receptors such as the lipid scavenger receptor CD36 (Ryeom et al., 1996), the integrin adhesion receptor αvβ5 (Finnemann et al., 1997; Miceli et al., 1997; Lin and Clegg, 1998) and the receptor tyrosine kinase Mer (MerTK) (D'Cruz et al., 2000; Nandrot et al., 2000). *In vitro* phagocytosis assays studying primary or permanent RPE in culture fed with isolated POS suggest that CD36 and MerTK participate in the engulfment step of the phagocytic process (Chaitin and Hall, 1983; Finnemann and Silverstein, 2001), while αvβ5 integrin promotes POS recognition/binding and initiates a downstream cytoplasmic signaling cascade in the RPE (Finnemann et al., 1997). However, the precise function of these receptors and their roles in the intact retina are so far only poorly understood. Most recently, we have begun to study phagocytosis and receptor activity in animal models that lack αvβ5 integrin or MerTK to determine how these different plasma membrane receptors of the RPE functionally interact to coordinate particle uptake.

*Margaret M. Dyson Vision Research Institute, [1]Department of Ophthalmology and [2]Department of Cell and Developmental Biology, Weill Medical College of Cornell University, Box 233, 1300 York Avenue, New York, NY10021, USA.

499

2. ROLE OF MERTK ACTIVATION IN RPE PHAGOCYTOSIS

Activity of the Mer tyrosine kinase receptor MerTK is essential for efficient engulfment of POS by RPE *in vivo* and *in vitro* (Mullen and LaVail, 1976; Edwards and Szamier, 1977) Despite its importance, mechanisms of MerTK activation and MerTK downstream signaling target proteins in RPE are still largely obscure. Retinal ligands of MerTK have not yet been conclusively identified. Moreover, we still do not know which RPE proteins serve as substrates for MerTK's kinase activity during RPE phagocytosis. However, both endogenous and overexpressed MerTK reveal a striking redistribution to the sites of internalized POS in *in vitro* phagocytosis assays suggesting that MerTK receptors may be components of the phagocytic machinery of the RPE (Feng et al., 2002; Finnemann, 2003). Furthermore, challenge with isolated POS causes increased phosphorylation at tyrosine residues of MerTK in RPE in culture (Feng et al., 2002; Finnemann, 2003). Although their mutual dependence has not been demonstrated directly, levels of MerTK tyrosine phosphorylation commonly serve to assess the extent of MerTK activity.

3. FOCAL ADHESION KINASE SIGNALING ACTIVATES MERTK DURING RPE PHAGOCYTOSIS

Our previous studies on phagocytic signaling in RPE suggest an important role for focal adhesion kinase (FAK) in MerTK activation. FAK is a cytoplasmic non-receptor tyrosine kinase that colocalizes with integrin receptors at focal contacts where it commonly transduces signaling pathways downstream of activated integrins (for a recent review on FAK see (Parsons, 2003). Reversible activation of FAK is critical for integrin functions that involve cytoskeletal reorganization (Ilic et al., 1995). We expressed a C-terminal fragment of FAK that competes with full-length endogenous FAK for cytoskeletal anchorage. This fragment has been shown to act as a dominant-negative inhibitor of endogenous FAK abrogating FAK downstream signal transduction. We showed that the rat derived RPE-J cell line, like primary wild-type rat RPE, utilizes endogenous MerTK to engulf POS. Importantly, expression of the dominant-negative FAK C-terminal fragment in RPE-J cells inhibited POS engulfment (Finnemann, 2003). Furthermore, it eliminated the increase in MerTK tyrosine phosphorylation that is elicited by phagocytic challenge of RPE cells in culture (Finnemann, 2003). On the contrary, RPE cultures derived from RCS rats retained similar FAK activation as wild-type Long Evans rat RPE cultures in response to OS challenge. These results identify a novel signal transduction pathway in which FAK acts upstream of MerTK to stimulate the internalization machinery of the RPE.

4. αvβ5 SIGNALING VIA FOCAL ADHESION KINASE ACTIVATES MERTK DURING RPE PHAGOCYTOSIS *IN VIVO* AND *IN VITRO*

Previous studies have shown that phosphorylation of FAK at tyrosine residue 861 promotes direct binding of FAK to the cytoplasmic face of αvβ5 integrin receptors (Eliceiri et al., 2002). In our studies, we observed increased levels of FAK in αvβ5 protein complexes isolated by immunoprecipitation from RPE-J cells during the early POS binding phase but loss of FAK from the integrin complex during the later POS internalization phase

(Finnemann, 2003). Residence of FAK in the complex correlated well with elevated phosphorylation of tyrosine 861, while phosphorylation of other tyrosine residues that indicate FAK enzymatic activity persisted beyond the time of FAK in the integrin complex. These results suggest that RPE cells activate FAK recruited to its apical αvβ5 surface receptors in response to POS phagocytic challenge *in vitro*.

To directly determine whether αvβ5 integrin receptors were required for FAK and MerTK activation in RPE, we tested FAK and MerTK activation during RPE phagocytosis by RPE cells of β5 integrin knockout mice that lack all αvβ5 integrin receptors. β5 null RPE cells in culture largely fail to phagocytose isolated POS (Nandrot et al., 2004). β5 null retina lacks the synchronized burst of RPE phagocytosis that characteristically follows early morning rod shedding in rodent retina (Nandrot et al., 2004). The detrimental effects of this abnormal timing of phagocytosis on retinal function in β5 integrin null mice of age are discussed in more detail in the chapter by Nandrot and Finnemann in this volume.

When we fed isolated POS to wild-type mouse RPE, we found robust FAK and MerTK activation confirming our earlier results using stable and primary rat RPE (shown for MerTK in Figure 69.1 a, β5+/+). In contrast, β5 null RPE cells in primary culture did not increase tyrosine phosphorylation of either FAK or MerTK in response to POS, although they expressed both proteins at normal levels (shown for MerTK in Figure 69.1 a, β5–/–). Fur-

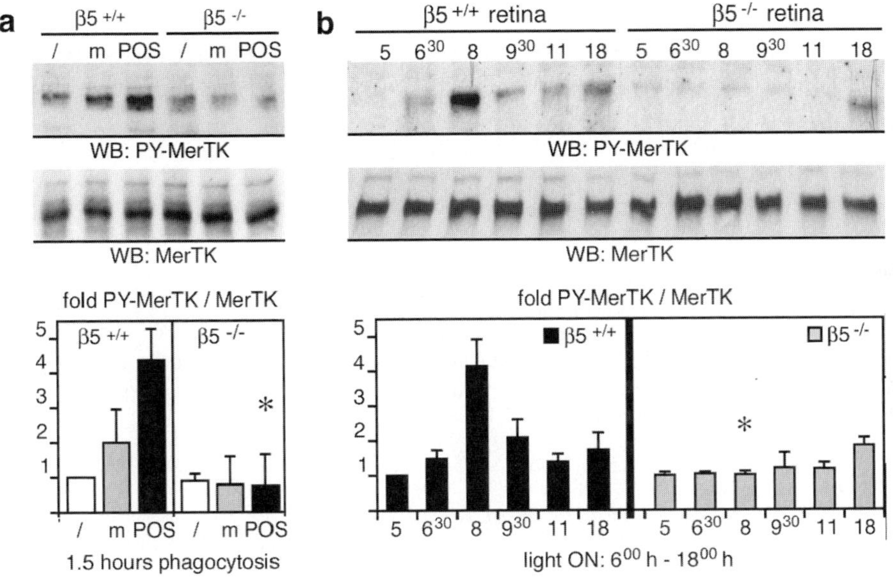

Figure 69.1. MerTK activation requires αvβ5 integrin. a. Primary RPE in culture from β5+/+ or β5–/– mice received isolated POS (POS) or assay medium alone (m) for 1.5 hours before lysis of cells. b. Eyecups were harvested from 3 week old strain-matched β5+/+ and β5–/– mice at different times of day as indicated and protein lysates prepared immediately. a and b. Lysates were analyzed by SDS-PAGE and immunoblotting for MerTK protein and tyrosine-phosphorylated MerTK (PY-MerTK). Band intensities were quantified to calculate relative levels of MerTK phosphorylation (= activation). Bars represent means ± SD, n = 3. Significant differences between equivalent β5+/+ and β5–/– values were determined by Student's *t*-test and are indicated by asterisks (P < 0.001 for a, P < 0.05 for b). Modified from Nandrot et al. (2004). (Reproduced from The Journal of Experimental Medicine, 2004, Vol. 200, pgs. 1539–1545 by copyright permission of The Rockcfeller University Press).

thermore, *in vivo* phagocytic signaling via FAK and MerTK was strongly and transiently stimulated following light onset in wild-type mouse retina but was absent in $\beta 5$ knockout mouse retina (shown for MerTK in Figure 69.1 b). These data provide conclusive evidence that $\alpha v\beta 5$ integrin signaling regulates rhythmic activation of FAK and MerTK during RPE phagocytosis in the intact retina.

5. PERSPECTIVE

The results discussed here are the first to describe signaling activities by phagocytic receptors in intact retina that precisely correlate temporally with POS shedding and uptake by RPE cells. The late onset retinal dysfunction of the $\beta 5$ knockout mouse as a consequence of lack of such phagocytic signaling emphasizes the importance of precise temporal regulation of POS phagocytosis by the RPE. In future studies, we will use similar experimental approaches exploring *in vivo* signaling in normal and mutant animal models to identify further components of the RPE phagocytic mechanism and to unravel their functional interactions.

6. ACKNOWLEDGMENTS

This work was supported by NIH grants EY13295 and EY14184, by a Karl Kirchgessner research grant, and by the Irma T. Hirschl/Monique Weill-Caulier Trust.

7. REFERENCES

Chaitin, M. H., and Hall, M. O., 1983, Defective ingestion of rod outer segments by cultured dystrophic rat pigment epithelial cells. *Invest. Ophthalmol. Vis. Sci.* **24**:812-820.

D'Cruz, P. M., Yasumura, D., Weir, J., Matthes, M. T., Abderrahim, H., LaVail, M. M., and Vollrath, D., 2000, Mutation of the receptor tyrosine kinase gene Mertk in the retinal dystrophic RCS rat. *Hum. Mol. Genet.* **9**:645-651.

Dowling, J. E., and Sidman, R. L., 1962, Inherited retinal dystrophy of the rat. *J. Cell Biol.* **14**:73-109.

Edwards, R. B., and Szamier, R. B., 1977, Defective phagocytosis of isolated rod outer segments by RCS rat retinal pigment epithelium in culture. *Science.* **197**:1001-1003.

Eliceiri, B. P., Puente, X. S., Hood, J. D., Stupack, D. G., Schlaepfer, D. D., Huang, X. Z., Sheppard, D., and Cheresh, D. A., 2002, Src-mediated coupling of focal adhesion kinase to integrin $\alpha v\beta 5$ in vascular endothelial growth factor signaling. *J. Cell Biol.* **157**:149-160.

Feng, W., Yasumura, D., Matthes, M. T., LaVail, M. M., and Vollrath, D., 2002, Mertk triggers uptake of photoreceptor outer segments during phagocytosis by cultured retinal pigment epithelial cells. *J. Biol. Chem.* **277**:17016-17022.

Finnemann, S. C., 2003, Focal adhesion kinase signaling promotes phagocytosis of integrin-bound photoreceptors. *EMBO J.* **22**:4143-4154.

Finnemann, S. C., Bonilha, V. L., Marmorstein, A. D., and Rodriguez-Boulan, E., 1997, Phagocytosis of rod outer segments by retinal pigment epithelial cells requires $\alpha v\beta 5$ integrin for binding but not for internalization. *Proc. Natl. Acad. Sci. U. S. A.* **94**:12932-12937.

Finnemann, S. C., and Rodriguez-Boulan, E., 1999, Macrophage and retinal pigment epithelium phagocytosis: apoptotic cells and photoreceptors compete for $\alpha v\beta 3$ and $\alpha v\beta 5$ integrins, and protein kinase C regulates $\alpha v\beta 5$ binding and cytoskeletal linkage. *J. Exp. Med.* **190**:861-874.

Finnemann, S. C., and Silverstein, R. L., 2001, Differential roles of CD36 and $\alpha v\beta 5$ integrin in photoreceptor phagocytosis by the retinal pigment epithelium. *J. Exp. Med.* **194**:1289-1298.

Gal, A., Li, Y., Thompson, D. A., Weir, J., Orth, U., Jacobson, S. G., Apfelstedt-Sylla, E., and Vollrath, D., 2000, Mutations in MERTK, the human orthologue of the RCS rat retinal dystrophy gene, cause retinitis pigmentosa, *Nat. Genet.* **26**:270-271.

Ilic, D., Furuta, Y., Kanazawa, S., Takeda, N., Sobue, K., Nakatsuji, N., Nomura, S., Fujimoto, J., Okada, M., and Yamamoto, T., 1995, Reduced cell motility and enhanced focal adhesion contact formation in cells from FAK-deficient mice. *Nature.* **377**:539-544.

Lin, H., and Clegg, D. O., 1998, Integrin αvβ5 participates in the binding of photoreceptor rod outer segments during phagocytosis by cultured human retinal pigment epithelium. *Invest. Ophthalmol. Vis. Sci.* **39**:1703-1712.

Miceli, M. V., Newsome, D. A., and Tate, Jr., D. J., 1997, Vitronectin is responsible for serum-stimulated uptake of rod outer segments by cultured retinal pigment epithelial cells. *Invest. Ophthalmol. Vis. Sci.* **38**:1588-1597.

Mullen, R. J., and LaVail, M. M., 1976, Inherited retinal dystrophy: primary defect in pigment epithelium determined with experimental rat chimeras. *Science.* **192**:799-801.

Nandrot, E., Dufour, E. M., Provost, A. C., Pequignot, M. O., Bonnel, S., Gogat, K., Marchant, D., Rouillac, C., Sepulchre de Conde, B., Bihoreau, M. T., Shaver, C., Dufier, J. L., Marsac, C., Lathrop, M., Menasche, M., and Abitbol, M. M., 2000, Homozygous deletion in the coding sequence of the c-mer gene in RCS rats unravels general mechanisms of physiological cell adhesion and apoptosis. *Neurobiol. Dis.* **7**:586-599.

Nandrot, E. F., Kim, Y., Brodie, S. E., Huang, X., Sheppard, D., and Finnemann, S. C., 2004, Loss of synchronized retinal phagocytosis and age-related blindness in mice lacking αvβ5 integrin. *J. Exp. Med.* **200**:1539-1545.

Parsons, J. T., 2003, Focal adhesion kinase: the first ten years. *J. Cell Sci.* **116**:1409-1416.

Ryeom, S. W., Sparrow, J. R., and Silverstein, R. L., 1996, CD36 participates in the phagocytosis of rod outer segments by retinal pigment epithelium. *J. Cell Sci.* **109**:387-395.

CHAPTER 70

PHOTORECEPTOR RETINOL DEHYDROGENASES
An attempt to characterize the function of Rdh11

Anne Kasus-Jacobi, David G. Birch, and Robert E. Anderson*

1. INTRODUCTION

Vertebrate vision begins with the absorption of light by visual pigments in photoreceptor cells. Visual pigments, or opsins, are seven membrane spanning, G protein-coupled receptors located in the membrane of the outer segment discs of rods and cones. In the dark, the light sensitive chromophore 11-*cis*-retinal is covalently attached to opsin through a Schiff base linkage to a specific lysine residue located in the center of the seventh transmembrane alpha helix. Light stimulation results in isomerization of 11-*cis*-retinal to all-*trans*-retinal, which causes a change in the conformation of rhodopsin. The resulting photoactivated metarhodopsin II interacts with the G protein transducin and triggers the phototransduction cascade leading to hyperpolarization of photoreceptors and ultimately to inhibition of neurotransmitter release at the synaptic terminus. After isomerization of 11-*cis*-retinal to the *trans* configuration, the Schiff base is hydrolyzed and the photolyzed chromophore separates from opsin. Whether all-*trans*-retinal is released in the lumen of the discs and subsequently transported to the cytosol by the retinal ATP-binding cassette transporter (ABCR)[1] or directly released into the cytosol[2] is controversial. Cytosolic all-*trans*-retinal is then reduced to all-*trans*-retinol by a retinol dehydrogenase (RDH) located in the membrane of the photoreceptor outer segment discs. This or these enzymes have not yet been identified. However, six distinct RDHs expressed in photoreceptors have recently been cloned (Table 70.1). Their functions, *in vivo,* are unknown, but all of them were shown to reduce all-*trans*-retinal *in vitro*.

Several lines of evidence suggest that reduction of all-*trans*-retinal in photoreceptor cells is crucial to maintain the functionality and integrity of the retina. This reaction is the first step of the biochemical pathway called the visual cycle, which is essential for a sus-

*Anne Kasus-Jacobi and Robert E. Anderson, University of Oklahoma Health Sciences Center, Dean A. McGee Eye Institute, Oklahoma City, OK 73104. David G. Birch, The Retina Foundation of the Southwest, Dallas, TX 75231. Corresponding author: A. Kasus-Jacobi, E-mail: anne-kasus-jacobi@ouhsc.edu.

505

Table 70.1. Photoreceptor retinol dehydrogenases. All listed RDHs are from human, except mouse Rdh11. Accession numbers are as follow: Rdh11 (AF474027); RDH12 (AAH25724); RDH13 (AAH09881); RDH14 (AAH09830); retSDR1 (O75911) and prRDH (AF229845). PR, photoreceptor; IS, inner segment; OS, outer segment; LCA, Leber congenital amaurosis.

Name	Localization	Activity, coenzyme (*in vitro* assay)	Disease	% Identity to Rdh11
Rdh11	PR (IS)	*trans*- and *cis*-retinal reductase, NADPH	–	100
RDH12	PR (?)	*trans*- and *cis*-retinal reductase, NADPH	LCA	70
RDH13	PR (IS)	None detected	–	38
RDH14	PR (OS)	*trans*- and *cis*-retinal reductase, NADPH	–	44
RetSDR1	Cone (OS)	all-*trans*-retinal reductase, NADPH	–	22
prRDH	PR (OS)	all-*trans*-retinal reductase, NADPH	–	22

tained phototransduction.[3] This pathway takes place in photoreceptor and retinal pigment epithelium (RPE) cells and allows the recycling of all-*trans*-retinal to 11-*cis*-retinal (see Figure 70.1). When all-*trans*-retinol is produced in photoreceptors from the reduction of all-*trans*-retinal, it is transported into the RPE where it is esterified by lecithin retinol acyl transferase (LRAT) and stored as all-*trans*-retinyl ester. All-*trans*-retinol is also supplied to the RPE by the choroidal vasculature, entering the RPE, in a receptor-mediated process involving a serum retinol-binding protein/transthyretin complex.[4] Retinyl esters stored in the RPE are the substrate for isomerohydrolase (IMH),[5] an enzyme proposed to catalyze the concerted hydrolysis of all-*trans*-retinyl ester and the isomerization to 11-*cis*-retinol. Oxidation of 11-*cis*-retinol to 11-*cis*-retinal by the 11-*cis*-retinol dehydrogenase RDH5 in the RPE completes the visual cycle. 11-*cis*-Retinal is transported back to the photoreceptors where it combines with opsin to regenerate photosensitive rhodopsin. The first step of the visual cycle is important because it generates all-*trans*-retinol, used to replenish the store of retinyl ester in the RPE. However, it is not the only source of all-*trans*-retinol since circulating all-*trans*-retinol can be used alternatively.

Reduction of all-*trans*-retinal is important because all-*trans*-retinal is a reactive molecule that can form toxic adducts like A2E,[6] mediate photodamage,[7] bind and activate opsin,[8] or inhibit photoreceptor ion channels,[9-11] These effects are triggered by light and are theoretically dependent on the rate of all-*trans*-retinal reduction, which is a slow process that takes tens of minutes in rods.[12] As pointed out above, the identity of enzyme(s) catalyzing this reaction in photoreceptors is unknown but the recently cloned photoreceptor RDHs are good candidates.

2. PHOTORECEPTOR RDHs

All six RDHs listed in Table 70.1 belong to the short-chain dehydrogenase/reductases (SDR) family of oxidoreductases. Members of this family are one-domain NAD(P)(H)-dependent enzymes of 250 to 300 amino acid residues. The family is highly divergent, with typically 15%-30% residue identity in pairwise comparisons. The criterion for SDR membership is the occurrence of conserved sequence motifs, arranged in a specific manner.[13] In humans, about 60 members of this family have been identified in the genome.

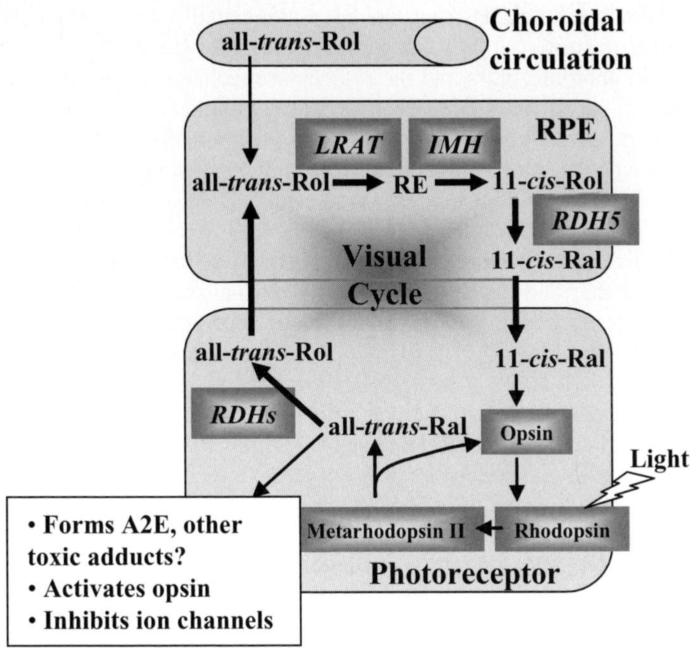

Figure 70.1. Retinoid metabolism in photoreceptor and RPE cells and effects of all-*trans*-retinal in photoreceptors. Reactions of the visual cycle are indicated with bolded arrows. Ral, retinal; Rol, retinol, RDH, retinol dehydrogenase; LRAT, lecithin retinol acyl transferase; IMH, isomerohydrolase; RE, retinyl ester.

Some of them have been associated with important functions and lead to various diseases if mutated.[14] The function of several members of this family, including photoreceptor RDHs, is still unknown.

Mouse Rdh11 has been cloned as a gene regulated by the transcription factors sterol regulatory element-binding proteins (SREBPs).[15] It is 85% identical to its human ortholog, RDH11, a protein that was first discovered as a gene that is expressed at very high levels in human prostate.[16] As revealed by immunofluorescence, Rdh11 is expressed in four layers of the mouse retina, including photoreceptor inner segments.[15] Absence of Rdh11 in the outer segment of photoreceptors was confirmed by fractionation of the retina on sucrose gradient, separating the outer segments from the rest of the retina, followed by immunoblotting (manuscript submitted for publication). Using a monoclonal antibody generated against human RDH11, immunofluorescence in monkey and bovine eye sections revealed a signal mostly located in the retinal pigment epithelium (RPE). Only a faint signal was detected in photoreceptor inner segments.[17] Human and mouse catalytic activities have been characterized *in vitro*. Both enzymes are able to reduce all-*trans*- and *cis*-retinal with low Km ranging from 0.1 to 1 μM, and specifically use NADPH as coenzyme.[15,18]

RDH12, 13, and 14 were first identified in nucleic acid and protein sequence databases by similarity with the previously identified RDH11 sequence.[17] Their localization in photoreceptors was shown by *in situ* hybridization (for RDH12) and immunofluorescence (for RDH13 and 14).[17] RDH12 is the gene the most closely related to RDH11; it is also the only gene among photoreceptor RDHs that has been associated with a retinopathy, the severe early-onset retinal dystrophy Leber Congenital Amaurosis (LCA).[19,20] The localization of RDH12 protein in photoreceptors and the molecular mechanism leading to the disease are unknown. However, given its similarity with RDH11, understanding the function of the latter, *in vivo*, might give some clues regarding RDH12 function.

RetSDR1, predominantly localized in cone outer segments, was first identified by searching an EST database from human retina with a DNA sequence corresponding to a conserved domain among RDHs.[21]

prRDH, localized in rod and cone outer segments, was identified from a cDNA library from bovine retina that had been simultaneously normalized and subtracted with bovine brain cDNA, in order to enrich it in genes expressed specifically in the retina.[2]

The subcellular localization of RDHs in photoreceptors is an important indication for their function. RDH14, retSDR1, and prRDH are located in the outer segment of photoreceptors, and therefore are likely candidates for the catalysis of the first step of the visual cycle. On the other hand, Rdh11, RDH13, and possibly RDH12, are located in the inner segment, which suggests a distinct function.

3. CHARACTERIZATION OF RDH11 KNOCKOUT MICE

Rdh11 is the first photoreceptor RDH to be studied *in vivo* (manuscript submitted for publication). Rdh11 knockout mouse was created by replacing *Rdh11* coding sequence with the *LacZ* reporter gene for expression profiling.[23] X-Gal staining of retinal section from *Rdh11*[+/−] mice confirmed an active transcription of this gene, only in photoreceptor cells. *Rdh11*[−/−] mice appeared normal and fertile, producing litters of normal size.

The visual phenotype of these mice was investigated by electroretinography (ERGs). These experiments revealed that the dark adaptation of knockout mice is delayed by a factor 2.5 to 3 compared to wild types. This result confirms that Rdh11 is involved in vision, more specifically during the process of dark adaptation. After illumination and return to the dark, a number of pathways are activated in photoreceptors to allow their return to the dark adapted state, which is the state of full sensitivity to light.[24] This relatively slow process comprises the regeneration of 11-*cis*-retinal through the visual cycle. However, none of the intermediates of the cycle was significantly changed in the Rdh11 knockout mice during dark adaptation, suggesting that the defect in dark adaptation is not due to a defect in the visual cycle (manuscript submitted for publication).

Rdh11 reduces all-*trans*-retinal *in vitro*; therefore, a disruption of Rdh11 is expected to create a delayed clearance of all-*trans*-retinal in inner segments during dark adaptation. However, it is difficult to demonstrate *in vivo*, because the portion of all-*trans*-retinal located in inner segments is small compared to the large amounts released in outer segments during illumination. A local change of all-*trans*-retinal in inner segments is not expected to significantly change the total amount. Indeed, there is no significant increase of all-*trans*-retinal amount in Rdh11 knockout mice during dark adaptation, at least when retinoids were extracted from whole eyes (see Figure 70.2).

In vitro experiments using inner segment membrane fractions collected from wild type and Rdh11 knockout retinas would be a more sensitive assay to confirm a decreased rate of all-*trans*-retinal reduction in Rdh11 knockout mice. If confirmed, such a difference could have significant effects leading to the delay of dark adaptation.

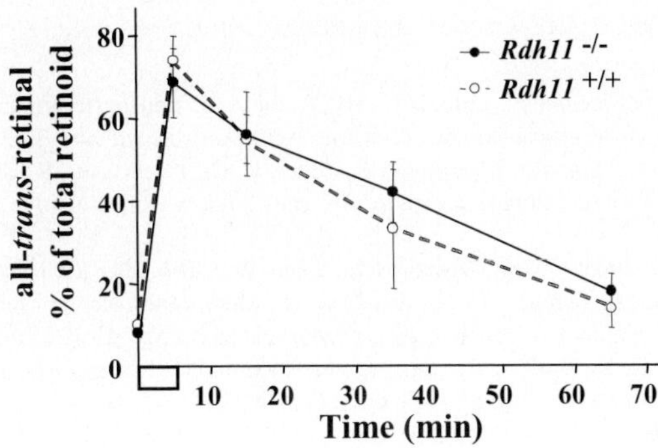

Figure 70.2. Changes in all-*trans*-retinal amount in whole eyes of living mice, before and after 5 min illumination (white box). All mice were first dark-adapted for a minimum of 12 hours. Each pair of eyes was enucleated and extracted before or after illumination, and at different times of recovery in the dark. All-*trans*-retinal amounts are shown as percent of total retinoid extracted from a pair of eyes. Error bars indicate the standard deviation (n = 5).

For example, light induced hyperpolarization of photoreceptor triggers the closure of L-type voltage-gated Ca^{2+} channels located in rod inner segments. In addition to the electrical signal, all-*trans*-retinal has been shown to directly inhibit these channels, at micromolar concentrations.[11] In the inner segment, Ca^{2+} regulates synaptic transmission, cell metabolism, cytoskeletal dynamics, gene expression and cell death[25]. Inhibition of such channels by all-*trans*-retinal, leading to a modification of Ca^{2+} homeostasis in the inner segment, could explain how a local increase of all-*trans*-retinal can change the kinetics of dark adaptation.

Ca^{2+} homeostasis can be tested in Rdh11 knockout mice and if a modification is found, it will demonstrate the importance of RDHs localized in photoreceptor inner segments. This hypothesis will be particularly interesting to consider in the case of the retinal dystrophy caused by RDH12 mutations, if in the future it is established that RDH12 is located in the inner segment of photoreceptors.

In summary, we have confirmed the localization of Rdh11 in photoreceptor inner segments and shown that a disruption of this gene leads to a delayed dark adaptation in mice. The molecular mechanism leading to this functional defect is currently under investigation.

4. ACKNOWLEDGMENTS

This work was supported by grants from the National Institute of Health (HL 20948, EY00871, EY04149, EY12190, EY015299, and RR17703), Research to Prevent Blindness, Foundation Fighting Blindness, Moss Heart Fund, and Perot Family Foundation. We thank Drs. Michael S. Brown, Joseph L. Goldstein, and Albert O. Edwards for their constant support and helpful discussions. We also thank Regeneron Pharmaceuticals, Inc. for the production of Rdh11 knockout mouse, and Kirsten G. Locke for the ERG analysis of the mice.

5. REFERENCES

1. Sun, H., and Nathans, J., Mechanistic studies of ABCR, the ABC transporter in photoreceptor outer segments responsible for autosomal recessive Stargardt disease, *J Bioenerg Biomembr*, **33**:523 (2001).
2. Schadel, S. A., Heck, M., Maretzki, D., Filipek, S., Teller, D. C., Palczewski, K., and Hofmann, K. P., Ligand channeling within a G-protein-coupled receptor. The entry and exit of retinals in native opsin, *J Biol Chem*, **278**:24896 (2003).
3. Rando, R. R., The biochemistry of the visual cycle, *Chem Rev*, **101**:1881 (2001).
4. Pfeffer, B. A., Clark, V. M., Flannery, J. G., and Bok, D., Membrane receptors for retinol-binding protein in cultured human retinal pigment epithelium, *Invest Ophthalmol Vis Sci*, **27**:1031 (1986).
5. Moiseyev, G., Crouch, R. K., Goletz, P., Oatis, J., Jr., Redmond, T. M., and Ma, J. X., Retinyl esters are the substrate for isomerohydrolase, *Biochemistry*, **42**:2229 (2003).
6. Mata, N. L., Tzekov, R. T., Liu, X., Weng, J., Birch, D. G., and Travis, G. H., Delayed dark-adaptation and lipofuscin accumulation in abcr +/− mice: implications for involvement of ABCR in age-related macular degeneration, *Invest Ophthalmol Vis Sci*, **42**:1685 (2001).
7. Boulton, M., Râozanowska, M., and Râozanowski, B., Retinal photodamage, *J Photochem Photobiol B*, **64**:144 (2001).
8. Jèager, S., Palczewski, K., and Hofmann, K. P., Opsin/all-trans-retinal complex activates transducin by different mechanisms than photolyzed rhodopsin, *Biochemistry*, **35**:2901 (1996).
9. Dean, D. M., Nguitragool, W., Miri, A., McCabe, S. L., and Zimmerman, A. L., All-trans-retinal shuts down rod cyclic nucleotide-gated ion channels: a novel role for photoreceptor retinoids in the response to bright light?, *Proc Natl Acad Sci U S A*, **99**:8372 (2002).
10. McCabe, S. L., Pelosi, D. M., Tetreault, M., Miri, A., Nguitragool, W., Kovithvathanaphong, P., Mahajan, R., and Zimmerman, A. L., All-trans-retinal is a closed-state inhibitor of rod cyclic nucleotide-gated ion channels, *J Gen Physiol*, **123**:521 (2004).
11. Vellani, V., Reynolds, A. M., and McNaughton, P. A., Modulation of the synaptic $Ca2^+$ current in salamander photoreceptors by polyunsaturated fatty acids and retinoids, *J Physiol*, **529**:333 (2000).
12. Tsina, E., Chen, C., Koutalos, Y., Ala-Laurila, P., Tsacopoulos, M., Wiggert, B., Crouch, R. K., and Cornwall, M. C., Physiological and microfluorometric studies of reduction and clearance of retinal in bleached rod photoreceptors, *J Gen Physiol*, **124**:429 (2004).
13. Kallberg, Y., Oppermann, U., Jèornvall, H., and Persson, B., Short-chain dehydrogenase/reductase (SDR) relationships: a large family with eight clusters common to human, animal, and plant genomes, *Protein Sci*, **11**:636 (2002).
14. Oppermann, U. C., Filling, C., and Jèornvall, H., Forms and functions of human SDR enzymes, *Chem Biol Interact*, **130-132**:699 (2001).
15. Kasus-Jacobi, A., Ou, J., Bashmakov, Y. K., Shelton, J. M., Richardson, J. A., Goldstein, J. L., and Brown, M. S., Characterization of mouse short-chain aldehyde reductase (SCALD), an enzyme regulated by sterol regulatory element-binding proteins, *J Biol Chem*, **278**:32380 (2003).
16. Lin, B., White, J. T., Ferguson, C., Wang, S., Vessella, R., Bumgarner, R., True, L. D., Hood, L., and Nelson, P. S., Prostate short-chain dehydrogenase reductase 1 (PSDR1): a new member of the short-chain steroid dehydrogenase/reductase family highly expressed in normal and neoplastic prostate epithelium, *Cancer Res*, **61**:1611 (2001).
17. Haeseleer, F., Jang, G. F., Imanishi, Y., Driessen, C. A., Matsumura, M., Nelson, P. S., and Palczewski, K., Dual-substrate specificity short chain retinol dehydrogenases from the vertebrate retina, *J Biol Chem*, **277**:45537 (2002).
18. Kedishvili, N. Y., Chumakova, O. V., Chetyrkin, S. V., Belyaeva, O. V., Lapshina, E. A., Lin, D. W., Matsumura, M., and Nelson, P. S., Evidence that the human gene for prostate short-chain dehydrogenase/reductase (PSDR1) encodes a novel retinal reductase (RalR1), *J Biol Chem*, **277**:28909 (2002).
19. Janecke, A. R., Thompson, D. A., Utermann, G., Becker, C., Hèubner, C. A., Schmid, E., McHenry, C. L., Nair, A. R., Rèuschendorf, F., Heckenlively, J., Wissinger, B., Nèurnberg, P., and Gal, A., Mutations in RDH12 encoding a photoreceptor cell retinol dehydrogenase cause childhood-onset severe retinal dystrophy, *Nat Genet*, **36**:850 (2004).
20. Perrault, I., Hanein, S., Gerber, S., Barbet, F., Ducroq, D., Dollfus, H., Hamel, C., Dufier, J. L., Munnich, A., Kaplan, J., and Rozet, J. M., Retinal dehydrogenase 12 (RDH12) mutations in leber congenital amaurosis, *Am J Hum Genet*, **75**:639 (2004).

21. Haeseleer, F., Huang, J., Lebioda, L., Saari, J. C., and Palczewski, K., Molecular characterization of a novel short-chain dehydrogenase/reductase that reduces all-trans-retinal, *J Biol Chem*, **273**:21790 (1998).

22. Rattner, A., Smallwood, P. M., and Nathans, J., Identification and characterization of all-trans-retinol dehydrogenase from photoreceptor outer segments, the visual cycle enzyme that reduces all-trans-retinal to all-trans-retinol, *J Biol Chem*, **275**:11034 (2000).

23. Valenzuela, D. M., Murphy, A. J., Frendewey, D., Gale, N. W., Economides, A. N., Auerbach, W., Poueymirou, W. T., Adams, N. C., Rojas, J., Yasenchak, J., Chernomorsky, R., Boucher, M., Elsasser, A. L., Esau, L., Zheng, J., Griffiths, J. A., Wang, X., Su, H., Xue, Y., Dominguez, M. G., Noguera, I., Torres, R., Macdonald, L. E., Stewart, A. F., DeChiara, T. M., and Yancopoulos, G. D., High-throughput engineering of the mouse genome coupled with high-resolution expression analysis, *Nat Biotechnol*, **21**:652 (2003).

24. Fain, G. L., Matthews, H. R., and Cornwall, M. C., Dark adaptation in vertebrate photoreceptors, *Trends Neurosci*, **19**:502 (1996).

25. Krizaj, D., and Copenhagen, D. R., Calcium regulation in photoreceptors, *Front Biosci*, **7**:d2023 (2002).

PIGMENT EPITHELIUM-DERIVED GROWTH FACTOR INHIBITS FETAL BOVINE SERUM STIMULATED VASCULAR ENDOTHELIAL GROWTH FACTOR SYNTHESIS IN CULTURED HUMAN RETINAL PIGMENT EPITHELIAL CELLS

Piyush C. Kothary*, Rhonda Lahiri, Lynn. Kee, Nitin Sharma, Eugene Chun, Angela Kuznia, and Monte A. Del Monte*

1. INTRODUCTION

The human retinal pigment epithelium (hRPE) is a monolayer of cells that is located between the photoreceptors and Bruch's membrane. Normally, it is mitotically inactive in adult eyes but sometimes it undergoes mitosis and cell division in pathologic states. Growth factors have been implicated in inducing proliferation and migration of hRPE (Kothary and Del Monte, 2003).

Fetal Bovine Serum (FBS) has been shown to stimulate proliferation in hRPE cells by stimulating intracellular protein synthesis of vascular endothelial growth factor (VEGF) (Lahiri et al., 2004) and fibroblast growth factor 2 (Kothary et al., 2001).

Pigment epithelium-derive growth factor (PEDF), which was first isolated from fetal RPE culture, is a 50 kDa antiagiogenic protein. Increased expression of PEDF induces regression in choroidal neovasculariztion in a mouse model (Mori et al., 2002). In this study we examined the effect of PEDF on VEGF synthesis in hRPE cells.

2. MATERIALS AND METHODS

2.1. Chemicals

PEDF was purchased from Bioproducts, Middletown, MD. Anti-VEGF was purchased from Upstate Biotech, Inc., Charlottesville, VA. 3H-thymidine and 14C-methionine were

---

*Department of Ophthalmology, University of Michigan Medical Center, Ann Arbor, MI 48105.

purchased from Amersham Corporation, Arlington Heights, IL. Ham's nutrient medium (F12), Dulbecco's minimum essential media (DMEM), Hank's balanced salt solution, fetal bovine serum (FBS), penicillin and streptomycin and trypsin were purchased from GIBCO BRL, Gaithersburg, MD.

2.2. Establishment and Maintenance of hRPE Cells Cultures

Primary cultures of hRPE were established from three human eyes obtained from Michigan Eye Bank as described previously (Del Monte et al., 1991; Kusaka et al., 1998). Briefly, the anterior segment, vitreous and the retina of eye were surgically removed. The posterior segment was then washed, treated with trypsin and the loosely adherent hRPE cells were detached by gentle brushing and hydrostatic pressure with a sterile Pasteur pipet. The cells were plated in 16-mm Primaria plates and incubated at 37 degrees C in a 95% air/5% CO_2 incubator. The medium-1, which contained Ham f12 + 14% FBS, was changed every three days until the cells were confluent.

2.3. Cellular Proliferation

Cellular proliferation of cultured hRPE cells was measured by tritiated thymidine incorporation and trypan blue exclusion methods as described previously (Kusaka et al., 1998). Briefly, hRPE cells at passage 4-8 were trypsinized and plated in 16-mm wells of 24-well plates at $1 \times 10,000$ cells per well in medium-1. Experimental reagents were added for 24-48 hours (after the cells were confluent.). The cells were then analysed using the trypan blue exclusion and ^{3}H-Thymidine incorporation methods separately.

2.4. Immunoprecipitation of 14-C-Methionine-VEGF

14C-methionine incorporation was measured as described previously (Bitar et al., 1998; Kothary et al., 2004). To measure intracellular VEGF synthesis, hRPE cells were labeled by 14-C-methionine and treated with experimental reagents. hRPE cells were then lysed with zwittergent 3-12 and precipitated with anti-VEGF specific antibody and protein-A.

2.5. Immunohistochemical Detection

Immunohistochemical analysis was carried out by the method previously described (Oncogene Science, Inc Brochure, Uniondale, NY). Briefly, to identify intracellular VEGF, hRPE cells which were grown on coverslips were treated with experimental reagents. The hRPE cells were then incubated with anti-VEGF and then with anti-rabbit-rhodamine. Coverslips then were dried and mounted on a slide using gel mount.

2.6. Statistical Analysis

All values, representing the mean $\pm$ SEM differences between two groups of data were tested by students' 't' test. A $p < 0.05$ was used to assess significant differences between two groups and are indicated by single or double or triple asterisks in each figure.

3. RESULTS

3.1. Effect of PEDF on hRPE Cell Number and ^{3}H-Thymidine Incorporation

FBS stimulates hRPE cell proliferation as determined by the trypan blue exclusion method as well as tritiated thymidine incorporation in a dose dependent manner (Data not shown). Figure 71.1 and Figure 71.2 show that PEDF inhibits FBS (10%) stimulated hRPE cells proliferation and ^{3}H-thymidine incorporation, also in a dose dependent manner.

3.2. Effect of FBS and PEDF on VEGF Synthesis in hRPE Cells

Figure 71.3 shows that FBS stimulates 14-C-methionine-VEGF in dose dependent manner. Figure 71.4 shows that PEDF (0.1 µg/ml) inhibits FBS (10%) stimulated 14-C-

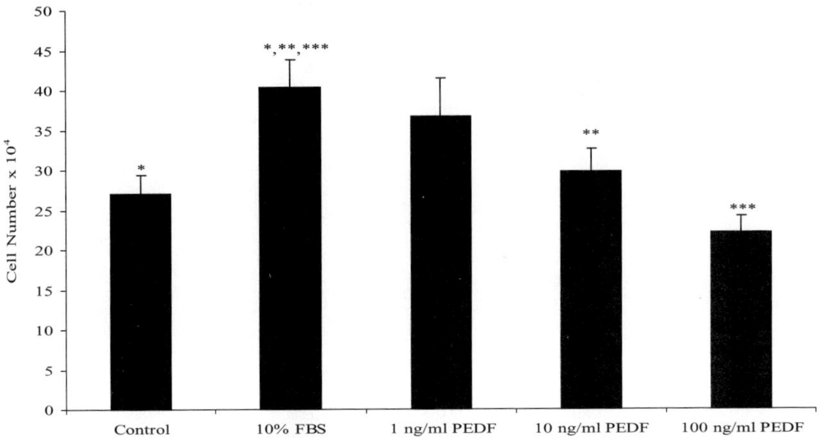

Figure 71.1. Effect of PEDF on FBS stimulated hRPE cell proliferation (*, **, *** = p < 0.05 when compared with each other.

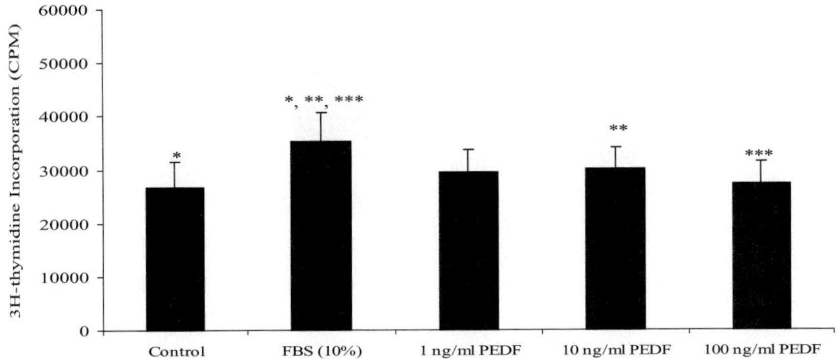

Figure 71.2. Effect of PEDF on FBS stimulated hRPE cell proliferation (*, **, *** = p < 0.05 when compared with each other).

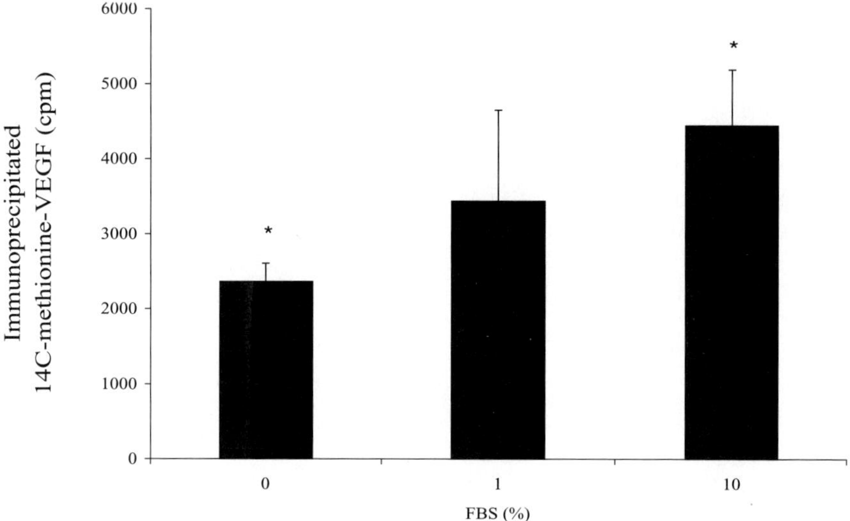

Figure 71.3. Effect of FBS on 14C-methionine-VEGF synthesis (* = p < 0.05 when compared with each other).

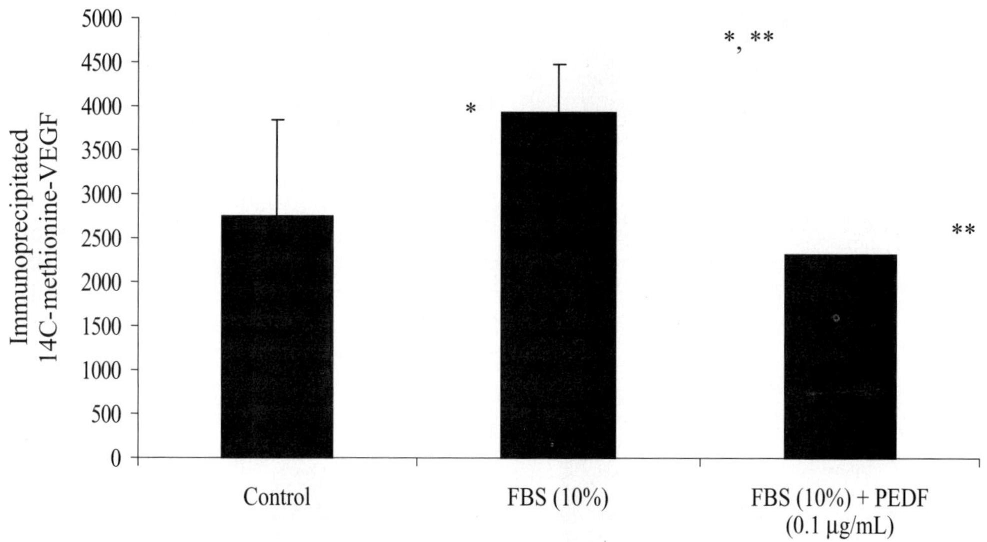

Figure 71.4. Effects of PEDF on FBS (10%) stimulated 14C-methionine-VEGF synthesis (*, ** = p < 0.05 when compared with each other).

Figure 71.5. The effect of PEDF (0.1 µg/ml) on FBS (10%) stimulated VEGF synthesis in cultured hRPE cells by immunohistochemical localization of VEGF (left vertical panel) and DIC microscopy (right vertical panel).

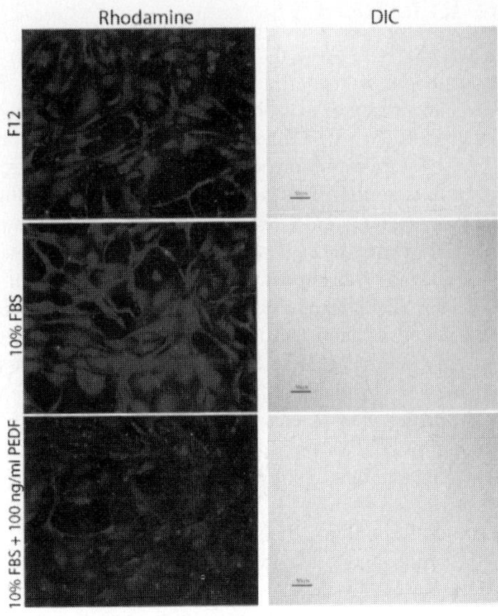

methionine VEGF. These findings were confirmed by immunohistochemical studies which showed less VEGF positive immunoreactivity in hRPE cells exposed to PEDF and FBS together than FBS alone (Figure 71.5).

4. DISCUSSION

In this study, we have demonstrated that PEDF inhibits FBS stimulated proliferation of hRPE cells. Our data also indicate that the inhibitory effect of PEDF is not caused by cytotoxicity, as remaining cells appear healthy and viable. Instead, the effect of PEDF on FBS induced hRPE cell proliferation was inhibitory. The concentrations of PEDF (10-100 ng/ml) that inhibited hRPE cell number and titrated thymidine incorporation arc in the same range as that of Duh et al. (2002) who showed that PEDF at the concentration of 2-20 nM partially suppressed the work of VEGF-induced retinal endothelial cell proliferation.

To characterize the biochemical events affected by PEDF, we examined the effects of PEDF on well known growth inducing factor, VEGF. We demonstrated that FBS stimulates VEGF synthesis in a dose dependent manner and PEDF (0.1 µg/ml) inhibits it in our hRPE cell cultures. The inhibitory effect of PEDF on FBS stimulated hRPE cell number and VEGF syntheses were almost equivalent on a percentage basis. This suggests that VEGF plays an important role in FBS mediated hRPE cell proliferation.

We have shown that PEDF inhibits FBS stimulated hRPE cell proliferation as well as intracellular VEGF synthesis. If we extend our observation, it is reasonable to postulate that PEDF may inhibit hRPE cell proliferation in vivo. Intravenous injection of PEDF has shown to inhibit retinal neovascularization in a mouse model (Duh et al., 2002). This suggests that PEDF may be of therapeutic value in proliferative eye disease.

5. REFERENCES

Bitar, K. N., Kothary, S., and Kothary, P. C., 1996, Somatostatin inhibits bombesin-stimulated Gi-protein via its own receptor in rabbit colonic smooth muscle cells, J. of Pharm & Exp. Therap. **276**:714-9.

Del Monte, M. A., Rabbani, R., Diaz, T. C., Lattimer, S. A., Nakamura, J., Brennan, M. C., and Greene, D. A., 1991, Sorbitol, myoinositol, and rod outer segment phagocytosis in cultured hRPE cells exposed to glucose. In vitro model of myoinositol depletion hypothesis of diabetic complications, Diabetes. **40**:1335-45.

Duh, E. J., Yang, H. S., Suzuma, I., Miyagi, M., Youngman, E., Mori, K., Katai, M., Yan, L., Suzuma, K., West, K., Davarya, S., Yong, P., Gehlbach, P., Pearlman, J., Crabb, J. W., Aieloo, L. P., Campochario, P., and Zack, D. J., 2002, Pigment epithelium-derived factor suppresses ischemia-induced retinal neovascularization and VEGF- induced migration and growth, Invest Ophthalmol Vis Sci. **43**:823-9.

Kothary, P. C., and Del Monte, M. A., 2003, in: Recent Res. Devel. Cell Biochem, edited by S. G. Pandalai, Transworld Research Network, Trivandrum (India), pp. 99-116.

Kothary, P. C., Singal, P., Patel, P., and Del Monte, M. A., 2001, in: Proceedings of World CongressNeuroinformatics, Part II, edited by Frank Rattay, (ARGESIN/ASIM- Verlag, Vienna (Austria), pp. 341-8.

Kothary, P. C., Paauw, J. D., Bansal, A. K., Grace, C. C., and Del Monte, M. A., 2004, in: Proceedings of 5th International Symposium on Ocular Pharmacology and Therapeutics, Medimond S.r.l., Bologna (Italy), pp.237-41.

Kusaka, K., Kothary, P. C., and Del Monte, M. A., 1998, Modulation of basic fibroblast growth factor effect by retinoic acid in cultured retinal pigment epithelium, Current Eye Research. **17**: 524-30.

Mori, K., Gehlbach, P., Ando, A., Mcvey, D., Wei, L., and Campochiaro, P. A., 2002, Regression of ocular neovascularization in response to increased expression of pigment epithelium-derived factor, Invest Ophthalmol Vis Sci. **43**:2428-3.

Lahiri, R., Kothary, P., and Del Monte, M. A., 2004, The effect of pigment epithelium derived factor (PEDF) on human retinal pigment epithelial cell proliferation and intracellular growth factor synthesis. Poster presentation at the ARVO Meeting, Ft. Lauderdale, Florida.

CHAPTER 72

THE RETINAL PIGMENT EPITHELIUM APICAL MICROVILLI AND RETINAL FUNCTION

Vera L. Bonilha, Mary E. Rayborn, Sanjoy K. Bhattacharya, Xiarong Gu, John S. Crabb, John W. Crabb, and Joe G. Hollyfield*

1. INTRODUCTION

The RPE performs highly specialized, unique functions essential for homeostasis of the neural retina. These include phagocytosis of photoreceptors shed outer segments, directional transport of nutrients into and removal of waste products from photoreceptor cells and visual pigment transport and regeneration. All of these functions involve the RPE apical microvilli.[1-4]

The RPE is a low cuboidal epithelium containing very long sheet-like apical microvilli that project into the interphotoreceptor matrix. The microvilli interact with the tips of the rod and cone photoreceptor outer segments extending from the outer retinal surface. The cone-RPE association is much less studied however, as many as 30-40 microvilli can be associated with a single cone. These vary in length with only a few reaching the outer segment. The RPE apical microvilli ensheath the outer segments of photoreceptor cells, extending for as long as half the outer segment.[5] A single RPE microvillous may completely surround the outer segment or multiple microvilli can encircle each other while surrounding the photoreceptor outer segments. Intracellular organelles are mostly absent from the cone-ensheathing microvilli while they are very abundant in the microvilli ensheathing the rod outer segments.

The RPE basal surface is highly infolded and interacts with the underlying Bruch's membrane,[1] an acellular layer separating the RPE from the choriocapillaris. The polarized organization of the RPE is essential for the vectorial transport of different molecules between the choriocapillaris and the neural retina and vice-versa. A unique characteristic of the RPE is the "reversed polarity" of select proteins such as the Na,K-ATPase pump, EMMPRIN and the adhesion molecule N-CAM. These proteins are found at the apical surface of the RPE, rather than at the basolateral surface as in other epithelia.[6-8]

---

*Vera L. Bonilha et al., Cole Eye Institute, Cleveland Clinic Foundation, OH 44195, USA.

2. RPE MICROVILLI STRUCTURE

The RPE microvillar structure has not been extensively studied. However, available information indicates that RPE microvilli possess an internal core bundle of densely packed actin filaments.[9] Myosin VIIa has been detected at the base of apical processes[10] while villin, fimbrin and myosin I have not been detected.[11,12] The entire length of the RPE microvilli has been shown to contain ezrin and EBP50.[11,13-15] The mouse RPE microvilli-enriched fraction, described below, contained several cytoskeletal components, among them various types of actin and tubulin, β-spectrin, ezrin, moesin, EBP50, and profilin.

A more complete definition of the protein composition of the RPE apical microvilli should provide insights into other biochemical processes occurring at this critical interface that are important for the support and maintenance of vision.

3. RPE MICROVILLI PROTEINS AND FUNCTION

Recently, we have improved a method to isolate RPE apical microvilli. The procedure relies on the binding of N-acetylglucosamine and sialic acid-containing glycoconjugates present in abundance on the RPE apical surface[16] to the WGA lectin conjugated to agarose beads. Mass interactions of the surface glycoconjugates with the immobilized lectin on the bead allow for the detachment of the RPE microvilli upon physical removal of the WGA beads. The RPE isolated microvilli are resolved by SDS-PAGE, in gel digested with trypsin, and peptides extracted and analyzed by mass spectrometry.[3,17] This procedure was done in mice eyecups with the RPE exposed and it has resulted in the identification of over 283 proteins, distributed over functional categories such as retinoid-metabolizing, cytoskeletal, enzymes, extracellular matrix components, membrane proteins and transporters, among others. A summary of selected proteins identified by this method is presented in Table 72.1 and has been recently described.[3,17]

The beads with the isolated RPE microvilli on their surface can be used for immuno-labeling experiments, morphological (light and electron microscopy) as well as in bio-chemical experiments. In Figure 72.1 beads with isolated mouse RPE were fixed in 4% paraformaldehyde, permeabilized in triton X100, reacted with both a rabbit antibody to (A) protein kinase A regulatory subunit II (PKA_{RII}) and (B) a mouse antibody to protein kinase A regulatory subunit I (PKA_{RI}). Parallel samples were processed for transmission electron microscopy and RPE microvilli are observed on the surface of the agarose beads (C and D).

Examples of proteins identified in the RPE microvilli by both mass spectrometry and other methods include Na,K-ATPase, Glut-1, monocarboxylate transporter, carbonic anhy-drase, basigin, and the chloride intracellular channel 6.[17] The cone and rod-associated matrix, present on top of the RPE apical surface, are firmly attached to the RPE apical surface. This is one of the reasons for the mass spectrometric identification of several novel extracellular matrix components such as fibromodulin, lumican, undulin 1, and neuroglycan C in the RPE isolated microvilli. The proteomic method therefore provides an unbiased account of pro-teins present in the RPE apical microvilli.

RPE apical microvilli play important roles in retinal attachment. Ensheathment of the outer segment tips by apical projections may contribute to adhesion by providing frictional or electrostatic interactions.[18] Any disruption of the relationship between cone and rod pho-toreceptors and the RPE will result in pathological consequences. A retinal detachment, for example, is a separation of the photoreceptor outer segments from its apical RPE microvilli. After clinical reattachment, return of normal vision depends, upon the restoration of a functional relationship between proteins present in the RPE apical surface and the photoreceptors outer segments.

Alterations in the proteins present in the RPE apical microvilli will likely impair vision as a consequence of disrupting the structural and functional nurturing of the photoreceptors by the RPE. Some of the RPE apical proteins identified in the RPE microvilli fraction have already been shown to be involved in retinal degenerations. The list of RPE apical proteins involved in retinal diseases is likely to grow as we learn more about the RPE proteome.

An interaction between cellular retinaldehyde-binding protein (CRALBP) and ERM-binding phosphoprotein 50 (EBP50) in RPE microsomes was recently described.[15] Our proteomic analyses was highly enriched in several retinoid processing proteins such as cellular retinaldehyde-binding protein, 11-*cis*-retinol dehydrogenase, cellular retinol-binding protein 1, interphotoreceptor retinoid-binding protein, EBP50, and ezrin. These results support the existence of a visual cycle protein complex in the RPE apical microvilli.[3] Several forms of retinitis pigmentosa are known to be caused by mutations in visual-cycle protein genes such as RPE65, CRALBP, IRBP.[19,20]

Macular edema resulting from pathologies such as uveitis, postoperative period following cataract extraction,[21] retinitis pigmentosa,[22] serpiginous choroiditis[23] and epiretinal membranes,[24] has been widely treated with carbonic anhydrase inhibitors.[25] Polarized dis-

Table 72.1. Selected proteins identified on WGA-beads after incubation with apical RPE.

Proteins	Accession Number[a]	Peptide Matches	Frequency[b]
Annexin A2	P07356	2	1
Annexin A5	P48036	3	1
Basigin	P18572	6	3
Carbonic anhydrase XIV	Q9WVT6	2	1
Chloride intracellular channel 6	Q96NY7	2	3
Cellular retinaldehyde-binding protein (CRALBP)	Q9Z275	6	3
ERM-binding phosphoprotein (EBP50)	Q9JJ19	1	1
Ezrin	P26040	4	2
Fibromodulin	P50608	4	2
Glucose transporter type 1 (Glut-1)	*P17809*	3	3
Interphotoreceptor retinoid-binding protein (IRBP)	P49194	5	3
L-lactate dehydrogenase A chain (LDH)	P06151	3	1
Lumican	P51885	9	3
Malate dehydrogenase	P14152	2	2
Membrane-associated adenylate kinase	Q9R0Y4	2	1
Monocarboxylate transporter 1	*AAC13720*	3	3
Neuroglycan C	Q9QY32	1	1
Retinol dehydrogenase, 11-cis (RDH5)	Q27979	3	3
Sodium/potassium-transporting ATPase alpha-1	P06685	6	3
Undulin 1	A40970	7	2
Vitronectin receptor α subunit (integrin αv)	P43406	3	2

[a] Swiss Protein database and NCBI (in italics) accession numbers are shown.
[b] Results from three independent experiments.

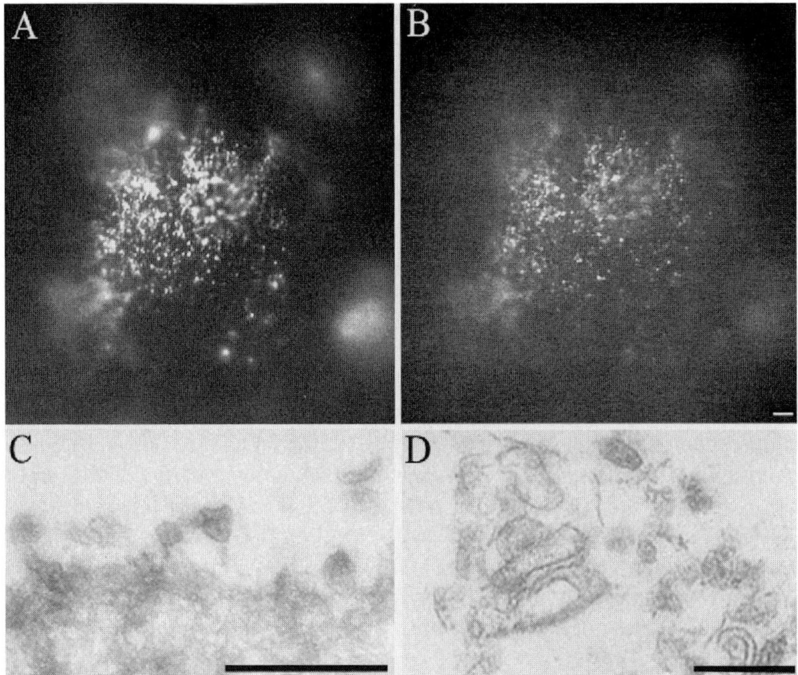

Figure 72.1. Morphological analysis of isolated WGA-beads with mouse RPE microvilli on their surface.
WGA beads scraped off the mouse eyecups were reacted with antibodies to protein kinase A regulatory subunit alpha II (PKA$_{RII}$) (A) and protein kinase A regulatory subunit I (PKA$_{RI}$) (B). Alternatively, the isolated beads were fixed in 2.5% glutaraldehyde, and processed for transmission electron microscopy (TEM). Low (C) and high (D) magnification of these beads revealed extensive surface areas covered by the RPE microvilli. Bars = 100 µm (A, B), 1 µm (C) and 0.5 µm (D).

tribution of carbonic anhydrase activity in the RPE apical surface has been reported. Carbonic anhydrase XIV was one of the proteins we identified by proteomic analysis in isolated RPE microvilli.[17]

Aging studies have shown a decrease in both the number and the length of epithelial microvilli, and a declined function of plasma membrane enzymes and receptors.[26-29] Specifically, a decrease in the activity of some of the enzymes detected in RPE microvilli like Na,K-ATPase, LDH, glutathione S-transferase, phosphoglycerate kinase, adenylate kinase[30] and catalase has been established in various epithelia.[31-33] Future studies involving these and other proteins may help to improve our understanding of aging diseases such as macular degeneration.

Most recently we have pursued proteomic analysis of rat RPE microvilli. One of the proteins consistently found in rat RPE microvilli is ceruloplasmin. The localization of ceruloplasmin in the RPE has been previously described.[34,35] Retinal degeneration has been reported in patients with the autosomal recessive disease called aceruloplasminemia, a deficiency in ceruloplasmin.[36]

4. CONCLUSIONS

Progress is being made in characterization of the RPE and its apical microvilli. The years to come will bring further definition of key proteins and pathways present in RPE microvilli as well as a better understanding of their function in vision.

5. ACKNOWLEDGMENTS

Supported by NIH grants EY06603, EY14239, EY014240 and an infrastructure grant EY015638, a Research Center grant from the Foundation Fighting Blindness, a grant from the National Glaucoma Research Program of American Health Assistance Foundation (G2004-047 to SKB), and funds from the Cleveland Clinic Foundation.

6. REFERNCES

1. K. M. Zinn, and J. V. Benjamin-Henkind., Anatomy of the human retinal pigment epithelium. *in: The Retinal Pigment Epithelium*, edited by K. M. Zinn, M. F. Marmor, (Harvard University Press; Cambridge, MA: 1979), pp:3-31.
2. D. Bok, The retinal pigment epithelium: a versatile partner in vision. *J Cell Sci Suppl* **17**:189-195 (1993).
3. V. L. Bonilha, S. K. Bhattacharya, K. A. West, J. S. Crabb, J. Sun, M. E. Rayborn, M. Nawrot, J. C. Saari, and J. W. Crabb., Support for a proposed retinoid-processing protein complex in apical retinal pigment epithelium. *Exp Eye Res* **79**:419-422 (2004).
4. T. D. Lamb, and E. N. Jr. Pugh, Dark adaptation and the retinoid cycle of vision. *Prog Retin Eye Res* **23**:307-380 (2004).
5. R. H. Steinberg, and I. Wood, The Relationship of the Retinal Pigment Epithelium to Photoreceptor Outer Segments in Human Retina. *in: The Retinal Pigment Epithelium*, edited by K. M. Zinn, M. F. Marmor (Harvard University Press; Cambridge, MA: 1979), pp:32-44.
6. D. Gundersen, J. Orlowski, and E. Rodriguez-Boulan, Apical polarity of Na,K-ATPase in retinal pigment epithelium is linked to a reversal of the ankyrin-fodrin submembrane cytoskeleton. *J Cell Biol* **112**:863-872 (1991).
7. D. Gundersen, S. K. Powell, and E. Rodriguez-Boulan, Apical polarization of N-CAM in retinal pigment epithelium is dependent on contact with the neural retina. *J Cell Biol* **121**:335-343 (1993).
8. A. D. Marmorstein, S. C. Finnemann, V. L. Bonilha, and E. Rodriguez-Boulan, Morphogenesis of the retinal pigment epithelium: toward understanding retinal degenerative diseases. *Ann N Y Acad Sci* **857**:1-12 (1998).
9. D. K. Vaughan, and S. K. Fisher, The distribution of F-actin in cells isolated from vertebrate retinas. *Exp Eye Res* **44**:393-406 (1987).
10. T. Hasson, M. B. Heintzelman, J. Santos-Sacchi, D. P. Corey, and M. S. Mooseker, Expression in cochlea and retina of myosin VIIa, the gene product defective in Usher syndrome type 1B. *Proc Natl Acad Sci U S A* **92**:9815-9819 (1995).
11. D. Hofer, and D. Drenckhahn, Molecular heterogeneity of the actin filament cytoskeleton associated with microvilli of photoreceptors, Muller's glial cells and pigment epithelial cells of the retina. *Histochemistry* **99**:29-35 (1993).
12. K. Owaribe, and G. Eguchi, Increase in actin contents and elongation of apical projections in retinal pigmented epithelial cells during development of the chicken eye. *J Cell Biol* **101**:590-596 (1985).
13. V. L. Bonilha, S. C. Finnemann, and E. Rodriguez-Boulan, Ezrin promotes morphogenesis of apical microvilli and basal infoldings in retinal pigment epithelium. *J Cell Biol* **147**:1533-1548 (1999).
14. V. L. Bonilha, and E. Rodriguez-Boulan, Polarity and developmental regulation of two PDZ proteins in the retinal pigment epithelium. *Invest Ophthalmol Vis Sci* **42**:3274-3282 (2001).
15. M. Nawrot, K. West, J. Huang, D. E. Possin, A. Bretscher, J. W. Crabb, and J. C. Saari, Cellular retinaldehyde-binding protein interacts with ERM-binding phosphoprotein 50 in retinal pigment epithelium. *Invest Ophthalmol Vis Sci* **45**:393-401 (2004).

16. N. G. Cooper, B. I. Tarnowski, and B. J. McLaughlin, Lectin-affinity isolation of microvillous membranes from the pigmented epithelium of rat retina. *Curr Eye Res* **6**:969-979 (1987).

17. V. L. Bonilha, S. K. Bhattacharya, K. A. West, J. Sun, J. W. Crabb, M. E. Rayborn, and J. G. Hollyfield, Proteomic characterization of isolated retinal pigment epithelium microvilli. *Mol Cell Proteomics* **3**:1119-1127 (2004)

18. M. F. Marmor. Mechanisms of Retinal Adhesion. *in: Progress in Retinal Research*, edited by N. Osborne, G. Chader (Pergamon Press, New York, NY: 1993), pp. 179-204.

19. M. A. Maw, B. Kennedy, A. Knight, R. Bridges, K. E. Roth, E. J. Mani, J. K. Mukkadan, D. Nancarrow, J. W. Crabb, and M. J. Denton, Mutation of the gene encoding cellular retinaldehyde-binding protein in autosomal recessive retinitis pigmentosa. *Nat Genet* **17**:198-200 (1997).

20. Q. Wang, Q. Chen, K. Zhao, L. Wang, and E. I. Traboulsi, Update on the molecular genetics of retinitis pigmentosa. *Ophthalmic Genet* **22**:133-154 (2001).

21. M. D. Farber, S. Lam, H. H. Tessler, T. J. Jennings, A. Cross, and M. M. Rusin, Reduction of macular oedema by acetazolamide in patients with chronic iridocyclitis: a randomised prospective crossover study. *Br J Ophthalmol* **78**:4-7 (1994).

22. G. A. Fishman, L. D. Gilbert, R. G. Fiscella, A. E. Kimura, and L. M. Jampol, Acetazolamide for treatment of chronic macular edema in retinitis pigmentosa. *Arch Ophthalmol* **107**:1445-1452 (1989).

23. J. C. Chen, F. W. Fitzke, and A. C. Bird, Long-term effect of acetazolamide in a patient with retinitis pigmentosa. *Invest Ophthalmol Vis Sci* **31**:1914-1918 (1990).

24. M. F. Marmor, Hypothesis concerning carbonic anhydrase treatment of cystoid macular edema: example with epiretinal membrane. *Arch Ophthalmol* **108**:1524-1525 (1990).

25. T. J. Wolfensberger. The role of carbonic anhydrase inhibitors in the management of macular edema. *Doc Ophthalmol* **97**:387-397 (1999).

26. I. Weisse, Changes in the aging rat retina. *Ophthalmic Res.* 1995;**27**:154-163.

27. T. Hirai, S. Kojima, A. Shimada, T. Umemura, M. Sakai, and C. Itakura, Age-related changes in the olfactory system of dogs. *Neuropathol Appl Neurobiol* **22**:531-539 (1996).

28. I. Jang, K. Jung, and J. Cho, Influence of age on duodenal brush border membrane and specific activities of brush border membrane enzymes in Wistar rats. Exp Anim **49**:281-287 (2000).

29. J. M. Serot, M. C. Bene, and G. C. Faure, Choroid plexus, aging of the brain, and Alzheimer's disease. *Front Biosci* **8**:s515-521 (2003).

30. M. Kadlubowsk, and P. S. Agutter, Changes in the activities of some membrane-associated enzymes during in vivo ageing of the normal human erythrocyte. *Br J Haematol* **37**:111-125 (1977).

31. P. Napoleone, E. Bronzett, and F. Amenta, Enzyme histochemistry of aging rat kidney. *Mech Ageing Dev* **61**:187-195 (1991).

32. L. Teillet, L. Preisser, J. M. Verbavatz, and B. Corman, Kidney aging: cellular mechanisms of problems of hydration equilibrium. *Therapie* **4**:147-154 (1999).

33. D. L. Schmucker, K. Thoreux, R. L. Owen, Aging impairs intestinal immunity. *Mech Ageing Dev* **122**:1397-1411 (2001).

34. P. Hahn, T. Dentchev, Y. Qian, T. Rouault, Z. L. Harris, and J. L. Dunaief JL, Immunolocalization and regulation of iron handling proteins ferritin and ferroportin in the retina. *Mol Vis.* **10**:598-607 (2004).

35. P. Hahn, Y. Qian, T. Dentchev, L. Chen, J. Beard J, Z. L. Harris, and J. L. Dunaief, Disruption of ceruloplasmin and hephaestin in mice causes retinal iron overload and retinal degeneration with features of age-related macular degeneration. *Proc Natl Acad Sci U S A* **101**:13850-13855 (2004). Epub 12004 Sep 13813.

36. K. Yamaguchi, S. Takahash, T. Kawanami, T. Kato, and H. Sasaki, Retinal degeneration in hereditary ceruloplasmin deficiency. *Ophthalmologica* **212**:11-14 (1998).

CHAPTER 73

UPREGULATION OF TRANSGLUTAMINASE IN THE GOLDFISH RETINA DURING OPTIC NERVE REGENERATION

Kayo Sugitani, Toru Matsukawa, Ari Maeda, and Satoru Kato*

1. SUMMARY

To elucidate the molecular involvement of transglutaminase (TG) in central nervous system (CNS) regeneration, we cloned a full-length cDNA for neural TG (TG_N) from axotomized goldfish retinas and produced a recombinant TG_N protein from this cDNA. The levels of TG_N mRNA and protein were increased at 10-30 days after optic nerve transection, and this increase in TG_N was only localized in the ganglion cells in goldfish retinas. In retinal explant cultures, the recombinant TG_N protein induced a drastic enhancement of neurite outgrowth, while TG_N-specific RNAi significantly suppressed this neurite outgrowth. Taken together, these data strongly indicate that TG_N is a key regulatory molecule for CNS regeneration.

2. INTRODUCTION

Transglutaminase (TG), a protein cross-linking enzyme, is widely distributed in mammalian cells and tissues. Neural TG (TG_N), which is expressed in neural tissue, rapidly increased in rat sciatic nerves[1] and superior cervical ganglia[2] after nerve injury. In the central nerve system, the TG_N activity of goldfish optic nerve increased, whereas that of rat optic nerve decreased after optic nerve crush.[3] Fish can successfully regenerate the optic axons and eventually function after nerve injury, whereas rat cannot regenerate their optic axons. Therefore, to elucidate a functional role of TG_N on CNS regeneration in genetic level, we first isolated a full-length cDNA clone for TG_N from a cDNA library prepared from axotomized goldfish retinas. In addition, we produced a recombinant TG_N protein and anti-TG_N

*Satoru Kato, Department of Molecular Neurobiology, Graduate School of Medicine, Kanazawa University, 13-1 Takara-machi, Kanazawa 920-8640, Japan, TEL: +81 76 265 2450; E-mail: satoru@med.kanazawa-u.ac.jp.

antiserum. We also investigated the expression and localization of the TG_N protein by immunohistochemical staining. Moreover, we made a TG_N specific small interference RNA to estimate the effect of TG_N on neurite outgrowth in explant culture system. In the present study, we showed a novel functional role of TG_N on axonal elongation of goldfish optic nerve after injury.

3. METHODS

3.1. Animals

Adult common goldfish (*Carassius auratus*; body length about 6-8cm) were used throughout this study. Goldfish were anesthetized with ice-cold water. The optic nerve was sectioned 1mm away from the posterior of the eyeball with scissors. After surgery the goldfish were kept in water tanks at $22°C \pm 1°C$ for 1-40 days.

3.2. Cloning of Goldfish Neural Transglutaminase (TG_N)

A cDNA library was constructed from poly $(A)^+$ RNA $(5\,\mu g)$ from goldfish retinas of which optic nerve had been transected 5 days before as described previously.[4] Tissue-type transglutaminase (tTG) cDNA from red sea bream (*Pagrus major*) liver[5] (gift from Dr. Yasueda, Ajinomoto Co.) was labeled with $[^{32}P]$ dCTP and 2×10^5 colonies were screened with this probe. Five positive clones were subcloned into pBK-CMV phagemid and sequenced usig DNA seqencer. Two independent clones were obtained. The 5' TG_N mRNA was cloned by the RACE method.

3.3. Purification of Recombinant TG_N

A full-length TG_N cDNA clone was inserted into the expression vector pFLAG-CMV-1, and the constructs were transfected into HEK 293 cells using Lipofectamine. For the control, only the pFLAG-CMV-1 vector was transfected to create mock cells. All cells were maintained in Dulbecco's MEM containing 10% fetal calf serum in a 5% CO2 humidified incubator for 48h at 37°C. The cells were then harvested, lysed and centrifuged for 30min at 15,000g. The FLAG-tagged enzyme was purified using ANTI-FLAG M2 affinity gel.

3.4. Immunohistochemistry

A rabbit antiserum against TG_N was obtained by subcutaneous injection of purified TG_N. Tissue fixation and cryosectioning were carried out as described previously.[6] Retinal sections were autoclaved at 121°C for 15min in 10mM citrate buffer. After washing and blocking, the sections were incubated with the rabbit polyclonal anti-TG_N antibody (1:100 dilution) overnight at 4°C. Following incubation with a biotinylated secondary antibody for 2h at room temperature, the bound antibodies were detected using horseradish peroxidase-conjugated streptavidin and 3-amino-9-ethylcarbazole.

3.5. Retinal Explant Culture

Retinal explant culture was performed according to a previous method of Matsukawa et al.[6] For siRNA synthesis, the in vitro Transcription T7 Kit for siRNA synthesis was used according to the manufacture's instruction. Transfections of siRNA to retinal explants were carried out using Lipofectamine 2000. For each transfection sample, 2 µl of Lipofectamine 2000 diluted in 98 µl of L-15 medium was mixed with 100 pmol of siRNA diluted in 100 µl of L-15 medium, incubated for 20 min at room temperature to allow complex formation and then added to 0.8 ml of the resuspended retinal culture. The retinal explants were gently mixed with the culture medium for 3 h and then divided into two 35-mm culture dishes. After incubation at 28°C overnight, 50 µl of fetal calf serum was added to each dish and the culture was continued.

4. RESULTS

4.1. TG$_N$ Expression in the Retina after Optic Nerve Transection

The expression of TG$_N$ protein levels in the goldfish retina was investigated after optic nerve transection using the anti-TG$_N$ antiserum. Weak signals for the TG$_N$ protein could be seen in the ganglion cell layers of control retina (Fig. 73.1a).

The immunoreactivity in the ganglion cell layer started to increase at 10 days and peaked at 20-30 days (Fig. 73.1b) and then decreased by 40 days after axotomy (Fig. 73.1c). The increase in TG$_N$ immunoreactivity was only localized in the ganglion cells and nerve layers (Fig. 73.1a,b,c). The expression pattern of TG$_N$ mRNA after optic nerve injury was the same as that of TG$_N$ protein (data not shown).

4.2. Moduration of Neurite Outgrowth by a Recombinant TG$_N$ Protein and RNAi in Retinal Explant Cultures

Addition of the recombinant TG$_N$ protein induced a large number of explants with long and thick neurites after 2 days (Fig. 73.2b,d), as compared with the control culture (Fig. 73.2a,d). The neurite outgrowth of the culture containing recombinant TG$_N$ protein was evoked in 50% of the explants during 2 days of culture whereas neurite outgrowth of the

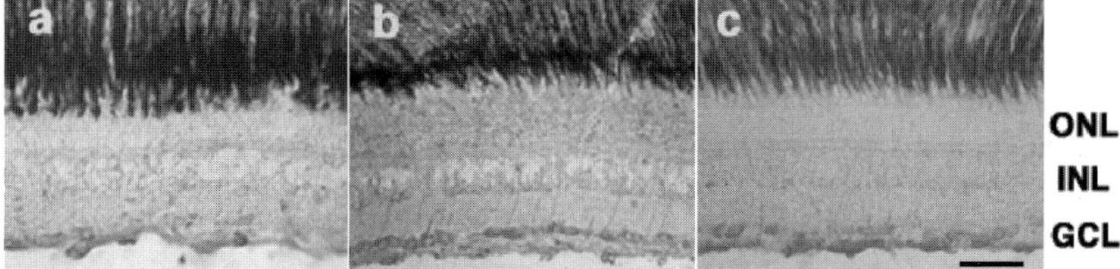

Figure 73.1. Immunohistochemical staining of goldfish retina with the anti-TG$_N$ antibody. (a) control retina, (b) at 20 days after optic nerve transection, (c) at 40 days after optic nerve transection. GCL, ganglion cell layer; INL, inner nuclear layer; ONL, outer nuclear layer. Scale bar = 40 µm.

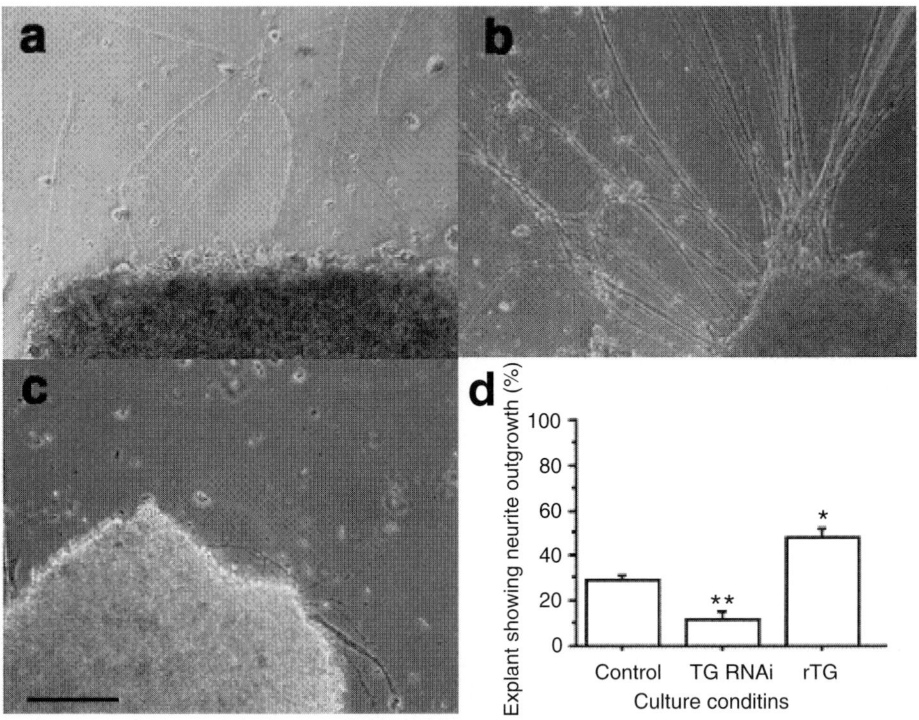

Figure 73.2. Explant culture of adult goldfish retinas treated with recombinant TG_N protein and TG_N-specific RNAi. (a) No addition. (b) Recombinant TG_N protein (4 μg/ml). (c) TG_N-specific siRNA (100 pmol/ml). (d) Graphical representation of the neurite outgrowth for 2 days of culture. Note a suppression of neurite outgrowth by TG_N-specific siRNA (**$p < 0.01$) and an enhancement of neurite outgrowth by recombinant TG_N (*$p < 0.01$) compared with control cultures. The values represent the mean ± SD in five independent experiments. Scale bar = 200 μm.

control culture was evoked in 30% of the explants (Fig. 73.2d). 21-bp siRNAs for TG_N-specific siRNA and random siRNA were chemically synthesized and transfected into retinal explants in culture using Lipofectamine 2000. Figure 73.2c shows the neurite outgrowth from retinal explants transfected with the TG_N-specific siRNA (100 pmol/ml) was clearly inhibited and only short neurites could be seen (Figure 73.2c,d). In contrast, retinal explants transfected with the random siRNA (100 pmol/ml) had no difference of neurite outgrowth of the control retina (data not shown).

5. DISCUSSION

5.1. Changes in TG_N Expression after Optic Nerve Transection

Eitan and Schwaltz reported that a crude neural TG (TG_N) enzyme preparation from injured goldfish optic nerves could partially regenerate injured rat optic nerves in vivo.[7] Hence, we used goldfish retinas for characterization of the TG_N in this study. First, to investigate the functional role of TG_N in CNS regeneration at a genetic level, we identified a full-length of TG_N cDNA clone using a retinal cDNA library from axotomized goldfish retinas.[4] The cDNA clone for TG_N encoded 678 amino acid residues with a molecular mass of 76 kDa. Levels of TG_N mRNA and protein started to increase in the goldfish retina at 10 days and peaked at 20-30 days after axotomy. Furthermore, we clearly showed that the increases in TG_N mRNA and protein was only localized in the retinal ganglion cells (RGCs) and nerve fiber layers. The period of 10-30 days for the upregulation of TG_N corresponds to a stage of axonal elongation in the goldfish visual system.[8] In our immunohistochemical study, TG_N protein in the rat RGCs rapidly decreased at 3 days after optic nerve injury (data not shown). These contrastive results of TG_N expression in the rat and goldfish retinas suggest that induction of TG_N is an important event for optic nerve regeneration in goldfish.

5.2. The Functional Role of TG_N during Optic Nerve Regeneration

TG catalyzes post-translational, covalent protein cross-linking reactions in diverse processes in nervous systems.[9] During development, TG_N activity is highest in the early postnatal stage of the mouse CNS.[10] In cerebellar granule neurons, TG inhibitors caused destabilization of neurites during the initial outgrowth period of cultured granule cells.[11] In our culture study, we clearly demonstrated that recombinant TG_N induced a drastic extension of long and thick neurites and that TG_N-specific RNAi significantly inhibited neurite outgrowth. The culture study shows that TG_N activity directly enhanced neurite outgrowth from RGCs after nerve injury via dimerization of bioactive peptides. Although the target proteins or substrates for TG_N have not yet been identified, midkine and galectin-3 are known to be present in embryonic mouse cerebellar granule neurons as putative substrates.[11,12] In a study on the partial regeneration of rat optic nerves mediated by goldfish TG_N, the authors described that interleukin-2 (IL-2) was a substrate for TG_N.[7,13] However in the current study on goldfish, the retinal explants do not contain any oligodendrocytes and therefore have no myelin inhibitory factors. Hence, cytotoxic IL-2 is not the substrate for TG_N in this goldfish optic nerve regeneration system.

6. REFERENCES

1. S. Shyne-Athwal, R.V. Riccio, G. Chakraborty, and N.A. Ingoglia, Protein modification by amino acid addition is increased in crushed sciatic but not optic nerves, *Science* **231**:603-605 (1986).
2. M. Ando, S. Kunii, T. Tatematsu, and Y. Nagata, Rapid and transient alterations in transglutaminase activity in rat superior cervical ganglia following denervation or axotomy, *Neurosci. Res.* **17**:47-52 (1993).
3. G. Chakaraborty, T. Leach, M.F. Zanakis, J.A. Sturman, and N.A. Ingoglia, Posttranslational protein modification by polyamines in intact and regenerating nerves, *J. Neurochem.* **48**:669-675 (1987).
4. S. Eitan, A. Solomon, V. Lavie, E. Yoles, D.L. Hirschberg, M. Belkin, and M. Schwartz, Recovery of visual response of injured adult rat optic nerves treated with transglutaminase, *Science* **264**:1764-1768 (1994).
5. H. Yasueda, K. Nakanishi, Y. Kumazawa, K. Nagase, M. Motoki, and H. Matsui, Tissue-type transglutaminase from red sea bream (Pagrus major). Sequence analysis of the cDNA and functional expression in Escherichia coli, *Eur. J. Biochem.* **232**:411-419 (1995).
6. T. Matsukawa, K. Sugitani, K. Mawatari, Y. Koriyama, Z. Liu, M. Tanaka, and S. Kato, Role of purpurin as a retinol-binding protein in goldfish retina during the early stage of optic nerve regeneration: its priming action on neurite outgrowth, *J. Neurosci.* **24**:8346-8353 (2004).
7. Z.W. Liu, T. Matsukawa, K. Arai, M. Devadas, H. Nakashima, M. Tanaka, K. Mawatari, and S. Kato, Na,K-ATPase alpha3 subunit in the goldfish retina during optic nerve regeneration, *J. Neurochem.* **80**:763-770 (2002).

8. T. Matsukawa, K. Arai, Y. Koriyama, Z. Liu, and S. Kato, Axonal regeneration of fish optic nerve after injury, *Biol. Pharm. Bull.* **27**:445-451 (2004).

9. M. Lesort, J. Tucholski, M.L. Miller, and G.V. Johnson, Tissue transglutaminase: a possible role in neurodegenerative diseases, *Prog. Neurobiol.* **61**:439-463 (2000).

10. M.J. Perry and L.W. Haynes, Localization and activity of transglutaminase, a retinoid-inducible protein, in developing rat spinal cord, *Int. J. Dev. Neurosci.* **11**:325-337 (1993).

11. S.A. Mahoney, M. Wilkinson, S. Smith, and L.W. Haynes, Stabilization of neurites in cerebellar granule cells by transglutaminase activity: identification of midkine and galectin-3 as substrates, *Neuroscience* **101**:141-155 (2000).

12. S. Kojima, T. Inui, H. Muramatsu, Y. Suzuki, K. Kadomatsu, M. Yoshizawa, S. Hirose, T. Kimura, S. Sakakibara, and T. Muramatsu, Dimerization of midkine by tissue transglutaminase and its functional implication, *J. Biol. Chem.* **272**:9410-9416 (1997).

13. S. Eitan and M. Schwartz, A transglutaminase that converts interleukin-2 into a factor cytotoxic to oligodendrocytes, *Science* **261**:106-108 (1993).

SURVIVAL SIGNALING IN RETINAL PIGMENT EPITHELIAL CELLS IN RESPONSE TO OXIDATIVE STRESS: SIGNIFICANCE IN RETINAL DEGENERATIONS

Nicolas G. Bazan*

1. SUMMARY

Photoreceptor survival depends on the integrity of retinal pigment epithelial (RPE) cells. The pathophysiology of several retinal degenerations involves oxidative stress-mediated injury and RPE cell death; in some instances it has been shown that this event is mediated by A2E and its epoxides. Photoreceptor outer segments display the highest DHA content of any cell type. RPE cells are active in DHA uptake, conservation, and delivery. Delivery of DHA to photoreceptor inner segments is mediated by the interphotoreceptor matrix. DHA is necessary for photoreceptor function and at the same time is a target of oxidative stress-mediated lipid peroxidation. It has not been clear whether specific mediators generated from DHA contribute to its biological properties. Using ARPE-19 cells, we demonstrated the synthesis of 10,17S-docosatriene [neuroprotectin D1 (NPD1)]. This synthesis was enhanced by the calcium ionophore A-23187, by IL-1β, or by supplying DHA. Added NPD1 (50 nM) potently counteracted H_2O_2/tumor necrosis factor-α oxidative stress-triggered apoptotic DNA damage in RPE. NPD1 also up-regulated the anti-apoptotic proteins Bcl-2 and Bcl-xL and decreased pro-apoptotic Bax and Bad expression. Moreover, NPD1 (50 nM) inhibited oxidative stress-induced caspase-3 activation. NPD1 also inhibited IL-1β-stimulated expression of COX-2. Furthermore, A2E-triggered oxidative stress induction of RPE cell apoptosis was also attenuated by NPD1. Overall, NPD1 protected RPE cells from oxidative stress-induced apoptosis. In conclusion, we have demonstrated an additional function of the RPE: its capacity to synthesize NPD1. This new survival signaling is potentially of interest in the understanding of the pathophysiology of retinal degenerations and in exploration of new therapeutic modalities.

---

*Neuroscience Center of Excellence and Department of Ophthalmology, Louisiana State University Health Sciences Center School of Medicine in New Orleans, 2020 Gravier Street, Suite D, New Orleans, LA 70112.

2. INTRODUCTION

Omega-3 fatty acids provided by the diet are necessary for retina, brain, and overall cellular functional integrity and human health (Simopoulos et al., 1999). Docosa-hexaenoic acid (22:6, n-3, DHA, found in fish oil and marine algae) is a quantitatively major omega-3 fatty acid highly concentrated in photoreceptors, brain, and retinal synapses (Bazan, 1990). Either DHA or its precursor 18:3, n-3 from the diet are initially taken up by the liver and then released into blood lipoproteins for distribution. DHA is in in high demand during photoreceptor cell biogenesis and synaptogenesis (Scott and Bazan, 1989) and has been shown to be critical for brain and retina development, excitable membrane function (Salem et al., 1986; Litman et al., 2001), memory (Catalan et al., 2002; Moriguchi and Salem, 2003), photoreceptor biogenesis and function (Wheeler et al., 1975; Stinson et al., 1991; Organisciak et al., 1996; Anderson et al., 2001; Bicknell et al., 2002; Anderson et al., 2002), and neuroprotection (Kim et al., 2000; Rotstein et al., 2003).

3. PHOTORECEPTOR RENEWAL AND THE SIGNIFICANCE OF THE RPE IN DHA CONSERVATION

In the outer segments of photoreceptors, rhodopsin is immersed in phospholipids endowed with the highest content of DHA of any cell type (Bazan, 1990; Choe and Anderson, 1990; Anderson et al., 2002). The RPE cells, which are in close contact with the photoreceptor tips, are the most active phagocytes of the body, and phagocytize the distal tips of photoreceptor outer segments in a daily process of rod outer segment renewal (Hu and Bok, 2001) that is completed by addition of new membrane to the base of the outer segments. DHA is conserved in photoreceptors by its retrieval through the interphotoreceptor matrix, which supplies the fatty acid for outer segment biogenesis (Bazan et al., 1985; Stinson et al., 1991; Gordon et al., 1992). This renewal is tightly regulated to maintain photoreceptor length and chemical composition, including that of their phospholipids. Most of the DHA in photoreceptor phospholipids is esterified in carbon-2 of the glycerol backbone, but DHA-containing molecular species of phospholipids also occupy both C1 and C2 positions of the glycerol backbone (Aveldano de Caldironi and Bazan, 1977; Wiegand and Anderson, 1983; Choe and Anderson, 1990). Retina and brain tenaciously retain DHA, even during very prolonged dietary deprivation of essential fatty acids of the omega-3 family. Dietary deprivation for more than one generation has been necessary to effectively reduce the content of DHA in retina and brain in rodents and even in non-human primates (Neuringer et al., 1986; Weisinger et al., 2002), conditions under which impairments of retinal function occur (*e.g.*, Wheeler et al., 1975; Neuringer et al., 1984).

RPE cells also participate in transport and reisomerization of bleached visual pigments, and contribute to the integrity of the blood-outer retinal barrier. Injury to the RPE, including retinal detachment or trauma, triggers cellular dysfunctions that lead to the onset and development of proliferative vitreoretinopathy.

Oxidative stress-mediated injury and cell death in RPE can in turn trigger photoreceptor death and impair vision, particularly when the macular RPE cells are affected. Oxidative stress leading to apoptosis of RPE cells is key in the pathophysiology of many retinal degenerations, such as age-related macular degenerations, including Stargardt's disease (Sieving et al., 2001; Radu et al., 2003; Sparrow et al., 2003).

Retinal DHA is a target of oxidative stress-mediated lipid peroxidation (Organisciak et al., 1996), which generates neuroprostanes from DHA through an enzyme-independent reaction (Roberts et al., 1998). In an intriguing contrast, some studies demonstrated DHA-mediated neuroprotection in photoreceptors (Politi et al., 2001; Rotstein et al., 2003) and brain (Kim et al., 2000). Does this neuroprotection result from the replenishment of DHA into membranes, or is there a selective neuroprotective signaling by a DHA-derived mediator?

4. IDENTIFICATION AND CHARACTERIZATION OF NPD1

We recently reported the isolation and structural characterization of 10,17*S*-docosatriene in ARPE-19 cells using tandem LC-PDA-ESI-MS-MS-based lipidomic analysis and ARPE-19 cells (Mukherjee et al., 2004). We termed the newly isolated dihydroxy-containing DHA derivative "neuroprotectin D1" (NPD1) (1) because of its *neuroprotective* properties in brain ischemia-reperfusion (Marcheselli et al., 2003) and in oxidative stress-challenged RPE cells (Mukherjee et al., 2004); (2) because of its potent ability to *in*-activate pro-apoptotic signaling (Mukherjee et al., 2004); and (3) because it is the first identified neuroprotective mediator of *D*HA. NPD1 synthesized by ARPE-19 cells is the same as that of the docosatriene found in human blood, glial cells, mouse brain (Hong et al., 2003), and during brain ischemia-reperfusion (Marcheselli et al., 2003). The biological activity of NPD1 seems to be exerted through potent inhibition of oxidative stress-induced apoptosis and of cytokine-triggered pro-inflammatory COX-2 gene-promoter induction (**Figure 74.1**).

5. PHOSPHOLIPASE A₂ AND NPD1 SYNTHESIS

NPD1 is formed through enzyme-mediated steps involving a phospholipase A_2 followed by a 15-lipoxygenase-like enzyme (**Figure 74.1**). Retina synthesizes mono-, di-, and trihydroxy derivatives of DHA, and certain lipoxygenase inhibitors block this synthesis, which suggests a lipoxygenase enzymatic process (Bazan et al., 1984). The availability of unesterified DHA is tightly regulated in RPE cells, as in brain, which we demonstrated in ARPE-19 cells, and which is supported by the observation that DHA pool size in retina and brain is negligible under basal, unstimulated conditions (Bazan, 1970; Aveldano and Bazan, 1974; Aveldano and Bazan, 1975). Therefore, the regulation of the phospholipase A_2 that releases free DHA is important in the pathway leading to the formation of NPD1. Ischemia or seizures elicit rapid activation of free DHA release in brain as well (Bazan, 1970; Aveldano and Bazan, 1975). The calcium ionophore A23187, or to a lesser extent, IL-1β, activates the synthesis of NPD1 in ARPE-19 cells. Under these conditions, there is a time-dependent increase in endogenous free DHA that is approximately 3- to 4-fold higher than the amount of NPD1 being synthesized.

DHA is highly concentrated as an acyl group of phospholipids in photoreceptor outer segment disc membranes (Bazan, 1990). The RPE cell actively recycles DHA from phagocytized disc membranes back to the inner segment of the photoreceptor cell (Bazan et al., 1985). In addition, the RPE cell takes up DHA from the blood stream through the choriocapillaris, and in turn supplies the fatty acid to photoreceptors through the interphoto-

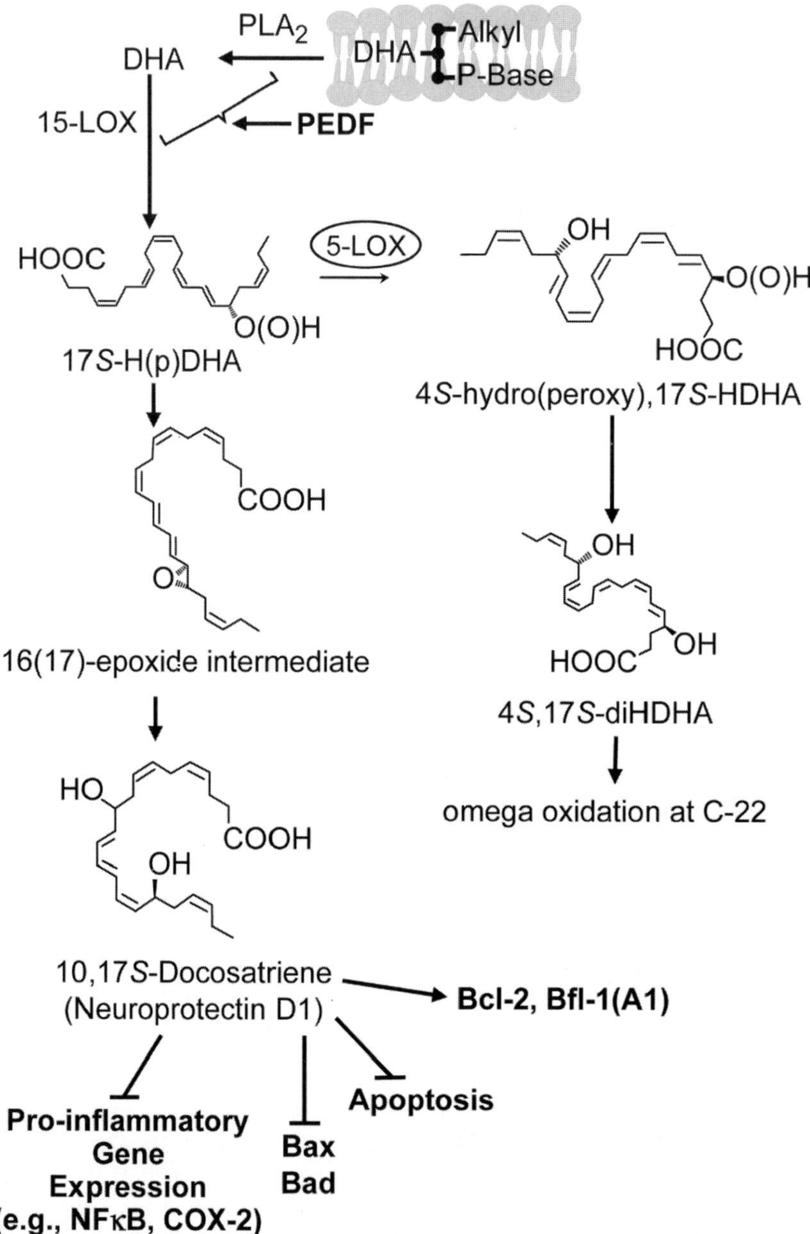

Figure 74.1. Proposed biosynthetic pathway for NPD1 synthesis. DHA esterified in membrane phospholipids is released through phospholipase A₂. PEDF (pigment epithelium-derived growth factor), a pleiotropic serpine-related growth factor, is one of the agonists recently identified (Bazan et al., 2005). DHA is then converted to 17-hydroperoxy-DHA (the 15-lipoxygenase action), which, through a possible 16(17)-epoxide intermediate, leads to the 10,17S-docosatriene, NPD1. Omega oxidation at C-22 may be a catabolic route. At the bottom are listed some of the bioactivities of NPD1.

receptor matrix (Bazan et al., 1985). This uptake is very active during early postnatal development, when photoreceptor outer segment biogenesis occurs (Scott and Bazan, 1989). In addition, active docosahexaenoyl CoA synthases in the RPE and the retina channel free DHA to acyltransferases that incorporate the fatty acid into membrane phospholipids (Reddy and Bazan, 1984). The RPE cell thus is very active in the uptake, conservation, and delivery of DHA to photoreceptors (Bazan, 1990; Bazan et al., 1985). We have demonstrated an additional function of the RPE cell; i.e., its capacity to synthesize NPD1 (Mukherjee et al., 2004). The biological activity of NPD1 may be elicited through a receptor, and in turn modulate signaling including induction of NF-κB and other transcription factors, and as a consequence, down-regulate pro-inflammatory genes (Marcheselli et al., 2003; Bazan, *in press*). NPD1 may act in autocrine fashion and/or may diffuse through interphotoreceptor matrix proteins and act in paracrine fashion on photoreceptor cells.

6. OXIDATIVE STRESS IN THE RETINAL PIGMENT EPITHELIUM

Oxidative stress triggers multiple signaling pathways, including some that are cytoprotective and others that contribute to cell damage and eventually cell death. Among these are the Bcl-2 family proteins. In fact, expression of pro- and anti-apoptotic Bcl-2 family proteins is altered by oxidative stress, and these proteins represent a major factor, insofar as the outcome of the apoptotic signaling, since cell survival reflects the predominance of one set of proteins over the other (Mattson and Bazan, *in press*). In the RPE and retina, oxidative stress, increased by several factors including retinal light exposure or reactive oxygen species, shifts the balance of the Bcl-2 family protein expression toward those that favor cell damage (e.g., Osborne et al., 1997; Liang et al., 2000). Since our results show oxidative stress-induced changes in the expression of Bcl-2 proteins (Mukherjee et al., 2004), they imply that the early RPE response to oxidative stress includes transcriptional, translational, and/or post-translational events upstream of the mitochondrial apoptotic step. In this connection we have observed that oxidative stress-triggered ARPE-19 cell damage includes changes in the expression of Bcl-2, Bcl-xL, Bax, and Bad. It is conceivable that oxidative stress, cytokines, and other intercellular signals (certain growth factors?) may activate NPD1 formation, in an effort to counteract the injury/pro-inflammatory response and restore homeostasis. Exogenous NPD1 (50 nM) promotes a differential modification in the expression of Bcl-2 family proteins under these conditions, up-regulating the protective Bcl-2 proteins and attenuating the expression of the proteins that challenge cell survival, particularly Bax and Bad (Mukherjee et al., 2004). These observations suggest a critical coordinate regulation of the availability of Bcl-2 proteins for subsequent downstream signaling. NPD1 may act at the level of signaling that regulates promoters of the genes encoding death repressors and effectors of the Bcl-2 family of proteins we studied (Mattson and Bazan, *in press*). On the other hand, translational or post-translational events may also integrate a concerted responsive machinery to counteract oxidative stress. The precise molecular mechanisms remain to be defined. The exploration of these events will provide an important insight into regulatory survival signaling. Bcl-2 family proteins regulate apoptotic signaling at the level of the mitochondrion and endoplasmic reticulum. As a consequence, cytochrome c is released from mitochondria and effector caspase-3 is activated (Mattson and Bazan, *in press*). In agreement with this sequence, we showed that oxidative stress activates caspase-3 in ARPE-19 cells and that the lipid mediator NPD1 (50 nM) decreases oxidative stress-activated caspase-3 (Mukherjee et al., 2004). Moreover, apoptosis was an outcome of TNFα/H$_2$O$_2$-induced oxidative stress in ARPE-19 cells. Interestingly, NPD1 was very effective in counteracting oxidative stress-induced apoptosis in ARPE-19 cells. This action was not mimicked by eicosanoids such as PGE$_2$, LTB$_4$, or even arachidonic acid (Mukherjee et al., 2004). This supports the selectivity of the new class of mediator, NPD1. Perhaps one of the most interesting observations is that DHA itself inhibited oxidative stress-induced apoptosis. Under those conditions, a remarkable, time-dependent formation of NPD1 occurred. Significantly, the potency of DHA for cytoprotection was much higher than that of added NPD1 (Mukherjee et al., 2004). This suggests that endogenously generated NPD1 may exert its action near the subcellular site of its synthesis. Alternatively, it may imply that other NPD1-like mediators may participate in promoting RPE cell survival (Bazan, *in press*). It is indeed possible that related NPD mediators are formed in an attempt to cope with the multiplicity of cellular signaling that has the potential of going awry in RPE or neurons when confronted with oxidative stress. In support of this possibility, brain does make a series of other potentially bioactive DHA-oxygenated derivatives, such as those generated in the presence of aspirin during ischemia-reperfusion (Marcheselli et al., 2003).

7. DHA AND RETINAL DEGENERATION

Until now it was thought that high DHA content in photoreceptors and RPE mainly endowed photoreceptor membrane domains with physical properties that contribute to the modulation of receptors (e.g., rhodopsin), ion channels, transporters, etc. In other cells, DHA modulates G-protein-coupled receptors and ion channels. Moreover, DHA has been suggested to regulate membrane function by maintaining its concentration in phosphatidylserine (Salem et al., 2001). As a target of oxidative stress, DHA is acted upon by reactive oxygen intermediates that generate DHA-peroxidation products that in turn participate in RPE and photoreceptor cell damage.

Rhodopsin mutations in rodent retinitis pigmentosa are associated with decreased photoreceptor DHA (Anderson et al., 2002), which can be interpreted as a retinal response to metabolic stress, where the strategy of decreasing the amount of the major target of lipid peroxidation, DHA, contributes to protection (Anderson et al., 2002). We propose that the retinal DHA pool size available for synthesis of neuroprotective docosanoids is compromised due to lipid peroxidation. Retinal degeneration induced by constant light promotes DHA loss from photoreceptors, and rats reared in bright cyclic light are protected, suggesting an unidentified adaptation or plasticity that may involve endogenous molecules (Li et al., 2001). Some of these may be lipid mediators, such as NPD1.

8. NPD1 PREVENTS APOPTOSIS AFTER A2E PHOTO-OXIDATION

N-retinylidene-*N*-retinylethanolamine (A2E) is the major hydrophobic fluorophore that accumulates in retinal pigment epithelial cells in Stargardt's disease and in age-related retinal degeneration (Radu et al., 2003). Blue light accelerates the production of A2E epoxides, which trigger apoptosis in RPE cells. NPD1 (50 nM) prevented A2E-induced cell death in ARPE-19 cells (Barreiro et al., 2005). We also monitored in parallel by LC-PDA-ESI-MS-MS the formation of A2E epoxides under these conditions. A2E epoxides produced oxidative stress and apoptosis in ARPE cells. NPD1 inhibited apoptosis when the A2E was added 15 minutes after blue light exposure. A2E alone, without light, did not trigger apoptosis at lower concentrations and short exposures. The accumulation of A2E in retinal pigment epithelial cells participates in apoptotic cell death and RPE responses in retinal degenerations such as Stargardt's disease. Moreover, blue light accelerates the oxidation of this compound into epoxides, and apoptosis can be prevented by NPD1 (Barreiro et al., 2005).

9. CONCLUSIONS AND EVOLVING QUESTIONS

Do growth factors contribute to the pro-survival actions of NPD1? For example, FGF2 induces bovine RPE cell survival in cultures through a sustained adaptive phenomenon that involves ERK2 activation by secreted FGF1 and ERK2-dependent Bcl-xL production (Bryckaert et al., 1999). Bcl-xL may play a key role in integrating and transmitting exogenous FGF2 signals for RPE cell survival. Very recently evidence has been provided that PEDF promotes NPD1 synthesis in human RPE cell primary cultures (Bazan et al., 2005). These issues present intriguing possibilities for future investigations.

RPE cell damage and apoptosis impair photoreceptor cell survival, a dominant factor in age-related macular degeneration (Hinton et al., 1998). In Stargardt's disease (a juvenile form of macular degeneration), oxidative stress mediated by the lipofuscin fluorophore A2E damages RPE, and caspase-3 is part of the damaging cascade; whereas Bcl-2 exerts cellular protection (Sparrow and Cai, 2001).

Thus NPD1, a DHA-derived mediator endogenously synthesized by neuro-epithelium-derived RPE cells, is a modulator of signaling pathways that promote cell survival (Bazan, *in press*). One pathway is the regulation of Bcl-2 family protein expression, a pre-mitochondrial apoptotic target of NPD1 under conditions of oxidative stress. Consequently, downstream signaling, including effector caspase-3 activation and DNA degradation, is attenuated (Mukherjee et al., 2004). NPD1 also potently counteracted cytokine-triggered pro-inflammatory COX-2 gene induction, another major factor in cell damage (Mukherjee et al., 2004). In ischemia-reperfusion-injured hippocampus and in neural progenitor cells stimulated by IL-1β, COX-2 expression seems to be related to NF-κB activation. NPD1 inhibits NF-κB and COX-2 induction under those conditions (Marcheselli et al., 2003). A similar regulatory mechanism may operate in RPE cells; *i.e.*, NPD1 down-regulation of cytokine-mediated NF-κB activation. Pro-inflammatory injury of the RPE can promote pathoangiogenesis and proliferative vitreoretinopathy, hallmarks of several diseases, including diabetic retinopathy. NPD1's neuroprotective bioactivity in brain ischemia-reperfusion includes decreased infarct size and inhibition of polymorphonuclear leukocyte infiltration (Marcheselli et al., 2003). The addition of DHA to the culture medium promoted strong cytoprotection when RPE cells were confronted with oxidative stress (Mukherjee et al., 2004). In vivo the active DHA supply to brain and retina from the liver through the blood stream is necessary for cell development and function, and may play a critical role in conditions where, due to enhanced oxidative stress, the polyunsaturated fatty acyl chains of membrane phospholipids are decreased as a consequence of lipid peroxidation, as occurs in aging, retinal degenerations, and neurodegenerations such as Alzheimer's disease (Scott and Bazan, 1989; Nourooz-Zadeh et al., 1999). A greater understanding of the signals that modulate NPD1 synthesis may provide windows for therapeutic intervention in neurodegenerative diseases. In addition, selective DHA-delivery systems to the retina and brain may promote neuroprotection. Moreover, NPD1 and its cellular activities might be manipulated by novel approaches that result in RPE cytoprotection and enhanced photoreceptor survival in retinal degenerations.

10. ACKNOWLEDGMENT

This work was supported by NIH grants EY05121 (NEI) and P20RR16816 (from the COBRE Program, NCRR), the Eye, Ear, Nose and Throat Foundation, and the American Health Assistance Foundation.

11. REFERENCES

Anderson, R. E., Maude, M. B., and Bok, D., 2001, Low docosahexaenoic acid levels in rod outer segment membranes of mice with rds/peripherin and P216L peripherin mutations. *Invest. Ophthalmol. Vis. Sci.* **42**:1715-1720.

Anderson, R. E., Maude, M. B., McClellan, M., Matthes, M. T., Yasumura, D., and LaVail, M. M., 2002, Low docosahexaenoic acid levels in rod outer segments of rats with P23H and S334ter rhodopsin mutations. *Mol. Vis.* **8**:351-358.

Aveldano, M. I., and Bazan, N. G., 1974, Displacement into incubation medium by albumin of highly unsaturated retina free fatty acids arising from membrane lipids. *FEBS Lett.* **40**:53-56.

Aveldano, M. I., and Bazan, N. G., 1975, Differential lipid deacylation during brain ischemia in a homeotherm and a poikilotherm. Content and composition of free fatty acids and triacylglycerols. *Brain Res.* **100**:99-110.

Aveldano de Caldironi, M. I., and Bazan, N. G., 1977, Acyl groups, molecular species, and labeling by 14C-glycerol and 3H-arachidonic acid of vertebrate retina glycerolipids. *Adv. Exp. Med. Biol.* **83**:249-256.

Barreiro, S. G., Marcheselli, V. L., and Bazan, N. G., 2005, Human retinal pigment epithelial cells protected by NPD1 after A2E-epoxide induction. ARVO abstract B224.

Bazan, N. G., 1970, Effects of ischemia and electroconvulsive shock on free fatty acid pool in the brain. *Biochim. Biophys. Acta* **218**:1-10.

Bazan, N. G., 1990, Supply of n-3 polyunsaturated fatty acids and their significance in the central nervous system. *Nutrition and the Brain, vol. 8*, R. J. Wurtman, and J. J. Wurtman, eds., Raven Press, Ltd., New York, pp. 1-24.

Bazan, N. G., *in press* Eicosanoids, docosanoids, platelet-activating factor, and inflammation. *Basic Neurochemistry 7th ed.* G. Siegel, R. W. Albers, S. Brady, and D. Price, eds. London, Elsevier.

Bazan, N. G., Birkle, D. L., and Reddy, T. S., 1984, Docosahexaenoic acid (22:6, n-3) is metabolized to lipoxygenase reaction products in the retina. *Biochem. Biophys. Res. Commun.* **125**:741-747.

Bazan, N. G., Birkle, D. L., and Reddy, T. S., 1985, Biochemical and nutritional aspects of the metabolism of polyunsaturated fatty acids and phospholipids in experimental models of retinal degeneration. *Retinal Degeneration: Experimental and Clinical Studies.* M. M. LaVail, R. E. Anderson, and J. Hollyfield, eds. Alan R. Liss, Inc., New York, pp. 159-187.

Bazan N. G., Marcheselli, V. L., Hu, J., Finley, J., Bok, D., and Chandamuri, B., 2005, Pigment epithelium-derived growth factor (PEDF) selectively up-regulates NPD1 synthesis and release throught the apical side of human RPE cells in primary cultures. ARVO abstract B141.

Bicknell, I. R., Darrow, R., Barsalou, L., Fliesler, S. J., and Organisciak, D. T., 2002, Alterations in retinal rod outer segment fatty acids and light-damage susceptibility in P23H rats. *Mol. Vis.* **8**:333-340.

Bryckaert, M., Guillonneau, X., Hecquet, C., Courtois, Y., and Mascarelli, F., 1999, Both FGF1 and bcl-x synthesis are necessary for the reduction of apoptosis in retinal pigmented epithelial cells by FGF2: role of the extracellular signal-regulated kinase 2. *Oncogene.* **18**:7584-7593.

Catalan, J., Moriguchi, T., Slotnick, B., Murthy, M., Greiner, R. S., and Salem, N. Jr., 2002, Cognitive deficits in docosahexaenoic acid-deficient rats. *Behav. Neurosci.* **116**:1022-1031.

Choe, H.-G., and Anderson, R. E., 1990, Unique molecular species composition of glycerolipids of frog rod outer segments. *Exp. Eye Res.* **51**:159-165.

Gordon, W. C., Rodriguez de Turco, E. B., and Bazan, N. G., 1992, Retinal pigment epithelial cells play a central role in the conservation of docosahexaenoic acid by photoreceptor cells after shedding and phagocytosis. *Curr. Eye Res.* **11**:73-83.

Hinton, D. R., He, S., and Lopez, P. F., 1998, Apoptosis in surgically excised choroidal neovascular membranes in age-related macular degeneration. *Arch. Ophthalmol.* **116**:203-209.

Hong, S., Gronert, K., Devchand, P. R., Moussignac, R. L., and Serhan, C. N., 2003, Novel docosanoids and 17S-resolvins generated from docosahexaenoic acid in murine brain, human blood, and glial cells: autacoids in anti-inflammation. *J. Biol. Chem.* **278**:14677-14687.

Hu, J., and Bok, D., 2001, A cell culture medium that supports the differentiation of human retinal pigment epithelium into functionally polarized monolayers. *Mol. Vis.* **7**:14-19.

Kim, H. Y., Akbar, M., Lau, A., and Edsall, L., 2000, Inhibition of neuronal apoptosis by docosahexaenoic acid (22:6n-3). Role of phosphatidylserine in antiapoptotic effect. *J. Biol. Chem.* **275**:35215-35223.

Li, F., Cao, W., and Anderson, R. E., 2001, Protection of photoreceptor cells in adult rats from light-induced degeneration by adaptation to bright cyclic light. *Exp. Eye Res.* **73**:569-577.

Liang, Y. G., Jorgensen, A. G., Kaestel, C. G., Wiencke, A. K., Lui, G. M., la Cour, M. H., Ropke, C. H., and Nissen, M. H., 2000, Bcl-2, Bax, and c-Fos expression correlates to RPE cell apoptosis induced by UV-light and daunorubicin. *Curr. Eye Res.* **20**:25-34.

Litman, B. J., Niu, S. L., Polozova, A., and Mitchell, D. C., 2001, The role of docosahexaenoic acid containing phospholipids in modulating G protein-coupled signaling pathways: visual transduction. *J. Mol. Neurosci.* **16**:237-242.

Marcheselli, V. L. Hong, S., Lukiw, W. J., Tian, X. H., Gronert, K., Musto, A., Hardy, M., Gimenez, J. M., Chiang, N., Serhan, C. N., and Bazan, N. G., 2003, Novel docosanoids inhibit brain ischemia-reperfusion-mediated leukocyte infiltration and pro-inflammatory gene expression. *J. Biol. Chem.* **278**:43807-43817. Erratum in: *J. Biol. Chem.* 2003, **278**:51974.

Mattson, M. P., and Bazan, N. G., *in press*, Apoptosis and necrosis. *Basic Neurochemistry 7th ed.* G. Siegel, R. W. Albers, S. Brady, D. Price, eds., London, Elsevier.

Moriguchi, T., and Salem, N. Jr., 2003, Recovery of brain docosahexaenoate leads to recovery of spatial task performance. *J. Neurochem.* **87**:297-309.

Neuringer, M., Connor, W. E., Van Petten, C., and Barstad, L., 1984, Dietary omega-3 fatty acid deficiency and visual loss in infant rhesus monkeys. *J. Clin. Invest.* **73**:272-276.

Neuringer, M., Connor, W. E., Lin, D. S., Barstad, L., and Luck, S., 1986, Biochemical and functional effects of prenatal and postnatal omega 3 fatty acid deficiency on retina and brain in rhesus monkeys. *Proc. Natl. Acad. Sci. U. S. A.* **83**:4021-4025.

Nourooz-Zadeh, J., Liu, E. H. C., Yhlen, B., Änggård, E. E., and Halliwell, B., 1999, F4-isoprostanes as specific marker of docosahexaenoic acid peroxidation in Alzheimer's disease. *J. Neurochem.* **72**:734-740.

Organisciak, D. T., Darrow, R. M., Jiang, Y. L., and Blanks, J. C., 1996, Retinal light damage in rats with altered levels of rod outer segment docosahexaenoate. *Invest. Ophthalmol. Vis. Sci.* **37**:2243-2257.

Osborne, N. N., Cazevieillem C., Pergandem G., and Wood, J. P., 1997, Induction of apoptosis in cultured human retinal pigment epithelial cells is counteracted by flupirtine. *Invest. Ophthalmol. Vis. Sci.* **38**:1390-1400.

Politi, L. E., Rotstein, N. P., and Carri, N. G., 2001, Effect of GDNF on neuroblast proliferation and photoreceptor survival: additive protection with docosahexaenoic acid. *Invest. Ophthalmol. Vis. Sci.* **42**:3008-3015.

Radu, R. A., Mata, N. L., Nusinowitzm S., Lium X., Sieving, P. A., and Travis, G. H., 2003, Treatment with isotretinoin inhibits lipofuscin accumulation in a mouse model of recessive Stargardt's macular degeneration. *Proc. Natl. Acad. Sci. USA* **100**:4742-4747.

Reddy, T. S., and Bazan, N. G., 1984, Synthesis of arachidonoyl coenzyme A and docosahexaenoyl coenzyme A in retina. *Curr. Eye. Res.* **3**:1225-1232.

Roberts, L. J. 2nd, Montine, T. J., Markesbery, W. R., Tapper, A. R., Hardy, P., Chemtob, S., Dettbarn, W. D., and Morrow, J. D., 1998, Formation of isoprostane-like compounds (neuroprostanes) in vivo from docosahexaenoic acid. *J. Biol. Chem.* **273**:13605-13612.

Rotstein, N. P., Politi, L. E., German, O. L., and Girotti, R., 2003, Protective effect of docosahexaenoic acid on oxidative stress-induced apoptosis of retina photoreceptors. *Invest. Ophthalmol. Vis. Sci.* **44**:2252-2259.

Salem, N. Jr, Kim, H. Y., and Yergey, J. A., 1986, Docoshexaenoic acid: membrane function and metabolism. *The Health Effects of Polyunsaturated Fatty Acids in Seafoods*, A. P. Simopoulos, R. R. Kifer, and R. Martin, eds., Academic Press, New York, NY, pp. 263-317.

Salem, N. Jr., Litman, B., Kim, H. Y., and Gawrisch, K., 2001, Mechanisms of action of docosahexaenoic acid in the nervous system. *Lipids.* **36**:945-959.

Scott, B. L., and Bazan, N. G., 1989, Membrane docosahexaenoate is supplied to the developing brain and retina by the liver. *Proc. Natl. Acad. Sci. U. S. A.* **86**:2903-2907.

Sieving, P. A., Chaudhry, P., Kondo, M., Provenzano, M., Wu, D., Carlson, T. J., Bush, R. A., and Thompson, D. A., 2001, Inhibition of the visual cycle in vivo by 13-cis retinoic acid protects from light damage and provides a mechanism for night blindness in isotretinoin therapy. *Proc. Natl. Acad. Sci. USA* **98**:1835-1840.

Simopoulos, A. P., Leaf, A., and Salem, N. Jr., 1999, Workshop on the essentiality of and recommended dietary intakes for omega-6 and omega-3 fatty acids. *J. Am. Coll. Nutr.* **18**:487-489.

Sparrow, J. R., and Cai, B., 2001, Blue light-induced apoptosis of A2E-containing RPE: involvement of caspase-3 and protection by Bcl-2. *Invest. Ophthalmol. Vis. Sci.* **42**:1356-1362.

Sparrow, J. R., Vollmer-Snarr, H. R., Zhou, J., Jang, Y. P., Jockusch, S., Itagaki, Y., and Nakanishi, K., 2003, A2E-epoxides damage DNA in retinal pigment epithelial cells. Vitamin E and other antioxidants inhibit A2E-epoxide formation. *J. Biol. Chem.* **278**:18207-18213.

Stinson, A. M., Wiegand, R. D., and Anderson, R. E., 1991, Recycling of docosahexaenoic acid in rat retinas during n-3 fatty acid deficiency. *J. Lipid Res.* **32**:2009-2017.

Weisinger, H. S., Armitage, J. A., Jeffrey, B. G., Mitchell, D. C., Moriguchi, T., Sinclair, A. J., Weisinger, R. S., and Salem, N. Jr., 2002, Retinal sensitivity loss in third-generation n-3 PUFA-deficient rats. *Lipids.* **37**:759-765.

Wheeler, T. G., Benolken, R. M., and Anderson, R. E., 1975, Visual membranes: specificity of fatty acid precursors for the electrical response to illumination. *Science.* **188**:1312-1314.

Wiegand, R. D., and Anderson, R. E., 1983, Phospholipid molecular species of frog rod outer segment membranes. *Exp. Eye Res.* **37**:159-173.

INDEX

Note: Page numbers followed by f indicate figures; t, tables.

A

A1E and A2E (RPE lipofuscin fluorophores)
cytotoxicity of, 63-67, 69-72
structures of, 64-65
UV spectrum of, 71-72
AAV. *See* Adeno-associated virus (AAV)
AAV5-mOPs-RS1 treatment, 286-288
ABCA4 (ABCR or rim protein), 465-469,
466f
Acetazolamide, 439-445
Acidification, retinal, 414, 439, 444
Activity-dependent mechanisms, 389,
392-394
Activity-independent mechanisms, 389-392
Actograms, 242f
Adeno-associated virus (AAV)
characteristics of, 236-238, 237t
derived vector, 248
disadvantages of, 256
mediated gene delivery by, 267-273
recombinant (rAAV), 240, 242-244, 270
Adenosine triphosphate (ATP) binding,
466-468
Adult-onset foveomacular dystrophy
(AOFMD), 35-39
Agarose beads, 520, 521t, 522f
Age-related macular degeneration (AMD)
animal models of, 111-114
clinical subtypes of, 41
and defective phagocytosis, 120-122
defined, 133
dry or wet exudative type, 109
genetic model of, 41-47
geographic atrophy type, 109
pathogenesis of, 69, 109-114
and phototherapy damage, 297
prevalence of, 377
resemblance to L-ORD, 47
risk factors for, 41-42, 110, 112, 114
and Stargardt-causing mutations, 465
Aging studies, 522
AIPL1. *See* Aryl hydrocarbon receptor
interacting protein like-1 (AIPL1)
All-*trans*-retinal
formation of, 70
removal in photoreceptors, 465-467, 469,
505-509
Alleles
in *trans*, 4, 6
variants in LCA, 11-13
αvβ5 integrin, 120-121, 500-501
Alphatransducin (Tα), 125-130
Alzheimer's disease, 110, 112
Amacrine cells, 379-380, 405-408
AMD. *See* Age-related macular degeneration
(AMD)
Amines, biogenic, 70
Amino-retinoid compounds, 69-73
formation of, 70
UV spectrum of, 72f
Amphibia, 390
Amphiphilic electrostatic attraction, 67
Amyotrophic lateral sclerosis (ALS), 147
Aneurysms. *See* Microaneurysms;
Neovascularization (NV)
Angiogenesis
in diabetic retinopathy, 187
genetic factors in, 35-39

and laser photocoagulation, 198-199
oxygen regulation of, 425
phototherapy (PDT) for, 297
therapies for, 192
VEGF and, 187-192, 303-306
Animal Enclosure Model-nursing facility
    (AEM-NF), 418
Animal models
  of AIPL1 in retina, 89-93
  amphibia as, 390
  APOE TR (targeted gene replacement) in,
    111-114
  β5 knockout mouse, 120-122
  for behavioral testing, 169-172
  chicken retina, 316-317
  crossbreeding in, 96
  Drosophila, 460-462
  fish retina
    characteristics of, 378
    goldfish (*Carrassius auratus*), 333-337,
      390-392, 394, 525-529
    zebrafish (*Danio rerio*), 201-206
  of HRGP-*cre*/R26R mice, 174-177
  light/dark adaptation in, 125-126, 129f,
    135f
  lizards (*Ctenophorus ornatus*), 392-394
  Long Evans rats, 500
  neonatal rats, 418-423, 447-450
  primates, 398, 401
  for retinal neovascularization, 163-167
  RGC survival in, 390
  Royal College of Surgeons (RCS) rats, 95,
    321-325, 368, 500
  for RP, 81-86
  of RPE cell mutations, 95-99
  Sprague-Dawley rats, 418-423
  transgenic, 141-145, 148, 167, 180,
    201-206, 368
  Xenopus laevis frog, 141-145, 219-222,
    456-457
Annexins (lipocortins)
  antibodies of, 76-78
  in Bruch's membrane and drusen, 75-78
  functions of, 76, 78
  on immunohistochemistry, 77f
Antibodies
  of annexin in Bruch's membrane and
    drusen, 76-78, 77f
  for immunostaining, 180-181, 212,
    334-335
  to localize Cre expression, 174
    neuron-specific β-tubulin III isoform
      (TUJI), 399-401
    polyclonal anti-opsin (1D4), 127, 128f
Antisense oligonucleotides, 304
AOFMD (adult-onset foveomacular
    dystrophy), 35-39
*APOE*4 TR mice retina, 109-114
Apoptosis, 276-277, 363-366
Aptamers, 304-305
Argon lasers, 196
Arl2 and Arl3, 487, 489
ARPE-19 cells, 63-67
Arrestin, 455-463
  bound to phospholipids, 459-460, 461f
  lysine mutants of, 457
  myc-tagged (EQKLISEEDL), 456, 459
  translocation of, 457-459
Arteries. *See specific arteries*
Artificial silicon retina (ASR™), 321-325
  design and surgical procedure for, 322
  mechanisms of, 321
Aryl hydrocarbon receptor interacting protein
    like-1 (AIPL1)
  activities of, 471-473
  biochemical function of, 89-93
  chaperone function of, 471-475
  holozyme stability in, 92-93
  knockdown, 91
  in photoreceptors, 472
Astrocytes, 211
Astronauts, 422-423
ATP (adenosine triphosphate) binding,
    466-468
Aura, epileptic, 24f
Autosomal dominant retinitis pigmentosa
    (adRP)
  genetic factors in, 3-7
  siRNA technology and, 233-238
Autosomal recessive optic atrophy (ROA1)
  candidate gene studies in, 25-26
  and chromosome 8q21-q22, 21-26, 24f
  ERG and VER in, 24f
  pedigree and haplotype in, 24-26

B
Bardet-Biedl syndrome, 13
Basic fibroblast growth factor (bFGF),
    197-199, 357-360
Bax protein, 334-335
BCD. *See* Bietti crystalline corneoretinal
    dystrophy (BCD)

BDNF. *See* Brain-derived neurotrophic factor
    (BDNF)
Behavioral testing
   circadian rhythm-based approach, 239-244
   of mouse model vision, 169-172
   for visual neural responses, 369
β5 null retina, 501
Beta2/NeuroD transcription factor, 155-160
   and photoreceptor cell loss, 157f
   relationship to *Bhlhb4*, 156
   role in pancreas, 159
   scotopic ERGs with, 158-159
BHLH (basic helix loop helix) transcription
    factor, 155-160
Bhlhb4 transcription factor
   and photoreceptor cell loss, 157f
   relationship to *Beta2/NeuroD*, 156
   role in rods, 159
   scotopic ERGs with, 158-159
Bietti crystalline corneoretinal dystrophy
    (BCD)
   ERG of, 51t
   genetic factors in, 49-52
   lipid metabolism and, 50-52
Blindness
   causes of, 133, 297, 303, 349
   congenital, 9-13, 15, 89, 206
   and POS phagocytosis, 499
Blood clots, 187
Blood vessels
   abnormal. *See* Choroidal
      neovascularization (CNV)
   aneurysms in. *See* Microaneurysms
   hemorrhage of, 196
   occlusions of, 310-313, 399f
   retinal vessel attenuation, 96-97
   See also specific vessels
Blotting techniques, 44-47, 250f
Bmi1 repressor loss, 209-214
Bone marrow-derived stem cells (MSCs),
    384-386
Brain-derived neurotrophic factor (BDNF)
   for neuroprotection, 447-449
   transplantation of, 363-366
   vitreous injections of, 297-301
Brain-derived stem cells (SCs), 384
Brain development, 211
Bruch's membrane
   annexins in, 75-78
   neuronal breaching of, 409f
   thickening of, 112-114

C
C-terminal segment, 285
Cadherins, 350-352f
CAIs. *See* Carbonic anhydrase inhibitors
    (CAIs)
Calbindin, 379-380
Candidate genes, 25-26
Capillary beds, 164-166, 165f, 167t
Carbonic anhydrase inhibitors (CAIs)
   applications for, 439
   dose-dependence of, 441
   effects of, 439-445, 442-443f
   for macular edema, 521-522
Caspase-3 activity, 333-336, 365, 449-450,
    531
Cathepsin D (CatD), 267-268, 270-273
CD36 (lipid scavenger receptor), 499
Cell membranes
   detergent-resistant, 491-495
   G90D mutation in, 125-130, 127f
   permeability of, 64-67
   retinoid transport across, 465-469
   transmembrane proteins in, 179
Cells
   acidification of, 414, 439, 444
   amacrine, 379-380, 405-408, 409f
   astrocytes, 211
   bipolar, 407-408
   culture reagents for, 269
   differentiation of, 155
   embryonic stem (ESs), 173, 206
   ganglion, 276-277, 405-406, 421
   membranes of. *See* Cell membranes
   Müller glial cells, 209, 277-278, 327,
     329-331, 405-407, 445
   neural and retinal stem cells, 210-211,
     377-380, 383
   neural progenitor cells (NPCs), 211
   programmed death of, 315
   signals of, 275-280, 284, 335-336, 531-538
   survival of, 6-7, 133, 150-151, 315-318,
     337-340, 390, 434
   toxic damage to, 63-67, 69-72, 291-294,
     425-435
   See also specific cells
Cellular acidification, 414, 439, 444
Cellular retinaldehyde-binding protein
    (CRALBP), 477-481
   characteristics of, 477
   domain structural model, 478-479
   and EBP50, 521

immunohistochemistry of, 212-214
ligand interactions in, 478-479
mutation-related diseases of, 477-478
protein interactions in, 479-481
Central polypurine tract (CPPT), 256
Central retinal artery occlusions (CRAOs),
310-313
Central retinal vein occlusions (CRVOs),
310-313
Cephalad fluid shifts, 417
Ceruloplasmin, 522
Chaperone function, 471-475
Chicken retina, 316-317
Choline aceytltransferase, 179, 183
Chondroitin sulfate, 365
Choroid
atrophy of, 57-60
neovascularization. *See* Choroidal
neovascularization (CNV)
neuronal escape into, 409f
Choroidal neovascularization (CNV)
in *APOE*4 TR-ch mice, 109-114
in diabetic retinopathy, 187
genetic factors in, 35-39
oxygen regulation of, 425
phototherapy for (PDT), 198-199, 297
VEGF and, 187-192, 303-306
Choroideremia
histopathology of, 57, 60
optical coherence tomography (OCT) of,
57-60, 59f
Chromophore isolation, 104-106
Chromosomes
8q21-q22 (ROA1 locus), 21-26, 24f,
471-475
Xp21.1, 29
Ciliary margin zone (CMZ), 378
Ciliary neurotrophic factor (CNTF)
characteristics of, 147, 150, 212
effects on photoreceptors, 292
neuroprotection from, 331
Circadian rhythms
acrophase of, 241-242, 243f
characteristics of, 411-412
defined, 240
photoentrainment of, 239-244
and susceptibility to light damage, 411-414
CME (cystoid macular edema), 310-313
CMZ (ciliary margin zone), 378
CNV. *See* Choroidal neovascularization
(CNV)

Cochlear dysfunction, 341-347
Collagen X, 47
Colour vision, 23f, 439
Cone loss, 133, 315-318
Cone outer layer (COS), 125, 129
Cone-rod homeobox transcription factor
(CRX), 150-151
Cones
CAIs effect on, 441-445
conditional gene knockout system in,
173-177
cone-rod dystrophy, 10, 13
Cre expression in, 175-177
defined, 133
degeneration of, 90-91, 93, 105-106, 141,
143-144
delayed photoresponse in, 133-138
distribution and function of, 133, 176-177,
315
gene expression of, 147-151
hyperoxic effects on, 427f, 428, 431, 434
light flash response of, 83-84
microvilli ensheathing of, 519
photosensitivity of, 102-103
pigment regeneration in, 101-106
transcriptional regulation in, 151
of zebrafish, 201
See also Photoreceptors; Rods
Congenital stationary night blindness
(CSNB), 125, 129, 159-160
Contrast sensitivity measurements, 170-172,
171f
Corneas, 49-52, 51t
Corneoretinal dystrophy, 49-52, 51t
Corticosteroids, 309-313
Cosmic rays, 422
CPPT (central polypurine tract), 256
CRALBP. *See* Cellular retinaldehyde-binding
protein (CRALBP)
CRAOs (central retinal artery occlusions),
310-313
Cre expression, 173-177
Cre/*lox* (conditional knockout) system,
173-177, 206
CRVOs (central retinal vein occlusions),
310-313
CRX (cone-rod homeobox) transcription
factor, 150-151
Crystal deposits, 49-52
CSNB (congenital stationary night
blindness), 125, 129, 159-160

Cyclic GMP phosphodierserase 6 (PDE6), 485-489
   function and localization in photoreceptors, 487-489
   interactions of, 485-487
   structure of, 487
Cyclohexamide (CHI), 456-458
Cystoid macular edema (CME), 310-313
Cysts, intraretinal, 313f
Cytokines
   effect on developing CNS, 279
   examples of, 279
   and retinal degeneration, 275-280
   suppressors of cytokine signaling (SOCS), 275-280
   See also specific cytokines
Cytoskeleton (actin), 351, 352f

D
Dark-adaptation, 125-126, 129f, 135f, 508-509
Deafness, 341, 349
Dendritic structure, 398-401
Detergent-resistant membranes (DRMs), 491-495
   absence of phosphatase in, 493
   fractionation and phosphorylation of, 492, 494f
DHA. *See* Docosa-hexaenoic acid (DHA)
Diabetes mellitus (DM), 159, 187-192
Diabetic retinopathy (DR)
   characteristics of, 195-196
   laser photocoagulation for, 195-199
   and neovascularization, 303-306
   pathogenesis of, 163-167, 187-192
Discoidin (DS) domain, 284f, 285
DNA
   injected for transgenesis, 203
   plasmid construction, 268-269, 363
   regulatory element of, 219-220
DNA tracings, 60f
Docosa-hexaenoic acid (DHA), 532-538
   endogenous free, 533-535
   for neuroprotection, 536-538
Dominant negative effect, 142-145
Dopamine, 70-71, 412-413
DRMs. *See* detergent-resistant membranes (DRMs)
Drosophila, 460-462
Drusen
   annexins in, 75-78
   and ApoE protein, 110, 112, 114

E
EAATs (excitatory amino acid transporters), 327
EBP50 (NHERF-1), 480-481
Edema
   carbonic anhydrase inhibitors (CAIs) for, 439, 521-522
   macular. *See* Macular edema
   in retinitis pigmentosa (RP), 309-313
EGF (epithelial growth factor), 357-360
Electron micrographs (EMs)
   of *APOE* TR mice retina, 113-114
   of GABAergic amacrine cell, 409f
   of myosin VIIa labelling, 342-344
   of normal versus abnormal RPE cells, 98f
Electroretinograms (ERGs)
   a-wave amplitudes in, 159, 286, 442-444
   with AAV5-mOPs-RS1 treatment, 286
   after BDNF injection and phototherapy, 301f
   b-wave amplitudes in, 135-136f, 158-159, 177f, 283, 286, 441-444
   of Bietti crystalline corneoretinal dystrophy (BCD), 51t, 52f
   CAI effect on, 439-445, 442-443f
   changes with ischemia on, 330-331
   of delayed photoresponse, 134-136
   flash testing on
      with CAIs onboard, 440, 442f
      with cell loss, 158f
      preparation for, 82
      with subretinal injection, 252f
   flicker-type, 102f, 440-441, 442f, 443
   fullfield and multifocal, 368
   with MerTK, 120
   normal response in, 24f
   of Rdh11 mice, 509f
   rod and cone responses on, 83-84, 91f, 102f
   scotopic
      of AMD, 122f
      of *BETA2/NeuroD* cell loss, 158f
      of rod and cone responses, 102f
   with subretinal implant, 322-323
Embryonic stem (ES) cells
   versus adult (ASCs), 382-383
   homologous gene-targeted, 173
   of zebrafish, 206
Endoplasmic reticulum (ER) membrane, 284, 489f
Enhanced GFP (EGFP) reporter, 204-205
Enhancer trapping, 204

ENU (N-ethyl-N-nitrosourea), 95-97
Enzymatic activity. *See specific enzymes*
EphA ephrins, 391-392
Episcleral vein, 399f
Epithelial growth factor (EGF), 357-360
ERGs. *See* Electroretinograms (ERGs)
ERM-binding phosphoprotein 50 (EBP50),
     480-481, 521
Excitatory amino acid transporters (EAATs),
     327
Exons, 29-32
Eye extract, 180-181, 183
Eyes
    adult human, 379-380
    differences between species, 377-378
    See also Animal models

F
FACS (fluorescence activated cell sorter)
     analysis, 269-270
FAK (focal adhesion kinase), 500-502
Farnesylation, 92
Fetal bovine serum (FBS), 513-517
Fetal retinal sheet transplants, 372-374
Filamin A (actin-binding protein), 351, 352f
Fish retina. *See under* Animal models
FITC (fluorescein isothiocyanate), 358
Flowcharts
    clinical for LCA, 17f
    decisional for LCA diagnosis, 19f
Fluorescein angiography
    of Bietti crystalline corneoretinal
         dystrophy (BCD), 51f
    of choroidal neovascularization, 37-38f
    of posterior pole atrophy, 58f
    of retinal perfusion, 188-192
    of vasoproliferative retinopathy, 188-192
Fluorescein isothiocyanate (FITC), 358
Fluorescence activated cell sorter (FACS)
     analysis, 269-270
Fluorescence micrographs, 271f
Fluorescent microscopy
    A1E and ARPE-19 cells on, 66f
    of Cre expression, 175t
    of myosin VIIa labelling, 342-343
Focal adhesion kinase (FAK), 500-502
Fovea centralis
    adult-onset foveomacular dystrophy
         (AOFMD), 35-39
    thickness of, 311-313
Functional titers, 258-259

Fundus changes (choroideremia), 57-60
Fundus photography
    of Bietti crystalline corneoretinal
         dystrophy (BCD), 51f
    of GFP expression, 262-263f
    of posterior pole atrophy, 58f
    of retinal vessel attenuation, 96-97

G
G protein-coupled receptor kinase 1 (GRK1),
     134-138
G90D (aspartic acid), 125-130
GABA-ergic inhibition, 409f
GAL4-VP16 transactivator, 272
Ganglion cells, 276-277, 405-406, 421
GDNF (glial cell line-derived neurotrophic
         factor), 212
Gel electrophoresis
    for molecular weight, 46f
    SDS-polyacrylamide (PAGE), 44-47
Gender, 6-7
Gene therapy (GT)
    assessing efficacy of, 239-244
    for LCA, 247-252
    for neovascularization, 305
    for retinoschisis, 285-288
    viral vectors for, 305
Genes
    AIPL1, 89-93
    APOE, 110-114
    β-*PDE*, 217-228
        and α-PDE, 225-228
        binding of, 221, 223
        expression in, 217-218
        functional testing of, 220-221
        promotor, 218-228
        transcriptional studies of, 218-225, 222f
        translational efficiency of, 225-228, 226f
    bFGF/FGF2, 198-199
    bHLH class B, 155-160
    C1QTNF5, 41-47
        assembly and stability testing of, 45-47
        domain constructs for, 42-43
        expression and purification of, 44
        molecular size/weight of, 43, 46f
    candidate, 25-26
    CYP4V2, 49-52
    exons of, 30-32
    expression changes after photocoagulation,
         198-199
    expression regulation in, 217-228

G90D (aspartic acid), 125-130
Gcgr (glucagon receptor gene), 170
GFAP, 198
Grk1, 134-138
GUCY2D, 16-18
HRGP-Cre, 173-177
hVEGF, 164-167
IMPDH1
    enzymatic activity of, 82, 84
    expression studies on, 82-85
    photoreceptor degeneration in, 81-86
    and retinal development, 84
LCA-related, 9-13, 15-19
Mertk, 95, 97
mutations of. *See* Genetic mutations
MYO7A, 341
PDGF, 198-199
peripherin/*rds*, 35-39
    mutations in, 141-145
    structure and function of, 145
and photoentrainment, 239-244
R26R (Cre-activatable *lacZ* reporter),
    173-177
Rd1-Bmi1, 209-214
RdCVF, 316-318
RDH12 , 507-508
retinal delivery efficacy of, 255-264
RHO (rhodopsin), 233-238
RP1, 3-7, 6t
RPE65
    and cone/rod responses, 102-103
    and LCA, 247-252
    mouse knockout, 240-244
    therapeutic level in retina, 251
    transgene expression, 249
RPGR mutations/deletions, 29-32
RS1, 284-288
    characteristics of, 283-285
    discoidin domain in, 284f, 285
silencing of, 233
siRNA design and screening, 234-235
thioredoxin-like 6 (TXNL6), 10-13
transcriptional regulation in, 217-228
transgene expression, 125-126, 249
USH1C, 341, 346
Usher 1 types of, 341-347
widely expressed essential, 176-177
Genetic heterogeneity
    of *C1QTNF5* gene, 42
    of LCA, 9, 15-19, 16t
    in proliferative retinopathy, 187-192

Genetic homozygosity, 23-24, 26
Genetic mutations
    of *AIPL1* gene, 89-93
    Arg224Pro and Asp226Asn, 81-86
    Arg677ter, 3-7
    BEMr15 (r15) and BEMr (r18), 95-97
    in *CYP4V2* gene, 49-52
    disease-linked missense, 284-285
    of *G90D*, 125-130, 127f
    lakritz mutant, 205
    LCA-related, 10-13, 15-19
    P23H, 103-104
    $R172W^{+/+}/rds^{+/+}$, 144
    of *RDS/peripherin* gene, 35-39
    and retinal degeneration, 169
    of RP1 and RPO1, 3-7
    Ser163Arg, 42
    types of, 30-31
    Tyr-141-Cys, 35-36, 39
    X-linked in XLRP, 29-32
GFAP (glial fibrillary acidic protein), 288
GFP. *See* Green fluorescent protein (GFP)
Glaucoma, 397-401
    carbonic anhydrase inhibitors (CAIs) for,
        439, 445
    prevalence of, 397
    vision loss from, 196
Glial cell line-derived neurotrophic factor
        (GDNF), 212
Glial cells, 209, 211-214, 379-380
Glial fibrillary acidic protein (GFAP), 288
Glutamate transport, 327-331
    ERG of, 330-331
    histological view of, 329f
Glutamine concentration, 406-407
Glycogen synthase kinase 3-β (GSK-3β),
        205
Goldfish (*Carrassius auratus*) retina,
        333-337, 390-392, 525-529
Gp130 (tyrosine kinase receptor), 147,
        150-151
Graded retinopathy, 190t
Grafts, retinal, 369-374, 382, 392
Green fluorescent protein (GFP)
    in ARPE-19 cells, 64-65
    expression after AAV injection, 270
    immunolabelled in subretinal space, 250f
    in RPE cells, 248-252
    in vivo, 260-261, 262f
GRK-1 (G protein-coupled receptor kinase
        1), 134-138

Growth factors
 basic fibroblast growth factor (bFGF),
  197-199, 357-360
 epithelial growth factor (EGF), 357-360
 PEDF. *See* Pigment epithelium-derived
  growth factor (PEDF)
 VEGF. *See* Vascular endothelial growth
  factor (VEGF)
GTPases, 487-489
Guanine nucleotides, 81

H
H_2O_2 treatment, 292-293
Haploinsufficiency, 142-145
Haplotypes
 and gender, 6-7
 for ROA1, 24-26
 in *trans*, 4, 6
Harmonin, 341-347
 isoforms of, 342f
 localization of, 344-347, 345-346f
 and USH proteins, 349-352
Head tracking responses, 370-371
Hemorrhage, subretinal, 196
Hereditary disorders. *See specific disorders*
HICD (hyperoxia-induced cell death),
  428-434
High fat diets, 110-114
High-performance liquid chromatography
  (HPLC), 70-72, 467f
HIV-based vectors, 256
HPLC (high-performance liquid
  chromatography), 70-72, 467f
Human retinal pigment epithelium (hRPE)
 cell proliferation in, 514, 515f
 VEGF synthesis in, 513-517
Hydrogen ions, 444
Hydrolysis, 468
Hyperlipoproteinemia, 112
Hyperoxia
 effects on cones and rods, 427f, 428, 431,
  434
 retinal toxicity of, 425-435
 sequence of events in, 428-430,
  432-433
 specificity to photoreceptors, 428-431,
  433-434
Hyperoxia-induced cell death (HICD),
  428-434
Hypertension, ocular. *See* Glaucoma
Hypoglycemia, 170-172

Hypoxia
 HIF-1 and HRE activation in, 303
 versus normoxia on RT-PCR, 181f, 184
 photoreceptor vulnerability to, 425-435
 retinal cell response to, 278-280
 and retinal neovascularization, 163
 and TJ protein expression, 179-184
 and vascular endothelial growth factor
  (VEGF), 198-199
Hypoxia-inducible factor (HIF-1), 303
Hypoxia response element (HRE), 303

I
Immunoblot analysis, 134, 137
Immunofluorescence microscopy
 with AAV5-mOPs-RS1 treatment, 287
 of myosin VIIa labelling, 342-343
 of opsin antibody-stained retina, 148-150,
  149f
 Rdh11 expression on, 507
Implants, subretinal, 321-325
Inflammation/inflammatory mediators,
  275-280, 537
Inner nuclear layer (INL)
 exposed to space environment,
  419-423
 thickness of, 157f
Inner plexiform layer (IPL), 405-406,
  420-422
Insulin
 and detergent-resistant membrane rafts,
  491-495
 and photoreceptor degeneration, 155
 role of, 159
 See also Diabetic retinopathy (DR)
Interferon gamma (IFNγ), 176f, 275-278
Interleukin-4 (IL-4), 277-278
Intraocular injection procedure, 260
Intraocular pressure (IOP), 397-401, 422
Intraperitoneal injections, 440
Intravitreal injections
 of AAV, 270
 of acetazolamide, 444
 of BDNF, 297-301
 flash testing on, 252f
 procedure for, 260
 of triamcinolone acetonide (TA), 309-313
IOP (intraocular pressure), 397-401, 422
IPL (inner plexiform layer), 405-406,
  420-422
Irvine-Gass syndrome, 310-313

Ischemic retinopathy
  aneurysm-related, 164-166, 165f, 167f, 309
  occlusion-related, 187, 310-313, 339f
  and retinal glutamate transport, 327-331
  retinal vessel attenuation, 96-97
  See also Neovascularization (NV)
Isoforms
  E2, E3, and E4 encoding, 110-114
  of harmonin, 341-342
  neuron-specific β-tubulin III isoform (TUJI), 399-401
Isomers, 71f, 72
Isorhodopsin, 105

J
JAK/STAT signal transduction pathway, 277f, 279
Janus kinase (JAK), 277-279

K
Kaplan-Meier survival analysis, 6-7
Knockout system
  β5 knockout mouse, 120-122
  Cre/*lox*, 173-177, 206
  of exons, 30-32
  versus knockdown in AIPL1, 91
  of photoreceptor PDE6, 489
  Rdh11 mice characterization, 508-509
  Rpe65$^{-/-}$ mouse model, 240-244
Krypton lasers, 196

L
Labelling and staining
  immunolabelling and staining
    with anti-TGN antibody, 527f
    with Dil, 399-400
    of GFP and *RPE65*, 250f
    occludins in, 181-184
    with TUJI antibodies, 399-401
    TUNEL (terminal deoxynucleotidyl transferase UTP nick end labelling) technique, 426-433
  photolabelling of CRALBP, 478
Laser photocoagulation
  of angiogenesis, 198-199
  characteristics of, 196
  for diabetic retinopathy (DR), 195-199
  gene expression changes after, 198-199
  histological changes after, 196-197
  long-term effects of, 199
  purpose of, 195
  risks of, 197
Laser (Light Amplification by Stimulated Emmision of Radiation) treatments, 196
Late-onset retinal degeneration (L-ORD), 41-47
LD. *See* Light damage (LD)
Leber congenital amaurosis (LCA)
  and AIPL1 function, 475
  clinical and molecular survey in, 15-19
  ERG responses and histopathology in, 91f
  flowcharts of, 17f
  gene therapy for, 247-252
  and *RDH12* gene, 507-508
  rod-derived cone viability variants in, 9-13
  role of AIPL1 in, 89-93
Lenses
  as depot for INFγ release, 275-276
  LIF expression in, 148
Lentivirus (LV)
  characteristics of, 255-257
  lentiviral vectors
    design and biosafety of, 255-256
    for gene therapy, 247-252
    production of, 257-258
    for retinal gene delivery, 255-264
    targeting of, 256-257
Lentivirus-derived control vector (LV-RO.8-GFP), 248-252
LF. *See* Lipofuscin (LF)
LIF (leukemia inhibitory factor), 147-151
Light
  adaptation studies, 125-126, 129f, 135f
  cycling of, 240-241
  low-light sensitivity, 371-372
  as phosphorylation trigger, 494-495
  photoentrainment, 240-244
  and phototransduction cascade, 505-506
Light-activated phosphorhodopsin (R*P), 459-460
Light damage (LD)
  circadian-dependent, 411-414
  longterm effects of, 407-408
  methodology for induced, 406
Light-dependent translocation, 128-129
Light flashes/flash testing, 82-84, 135, 136f
Light-induced retinal degeneration (LIRD) model, 411-413

Light micrographs
  of abnormal morphology, 183f
  of abnormal pigmentation, 97f
  after BDNF injection and phototherapy,
      298, 299f
  of *APOE* TR mice retinas, 113-114
  of *BETA2/NeuroD* cell loss, 157f
  of cones, 84f
  of photoreceptor loss, 97f
  with subretinal implant, 324
Lipid metabolism
  and ApoE protein, 110
  and corneoretinal dystrophy, 50-52
Lipid rafts ("caveolae"), 491-495
Lipocortins. *See* Annexins (lipocortins)
Lipofuscin (LF)
  composition of, 79
  in RPE (RPE LF), 63-67, 72-73
  and RPE function, 120
LIRD (light-induced retinal degeneration)
      model, 411-413
Lizard (*Ctenophorus ornatus*) retina, 392-394
Loki analysis, 5f
Luciferase, 219, 222, 224
Luciferase assay system, 269-270, 272
LV. *See* Lentivirus (LV)
Lysine mutants, 457

M
Macrodeletion, 31-32
Macula lutea
  adult-onset foveomacular dystrophy
      (AOFMD), 35-39
  atrophy of, 283
  degeneration of. *See* Macular degeneration
  edema of. *See* Macular edema
  OCT through, 59f
  in retinitis pigmentosa (RP), 309-313
Macular degeneration
  genetic factors in, 63
  X-linked retinoschisis (RS), 283-288
  See also Age-related macular degeneration
      (AMD)
Macular dystrophy (MD)
  pathogenesis of, 144
  and peripherin/*rds* mutations, 141, 143-144
  and R172W, 142-143
Macular edema
  carbonic anhydrase inhibitors (CAIs) for,
      439, 521-522
  clinically significant (CSME), 309-313

cystoid (CME), 310-313
  in retinitis pigmentosa (RP), 309-313
Magnocellular pathways, 397-398, 401
Mean foveal thickness, 311f
Melatonin, 412
Membrane blebbing, 64-67
Membranous whorls, 97, 98f
Mer tyrosine kinase (MerTK)
  receptors of, 120
  in vitro phagocytosis, 499-500, 508
  in vivo phagocytosis, 499-502, 501f
Metabolism, retinal, 413-414
Methazolamide, 444
Mice. *See under* Animal models
Microaneurysms, 164-166, 165f, 167t, 309
Microarray analysis, 377-380
Microgravity, 419-423
Microneuromas, 406-408
Microvilli, apical, 99, 119, 519-523
Mission STS-72, 418-419
Mitochondrial RNA (ribonucleic acid)
  extraction of, 358-359
  injected for transgenesis, 205
  neurofilaments of, 401
  in opsin, 413
Molecular size/weight determination, 43, 45,
      46f
Müller glial cells
  buffering effect of, 445
  characteristics of, 209
  inflammatory stimulation of, 277-278
  and retinal glutamate transport, 327,
      329-331
  seal formation of, 405-407
Mutations. *See* Genetic mutations
Myc-tagged (EQKLISEEDL) arrestin, 456,
      459
Myosin VIIa
  in apical processes, 520
  binding properties of, 351, 352f
  and harmonin, 346
  localization of, 343-344, 347
  roles in Usher syndrome, 341-344

N
N-ethyl-N-nitrosourea (ENU), 95-97
N-retinylidene-*N*-retinylethanolamine (A2E),
      536-537
NEDD8 ultimate buster (NUB1), 92-93,
      472-474
Neonates, 449-450

Neovascularization (NV)
  animal models for, 163-167
  in *APOE*4 TR-ch mice, 109-114
  choroidal. *See* Choroidal
      neovascularization (CNV)
  defined, 109, 163
  and diabetic retinopathy (DR), 187-192,
      303-306
  histological views of, 191f
  therapies against, 303-306
  and VEGF, 180-184
Neural cell markers, 359-360
Neural plasticity, 405-409
Neural progenitor cells (NPCs), 211
Neural retinal leucine (*Nrl*) zipper, 133-138,
      148, 150-151
Neural stem cells (NSCs), 210-211
Neurite outgrowth, 528f
Neuron-specific β-tubulin III isoform (TUJI),
      399-401
Neuronal migration, 407-408
Neuroprotectin D1 (NPD1)
  identification and characterization of, 533,
      537
  neuroprotection of, 536-537
  synthesis of, 531, 533-534, 537
Neuroprotection
  with BDNF injections and transplantations,
      297-301, 363-366, 447-449
  and Bmil loss, 209-214
  with CNPD1, 531-537
  with CTNF, 331
  by glutamate transport modulation,
      327-331
  by light activation of insulin receptor, 494
  molecular mechanisms of, 291-294
  PEDF for, 292-294
  by preconditioning, 435
  with RdCVFs, 316-318
  by SOCS proteins, 275-280
  with subretinal implants, 321-325
Neurotrophic factors/neurotrophins
  in retinal cell differentiation, 357-360
  and retinal ganglion cells (RGCs), 447-449
Night blindness. *See* Nyctalopia (night
      blindness)
NIH.R3 Life Science Payload, 418
NMDA and AMPA mediation, 392-393
Norepinephrine, 70-72
NPCs (neural progenitor cells), 211
NPD1. *See* Neuroprotectin D1

Nr2e3 (nuclear receptor) transcription factor,
      151
NRL (neural retina leucine) transcription
      factor, 133-138, 148, 150-151
NUB1 (NEDD8 ultimate buster), 92-93,
      472-474
NV. *See* Neovascularization (NV)
Nyctalopia (night blindness)
  in choroideremia, 57
  congenital (CNB), 125, 129, 159-160
  in corneoretinal dystrophy, 49
  in retinitis pigmentosa, 4

O
OA. *See* Optic atrophies (OAs)
Observational (OB) protocol, 169-171
  observer bias in, 170-172
Occludins, 179, 181-184
Occlusions. *See* Vein occlusions
Oligonucleotide therapy, 304-305
Omega-3 fatty acids, 532
ONL. *See* Outer nuclear layer (ONL)
Opsins (visual pigments)
  function in cones, 137
  M-opsin expression, 151, 175, 177
  polyclonal anti-opsin (1D4), 127, 128f
  rhodopsin function, 102-106, 104f
  in rods, 125-130
  S-opsin expression, 175, 177
  UV opsin promoter, 204
Optic atrophies (OAs)
  autosomal dominant (DOA), 21
  autosomal recessive (ROA1), 21-26
Optic axons, 525
Optic nerve (ON)
  cell survival after injury (crush) to,
      333-337, 447, 449
  optical coherence tomography (OCT)
      through, 59f
  regeneration of, 389-394
    molecular prerequisites for, 391-392
    role of training in, 392-394, 393f
    upregulation of TGN in, 525-529
Optical coherence tomography (OCT)
  after TA injection, 310-313
  of choroideremia, 57-60, 59f
  of exon knockout, 32f
  through optic nerve and macula, 59f
Optokinetic testing, 128f
Optomotor responses, 169-172
Outer membrane protein 25 (OMP25), 346

Outer nuclear layer (ONL)
exposed to space environment, 419-423
of rods, 127-128
thickness of, 155-157, 287f, 299-301, 431
Oxidative stress, 531-538
and cell survival, 411, 535-537
and DHA, 532-533
Oxygen
and bronchopulmonary epithelium, 425
oxidative phosphorylation, 291-292
oxidative stress, 411, 531-538
See also Hyperoxia; Hypoxia
Oxygen toxicity hypothesis, 426

P
Pancreas, 159, 187-192
Pars plana and *pars plicata*, 378
Particle titers, 258-260
PCR. *See* Polymerase chain reaction (PCR)
PDE (phosphodiesterase) holoenzyme,
92-93
PDE6 (cyclic GMP phosphodierserase 6),
485-489
PDT. *See* Phototherapy (PDT)
PDZ domains, 350-352, 480-481
PDZ proteins, 341-342, 344, 346, 350
PEDF. *See* Pigment epithelium-derived
growth factor (PEDF)
Peripheral nerve grafting, 392
Phagocytes, 119-120
Phagocytosis
in photoreceptor outer segments (POS),
119-122
process of, 120-121
in vitro RPE, 97, 499-500, 508
in vivo RPE, 499-502
Phenotype-genotype correlations, 17-18
Phenotypes
function-related, 205-206
rescue of, 205
in transgenic mice, 144
Phospho-Akt (p-Akt), 333-337
Phospho-Bad (p-Bad), 333, 335-336
Phosphoinositide 3-kinase (P13K), 492
Phospholipids, 459-460, 461f
Phosphorylation
characteristics of, 103
and GRK1, 137-138
light-dependent, 135
oxidative, 291-292
patterns of, 413

in rod degeneration, 105
tyrosine to ROS, 492-495
Photodynamic therapy (PDT), 297-301
See also Laser photocoagulation
Photoperiod reversal, 241-242
Photoreceptor outer segments (POSs)
DHA levels in, 531
exposed to space environment, 419-422
phagocytosis in, 119-122, 499-502
Photoreceptors
age-dependency of, 136-138
AIPL1 in, 472-473
arrestin translocation in, 455-463, 458f
and CAIs, 439-440
conditional gene knockout system in,
173-177
counts after subretinal implant, 324-325
Cre expression in, 175-177
damage to, 192
death of
after return to room air, 429-431, 430f
characteristics of, 364, 365f
and macular deposits, 41-42
oxygen-induced, 425-435
sequence of, 428-430, 432-433, 435
degeneration of, 81-86, 95-99, 155-160
development of, 89
distribution and function of, 176-177
exposed to space environment, 417-423
G90D mutant opsin in, 125-130
gene therapy for, 283-288
hyperoxic specificity to, 428-431,
433-434
and hypoxia, 163, 278-280, 425-435
and laser photocoagulation, 198-199
light-induced lesions in, 405-409
neuroprotection of. *See* Neuroprotection
nuclear layer of, 90
outer segment phagocytosis in, 119-122,
499-502
phagocytosis in, 499-502
and phototransduction cascade, 505-506
with *Rd1-Bmi1* loss, 209-214
retinal hydrogenases in, 505-509, 506t
role of ABCA4 in, 469
stem cell differentiation for, 383-386
transduction window for, 264
transplantation of, 316
and VEGF, 187-192
See also Cones; Rods
Photosensitivity cells, 101

Phototherapy (PDT), 297-301, 299f, 310-313
    damage from, 406-414
Phototransduction cascade, 505-506
Phototransduction proteins, 413
Pigment epithelium-derived growth factor
        (PEDF), 513-517
    for neuroprotection, 292-294
    and VEGF synthesis in hRPE cells,
        515f-516f
Pigment/pigmentation, retinal, 101-106
    depigmented patches on RPE, 96f
    regeneration of, 101-106
    on retina, 3-7
    rhodopsin. *See* Rhodopsin
    See also Retinitis pigmentosa (RP)
Plasmids, 268-269, 363
Polymerase chain reaction (PCR)
    with occludin and ZO-1 labelling, 181-182
    quantitative reverse transcriptase (QRT-
        PCR), 259-261
    real time, 235-236
    semi-quantitative RT, 148-150, 174
POS. *See* Photoreceptor outer segments
        (POSs)
Posterior pole atrophy, 58f
Prenylation, 92
Primer sequences, 359t
Promotors, tissue/cell-specific, 204-206,
        267-273
Protein kinases
    A (PKA), 205, 328, 330-331
    C (PKC), 328, 330-331
Protein misfolding, 82, 85-86, 103-104
Protein synthesis inhibition, 457-458
Putrescine, 70-71

Q
Quantitative reverse transcriptase (QRT-
        PCR), 259-260

R
Rab escort protein (REP-1), 57, 59
Rat retina. *See under* Animal models
RCCI-like domain (RLD), 29-32, 30f
RDHs. *See* Retinol dehydrogenases (RDHs)
Reactive oxygen species, 291
Receptors
    αvβ5 integrin, 120-121, 500-501
    Mer tyrosine kinase (MerTK), 120,
        499-502
    scavenger CD36, 120

Recoverin, 379-380, 413
Regulator of chromatin condensation (RCCI),
        29-32
Retina
    acidification of, 414, 439, 444
    AIPL1 expression in, 90-93, 472-473
    caspase-3 activity in, 334, 365, 449-450,
        531
    crystal deposits on, 49-52
    degeneration of. *See* Retinal degeneration
    development of, 90-93
    differences between species, 377-378
    exposed to space environment, 419-423
    function of, 519-523
    NUB1 expression in, 472-473
    phototherapy damage to, 297-301
    pigmentation on, 3-7, 96f, 101-106
    pure-cone type, 137
    regeneration of, 377-378
    RS1 role in, 284-288
    supramolecular Usher protein complexes
        in, 349-352
    thickness defined, 164, 166
    Usher 1 proteins in, 341-347
    VEGF overexpression in, 179-184
    See also specific retina structures
Retinal
    9-*cis* type, 105
    11-*cis*, 125, 505
    all-*trans*-retinal, 70, 465-467, 466f, 505
Retinal degeneration
    age-related/late-onset, 41-47
    animal models of, 79-228
    assessment of progression, 169-172
    basic underlying science of, 455-538
    and bHLH transcription factors, 155-160
    causes of, 417
    cytokine-induced, 275-280
    diagnostic, clinical, cytopathological, and
        physiologic aspects of, 55-78, 169
    and docosa-hexaenoeic acid (DHA),
        536-537
    gene therapy and neuroprotection for,
        231-337
    and hypoglycemia, 170-172
    induced, 405-450
    inflammation/inflammatory mediators and,
        275-280
    LIF expression and, 150-151
    molecular genetics and candidate genes in,
        3-52

and oxidative stress, 535-536
  phases of, 405
  process of, 405-409
  in space flight environments, 417-423
  stem cell repair of, 381-386
  and survival signalling, 531-538
  in Usher syndrome, 341-352
Retinal detachment, 163, 167t, 196
Retinal ganglion cells (RGCs)
  death of, 397, 399, 401, 449-450
  Dil-labelled, 399-400
  function of, 389
  injury-induced loss of, 447-450
  and neurotrophins, 447-449
  remodelling in, 397-401
  survival after injury (crush) to, 333-337
Retinal perfusion, 188-192
Retinal pigment epithelium (RPE)
  amino-retinoid compounds in, 69-73
  animal model mutations for, 95-99, 180
  apical microvilli of, 519-523
  apoptosis of cells in, 363-366
  atrophy of, 49, 57-60
  culture of, 357-358
  differentiation of, 359-360
  dysfunction and death of, 41-42, 364, 365f
  exposed to space environment, 419-423
  function of, 499, 532
  human (hRPE), 513-517
  inflammatory processes and, 537
  lipofuscin in (RPE LF), 63-67, 72-73
  neural differentiation of, 357-360
  as outer blood-retinal barrier, 179
  phagocytosis in, 97, 119-122, 499-502
  phototherapy damage to, 297-301
  pigment regeneration in, 101-106
  retinoid-processing proteins in, 480-481
  specific promoter for, 247-252
  suitable for transplantation, 366
  support of photoreceptors, 95
  survival signalling in, 531-538
  targeting of, 267-273
  tight junction (TJ) proteins in, 179-184
  vacuoles in, 96, 98f, 112-113
Retinal stem cells (RSCs), 210, 377-380
Retinal transplantation, 367-374
  of fetal retinal sheets, 372-374
  mechanical support for, 373-374
  pretesting for, 369
  procedure for, 369, 373
  results of, 369-372
    trophic effects and permanence of, 373
    visual responses after, 370-372
Retinal vein occlusions, 187, 310-313
Retinal vessel attenuation, 96-97
Retinitis pigmentosa (RP)
  autosomal dominant. *See* Autosomal
    dominant retinitis pigmentosa (adRP)
  and C214S, 142-143
  clinical flowchart of, 17f
  defined, 133
  and EBP50, 521
  GTPase Regulator (RPGR), 29-32
  and L185P/Rom-1, 142-143
  and macular edema, 309-313
  night blindness in, 4
  onset times of, 349
  and P216L, 142-143
  pathogenesis of, 142-143, 349, 351
  and peripherin/*rds* mutations, 141-142
  prevalence of, 377
  TA injection for, 310-313
  treatment of, 147
  X-linked (XLRP) mutations in, 29-32
Retinoblastoma (Rb) protein, 214,
  219-222
Retinoids
  binding properties of, 466-469
  transport and metabolism of, 465-469,
    507f
Retinol dehydrogenases (RDHs)
  function of Rdh11, 505-509, 506t
  interaction with CRALBP, 479-480
  Rdh12 expression on, 507-508
  RetSDR1 and prRDH, 508
  SDR family of, 506-507
Retinopathy of prematurity (ROP), 425
Retinoschisis (RS)
  disease mechanisms in, 283-285
  gene therapy for, 285-288
Retinyl ester, 101
RGC. *See* Retinal ganglion cells (RGCs)
Rhodopsin, 102-106
  and autosomal dominant RP (ADRP),
    233-238
  formation of, 125-130
  GRK1 inactivation of, 138
  isorhodopsin, 105
  levels of, 413
  light-activated phosphorhodopsin (R*P),
    459-460
  misrouting of, 405

P23H
    design and screening with, 234-235
    long-term suppression of, 236-238
    siRNA effects on, 235-236
rhodopsin kinase (GRK1), 488-489
RhoGDI, 487-488
Rhodopsin mutations, 81, 86
RNA (ribonucleic acid)
    induced silencing complex (RISC) of, 233,
        305-306
    interference (RNAi), 233-238, 305-306
    mitochondrial (mRNA), 205, 358-359, 401
    short-interfering (siRNA), 233-238,
        527-528
RO.8 promotor, 248-249, 251
ROAI. *See* Autosomal recessive optic atrophy
        (ROA1)
Rod-dependent cone viability factor
        (RdCVF), 316-318
Rod-derived cone viability factor (RdCVF), 10
Rod inner segments (RIS), 128f, 130, 455,
        459
Rod outer segments (ROS)
    detergent-resistant membrane rafts of,
        491-495
    development of, 126-128
    disrupted structure of, 145
    function of, 455, 459
    microvilli ensheathing of, 519
    phagocytosis of, 69
    shortened, 489
    tyrosine phosphorylation of, 492-495
Rods
    arrestin translocation in, 455-463, 458f
    CAIs effect on, 441-445
    cGMP-phosphodiesterase β-subunit gene
        in, 217-228
    cone survival and, 434
    defined, 133
    degeneration of, 90-91, 93, 105, 141-145
    EGFP in, 204-205
    function and loss of, 133
    hyperoxic effects on, 427f, 428, 431, 434
    light flash response of, 83-84
    photosensitivity of, 102-103
    pigment regeneration in, 101-106
    rapid loss of, 368
    rod a-waves, 159
    rod-cone dystrophy, 10
    sensitivity depression in, 125-130
    See also Cones; Photoreceptors

ROS. *See* Rod outer segments (ROS)
Royal College of Surgeons (RCS) rat, 95,
        321-325, 368, 500
RP. *See* Retinitis Pigmentosa (RP)
R*P (light-activated phosphorhodopsin),
        459-460
RPE. *See* Retinal pigment epithelium (RPE)
RPE65 proteins
    as cell markers, 379-380
    defined, 240
    See also Genes
RS. *See* Retinoschisis (RS)
RS1 protein (retinoschisin)
    characteristics of, 283-284
    structure and disease mechanisms of,
        284-285

S
SCs. *See* Stem cells (SCs)
SDS-polyacrylamide gel electrophoresis
        (PAGE), 44-47
Serotonin, 70-71
Short chain dehydrogenase/reductases (SDR),
        506-507
Short interfering RNA (siRNA), 233-238,
        234-235
Signal peptides, 284
Signal transducers and activators of
        transcription (STAT) pathway,
        277-279
Sodium bicarbonate transporter (NBC3), 350
Space flight environments
    ionizing nuclei and cosmic rays, 422
    microgravity, solar radiation, and O_2
        fluctuations in, 423
    orbital data of, 418
    retinal degeneration in, 417-423
    Sprague-Dawley rats in, 418-423
Spermine/spermidine, 70-71
Sprague-Dawley rats, 418-423
Staining. *See* Labelling and staining
Stargardt disease (STGD)
    and all-*trans* retinal, 465-467, 469
    genetic factors in, 10-13, 63
STAT pathway, 277-279
Stem cells (SCs)
    bone marrow-derived (MSCs), 384-386
    brain-derived, 384
    characteristics of, 377-380
    controversy over use, 382-384
    defined, 381-382

differentiation of, 382
embryonic. *See* Embryonic stem (ES)
  cells
neural (NSCs), 210-211
retinal (RSCs), 210, 377-380, 383
for retinal repair, 381-386
STGD. *See* Stargardt disease (STGD)
Subretinal implantation, 321-325
Subretinal space
  GFP expression in, 250f
  hemorrhage in, 196
  histological neovascularization in, 190,
    191f
  injected with rAAV, 267-273
  membranous whorls in, 97
  retinal implants in, 321-325
  "ubiquitous" promoters in, 264
Sucrose gradient sedimentation, 43, 45
Superior colliculus (SC), 369-372,
    447-449
Suppressors of cytokine signaling (SOCS),
    275-280
Survival signalling, 531-538
Synaptic terminals
  localization of, 346-347
  and neural activity, 351, 352f, 392-393
  remodelling of, 406-407
Syntaxin, 379-380

T
Targeted gene replacement (TR), 111-114
Tectum, 389-393
Tetramers/tetramerization, 143
TFIIB-TBP binding complex, 221, 223
Threshold measurements, 170-171
Thrombi/emboli, 187
Thyroid hormone receptor beta 2 (TRβ2),
    150-151
Tight junction (TJ) proteins, 179-184
Tol2 transposase based systems, 203-204
Topography/topographic maps (retinotectal)
  construction of, 390-391
  in lizard models, 392-394
  restoration of, 389-394, 391-393
  upregulation in, 391-392
Toxicity, retinal, 63-67, 69-72, 291-294,
    425-435
Transcription factors
  and β-PDE promoter, 218-228
  Beta2/NeuroD, 155-160
  bHLH (basic helix loop helix), 155-160

Bmi1 repressor, 209-214, 210f
CRX (cone-rod homeobox), 150-151
GAL4, 268-270, 272-273
gene regulation by, 217-228
Nr2e3 (nuclear receptor), 151
NRL (neural retina leucine), 133-138, 148,
    150-151
Sp1, Sp3, and Sp4, 218-219, 223-225,
    227
Transducin, 413
Transducin alpha subunit (Tα), 125-126,
    129f
Transgene expression, 125-126, 249
Transgene manipulation, 188
Transgenesis/transgenetic technology,
    201-206
  limitations and future of, 205-206
  in zebrafish, 202-203
Transglutaminase, neural (TGN), 525-529
  antiserum production, 525-526
  and optic nerve regeneration, 529
Transplantation
  of BDNF, 363-366
  pretesting for, 369
  safety of, 385-386
  timing of, 382-383
  See also Retinal transplantation; Stem cells
    (SCs)
Triamcinolone acetonide (TA), 309-313
  procedure for injection, 311
Trophic factors, 373
Tryptamine, 70-71
TUJI. *See* Neuron-specific β-tubulin III
    isoform (TUJI)
TUNEL (terminal deoxynucleotidyl
    transferase UTP nick end labelling)
    technique, 426-433, 449-450
  quantification of, 427-428
Two-alternative forced choice (TAFC)
    protocol, 169-172
Tyramine, 70-71
Tyrosine, 492-495

U
Ultraviolet (UV) spectrum, 71-72
Uncoupling protein UCP2, 291-292
Unfolded protein response (UPR), 85-86
Upregulation
  of FGF2, GFAP, and c-Jun expression, 427,
    431-432
  in optic nerve, 525-529

in topographic maps (retinotectal),
391-392
USH proteins
molecular analysis of, 349-352
supramolecular complexes of, 350-351,
352f
and Usher syndrome, 346
Usher syndrome (USH), 341-347, 349

V
Vacuoles in RPE, 96, 98f, 112-113
Vascular endothelial growth factor (VEGF)
controlling expression of, 303-306
and hypoxia, 163-164, 167, 198-199
and neovascularization, 180-184
retinal overexpression of, 187-192
synthesis inhibition in, 513-517
tight junction (TJ) proteins in, 182-184
Vasoproliferative retinopathy, 187-192
Vector titers, 258-261
VEGF. *See* Vascular endothelial growth factor
(VEGF)
Vein occlusions, 187, 310-313
Veins. *See specific veins*
VER (visual evoked response), 24f
Verteporfin (Visudyne), 36, 297-301
Vimentin, 379-380

Vision
assessment in mice, 169-172
central, 196, 283
color vision, 23f, 439
night blindness (nyctalopia), 4, 23, 49
peripheral, 283
sudden loss of, 196
Visual acuity
after TA injection, 312f
measurement of, 170-172, 171f
Visual cycle physiology, 47, 481
Visual evoked response (VER), 24f
Visual fields, 23f
Vitreous
hemorrhage into, 196
See also Intravitreal injections

W
Western blot of *RPE65* expression, 250f

X
X-linked retinitis pigmentosa (XLRP), 29-32
Xenopus laevis, 141-145, 219-222, 456-457

Z
Zebrafish (*Danio rerio*) retina, 201-206
ZO-1 labelling, 181-184